AF869034

I CAN Learn®
EDUCATION SYSTEMS

JRL Enterprises, Inc.
New Orleans, Louisiana

Pre-Algebra Textbook 5th Edition

ISBN-10: 0-9789017-6-2

ISBN-13: 978-0-9789017-6-9

Protected by U.S. Patents No. 5267865, 5441415, 5788508, 6064856, 6758674 and Des. 385431 and European Patent No. 0656139.

JRL Enterprises, Inc.

400 Poydras Street, Suite 1000

New Orleans, LA 70130

www.icanlearn.com

Printed in Canada

Table of Contents

MPA-001: Identifying, Comparing, and Ordering Whole Numbers Through Billions

There are many things you can do with numbers, as you will find out throughout these lessons. One thing you do all the time without thinking about is to compare numbers. For example, let's say you and a friend have jobs that pay an hourly wage. You make $7.00 per hour and your friend makes $7.25 per hour. You could describe this in two ways. One, you make *less* than your friend. Another is that your friend makes *more* than you do. Using math symbols, you would write:

$7.25 > $7.00, which would be read as "$7.25 is greater than $7.00"

or

$7.00 < $7.25, which would be read as "$7.00 is less than $7.25"

If you and your friend made the same amount, you would say these amounts are *equal.*

$7.25 = $7.25

You also compare numbers when buying things. If the item costs $3.57, then you must give the clerk at least this amount. You could write this as:

amount given ≥ $3.57, which would be read as "amount given is greater than or equal to $3.57"

Note: Notice the line under the greater than sign. It means that $3.57 could equal the amount given.

You even compare numbers when watching a baseball game. You know who wins the game by comparing the final scores. The team with *more* runs wins.

You compare numbers by comparing their **place values**. Let's look at the number "23" to help understand place value. You know this number is read "twenty-three," but think about what this really means. Twenty-three is *one* number but it has *two* digits. The digits in the number system are the **whole numbers**: 0, 1, 2, 3, 4, 5, 6, 7, 8, 9. Each digit has a specific value depending on where it is placed in the number.

millions	hundred thousands	ten thousands	thousands	hundreds	tens	ones
					2	3

Notice the number 23 in the place value chart above. The "2" is in the "tens" place. This means you have 2 sets of 10, or, in other words, "20." The "3" is in the "ones" place. It means you have 3 sets of 1 or "3." Put these two values together and that is what "23" really means — two groups of 10 and 3 groups of 1.

Let's look at a larger number like 52,609. Going backwards, the "9" means 9 sets of 1 or "9". The "0" means there are no sets of 10. The "6" means 6 sets of 100 or "600". The "2" means 2 sets of 1000 or "2000". Finally, the "5" means 5 sets of 10,000 or "50,000." When you add these individual values together, you get the number "52,609."

millions	hundred thousands	ten thousands	thousands	hundreds	tens	ones
		5	2,	6	0	9

Note: Notice the comma between the 2 and the 6. This helps you read the number by sectioning the digits into ***periods****. The set of 3 digits all the way to the right is called the hundreds period. Moving from* right *to* left, *the next set of 3 digits is called the thousands period, followed by the million and billion periods. Separate these periods using commas. For example, 9240728 becomes 9,240,728.*

Use place value to compare or order numbers. Follow these steps:

Example 1 Compare 4,126,038,745 and 4,126,145,219

4,126,038,745 4,126,145,219	**Step 1:** Vertically align the numbers so that digits in the same place value (ones, tens, and so on) match up.
4,126,038,745 **4**,126,145,219	**Step 2:** Compare digits beginning in the highest place value; in this case, it is the billions place. Think of this in terms of win, lose, or tie. In this case you have a tie because both digits are 4's.
4,**1**26,038,745 4,**1**26,145,219	**Step 3:** The digits in the hundred millions place is a tie, so you keep going until you find a place value in which the digits are different.
4,126,**0**38,745 4,126,**1**45,219	**Step 4:** The digits are different in the hundred thousands place. The 1 "wins" and the 0 "loses," so the second number is greater overall; 1 > 0 or 0 < 1.
4,126,038,745 < 4,126,145,219 or 4,126,145,219 > 4,126,038,745	**Answer:** You can write this comparison two ways.

You can also use these steps to list the numbers in order either **ascending**, from least to greatest, or **descending**, from greatest to least.

Example 2 The net incomes in 1999 for five companies were:

Company A $311,870,792

Company B $98,937,765

Company C $312,458,531

Company D $1,071,845,292

Company E $312,458,777

Rank these companies according to their net incomes in descending order, from greatest to least.

$311,870,792 $98,937,765 $312,458,531 $1,071,845,292 $312,458,777	**Step 1:** Line up the numbers so that digits with the same place value are aligned.
Company A $311,870,792 Company B $98,937,765 Company C $312,458,531 #1 Company D $**1**,071,845,292 Company E $312,458,777	**Step 2:** Look for the highest place value for all the numbers. In this case, Company D has a digit in the billions place. Because none of the others do, Company D has the highest value.
Company A $**3**11,870,792 Company B $98,937,765 Company C $**3**12,458,531 #1 Company D $1,071,845,292 Company E $**3**12,458,777	**Step 3:** Look for the next highest place value. Companies A, C, and E all have digits in the hundred millions place. All three digits also happen to be "3." You need to think in terms of win, lose, and tie again. Right now you have a tie, so you need to keep going until you have a "winner."

#4	Company A	$311,870,792
	Company B	$98,937,765
#3	Company C	$312,458,531
#1	Company D	$1,071,845,292
#2	Company E	$312,458,777

Step 4: Comparing the 3 digits in the millions place, Company A loses, so you look at Companies C and E. It is in the hundreds place that you find a "winner" — Company E. Comparing these three companies, E has the highest value, followed by Company C, then A.

#4	Company A	$311,870,792
	Company B	$98,937,765
#3	Company C	$312,458,531
#1	Company D	$1,071,845,292
#2	Company E	$312,458,777

Step 5: From reading the chart, you can see Company B has the lowest net income, because it is the only one left.

#1	Company D	$1,071,845,292
#2	Company E	$312,458,777
#3	Company C	$312,458,531
#4	Company A	$311,870,792
#5	Company B	$98,937,765

Answer: The companies ranked from highest to lowest net income are Company D, E, C, A, and B.

Problem Set

1. Choose >, <, or = to complete the expression.
895,012__895,212

2. Choose >, <, or = to complete the expression.
5,178__5,178

3. Choose >, <, or = to complete the expression.
1,203,584,354__1,203,584,359

4. Choose >, <, or = to complete the expression.
863,029,874,389__863,025,468,901

5. Choose >, <, or = to complete the expression.
1,398,730,989__1,489,575,521

6. Place the following numbers in descending order.
682; 579; 702; 683; 568

7. Place the following numbers in ascending order.
2,650; 2,498; 2,402; 2,561; 2,489

8. Which class collected the second highest number of pennies?

Class	Pennies
Seniors	178,902
Juniors	227,628
Sophomores	187,523
Freshmen	107,204

9. The following people ran for governor. Who received the greatest number of votes?

	Votes
Livingston	1,178,902
Johnson	1,127,628
Ackerman	1,187,523
Carlson	192,204

10. The following people ran for governor. Who received the least number of votes?

	Votes
Livingston	1,178,902
Johnson	1,127,628
Ackerman	1,187,523
Carlson	192,204

11. Identify the digit in the thousands place: 578,621

12. Find the place value for the digit 5: 578,621

hundreds　　hundred thousands

thousands　　ten thousands

13. Identify the digit in the hundreds place: 79,345

14. Find the place value for the digit 4: 79,345

tens　　ten thousands

hundreds　　thousands

15. Identify the digit in the hundred thousands place: 4,578,621

16. Find the place value for the digit 7: 4,578,621

hundreds　　hundred thousands

ten thousands　　tens

17. Identify the digit in the hundred thousands place: 4,019,357

18. Identify the digit in the hundreds place: 70,957

19. Find the place value for the digit 9: 643,791

thousands　　hundreds

ten thousands　　tens

20. Find the place value for the digit 2: 39,427

thousands　　ten thousands

hundreds　　tens

MPA-002: Adding, Subtracting, Multiplying, and Dividing Whole Numbers

There are four basic operations you can perform on numbers: addition, subtraction, multiplication, and division. You do them every day when you balance your checkbook, buy groceries, drive a car, cook a meal, and so on. In this lesson, you will review the procedure for each operation.

Addition

Sum	The answer to an addition problem.

Example 1 16,373 + 948 + 134,118 + 4,175

	16,373 948 134,118 + 4,175	**Step 1:** Write the numbers under each other so that the place values match up, similar to comparing numbers.
The 2 was "carried." The total of the ones column is 24.	*2* 16,37**3** 94**8** 134,11**8** + 4,17**5** 4	**Step 2:** Beginning with the ones column, add the digits in that column. If the sum is more than 10, you must regroup. Put the "ones" digit of your answer under the ones column and "carry" the tens value to the top of the tens column to repeat the process.
These numbers have been carried. →	*1 1 2 2* 16,373 948 134,118 + 4,175 155,614	**Step 3:** Now add the tens column, including the 2 that was carried from the ones column. If necessary, write the ones digit of your total under the tens column and carry the tens value to the top of the hundreds column. Continue until all columns have been added.

Answer: The sum is 155,614

There are **properties** of addition that allow you to do certain things when adding numbers.

Identity Property: Suppose that you have the number 3. You want to add something to it but you don't want to change its "identity" (value). What is that number? Zero! The Identity Property states that you can add zero to any number and that number will remain the same.

Identity Property of Addition	The Identity Property of Addition states that when zero is added to any number, the result is the same as that number.

Commutative Property: The word to concentrate on is "commute." Commute means to go back and forth. You commute to school. When you relate it to addition, it means that you can take two numbers and move their positions in the problem. For example, 2 + 3 = 3 + 2. The numbers can change position and the sum remains the same.

Commutative Property of Addition	The Commutative Property states that the order in which you add numbers does not affect the sum.

Associative Property: Think about numbers associating with one another. For example, 2 + 3 + 4. The 3 in the middle either "associates" with the 2 first, giving you a sum of 5, then add 4 for a total of 9. It can also associate with the 4 first, giving you a sum of 7, then add 2 for a total of 9. Notice that you get the same answer either way. The results are the same, and by regrouping, you can use combinations of numbers that are easy to solve.

Associative Property of Addition	The Associative Property states that changing the grouping of numbers when you add two or more numbers does not change the sum.

Subtraction

The order in which you subtract matters! This operation is **not** commutative or associative.

Difference	The answer to a subtraction problem.

Example 2 Subtract: 21,192 – 4,358

Note:Be careful to subtract the bottom digit from the top digit in each column. For example, in the ones column, you must take 8 from 2, not the other way around. Regroup, or "borrow" when needed.

```
  21,192
–  4,358
```

Step 1: Write the numbers under each other so that the place values match up.

```
       8
  21,1̸92
–  4,358
       4
```

Step 2: You will begin with the ones column. In this case, you must take 8 ones *away* from 2 ones. This is not possible. Now go to the 9 in the tens column (remember that this 9 actually means 90) and "borrow" 1 ten (or 10) to add to the 2 that is in the ones column for a total of 12. There is now an 8 in the tens column, because you borrowed 1 ten. Now you can subtract 8 from 12.

```
       8
  21,192
–  4,358
      34
```

Step 3: Now move to the tens column. You need to take 5 away from 8. You can do this, so there is no need to borrow.

```
  10   8
  21,192
–  4,358
  16,834
```

Step 4: Continue subtracting the columns from right to left, borrowing when needed.

Answer: The difference is 16,834.

Multiplication

Product	The answer to a multiplication problem.

Properties exist for multiplication as well.

Associative Property: Similar to addition, if you have 3 × 4 × 5, you can multiply the 4 by the 3, then the 5, or multiply the 4 by the 5, then the 3. Either way the product is 60. Try it!

Associative Property of Multiplication	The Associative Property states that changing the grouping of numbers when you multiply two or more numbers does not change the product.

Commutative Property: Similar to addition, you can switch the order of two numbers and the product will be the same. For example, $3 \times 4 = 4 \times 3$.

Commutative Property of Multiplication	The Commutative Property of Multiplication states that the order in which you multiply numbers does not affect the product.

Identity Property: What can you multiply 4 by and not change its identity (value)? One! You can multiply any number by one and the number will remain the same.

Identity Property of Multiplication	The Identity Property of Multiplication states that when any number is multiplied by one, the result is the same as that number.

Zero Property: This is a special property just for multiplication. When you multiply any number by zero, the product will be zero.

Zero Property of Multiplication	The Zero Property of Multiplication states that when any number is multiplied by zero, the result is zero.

Example 3 Multiply: $24 \times 13{,}745$

$\begin{array}{r} 13{,}745 \\ \times \quad 24 \\ \hline \end{array}$		**Step 1:** Write the problem so that the place values match up. Write the number with more digits on top (this is not necessary, but will make multiplication easier).
$\begin{array}{r} \mathit{2} \\ 13{,}7\not{4}5 \\ \times \quad 24 \\ \hline 0 \end{array}$	$5 \times 4 = 20$	**Step 2:** You must multiply the ones digit in the second number by *every* place value in the first number, from right to left. In this case, write the 0 under the ones column and carry the 2.
$\begin{array}{r} \mathit{12\ 12} \\ \mathbf{13{,}745} \\ \times \quad \mathbf{24} \\ \hline 54\,980 \end{array}$		**Step 3:** Now multiply four times four and add the two that was carried. Numbers that are carried are always *added*. Continue to the left. This is called a partial product.
$\begin{array}{r} 13{,}745 \\ \times \quad 24 \\ \hline 54\,980 \\ 0 \end{array}$		**Step 4:** Next, multiply by the 2 in the tens place. Remember that this actually means you will multiply by 20. For this reason, always begin the second partial product by placing a zero in the ones column.
$\begin{array}{r} 13{,}745 \\ \times \quad 24 \\ \hline 54\,980 \\ 274\,900 \end{array}$	Begin with 5×2.	**Step 5:** Now multiply all the digits of the first number by the tens digit of the second number, carrying when necessary. This is the second partial product. *Note: If the second number had a hundreds digit, you would begin the third partial product by placing* ***two*** *zeroes at the end.*
$\begin{array}{r} 13{,}745 \\ \times \quad 24 \\ \hline 54\,980 \\ +274\,900 \\ \hline 329{,}880 \end{array}$		**Step 6:** Add the partial products.

Answer: The product is 329,880.

Division

Division is **not commutative and not associative.** Be careful when writing your division problem. For example, $6 \div 3$ does not mean the same thing as $3 \div 6$.

Quotient	The answer to a division problem.
Divisor	The number that you divide by.
Dividend	The number that is being divided.

Some ideas to think about…

- When the divisor is 1, the quotient will be the dividend; for example, $82 \div \mathbf{1} = 82$.
- When the dividend is 0, the quotient will also be 0; for example, $\mathbf{0} \div 45 = 0$.
- Zero can *never* be the divisor. You cannot divide by 0! For example, it is not possible to divide 6 objects into 0 groups.
- When the divisor and the dividend are the same number (not 0), the quotient will be 1, for example, $90 \div 90 = 1$.

Example 4 Divide: $211{,}644 \div 9$

$9\overline{)211{,}644}$

Step 1: Rewrite the problem using the long division symbol.

$$\begin{array}{r} 2\phantom{0{,}000} \\ 9\overline{)211{,}644} \end{array}$$

Step 2: Divide the first digit of the dividend by the divisor. If this is not possible, try the number formed by the first two digits or perhaps the first three digits. In this case, the 2 cannot be divided by 9, but 21 can be.

Note: Notice that the 9 divides into the number 21 twice, and the 2 is written above the "1" in "21," not above the 2.

$$\begin{array}{r} 2\phantom{0{,}000} \\ 9\overline{)211{,}644} \\ \underline{-18}\phantom{1{,}644} \\ 3\phantom{1{,}644} \end{array}$$

Step 3: Multiply the 2 in the quotient by the divisor, in this case the 9, and write the product under the 21. Subtract.

$$\begin{array}{r} 23{,}516 \\ 9\overline{)211{,}644} \\ \underline{-18}\phantom{1{,}644} \\ 31\phantom{{,}644} \\ \underline{-27}\phantom{{,}644} \\ 46 \\ \underline{-45} \\ 14 \\ \underline{-9} \\ 54 \\ \underline{-54} \\ 0 \end{array}$$

Step 4: Bring down the next digit in the dividend and continue this process (Divide, Multiply, Subtract, Bring Down) until each digit in the dividend has been used.

Note: If there is a remainder, write it after the quotient using a capital letter "R." For example, 182 R 3.

Answer: 23,516

Problem Set

1. Add: 96,486 + 36,531

2. Add: 9,643 + 7,361

3. Subtract: 9,683 – 3,643

4. Subtract: 76,384 – 27,426

5. Divide: 4,863 ÷ 28

6. Ashley decided to make a record of how many seconds it takes her to complete a set of tasks over an entire day. Her results are as follows:

Task	Seconds
Sleep	28,800
School	3,600
Ready for School	25,200
Homework	10,800
Television	18,000

How many total seconds does it take her to do her tasks over the course of a day?

7. Add: 9,875,254 + 29,785,548

8. Subtract: 875,124,321 – 1,256,782

9. Multiply: 786,851 × 782

10. Divide: 2,984 ÷ 17

11. Divide: 61,751 ÷ 751

12. Find the sum of the following numbers:

891,781
791,751
12,210
1,854,655

13. If 36,486,587 people donated blood in 1995 and 36,486,593 people were donors in 1996, what was the difference in the number of donors between 1995 and 1996?

14. Margie is studying the solar system in school. Her teacher gave her the following chart to help her remember the distance of each planet from the sun:

Planet	Distance from the sun
Mercury	36,000,000 mi
Venus	67,000,000 mi
Earth	93,000,000 mi
Mars	142,000,000 mi
Jupiter	483,000,000 mi
Saturn	886,000,000 mi
Uranus	1,790,000,000 mi
Neptune	3,000,000,000 mi
Pluto	5,000,000,000 mi

What is the sum of the distances of the first five planets from the sun in the solar system?

15. If Darren works 8 hours a day, five days a week, for fifty-two weeks a year, how many hours will he have worked after 15 years?

16. Multiply: 3,126 × 4

17. Multiply: 363 × 11

18. Multiply: 950 × 9

19. Multiply: 594 × 5

20. Divide: 3,702 ÷ 3

MPA-003: Using Four-Step Plan for Problem Solving

You solve problems all the time. If you need to be at an appointment by 3:00, you probably take a moment or two to figure out how long it will take you to get to the appointment based on where you will be until then. Suppose you are going out of town and cannot take your pet with you. What you will do with your pet while you are gone becomes a problem. Perhaps you will ask a neighbor to watch your cat or maybe you will take the dog to a kennel. Solving a problem might be as simple as deciding what you will have for dinner tonight. Do you have to go to the store? How many people are you serving?

Whatever the problem is—and of course you will be focusing on math problems—there are steps you can take to increase your chances of finding a successful solution.

Four-Step Plan for Problem Solving

Step 1: The first step is that you must **understand** the problem. This might include reading the problem more than once (maybe even two, three, or more times), or asking someone to repeat the problem to you. This step is extremely important. You could end up solving the wrong problem if you don't really understand what is being asked of you.

Step 2: In the second step, you come up with a **plan** to solve the problem. When you are solving problems involving numbers you might try "guess and check," working backwards, making a chart or table, finding a pattern, drawing a picture, or some other strategy. This step may take the most time.

Step 3: Actually **solving** the problem is the third step. This is where you carry out the plan you made in Step 2. If the plan is well thought out, then solving the problem should be very easy.

Step 4: Finally, you want to take a moment to **check** your work to be sure you found the correct answer to the question you were asked.

Example 1 Justin has 6 errands to run in $3\frac{1}{2}$ hours. He also needs to take a $\frac{1}{2}$ hour lunch break. How much time will he have to run each errand?

Step 1: Understand: Justin has $3\frac{1}{2}$ hours to run 6 errands. Within that time, he will take a $\frac{1}{2}$ hour lunch break. This will take away from the time he has to run the errands.

Step 2: Plan: Because the lunch break will take away from the amount of time Justin has to run errands, you will *subtract* $\frac{1}{2}$ hour from the total time he has, $3\frac{1}{2}$ hours. This will give us the amount of time Justin really has. Then you can *divide* the remaining time by 6 because you need to find the amount of time he has for *each* errand. Using minutes instead of hours will make the work easier. There are 60 minutes in one hour, so there are 210 minutes in $3\frac{1}{2}$ hours.

Step 3: Solve. Write the values for the plan.

$$3\frac{1}{2} \text{ hours} - \frac{1}{2} \text{ hour} = 210 \text{ minutes} - 30 \text{ minutes}$$
$$= 180 \text{ minutes}$$

Now divide 180 minutes by 6 errands.

$$\frac{180 \text{ minutes}}{6 \text{ errands}} = 30 \text{ minutes per errand}$$

This means Justin will have about 30 minutes to run each errand.

Step 4: Check:

$$(30 \text{ minutes} \times 6 \text{ errands}) + 30 \text{ minutes for lunch}$$
$$= 180 + 30$$
$$= 210 \text{ minutes}$$
$$= 3.5 \text{ or } 3\frac{1}{2} \text{ hours}$$

Answer: Justin will have about 30 minutes to run each errand.

Example 2

For Halloween, Mrs. Malone wants to give each of the 6 children in her neighborhood 4 pieces of candy. If she has 30 pieces total, how many will she have left over?

Step 1: Understand: Mrs. Malone has 30 pieces of candy. If she gives 4 pieces each to 6 children, how much candy will she have left?

Step 2: Plan: You will *multiply* 4×6 to give us the total amount of candy she will need. You may also *draw a picture* of 6 groups of 4 pieces of candy. You can then *subtract* this total to be given to the children from the total amount of candy that Mrs. Malone has to find how much is left over.

Step 3: Solve: Mrs. Malone needs 24 pieces of candy: $4 \times 6 = 24$. She will have 6 pieces left over: $30 - 24 = 6$.

Step 4: Check:

$$(4 \times 6) + 6 = 30$$
$$24 + 6 = 30$$
$$30 = 30$$

Answer: She will give 24 pieces of candy to the children, and will have 6 pieces left over.

Example 3 Find two consecutive odd numbers whose product is 1,155.

Step 1: Understand: There are some words in this problem you need to make sure you understand. The word consecutive means you are looking for two numbers that follow each other in the number system, such as 2 and 3 or 56 and 57. In this case, you are looking for consecutive *odd* numbers, so you are looking for a pair of numbers like 1 and 3 or 11 and 13. Product means you will be multiplying.

Step 2: Plan: Multiply two consecutive numbers and see if the product is 1,155. If the product is below 1,155, try two more consecutive numbers higher than the original two; if the product is above 1,155, try two more numbers lower than the original two. Keep using this process until the product is 1,155.

Step 3: Solve the problem.

$29 \times 31 = 899$ too low
$31 \times 33 = 1{,}023$ too low
$33 \times 35 = 1{,}155$ just right

Step 4: Check. $33 \times 35 = 1{,}155$. You might want to check this problem with a calculator if you solved the problem using paper and pencil.

Answer: The two consecutive odd numbers whose product is 1,155 are 33 and 35.

Problem Set

1. Jeffrey owns an ice cream shop that sells three different flavors: chocolate, vanilla and strawberry. On the first day he sold 24 ice cream cones. On the second day he sold 20 ice cream cones, and on the third day he sold 34 ice cream cones. If each ice cream cone costs $2 and he sold equal amounts of all, how much money did each flavor make?

2. Glenda is in charge of scheduling the games for the ping pong tournament at her school. If a person loses a game, he is eliminated from the competition. There are 68 people participating in the tournament. How many games will need to be scheduled to determine a champion?

3. Devon wants a motorcycle that costs $9,000. His father is going to pay for half of the motorcycle. Devon will receive a paycheck for $1,385 before buying the motorcycle. How much money does Devon need to buy the motorcycle?

4. Germaine has $19 in the morning. He eats breakfast at Shawney's and spends $5.50; then he eats a school lunch that costs him $2.75. His friend repays him $3, and then he buys a cola and cookie for a combined $1.50. How much money does Germaine have at the end of the day?

5. Eddy and his friend went shrimping in Lake Pontchartrain. Eddy caught 7 pounds on the first round. He then caught another 13 pounds. Eddy's friend caught 3 pounds more than twice what Eddy caught. How many pounds did they catch together?

6. A flock of pigeons begins with 23 birds. After a month, 4 pigeons are born. The next month, 7 pigeons are born, but 3 leave. When winter comes, 12 birds leave, 2 pigeons are born, and 8 more pigeons join the flock. How many pigeons are left?

7. Brian cooked a platter of chicken wings. Trey took 17 chicken wings from the platter and ate all of them. Matt took 13, but put 7 back on the platter. Brian ate 12 chicken wings that he originally took and the 7 that Matt put back. If there were no chicken wings left, how many chicken wings were originally on the platter?

8. Mr. Thomas tells his math class that when two consecutive numbers are multiplied, their product is 650. What are the two numbers?

25 and 26	29 and 30
20 and 21	34 and 35

9. Shawntelle works at a Chinese restaurant. On Monday she worked from 3 p.m. to 10:08 p.m. On Tuesday she worked from 11:37 a.m. to 8:37 p.m. and on Wednesday she worked from 1:34 p.m. to 9:12 p.m. How long did Shawntelle work in those three days?

10. John has $1 bills, $5 bills, and $20 bills. He has 15 bills that total $138. How many $5 bills does he have?

11. Truman has 8 bananas in his tree house. He goes to search for more bananas, and along the way he finds 10 more. While swinging from a vine he drops 4 of them, but when he lands on the tree he finds 9 more. At the final banana tree near his tree house, he comes across 7 more bananas. Unfortunately, while swinging back to his house, he drops 6 of them. How many bananas will Truman have at home?

12. Two consecutive even numbers multiplied together have a product of 5,928. What are the two numbers?

76 and 78	82 and 84
72 and 74	86 and 88

13. Sylvia found 31 cents in her pocket and 93 cents in her car. How much money did she find altogether?

14. Donald bought 23 shares of stock A and 44 shares of stock B. How many shares of stock A and B did Donald buy?

15. Mandy allotted $400.00 for gift buying this year. If she allows herself to spend no more than $25.00 per gift, how many gifts will she be able to buy?

16. Lisa bought 5 raffle tickets to her school's fair. Each ticket cost $2. How much did she spend on the tickets?

17. Theodore needs spark plugs for his truck. If he needs 6 spark plugs for his truck and each plug costs $1.50, how much, excluding tax, will the spark plugs cost him?

18. Tameka is working part-time at an advertising agency. This week she worked 25 hours. If her hourly wage is $9.15, how much will she receive before taxes?

19. Phillip and Sonya went out to eat. Phillip's meal cost $9.73 and Sonya's meal cost $8.49. What was the total cost of their meals?

20. At a local high school track meet, Amanda's time for the 400-meter dash was 58 seconds, while Sherial's time was 49 seconds. How much faster was Sherial's time than Amanda's time?

MPA-004: Using Rounding to Estimate

You estimate all the time. How much time will it take to get to work? How much will it cost to go on vacation? How many pizzas are needed for the birthday party? The reason you estimate is because you cannot, or choose not, to calculate an exact answer. If you estimate that you will pay $1,000 for taxes this year, and you actually owe $1,200, you had a pretty good estimate. But if you end up owing $5,000 in taxes, either your estimate was way off or a mistake was made in figuring out your taxes. You will probably check to see if the calculations were correct.

We can use a process called *rounding* to estimate. Let's see how this works.

Rounding Numbers

To round numbers, it is important that you understand place value. Go back to the lesson on comparing and ordering numbers if you need to review place value.

Let's take the number **23,826.**

If you want to round this number to the nearest *thousand*, you need to focus on the digit in the thousands place, 23,826. The digit is 3. Think about this for a second. This number must be between 23,000 and 24,000. The real question is, which of these is 23,826 closer to? The digit to the right of the thousands digit will tell us. In this case, the digit to the right of the thousands place is 8. Here are the rules for rounding:

Rules for Rounding

Step 1: If the digit to the right is 5 or greater, then round up to the next thousand, hundred, ten, etc.

Step 2: If the digit to the right is less than 5, then round down to the nearest thousand, hundred, ten, etc.

In this example, the digit to the right is 8, so you will round up to 24,000.

Let's take the same number and round to the nearest *hundred*. We now need to focus on the digit in the hundreds place, in this case, 8. This number will fall between 23,800 and 23,900. Using the same steps from before, look at the digit to the right. It is 2, so you will round down to 23,800.

> *Note: Notice that you rounded the same number to different place values. When you rounded to the nearest hundred, it was closer to the original number than when you rounded to the nearest thousand. The higher the place value that the number is rounded to, the less accurate the rounded number will be.*

The rest of this lesson concentrates on using rounding to estimate sums and differences.

Estimates of Sums and Differences

Step 1: If you are told to round to a specific place value, then do so. If not, then round all the numbers in that problem to the highest common place value.

Step 2: Find the sum or difference of the rounded numbers.

Example 1	Round 3,761,167 to the nearest ten thousand.
3,7**6**1,167	**Step 1:** Concentrate on the ten thousands place, 3,7**6**1,167. This number is between 3,760,000 and 3,770,000.
3,76**1**,167	**Step 2:** The number to the right of the ten thousands place, 1, tells us to round down to 3,760,000.
3,760,000	**Answer:** The answer is 3,760,000.

Example 2 On a road trip, Henry drove from New Orleans to Atlanta to New York and then to Chicago. From New Orleans to Atlanta he drove 473 miles, from Atlanta to New York he traveled 870 miles, and from New York to Chicago he traveled 820 miles. Henry drove 929 miles from Chicago to his home in New Orleans. Estimate Henry's total mileage.

Note: We are not told which place value to round to, so round each number to the hundreds place, since this is the highest place value that all three numbers have in common.

473	The 7 to the right tells us to round up.	500
870	The 7 to the right tells us to round up.	900
820	The 2 to the right tells us to round down.	800
+ **9**29	The 2 to the right tells us to round down.	+ 900
		3,100

Answer: The estimate is 3,100 miles.

Go back and add the actual mileage. How close was the estimate?

Example 3 Three companies are planning to merge into one larger company. The net value of the first company is $160,534,000. The net value of the second company is $1,514,195,155. The net value of the last company is $783,391,616. Approximate the net value of the merged company to the hundred thousands place.

This time you are given a place value, the hundred thousands place.

160,**5**34,000	The 3 to the right tells us to round down.	160,500,000
1,514,**1**95,155	The 9 to the right tells us to round up.	1,514,200,000
+ 783,**3**91,616	The 9 to the right tells us to round up.	+ 783,400,000
		2,458,100,000

Answer: The net value of the merged company would be *about* $2,458,100,000.

Example 4 In Example 3, what is the difference between the highest and lowest net values? Estimate to the nearest hundred million.

1,**5**14,195,155	The 1 tells us to round down.	1,500,000,000
− **1**60,534,000	The 6 tells us to round up.	− 200,000,000
		1,300,000,000

Answer: The difference between the highest and lowest net values is $1,300,000,000.

Problem Set

1. Round the following items to the greatest place value, then add the rounded numbers together: 1,650 + 2,501

2. Round the following items to the greatest place value, then add the rounded numbers together: 1,358 + 2,150

3. Round the following items to the greatest place value, then add the rounded numbers together: 1,758 + 2,201

4. Round the following numbers to the nearest hundreds, then perform the indicated operation: 1,797 + 5,451

5. Round the following numbers to the nearest thousands, then perform the indicated operation: 1,650 + 2,501 + 12,958

6. Round the following numbers to the nearest thousands, then perform the indicated operation: 25,735 + 5,981 + 4,349

7. Round the following numbers to the nearest thousands, then perform the indicated operation: 52,472 – 39,852 + 6,358

8. Juan wants to buy four CDs that are on sale. The sale prices, which include sales tax, are as follows: $7.95, $11.90, $12.30, $9.75. Use rounding to the nearest dollar to determine whether he has enough money to purchase the CDs if he has $45. Answer yes or no.

9. Round the following numbers to the nearest thousands, then perform the indicated operation: 4,235 + 5,651

10. Round the following numbers to the nearest thousands, then perform the indicated operation: 4,235 + 2,875

11. Round the following numbers to the nearest thousands, then perform the indicated operation: 4,235 + 4,502

12. Round the following numbers to the nearest thousands, then perform the indicated operation: 4,235 + 5,499

13. Round the following numbers to the nearest thousands, then perform the indicated operation: 46,549 – 18,345

14. Round the following numbers to the nearest thousands, then perform the indicated operation: 26,549 – 12,458

15. Round the following numbers to the nearest thousands, then perform the indicated operation: 7,459 + 21,501

16. Round the following numbers to the nearest thousands, then perform the indicated operation: 7,459 + 12,624

17. Round the following numbers to the nearest thousands, then perform the indicated operation: 24,876 – 8,798

18. In preparing a budget, Jules made a list of items that he is responsible for each month. They are:

Car insurance	$125.00
Rent	$770.00
Utilities (phone, electric, etc.)	$375.00
Gasoline	$150.00

Jules earns $1,800 per month. Rounding the amount he owes for each expense to the nearest ten, estimate the amount he would have left.

19. Round to the greatest common place value, then multiply: 812×24

20. Round to the greatest common place value, then divide: $7,829 \div 31$

MPA-005: Estimating Products and Quotients Using Patterns

In a previous lesson, you learned rules for rounding numbers to estimate answers. Specifically, you concentrated on estimates of sums and differences. In this lesson, you will use the same rules and study examples involving products and quotients.

First, let's review the rules you use to round numbers.

If you want to round the number 4,729 to the nearest ten, you would first find the digit in the tens place. In this case, it is the 2. This number will be between 4,720 and 4,730. Next, look at the digit to the right of this place value and use the following guidelines:

Rules for Rounding

Step 1: If the digit to the right is 5 or greater, then round up.

Step 2: If the digit to the right is less than 5, then round down.

In this example, the digit to the right of the tens place is a 9. This means you will round up to 4,730.

Estimates of Products and Quotients

The rules are the same as when estimating sums and differences.

Step 1: If you are told which place value to round to, then do so. If not, then round to the highest common place value.

Step 2: Multiply or divide the rounded numbers.

Since we are estimating we should use the symbol, ≈, which means "is approximately equal to," instead of the equal sign in our answer.

Example 1 Estimate $6,829 \times 2,701$

	The highest common place value is the thousands place. Therefore, you will focus on the digits in the thousands place.	
6,829	The 8 to the right tells you to round up.	7,000
× **2**,701	The 7 to the right tells you to round up.	× 3,000
		21,000,000

Answer: $6,829 \times 2,701 \approx 21,000,000$

Note: Go back and solve the actual problem now. How close is the estimate to the actual product?

Example 2 Estimate $62\overline{)8824}$

$60\overline{)8820}$

The highest common place value is the tens place. You will focus on the tens place in each number. The divisor, 62, is between 60 and 70. The digit to the right, 2, tells you to round down to 60. The dividend, 8,824, is between 8,820 and 8,830. The digit to the right is 4. This means you will round this number to 8,820. To get an estimate, simply divide these rounded numbers.

$$\begin{array}{r} 147 \\ 60\overline{)8820} \\ \underline{-60} \\ 282 \\ \underline{-240} \\ 420 \\ \underline{420} \\ 0 \end{array}$$

Answer: The estimated quotient for this division problem is 147.

Note: Go back and solve the actual problem now. How close is the estimate and actual product?

Example 3 Mark plans to work this summer. He will earn $206 for each week that he works. Estimate how much money he will make if he works for 12 weeks.

$$\begin{array}{r} 206 \\ \underline{\times 12} \end{array}$$

This problem requires multiplication, so you will estimate the product to solve.

$$\begin{array}{r} 210 \\ \underline{\times\ 10} \\ 2100 \end{array}$$

The highest common place value is the tens place. The estimated product is 2,100.

Answer: Mark will make about $2,100 this summer.

Note: Go back and solve the actual problem now. How close is the estimate to the actual product?

Problem Set

1. Estimate the product: 421 × 93

2. Estimate the product: 3,150 × 95

3. Estimate the product: 75 × 91

4. Estimate the quotient: 36,024 ÷ 789

5. Estimate the quotient: 5,842,618 ÷ 81,345

6. If the Bobcats pick 21 tailbacks in the draft each year, about how many tailbacks will they have drafted after 68 years?

7. If You-Rent-Em Car Company purchased 29 cars at $19,995 each, about how much will the company have spent?

8. If Diedra sold 102 stuffed animals at $9 each, about how much money will she make?

9. If Rachel wrote 23 books in her career, which lasted 19 years, about how many books did she average each year?

10. If Orinthall scores 4,998 points in 18 seasons, about how many points does he score each season?

11. Estimate the product $7{,}836 \times 491$ by first rounding to the nearest hundred.

12. Estimate the quotient $7{,}557 \div 419$ by first rounding to the nearest ten.

13. Estimate the product 381×215 by first rounding to the nearest hundred.

14. Estimate the product 219×82 by first rounding to the nearest ten.

15. Estimate the quotient $877 \div 113$ by first rounding to the nearest ten.

16. Estimate the quotient $285{,}364 \div 3{,}279$ by first rounding to the nearest ten.

17. Estimate the product 248×114 by first rounding to the nearest ten.

18. Estimate the product $4{,}018 \times 877$ by first rounding to the nearest hundred.

19. Estimate the quotient $82{,}481 \div 3{,}256$ by first rounding to the nearest hundred.

20. Estimate the quotient $740{,}931 \div 2{,}843$ by first rounding to the nearest thousand.

MPA-006: Determining Reasonableness of Answers and Appropriate Method of Computation

Every time you solve a problem that involves numbers, there are some questions you need to ask yourself.

Methods of computation

Before you solve any problem, you need to know *how* you will compute the answer. The first question you will ask is **"Do you need an exact answer?"** In other words, do you need the actual answer or just an estimate? If you do not need an exact answer, then you can estimate the answer. If you *do* need an actual answer, then you have to ask yourself another question, **"Are the numbers small and easy to work with?"** If yes, then use paper and pencil. If no, then use a calculator. Use this diagram to help you.

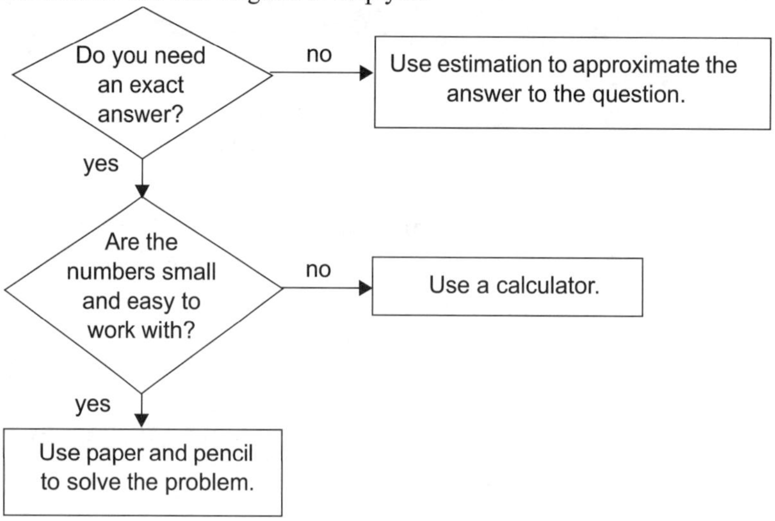

Is the answer *reasonable*?

After you solve any problem, you want to be sure that you answered the question. First, ask yourself, "Is the answer reasonable?" If it does not seem to make sense, then you need to rework the problem. If the answer does seem to make sense, then you need to ask yourself, "Did I need an exact answer?" If not, use the previous estimate as your answer. If you need an exact answer, then check your answer by working backwards, using inverse operations, or reworking the problem. Use the following diagram to help you.

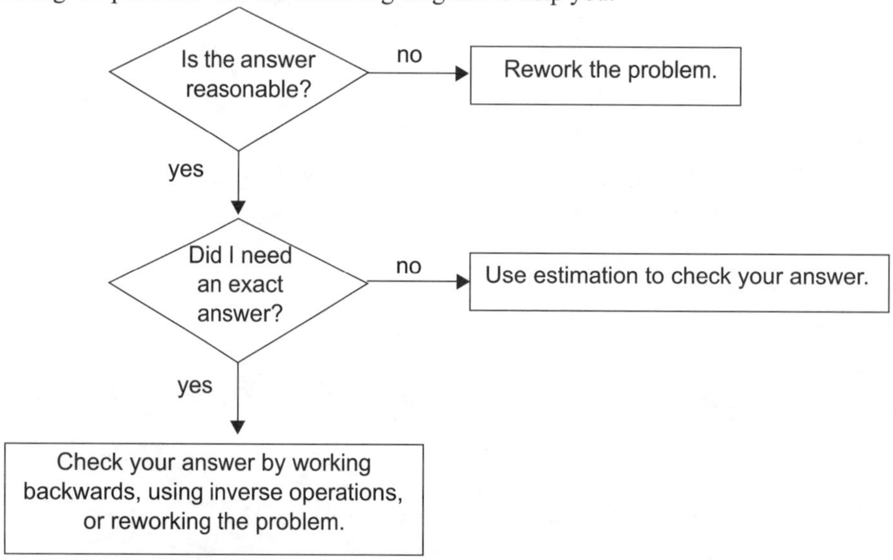

Example 1 A carpet-cleaning company charges $59 for the basic five-room cleaning service, plus $18 for each additional room. If there were 4,363 service calls last year for the basic five-room cleaning, about how much money did they make from the basic cleaning charges?

$59

4,363 calls

Using the four-step plan to solve problems:

Step 1: Understand the problem. A carpet-cleaning company charges $59 for the basic five-room cleaning service. There were 4,363 service calls for this service last year. (The $18 is not important, because the question only asks about calls for the basic service.)

Step 2: Develop a plan. Let's first look at the "Methods of Computation" diagram.

- *Do you need an exact answer?* The word *about* tells you that you do not need an exact answer. An approximation will be fine. So you will use an estimate to answer the question.
- You will use multiplication to solve the problem. You will estimate using rounding rules because you do not need an exact answer. You will round to the nearest ten because that is the highest place value in both numbers.

$$\begin{array}{r} 4{,}363 \\ \times\ 59 \\ \hline \end{array} \qquad \begin{array}{r} 4{,}360 \\ \times\ 60 \\ \hline \end{array}$$

Step 3: Solve the problem.

4,36**3**	The 3 tells you to round down.	4,360
× 5**9**	The 9 tell you to round up.	× 60
		261,600

Step 4: Check. This is where you use the second diagram to see if the answer is reasonable. Does this amount seem too high, too low, or just right? Because you didn't need an exact answer, you can use your estimate as the answer.

Answer: The carpet-cleaning company made approximately $261,600 from the basic cleaning service.

Example 2 Tommy is shopping for new CDs and has $50 to spend. He has chosen three items that cost $11, $16.95, and $18.95. What will be the total cost of the three items? Does he have enough money to buy all three CDs?

Step 1: Understand the problem. Tommy wants to buy three CDs that cost $11, $16.95, and $18.95. You need to calculate the total.

Step 2: Develop a plan. Using the first diagram, "Do you need an exact answer?" Yes, you do, because a total cost for the three items is required to determine if Tommy has enough money to buy them. The word *about* does not appear in the problem, so you are probably not looking for an estimate. You will add the cost of the three items to find the total cost. The second question in the first diagram asks, "Are the numbers small and easy enough to work with?" Yes, they are, so you will use paper and pencil to solve the problem.

Step 3: Solve the problem.

$$\begin{array}{r} 11.00 \\ 16.95 \\ +\ 18.95 \\ \hline 46.90 \end{array}$$

The total cost of the three CDs is $46.90. Because Tommy has $50 to spend, he can buy all CDs.

Step 4: Check. Use the second diagram. Does this answer seem reasonable? Does the total seem too high, too low, or just right? Because you needed an exact answer, you will check your answer by reworking the problem to be sure you added correctly.

$$\begin{array}{r} 11.00 \\ 16.95 \\ +\ 18.95 \\ \hline 46.90 \end{array}$$

Answer: The answer checks, so $46.90 is the total amount Tommy must pay. Since he has $50, he has enough money to buy all three CDs.

Example 3 Lisa was one of three winners in last night's lottery. If the jackpot was 55.8 million dollars, how much money will Lisa get?

Step 1: Understand the problem. Lisa was one of three winners of the 55.8 million-dollar jackpot. This means that the jackpot will be split into three equal parts.

Step 2: Develop a plan. *"Do you need an exact answer?"* I would if I were Lisa! You will divide the jackpot by three. *"Are the numbers small and easy enough to work with?"* In this case, you might say no and decide to use a calculator.

Step 3: Solve the problem.

On the calculator divide 55.8 by 3. The answer should be 18.6. In other words, Lisa's share of the jackpot is 18.6 million dollars.

Step 4: Check. "Does the answer seem reasonable?" Because you needed an exact answer, you will check your answer using the inverse operation. Because you were dividing, you will now multiply to check your answer. Does

$18.6 \times 3 = 55.8$? Check on your calculator.

Answer: Lisa's share of the jackpot is 18.6 million dollars.

Problem Set

1. Your family is going to a water park. Should you estimate or determine an exact amount of money to bring? Answer estimate or exact.
2. Your family is planning a trip to the zoo. Is it reasonable to overestimate the amount of money to bring? Answer yes or no.
3. If you went to a restaurant, would you pay for the meal with an estimate or an exact amount? Answer estimate or exact.
4. You are going to the mall to buy shoes. Is it better to bring an exact amount or estimate? Answer estimate or exact.
5. Your class is voting to choose a class pet. You are in charge of counting the votes. Should you estimate or determine an exact number? Answer estimate or exact.
6. Little Susie asks how many times you blink in a day. Would the answer you give her most likely be an estimate or an exact amount? Answer estimate or exact.

7. $16{,}869 - 8{,}762 \approx 8{,}000$. Is this the exact answer? Answer yes or no.

8. $4{,}692 - 86 \approx 4{,}600$. Is this answer reasonable? Answer yes or no.

9. You are going to a concert. Your ticket costs $18.50. You want a souvenir T-shirt for $7.56 and a snack that costs about $3.85. Should you bring $25 or $50?

10. You are going shopping for new shoes and noticed they are on sale for $39.99. How much money should you take with you, $39.99 or $45?

11. Would you need an estimate or an exact answer if planning refreshments for the number of people attending a wedding? Answer estimate or exact.

12. Would you need an estimate or an exact answer if planning refreshments for the number of people attending a birthday party? Answer estimate or exact.

13. Would you need an estimate or an exact answer to know the amount of money that is in your savings account when transferring money from your savings account to your checking account? Answer estimate or exact.

14. Would you need an estimate or an exact answer to pay a bill with a check? Answer estimate or exact.

15. Is this statement reasonable? Answer yes or no. 64 + 51 is approximately 110.

16. Raffle tickets cost $5.00 each. Approximately how many tickets must be sold for the raffle to break even if there are three prizes costing $212.00, $489.00, and $311.00 to be given away?

42	1,000
200	5,000

17. You would like to purchase a computer system. The computer cost is $119.00. The monitor cost is $279.00. The printer cost is $67.00. Approximate to the nearest ten dollars and decide how much money you will need to buy the system.

18. The Humber Estuary Bridge in Great Britain is 4,626 feet long. The Akashi-Kaikyo Road Bridge in Japan is 6,528 feet long. Approximate to the nearest hundred feet and estimate how much longer the Japanese bridge is than the Great Britain bridge.

19. Rokyo Products produced 5,018 radio components in their first quarter of business. If Rokyo Products sells each component for $15.29, approximately how much revenue will the company generate?

20. Virlomi drove her new motorized bicycle 797 miles on a full tank of gas. The tank holds 9.89 gallons of gas. Estimate how many miles Virlomi could drive on one gallon of gas.

MPA-007: Solving Problems Using Logical Reasoning Skills

Read the following statements:

- Every weekend has been beautiful and sunny for the last five weeks. This weekend will be beautiful and sunny.
- Every time I go to the grocery store, I see Susan. When I go to the grocery store today, I will probably see Susan.
- Everyone in class is speaking Spanish. I must be in Spanish class.

What is happening in these situations? Conclusions are being made based on observations. This type of reasoning is called **inductive reasoning**.

Inductive reasoning	A type of reasoning in which a conclusion is made based on a group of facts or observations.

Now read the following statements:

- A foul shot is worth one point. Bill made a foul shot, so the team earned one point.
- Joey will get grounded if he gets home after 9 p.m. Joey comes home at 10 p.m; therefore, he gets grounded.
- Everything in the store costs $1.00. If I purchase 10 items, then the total cost is $10.

What is happening in these situations? Conclusions are being made according to a rule. This type of reasoning is called **deductive reasoning**.

Deductive reasoning	A type of reasoning in which a conclusion is made based on a rule.

Think of it this way: inductive reasoning creates a rule based on a set of facts and deductive reasoning uses a rule to apply to a situation to make conclusions. Both types of reasoning are used to solve problems.

Use the following steps when solving problems using logical reasoning:

Step 1: Read and understand the question.

Step 2: Develop a plan.

- When the data will lead you to a simple conclusion, organize the data in a simple chart.
- When the data will lead you to a complex conclusion, organize the data in a table.

Step 3: Solve the problem.

Step 4: Check your work.

Now see if you can solve the following examples.

Example 1 Rick, John, Ed, and Pat are lining up for a picture. Pat stands on the right of John, Ed stands on the left of Rick, and Rick stands on the left of John. What is the order from left to right?

Put four people in order from left to right based on some clues.

Step 1: Read and understand the problem.

You need to find the correct order of these four people. To do so, use a simple chart.

Step 2: Develop a plan.

Step 3: Create a chart that shows the four people standing next to each other. First, draw the chart with four spaces for the number of people.

John	Pat		

Step 4: Begin filling in the information from the problem. The first clue states that Pat stands on the right of John.

John	Pat	Ed	Rick

Step 5: The next clue indicates that Ed stands on the left of Rick.

John	Pat		

		John	Pat

	Rick	John	Pat

Ed	Rick	John	Pat

Step 6: Finally, Rick stands on the left of John. However, this presents a problem with your chart. Rick is not standing to the left of John on the chart that is currently drawn. You must rearrange the order of the chart. You know that Pat must be on the right of John, and Rick must be on the left. So, the order for these three people must be Rick, John, Pat. Because Ed stands on the left of Rick, you can put Ed first, then Rick, John and Pat.

Ed	Rick	John	Pat

Answer: On the left is the completed chart. Check your work by rereading the problem to make sure that all the requirements are met. In this case, all requirements are met in the chart.

Example 2 Lisa, Miki, Lynn, and Amy each rented a movie to watch at their slumber party. They each rented a different type of movie: Horror, romance, drama and comedy. Using the information given here, figure out what type of movie each person rented.

Lisa came to the slumber party with the girl who brought the horror movie.

Miki and Lynn hate romantic movies.

Amy does not like horror or romantic movies.

Miki loves to laugh.

You must match each person with the type of movie that she brought based on a set of clues.

Step 1: Read and understand the problem.

This problem is complex and will require a logic matrix.

Step 2: Develop a plan.

	Lisa	Miki	Lynn	Amy
horror				
romance				
comedy				
drama				

Step 3: Create a table to solve the problem. First, draw the matrix to include the number of people and the number of movie types.

	Lisa	Miki	Lynn	Amy
horror	X			
romance				
comedy				
drama				

Step 4: Fill in the table based on the clues given in the problem. Put an X in the box to indicate that a person did NOT bring that type of movie. An O means that a person DID bring that type of movie. The first clue tells you that Lisa did not bring the horror movie, so put an X in the box under "Lisa" and next to "horror."

	Lisa	Miki	Lynn	Amy
horror	X			
romance		X	X	
comedy				
drama				

Step 5: The second clue tells you that Miki or Lynn probably did not bring the romantic movie. Put an X in the boxes under "Miki" and under "Lynn" and across from "romance."

	Lisa	Miki	Lynn	Amy
horror	X			X
romance	O	X	X	X
comedy	X			
drama	X			

Step 6: The third clue tells you that Amy probably did not bring a romantic or horror movie. Put an X in the boxes under "Amy" and across from "romance" and "horror."

Note: Notice that the row next to "romance" has only one empty box in it. This must be an O. *You now know that Lisa brought the romantic movie, so you can put an* X *in each box in the rest of the column.*

	Lisa	Miki	Lynn	Amy
horror	X	X	O	X
romance	O	X	X	X
comedy	X	O	X	X
drama	X	X	X	O

Step 7: The fourth clue tells you that Miki probably brought the comedy. Put an O under "Miki" and next to "comedy." Next, put an X in each remaining box under "Miki."

Note: Notice that only one box is left in the "horror" row. Put an O *in this box. This means that Lynn brought the horror movie. Then fill in the rest of Lynn's column with* X*'s. Now, the row next to "drama" has only one empty box next to it. Put an* O *in it and an* X *in the other box in that column.*

Lisa = romantic
Miki = comedy
Lynn = horror
Amy = drama

Answer: Lisa brought the romantic movie.
Miki brought the comedy.
Lynn brought the horror movie.
Amy brought the drama.

Problem Set

Answer the following questions:

1. Refer to the following statement and decide the type of reasoning that was used to determine the conclusion.

$3 \div 3 = 1$

$4 \div 4 = 1$

$5 \div 5 = 1\ldots$

Therefore, any number divided by itself is one.

2. Refer to the following statement and decide the type of reasoning that was used to determine the conclusion.

$1 + 0 = 1$

$2 + 0 = 2$

$3 + 0 = 3\ldots$

Therefore, zero added to any number is that number.

3. Refer to the following statement and decide the type of reasoning that was used to determine the conclusion.

$5 - 5 = 0$

$3 - 3 = 0$

$1 - 1 = 0\ldots$

Therefore, any number subtracted from itself is zero.

4. Refer to the following statement and decide the type of reasoning that was used to determine the conclusion.

All even numbers are divisible by two. 16 is an even number. Therefore, 16 is divisible by two.

5. Refer to the following statement and decide the type of reasoning that was used to determine the conclusion.

All odd numbers are not divisible by two. 19 is an odd number. Therefore, 19 is not divisible by two.

6. Refer to the following statement and decide the type of reasoning that was used to determine the conclusion.

All numbers whose last digit is zero are divisible by five. 230 is a number whose last digit is zero. Therefore, 230 is divisible by five.

7. Fill in the blank:

Mary is older than Ann. Therefore, Ann is _____________ than Mary.

8. Fill in the blank:

Mary is shorter than Elizabeth. Therefore, Elizabeth is _____________ than Mary.

9. Fill in the blank:

The city of LaPlace is smaller than Baton Rouge. The city of Baton Rouge is smaller than New Orleans. Therefore, the city of LaPlace is _____________ than New Orleans.

10. Fill in the blank:

Walking is slower than jogging. Therefore, jogging is _____________ than walking.

11. Fill in the blank:

John weighs more than Mary. Therefore, Mary weighs _____________ than John.

12. Fill in the blank:

A pound of feathers weighs the same as a pound of lead. Therefore, a pound of lead is _____________ weight as a pound of feathers.

13. Fill in the blank:

A gallon of water weighs more than a gallon of air. Therefore, a gallon of air weighs _____________ than a gallon of water.

14. Fill in the blank:

A dollar bill has the same value as 100 pennies. Therefore, 100 pennies has _____________ value as a dollar bill.

15. Fill in the blank:

Ten can be factored into 2×5. Therefore, 5 is ______________ 10.

16. Marcus, Duane, and Charles are friends. Marcus is taller than Duane, and Duane is taller than Charles. Who is the tallest among them?

17. Marcus, Duane, and Charles are friends. Marcus is taller than Duane, and Duane is taller than Charles. Who is the shortest among them?

18. Jessica, Jennifer, and Jane are friends. Jane is taller than Jennifer and Jessica is shorter than Jane. Who is the tallest among them?

19. Jessica, Jennifer, and Jane are friends. Jane is taller than Jennifer and Jessica is shorter than Jane. Who is the shortest among them?

20. There are four students in Mrs. Brown's class who are competing for the best student of the year award: John, Roberta, Bob, and Georgette. Consider the following clues and decide who will be the best student of the year in Mrs. Brown's class according to their grade point averages.

Clues:

1. Georgette has a higher grade point average than Bob, but not as high as Roberta.

2. Bob has a higher grade point average than John, but not as high as Georgette.

21. There are four students in Mrs. Brown's class that are competing for the best student of the year award: John, Roberta, Bob and Georgette. Consider the following clues and decide who will be the best student of the year in Mrs. Brown's class according to their grade point averages.

Clues:

1. Roberta has a higher grade point average than Bob, but not as high as John.

2. Georgette has a higher grade point average than Roberta, but not as high as John.

22. Each student is enrolled in a different elective. Use the information provided in the given matrix to answer the question. Remember "O" represents a true statement and "X" represents a false statement.

	Chorus	Band	Art	Photography	Drama
Julio	X		X		X
Lam	X	X	X		
Paul	O	X	X	X	X
Jeff	X		X	X	
Ann	X	X	O	X	X

Which student is enrolled in Photography?

23. Each student is enrolled in a different elective. Use the information provided in the given matrix to answer the question. Remember "O" represents a true statement and "X" represents a false statement.

	Chorus	Band	Art	Photography	Drama
Julio	X		X		X
Lam	X	X	X		
Paul	O	X	X	X	X
Jeff	X		X	X	
Ann	X	X	O	X	X

Which student is enrolled in Drama?

24. Each woman drives a differently colored car. Use the information provided in the given matrix to answer the question. Remember "O" represents a true statement and "X" represents a false statement.

	Red	Blue	White	Tan	Black
Sami	X	X	X	O	X
Jen	X	X		X	
Juanita	X		X	X	
Shequita	X			X	X
Heidi	O	X	X	X	X

Which woman drives the tan car?

25. Each woman drives a differently colored car. Use the information provided in the given matrix to answer the question. Remember "O" represents a true statement and "X" represents a false statement.

	Red	Blue	White	Tan	Black
Sami	X	X	X	O	X
Jen	X	X		X	
Juanita	X		X	X	
Shequita	X			X	X
Heidi	O	X	X	X	X

Which woman drives the blue car?

26. Jeff, John, and Paul each have a different coin; a nickel, a dime, and a quarter. Use the information provided in the given matrix to answer the question. Remember "O" represents a true statement and "X" represents a false statement.

	Jeff	John	Paul
Nickel	X	X	O
Dime	O	X	X
Quarter	X	O	X

Who has the dime?

27. Jeff, John, and Paul each have a different bill; a five-dollar bill, a one-dollar bill, and a ten-dollar bill. Use the information provided in the given matrix to answer the question. Remember "O" represents a true statement and "X" represents a false statement.

	Jeff	John	Paul
One	O	X	X
Five	X	X	O
ten	X	O	X

Who has the five-dollar bill?

28. Each student chose a different study group. Based on the information provided in the given matrix select the correct statement. Remember "O" represents a true statement and "X" represents a false statement.

Group	Donne	Milton	Blake	James	Joyce
1	X	X	X	O	X
2				X	
3				X	
4				X	
5				X	

James chose Group 1.

Blake chose Group 3.

James and Joyce did not choose Group 1.

Blake did not choose Group 2.

29. Each student chose a different study group. Based on the information provided in the given matrix select the correct statement. Remember "O" represents a true statement and "X" represents a false statement.

Group	Donne	Milton	Blake	James	Joyce
1	X	X	X	O	X
2	X	X	O	X	X
3			X	X	
4			X	X	
5			X	X	

Joyce chose group 2.

James chose Group 1 and Blake chose Group 2.

James and Joyce did not choose Group 1.

Milton did not choose Group 3.

30. Each student chose a different study group. Based on the information provided in the given matrix select the correct statement. Remember "O" represents a true statement and "X" represents a false statement.

Group	Donne	Milton	Blake	James	Joyce
1	X	X	X	O	X
2	X	X	O	X	X
3			X	X	
4			X	X	
5			X	X	

Who chose Group 3?

James and Joyce

Blake

Milton

There is not enough information to determine an answer.

31. Writers: Willa Cather, Susan Glaspell, Marita Bonner, Mary E. Wilkins Freeman, Louisa May Alcott.

Stories: *Hate is Nothing, Louisa, A Jury of Her Peers, A Whisper in the Dark, A Wagner Matinee*

Information:

A. Neither Glaspell nor Cather wrote *Hate is Nothing*, but one wrote *A Jury of Her Peers* and the other wrote *A Wagner Matinee*.

B. Alcott did not write the story with a title containing her name, but either Glaspell or Freeman did.

C. Cather wrote either *Louisa* or *A Wagner Matinee*.

D. *A Whisper in the Dark* was not written by Bonner.

Which story was written by Marita Bonner?

32. Writers: Willa Cather, Susan Glaspell, Marita Bonner, Mary E. Wilkins Freeman, Louisa May Alcott.

Stories: *Hate is Nothing, Louisa, A Jury of Her Peers, A Whisper in the Dark, A Wagner Matinee*

Information:

A. Neither Glaspell nor Cather wrote *Hate is Nothing*, but one wrote *A Jury of Her Peers* and the other wrote *A Wagner Matinee*.

B. Alcott did not write the story with a title containing her name, but either Glaspell or Freeman did.

C. Cather wrote either *Louisa* or *A Wagner Matinee*.

D. *A Whisper in the Dark* was not written by Bonner.

Who wrote *A Whisper in the Dark*?

33. Authors: V. Flourney, A. Diamond, A. Say, L. Yep, M. Lee.

Books: *The Lost Lake, Smallpox and the American Indian, The Star Fisher, The Patchwork Quilt, Finding My Voice.*

Information:

A. *The Patchwork Quilt* was not written by Say or Lee, but one did write *The Lost Lake*.

B. Of Say and Diamond, one wrote *The Lost Lake* and the other wrote *Smallpox and the American Indian*.

C. *The Star Fisher* was not written by Flourney or Lee, but one did write *The Patchwork Quilt*.

Who wrote *The Patchwork Quilt*?

34. Authors: T. Feelings, S. Begay, S. Hamanaka, C. Garza, A. Dorros.

Books: *Abuela, Family Pictures, The Journey, Navajo: Visions and Voices Across the Mesa, Soul Looks Back in Wonder.*

Information:

A. Of Garza and Dorros, one wrote *Abuela* and the other wrote *Family Pictures*.

B. *The Journey* was not written by Begay or Garza, but one did write *Navajo: Visions and Voices Across the Mesa*.

C. *Abuela* was not written by Garza or Feelings, but *Soul Looks Back in Wonder* was written by one of them.

Who wrote *The Journey*?

35. <u>Authors:</u> L. Erdrich, L. Hogan, C.R. Lee, T. Morrison, G. Naylor.

<u>Books:</u> *Mean Spirit, Tracks, Beloved, Mama Day, Native Speaker.*

<u>Information:</u>

A. Of Hogan and Naylor, one wrote *Mean Spirit* and the other wrote *Mama Day.*

B. *Tracks* was not written by Lee or Naylor, but one did write *Mama Day.*

C. *Beloved* was not written by Erdrich or Lee.

Who wrote *Tracks*?

36. <u>Authors:</u> J. Okimoto, G. Paulsen, C. Potok, S. Power, G. Soto

<u>Books:</u> *Grass Dancer, The Chosen, Jesse, Talent Night, Hatchet*

<u>Information:</u>

A. Of Okimoto and Soto, one wrote *Talent Night* and the other wrote *Jesse*.

B. *Talent Night* was not written by Soto or Power, but one did write *Grass Dancer.*

C. Potok did not write *Hatchet*.

Who wrote *The Chosen*?

37. <u>Authors:</u> A.&J. Parker, S. Banks, P. Yee, G. Soto, E. Brunting

<u>Books:</u> *Kwanza, Local News, Tales from the Gold Mountain, Remember My Name, Going Home*

<u>Information:</u>

A. Of Banks and Soto, one wrote *Local News* and the other wrote *Remember My Name*.

B. *Tales from the Gold Mountain* was not written by Soto or Brunting, but one did write *Going Home*.

C. *Kwanza* was not written by Yee or Banks, but *Remember My Name* was written by one of them.

Who wrote *Remember My Name*?

38. <u>Authors:</u> W.D. Meyers, J. Bruchac, Y. Uchida, L. Yep, H. Arrow

<u>Books:</u> *The First Strawberries, The Bracelet, The Dollmaker, Fallen Angels, The Rainbow People*

<u>Information:</u>

A. Neither Meyers, Uchida, nor Yep wrote *The First Strawberries*, although one did write *Fallen Angels*.

B. *The Dollmaker* was written by either Arrow or Meyers.

C. Yep did not write *The Bracelet* or *Fallen Angels*.

D. Of Bruchac and Uchida, one wrote *The First Strawberries* and the other wrote *The Bracelet.*

Who wrote *The Dollmaker*?

39. <u>Writers:</u> Rebecca Davis, Harriet Jacobs, Ann Petry, Charlotte Gilman, Caroline Kirkland

<u>Stories:</u> *Incidents in the Life of a Slave Girl, Life in the Iron Mills, The Yellow Wallpaper, The Schoolmaster's Progress, The Bones of Louella Brown*

<u>Information:</u>

A. Neither Jacobs nor Kirkland wrote *Life in the Iron Mills*, but one wrote *The Schoolmaster's Progress* and the other wrote *Incidents in the Life of a Slave Girl*.

B. Petry wrote either *The Yellow Wallpaper* or *The Bones of Louella Brown*.

C. *The Yellow Wallpaper* was written by someone whose first name begins with the letter "C", but *Incidents in the Life of a Slave Girl* was not written by someone whose first name begins with the letter "C".

Who wrote *Life in the Iron Mills*?

40. <u>Writers:</u> Willa Cather, Susan Glaspell, Marita Bonner, Mary E. Wilkins Freemont, Louisa May Alcott

<u>Place of Birth:</u> Randolph, Massachusetts; Concord, Massachusetts; Davenport, Iowa; Boston, Massachusetts; Winchester, Virginia.

<u>Information:</u>

A. Alcott, Bonner, and Freeman were from the same state.

B. Neither Alcott nor Cather was born in Boston, but one was born in Concord and the other was born in Winchester.

C. The first letter of Bonner's last name is the same as the first letter of the city where she was born.

Who was born in Boston, Massachusetts?

MPA-008: Order of Operations

Look at the following mathematical expression: $3 + 2 \cdot 6$. If you were to *simplify* the expression, or, in other words, do the operations and come up with a single answer, what would it be?

One person might add the 3 and 2 first and then multiply by 6, and obtain an answer of 30. Another person might multiply the 2 and 6 first and then add the 3, and obtain an answer of 15.

Who is correct? Can they both be right?

The answer to that question is no; only one answer can be correct. You must know which operation to perform first when a problem contains more than one operation.

Here is the correct order of operations:

- Parentheses
- Exponents
- Multiplication and division
- Addition and subtraction

This sequence is called the **order of operations**.

Order of Operations	A series of rules that determine the order, or sequence, in which you perform operations. If a particular operation is not included in the problem, go to the next step. Step 1: Perform all operations in grouping symbols, starting with the innermost group of symbols. Examples of grouping symbols are ( ), [ ], { }, and the fraction bar. Step 2: Simplify all exponents. Step 3: Multiply or divide, in order, from left to right. Step 4: Add or subtract, in order, from left to right.

The sentence "Please Excuse My Dear Aunt Sally" will help you remember the order of operations. The first letter of each word reminds you of the order of the steps:

P: Simplify inside parentheses ( ), brackets [ ], and braces {}.

E: Simplify any exponents.

M D: Perform all multiplication and division from left to right as they occur.

A S: Perform all addition and subtraction from left to right as they occur.

Example 1 Simplify: $6(2 + 5) - 10$

$6(7) - 10$	**Step 1:** Since you have an operation within parentheses (addition), you will do this first.
$6(7) - 10$ $42 - 10$	**Step 2:** There are no exponents, so now perform multiplication. There is no division.
$42 - 10$ 32	**Step 3:** Perform addition and subtraction. In this problem, only subtraction is left.
	Answer: 32

Example 2 Simplify: $16 \div 4 + 3 \cdot 3 - 21 \div 3$

$4 + 3 \cdot 3 - 21 \div 3$ $4 + 9 - 21 \div 3$ $4 + 9 - 7$	**Step 1:** There are no parentheses or exponents, so perform any multiplication and division. Remember to perform all multiplication and division from left to right as they occur. $16 \div 4 = 4$ and $3 \cdot 3 = 9$ and $21 \div 3 = 7$.
$4 + 9 - 7$ $13 - 7$ 6	**Step 2:** The only operations left are addition and subtraction. Because addition comes first in the problem, do the addition first, followed by subtraction.
	Answer: 6

Example 3 Simplify: $6(8 - 3 \times 2) + 5$

$6(8 - 3 \times 2) + 5$ $6(8 - 6) + 5$ $6(2) + 5$	**Step 1:** Perform the operations in parentheses first, following the order of operations. This means you will multiply, then subtract.
$6(2) + 5$ $12 + 5$	**Step 2:** Now that you have simplified the operations inside the parentheses, the next operation is multiplication.
$12 + 5$ 17	**Step 3:** The last operation is addition.
	Answer: 17

Example 4 Simplify: $5^2 - (9 + 11) + 1^3$

$5^2 - (9 + 11) + 1^3$ $5^2 - (20) + 1^3$	**Step 1:** Operations in the parentheses are done first. In this case, perform the addition.
$5^2 - (20) + 1^3$ $25 - 20 + 1^3$ $25 - 20 + 1$	**Step 2:** Next, simplify the exponents. Be sure to solve the problem one step at a time.
$25 - 20 + 1$ $5 + 1$ 6	**Step 3:** Because there is no multiplication or division, skip to addition and subtraction. Subtraction is first, so you will subtract, then add.
	Answer: 6

Example 5 Simplify: $4[3^2 - (2^3 - 4) + 7]$

$4[3^2 - (2^3 - 4) + 7]$ $4[3^2 - (8 - 4) + 7]$ $4[3^2 - (4) + 7]$	**Step 1:** This problem has parentheses *and* brackets. Always do the operations in the innermost set first. Remember to follow the order of operations within the parentheses and brackets. Here, you will simplify the exponent, followed by subtraction.
$4[3^2 - (4) + 7]$ $4[9 - 4 + 7]$	**Step 2:** Perform operations within the brackets following the correct order. In this case, simplify the exponent first.
$4[9 - 4 + 7]$ $4[5 + 7]$ $4[12]$	**Step 3:** Subtraction is first, therefore you will subtract, then add.
$4[12]$ 48	**Step 4:** Finally, you will multiply.
	Answer: 48

Problem Set

Evaluate:

1. $7 - 3 + 5 + 6$
2. $8 \times 3 - 6 + 3$
3. $4 - (2 \times 1) + 6$
4. $(1 \times 3) + 6 + 2$
5. $(5 \times 2) - (4 \times 2)$
6. $5 - 3 + (2 \times 3)$
7. $8 - (3 \times 2) + 6$
8. $3 \times (3 \times 5) - 6$
9. $3 \times (7 + 2) - 6$
10. $2 + (3 \times 2) + 3$
11. $4 \times (32 \div 8) + 16$
12. $8 + (4 \times 2) + 2$
13. $(8 \times 4) \div (2 + 2)$
14. $14 - (7 \times 2) + 6$
15. $4[4 \times (3 - 2) + 6(2 - 0)]$

16.

Used Beepers

Black	Green	Red	Blue	Marble
$4	$6	$7	$9	$2

Jeanne buys 3 green beepers and 4 blue ones. How much does she spend?

17.

Barbara's Birthday Supplies

Hats	Balloons	Cake	Wrapping Paper	Noise Makers
$1	$4	$12	$8	$3

Joseph buys 7 hats and 4 balloons. How much does he spend?

18.

The Lunch Counter Cafe

Bologna	Ham	Roast Beef	Turkey	Salami	Steak
$2 per lb.	$6 per lb.	$5 per lb.	$4 per lb.	$1 per lb.	$9 per lb.

Stacy buys 3 lbs. of turkey and 2 lbs. of salami. How much money does she spend?

19.

Super CDs Sale

Rap	Rock	Classical	Techno	Opera
$8	$10	$2	$5	$3

Natoya buys 6 techno CDs and 4 rap CDs. How much does she spend?

20.

Serina's School Supplies

Pack of Pencils	Marker Set	Notebook	Calculator	Zipper Pencil case
$1	$2	$5	$7	$6

Anna buys 1 zipper pencil case and 3 notebooks. How much does she spend?

MPA-009: Solving One-Step Equations Using a Box

You have saved \$300 for your vacation, which will cost \$1,000. How much more money do you need?

In this situation, you do not know the amount of money you still need. This is called the unknown. One way to write this using numbers and symbols is $\$300 + \square = \$1{,}000$. Notice that the box takes the place of the unknown. Before solving this problem, you must understand the word "equation."

Equation	An equation is simply a mathematical sentence that has an equal sign. Examples: $2 + 2 = 4$ or $5 \times 6 = 30$

In an equation, what is on one side of the equal sign is equal to what is on the other side of the equal sign. Look at the following statement:

$$2 + 3 = 5$$

Of course, you know this is true. However, did you ever think about this? If you have 2 people and 3 more arrive, *it is the same as* having 5 people. In other words, even though "2 + 3" and "5" *look* different, they have the same value. The equal sign separates the left side of the equation from the right side. You can think of this visually as a balance:

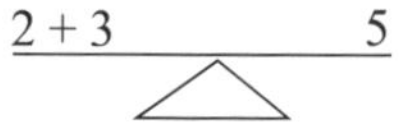

The value of the left side must balance the value of the right side. When you solve equations like the one above, $\$300 + \square = \$1{,}000$, you must find a value for the box that "balances" the equation. Solving the problem, you find that the box equals \$700. To check this, replace the box with \$700 and make sure that the left side of the equation has the same value as the right side:

$$\begin{aligned} \$300 + \square &= \$1{,}000 \\ \$300 + \$700 &= \$1{,}000 \\ \$1{,}000 &= \$1{,}000 \end{aligned}$$

This equation can also be described as a **one-step equation**, which means that there is only one operation (addition, subtraction, multiplication, or division) in the problem. Other examples of one-step equations include:

$$\begin{aligned} 6 \cdot \square &= 18 \text{ multiplication} \\ 23 - \square &= 15 \text{ subtraction} \\ 30 \div \square &= 30 \text{ division} \end{aligned}$$

To solve one-step equations:

1. Translate the problem into an equation using numbers and symbols. Use a box to represent the unknown.
2. Find a value for the box that balances the equation.
3. Check the problem.

Example 1 Adam has \$950 in the bank. He must make a \$500 car payment. How much will he have in the bank after the payment?

$\$500 + \square = \$950$

$\$500 + \$450 = \$950$

From reading the problem, you know that the \$500 car payment and the amount left in the account after the payment must equal \$950. The equation can be written as $\$500 + \square = \$950$. You must find a value for the box that will balance the equation.

Answer: He will have \$450 left in his account.

Check:

$$\$500 + \square = \$950$$
$$\$500 + \$450 = \$950$$
$$\$950 = \$950$$

Example 2 At registration, 150 people sign up to join a softball league. There are 15 people on each team. How many teams are there?

$15 \cdot \square = 150$

$15 \cdot 10 = 150$

You know that there is a total of 150 people and 15 people on *each* team. Therefore, 15 times some number must equal 150. You can write the equation as $15 \cdot \square = 150$, and find the value for the box that balances the equation.

Answer: 10 teams will be formed.

Check:

$$15 \cdot \square = 150$$
$$15 \cdot 10 = 150$$
$$150 = 150$$

Problem Set

Solve:

1. Solve: $\square + 2 = 7$

2. Solve: $\square + 13 = 15$

3. Solve: $64 \times \square = 256$

4. Solve: $\frac{\square}{3} = 4$

5. Solve: $\square - 11 = 9$

6. Michael's goal is to deliver 251 papers by the end of the day. If he has 30 left, how many papers has he already delivered?

7. Josh spent \$5.50 of his allowance on baseball cards. He has \$17.50 left. How much allowance did he receive?

8. Robbie ran 4 laps around the track for a total of 2,400 meters. How many meters are in each lap?

9. Is 70 a solution of the equation?

Write Yes or No.

$\square - 7 = 75$

10. Is 12 a solution of the equation?

Write Yes or No.

$14 + \square = 22$

11. Is 17 a solution of the equation?

Write Yes or No.

$39 = \square - 12$

12. Is 33 a solution of the equation?

Write Yes or No.

$4 + \square = 27$

13. Which of the numbers is a solution, 39 or 57?

$48 = \square + 9$

14. Which of the numbers is a solution, 100 or 60?

$\square \div 12 = 5$

15. Is 117 a solution of the equation?

Write Yes or No.

$\square - 7 = 100$

16. Is 0 a solution of the equation?

Write Yes or No.

$12 + \square = 12$

17. Is 4 a solution of the equation?

Write Yes or No.

$32 = 8 \times \square$

18. Solve for the missing value:

The sum of $\square$ and three is sixteen.

19. Solve for the missing value:

$\square$ decreased by five is five.

20. Solve for the missing value:

Eight increased by seven is $\square$.

MPA-010: Solving One-Step Equations of Whole Numbers Using Addition and Subtraction

The goal of all addition, subtraction, multiplication, and division operations is to solve equations. For most questions in algebra, solving an equation will produce the numerical answer you're looking for.

The value for which you solve—the unknown—will be represented by a letter called a **variable**.

Variable	A letter or symbol representing an unknown quantity.

A variable can represent any quantity, but a variable in an equation is usually limited to only one value. One-step equations are used to find that value. To discover the value of the variable, the variable must be isolated on one side of the equation. Unfortunately, variables are usually not isolated on one side of the equation. In such cases, you must perform **inverse operations**.

Inverse Operation	An operation that reverses, or "undoes," another operation.

Subtraction is the inverse of addition, and addition is the inverse of subtraction. Division is the inverse of multiplication and multiplication is the inverse of division. If a number is added to a variable, subtract it from both sides of the equation. If a number is subtracted from a variable, add it to both sides of the equation. Here are some examples:

Example 1 Solve: $x + 3 = 7$

$$x + 3 = 7$$
$$x + 3 - 3 = 7 - 3$$
$$x = 4$$

Step 1: Since 3 is added to the variable, x, subtract 3 from both sides of the equation, thereby isolating the variable x.

$$4 + 3 = 7$$
$$7 = 7$$

Step 2: Check your work by substituting 4 for x in the original equation.

Answer: $x = 4$

Example 2 Solve: $y + 2 = 5$

$$y + 2 = 5$$
$$y + 2 - 2 = 5 - 2$$
$$y = 3$$

Step 1: Since 2 is added to the variable, y, subtract 2 from both sides of the equation to isolate the variable.

$$3 + 2 = 5$$
$$5 = 5$$

Step 2: Check your work by substituting 3 for y in the original equation.

Answer: $y = 3$

Example 3 Solve: $c - 26 = 20$

$$c - 26 = 20$$
$$c - 26 + 26 = 20 + 26$$
$$c = 46$$

Step 1: Since 26 is subtracted from the variable, *c*, add 26 to both sides of the equation to isolate the variable.

$$46 - 26 = 20$$
$$20 = 20$$

Step 2: Check your work by substituting 46 for *c* in the original equation.

Answer: $c = 46$

Example 4 Solve: $d - 4 = 6$

$$d - 4 = 6$$
$$d - 4 + 4 = 6 + 4$$
$$d = 10$$

Step 1: Since 4 is subtracted from the variable, *d*, add 4 to both sides of the equation to isolate the variable.

$$10 - 4 = 6$$
$$6 = 6$$

Step 2: Check your work by substituting 10 for *d* in the original equation.

Answer: $d = 10$

Example 5 Solve: $a - 14 = 98$

$$a - 14 = 98$$
$$a - 14 + 14 = 98 + 14$$
$$a = 112$$

Step 1: Since 14 is subtracted from the variable, *a*, add 14 to both sides of the equation to isolate the variable.

$$112 - 14 = 98$$
$$98 = 98$$

Step 2: Check your work by substituting 112 for *a* in the original equation.

Answer: $a = 112$

Problem Set

1. Solve: $x + 9 = 63$

2. Solve: $f - 71 = 16$

3. A new scientific calculator is on sale. The sale price is \$22. It was marked down \$6. Let *p* represent the original price. How much was the calculator before the sale?

4. You have 13 right-handed people and enough left-handed people to total 25 people for your experiment. How many left-handed people do you need?

5. Solve: $d - 236 = 600$

6. Pappio's Pizza is mailing coupons to local residents. 685 of the total coupons they mailed were sent to adults. 185 coupons were sent to teens. How many coupons were sent to adults and teens?

7. Lana has 95 hair clips. All but 7 have butterflies. How many hair clips have butterflies?

8. Solve: $c - 1{,}536 = 3{,}135$

9. Solve: $d + 4{,}126 = 9{,}230$

10. Which equation would you use to solve this problem? Batrika wants to buy a car. She only has \$9,716. The car she wants is \$892 more than she has now. What is the cost of the car Batrika wants?

none of these	$892 - p = 9{,}716$
$9{,}716 - 892 = p$	$p - 892 = 9{,}716$

11. State the inverse operation of adding 33.

12. State the inverse operation of adding 77.

13. Which operation should be performed to solve the equation? $x + 22 = 57$

multiplication	subtraction
division	addition

14. Which operation should be performed to solve the equation? $65 = x + 42$

subtraction	multiplication
division	addition

15. What is the inverse operation of subtracting 544?

16. What is the inverse operation of subtracting 99?

17. What is the inverse operation of subtracting 301?

18. Which operation should be performed to solve the equation? $x - 99 = 21$

subtraction	division
addition	multiplication

19. Which operation should be performed to solve the equation? $w - 9 = 14$

multiplication	subtraction
addition	division

20. Which operation should be performed to solve the equation? $104 = x - 57$

multiplication	addition
subtraction	division

MPA-011: Solving One-Step Equations of Whole Numbers Using Multiplication and Division

In a previous lesson, you learned how to solve one-step equations using addition and subtraction. In this lesson, you will expand on what you learned by following the same rules to solve one-step equations of whole numbers using multiplication and division. Recall that division is the inverse of multiplication and multiplication is the inverse of division.

Solve One-Step Equations of Whole Numbers Using Division

When solving such equations, keep in mind the Division Property of Equality, which states that if you divide each side of an equation by the same non-zero number, then both sides of the equation will remain equal to each other. In other words, if you divide one side of an equation by a number, then you must also divide the other side of the equation by the same number.

Division Property of Equality	If you divide each side of an equation by the same non-zero number, then the two sides of the equation will remain equal.

The examples below demonstrate how to use the Division Property of Equality.

Example 1 Lonna works for a dog-walking company. Today, she has 48 dogs to walk. Because the dogs are so active, she can only walk 4 at a time. How many walks will she have to take today?

$4w = 48$

Step 1: Set up an equation. Use w to represent the number of walks. 4 (the number of dogs per walk) times w will equal the total number of dogs, 48.

$\frac{\cancel{4}w}{\cancel{4}} = \frac{48}{4}$

$w = 12$

Step 2: To isolate the variable, divide both sides of the equation by 4.

$4(12) = 48$

$48 = 48$

Step 3: Check your answer by substituting 12 for w in the original equation. Your answer is correct.

Answer: 12

Example 2 Roshonda is organizing the annual 12-kilometer race for her community. The runners will be in teams of 4, and each team member will run the same distance as other members of the team. How far will each team member run if 4 people are on each team and the race is 12 kilometers long?

$4y = 12$

Step 1: Set up the equation. Let y represent the distance each person will run. Multiply 4, the number of people in each team, by y. The product should be 12, the distance of the race.

$\frac{\cancel{4}y}{\cancel{4}} = \frac{12}{4}$

$y = 3$

Step 2: Solve by isolating the variable. Divide both sides of the equation by 4.

$4(3) = 12$

$12 = 12$

Step 3: Check your answer by substituting 3 kilometers for y in the original equation.

Answer: 3 km

Solve One-Step Equations of Whole Numbers Using Multiplication

The Multiplication Property of Equality is similar to the Division Property of Equality. The Multiplication Property states that in an equation involving whole numbers, if you multiply one side of an equation by a number, then you must perform the same action on the other side of the equation. In other words, to keep the two sides of an equation equal, you must multiply both sides of an equation by the same non-zero number.

Multiplication Property of Equality	If you multiply each side of an equation by the same non-zero number, then the two sides of the equation will remain equal.

The next example demonstrates how to use the Multiplication Property of Equality.

Example 3 The drama coach has asked you to help him arrange chairs in the auditorium for tomorrow night's performance of the new school play. If the auditorium can fit 27 rows of 9 chairs in it, how many chairs will you need?

$$\frac{c}{27} = 9$$

Step 1: Set up the equation. Let c equal the number of chairs. The total number of chairs (represented by c), divided by the number of rows (27), should equal the number of chairs per row (9).

$$\frac{c}{\cancel{27}} \cdot \cancel{27} = 9 \cdot 27$$
$$c = 243$$

Step 2: To solve, multiply both sides by 27.

$$\frac{243}{27} = 9$$
$$9 = 9$$

Step 3: Check your answer by substituting 243 for c in the original equation.

Answer: 243 chairs

Try solving the following problem using both properties of equality.

Example 4 Solve for x: $\frac{7x}{14} = 3$

$$\frac{7x}{\cancel{14}} \cdot \cancel{14} = 3 \cdot 14$$
$$7x = 42$$

Step 1: To isolate the variable, first multiply both sides of the equation by 14.

$$\frac{\cancel{7}x}{\cancel{7}} = \frac{42}{7}$$
$$x = 6$$

Step 2: Then, divide both sides of the equation by 7 using the Division Property of Equality to isolate the variable x.

$$\frac{7(6)}{14} = 3$$

Step 3: Check your answer by substituting 6 for x in the original equation.

$$\frac{42}{14} = 3$$

$$3 = 3$$

Answer: $x = 6$

Problem Set

1. Katrina is saving money for a new CD player. If she saves $13 a week and the CD player costs $169, how many weeks will it take her to save up enough money to buy the CD player?

2. Solve: $2m = 18$

3. Solve: $28 = 4m$

4. There are 1,425 seventh graders interested in playing in the summer soccer league. If there are 25 people to a team, how many teams would need to be created to accommodate all of the seventh graders?

5. The product of 3 and a number is 207. Find the number. Use the equation: $3n = 207$.

6. The quotient of a number and 4 is 36. Find the number. Use the equation: $\frac{n}{4} = 36$

7. The quotient of a number and 22 is 7. Find the number.

8. The product of a number and 22 is 198. Find the number.

9. The high school band is selling candy to raise money for new uniforms. The top 5 sellers were Nateka, Corrina, Kenneth, Guillaume, and Anthony. Guillaume sold three times as much as Corrina. If Guillaume sold 132 candy bars, how many did Corrina sell?

10. Solve: $19d = 76$

    95 4
    76 none of these

11. State the inverse operation of multiplying by 13.

12. What is the inverse operation of multiplying by 108?

13. What is the inverse operation of multiplying by 917?

14. Which operation should be performed to solve the equation? $3x = 87$

    subtract 3 add 3
    divide by 3 multiply by 3

15. Which operation should be performed to solve the equation? $102w = 204$

    multiply by 102 add 102
    subtract 102 divide by 102

16. Which operation should be performed to solve the equation? $97w = 582$

    subtract 97 add 97
    divide by 97 multiply by 97

17. Which operation should be performed to solve the equation?

$$\frac{x}{5} = 3$$

add 5   divide by 5

multiply by 5   divide by 3

18. Which operation solves the equation?

$$\frac{r}{2} = 250$$

multiply by 2   add 2

subtract 2   divide by 2

19. Which operation solves the equation?

$$4 = \frac{y}{3}$$

subtract 3   multiply by 3

add 3   divide by 3

20. What is the inverse operation of dividing by 44?

MPA-012: Solving One-Step Equations of Whole Numbers Using All Operations

In this lesson, you will solve one-step equations using the four basic operations together. This lesson is just a review, since you've had practice with both addition/subtraction and multiplication/division inverse operations. Remember:

- Subtraction is the inverse operation of addition.
- Addition is the inverse operation of subtraction.
- Division is the inverse operation of multiplication.
- Multiplication is the inverse operation of division.

Subtraction Property of Equality	Subtract the same number from both sides of the equation and the two sides of the equation remain equal. $a = c$ $a - b = c - b$
Addition Property of Equality	Add the same number from both sides of the equation and the two sides of the equation remain equal. $a = c$ $a + b = c + b$
Division Property of Equality	Divide both sides of the equation by the same non-zero number and the two sides of the equation remain equal. $ab = c$ $\frac{ab}{b} = \frac{c}{b}$ $a = \frac{c}{b}, b \neq 0$
Multiplication Property of Equality	Multiply both sides of the equation by the same number and the two sides of the equation remain equal. $a = b$ $ac = bc$

Example 1 Solve: $d + 93 = 109$

$$d + 93 = 109$$
$$d + 93 - 93 = 109 - 93$$
$$d = 16$$

Step 1: Because 93 is added to the variable, *d*, subtract 93 from both sides of the equation to isolate the variable *d*.

$$16 + 93 = 109$$
$$109 = 109$$

Step 2: Check your work by substituting 16 for *d* in the original equation.

Answer: After simplification, *d* equals 16.

Example 2 Solve: $p - 24 = 15$

$$p - 24 = 15$$
$$p - 24 + 24 = 15 + 24$$
$$p = 39$$

Step 1: Because 24 is subtracted from the variable, p, add 24 to both sides of the equation to isolate the variable p.

$$39 - 24 = 15$$
$$15 = 15$$

Step 2: Check your work by substituting 39 for p in the original equation.

Answer: $p = 39$

Example 3 Solve: $h \times 8 = 64$

$$h \times 8 = 64$$
$$\frac{h \times 8}{8} = \frac{64}{8}$$
$$h = 8$$

Step 1: Because the variable, h, is multiplied by 8, divide both sides of the equation by 8 to isolate the variable h.

$$8 \times 8 = 64$$
$$64 = 64$$

Step 2: Check your work by substituting 8 for h in the original equation.

Answer: $h = 8$

Example 4 Solve: $\frac{q}{5} = 4$

$$\frac{q}{5} = 4$$
$$\left(\frac{q}{5}\right)5 = (4)5$$
$$q = 20$$

Step 1: Because the variable, q, is divided by 5, multiply both sides of the equation by 5 to isolate the variable q.

$$\frac{20}{5} = 4$$
$$4 = 4$$

Step 2: Check your work by substituting 20 for q in the original equation.

Answer: $q = 20$

Problem Set

1. Solve: $z - 630 = 1{,}000$

2. Conga received a cell phone for his birthday. He cannot remember how many minutes he is allowed to use each month. When he called the cell phone company, they informed him that he already used 283 minutes, and has 117 minutes left on his plan this month. How many total minutes is Conga allowed to use his cell phone each month if he intends to stay within his plan?

3. Solve: $\frac{a}{19} = 1{,}064$

4. John and Lisa are planning a party. John decides that each person should be able to eat 10 pieces of candy at the party. Twenty-five people are invited to the party (including John and Lisa). How many pieces of candy must John and Lisa provide at the party so that each guest may have 10 pieces?

5. Solve for y: $25 + y = 628$

6. Teresa and Michael are saving money to buy a tent. Together, they have saved \$45.00. Michael has saved \$16.00. How much has Teresa saved?

7. Suzanne is an artist. After selling 50 paintings in one month, she had 20 left. How many paintings did Suzanne have at the beginning of the month?

8. Jeff is cooking pizza for a party to which 220 people are invited. Jeff figures that each person will eat 3 slices of pizza. How many slices of pizza must Jeff make for the party so that each guest will be able to eat 3 slices? Let p = the number of slices of pizza Jeff must make.

9. Solve for m: $\frac{m}{3} = 52$

10. Solve for b: $153 = 17b$

11. Is the given number a solution to the equation?
    True or False: $4 + y = 7;\ y = 3$

12. Is the given number a solution to the equation?
    True or False: $h - 13 = 66;\ h = 79$

13. Is the given number a solution to the equation?
    True or False: $78 - w = 34;\ w = 34$

14. Is the given number a solution to the equation?
    True or False: $12g = 48;\ g = 4$

15. Is the given number a solution to the equation?
    True or False: $\frac{75}{v} = 15;\ v = 5$

16. Is the given number a solution to the equation?
    True or False: $\frac{g}{12} = 5;\ g = 60$

17. Is the given number a solution to the equation?
    True or False: $y + 18 = 54;\ y = 26$

18. Is the given number a solution to the equation?
    True or False: $14 + p = 37;\ p = 33$

19. Is the given number a solution to the equation?
    True or False: $\frac{q}{4} = 96;\ q = 284$

20. Is the given number a solution to the equation?
    True or False: $\frac{426}{m} = 71;\ m = 6$

MPA-013: Using Powers and Exponents in Expressions

Think about this for a minute: $2 + 2 + 2 + 2 + 2 = 10$. Another way of writing this is $5 \times 2 = 10$. Multiplication is really just a shortcut for writing and doing addition when you are adding the same number over and over.

Now look at this: $2 \times 2 \times 2 \times 2 \times 2 = 32$. This time you are *multiplying* the 2 over and over. You have a shortcut for writing this as well.

Power

2^5 — Exponent

Base

Exponential notation	A shortcut for repeated multiplication of the same factor.
Exponent	A number representing how many times the base is multiplied.
Base	The number or factor that is being multiplied over and over. A number or variable that has an exponent attached to it.
Power	An expression with a base and an exponent, such as 2^5 (read as "2 to the 5th power").

Reading Exponential Notation

You read exponential notation as the base raised to some *power*. You would read the example above as "2 to the fifth power." There are two exceptions to the way these are read:

- If a number is multiplied twice, in other words, to the second power, you can read it as being *squared*. For example, 4^2 can be read as "four squared."
- If a number is multiplied three times, in other words, to the third power, you can read it as being *cubed*. For example, 7^3 can be read as "seven cubed."

Special exponents

The value of any number raised to the first power (where the exponent is 1) *is that number*. For example, $8^1 = 8$.

Any number (except for zero) raised to the zero power is equal to 1. For example, $8^0 = 1$. Here is why…

Look at this pattern:

$2 \times 2 \times 2 \times 2 \times 2 = 2^5 = 32$
$2 \times 2 \times 2 \times 2 = 2^4 = 16$
$2 \times 2 \times 2 = 2^3 = 8$
$2 \times 2 = 2^2 = 4$
$2 = 2^1 = 2$
$2^0 = ?$

Notice that as the exponent decreases by 1, the values 32, 16, 8, and so on, are being divided by 2 each time. If the pattern continues, the question mark must be 1. Try this for another example with a different base but the same exponents, and it will happen again!

Let's look at some examples:

Example 1 Write $5 \times 5 \times 5 \times 5 \times 5 \times 5 \times 5 \times 5 \times 5 \times 5$ using exponential notation.

Answer: 5^{10}

Example 2 What does 3^4 mean?

Answer: $3 \times 3 \times 3 \times 3$

Example 3 Simplify 6^3

Answer: $6 \times 6 \times 6 = 216$

Example 4 Simplify 99^0

Answer: 1

Example 5 Simplify 13^1

Answer: 13

Example 6 How do you read the following problems?

a) 10^2

b) 2^3

c) 1^{11}

Answer: a) "ten squared"

Answer: b) "two cubed"

Answer: c) "one to the eleventh power"

Variable	A variable is a symbol, usually a letter, used to represent a number in a mathematical expression.

Example 7 Write this power as a factor: p^4

$p \times p \times p \times p$	**Step 1:** Since p is the base and 4 is the exponent, write the power as repeated multiplication.
	Answer: $p^4 = p \times p \times p \times p$

Example 8 Evaluate: n^6 for $n = 1$

1^6	**Step 1:** Substitute 1 for n in the expression.
$1^6 = 1 \times 1 \times 1 \times 1 \times 1 \times 1$ $= 1$	**Step 2:** Write the power as repeated multiplication and multiply.
	Answer: 1

Problem Set

1. Simplify: 5^3

2. Evaluate: c^3 for $c = 15$

3. Evaluate: b^2 for $b = 23$

4. Evaluate: m^3 for $m = 40$

5. Simplify: 14^2

6. Identify the base of the number 4^5.

7. Identify the exponent of the number 101^7.

8. Write the number using exponents: (11)(11)(11)(11)

9. Write the number using exponents: (13)(13)(13)(13)(13)

10. What would the exponent of the number (107)(107)(107)(107)(107)(107) be if the number was written in exponential form?

11. Which number is the exponent for 516^{615}?

12. Which number is the exponent for 987^{789}?

13. Which number is the base of the number 876^{678}?

14. Which number is the base of the number 25^{89}?

15. Which number is the base of the number 516^{615}?

16. Simplify the number $9{,}871^1$

17. Simplify the number: $6{,}543^0$

18. Write the number in expanded form and simplify: 5^3

19. Evaluate: y^5 for $y = 4$

20. Evaluate: y^4 for $y = 6$

MPA-014: Evaluating Expressions for Given Variables

You have already learned that letters are sometimes used to replace numbers you have not found yet; these letters are called *variables*.

For instance, your uncle tells you that he will give you $10 a week in addition to any money you receive from your job. Since the amount you receive from your job changes every week, use "x" to represent that amount. The $10 is called a **constant**, because that amount does not change every week.

Using numbers and symbols, you can write $x + 10$ to show the total amount of money you will receive. This is called an **expression**. An expression contains numbers, variables, or a combination of both, and at least one operation (addition, subtraction, multiplication, and/or division).

Numerical expression	A mathematical expression that contains only numerals and at least one operation symbol. Examples: 2 + 3, 9 – 2, 3 x 4, 10 ÷ 2
Algebraic expression	A mathematical expression formed by a combination of variables, numbers, and at least one operation. Examples: $a + 6$, $x + y - z$, xy

How much money did you make according to the expression $x + 10$? You won't know until you determine the amount you receive from your job. If you make $200, you can substitute this amount into the expression, which becomes $200 + $10 = $210. If you make $245, the expression then becomes $245 + $10 = $255. As you can see, the expression has no meaning until you know the value of the variable.

When you simplify expressions, remember the sentence that helps you with the order of operations: "Please Excuse My Dear Aunt Sally" (PEMDAS).

P	Simplify inside parentheses ( ), brackets [ ], and braces {}.
E	Simplify any exponents.
M D	Perform all multiplication and division from left to right as they occur.
A S	Perform all addition and subtraction from left to right as they occur.

Example 1 Simplify the numerical expression $(3 + 4)^2 \div 7 \times 2 - 6$.

$(7)^2 \div 7 \times 2 - 6$	**Step 1:** The "P" means perform operations in the parentheses first. If there is more than one operation, you must use the same order of operations *within* the parentheses. In this case, add 3 + 4 and replace it with 7.
$49 \div 7 \times 2 - 6$	**Step 2:** The "E" means exponents. This problem has only one term with an exponent, 7^2. Replace it with the value 49.
$7 \times 2 - 6$	**Step 3:** The "M" means multiplication and the "D" means division. Perform the operation that comes first; if division comes first in the problem, perform the division, then perform the next multiplication or division operation. In this problem, you will divide 49 by 7, and replace it with 7. Then multiply 7 by 2, and replace it with 14.
$14 - 6$	**Step 4:** The "A" means addition and the "S" means subtraction. In this problem, subtraction is the only operation left. Subtract 6 from 14.
8	**Answer:** When the expression is simplified, its value is 8.

What happens when you have an algebraic expression such as $x^2 - (yz + 6) \div 2$? When you were calculating the amount of money you made each week, you had to know the amount you made from your job (the value of x). This example is similar; you must know the value of each variable before you can simplify.

Example 2 Evaluate the algebraic expression $x^2 - (yz + 6) \div 2$ if $x = 5$, $y = 2$, and $z = 4$.

$(5)^2 - (2(4) + 6) \div 2$	**Step 1:** Substitute the variables with their actual values. *Note: Put parentheses around the values that you substitute.*
$(5)^2 - (8 + 6) \div 2$	**Step 2:** Now use the order of operations. Begin with parentheses. The first set of parentheses has no operation, so move to the next set of parentheses. Here you have $((2)(4) + 6)$. Multiply first, then add.
$(5)^2 - (14) \div 2$	**Step 3:** Next, you will do exponents. The number being squared is the 5.
$25 - (14) \div 2$	**Step 4:** Look for any multiplication or division. Now divide.
$25 - 7 = 18$	**Step 5:** Finally, you will subtract.

Answer: $x^2 - (yz + 6) \div 2 = 18$ if $x = 5$, $y = 2$, and $z = 4$.

Example 3 Evaluate $5x - 2yz$ if $x = 7$, $y = 2$, and $z = 6$.

$5(7) - 2(2)(6)$	**Step 1:** Substitute the variables with the values given.
$35 - 2(2)(6)$ $35 - 4(6)$	**Step 2:** Follow the order of operations. There are no operations to perform *within* the parentheses, and there are no exponents, so you will perform multiplication first.
$35 - 24 = 11$	**Step 3:** Then, subtract.

Answer: $5x - 2yz = 11$ if $x = 7$, $y = 2$, and $z = 6$.

Example 4 Evaluate $\dfrac{b + d + 3}{c}$ if $b = 15$, $c = 9$, and $d = 18$.

Note: Pretend that the numerator and denominator of a fraction have parentheses around them. In other words, even though parentheses do not exist, simplify the numerator first, followed by the denominator, then divide.

$\dfrac{(15) + (18) + 3}{(9)}$	**Step 1:** Substitute the variables with the values given. Then, follow the order of operations.
$\dfrac{33 + 3}{9}$	**Step 2:** Simplify the numerator.
$\dfrac{36}{9} = 4$	**Step 3:** Now divide.

Answer: $\dfrac{b + d + 3}{c} = 4$ if $b = 15$, $c = 9$, and $d = 18$.

Example 5 Evaluate $v^2 - qr^3$ if $v = 3$, $q = 2$, and $r = 1$.

$(3)^2 - (2)(1)^3$	**Step 1:** Substitute the variables with the values given. Follow the order of operations.
$9 - (2)(1)^3$	**Step 2:** There are no operations to perform *within* the parentheses, so move to the exponents. Note that only the 1 is being cubed.
$9 - (2)(1)$	**Step 3:** Next, perform multiplication.
$9 - 2 = 7$	**Step 4:** Finally, subtract.

Answer: $v^2 - qr^3 = 7$ if $v = 3$, $q = 2$, and $r = 1$.

Problem Set

Evaluate:

1. $x + 7$ if $x = 4$

2. $x + 5$ if $x = 2$

3. $(a)3$ if $a = 2$

4. $(a) \div 4$ if $a = 16$

5. $7x^2 + 7$ if $x = 2$

6. $5(g + h)^2$ if $g = 4$ and $h = 6$

7. $3s \div g^2$ if $s = 8$ and $g = 2$

8. $(a + g)^2 \div (10 + c)$ if $a = 3$, $c = 10$, and $g = 7$

9. $3[s + (a - c)^2]$ if $a = 3$, $c = 2$, and $s = 4$

10. $6[m + (n - p)^2]$ if $m = 4$, $n = 7$, and $p = 5$

Simplify:

11. $9 \times 5 - 6$

12. $2^3 - 6 + 5 \times 3$

13. $6 \div 2 - 1 + 7 \times 2$

14. $5 + 6 - 18 \div 9 + 3^2$

15. $6^2 \div 4 + 6 + 3^2$

16. $6^2 - 4^2 + 7$

17. $25 \div 5 + 6 - 3$

18. $7^2 - 15 \div 3$

19. $4 \times 7 + 9^2$

20. $4 \times 18 - 115^0$

MPA-015: Identifying the Place Value of Decimals Through Thousandths

A **decimal** is another way of writing a fraction. Writing monetary amounts is a common way of using decimals. For example, in $6.39, 39 cents is the decimal because it is part of a dollar. It is $\frac{39}{100}$ of a dollar.

You should be familiar with the term **place value** of whole numbers. As a quick review, the whole number 4,517 and the place value of each digit follow:

Thousands	Hundreds	Tens	Ones
4	5	1	7

Thousands, hundreds, tens, and ones are examples of place values for whole numbers that come *before* a **decimal point**. However, the term place value also applies *after* a decimal point. When you move to the right of a decimal, the value decreases by a power of ten. Consider the number 4,517.368. Notice that a decimal point and three digits have been added to the number from the previous example.

Thousands	Hundreds	Tens	Ones		Tenths	Hundredths	Thousandths
4	5	1	7	.	3	6	8

One place after the decimal point stands for tenths: $0.3 = \frac{3}{10} = 3$ tenths

Two places after the decimal point stands for hundredths: 0.36 = 36 hundredths

Three places after the decimal point stands for thousandths: 0.368 = 368 thousandths

Example 1 Read these decimals.

0.4	four tenths
0.217	two hundred seventeen thousandths
0.56	fifty-six hundredths
0.003	three thousandths
0.021	twenty-one thousandths

Note: All place values after the decimal point end with "ths" (tenths, hundredths, thousandths).

A mixed decimal contains a whole number and a decimal. The decimal number 13.52 is a mixed decimal. The decimal point separates the whole number (13) from the decimal (0.52). The decimal is read as **"and"** or "**point**." For example, the number 13.52 is read "thirteen and fifty-two hundredths" or "thirteen point fifty-two."

Let's practice identifying numbers using your knowledge of place value following a decimal point—tenths, hundredths, and thousandths.

Example 2 Read these mixed decimals.

7.3	Seven and three tenths
10.02	Ten and two hundredths
914.006	Nine hundred fourteen and six thousandths
85.112	Eighty-five and one hundred twelve thousandths

It is important to be able to write numbers in decimal form. Read the words carefully, remembering that **"and"** means the decimal point, and **"ths"** (tenths, hundredths, thousandths) means that the number includes a fractional or decimal part.

Example 3 Write these words as decimals.

Three hundred seven and nine thousandths	307.009
Fifteen and sixty-two hundredths	15.62
Seven and one tenth	7.1
Six thousand two and four hundredths	6,002.04

Problem Set

1. Express the following number as a decimal: Twenty-eight hundredths.

2. Express the following number as a decimal: Nine tenths

3. Express the following number as a decimal: Eighty-five thousandths

4. Express the following number as a decimal: Five and nine hundredths.

5. Write the word form for the following number: 0.38

6. Write the word form for the following number: 101.01

7. Write the following number in decimal form: $58\frac{2}{10}$

8. Write the following number in decimal form: $\frac{48}{100}$

9. Write the following number in decimal form: $\frac{892}{1,000}$

10. Write the following number in decimal form: $\frac{1,000}{1,000}$

11. What is the place value of the digit 7 in the following number? 4.794

12. What is the place value of the digit 1 in the following number? 3.512

13. What is the place value of the digit 2 in the following number? 3.512

14. What is the place value of the digit 2 in the following number? 5.132

15. Write the following number in decimal form: $\frac{11}{100}$

16. Write the following number in decimal form: $\frac{532}{1,000}$

17. Express seventy-five thousandths as a decimal.

18. Express eight and seven thousandths as a decimal.

19. Express twelve and eighty-five thousandths as a decimal.

20. Express one and thirty-five thousandths as a decimal.

MPA-016: Comparing and Ordering Decimals

The name of a decimal depends upon the number of digits in the decimal. For example, if there is only one digit after the decimal (0.4), its name is **tenths**. If there are two digits after the decimal (0.12), its name is **hundredths**, and a decimal number with three digits after the decimal (0.267) has the name **thousandths**.

Several methods can be used to compare and order decimals. The two ways that will be discussed in this lesson are:

Number Line	A number line can be used to compare small numbers. Remember that numbers on the right of the number line are larger than numbers on the left.
Place Value Chart	A place value chart can be used to compare decimal numbers from least to greatest or greatest to least.

Let's look at some examples using each of these methods for comparing and ordering decimals. We'll use the > (greater than), < (less than), and = (equal to) symbols to compare decimals.

Example 1 Compare 3.6 and 3.8 using the number line.

Step 1: Graph each decimal on a number line. Since both numbers have a 3 in the ones place we know the numbers will fall between 3 and 4 on the number line.

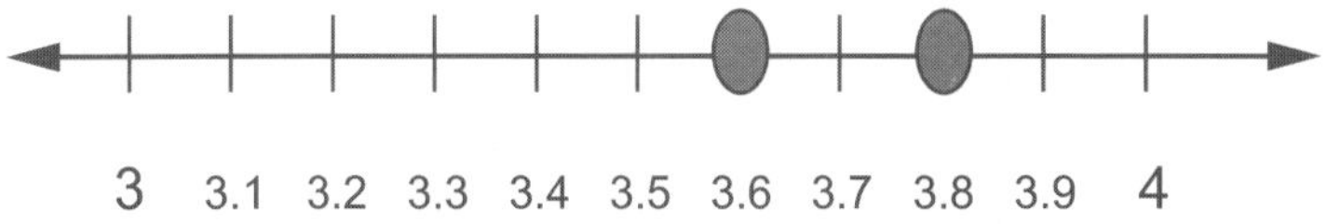

Answer: Since 3.8 is to the right of 3.6, 3.8 is greater than 3.6. Therefore 3.6 < 3.8.

Example 2 Compare 0.127 and 0.326 using a place value chart.

Ones	.	Tenths	Hundredths	Thousandths
0	.	1	2	7
0	.	3	2	6

Step 1: Place the numbers in a place value chart.

Step 2: Compare the tenths column. The 3 is greater than the 1.

Answer: 0.127 < 0.326

Example 3 A centigram is 0.01 of a gram, and a decigram is 0.1 of a gram. Which is larger?

Ones	.	Tenths	Hundredths	Thousandths
0	.	0	1	
0	.	1	0	

Step 1: Place the numbers in a place value chart.

0.01 < 0.10

Step 2: Compare the tenths column. The 1 is greater than the 0.

Answer: A decigram is larger than a centigram.

Example 4 Order these decimals from least to greatest: 1.062, 1.103, 1.115, 1.045

Ones	.	Tenths	Hundredths	Thousandths
1	.	0	6	2
1	.	1	0	3
1	.	1	1	5
1	.	0	4	5

Step 1: Place the numbers in a place value chart.

Ones	.	Tenths	Hundredths	Thousandths
1	.	0	6	2
1	.	0	4	5

Step 2: Look at the ones column. All the numbers have a one in the ones column. Next, look at the tenths column. Two numbers have a zero and two numbers have a one. Look at only the numbers that have a zero in the tenths column, 1.045 and 1.062. In the hundredths place one number has a 6 and one has a 4. Since 6 is greater than the 4, 1.045 < 1.062.

Ones	.	Tenths	Hundredths	Thousandths
1	.	1	0	3
1	.	1	1	5

Step 3: Look at the two remaining numbers that have a one in the tenths column, 1.103 and 1.115. In the hundredths column there is a 1 and 0. Therefore, $1.103 < 1.115$.

Answer: Place all of the numbers in order from least to greatest:
1.045, 1.062, 1.103, 1.115

Example 5 Order the list of decimal numbers from greatest to least: 0.272, 0.7, 0.27, 0.702, 0.027

Ones	.	Tenths	Hundredths	Thousandths
0	.	2	7	2
0	.	7	0	0
0	.	2	7	0
0	.	7	0	2
0	.	0	2	7

Step 1: Place the numbers in a place value chart. Make each number the same number of decimal places by filling in any blank decimal places with a zero.

Step 2: Look at the tenths column. Two numbers have a 2, two numbers have a 7, and one number has a 0 in the tenths column. Therefore, the smallest number is 0.027.

Step 3: Look at the two numbers that have a 2 in the tenths column, 0.272 and 0.270. Both numbers also have a 7 in the hundredths column. Look at the thousandths column. One number has a 2 and the other has a 0. Therefore, $0.270 < 0.272$.

Step 4: Now, look at the two numbers that have a 7 in the tenths column, 0.700 and 0.702. Both numbers also have a 0 in the hundredths column so look at the thousandths column. One number has a 0 and the other has a 2. Therefore, $0.700 < 0.702$.

Answer: Ordering the numbers from greatest to least:
0.702, 0.7, 0.272, 0.27, 0.027

Problem Set

Use >, <, or = to complete the statements:

1. 8.4____4.8 **2.** 7 ___ 7.00 **3.** 0.05 ___ 0.5 **4.** 912 ___ 912.4

In questions 5-10, which set of decimals is ordered from least to greatest?

5. (0.263; 0.262; 0.2632; 0.26; 0.2)
(0.263; 0.2; 0.262; 0.26; 0.2632)
(0.26; 0.2; 0.263; 0.2632; 0.262)
(0.2; 0.26; 0.262; 0.263; 0.2632)

6. ($66.33; $36.36; $36.63; $33.66; $63.36)
($33.66; $36.36; $36.63; $63.36; $66.33)
($66.33; $63.36; $36.63; $36.36; $33.66)
($36.36; $33.66; $66.33; $63.36; $36.63)
($63.36; $36.63; $36.36; $66.33; $33.66)

7. (0.473; 0.472; 0.4732; 0.47; 0.4)
(0.473; 0.4; 0.472; 0.47; 0.4732)
(0.4; 0.47; 0.472; 0.473; 0.4732)
(0.47; 0.4; 0.473; 0.4732; 0.472)

8. ($4.30; $6.72; $14.22; $14.30; $24.40)
($14.22; $4.30; $24.40; $14.30; $6.72)
($24.40; $14.30; $14.22; $6.72; $4.30)
($6.72; $4.30; $14.22; $14.30; $24.40)
($4.30; $24.40; $14.30; $14.22; $6.72)

9. (6.26262; 62.6262; 626.262; 6,262.62; 62,626.2)
(62.6262; 6.26262; 626.262; 6,262.62; 62,626.2)
(6.26262; 626.262; 62.6262; 6,262.62; 62,626.2)
(62,626.2; 6,262.62; 626.262; 62.6262; 6.26262)

10. (0.074; 0.07421; 0.0744; 0.0724; 0.071)
(0.074; 0.0744; 0.0724; 0.071; 0.07421)
(0.07421; 0.071; 0.0724; 0.0744; 0.074)
(0.071; 0.0724; 0.074; 0.07421; 0.0744)

Solve:

11. Janet ran 100 meters in 14.2332 seconds, while Kate ran the same distance in 14.2337 seconds. If they were racing, who would win the race?

12. Select the symbol that makes the statement true.
282.6____282.666

> < =

13. Select the symbol that makes the statement true.
36____36.00

> < =

14. Select the symbol that makes the statement true.
0.11____0.111

> < =

15. Which set of decimals is ordered from least to greatest?

(0.24; 0; 0.21; 0.401; 0.241)
(0.401; 0.241; 0.24; 0.21; 0)
(0; 0.21; 0.24; 0.241; 0.401)
(0.401; 0; 0.21; 0.241; 0.24)

16. Which set of decimals is ordered from least to greatest?

($64.75; $64; $64.05; $64.95; $64.15)
($64.05; $64.75; $64.95; $64.15; $64)
($64.95; $64.75; $64.15; $64.05; $64)
($64; $64.05; $64.15; $64.75; $64.95)

17. Susan, Shaquita, and Carlos have just received their allowances from their parents. Carlos received $10.20, Shaquita received $12.00, and Susan received $7.21. According to their allowances, who received the most money?

18. Choose the largest number:

65 hundredths 65 thousandths
6 tenths 54 thousandths

19. Choose the smallest number:

6 tenths 654 thousandths
65 hundredths 6 hundredths

20. Choose the smallest number:

65 hundredths 54 thousandths
65 thousandths 6 tenths

MPA-017: Rounding Decimals and Estimating Computations Using Decimals

Some decimal numbers can be quite long in length like 0.13333333… or 0.6666666… Both of these decimals are called **repeating decimals** because they don't terminate. We can rewrite the decimals using a shortcut in which a bar is placed above the number or numbers that continue to repeat. For example, $0.1\overline{3}$ and $0.\overline{6}$. But sometimes we can also use rounding or estimation when we don't need an exact answer. One estimation method would be to **truncate** the numbers.

Truncate	To cut off at a certain place value position, eliminating the digits that follow.

If you wanted to round the number to the thousandths place you would just drop off all of the numbers after the thousandths place. Therefore 0.133333333… would become 0.133 and 0.666666666… would become 0.666. However, the preferable method would be **rounding** because this will give you a better **approximation** for the answer.

Rounding	Step 1: Find the place you are rounding to. Step 2: Look at the digit to its right. If the digit is less than five, round down. If the digit is five or more, round up. Step 3: All digits to the right of the place you are rounding become zeros, which you may drop.

Approximation	An estimate, a guess, or a mathematical result that is approximately correct or close enough to exactness for a particular purpose. The symbol " ≈ " is used instead of "=".

Using the rounding method to the nearest thousandths place on 0.133333333… and 0.666666666… you would first need to look at the ten thousandths place to determine whether you need to round up or round down. For 0.13<u>3</u>3... there is a 3 in the ten thousandths place, which is less than five, so you round down to 0.133. For 0.66<u>6</u>6... there is a 6 in the ten thousandths place, which is greater than five, so you round the thousandths place up to a 7 resulting in 0.667. Notice this is a different value than the truncating method.

Example 1 Round 8.547 to the nearest tenths place.

Ones	.	Tenths	Hundredths	Thousandths
8	.	5	4	7

Step 1: Find the place value you are rounding to. The 5 is in the tenths place.

8.<u>5</u>47 ↑

Step 2: Look at the digit to the right, the hundredths place. There is a 4 which is less than 5, so round down.

8.500

Step 3: All digits to the right of the tenths place become zero, which may be dropped.

Answer: $8.547 \approx 8.5$

Example 2 Round each number to the nearest whole number to estimate the sum: $13.6 + 57.2 + 8.9$

$13.6 \rightarrow 14$ $57.2 \rightarrow 57$ $8.9 \rightarrow 9$	Step 1: Round each number to the nearest whole number. Look to the tenths place to determine if the number should be rounded up or down.
$13.6 \rightarrow 14$ $57.2 \rightarrow 57$ $+8.9 \rightarrow +9$ 80	Step 2: Add the rounded values.
	Answer: $13.6 + 57.2 + 8.9 \approx 80$

Example 3 Round each number to nearest whole number to estimate the product: 5.4×8.6

$5.4 \rightarrow 5$ $8.6 \rightarrow 9$	Step 1: Round each number to the nearest whole number. Look to the tenths place to determine if the number should be rounded up or down.
$5 \times 9 = 45$	Step 2: Multiply the rounded values.
	Answer: $5.4 \times 8.6 \approx 45$

Another estimation technique you can use is called **clustering**.

Clustering	When numbers in a set have approximately the same value, or "cluster" around the same number, that central number can be used to make an estimate.

Example 4 Use clustering to the nearest multiple of ten to estimate: $11.24 + 9.89 + 10.43 + 10.04 + 8.67$

$11.24 + 9.89 + 10.43 + 10.04 + 8.67$	Step 1: Determine what multiple of ten all of the numbers are clustered around. All of these values are clustered around 10.
$10 + 10 + 10 + 10 + 10 \approx 50$	Step 2: Change each value to 10 and add.
	Answer: $11.24 + 9.89 + 10.43 + 10.04 + 8.67 \approx 50$

Example 5 Paul has $64.76 to buy DVDs that cost $12.95 each. Rounding to the nearest dollar, about how many DVDs can he buy?

Paul has $64.76 ⟶ $65 Each DVD costs $12.95 ⟶ $13	Step 1: Round each value to the nearest dollar. Look at the tenths place to decide whether to round up or down.
$13\overline{)65}$ = 5	Step 2: Divide the two rounded values.

Answer: Paul will be able to purchase 5 DVDs.

Problem Set

1. Round 9.678 to the nearest hundredth.

2. Round 3.1737 to the nearest thousandth.

3. Round 6.2915 to the nearest thousandth.

4. Estimate to the nearest whole number:
3.79 + 18.64 + 82.11

5. Use clustering to the nearest multiple of ten to estimate: 21.29 + 20.43 + 18.73 + 17.61 + 21.16

6. Round 7.318 to the nearest tenth.

7. Round 26.178 to the nearest tenth.

8. Round 6.9316 to the nearest thousandth.

9. Round 19.67341 to the nearest thousandth.

10. Round to the nearest whole number to estimate the sum:
37.94 + 29.67 + 89.19

11. 92.19783 rounded to the nearest thousandth is 92.198. True or false?

12. 91.01112 rounded to the nearest tenth is 91.0. True or false?

13. 45.6789 rounded to the nearest hundredth is 45.678. True or false?

14. A reasonable estimate of 2.7 + 3.8 + 11.3 is 18. True or false?

15. Using clustering to the nearest multiple of ten, the estimated sum of 38.1 + 42.6 + 39.6 + 40.2 is 160. Is this statement True or False?

16. Bernice has $50 to buy CDs. Each CD cost $9.75. Rounding to the nearest dollar, about how many CDs can she buy?

17. Sally has $80.29 to buy CDs that cost $10.43 each. Rounding to the nearest dollar, about how many CDs can she buy?

18. Cory is making renovations to his home. He measures the length of a wall to be 12.2791 ft. When buying wallpaper for the wall, he rounds the length to the nearest thousandth. What is the length in feet of the wall that the paper will cover?

19. Romy works as a gift wrapper during the holidays. On average, she wraps 4.3 gifts per hour. Rounding to the nearest whole number, about how many gifts would she wrap in 2.8 hours?

20. Bernard the lumberjack chops down an average of 9.3 trees per hour. Rounding to the nearest whole number, about how many trees would he chop down in 6.7 hours?

MPA-018: Adding and Subtracting Decimals

When adding decimal numbers, it is important to know that only tenths are added to tenths, hundredths to hundredths, thousandths to thousandths, and so on. Therefore, it is necessary to **keep the decimal points in all the addends in a column**. When the addends do not have the same number of digits behind the decimal point, it is helpful to add zeros so that all the addends have the same number of digits after the decimal point. Once you have set up the problem correctly, you add as usual, remembering to regroup when necessary. The final rule to remember when adding decimals is that **the decimal point in the sum must be written directly below the decimal points in the addends**.

Example 1 Add 6.32, 1.4, and 7.835.

$$\begin{array}{r} 6.32 \\ 1.4 \\ +7.835 \\ \hline \end{array}$$

Step 1: Write the numbers vertically with decimals directly under each other.

$$\begin{array}{r} 6.320 \\ 1.400 \\ +7.835 \\ \hline \end{array}$$

Step 2: Add zeroes so that all the addends have the same number of digits behind the decimal point.

$$\begin{array}{r} {}^{1}6.320 \\ 1.400 \\ +7.835 \\ \hline 15.555 \end{array}$$

Step 3: Add as usual. Place the decimal point in the sum directly below the decimal points in the addends.

Answer: 15.555

Example 2 Add 6.5 and 15.87.

$$\begin{array}{r} 6.5 \\ +15.87 \\ \hline \end{array}$$

Step 1: Write numbers vertically.

$$\begin{array}{r} 6.50 \\ +15.87 \\ \hline \end{array}$$

Step 2: Add zeroes.

$$\begin{array}{r} {}^{1\ 1}6.50 \\ +15.87 \\ \hline 22.37 \end{array}$$

Step 3: Add as usual. Place decimal point in the sum.

Answer: 22.37

The alignment of the problem is also important when subtracting decimals. Write the problem with the decimal points in the **minuend** and **subtrahend** directly under each other. Again, it is better to add zeroes so that both the minuend and subtrahend have the same number of digits behind the decimal point. Then, subtract as you normally would, regrouping whenever necessary. The last thing you must remember is to place the decimal point in the answer (**difference**) directly below the other decimal points in the subtraction problem.

Example 3 Subtract 6.5473 from 7.3

$$\begin{array}{r} 7.3 \\ -\ 6.5473 \\ \hline \end{array}$$

Step 1: Write the numbers vertically with the decimal points in a column.

$$\begin{array}{r} 7.3000 \\ -\ 6.5473 \\ \hline \end{array}$$

Step 2: Add zeroes so that both numbers have the same number of digits behind the decimal point.

$$\begin{array}{r} {\scriptstyle 6\ 2\,9\,9\,1} \\ \not{7}.\not{3}\not{0}\not{0}0 \\ -\ 6.5473 \\ \hline 0.7527 \end{array}$$

Step 3: Subtract using regrouping. Place the decimal point in the answer directly below the other decimal points.

Answer: 0.7527

Example 4 Subtract 115.684 from 525.6

$$\begin{array}{r} 525.6 \\ -115.684 \\ \hline \end{array}$$

Step 1: Write numbers vertically.

$$\begin{array}{r} 525.600 \\ -115.684 \\ \hline \end{array}$$

Step 2: Add zeroes.

$$\begin{array}{r} {\scriptstyle 14\ 5\,9\,1} \\ 5\not{2}\not{5}.\not{6}\not{0}0 \\ -115.684 \\ \hline 409.916 \end{array}$$

Step 3: Subtract using regrouping. Place decimal point in the answer.

Answer: 409.916

It sometimes becomes confusing when working with word problems involving money. This confusion lies in setting up the problem. You must remember to include a decimal point when setting up the problem even when the amount is expressed differently, such as:

A five dollar bill = \$ 5.00

Fifteen cents = \$ 0.15

A twenty dollar bill = \$ 20.00

Example 5 Anne bought a bracelet for \$14.95. How much change would she get if she paid with a twenty dollar bill?

$$\begin{array}{r} \$20.00 \\ -14.95 \\ \hline \end{array}$$

Step 1: Write the amounts, placing the decimals directly under each other.

$$\begin{array}{r} \overset{1\,9\ 9\,1}{\$\not{2}\not{0}.\not{0}0} \\ -14.95 \\ \hline \$5.05 \end{array}$$

Step 2: Subtract, using regrouping. Place decimal point and dollar sign in the answer.

Answer: $5.05

Example 6 Billy ordered a hamburger ($2.49), fries ($1.26) a drink ($0.85) and an ice cream cup ($1.09). How much did Billy's order add up to, excluding tax?

$$\begin{array}{r} \$2.49 \\ 1.26 \\ .85 \\ +1.09 \\ \hline \end{array}$$

Step 1: Write all the amounts in a column using decimals.

$$\begin{array}{r} \overset{1\ 2}{\$2.49} \\ 1.26 \\ .85 \\ +1.09 \\ \hline \$5.69 \end{array}$$

Step 2: Add as usual. Place decimal point and dollar sign in the answer.

Answer: $5.69

Problem Set

1. Find the sum: 34.124 + 76.79
2. Find the sum: 543 + 17.05
3. Find the sum: 1.53 + 4.25 + 10.64
4. Find the difference: 15.06 – 1.83
5. Find the difference: 28 – 17.21
6. Find the difference: 29.004 – 8
7. Find the difference: 19.702 – 14.191
8. Find the sum: 13.2 + 5.8
9. Find the sum: 17.6 + 2.4
10. Find the sum: 21,208.8 + 1,610.1
11. Find the difference: 51.99 – 5.99
12. Find the sum or difference as indicated: 14.08 – 10.83
13. Find the sum or difference as indicated: 0.670 + 1.253
14. Find the sum or difference as indicated: 278.5 + 3.2
15. Katherine is buying accessories for her prom gown. The red leather shoes are on sale for $129.65. A red leather purse is $63.99 and hosiery is only $3.49. What is the total cost for her accessories?
16. Find the sum or difference as indicated: 18.46 + 2.54
17. Roberto swam 430.53 meters and Quinn swam 389.6 meters. How many more meters did Roberto swim than Quinn?
18. Kyle brought 3.5 gallons of punch and Jacobe brought 2.75 gallons to the party. How much punch did the 2 boys bring to the party?
19. Keith read 74.25 pages of a book. The book has 125 pages. How many more pages must Keith read to complete the book?
20. Angie needs 15 pounds of sugar to make a large batch of her delicious pralines. She has 10.375 pounds. How much more sugar does she need?

MPA-019: Multiplying Decimals

In a multiplication problem you multiply two **factors**, the **multiplicand** and the **multiplier**. You multiply by one number at a time to get the **partial products**. Next, you add the partial products to get the final answer or **product**. For example:

factors →	982	multiplicand
↘	× 56	multiplier
	5,892	partial product
	+49,100	partial product
	54,992	product

When multiplying decimals, multiply as usual. There is one more consideration to make, though, since a decimal or decimals are involved. To know where the decimal point goes in the answer, you first count the number of decimal places in both factors. That total is the number of places that must be marked off in the answer. An illustration with decimals would be:

factors →	9.82	2 places
↘	×5.6	+ 1 place
	5892	3 places in the answer
	+49100	
	54.992	3 places marked off in the answer

Let's do a few examples to see the different kinds of decimal problems you might be asked to multiply.

Example 1

Multiply a decimal by a whole number.

0.97	2 places
× 63	+ 0 places
291	2 places in the answer
5820	
61.11	2 places marked off in the answer

Example 2

Multiply a decimal by a decimal.

16.03	2 places
× 0.137	+ 3 places
11221	5 places in the answer
48090	
160300	
2.19611	5 places marked off in the answer

Example 3

Multiply two decimals with the same place values.

```
    1.38      2 places
  × 2.04    + 2 places
  ------    ----------
     552      4 places in the answer
     000
+  27600
  ------
  2.8152      4 places marked off in the answer
```

Note: When the two factors have the same number of places, you still add the total number of places to mark off in the answer. Multiplying is not to be confused with adding, where the decimal points would be written directly under each other.

Sometimes when you add up the decimal places in the two factors, you end up with more places than you have in the answer. When this happens, you must mark off the correct number of places. Let's see how this is done.

Example 4

```
   0.007      3 places
   × 0.3    + 1 place
  ------    ---------
  0.0021      4 places in the answer
```

Answer: 0.0021

You need to mark off four places. Add zeroes in front of the 21, not after (0.0021, not 0.2100).

Example 5

```
     0.006      3 places
   × 0.138    + 3 places
  --------    ----------
        48      6 places in the answer
       180
     + 600
  --------
  0.000828      6 places marked off in the answer
```

Answer: 0.000828

You need to mark off six places. Add three zeroes in front of 828 and write 0.000828.

Problem Set

1. Multiply: 8.1×44

2. Multiply: 28.4×13

3. Multiply: 8.7×9

4. Multiply: 0.06×3

5. Multiply: 9.3×6

6. Multiply: 7.22×3.48

7. Multiply: 68.3×4.87

8. Janet is cooking baked macaroni for a party. If she needs to buy 2.3 pounds of cheese that costs \$3.24 per pound, how much money will the cheese cost her? Round your answer to the nearest cent.

9. The football team wants to buy new jerseys. The jerseys cost \$17.49 each. They need 33 jerseys. How much will they cost? Round your answer to the nearest cent.

10. Multiply: 2.34×6

11. Write the product with the decimal in the proper place: $21.6 \times 57 = 12312$

12. Write the product with the decimal in the proper place: $10 \times 0.245 = 2450$

13. Write the product with the decimal in the proper place: $7.42 \times 3 = 2226$

14. Write the product with the decimal in the proper place: $409 \times 0.80 = 32720$

15. Write the product with the decimal in the proper place: $91.4 \times 0.8 = 7312$

16. Write the product with the decimal in the proper place: $3.01 \times 0.46 = 13846$

17. Multiply: 4.81×38.6

18. Darren purchased 10.2 pounds of onions at the grocery store. If the onions cost \$0.37 per pound, what was his total cost? Round your answer to the nearest cent.

19. Jack is a chef at a seafood restaurant. He is calling his suppliers to place orders for the day. He orders 53.3 pounds of shrimp at \$0.96 per pound. How much does the restaurant spend on shrimp? Round your answer to the nearest cent.

20. Liz wants to buy 7.3 pounds of chicken. The cost of the chicken is \$0.24 per pound. How much will it cost her? Round your answer to the nearest cent.

MPA-020: Multiplying Decimals by Powers of Ten

Before actually multiplying decimals by powers of ten, you must have a clear understanding about what the term "power of ten" means. To write a power of ten, you have 10 as a **base** and you have an **exponent** (a little number up and to the right of the base 10). The exponent tells you how many times 10 is used as a factor. For instance:

10^1 means 10 is used as a factor 1 time: 10.

10^2 means 10 is used as a factor 2 times: $10 \times 10 = 100$.

10^3 means 10 is used as a factor 3 times: $10 \times 10 \times 10 = 1000$.

When working with powers of 10, you can use a shortcut to writing any power of 10. Any power of 10 can be written by first writing a one, followed by as many zeroes as the number of the exponent.

Example 1 Simplify 10^5.

1	**Step 1:** Write a 1.
100000 12345	**Step 2:** The exponent is 5, so write five zeroes after the 1.
100,000	**Answer:** Therefore, 10^5 means one hundred thousand.

Example 2 Simplify 10^7.

1	**Step 1:** Write a 1.
10000000 1234567	**Step 2:** The exponent is 7, so write seven zeroes after the 1.
10,000,000	**Answer:** Therefore, 10^7 means ten million.

When multiplying or dividing decimals by powers of 10, you can use your knowledge about powers of 10 to do these problems quicker than by using long multiplication or division. To use this quicker method, you must remember certain rules for multiplying or dividing by powers of 10. To **multiply** by a power of 10, move the decimal point to the **right** the same number of spaces as the exponent of the number, or zeroes in the multiple of 10. To **divide** by a power of 10, move the decimal point to the **left** the correct number of spaces. It is also important to note that it may sometimes be necessary to add a zero in the spaces when writing your final solution.

Multiplying decimals by powers of ten	To multiply a decimal by a power of ten, move the decimal point to the right the same number of spaces as the exponent.
Dividing decimals by powers of ten	To divide a decimal by a power of ten, move the decimal point to the left the same number of spaces as the exponent.

Let's use the shortcut method to work some problems involving multiplying decimals by powers of 10.

Example 3 Simplify 0.5627×10^3

0.5627×10^3	**Step 1:** $10^3 = 10 \times 10 \times 10 = 1{,}000$ (3 zeroes)

0.5627 1 2 3 ⟶	**Step 2:** Move the decimal point three spaces to the right.
562.7	**Answer:** Write the solution with the decimal point in its new position.

Example 4 Simplify 49.7×10^4

49.7×10^4	**Step 1:** $10^4 = 10 \times 10 \times 10 \times 10 = 10{,}000$ (4 zeroes)
49.7 1 2 3 4 ⟶	**Step 2:** Move the decimal point four spaces to the right.
497,000	**Step 3:** Add zeroes to hold the place value of the spaces to the right.
497,000.	**Answer:** Write the solution with the decimal point in its new position. *Note: A whole number has an assumed decimal point.*

Example 5 Simplify 16.57×10 .

16.57×10	**Step 1:** $10 = 10^1 = 10$ (1 zero)
16.57 1 ⟶	**Step 2:** Move the decimal point one space to the right.
165.7	**Answer:** Write the solution with the decimal point in its new position.

Now let's work through a few division problems involving decimals and powers of 10. Remember that since you moved to the right when multiplying, you move to the left when dividing.

Example 6 Simplify $41.6 \div 10^2$.

$41.6 \div 10^2$	**Step 1:** $10^2 = 10 \times 10 = 100$ (2 zeroes)
41.6 2 1 ⟵	**Step 2:** Move the decimal point two spaces to the left.
0.416	**Answer:** Write the solution with the decimal point in its new position.

Example 7 Simplify $15.6 \div 10^4$.

$15.6 \div 10^4$	**Step 1:** $10^4 = 10 \times 10 \times 10 \times 10 = 10{,}000$ (4 zeroes)
15.6 4 3 2 1 ⟵	**Step 2:** Move the decimal point four spaces to the left.
00156	**Step 3:** Add zeroes to hold the place value of the spaces to the left.
0.00156	**Answer:** Write the solution with the decimal point in its new position.

Example 8	Simplify $17 \div 10^3$.
$17 \div 10^3$	**Step 1:** $10^3 = 10 \times 10 \times 10 = 1{,}000$ (3 zeroes)
17 (3 2 1 ←)	**Step 2:** Move the decimal point three spaces to the left. *Note: There is an assumed decimal point after a whole number.*
017	**Step 3:** Add a zero to hold the place value of the spaces to the left.
0.017	**Answer:** Write the solution with the decimal point in its new position.

Problem Set

Multiply:

1. 2.6871×100
2. 20.4681×100
3. 438.65×100
4. 8.9541×100
5. 78.5213×100
6. $9.49526 \times 1{,}000$
7. $54.674 \times 1{,}000$
8. $856.97 \times 1{,}000$
9. $0.09495 \times 1{,}000$
10. 0.02046×100
11. $0.9798 \times 1{,}000$
12. 479.581×10
13. 0.68543×10
14. 154.491×10
15. 9.45065×10

Solve:

16. A children's library has 100 bookshelves. Each bookshelf holds 247 books. How many books can the bookshelves hold?

17. Jemeka has 10 CD cases. Each CD case holds 6 CDs. How many CDs does Jemeka have?

18. A paper manufacturing plant produces 1,000 packs of paper in an hour. Each pack of paper has 250 sheets. How many sheets of paper can this plant produce in an hour?

19. Jason won a radio give-away contest. After taxes he should receive $2,156 per month for 10 months. How much money did Jason win?

    $215,600 $215.60
    $2,156,000 $21,560

20. Carson just purchased a new sofa. He put no money down, but he must make a payment of $100 per month for the next 10 months. How much did Carson's new sofa cost?

    $1,000 $10
    $100,000 $10,000

MPA-021: Converting Between Standard and Scientific Notation

The world has become connected by numerous cooperative and joint projects that involve many different countries. For scientists and mathematicians to share their information and findings, they must be able to understand each other's data, regardless of its country of origin. For this reason, scientists and mathematicians developed scientific notation. Scientific notation allows you to easily express extremely large and small numbers. When a number is written using scientific notation, the value of the number does not change. It is merely another way of writing these complicated numbers. For instance, it is time consuming and difficult to read such numbers as: 6,723,000,000,000,000,000,000 or 0.00000000000000000000023, but it is not as difficult to read them as 6.723×10^{21} or 2.3×10^{-22}.

Scientific Notation	A method of writing large and small numbers in the form $a \times 10^n$ where $1 \le n \le 10$ and n is an integer.

Let's learn to convert large and small numbers using scientific notation.

When considering very large numbers, you must first represent the number as a decimal between 1 and 10. Then, you must multiply that new decimal number by a power of 10, which is found by counting the number of places the decimal point has moved.

Note: When writing large numbers in scientific notation, the exponent of the base 10 is always ***positive.***

In the above example of 6,723,000,000,000,000,000,000, you can represent the number as a decimal between 1 and 10. Therefore, 6.723 is between 1 and 10, and is written as $1 < 6.723 < 10$. The next step is to count the number of places the decimal has moved. In this example of 6.723, the decimal moved 21 places or 10^{21}.

Example 1 Write 2,496,000,000,000 in scientific notation.

2.496 $1 < 2.496 < 10$	**Step 1:** Represent the number as a decimal between 1 and 10.
2,496,000,000,000. ← 12 places	**Step 2:** Count the number of places the decimal point has been moved (whole numbers have an assumed decimal point).
2.496×10^{12}	**Step 3:** Write the decimal between 1 and 10 multiplied by 10^{12}, because the decimal point moved 12 places.
	Answer: 2.496×10^{12}

Example 2 Write 859,000,000.00 in scientific notation.

8.59 $1 < 8.59 < 10$	**Step 1:** Write as a decimal between 1 and 10.
859,000,000.00 ← 8 places	**Step 2:** Count the places the decimal point moved.
8.59×10^8	**Step 3:** Write the decimal between 1 and 10 multiplied by 10^8.
	Answer: 8.59×10^8

Example 3	Write 1.57×10^{3} in scientific notation.
10^{3}	**Step 1:** The exponent on the power of ten is a positive; therefore this will be a number greater than 10.
1.57 → 3 places	**Step 2:** Move the decimal 3 places to the right and add a zero.
	Answer: 1,570

Now let's consider very small numbers. The rules for writing extremely small numbers are similar to what you have been doing, with one exception. Let's look at the example of 0.00000000000000000000023. The number is represented as a decimal between 1 and 10. Therefore, 2.3 is between 1 and 10, and is written as $1 < 2.3 < 10$. The next step is to count the number of places the decimal has moved. In this example of 2.3, the decimal moved 22 places or 10^{-22}. Remember that when you express such small numbers in scientific notation, expressing decimals as whole numbers, the exponent of base 10 will be a **negative** exponent. Let's work a few examples.

Example 4	Write 0.000000000407 in scientific notation.
4.07 $1 < 4.07 < 10$	**Step 1:** Write as a decimal between 1 and 10.
.000000000407 → 10 places	**Step 2:** Count the places the decimal point moved.
4.07×10^{-10}	**Step 3:** Write the decimal between 1 and 10 times a negative power of 10.
	Answer: 4.07×10^{-10}

Example 5	Write 0.000567 in scientific notation.
5.67 $1 < 5.67 < 10$	**Step 1:** Write as a decimal between 1 and 10.
.000567 → 4 places	**Step 2:** Count the places the decimal point moved.
5.67×10^{-4}	**Step 3:** Write the decimal between 1 and 10 multiplied by a negative power of 10.
	Answer: 5.67×10^{-4}

Example 6	Write 2.9×10^{-13} in scientific notation.
10^{-13}	**Step 1:** The exponent on the power of ten is a negative; therefore this will be a number less than 1 but greater than 0.
2.9 ← 13 places	**Step 2:** Move the decimal 13 places to the left and add zeros.
	Answer: 0.00000000000029

Scientific notation is a convenient way of expressing very large or very small numbers. Remember these steps:

- For both large or small numbers, write them as a decimal that falls between 1 and 10.

- For both large or small numbers, count the number of spaces the decimal point moved from the original position to the new position.
- If you started with a large number, the exponent of 10 is positive. If you started with a small number, the exponent of 10 is negative.

Problem Set

1. Multiply: 3.40×100

2. Choose the number that is written in scientific notation:

4.29×10^7 25.1×10^2

0.39×10^2 42.9×10^8

3. Choose the number that is written in scientific notation:

42.9×10^6 1.2×10^8

0.64×10^9 28×10^4

4. Choose the number that is written in scientific notation:

48.3×10^2 52.1×10^7

2.6×10^4 0.16×10^6

5. Choose the number that is NOT written in scientific notation:

3.71×10^6 0.16×10^6

8.42×10^4 2.1×10^2

6. Write this number in standard notation: 7.87×10^4

7. Write this number in standard notation: 2.15×10^5

8. Shamble likes to stargaze. He knows that the Sun is about 4,600,000,000 years old. How would Shamble express this number in scientific notation?

9. Neptune is 2,796,000,000 miles from the Sun. How would you express this in scientific notation?

10. Lourdes is a scientist. She figures that the first humans were on Earth 3,000,000 years ago. How would Lourdes express this in scientific notation?

11. Multiply: 6.25×100

12. Multiply: $119.26 \times 1{,}000$

13. Multiply: $7.866 \times 10{,}000$

14. Multiply: $224.99 \times 1{,}000$

15. Choose the number that is written in scientific notation:

0.421×10^3 4.21×10^3

42.1×10^3 0.36×10^2

16. Write this number in scientific notation: 55,800,000

17. Write this number in scientific notation: 899,000,000

18. Write this number in standard notation: 8.7×10^8

19. Latoya likes to stargaze. She knows that the planet Venus is 6.7×10^7 miles from the Sun. Write this number of miles in standard notation.

20. Uranus has a diameter of 32,000 miles. The diameter of Mars is 27,800 miles. How would you express the difference in diameter of these two planets in scientific notation?

MPA-022: Dividing Decimals

When dividing decimals, always remember to consider the decimal point. Where does the decimal point go in the quotient? This question is easily answered when you divide a decimal number by a whole number. To do so, first bring the decimal point up from the dividend to the quotient, then divide as usual.

Example 1 Divide 94.5 by 3.

$$\begin{array}{r} 31.5 \\ 3\overline{)94.5} \\ \underline{-9}\ \downarrow\downarrow \\ 04\ \downarrow \\ \underline{-3}\ \downarrow \\ 15 \\ \underline{-15} \\ 0 \end{array}$$

Step 1: Bring the decimal point up from the dividend to the quotient and divide as usual.

$31.5 \times 3 = 94.5$

Step 2: Check your answer by multiplying the quotient by the divisor.

Answer: 31.5

Go a step further and consider what must be done when the divisor is not a whole number but is instead a decimal number. It is easier to divide if the decimal point in the divisor is eliminated. This division can be easily accomplished if you move the decimal point to the right as many spaces as necessary so that the decimal becomes a whole number.

Remember: If you move the decimal point in the divisor, then you must also move the decimal point in the dividend. Both decimal points must be moved the same number of spaces.

A moved decimal point is represented by a caret (^). When a decimal point has been moved, place a decimal point in the quotient directly above the caret then divide as usual.

Example 2 Divide 281.47 by 0.7.

$.7_{\wedge}\overline{)281.47}$

Step 1: Eliminate the decimal in the divisor by moving it over one space and placing a caret.

$7_{\wedge}\overline{)281.4_{\wedge}7}$

Step 2: Move the decimal one space in the dividend, and place a caret. Remember that the decimal must be moved the same number of spaces in the divisor and in the dividend.

$\begin{array}{r} 402.1 \\ .7_{\wedge}\overline{)281.4_{\wedge}7} \\ \underline{-28}\downarrow\downarrow \\ 01\downarrow\downarrow \\ \underline{-0}\downarrow\downarrow \\ 14\downarrow \\ \underline{-14}\downarrow \\ 07 \\ \underline{-7} \\ 0 \end{array}$	**Step 3:** Place the decimal point in the quotient directly above the caret and divide as usual.

Answer: $281.47 \div 0.7 = 402.1$

Example 3 Divide 1.862 by 0.98.

$.98_{\wedge}\overline{)1.862}$	**Step 1:** To eliminate the decimal in the divisor, move it over two spaces and place a caret.
$.98_{\wedge}\overline{)1.86_{\wedge}2}$	**Step 2:** Now, move the decimal two spaces in the dividend, and place a caret.
$\begin{array}{r} 1.9 \\ .98_{\wedge}\overline{)1.86_{\wedge}2} \\ \underline{-98}\ \downarrow \\ 882 \\ \underline{-882} \\ 0 \end{array}$	**Step 3:** Place the decimal point in the quotient directly above the caret and divide as usual.

Answer: $1.862 \div 0.98 = 1.9$

Example 4 Divide 65.05 by 0.5.

$.5_{\wedge}\overline{)65.05}$	**Step 1:** To eliminate the decimal in the divisor, move it over one space and place a caret.
$.5_{\wedge}\overline{)65.0_{\wedge}5}$	**Step 2:** Now, move the decimal one space in the dividend and place a caret.

```
      130.1
.5‸)65.0‸5
   −5↓↓
    15↓
   −15↓
     00
     −0
      05
      −5
       0
```

Step 3: Place the decimal point in the quotient directly above the caret and divide as usual.

Answer: $65.05 \div 0.5 = 130.1$

Sometimes, you must add zeroes in the dividend when moving the decimal point. Remember that the purpose of adding zeroes is to eliminate a decimal point in the divisor. Always move the decimal point the same number of spaces in the dividend as you did in the divisor. If needed, add zeros to get the correct number of spaces. Let's work through an example where this is necessary.

Example 5 Divide 787.5 by 0.125.

```
.125‸)787.5
```

Step 1: Move the decimal point in the divisor three spaces.

```
.125)787.500‸
```

Step 2: The dividend has only one space to move, but you must move the decimal three spaces. Add two zeroes to the end of the dividend and place the caret.

```
          6300.
.125‸) 787.500‸
      −750↓↓↓
        375↓↓
       −375↓↓
          00↓
          −0↓
           00
```

Step 3: Place the decimal point in the correct position in the quotient and divide.

Answer: $787.5 \div 0.125 = 6,300$

Problem Set

1. Divide and Check: $541.8 \div 9$

2. Divide and Check: $56.49 \div 7$

3. What is the 3 called in $75 \div 3$?

   divisor — quotient
   none of these — dividend

4. In $12 \div 4 = 3$, 3 is the ________.

   divisor — dividend
   quotient — addend

5. Divide and Check: $37.17 \div 6.3$

6. Divide and Check: $48.858 \div 3.4$

7. Divide and Check: $573.6 \div 12$

8. Bill, Jane, Bou, and Sue work for Quickburger. Last night they made \$105.20 altogether. If they all made the same amount, how much money did each worker earn?

9. Jill has 24.6 ounces of milk. This afternoon she is giving her nieces and nephews milk and cookies. If she divides the milk 6 ways, how much milk will each child get?

10. Burgers cost \$1.75 each. The Johnson twins have \$28. How many burgers can they buy?

11. Divide and Check: $137.34 \div 3$

12. Divide and Check: $226.59 \div 13$

13. Divide and Check: $554.3 \div 241$

14. Divide and Check: $60 \div 1.0$

15. Divide and Check: $592.8 \div 5.7$

16. Divide and Check: $7{,}250.4 \div 212$

17. Peg gets paid at the end of every 7.5 hour shift. If she earns \$82.50 nightly, how much does she get paid per hour?

18. Divide: $587.1 \div 0.05$

19. Divide: $968.1 \div 0.06$

20. Divide: $162.9 \div 0.09$

MPA-023: Rounding Quotients Involving Decimals

You have already learned how to round numbers. In this lesson, you will learn how to round quotients that involve decimals. The rules for rounding decimals are the same. To round any number, you must know place values and what place value is being rounded.

To work with this decimal number, 437.2463, you must know the place values of each digit.

Hundreds	Tens	Ones		Tenths	Hundredths	Thousandths	Ten Thousandths
4	3	7	.	2	4	6	3

Suppose that you were asked to round 437.2463 to the nearest tenth. This would involve the place values for 437.2. There are only two possible answers: 437.2 or 437.3. To find the solution, look at the number next to the tenths place, a 4. This is not enough to round up one more number, so the answer remains 437.2.

Now, round 437.2463 to the nearest hundredth. The answer will be either 437.24 or 437.25. The next digit, 6, is in the thousandths place. This is enough to round up one more number, so the solution is 437.25.

Round 437.2463 to the nearest thousandth. The answer will be either 437.246 or 437.247. The next digit, 3, is in the ten thousandths place. This is not enough to round up, so the answer remains 437.246.

The following procedure will help you round quotients involving decimals. Remember to set up the problem properly.

Rounding quotients

1. Determine the place value to which you will round.
2. Divide using the rules for long division, dividing out to one place past the place to which you will round.
3. Look at the digit immediately to the right of the digit in the place to which you will round. If the digit to the right is 5 or greater, round the preceding digit up by 1. If the digit is less than 5, then the preceding digit stays the same.

Example 1 Divide 24.6147 by 3.3, and round to the nearest tenth.

3.3‸)24.6‸147

Step 1: Move the decimal point over one space in the divisor and the dividend.

```
        7.4
3.3‸)24.6‸147
    −231 ↓
      151
     −132
       19
```

Step 2: Place the decimal point in the quotient directly above the caret and divide.

```
        7.45
3.3‸)24.6‸147
    −231↓↓
      151↓
     −132↓
        194
       −165
```

Step 3: Divide one digit past the tenths place. There is no need to continue. The digit in the hundredths place will determine your answer.

$7.45 \approx 7.5$

Step 4: The digit, 5, is enough to round up one more number.

Answer: 7.5

Example 2 Divide 0.5047 by 0.7 to the nearest hundredth.

$.7_{\wedge}\overline{).5_{\wedge}047}$

Step 1: Move the decimal point one space over in the divisor and dividend.

$$\begin{array}{r} . \\ .7_{\wedge}\overline{).5_{\wedge}047} \end{array}$$

Step 2: Place the decimal point in the quotient directly above the caret.

$$\begin{array}{r} .721 \\ .7_{\wedge}\overline{).5_{\wedge}047} \\ \underline{-49}\downarrow\downarrow \\ 14\downarrow \\ \underline{-14}\downarrow \\ 07 \end{array}$$

Step 3: Divide one digit past the hundredths place. This digit lets you know whether your answer is 0.72 or 0.73.

$0.721 \approx 0.72$

Step 4: The digit, 1, is not enough to round up, so your answer remains as it is.

Answer: 0.72

Example 3 Divide 37.9304 by 6.8 to the nearest tenth.

$6.8_{\wedge}\overline{)37.9_{\wedge}304}$

Step 1: Move the decimal point one space over in the divisor and dividend.

$$\begin{array}{r} . \\ 6.8_{\wedge}\overline{)37.9_{\wedge}304} \end{array}$$

Step 2: Place a decimal point in the quotient directly above the caret and divide.

$$\begin{array}{r} 5.57 \\ 6.8_{\wedge}\overline{)37.9_{\wedge}304} \\ \underline{-340} \\ 393 \\ \underline{-340} \\ 530 \\ \underline{-476} \end{array}$$

Step 3: Divide one digit past the tenths place. This digit lets you know whether your answer is 5.5 or 5.6.

$5.57 \approx 5.6$

Step 4: The digit, 7, is enough to round up.

Answer: 5.6

Problem Set

1. Divide. Round the answer to the nearest tenth.

$51 \div 4$

2. Divide. Round the answer to the nearest tenth.

$12\overline{)45}$

3. Divide. Round the answer to the nearest tenth.

$52\overline{)87}$

4. Divide. Round the answer to the nearest hundredth.

$67\overline{)878}$

5. Divide. Round the answer to the nearest hundredth.

$7\overline{)19}$

6. Divide. Round the answer to the nearest hundredth.

$8\overline{)21}$

7. Divide. Round the answer to the nearest whole number.

$11\overline{)98}$

8. If the answer is rounded to the nearest whole number, then $254 \div 54 \approx 5$.

True False

9. Mike spends $8.16 on five bars of soap. How much does each bar of soap cost, rounded to the nearest penny?

10. Tricia buys 12 gallons of milk to make ice cream for her birthday party. Her total for the milk came to $37.09. How much does each gallon cost, rounded to the nearest penny?

11. Ken bought 18 tomatoes for $3.98. How much did each tomato cost, rounded to the nearest penny?

12. Octavio bought 9 markers. He spent $7.62 on them. How much did each marker cost, rounded to the nearest penny?

13. Timothy buys 14 steaks to feed his family at the family reunion. He pays a total of $61.83. How much did each steak cost, rounded to the nearest penny?

14. Divide, then round the answer to the nearest hundredth:

$9.9 \div 8$

15. Divide, then round the answer to the nearest hundredth:

$6.7 \div 4$

16. Divide, then round the answer to the nearest hundredth:

$5.5 \div 4$

17. Divide, then round the answer to the nearest hundredth:

$9.7 \div 4$

18. Divide, then round the answer to the nearest hundredth:

$9.7 \div 8$

19. Divide, then round the answer to the nearest hundredth:

$7.3 \div 4$

20. Divide, then round the answer to the nearest hundredth:

$7.4 \div 8$

MPA-024: Using Divisibility Rules

Often, it is helpful to know if a number is *divisible* by another number. This means that there is no remainder left after the division has been done. Usually, to find out if a number is divisible by another, you actually have to do the division and see if a remainder exists or not. For example, if you divide 100 by 25, the quotient is 4, so we would say 100 is divisible by 4. If you divide 100 by 33, the quotient is 3 R 1, so we would say 100 is *not* divisible by 33.

If you need to know if a number is divisible by 2, 3, 4, 5, 6, 9, or 10, rules exist to help you. If you memorize these rules, it will save you time so you won't have to do the actual division in some cases.

A number is divisible by...	if...
2	the digit in the ones place is even (ends in 0, 2, 4, 6, or 8).
3	the sum of the digits is divisible by 3.
4	the number formed by the last 2 digits is divisible by 4.
5	the digit in the ones place is a 0 or 5.
6	the number is divisible by both 2 and 3.
9	the sum of the digits is divisible by 9.
10	the digit in the ones place is a 0.

Let us look at an example of each rule.

Example 1 Is 337 divisible by 2?

Answer: Is the digit in the ones place even? The digit in the ones place is 7, so 337 is *not* divisible by 2.

Example 2 Is 747 divisible by 3?

Answer: Is the sum of the digits divisible by 3? When we add 7 + 4 + 7, the sum is 18. Eighteen is divisible by 3, so 747 is divisible by 3.

Example 3 Is 71,976 divisible by 4?

Answer: Is the number formed by the last two digits divisible by 4? In this case, is 76 divisible by 4? Try this with paper and pencil or a calculator. The quotient is 19 with no remainder, so 71,976 is divisible by 4.

Note: You may still need to divide, but it will be easier than dividing the entire number by 4.

Example 4 Is 751 divisible by 5?

Answer: Is the ones digit 0 or 5? The ones digit in this number is 1; therefore, 751 is *not* divisible by 5.

Example 5 Is 3,642 divisible by 6?

Answer: This is a two part test. Is 3,642 divisible by 2 (is the ones digit even)? Yes, the ones digit is 2.

Is 3,642 divisible by 3 (is the sum of the digits divisible by 3)? Yes, $3 + 6 + 4 + 2 = 15$, which is divisible by 3.

Because 3,642 is divisible by both 2 and 3, it is also divisible by 6.

Example 6 Is 123,462 divisible by 9?

Answer: Is the sum of the digits divisible by 9?

$1 + 2 + 3 + 4 + 6 + 2 = 18$, which is divisible by 9; therefore, 123,462 is divisible by 9.

Example 7 Is 6,340 divisible by 10?

Answer: Is the ones digit 0? The ones digit is 0, so 6,340 is divisible by 10.

You may be curious why rules do not exist for divisibility by 7 and 8. Rules do exist, but are so complicated that it is actually easier to perform the division than it is to use the rules.

Problem Set

1. Answer the following question using your divisibility rules: Which number can be divided evenly into 670?

6 5
3 9

2. Answer the following question using your divisibility rules: Which number can be divided evenly into 330?

9 4
7 6

3. Answer the following question using your divisibility rules: Which number can be divided evenly into 303?

10 3
8 2

4. Answer the following question using your divisibility rules: Which number can be divided evenly into 632?

6 9
2 5

5. State whether the first number is evenly divisible by the second number. Use your divisibility rules. Do not divide the number. 844; 4

No Yes

6. State whether the first number is evenly divisible by the second number. Use your divisibility rules. Do not divide the number. 715; 10

Yes No

7. State whether the first number is evenly divisible by the second number. Use your divisibility rules. Do not divide the number. 52,080; 10

Yes No

8. Which number is evenly divisible by 10?

6,402 7,650
2,362 6,443

9. Which number is evenly divisible by 9?

7,699 6,642
3,425 2,000

10. Is 520 evenly divisible by the following number? 6

Yes No

11. Is 5,624 evenly divisible by the following number? 5

Yes No

12. Which number can be divided evenly into 160?

9 5
7 3

13. Which number can be divided evenly into 685?

5 3
7 2

14. Is 650 evenly divisible by the following number? 2

Yes No

15. Is 7,512 evenly divisible by the following number? 3

Yes No

16. Is 771 evenly divisible by the following number? 5

No Yes

17. Is 1,095 evenly divisible by the following number? 5

Yes No

18. Is 981 evenly divisible by the following number? 4

Yes No

19. An astronomer looking through a telescope counts 639 stars in the sky on a clear night. The astronomer wants to divide the night sky into sectors with equal numbers of stars. Using your divisibility rules, decide how many sectors there will be if each sector has the same number of stars.

20. There are 78 beach balls to be put on display at a swimming pool equipment store. The balls will be displayed in bins containing equal numbers of balls. Using your divisibility rules, determine how many bins are needed.

MPA-025: Identifying Prime and Composite Numbers

As you are finding out, mathematics has its own special vocabulary. For example, words such as add, subtract, multiply, and divide all tell you to do very specific things with numbers. As you increase your vocabulary, the easier it will be to understand the lessons presented to you in the videos and in this book.

Mathematicians are always looking for patterns, because once a pattern is found, a rule or definition can be written. For instance, a pattern you are familiar with is even and odd numbers. Numbers that are divisible by 2 are called *even*, and numbers that are not are called *odd*.

This lesson concentrates on numbers that are divisible (no remainder) by only two numbers (1 and the number itself) and numbers that are divisible by more than two numbers.

Factor	The numbers that are multiplied to get a product.

For example, $3 \times 4 = 12$. The *factors* are 3 and 4, and 12 is the *product*.

Let's look at the number 10. What are some possible factors of 10?

Since $1 \times 10 = 10$, 1 and 10 are factors of 10.

We also have $2 \times 5 = 10$, so 2 and 5 are also factors of 10.

We can say that 1, 2, 5, and 10 are factors of 10.

How about the number 24? Can you think of all its factors?

1×24 $\quad$ 3×8 $\quad$ 2×12 $\quad$ 4×6

So your list of factors for 24 is 1, 2, 3, 4, 6, 8, 12, 24.

Try the number 36.

1×36 $\quad$ 3×12 $\quad$ 6×6 $\quad$ 2×18 $\quad$ 4×9

The factors are 1, 2, 3, 4, 6, 9, 12, 18, and 36.

Did you notice that the first pair of factors is always 1 and the number itself?

When these are the *only* factors of a number, you call the number **prime**. When there are *more* than just these two factors, you call the number **composite**.

Prime number	A whole number greater than one with exactly two factors, one and itself.
Composite number	A whole number greater than one with more than two factors.

The numbers above are composite because they all have more than two factors. The first prime number is 2. The only factors of 2 are 1 and 2.

Note: One is neither a prime nor a composite number. It does not fit either definition, so consider this number to be a special case.

Use the divisibility rules you learned to help you decide if a number is prime or composite. Of course, numbers can be divisible by numbers other than 2, 3, 4, 5, 6, 9, and 10, but this is a good place to start. As soon as you find that a number is divisible by more than 1 and itself, you know that it is composite.

Note: For example, notice that 2 is the only even prime number. All other even numbers will be divisible by 2 (think of the divisibility rule for numbers divided by 2) and therefore composite.

Let's look at some examples.

Example 1 Is the number 57 prime or composite?

Answer: Using the divisibility rules, you know this number is divisible by 3 because the sum of its digits (5 + 7 = 12) is divisible by 3. Because you know there are other factors of 57 besides 1 and 57, you know this number is *composite*.

Example 2 Is the number 67 prime or composite?

Answer: Again, using the divisibility rules, you find that 67 is not divisible by 2, 3, 4, 5, 6, 9, or 10. Check to see if the number is divisible by 7 or 8 by actually dividing. Because 67 is not divisible by any other numbers besides 1 and 67, the number is *prime*.

Problem Set

1. Is 9 prime, composite, or neither?

2. How many prime number(s) occur between 1 and 10, including the numbers 1 and 10?

3. A prime number is _______________.

a number that has only one factor
a number that has more than two factors
a number that has no factors
a number that has exactly two factors

4. A composite number is _______________.

a number that has only one factor
a number that has more than two factors
a number that has no factors
a number that has exactly two factors

5. What number(s) between 21 and 30 is/are composite, including the numbers 21 and 30?

6. Which number is a prime number?

1 35
21 37

7. Which number is a composite number?

11 29
39 5

8. Is 49 prime, composite, or neither?

9. All odd numbers are prime.

True False

10. Tameeca wishes to plant 10 rows of corn. She wants each row to have the same number of stalks. Will the total number of stalks be prime, composite, or neither?

11. Two prime numbers multiplied together equal fifteen. What are the numbers?

12. A number that has exactly two factors is a(n) _______ number.

even prime
composite odd

13. A prime number has more than 2 factors.

True False

14. A composite number is a number with more than two factors.

True False

15. Which set contains all prime numbers?

{49, 42} {38, 33, 36}
{19, 29} {27, 21, 24}

16. All numbers divisible by numbers other than one and themselves are ________.

neither composite
none of these prime

17. 1 is the smallest prime number.

True False

18. 4 is the lowest composite number.

True False

19. Are all multiples of 8 prime, composite, or neither?

20. Sergeant Carpenter can line his cadets in 3 even rows of 4. Does he have a prime or composite number of cadets?

MPA-026: Using Prime Factorization

In a previous lesson, you learned about prime and composite numbers. In this lesson, you will see that all composite numbers can be written as a product of prime numbers. The process you will use is called **prime factorization**.

Prime factorization	A process that allows you to write a composite number as a product of two or more prime numbers.

If you think about it, the words "prime factorization" tell you what you need to do. "Prime" reminds you that you are looking for prime numbers only, and the key word in "factorization" is factor, which means you will be multiplying.

We will break down composite numbers into a product of prime numbers using **factor trees**. A factor tree is just a diagram that will help you to do prime factorization.

Example 1 Find the prime factorization of 10.

10

Step 1: You will begin by writing 10 at the top of the "tree."

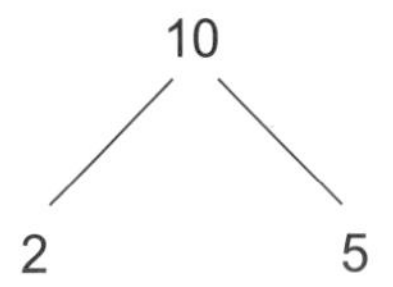

Step 2: Now think of *any* two factors of 10 and "branch" these numbers off the 10. In this case the only factors of 10 besides 1 and 10 are 2 and 5, so you will use 2 and 5.

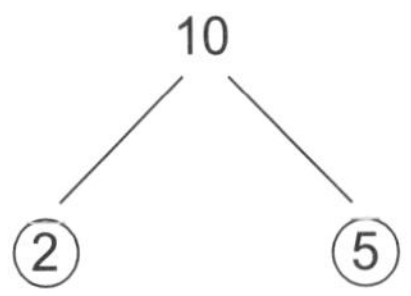

Step 3: Now, check if either or both of these factors are prime. Circle any prime factors. In this case, both are prime, so you are done.

Answer: The prime factorization of 10 is 2×5.

Example 2 Find the prime factorization of 72.

72

Step 1: 72 will be the top of your "tree."

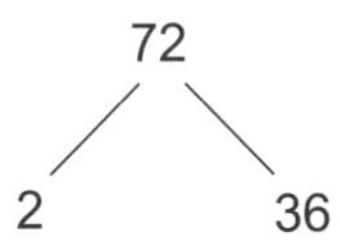

Step 2: Think of any two factors of 72. For example, 72 is even, so it must be divisible by 2, and $2 \times 36 = 72$, so 2 and 36 will be the "branches."

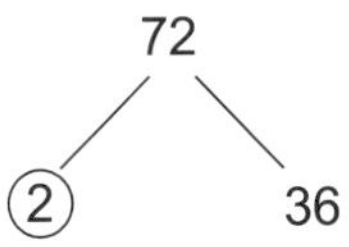

Step 3: Stop and think. Is either factor prime? The 2 is prime, so you will circle it.

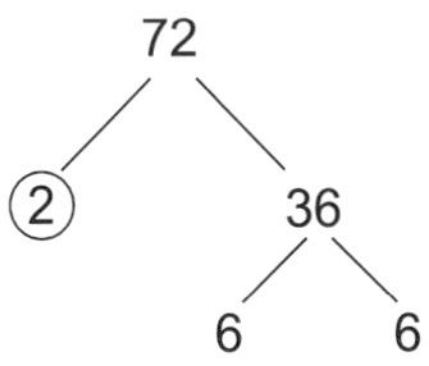

Step 4: Since 36 is not prime you must now think of two factors of 36. Any two factors of 36 would be fine. For this example, you will use 6×6 and "branch" these factors off the 36.

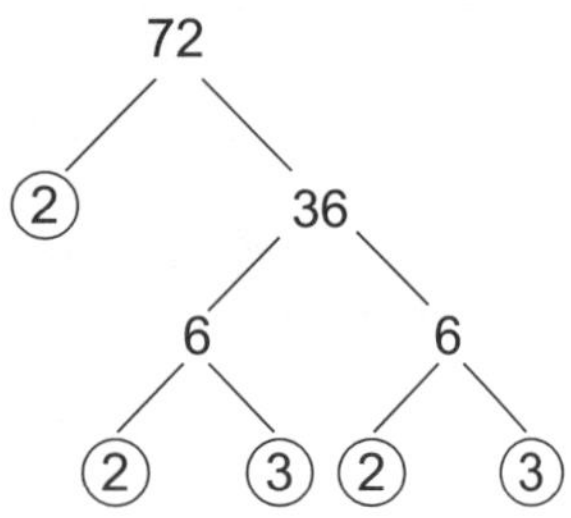

Step 5: Because 6 is not prime, you cannot circle anything. So you now find factors of 6 and "branch" these factors off *each* 6. Because 2 and 3 are prime, you will circle each one. Because all the branches are now prime, you are done.

$2 \times 2 \times 2 \times 3 \times 3$

Answer: Write all the circled numbers as a product. Write them in numerical order—remember that multiplication is commutative, so you can write the factors in any order you like. Check to be sure the product is 72.

Note: We can also write this using exponents: $2^3 \times 3^2 = 72$

Problem Set

1. Find the prime factorization of 36.

2. Find the prime factorization of 25.

3. Find the prime factorization of 48.

4. Find the prime factorization of 74.

5. Find the prime factorization of 30.

6. Find the prime factorization of 84.

7. Find the prime factorization of 39.

8. Find the prime factorization of 200.

9. Find the prime factorization of 83.

10. Find the prime factorization of 64.

11. Find the prime factorization of 180.

12. Find the prime factorization of 85.

13. List all of the factors for the given number: 50

14. List all of the factors for the given number: 78

15. Find the prime factorization of 1,000.

16. Find the prime factorization of 150.

17. Find the prime factorization of 105.

18. Find the prime factorization of 325.

19. Find the prime factorization of 170.

20. Find the prime factorization of 1,250.

MPA-027: Finding the Greatest Common Factor

How do you work with a fraction that contains large numbers in both the numerator and the denominator? Suppose, for instance, that you encounter a fraction whose numerator is 1,746 and whose denominator is 3,492. How would you even begin to work with this fraction? You can reduce large fractions to simplest form by finding the *greatest common factor* of the numerator and the denominator.

The greatest common factor (GCF) of two numbers is defined as the largest factor that two numbers have in common. Let's practice finding the GCF of two numbers by listing the factors and by using prime factorization.

Greatest Common Factor (GCF)	The largest factor that two or more numbers have in common.

Listing Factors

To find the GCF of two numbers, list the factors, and then identify the GCF.

Example 1 Find the GCF of 18 and 36.

$18 \rightarrow 1, 2, 3, 6, 9, 18$ $36 \rightarrow 1, 2, 3, 4, 6, 9, 12, 18, 36$	**Step 1:** List the factors of 18 and 36.
$18 \rightarrow 1, 2, 3, 6, 9, (18)$ $36 \rightarrow 1, 2, 3, 4, 6, 9, 12, (18), 36$	**Step 2:** Identify and circle the greatest factor that is common to both 18 and 36.
GCF = 18	**Answer:** The greatest common factor of 18 and 36 is 18.

Prime Factorization

You can also use prime factorization to find the GCF of two numbers. This method involves first writing each number as the product of prime numbers. Then, multiply the common prime numbers. The product is the GCF.

Example 2 Use prime factorization to find the GCF of 27 and 45.

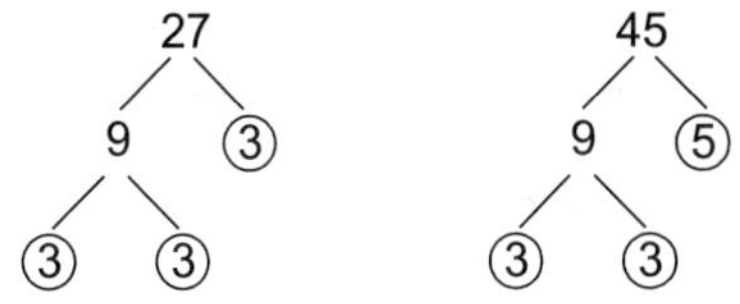

(factor trees above)	**Step 1:** Find the prime factorization of both 27 and 45.
$27 = 3 \times 3 \times 3$ $45 = 3 \times 3 \times 5$	**Step 2:** Write both 27 and 45 as the product of prime numbers and circle the common factors.
$3 \times 3 = 9$	**Step 3:** Multiply the common prime numbers. The product is the GCF.
GCF = 9	**Answer:** The greatest common factor of 27 and 45 is 9.

Now that you are familiar with the two methods used to find the GCF, you should feel more confident about simplifying fractions. For now, practice finding the GCF using the methods you just learned.

Example 3 Find the GCF of 12 and 24 by listing the factors.

12 → 1, 2, 3, 4, 6, 12 24 → 1, 2, 3, 4, 6, 8, 12, 24	**Step 1:** List the factors of 12 and 24.
12 → 1, 2, 3, 4, 6, (12) 24 → 1, 2, 3, 4, 6, 8, (12), 24	**Step 2:** Identify and circle the greatest factor that is common to both 12 and 24.
GCF = 12	**Answer:** The greatest common factor of 12 and 24 is 12.

Example 4 Find the GCF of 45 and 64 by listing the factors.

45 → 1, 3, 5, 9, 15, 45 64 → 1, 2, 4, 8, 16, 32, 64	**Step 1:** List the factors of 45 and 64.
45 → (1), 3, 5, 9, 15, 45 64 → (1), 2, 4, 8, 16, 32, 64	**Step 2:** Identify and circle the greatest factor that is common to both 45 and 64.
GCF = 1	**Answer:** The greatest common factor of 45 and 64 is 1.

Example 5 Use prime factorization to find the GCF of 80 and 60.

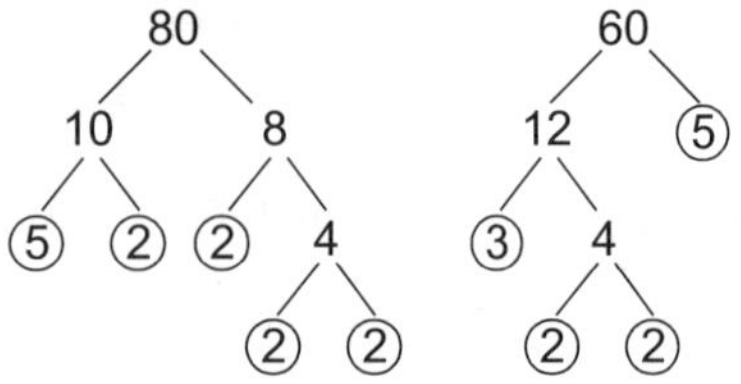

Step 1: Find the prime factorization of both 80 and 60.

$80 = 2 \times 2 \times 2 \times 2 \times 5$ $60 = 2 \times 2 \times 3 \times 5$	**Step 2:** Write both 80 and 60 as the product of prime numbers. Circle the prime numbers that the two numbers have in common.
$2 \times 2 \times 5 = 20$	**Step 3:** Multiply the common prime numbers. The product is the GCF.
GCF = 20	**Answer:** The greatest common factor of 80 and 60 is 20.

Example 6 Use prime factorization to find the GCF of 40 and 56.

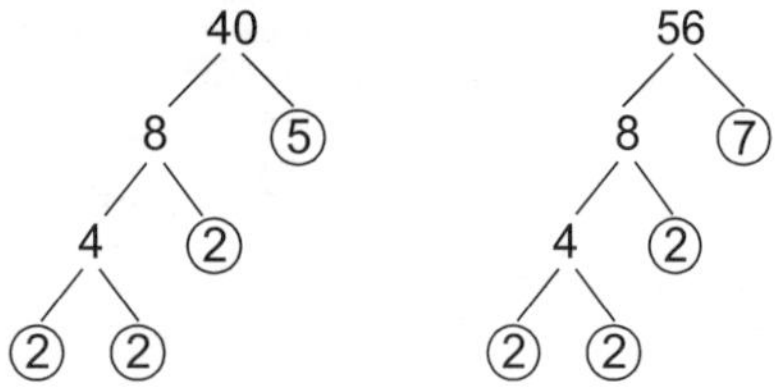

Step 1: Find the prime factorization of both 40 and 56.

$40 = 2 \times 2 \times 2 \times 5$ $56 = 2 \times 2 \times 2 \times 7$	**Step 2:** Write both 40 and 56 as the product of prime numbers. Circle the prime numbers that the two numbers have in common.

$2 \times 2 \times 2 = 8$	**Step 3:** Multiply the common prime numbers. The product is the GCF.
GCF = 8	**Answer:** The greatest common factor of 40 and 56 is 8.

Problem Set

1. Find the GCF (Greatest Common Factor) of 12 and 24 by listing the factors of each number.

2. Use prime factorization to find the GCF (Greatest Common Factor) of 20 and 30.

3. Find the GCF (Greatest Common Factor) of 3 and 6 by listing the factors of each number.

4. Use prime factorization to find the GCF (Greatest Common Factor) of 40 and 10.

5. Find the GCF (Greatest Common Factor) of 3 and 12 by listing the factors of each number.

6. Find the GCF (Greatest Common Factor) of 2 and 9 by listing the factors of each number.

7. Use prime factorization to find the GCF (Greatest Common Factor) of 10 and 100.

8. Use prime factorization to find the GCF (Greatest Common Factor) of 18 and 8.

9. Use prime factorization to find the GCF (Greatest Common Factor) of 24 and 14.

10. Find the GCF (Greatest Common Factor) of 42 and 35 by listing the factors of each number.

11. Find the GCF (Greatest Common Factor) of 14 and 49 by listing the factors of each number.

12. There are 24 bags of candy and 12 bags of apples to be packaged. Each container should have only one type of each item. All containers must have the same number of bags in them. What is the greatest number of bags that can be placed in each container?

13. There are 27 pieces of grape nut bread and 36 pieces of potato bread that need to be put into containers. Each container should have only 1 type of bread. All containers must have the same amount of pieces of bread in them. What is the greatest number of pieces of bread that can be placed in each of the containers?

14. There are 4 large folders and 6 small folders that need to be packaged. Each package should hold only one size folder. All packages must have the same number of folders. What is the greatest number of folders that need to be placed in each package?

15. There are 48 pads of yellow paper and 36 pads of white paper that need to be boxed. Each box should contain only one type of color paper. All boxes must have the same number of pads of paper. What is the greatest number of pads of paper that need to be placed in each box?

16. There are 32 black plastic bags and 24 white plastic bags that need to be boxed. Each box should have only one type of each kind of bag. All boxes must have the same number of plastic bags. What is the greatest number of bags that can be placed in each box?

MPA-028: Reducing Fractions to Lowest Terms / Simplest Form

Now that you know how to find the greatest common factor (GCF) for a set of numbers, you can simplify fractions to lowest terms. In fact, once you find the GCF of the numerator and denominator of a fraction, you only need to complete one more step to reduce the fraction to simplest form.

The numerator and denominator of any fraction always have at least one common factor (the number 1). Your goal when simplifying a fraction is to make the number 1 the only common factor. The easiest way to remove all the other common factors is to divide the numerator and denominator by the GCF for the two numbers.

A fraction, reduced to its simplest form, is equivalent to the original fraction. Reducing a fraction only makes it simpler; it does not change the value of the fraction. Let's look at some examples.

Example 1 Reduce $\frac{3}{6}$ to simplest form.

Step 1: Find the GCF, either by listing factors or by prime factorization.

Listing factors:

$3 \rightarrow 1, \mathbf{3}$

$6 \rightarrow 1, 2, \mathbf{3}, 6$

GCF = 3

Prime factorization:

3
(already prime)

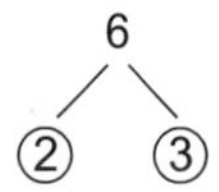

$3 = 3$

$6 = 2 \times 3$

GCF = 3

Step 2: Divide the numerator and denominator by their GCF, 3. Dividing both the numerator and denominator by 3 is the same as $\frac{3}{6} \div \frac{3}{3}$. Also notice that the GCF of 1 and 2 is 1, which means that $\frac{1}{2}$ is the simplest form; you cannot reduce it any further.

$$\frac{3}{6} = \frac{3 \div 3}{6 \div 3} = \frac{1}{2}$$

Answer: Therefore, $\frac{3}{6}$ reduced to simplest form is $\frac{1}{2}$.

$$\frac{3}{6} = \frac{1}{2}$$

Example 2 Reduce $\frac{6}{18}$ to simplest form.

Step 1: Find the GCF by listing factors or by prime factorization.

Listing factors:

$6 \rightarrow 1, 2, 3, \mathbf{6}$

$18 \rightarrow 1, 2, 3, \mathbf{6}, 9, 18$

GCF = 6

Prime factorization:

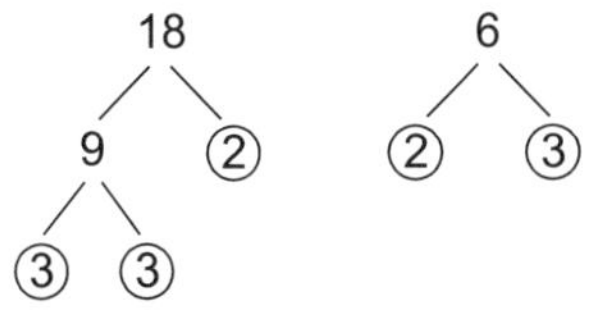

$6 = 2 \times 3$

$18 = 2 \times 3 \times 3$

$2 \times 3 = 6$

GCF = 6

Step 2: Divide the numerator and the denominator by their GCF, 6.

$$\frac{6}{18} = \frac{6 \div 6}{18 \div 6} = \frac{1}{3}$$

$$\frac{6}{18} = \frac{1}{3}$$

Answer: $\frac{6}{18}$ reduced to simplest form is $\frac{1}{3}$.

Example 3 Reduce $\frac{21}{28}$ to simplest form.

Step 1: Find the GCF by listing factors or by prime factorization.

Listing factors:

$21 \rightarrow 1, 3, \mathbf{7}, 21$

$28 \rightarrow 1, 2, 4, \mathbf{7}, 14, 28$

GCF = 7

Prime factorization:

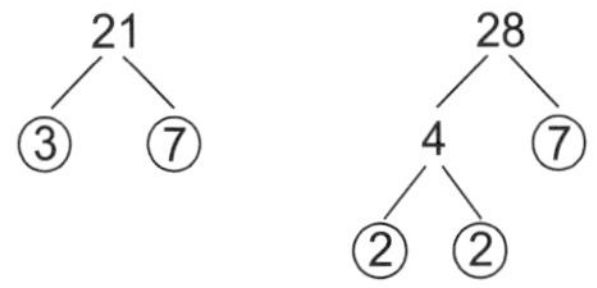

$21 = 3 \times 7$

$28 = 2 \times 2 \times 7$

GCF = 7

Step 2: Divide the numerator and the denominator by their GCF, 7.

$$\frac{21}{28} = \frac{21 \div 7}{28 \div 7} = \frac{3}{4}$$

$$\frac{21}{28} = \frac{3}{4}$$

Answer: $\frac{21}{28}$ reduced to simplest form is $\frac{3}{4}$.

Example 4 Reduce $\frac{35}{45}$ to simplest form.

Listing factors:

$35 \rightarrow 1, \mathbf{5}, 7, 35$

$45 \rightarrow 1, 3, \mathbf{5}, 9, 15, 45$

GCF = 5

Step 1: Find the GCF by listing factors or by prime factorization.

Prime factorization:

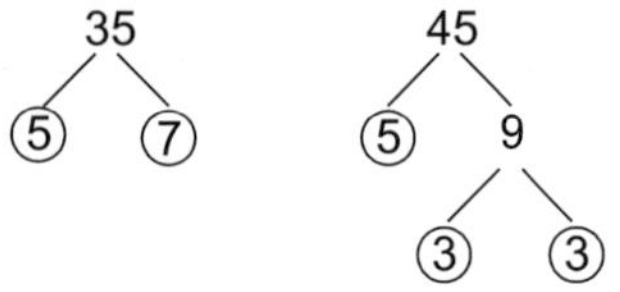

$35 = 5 \times 7$

$45 = 3 \times 3 \times 5$

GCF = 5

$\frac{35}{45} = \frac{35 \div 5}{45 \div 5} = \frac{7}{9}$

Step 2: Divide the numerator and the denominator by their GCF, 5.

$\frac{35}{45} = \frac{7}{9}$

Answer: $\frac{35}{45}$ reduced to simplest form is $\frac{7}{9}$.

Example 5 Reduce $\frac{72}{108}$ to simplest form.

Listing factors:

$72 \rightarrow 1, 2, 3, 4, 6, 8, 12, 18, 24, \mathbf{36}, 72$

$108 \rightarrow 1, 2, 3, 4, 6, 9, 12, 18, 27, \mathbf{36}, 54, 108$

GCF = 36

Step 1: Find the GCF by listing factors or by prime factorization.

Prime factorization:

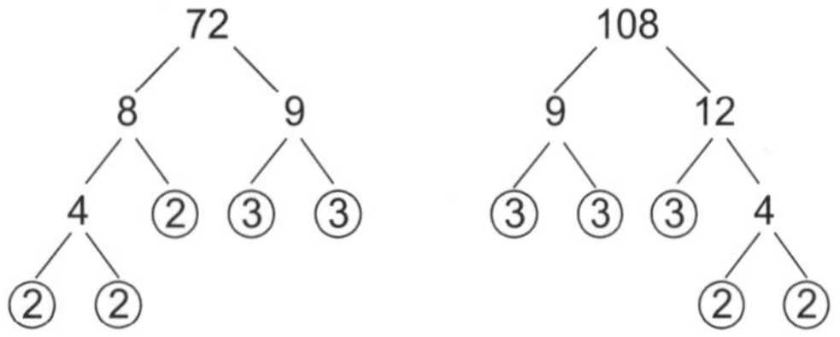

$72 = 2 \times 2 \times 2 \times 3 \times 3$

$108 = 2 \times 2 \times 3 \times 3 \times 3$

GCF = 36

$\frac{72}{108} = \frac{72 \div 36}{108 \div 36} = \frac{2}{3}$

Step 2: Divide the numerator and the denominator by their GCF, 36.

$\frac{72}{108} = \frac{2}{3}$

Answer: $\frac{72}{108}$ reduced to simplest form is $\frac{2}{3}$.

Problem Set

1. Reduce $\frac{2}{4}$ to simplest form.

2. Reduce $\frac{3}{9}$ to simplest form.

3. Reduce $\frac{2}{8}$ to simplest form.

4. Reduce $\frac{3}{6}$ to simplest form.

5. Reduce $\frac{5}{35}$ to simplest form.

6. Reduce $\frac{3}{7}$ to simplest form.

7. Reduce the following fraction to simplest form.

   $\frac{18}{63} = \square$

8. Reduce the following fraction to simplest form.

   $\frac{22}{48} = \square$

9. Reduce the following fraction to simplest form.

   $\frac{17}{51} = \square$

10. Choose the number that reduces to $\frac{1}{2}$.

    $\frac{1}{4}$ $\frac{17}{34}$ $\frac{2}{6}$ $\frac{15}{35}$

11. Choose the number that reduces to $\frac{3}{4}$.

    $\frac{20}{15}$ $\frac{75}{125}$ $\frac{12}{24}$ $\frac{12}{16}$

12. Jennifer is writing a cookbook. A recipe calls for $\frac{4}{24}$ cup of sugar. For clarity, she wants to reduce the fraction to simplest form. Which fraction should she use?

    $\frac{1}{3}$ cup $\frac{1}{6}$ cup $\frac{2}{12}$ cup $\frac{1}{4}$ cup

13. Reduce $\frac{40}{80}$ to simplest form.

14. Reduce $\frac{2}{14}$ to simplest form.

15. Reduce $\frac{18}{19}$ to simplest form.

16. Reduce $\frac{15}{30}$ to simplest form.

17. Reduce $\frac{16}{80}$ to simplest form.

18. Reduce the following fraction to simplest form.

    $\frac{80}{100} = \square$

19. Choose the fraction that reduces to $\frac{1}{3}$.

    $\frac{1}{30}$ $\frac{7}{10}$ $\frac{6}{2}$ $\frac{4}{12}$

20. Becca noticed that 75%, or $\frac{75}{100}$, students passed the math test last Monday. She wishes to reduce this fraction to simplest form. Which fraction should she choose?

    $\frac{1}{25}$ $\frac{3}{4}$ $\frac{2}{3}$ $\frac{1}{4}$

MPA-029: Converting Fractions and Decimals

How is it that a huge fraction sometimes is the same as a rather small decimal, like $\frac{473}{500} = 0.946$? Or, how is it that a very long decimal sometimes is the same as a rather small fraction, like $0.42857143 \approx \frac{3}{7}$?

It's actually easy to see. After all, what is a fraction? It's nothing but an expression of division. You can understand how large fractions are by converting them to decimals. And you can see what fractions compare to decimals by converting decimals to fractions. Division is key.

To convert a **fraction to a decimal**:

- Divide the numerator by the denominator.
- Round if indicated.

Example 1 Convert $\frac{3}{4}$ to a decimal.

$$\begin{array}{r} 0.75 \\ 4\overline{)3.00} \\ \underline{28} \\ 20 \\ \underline{20} \\ 0 \end{array}$$

Step 1: The numerator, which is also the dividend, is 3. The denominator, which is also the divisor, is 4. Divide the numerator 3 by the denominator 4.

Answer: $\frac{3}{4} = 0.75$

Unfortunately, it's not always that easy. Sometimes the decimals don't end. If you continued dividing these types of fractions, you'd never finish. However, you would find a pattern in the decimal. This pattern is unique for every **repeating decimal**.

Repeating Decimal	A decimal number with a pattern behind the decimal point that repeats infinitely. When you write repeating decimals, place a bar over the digits that repeat. For example, if $\frac{3}{7} = 0.42857142\ldots$, you would write $0.\overline{428571}$.

In general, thirds, sixths, sevenths, ninths, elevenths, thirteenths, and more, correspond to repeating decimals, but repeating decimals are not limited to these fractions.

Example 2 Convert $\frac{1}{6}$ to a decimal.

$$\begin{array}{r} 0.1666 \\ 6\overline{)1.0000} \\ \underline{-6} \\ 40 \\ \underline{-36} \\ 40 \\ \underline{-36} \\ 40 \\ \underline{-36} \\ 4 \end{array}$$

Step 1: The numerator, which is also the dividend, is 1. The denominator, which is also the divisor, is 6. Divide the numerator by the denominator. Stop dividing when you notice a pattern.

$\frac{1}{6} = 0.1666\ldots = 0.1\overline{6}$

Answer: Write the decimal, but write the repeated pattern only once with a bar over only one number of the pattern.

Now, to convert a **decimal to a fraction**:

- Place the numbers behind the decimal in the numerator.
- Use the place value of the decimal to determine the denominator.
- Reduce the fraction to simplest form.

Example 3 Convert 0.625 to a fraction.

$\frac{625}{}$

Step 1: Write a fraction with the numbers behind the decimal as the numerator.

$\frac{625}{1000}$

Step 2: Use the last place value, thousandths, as the denominator.

$\frac{625 \div 125}{1000 \div 125} = \frac{5}{8}$

Step 3: Reduce the fraction to lowest terms by dividing both the numerator and denominator by the GCF, 125.

Answer: $\frac{5}{8}$

Now try some examples.

Example 4 Convert $\frac{13}{20}$ to a decimal.

$$\begin{array}{r} 0.65 \\ 20\overline{)13.00} \\ \underline{120} \\ 100 \\ \underline{100} \\ 0 \end{array}$$

Step 1: The numerator, which is also the dividend, is 13. The denominator, which is also the divisor, is 20. Divide the numerator by the denominator.

Answer: $\frac{13}{20} = 0.65$

Example 5 Convert $\frac{5}{11}$ to a decimal.

$$
\begin{array}{r}
0.454545 \\
11\overline{)5.000000} \\
\underline{44} \\
60 \\
\underline{-55} \\
50 \\
\underline{-44} \\
60 \\
\underline{-55} \\
50 \\
\underline{-44} \\
60 \\
\underline{-55} \\
5
\end{array}
$$

Step 1: The numerator, which is also the dividend, is 5. The denominator, which is also the divisor, is 11. Divide the numerator by the denominator. Stop dividing when you notice a pattern.

$\frac{5}{11} = 0.454545\ldots 0.\overline{45}$

Step 2: Write the decimal, but write the repeated pattern only once with a bar over only one of the patterns.

Answer: $\frac{5}{11} = 0.\overline{45}$

Example 6 Convert 0.12 to a fraction.

$\frac{12}{}$

Step 1: Write a fraction with the numbers behind the decimal as the numerator.

$\frac{12}{100}$

Step 2: Use the last place value, hundredths, as the denominator.

$0.12 = \frac{12 \div 4}{100 \div 4} = \frac{3}{25}$

Step 3: Reduce the fraction to lowest terms by dividing both the numerator and denominator by the GCF, 4.

Answer: $\frac{3}{25}$

Example 7 Convert 0.8125 to a fraction.

$\frac{8{,}125}{}$

Step 1: Write a fraction with the numbers behind the decimal as the numerator.

$\frac{8{,}125}{10{,}000}$

Step 2: Use the last place value, ten thousandths, as the denominator.

$0.8125 = \frac{8{,}125 \div 625}{10{,}000 \div 625} = \frac{13}{16}$

Step 3: Reduce the fraction to lowest terms by dividing both the numerator and denominator by the GCF, 625.

Answer: $\frac{13}{16}$

Problem Set

1. Express $\frac{1}{6}$ as a decimal. Round to the nearest hundredth.

2. Express $\frac{3}{10}$ as a decimal.

3. Express $\frac{7}{10}$ as a decimal.

4. Express $\frac{4}{10}$ as a decimal.

5. Express 0.48 as a fraction in simplest form.

6. Express 0.7 as a fraction in simplest form.

7. Julie bought $\frac{7}{8}$ of a pound of peanuts. What is this value expressed as a decimal?

8. Cheryl bought 0.8 of a pound of candy. How much is this when expressed as a fraction in simplest form?

9. Express $\frac{75}{100}$ as a decimal.

10. Express $\frac{301}{1,000}$ as a decimal.

11. Express $\frac{9}{36}$ as a decimal.

12. Express 0.11 as a fraction reduced to its simplest form.

13. Express 0.21 as a fraction in simplest form.

14. Express $\frac{63}{84}$ as a decimal.

15. Express 0.255 as a fraction in simplest form.

16. Express 0.7 as a fraction.

17. Express 0.69 as a fraction.

18. Express $\frac{2}{3}$ as a decimal. Round to the nearest hundredth.

19. Identify which of the following fractions is a non-repeating decimal.

    $\frac{2}{15}$ $\quad$ $\frac{5}{15}$

    $\frac{3}{15}$ $\quad$ $\frac{1}{15}$

20. Mrs. Harrison asked her chemistry class to mix $\frac{1}{2}$ cup water with 0.3 ounces of salt. The students were given three packs of salt, each containing $\frac{1}{6}$ ounce of salt. Determine whether each pack contains 0.3 ounces of salt.

    Answer Yes or No.

MPA-030: Finding Least Common Multiple of Two or More Numbers

To find the multiples of a number, start multiplying the number by 1, 2, 3, 4, and so on. The answers you find from this process are called the **multiples** of that number. In this lesson, you'll learn how to find the **least common multiple** of two numbers.

Least Common Multiple (LCM)	The smallest non-zero number that is common between the multiples or products of two or more numbers.

Two methods will help you find the least common multiple (LCM) of numbers. The first is listing the multiples. To find the LCM by listing the multiples:

- Write the multiples of the numbers in a list.
- List the multiples until you find a multiple common to the lists.

Find the LCM in the following problems by listing the multiples.

Example 1 Find the LCM of 6 and 8.

$6 \rightarrow 6, 12, 18, \mathbf{24}, 30, 36, 42, 48$	**Step 1:** List the multiples of 6.
$8 \rightarrow 8, 16, \mathbf{24}$	**Step 2:** List the multiples of 8 until you find a common multiple.
	Answer: The LCM is 24.

Example 2 Find the LCM of 7 and 9.

$7 \rightarrow 7, 14, 21, 28, 35, 42, 49, 56, \mathbf{63}$	**Step 1:** List the multiples of 7.
$9 \rightarrow 9, 18, 27, 36, 45, 54, \mathbf{63}$	**Step 2:** List the multiples of 9 until you find a multiple common to both.
	Answer: The LCM is 63.

The second method of finding the LCM is to factor the numbers into primes. To find the LCM using factors:

- Create a factor tree to find the prime factorization of the numbers. Be careful that you do not confuse finding the least common multiple with finding the greatest common factor (GCF).
- Multiply all the unique factors and one of each of the common factors.

In the following examples, you'll use factoring trees to find the LCM.

Example 3 Find the LCM of 9 and 12.

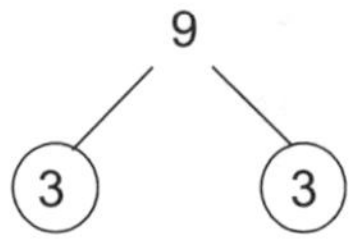

Step 1: Find the prime factorization of 9.

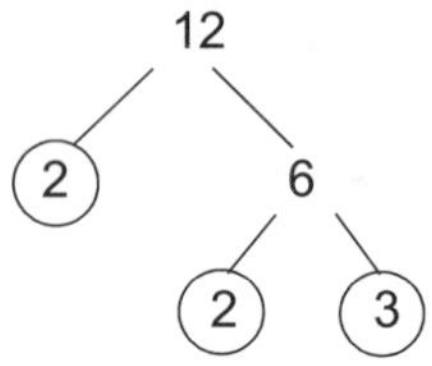

Step 2: Find the prime factorization of 12.

$9 = 3 \times 3$ $12 = 2 \times 2 \times 3$ $2 \times 2 \times 3 \times 3 = 36$	**Step 3:** Multiply all the factors except those that occur in both lists. The factors that occur in both lists should only be used once.
	Answer: The LCM is 36.

Example 4 Find the LCM of 16 and 20.

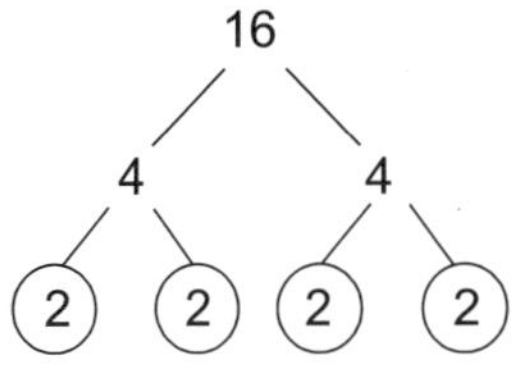

Step 1: Find the prime factorization of 16.

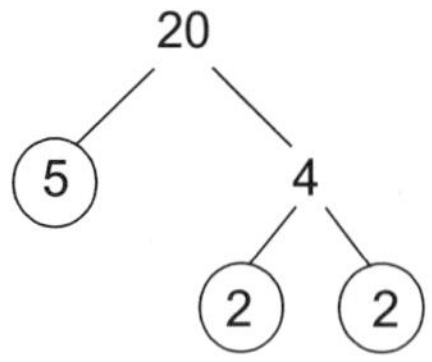

Step 2: Find the prime factorization of 20.

$16 = 2 \times 2 \times 2 \times 2$

$20 = 2 \times 2 \times 5$

$2 \times 2 \times 2 \times 2 \times 5 = 80$

Step 3: Multiply all the factors except those that occur in both lists. Use those only once.

Answer: The LCM is 80.

Example 5 Find the LCM of 32 and 62.

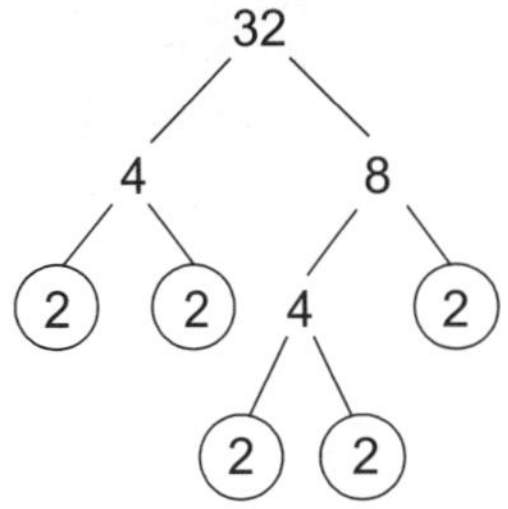

Step 1: Find the prime factorization of 32.

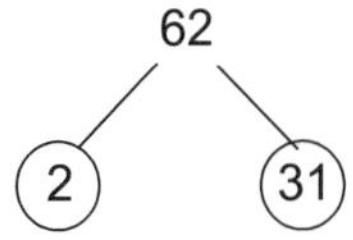

Step 2: Find the prime factorization of 62.

$32 = 2 \times 2 \times 2 \times 2 \times 2$

$62 = 2 \times 31$

$2 \times 2 \times 2 \times 2 \times 2 \times 31 = 992$

Step 3: Multiply all the factors except those that occur in both lists. Use those only once.

Answer: The LCM is 992.

Problem Set

1. Find the LCM of 5 and 7 by listing the multiples of each number.

2. LCM stands for:

   Least Common Multiple
   Least Common Mean
   Least Common Factor
   Greatest Common Factor

3. Use prime factorization to find the LCM of 9 and 6.

4. Use prime factorization to find the LCM of 9 and 81.

5. Use prime factorization to find the LCM of 12, 18 and 36.

6. Find the LCM of 6, 10 and 15 by listing the multiples of each number.

7. Find the LCM of 2, 5 and 8. (Use any method.)

8. Albert and Milton love to jog laps at the park. Albert makes a lap in 8 minutes and Milton makes one in 6 minutes. If they start at the same time, how many minutes will it take for them to meet up?

9. Michelle, Robert, and Danny are running a race. Michelle takes 20 seconds to complete one lap, Robert takes 40 seconds, and Danny takes 60 seconds. If Michelle and Danny start at the same time, in how many seconds will they cross the starting line together?

10. It takes Tom 45 seconds to complete one lap of the track. Brad takes 55 seconds. How many seconds will pass before they finally meet up?

11. Raymond's toy car takes 15 seconds to complete one lap. Sergio's car takes 25 seconds to complete the same lap. How many seconds will it take for the two cars to be side by side?

12. Jack, Jill, and Fred are running a race. It takes Jack 12 minutes to lap the track. Jill takes 16 minutes. Fred completes the lap in 18 minutes. After how many minutes will Jack and Fred first meet on the track?

13. Alexis was asked to subtract $\frac{1}{10}$ from $\frac{7}{12}$. In order to do this, she needed to find the least common multiple of the denominators for 10 and 12. What is the LCM for 10 and 12?

14. The LCM of 2 given numbers is the smallest number that the 2 given numbers will divide into evenly.

    False True

15. Find the LCM of 3 and 7 by listing the multiples of each number.

16. Find the LCM of 4 and 13. (Use any method.)

MPA-031: Comparing and Ordering Fractions and Decimals

How do you know that a half dollar is worth more than forty cents? Did you memorize that? Did you compare a fraction and a decimal without knowing it?

A number line will show how any two numbers compare. First, a half dollar is $\frac{1}{2}$ of a dollar and forty cents is 0.40 of a dollar. One is a fraction, and the other is a decimal. Plotted on a number line, they look like this:

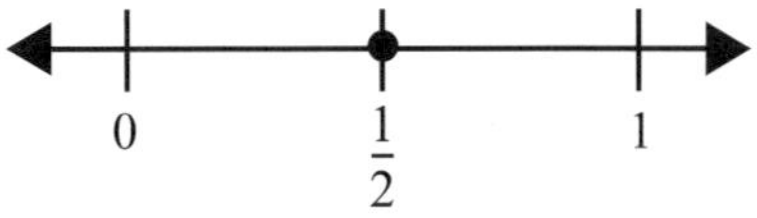

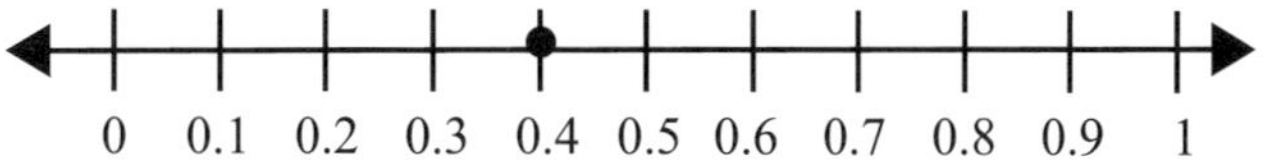

Superimposed, the number line looks like this:

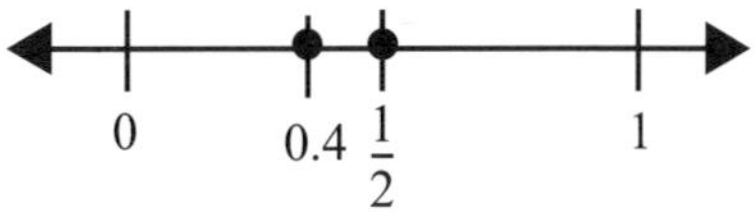

Because $\frac{1}{2}$ is closer to 1, and 0.4 is closer to 0, $\frac{1}{2}$ is greater than 0.4. This can be written as $\frac{1}{2} > 0.4$ or $0.4 < \frac{1}{2}$.

A number line will always help you compare two numbers, but sometimes numbers are very close together, such as $\frac{5}{8}$ and 0.627. It would be almost impossible to see the tiny gap between the two on a number line like the one above, so you need a different method.

To compare a decimal and a fraction, first convert the fraction to a decimal. You will then have two decimals to compare. Simply compare the place values.

Example 1 Compare 0.627 and $\frac{5}{8}$.

$\begin{array}{r} 0.625 \\ 8\overline{)5.000} \\ \underline{48} \\ 20 \\ \underline{16} \\ 40 \\ \underline{40} \\ 0 \end{array}$	**Step 1:** Convert $\frac{5}{8}$ to a decimal by division. Divide the denominator, 8, into the numerator, 5.
0.627 ___ 0.625	**Step 2:** Compare place values: 6 = 6 and 2 = 2. However, 7 > 5.
0.627 > 0.625	**Step 3:** Therefore, 0.627 > 0.625
$0.627 > \frac{5}{8}$	**Answer:** $0.625 = \frac{5}{8}$, so $0.627 > \frac{5}{8}$.

Another way to compare numbers is to convert the decimal to a fraction. Then, both numbers will be fractions. You can compare the numbers by cross-multiplication or by converting to equivalent fractions with a common denominator. The following example shows the cross-multiplication method.

Example 2 Compare 0.627 and $\frac{5}{8}$ by cross-multiplication.

$0.627 = \frac{627}{1,000}$

Step 1: Convert 0.627 to a fraction. 627 thousandths is the same as 627 over 1,000.

5,016 5,000

$\frac{627}{1,000} \times \frac{5}{8}$

Step 2: Cross-multiply. Put the answers just above their numerators, then compare.

$5,016 > 5,000$

$\frac{627}{1,000} > \frac{5}{8}$

Step 3: 5,016 is greater than 5,000, so $\frac{627}{1,000}$ is greater than $\frac{5}{8}$.

$0.627 > \frac{5}{8}$

Step 4: $\frac{627}{1,000} = 0.627$, so $0.627 > \frac{5}{8}$.

Answer: $0.627 > \frac{5}{8}$.

The following example shows how to compare two numbers by finding the common denominator.

Example 3 Compare 0.627 and $\frac{5}{8}$ by finding a common denominator.

$0.627 = \frac{627}{1,000}$

Step 1: Convert 0.627 to a fraction. 627 thousandths is the same as 627 over 1,000.

$\frac{5}{8} = \frac{5 \times 125}{8 \times 125} = \frac{625}{1,000}$

Step 2: Use 1,000 as the common denominator between 8 and 1,000.

$\frac{627}{1,000} > \frac{625}{1,000}$

Step 3: 627 is greater than 625, so $\frac{627}{1,000}$ is greater than $\frac{625}{1,000}$.

$0.627 > \frac{5}{8}$

Step 4: $\frac{627}{1,000} = 0.627$ and $\frac{625}{1,000} = \frac{5}{8}$, so $0.627 > \frac{5}{8}$.

Answer: $0.627 > \frac{5}{8}$.

Note that in the examples above, any method yields the same result: $0.627 > \frac{5}{8}$.

Now, let's compare numbers using a number line.

Example 4 Compare 0.3 and $\frac{1}{4}$ using a number line.

Step 1: Plot 0.3 on a number line.

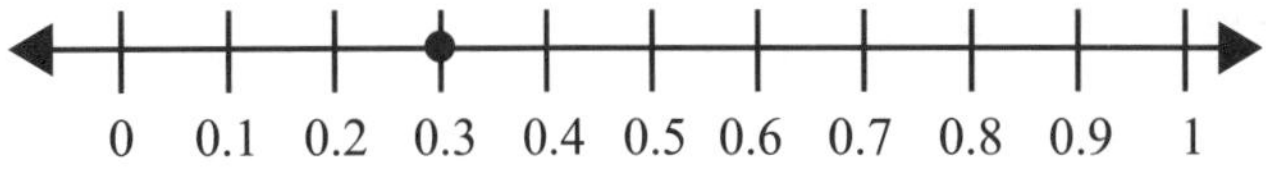

Step 2: Plot $\frac{1}{4}$ on a number line.

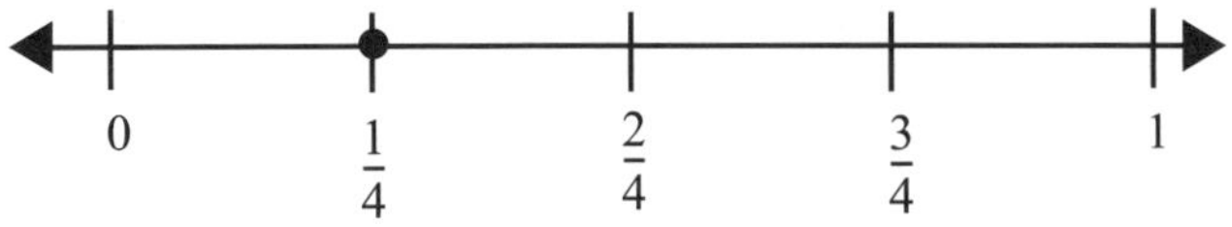

Step 3: Put one number line on top of the other.

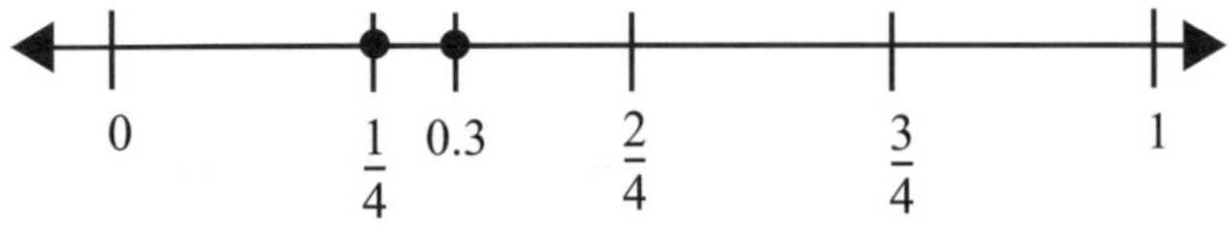

$0.3 > \frac{1}{4}$

Step 4: Because 0.3 is closer to 1 than $\frac{1}{4}$ is, 0.3 is greater than $\frac{1}{4}$.

Answer: $0.3 > \frac{1}{4}$

In the following example, convert the fraction to a decimal, then compare decimals.

Example 5 Compare $\frac{5}{16}$ and 0.36 by converting to decimals.

$$\begin{array}{r} 0.3125 \\ 16\overline{)5.0000} \\ \underline{48} \\ 20 \\ \underline{16} \\ 40 \\ \underline{32} \\ 80 \\ \underline{80} \\ 0 \end{array}$$

Step 1: Convert $\frac{5}{16}$ to a decimal by division.

0.3125 ___ 0.36

Step 2: Compare place values: 3 = 3, but 1 < 6.

0.3125 < 0.36

Step 3: Therefore, 0.3125 < 0.36.

$\frac{5}{16} < 0.36$

Step 4: $0.3125 = \frac{5}{16}$, so $\frac{5}{16} < 0.36$.

Answer: $\frac{5}{16} < 0.36$

See if you can solve the following examples.

Example 6 Compare 0.22 and $\frac{2}{9}$ by cross-multiplying.

$0.22 = \frac{22}{100} = \frac{22 \div 2}{100 \div 2} = \frac{11}{50}$

Step 1: Convert 0.22 to a fraction: 22 hundredths is the same as 22 over 100. Reduce the fraction to lowest terms.

$$\overset{99}{\frac{11}{50}} \times \overset{100}{\frac{2}{9}}$$

Step 2: Cross-multiply.

$\frac{11}{50} < \frac{2}{9}$

Step 3: 99 is less than 100, so $\frac{11}{50}$ is less than $\frac{2}{9}$.

$0.22 < \frac{2}{9}$

Answer: $\frac{11}{50} = 0.22$, so $0.22 < \frac{2}{9}$.

Example 7 Compare $\frac{19}{40}$ and 0.47 by finding a common denominator.

$0.47 = \frac{47}{100}$

Step 1: Convert 0.47 to a fraction: 47 hundredths is the same as 47 over 100.

$\frac{19}{40} = \frac{19 \times 10}{40 \times 10} = \frac{190}{400}$

$\frac{47}{100} = \frac{47 \times 4}{100 \times 4} = \frac{188}{400}$

Step 2: Use 400 as the common denominator between 40 and 100.

$\frac{190}{400} > \frac{188}{400}$

Step 3: 190 is greater than 188, so $\frac{190}{400}$ is greater than $\frac{188}{400}$.

$\frac{19}{40} > 0.47$

Answer: $\frac{19}{40} = \frac{190}{400}$ and $\frac{188}{400} = 0.47$, so $\frac{19}{40} > 0.47$.

Problem Set

1. Use <, >, or = to fill in the blank.

$\frac{5}{12}$ ___ $\frac{10}{12}$

2. Use <, >, or = to fill in the blank.

$\frac{7}{9}$ ___ $\frac{3}{9}$

3. Use <, >, or = to fill in the blank.

0.7 ___ $\frac{1}{2}$

4. Use <, >, or = to fill in the blank.

3.09 ___ 1.92

5. Use <, >, or = to fill in the blank.

1.12 ___ 2.09

6. Use <, >, or = to fill in the blank.

$\frac{3}{4}$ ___ $\frac{2}{6}$

7. Use <, >, or = to fill in the blank.

$\frac{6}{10}$ ___ $\frac{4}{5}$

8. Use <, >, or = to fill in the blank.

$\frac{6}{12}$ ___ $\frac{2}{6}$

9. Order from least to greatest.

$\frac{1}{8}, \frac{5}{8}, \frac{7}{8}, \frac{3}{8}$

10. Order from least to greatest.

$\frac{2}{4}, \frac{6}{10}, \frac{3}{20}, \frac{1}{25}$

11. Order from least to greatest.

$\frac{2}{6}, \frac{6}{8}, \frac{5}{12}, \frac{1}{4}$

12. Mona has $10.38 and Henry has $10.29. Who has more money, Mona or Henry?

13. Mrs. Quin's honors English class was required to read five books. Tabitha read $\frac{2}{5}$ of the books and Samantha read $\frac{1}{5}$ of the books. Who read more books, Tabitha or Samantha?

14. Spencer ate $\frac{4}{10}$ of his pizza. Stacey ate $\frac{5}{10}$ of her pizza. Who ate more pizza?

15. Joseph grew four plants for a science project. Over the four weeks of the project, plant A grew 3.02 inches, plant B grew 3.16 inches, plant C grew 3.7 inches, and plant D grew 3.59 inches. Order the plants by their growth from least to greatest.

16. At the Speedway 2,000 racetrack, the computers went down and the times got scrambled. Car #209 had a time of 6.325, car #213 had a time of 6.293, and car #301 had a time of 6.589. Help Mr. Conrad order the car's race times from least to greatest.

17. Order the following from least to greatest.

$\frac{4}{9}, \frac{13}{18}, \frac{1}{3}, \frac{3}{6}$

18. Chloe needs to purchase pecans. The two brands of pecans cost the same amount. Goody's bag contains 0.3 pounds and White's contains $\frac{1}{4}$ of a pound. Which brand offers more pecans for the money?

19. Shenita needs to buy school supplies. Bell's Discount Store is 0.37 miles from her house. Stone's Discount Store is $\frac{3}{10}$ of a mile from her house. Which store is closer?

20. Ashley can run $\frac{3}{4}$ of a mile in the same time it takes Haley to run 0.75 miles. Who runs the greater distance in that time?

MPA-032: Converting Improper Fractions and Mixed Numbers

Fractions are interesting numbers. A million different fractions can all mean the same thing. How is it that $1\frac{1}{2}$ is the same as $\frac{3}{2}$, and $17\frac{9}{19}$ means the same thing as $\frac{332}{19}$? These are all equivalent fractions, yet they're different. One in each pair is a **mixed number**, and one in each pair is an **improper fraction**.

Proper Fraction	A fraction in which the numerator is less than the denominator. A proper fraction is less than one.
Improper Fraction	A fraction in which the numerator is greater than or equal to the denominator. An improper fraction is greater than or equal to one.
Mixed Number	A combination, or mix, of a whole number and a fraction. A mixed number is greater than one.

Here are some examples of these kinds of numbers:

$\frac{4}{7}$	proper fraction	$4 < 7$	The numerator 4 is less than the denominator 7.
$\frac{17}{3}$	improper fraction	$17 > 3$	The numerator 17 is greater than the denominator 3.
$1\frac{1}{6}$	mixed number	$1\frac{1}{6} > 1$	The number $1\frac{1}{6}$ is greater than one and includes a fraction ($\frac{1}{6}$).

Fractions may be in simplest (reduced) form or they may not be. The numerator and denominator of fractions in simplest form share no factors except one. Improper fractions are never in simplest form. To simplify improper fractions, convert them to mixed numbers.

Examples of fractions in simplest form: $\frac{2}{3}, \frac{7}{9}, 2\frac{4}{5}, 7\frac{5}{9}$

Examples of fractions not in simplest form: $\frac{6}{9}, \frac{12}{15}, \frac{21}{7}, \frac{18}{11}, 2\frac{4}{12}$

Follow these steps to convert an **improper fraction to a mixed number:**

- Divide the numerator by the denominator.
- The whole number answer (without the remainder) is the whole number of the mixed number.
- The remainder is the new numerator.
- The denominator is the same.

A mixed number may be easier to understand visually:

$$\text{Quotient}\frac{\text{Remainder}}{\text{Denominator}}$$

Example 1 Simplify $\frac{18}{7}$.

$7\overline{)18}$ quotient $2r4$; -14; 4	**Step 1:** Divide the numerator, 18, by the denominator, 7.
2	**Step 2:** The whole number part of the quotient is the whole number part of the mixed number.
$2\frac{4}{\ }$	**Step 3:** The remainder of the quotient is the new numerator.
$2\frac{4}{7}$	**Step 4:** The denominator is the same as in the original improper fraction.

Answer: $\frac{18}{7} = 2\frac{4}{7}$

To convert a **mixed number to an improper fraction:**

1. Multiply the whole number by the denominator.
2. Add the result to the numerator.
3. Write a new fraction. Use the result of step two as the new numerator, and use the original denominator.
4. An improper fraction may be easier to understand visually:

$$\frac{\text{(whole number)(denominator)} + \text{original numerator}}{\text{original denominator}}$$

Example 2 Convert $3\frac{4}{5}$ to an improper fraction.

$3 \times 5 = 15$	**Step 1:** Multiply the whole number, 3, by the denominator, 5.
$15 + 4 = 19$	**Step 2:** Add that product to the original numerator.
$\frac{19}{\ }$	**Step 3:** The result is the new numerator.
$\frac{19}{5}$	**Step 4:** The denominator is the same as in the original mixed number.

Answer: $3\frac{4}{5} = \frac{19}{5}$

Example 3 Simplify $\frac{203}{15}$.

$15\overline{)203}$ = 13r8; 15; 53; 45; 8	**Step 1:** Divide the numerator, 203, by the denominator, 15.
13	**Step 2:** The whole number part of the quotient is the whole number part of the mixed number.
$13\frac{8}{\ }$	**Step 3:** The remainder of the quotient is the new numerator.
$13\frac{8}{15}$	**Step 4:** The denominator is the same as in the original improper fraction.

Answer: $\frac{203}{15} = 13\frac{8}{15}$

Example 4 Convert $25\frac{9}{17}$ to an improper fraction.

$25 \times 17 = 425$	**Step 1:** Multiply the whole number, 25, by the denominator, 17.
$425 + 9 = 434$	**Step 2:** Add that product to the original numerator.
$\frac{434}{\ }$	**Step 3:** The result is the new numerator.
$\frac{434}{17}$	**Step 4:** The denominator is the same as in the original mixed number.

Answer: $25\frac{9}{17} = \frac{434}{17}$

Problem Set

1. Change $1\frac{2}{5}$ to an improper fraction.

2. Change $3\frac{3}{4}$ to an improper fraction.

3. Write $\frac{42}{7}$ in simplest form.

4. Write $\frac{12}{12}$ in simplest form.

5. Write $\frac{38}{8}$ in simplest form.

6. Change $11\frac{5}{9}$ to an improper fraction.

7. Change 12 to an improper fraction.

8. Identify $\frac{11}{9}$ as a proper fraction, an improper fraction, or a mixed number.

9. Identify $\frac{11}{12}$ as a proper fraction, an improper fraction, or a mixed number.

10. Identify $\frac{11}{11}$ as a proper fraction, an improper fraction, or a mixed number.

11. Choose the correct symbol (<, >, or =) to make the statement true:

$1 \_\_ \frac{90}{2}$

12. Choose the correct symbol (<, >, or =) to make the statement true:

$\frac{13}{16} \_\_ 1$

13. Write $\frac{47}{5}$ in simplest form.

14. Write $\frac{9}{5}$ in simplest form.

15. Change $3\frac{7}{8}$ to an improper fraction.

16. Write $\frac{15}{2}$ as a mixed number in simplest form.

17. Write $\frac{21}{4}$ as a mixed number in simplest form.

18. Write $\frac{23}{6}$ as a mixed number in simplest form.

19. Choose the correct symbol (<, >, or =) to make the statement true:

$\frac{7}{5} \_\_ 1$

20. Choose the correct symbol (<, >, or =) to make the statement true:

$3\frac{1}{4} \_\_ 1$

MPA-033: Estimating Computations with Fractions and Mixed Numbers

What's the easiest way to add, subtract, multiply, or divide two numbers? When you don't need exact values to answer a question, estimation is not only easy, but it's quick. You can round numbers to make them easy to calculate. When you round two numbers so that the math is easier, you're making them **compatible numbers**. After you round the numbers, it's much easier to add, subtract, multiply, and divide.

Compatible Numbers	Numbers that have been rounded so that you can calculate easily.

To make proper fractions compatible, round them to 0, $\frac{1}{2}$, or 1—numbers that are easy to calculate.

Example 1 Subtract: $\frac{7}{8}-\frac{1}{3}$

$\frac{7}{8}-\frac{1}{3}\approx 1-\frac{1}{2}$ — **Step 1:** Round $\frac{7}{8}$ to 1 and $\frac{1}{3}$ to $\frac{1}{2}$.

$1-\frac{1}{2}=\frac{1}{2}$ — **Step 2:** Since 1 and $\frac{1}{2}$ are compatible numbers, you can subtract them.

Answer: $\frac{1}{2}$

If a question asks you to compute mixed numbers, it may be easier to round to the nearest whole number. The following example shows how easy it is to multiply two numbers when you round them to the nearest whole number.

Example 2 Multiply: $5\frac{6}{11}\times 9\frac{5}{6}$

$5\frac{6}{11}\times 9\frac{5}{6}\approx 6\times 10$ — **Step 1:** Round to 6 and to 10.

$6\times 10=60$ — **Step 2:** Now that you have compatible numbers, multiply them.

Answer: 60

If an expression already contains a compatible number, you do not need to round it. Sometimes you'll want to round more or less than just to the nearest whole number. Study the following example.

Example 3 Multiply: $\frac{1}{4}\times 97\frac{5}{7}$

$\frac{1}{4}\times 97\frac{5}{7}$ — **Step 1:** Leave $\frac{1}{4}$ as it is and round $97\frac{5}{7}$ to 100.

$\frac{1}{4}\times 97\frac{5}{7}\approx\frac{1}{4}\times 100$ — **Step 2:** Now that you have compatible numbers, multiply them.

$\frac{1}{4}\times 100 = 100\div 4 = 25$ — **Answer:** 25

Be careful when using the number zero. Multiplying by zero yields zero. For example, if you had rounded $\frac{1}{4}$ to 0, the final estimate would have been 0 instead of the more accurate 25. Division is another operation that doesn't work well with an estimated zero. Division by zero is impossible. When you estimate multiplication or division, don't round to the nearest whole number if the rounding will produce zero; round to compatible numbers instead.

Now try a few problems with rounding.

Example 4 Add: $\frac{4}{9}+\frac{8}{17}$

$\frac{4}{9}+\frac{8}{17}\approx\frac{1}{2}+\frac{1}{2}$	**Step 1:** Round $\frac{4}{9}$ to $\frac{1}{2}$ and $\frac{8}{17}$ to $\frac{1}{2}$.
$\frac{1}{2}+\frac{1}{2}=1$	**Step 2:** Now that you have compatible numbers, add them.
	Answer: 1

Example 5 Subtract: $18\frac{1}{3}-9\frac{2}{7}$

$18\frac{1}{3}-9\frac{2}{7}\approx 18-9$	**Step 1:** Round to the nearest whole numbers. $18\frac{1}{3}$ rounds to 18, and $9\frac{2}{7}$ rounds to 9, because 18 and 9 are compatible numbers.
$18-9=9$	**Step 2:** Subtract.
	Answer: 9

Example 6 Multiply: $9\frac{3}{7}\times 212\frac{3}{4}$

$9\frac{3}{7}\times 212\frac{3}{4}\approx 10\times 210$	**Step 1:** Round $9\frac{3}{7}$ to 10 and $212\frac{3}{4}$ to 210, because 10 and 210 are compatible numbers.
$10\times 210=2{,}100$	**Step 2:** Multiply.
	Answer: 2,100

Example 7 Divide: $7\frac{8}{9}\div\frac{2}{19}$

$7\frac{8}{9}\div\frac{2}{19}\approx 8\div\frac{1}{10}$	**Step 1:** Round $7\frac{8}{9}$ to the nearest whole number, 8, and round $\frac{2}{19}$ to $\frac{1}{10}$. *Note: Do not round to zero, because division by zero is impossible.*
$8\div\frac{1}{10}=8\times 10=80$	**Step 2:** Divide. Dividing by $\frac{1}{10}$ is the same as multiplying by 10.
	Answer: 80

Problem Set

1. Round to the nearest $\frac{1}{2}$ or whole number to estimate the answer:

$\frac{7}{9} + \frac{3}{8}$

2. Round to the nearest whole number to estimate the answer:

$12\frac{7}{15} - 6\frac{14}{15}$

3. Round to the nearest whole number to estimate the answer:

$9\frac{3}{5} + 8\frac{2}{10}$

4. Round to the nearest whole number to estimate the answer:

$7\frac{8}{9} + 7\frac{1}{9}$

5. Round to the nearest whole number to estimate the answer:

$24\frac{8}{10} - 16\frac{3}{9}$

6. Round to the nearest $\frac{1}{2}$ or whole number to estimate the answer:

$\frac{7}{8} + \frac{3}{5}$

7. Use compatible numbers to estimate:

$\frac{1}{2} \times 17$

8. Round to the nearest whole number to estimate the answer:

$9\frac{8}{10} \div 1\frac{4}{5}$

9. Round to the nearest whole number to estimate the answer:

$8\frac{1}{6} \times 4\frac{1}{5}$

10. Round to the nearest whole number to estimate the answer:

$6\frac{2}{3} \times 6\frac{1}{10}$

11. Round to the nearest whole number to estimate the answer:

$8\frac{5}{6} - 2\frac{1}{4}$

12. Round to the nearest whole number to estimate the answer:

$1\frac{3}{4} + 2\frac{6}{7} + 3\frac{3}{8}$

13. Round to the nearest whole number to estimate the answer:

$12\frac{11}{13} \times 4\frac{20}{22}$

14. Round to the nearest whole number to estimate the answer:

$10\frac{3}{5} - 3\frac{4}{7}$

15. Round to the nearest whole number to estimate the answer:

$19\frac{3}{4} + 18\frac{5}{7}$

16. Round to the nearest whole number to estimate the answer:

$9\frac{1}{7} \div 3\frac{1}{3}$

17. By rounding to the nearest whole number, is the following estimate reasonable? Answer yes or no.

$5\frac{1}{9} \times 4\frac{7}{9} = 21$

18. About 705 people go to the mall each day. One-fourth of these people visit a pet store. How many people visit this pet store each day? (Use compatible numbers to estimate.)
________ people

19. Round to the nearest whole number to estimate the answer:

$22\frac{3}{7} + 11$

20. Round to the nearest whole number to estimate the answer:

$8\frac{8}{9} \div 2\frac{4}{5}$

MPA-034: Adding and Subtracting Fractions

You know what fractions are, and you know what they represent. They're just numbers. They have values just like integers. So, can you add and subtract fractions? Sure you can. In fact, it's fairly simple.

If the fractions have a common denominator, add or subtract the numerators. If the fractions do not have a common denominator, find their least common denominator and convert, so that the fractions have the same denominator. After the fractions have the same denominator, add or subtract the new numerators. Always reduce the final sum or difference to lowest terms.

Here are two examples, one for addition and one for subtraction:

Example 1 Add: $\frac{3}{5}+\frac{4}{5}$

$\frac{3}{5}+\frac{4}{5}=\frac{7}{5}$

Step 1: The denominators are the same, so add the numerators.

$\frac{3}{5}+\frac{4}{5}=\frac{7}{5}=1\frac{2}{5}$

Step 2: Reduce $\frac{7}{5}$ to $1\frac{2}{5}$

Answer: $1\frac{2}{5}$

Example 2 Subtract: $\frac{4}{9}-\frac{2}{9}$

$\frac{4}{9}-\frac{2}{9}=\frac{2}{9}$

Step 1: The denominators are the same, so subtract the numerators.

Answer: $\frac{2}{9}$

In the following examples, the fractions do not have common denominators. Find the least common denominator then convert the fractions. Then, you can add or subtract.

Example 3 Add: $\frac{5}{8}+\frac{1}{12}$

Step 1: Find the least common denominator. The LCD of 8 and 12 is 24.

$\frac{5}{8}=\frac{15}{24}$

$\frac{1}{12}=\frac{2}{24}$

Step 2: Convert $\frac{5}{8}$ to $\frac{15}{24}$ and $\frac{1}{12}$ to $\frac{2}{24}$.

Note: $\frac{5}{8}=\frac{5\cdot 3}{8\cdot 3}=\frac{15}{24}$ and $\frac{1}{12}=\frac{1\cdot 2}{12\cdot 2}=\frac{2}{24}$

$\frac{15}{24}+\frac{2}{24}=\frac{17}{24}$

Step 3: Add the numerators. $\frac{17}{24}$ is already in lowest terms.

Answer: $\frac{17}{24}$

Example 4 Subtract: $\frac{69}{75} - \frac{1}{50}$

Step 1: The denominators are not the same, so find the LCD.

$\frac{69}{75} = \frac{138}{150}$

$\frac{1}{50} = \frac{3}{150}$

Step 2: The LCD of 75 and 50 is 150, so convert $\frac{69}{75}$ to $\frac{138}{150}$ and $\frac{1}{50}$ to $\frac{3}{150}$.

Note: $\frac{69}{75} = \frac{69 \cdot 2}{75 \cdot 2} = \frac{138}{150}$ and $\frac{1}{50} = \frac{1 \cdot 3}{50 \cdot 3} = \frac{3}{150}$

$\frac{138}{150} - \frac{3}{150} = \frac{135}{150}$

Step 3: Subtract the numerators.

$\frac{135}{150} = \frac{9}{10}$

Step 4: Reduce $\frac{135}{150}$ to $\frac{9}{10}$.

Answer: $\frac{9}{10}$

Problem Set

1. Combine and express the answer in simplest form.
$\frac{13}{21} + \frac{5}{21}$

2. Combine and express the answer in simplest form.
$\frac{7}{9} + \frac{1}{9}$

3. Combine and express the answer in simplest form.
$\frac{1}{2} + \frac{2}{2}$

4. Combine and express the answer in simplest form.
$\frac{7}{16} + \frac{7}{16}$

5. Combine and express the answer in simplest form.
$\frac{14}{25} + \frac{13}{25}$

6. Combine and express the answer in simplest form.
$\frac{2}{19} + \frac{7}{19} + \frac{10}{19}$

7. Combine and express the answer in simplest form.
$\frac{14}{22} + \frac{6}{22} + \frac{1}{22}$

8. Combine and express the answer in simplest form.
$\frac{12}{15} + \frac{2}{15}$

9. Combine and express the answer in simplest form.
$\frac{7}{11} - \frac{6}{11}$

10. Simplify completely.
$\frac{7}{12} - \frac{4}{15}$

11. Simplify completely.

$\frac{2}{15}+\frac{7}{12}$

12. Carol surveyed her class to determine how many hours the students spend doing homework each day. What fraction of her class spends one or two hours on their homework, if $\frac{1}{6}$ of the class spends one hour and $\frac{2}{9}$ of the class spends two hours?

13. Paul's recipe calls for $\frac{3}{4}$ cup of flour, and he already has $\frac{1}{4}$ cup. How much more flour does he need?

Solve and express the final answer in simplest form.

14. Kelly glues two pieces of wood together for shop. One is $\frac{3}{4}$ of a foot and the other is $\frac{3}{8}$ of a foot long.

How long is the combined piece?

Solve and express the final answer in simplest form.

15. Simplify completely.

$\frac{5}{8}+\frac{1}{2}$

16. Simplify completely.

$\frac{1}{2}-\frac{1}{8}$

17. Simplify completely.

$\frac{1}{2}+\frac{3}{8}$

18. Simplify completely.

$\frac{1}{4}-\frac{1}{12}$

19. Simplify completely.

$\frac{7}{8}-\frac{1}{10}$

20. Simplify completely.

$\frac{9}{10}+\frac{7}{8}$

MPA-035: Adding and Subtracting Mixed Numbers with Unlike Denominators

In this lesson, you'll review how to add and subtract mixed numbers with unlike denominators. To add mixed numbers with unlike denominators:

- Find the LCD of both fractions.
- Rewrite the fractions with their LCD as the denominator.
- Add the numerators and keep the LCD as the denominator. You may have to add one whole number to the whole number.
- Add the whole numbers.
- Express the combined whole number-fraction answer in simplest form.

To subtract mixed numbers with unlike denominators:

- Find the LCD of both fractions.
- Rewrite the fractions with their LCD as the denominator.
- Subtract the numerators and keep the LCD as the denominator. You may have to borrow one whole number from the whole number.
- Subtract the whole numbers.
- Express the combined whole number-fraction answer in simplest form.

Here are a few examples. The first example shows addition.

Example 1 Add: $5\frac{1}{4}+2\frac{3}{8}$

$5\frac{1}{4}+2\frac{3}{8} = 5\frac{2}{8}+2\frac{3}{8}$	**Step 1:** Express the fractions using 8 as the denominator, which is the LCD of 4 and 8.
$\frac{2}{8}+\frac{3}{8} = \frac{5}{8}$	**Step 2:** Add the fractions.
$5+2=7$	**Step 3:** Add the whole numbers.
$5\frac{1}{4}+2\frac{3}{8} = 7\frac{5}{8}$	**Step 4:** The whole number and fraction combined is a mixed number already in simplest form.
	Answer: $7\frac{5}{8}$

The following example shows subtraction.

Example 2 Subtract: $8\frac{2}{5}-3\frac{7}{9}$

$8\frac{2}{5}-3\frac{7}{9} = 8\frac{18}{45}-3\frac{35}{45}$	**Step 1:** Express the fractions using 45 as the denominator, which is the LCD of 5 and 9.
$\frac{18}{45}-\frac{35}{45} = ?$	**Step 2:** Since we cannot subtract the fractions, we borrow one whole number from 8.

$8\frac{18}{45} = 7 + \frac{45}{45} + \frac{18}{45}$ $= 7\frac{63}{45}$	**Step 3:** The one whole number borrowed from 8 is written as $\frac{45}{45}$. We add $\frac{45}{45}$ and $\frac{18}{45}$.
$8\frac{18}{45} - 3\frac{35}{45} = 7\frac{63}{45} - 3\frac{35}{45}$	**Step 4:** Now we write the problem.
$\frac{63}{45} - \frac{35}{45} = \frac{28}{45}$	**Step 5:** Subtract the fractions.
$7 - 3 = 4$	**Step 6:** Subtract the whole numbers. Remember that you borrowed one whole from 8, so the first whole number is now 7.
$4\frac{28}{45}$	**Step 7:** The whole number and fraction combined is a mixed number already in simplest form.
	Answer: $4\frac{28}{45}$

Try these examples now.

Example 3 Add: $7\frac{15}{16} + 1\frac{3}{4}$

$7\frac{15}{16} + 1\frac{3}{4} = 7\frac{15}{16} + 1\frac{12}{16}$	**Step 1:** Express the fractions using 16 as the denominator, which is the LCD of 16 and 4.
$\frac{15}{16} + \frac{12}{16} = \frac{27}{16}$	**Step 2:** Add the fractions.
$\frac{27}{16} = 1\frac{11}{16}$ $7 + 1 = 8$	**Step 3:** $\frac{27}{16}$ simplified is $1\frac{11}{16}$, so add this to the first whole number, 7.
$8 + 1 = 9$	**Step 4:** Add the whole numbers. Remember that you added one whole number to 7, so the first whole number is now 8.
$9\frac{11}{16}$	**Step 5:** The whole number and fraction combined is a mixed number already in simplest form.
	Answer: $9\frac{11}{16}$

Example 4 Add: $5\frac{1}{6} + 2\frac{8}{15}$

$5\frac{1}{6} + 2\frac{8}{15} = 5\frac{5}{30} + 2\frac{16}{30}$	**Step 1:** Express the fractions using 30 as the denominator, which is the LCD of 6 and 15.
$\frac{5}{30} + \frac{16}{30} = \frac{21}{30}$	**Step 2:** Add the fractions.

$5 + 2 = 7$	**Step 3:** Add the whole numbers.
$7\frac{21}{30} = 7\frac{7}{10}$	**Step 4:** Combine the whole number and fraction as a mixed number and simplify the fraction.

Answer: $7\frac{7}{10}$

Example 5 Subtract: $23\frac{8}{11} - 17\frac{3}{5}$

$23\frac{8}{11} - 17\frac{3}{5} = 23\frac{40}{55} - 17\frac{33}{55}$	**Step 1:** Express the fractions using 55 as the denominator, which is the LCD of 11 and 5.
$\frac{40}{55} - \frac{33}{55} = \frac{7}{55}$	**Step 2:** Subtract the fractions.
$23 - 17 = 6$	**Step 3:** Subtract the whole numbers.
$6\frac{7}{55}$	**Step 4:** The whole number and fraction combined is a mixed number already in simplest form.

Answer: $6\frac{7}{55}$

Example 6 Subtract: $46\frac{3}{10} - 28\frac{5}{6}$

$46\frac{3}{10} - 28\frac{5}{6} = 46\frac{9}{30} - 28\frac{25}{30}$	**Step 1:** Express the fractions using 30 as the denominator, which is the LCD of 10 and 6.
$\frac{9}{30} - \frac{25}{30} = ?$	**Step 2:** Since we cannot subtract the fractions, we borrow one whole number from 46.
$46\frac{9}{30} = 45 + \frac{30}{30} + \frac{9}{30}$ $= 45\frac{39}{30}$	**Step 3:** The one whole number borrowed from 46 is written as $\frac{30}{30}$. We then add $\frac{30}{30}$ and $\frac{9}{30}$.
$46\frac{9}{30} - 28\frac{25}{30} = 45\frac{39}{30} - 28\frac{25}{30}$	**Step 4:** Now we rewrite the problem.
$\frac{39}{30} - \frac{25}{30} = \frac{14}{30}$	**Step 5:** Subtract the fractions.
$45 - 28 = 17$	**Step 6:** Subtract the whole numbers. Remember that you borrowed one whole number from 46, so the first whole number is now 45.

$17\frac{14}{30} = 17\frac{7}{15}$

Step 7: Combine the whole number, 17, and fraction, $\frac{14}{30}$, as a mixed number and simplify the fraction.

Answer: $17\frac{7}{15}$

Problem Set

1. Add: $5\frac{1}{3} + 1\frac{1}{2}$

2. Add: $2\frac{1}{5} + 4\frac{3}{7}$

3. Add: $7\frac{3}{14} + 1\frac{2}{12}$

4. Add: $1\frac{13}{18} + 1\frac{6}{11}$

5. Add: $2\frac{2}{7} + 3\frac{13}{15}$

6. Add: $2\frac{3}{17} + 1\frac{9}{10}$

7. Add: $8\frac{4}{9} + 3\frac{1}{6}$

8. Add: $3\frac{1}{2} + 2\frac{4}{9}$

9. Subtract: $8\frac{2}{3} - 3\frac{3}{5}$

10. Subtract: $7\frac{6}{7} - 5\frac{3}{8}$

11. Subtract: $6\frac{5}{6} - 2\frac{5}{7}$

12. Subtract: $7\frac{1}{4} - 3\frac{3}{5}$

13. Subtract: $18\frac{3}{4} - 9\frac{7}{8}$

14. Nadine is on the swim team. At the last meet, she practiced by swimming a few laps. She swam her first lap in $1\frac{6}{10}$ minutes. She swam her last lap in $1\frac{1}{4}$ minutes. How long did it take her to swim the two laps?

15. Mohammed's parents live $102\frac{1}{2}$ miles away. He is driving there to visit with them. After an hour he had $54\frac{3}{5}$ miles left to drive. How far did Mohammed drive the first hour?

16. Jason and his dad are building a dog house for their puppy, Cleatas. They have boards that measure $26\frac{3}{4}$ feet long. They need the wood to be $5\frac{1}{2}$ feet long. How much do Jason and his dad have to cut off?

17. Sheryl is painting the house. She bought a can that contains $1\frac{1}{2}$ gallons of paint. It takes $6\frac{3}{4}$ gallons of paint to paint one room. How many more gallons of paint does Sheryl need to paint the room?

18. Subtract: $7\frac{15}{16} - 3\frac{4}{8}$

19. Subtract: $7\frac{20}{27} - 3\frac{20}{63}$

20. Subtract: $18\frac{23}{25} - 7\frac{31}{40}$

MPA-036: Multiplying Fractions and Mixed Numbers and Simplifying

Although you may not realize it, you are already multiplying fractions in your everyday life. For instance, a cookie recipe may require you to add $\frac{1}{2}$ cup of flour 3 times to the mixing bowl. To save time and effort, you quickly figure out that 3 times $\frac{1}{2}$ equals $1\frac{1}{2}$. Thus, you can measure out $1\frac{1}{2}$ cups of flour all at once.

In examples like the one above, the answer seems clear because you are multiplying a fraction by a whole number. However, if a recipe calls for you to add, for example, $\frac{3}{7}$ of a cup of sugar $4\frac{2}{8}$ times to the mix, then you may not be able to calculate the product mentally. To solve problems such as this one, you must become familiar with the rules involved in multiplying **mixed numbers, proper fractions,** and **improper fractions.** You should also know the steps involved in reducing a fraction to **simplest terms.**

Mixed Number	A number consisting of two parts: A whole number and a proper fraction.
Proper Fraction	A fraction whose numerator is less than the denominator.
Improper Fraction	A fraction whose numerator is greater than or equal to the denominator.
Simplest Terms	Improper fractions that have been converted to mixed numbers. Also, all fractions that are proper fractions in lowest terms.

Multiplying Fractions

You can use one of two approaches when multiplying fractions: The Greatest Common Factor Method, and the Cross-Canceling Method.

Greatest Common Factor Method

- Multiply fractions first: $\dfrac{\text{numerator} \times \text{numerator}}{\text{denominator} \times \text{denominator}}$
- Then, reduce the resulting fraction to simplest terms by dividing both the numerator and the denominator by the greatest common factor (GCF).

Example 1 Multiply $\frac{3}{2} \times \frac{5}{30}$ using the GCF Method.

$\frac{3}{2} \times \frac{5}{30} = \frac{15}{60}$

Step 1: Multiply the first numerator, 3, by the second, numerator, 5, and then the first denominator, 2, by the second denominator, 30.

$\frac{3}{2} \times \frac{5}{30} = \frac{15}{60} = \frac{1}{4}$

Step 2: Reduce the resulting fraction, $\frac{15}{60}$, to simplest terms by dividing the numerator and denominator by the GCF, 15.

Answer: $\frac{1}{4}$

Cross-Canceling Method

- First, reduce the fractions by cross-canceling and cancel out any common factors between the numerator and denominator
- Next, multiply the fractions: $\dfrac{\text{numerator} \times \text{numerator}}{\text{denominator} \times \text{denominator}}$

Example 2 Multiply $\frac{25}{4} \times \frac{12}{10}$ using cross-canceling.

$\frac{\not{25}^{5}}{{}_{1}\not{4}} \times \frac{\not{12}^{3}}{\not{10}_{2}} = \frac{5}{1} \times \frac{3}{2}$	**Step 1:** First, reduce the fractions by cross-canceling.
$\frac{5}{1} \times \frac{3}{2} = \frac{15}{2}$	**Step 2:** Multiply the first numerator, 5, by the second numerator, 3, and then the first denominator, 1, by the second denominator, 2.
$\frac{25}{4} \times \frac{12}{10} = \frac{15}{2}$	**Answer:** $\frac{15}{2}$

Multiplying Mixed Numbers

Now that you know how to convert mixed numbers and multiply fractions, you should be able to put these two skills together when asked to multiply mixed numbers.

1. Begin by converting the mixed number to an improper fraction.
2. Use either the GCF method (multiply the improper fractions and reduce to simplest terms using the GCF), or the cross-canceling method (reduce the improper fraction by cross-canceling and then multiply the fractions) to multiply the fractions.
3. Make sure that your answer is expressed in simplest terms.

Example 3 Multiply $3\frac{1}{3} \times 2\frac{2}{5}$ using the GCF method. Remember to multiply the improper fractions and then reduce to simplest terms using the GCF.

$3\frac{1}{3} = \frac{10}{3}, 2\frac{2}{5} = \frac{12}{5}$ $3\frac{1}{3} \times 2\frac{2}{5} = \frac{10}{3} \times \frac{12}{5}$	**Step 1:** Convert the mixed numbers to improper fractions.
$\frac{10 \times 12}{3 \times 5} = \frac{120}{15}$	**Step 2:** Multiply the fractions. Start by multiplying the first numerator, 10, times the second numerator, 12. Then, multiply the first denominator, 3, times the second denominator, 5.
$\frac{120}{15} = \frac{24}{3} = 8$	**Step 3:** Use the GCF, 5, to simplify the fraction. Divide both the numerator and denominator by 5. Express your answer in simplest terms.
$3\frac{1}{3} \times 2\frac{2}{5} = 8$	**Answer:** 8

Example 4 Multiply $1\frac{3}{4} \times 5\frac{2}{14}$ using the cross-canceling method. Remember to reduce the improper fractions by cross-canceling and then multiply the fractions.

$1\frac{3}{4} = \frac{7}{4}$, $5\frac{2}{14} = \frac{72}{14}$ $1\frac{3}{4} \times 5\frac{2}{14} = \frac{7}{4} \times \frac{72}{14}$	**Step 1:** Convert the mixed numbers to improper fractions.
$\frac{\cancel{7}^{1}}{\cancel{4}_{1}} \times \frac{\cancel{72}^{18}}{\cancel{14}_{2}} = \frac{1 \times 18}{1 \times 2} = \frac{18}{2} = 9$	**Step 2:** Use cross-canceling to multiply the fractions together. First, reduce each fraction to simplest terms. Then, multiply the first numerator, 1, by the second numerator, 18. Then, multiply the first denominator, 1, by the second denominator, 2. Then express the result in simplest terms.
$1\frac{3}{4} \times 5\frac{2}{14} = 9$	**Answer:** 9

Remember everything you have learned about mixed numbers, proper fractions, improper fractions, and simplest terms as you review the following examples.

Example 5 Multiply $\frac{3}{8} \times \frac{9}{12}$ using the GCF method. Remember to multiply the fractions and then reduce to simplest terms using the GCF.

$\frac{3 \times 9}{8 \times 12} = \frac{27}{96}$	**Step 1:** Multiply the first numerator, 3, by the second numerator, 9. Then, multiply the first denominator, 8, by the second denominator, 12.
$\frac{\cancel{27}^{9}}{\cancel{96}_{32}} = \frac{9}{32}$	**Step 2:** Use the GCF to express your answer in simplest terms. Divide the numerator and denominator by the GCF, 3. The result is already in simplest terms.
$\frac{3}{8} \times \frac{9}{12} = \frac{9}{32}$	**Answer:** $\frac{9}{32}$

Example 6 Multiply $\frac{2}{22} \times \frac{11}{30}$ using the cross-canceling method. Remember to reduce the fractions by cross-canceling and then multiply the fractions.

$\frac{\cancel{2}^{1}}{\cancel{22}_{2}} \times \frac{\cancel{11}^{1}}{\cancel{30}_{15}}$	**Step 1:** Reduce the fractions by cross-canceling.
$\frac{1 \times 1}{2 \times 15} = \frac{1}{30}$	**Step 2:** Multiply the first numerator, 1, by the second numerator, 1. Then, multiply the first denominator, 2, by the second denominator, 15.
$\frac{2}{22} \times \frac{11}{30} = \frac{1}{30}$	**Answer:** $\frac{1}{30}$

Example 7 Multiply $4\frac{1}{5} \times 1\frac{3}{7}$ using the GCF method. Remember to multiply the improper fractions and then reduce to simplest terms using the GCF.

$4\frac{1}{5} \times 1\frac{3}{7} = \frac{21}{5} \times \frac{10}{7}$

Step 1: Convert the mixed numbers to improper fractions.

$\frac{21 \times 10}{5 \times 7} = \frac{210}{35}$

Step 2: Multiply the first numerator, 21, by the second numerator, 10. Then, multiply the first denominator, 5, by the second denominator, 7.

$\frac{210}{35} = \frac{30}{5} = 6$

Step 3: Reduce answer to simplest terms, using the GCF, 7. Divide the numerator and the denominator by 7, and simplify.

$4\frac{1}{5} \times 1\frac{3}{7} = 6$

Answer: 6

Example 8 Multiply $5\frac{3}{9} \times 3\frac{3}{8}$ using cross-canceling. Remember to reduce the improper fractions by cross-canceling then multiply the fractions.

$5\frac{3}{9} = \frac{48}{9}$, $3\frac{3}{8} = \frac{27}{8}$

$5\frac{3}{9} \times 3\frac{3}{8} = \frac{48}{9} \times \frac{27}{8}$

Step 1: Convert the mixed numbers to improper fractions.

$\frac{\cancel{48}^{6}}{\cancel{9}_{1}} \times \frac{\cancel{27}^{3}}{\cancel{8}_{1}} = \frac{6}{1} \times \frac{3}{1}$

Step 2: Reduce the fractions by cross-canceling.

$\frac{6}{1} \times \frac{3}{1} = \frac{18}{1} = 18$

Step 3: Multiply the first numerator, 6, by the second numerator, 3. Then, multiply the first denominator, 1, by the second denominator, 1. Make sure that your answer is in lowest terms.

$5\frac{3}{9} \times 3\frac{3}{8} = 18$

Answer: 18

Example 9 What is $\frac{1}{4}$ of $\frac{12}{15}$?

$\frac{1}{4} \times \frac{12}{15}$

Step 1: The word "of" means multiply. Change the problem using the symbol for multiplication.

$\frac{1 \times 12}{4 \times 15} = \frac{12}{60}$

Step 2: Multiply the numerators, 1×12, and multiply the denominators, 4×15.

$\frac{\cancel{12}^{1}}{\cancel{60}_{5}}$

Step 3: Reduce the fraction to lowest terms.

Answer: $\frac{1}{5}$

Problem Set

1. Multiply: $\frac{2}{1} \times \frac{8}{4}$

Write the answer in simplest form.

2. Multiply: $\frac{7}{9} \times \frac{11}{12}$

Write the answer in simplest form.

3. Multiply: $\frac{5}{14} \times \frac{7}{11}$

Write the answer in simplest form.

4. Multiply: $\frac{1}{8} \times \frac{9}{8}$

Write the answer in simplest form.

5. Multiply: $\frac{2}{3} \times \frac{3}{6}$

Write the answer in simplest form.

6. Multiply: $\frac{3}{6} \times \frac{5}{12}$

Write the answer in simplest form.

7. Multiply: $\frac{11}{12} \times \frac{1}{11}$

Write the answer in simplest form.

8. Multiply: $\frac{3}{8} \times 3\frac{4}{14}$

Write the answer in simplest form.

9. Multiply: $1\frac{1}{8} \times \frac{7}{8}$

Write the answer in simplest form.

10. Multiply: $\frac{1}{2} \times \frac{2}{5}$

Write the answer in simplest form.

11. Multiply: $\frac{3}{4} \times \frac{2}{3}$

Write the answer in simplest form.

12. What is the area in simplest form?

________ ft^2

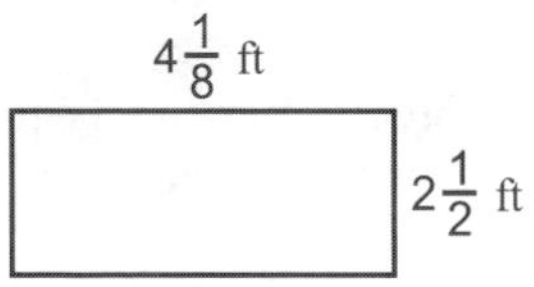

13. What is $\frac{4}{5}$ of 2?

14. What is $\frac{1}{3}$ of 8?

15. What is $\frac{3}{12}$ of $\frac{8}{11}$?

16. What is $\frac{1}{9}$ of $\frac{6}{7}$?

17. Multiply: $98 \times \frac{1}{98}$

Write the answer in simplest form.

18. Multiply: $5 \times \frac{1}{5}$

Write the answer in simplest form.

19. Multiply: $\frac{1}{9} \times \frac{5}{14}$

Write the answer in simplest form.

20. Multiply: $\frac{1}{8} \times 3$

Write the answer in simplest form.

MPA-037: Dividing Fractions and Mixed Numbers and Simplifying

Suppose that you have three friends and you want to split half of a pizza between the four of you. To split it evenly, you will have to divide.

What if you must read sixteen and a quarter pages in your history textbook before the test, and you have two and a half days left? To find out how many pages you have to read each day, you must know how to divide fractions and mixed numbers.

Before you can divide fractions, you must know what a reciprocal is.

Reciprocal	Any two numbers whose product is 1.

Look at the fraction $\frac{1}{8}$. The reciprocal of $\frac{1}{8}$ is a number that you can multiply by $\frac{1}{8}$ and obtain the number 1. The reciprocal of $\frac{1}{8}$ is $\frac{8}{1}$ because $\frac{1}{8} \times \frac{8}{1} = \frac{8}{8} = 1$.

Another thing to remember is that you must place any whole number over a denominator of 1 if you want to write the whole number as a fraction. For example, 3 can be written as the fraction $\frac{3}{1}$.

Dividing fractions

Step 1: Convert any mixed numbers to improper fractions.

Step 2: Rewrite the problem as the first fraction (dividend) multiplied by the reciprocal of the second fraction (divisor).

Step 3: Multiply. Check to see if you can cross-cancel any numbers, so that you do not need to reduce later.

Step 4: Write the answer in lowest terms.

Example 1 Simplify: $\frac{2}{4} \div \frac{1}{3}$

$\frac{2}{4} \div \frac{1}{3} = \frac{2}{4} \times \frac{3}{1}$	**Step 1:** Rewrite the problem as the first fraction multiplied by the reciprocal of the second.
$\frac{2}{4} \times \frac{3}{1} = \frac{1}{2} \times \frac{3}{1}$	**Step 2:** Simplify the fractions to lowest terms.
$\frac{1}{2} \times \frac{3}{1} = \frac{1 \times 3}{2 \times 1}$	**Step 3:** Multiply.
$\frac{3}{2} = 1\frac{1}{2}$	**Step 4:** Convert the improper fraction to a mixed number.

Answer: $\frac{2}{4} \div \frac{1}{3} = 1\frac{1}{2}$

Example 2 Simplify: $1\frac{3}{8} \div 2\frac{1}{2}$

$1\frac{3}{8} = \frac{11}{8}$ $2\frac{1}{2} = \frac{5}{2}$	**Step 1:** Convert the mixed numbers to improper fractions.
$\frac{11}{8} \div \frac{5}{2} = \frac{11}{8} \times \frac{2}{5}$	**Step 2:** Rewrite the problem as the first fraction multiplied by the reciprocal of the second.
$\frac{11}{{}_4\not{8}} \times \frac{\not{2}^1}{5} = \frac{11}{4} \times \frac{1}{5}$	**Step 3:** Use cross-canceling to simplify the multiplication problem.
$\frac{11}{4} \times \frac{1}{5} = \frac{11}{20}$	**Step 4:** Multiply.

Answer: $1\frac{3}{8} \div 2\frac{1}{2} = \frac{11}{20}$

Example 3 Jack has $2\frac{1}{3}$ pounds of dog food to feed to his 4 dogs. How many pounds of food will each dog get?

$2\frac{1}{3} \div 4 = \frac{7}{3} \div \frac{4}{1}$	**Step 1:** Convert the mixed number to an improper fraction.
$\frac{7}{3} \div \frac{4}{1} = \frac{7}{3} \times \frac{1}{4}$	**Step 2:** Rewrite the problem as the first fraction multiplied by the reciprocal of the second.
$\frac{7}{3} \times \frac{1}{4} = \frac{7}{12}$	**Step 3:** Multiply.

Answer: $2\frac{1}{3} \div 4 = \frac{7}{12}$ lbs.

Problem Set

1. Divide: $\frac{8}{9} \div \frac{8}{3}$

Express your answer in simplest form.

2. Divide: $\frac{5}{16} \div \frac{1}{4}$

Express your answer as a mixed number in simplest form.

3. Divide: $\frac{5}{6} \div \frac{3}{4}$

Express your answer as a mixed number in simplest form.

4. Divide: $\frac{1}{2} \div \frac{1}{4}$

Express your answer in simplest form.

5. Divide: $\frac{11}{12} \div \frac{1}{2}$

Express your answer in simplest form.

6. John, Laura, and Michael have ordered a large pepperoni pizza. The pizza is cut into 9 slices. Michael cuts each slice in half. How many slices are there after this?

7. Divide: $3\frac{1}{4} \div \frac{1}{8}$

Express your answer in simplest form.

8. John and Edward are making lasagna. The recipe they are using calls for $2\frac{1}{2}$ cups of mozzarella cheese. The bags of shredded cheese at the grocery store hold $\frac{1}{2}$ cup each. How many bags will John and Edward have to buy to make lasagna?

9. Leah has ordered 10 packages of paper to use on her engineering projects. Each project will use approximately $1\frac{1}{2}$ packages of paper. According to these measurements, how many projects can Leah complete with her 10 packages of paper? (Express your answer as a whole number.)

10. Divide: $4\frac{1}{2} \div 2\frac{1}{4}$

Express your answer in simplest form.

11. Divide: $\frac{1}{8} \div \frac{1}{17}$

Express your answer in simplest form.

12. Divide: $8\frac{3}{4} \div 6\frac{1}{4}$

Express your answer in simplest form.

13. In the following problem, what is the reciprocal of the divisor?

$\frac{1}{8} \div \frac{1}{2}$

14. Divide: $\frac{11}{12} \div \frac{1}{4}$

Express your answer in simplest form.

15. Divide: $\frac{3}{32} \div \frac{3}{8}$

Express your answer in simplest form.

16. Divide: $\frac{1}{36} \div \frac{4}{9}$

Express your answer in simplest form.

17. Rhonda is an engineer. She will need 8 reams of graph paper for her next project. Graph paper is packaged in packages of $\frac{1}{4}$ reams each. How many packages of graph paper will Rhonda need for her next project?

18. Divide: $6 \div \frac{1}{2}$

Express your answer in simplest form.

19. Divide: $\frac{2}{3} \div 8$

Express your answer in simplest form.

20. Divide: $\frac{6}{7} \div 9$

Express your answer in simplest form.

MPA-038: Solving One-Step Equations with Fractions Using Addition and Subtraction

You have learned how to work with one-step equations as well as how to work with fractions. In this lesson, you will learn how to combine that knowledge to solve one-step equations with fractions using addition and subtraction.

When working with such equations, you must remember to use the Addition and Subtraction Properties of Equality.

Addition Property of Equality	If you add one number to one side of an equation, then you must add that same number to the other side of that equation.
Subtraction Property of Equality	If you subtract a number from one side of an equation, then you must subtract that same number from the other side of that equation.

To solve equations that contain fractions and mixed numbers, follow these four steps:

- Set up the equation.
- Use either the Addition or Subtraction Property of Equality to isolate the variable.
- Add or subtract the fraction parts and then the whole numbers. Regroup when necessary.
- Be sure to express the answer in simplest form.

Example 1 After trick-or-treating for Halloween, Seth collected $\frac{12}{16}$ pound of candy. If he has only $\frac{5}{16}$ pound of candy left after one week, how much has he eaten?

n = Number of pounds of candy Seth has eaten

$$n + \frac{5}{16} = \frac{12}{16}$$

Step 1: Set up the equation.

$$n + \frac{5}{16} - \frac{5}{16} = \frac{12}{16} - \frac{5}{16}$$

Step 2: Using the Subtraction Property of Equality, subtract from both sides of the equation to isolate the variable.

$$n = \frac{7}{16}$$

Answer: Seth ate $\frac{7}{16}$ pound of candy.

Example 2 Blake is making a cake that requires $2\frac{2}{3}$ teaspoons of butter. He has already added $1\frac{1}{3}$ teaspoons of butter to the mix. How much more does he need to add?

x = Number of teaspoons of butter left to add

$$2\frac{2}{3} = x + 1\frac{1}{3}$$

Step 1: Set up the equation.

$$2\frac{2}{3} - 1\frac{1}{3} = x + 1\frac{1}{3} - 1\frac{1}{3}$$

Step 2: Use the Subtraction Property of Equality to subtract from both sides of the equation and isolate the variable.

$$x = 1\frac{1}{3} \text{ teaspoons}$$

Answer: Blake still must add $1\frac{1}{3}$ teaspoons of butter to the mix.

Example 3 Dion bought her dog JuJu a bag of dog food. JuJu ate $2\frac{1}{5}$ pounds of the food in one day, and $3\frac{2}{5}$ pounds are left. How much food was in the bag to begin with?

d = Total weight of the bag of dog food

Step 1: Set up the equation.

$d - 2\frac{1}{5} = 3\frac{2}{5}$

Step 2: Isolate the variable using the Addition Property of Equality, which states that you must add to both sides of the equation.

$d - 2\frac{1}{5} + 2\frac{1}{5} = 3\frac{2}{5} + 2\frac{1}{5}$

$d = 5\frac{3}{5}$ pounds

Answer: The total weight of the bag of dog food is $5\frac{3}{5}$ pounds.

Example 4 You want to help your father make an Italian dish. He says that he already added $\frac{7}{9}$ cup of parsley, and that you can add the remaining $\frac{1}{9}$ cup of parsley. How much parsley do you need in total?

p = Amount of parsley needed to complete the recipe

Step 1: Set up the equation.

$p - \frac{7}{9} = \frac{1}{9}$

Step 2: Follow the Addition Property of Equality, and add $\frac{7}{9}$ to both sides of the equation to isolate the variable.

$p - \frac{7}{9} + \frac{7}{9} = \frac{1}{9} + \frac{7}{9}$

$p = \frac{8}{9}$

Answer: The recipe calls for $\frac{8}{9}$ cup of parsley.

Example 5 LaTonia used up $4\frac{1}{6}$ feet of closet space and only has $2\frac{2}{6}$ feet of space left. How much total closet space did she have to begin with?

s = Total amount of closet space

Step 1: Set up the equation.

$s - 4\frac{1}{6} = 2\frac{2}{6}$

Step 2: Using the Addition Property of Equality, isolate the variable by adding to both sides of the equation.

$s - 4\frac{1}{6} + 4\frac{1}{6} = 2\frac{2}{6} + 4\frac{1}{6}$

$s = 6\frac{3}{6} = 6\frac{1}{2}$

Answer: LaTonia started with $6\frac{1}{2}$ feet of total closet space.

Problem Set

Solve:

1. Solve: $4\frac{1}{2} + z = 10\frac{3}{5}$

2. Solve: $n + 6\frac{3}{8} = 7\frac{3}{4}$

3. Solve: $x - \frac{2}{6} = \frac{11}{12}$

4. Solve: $x - \frac{4}{5} = 1\frac{1}{7}$

5. Ayesha started with $12\frac{17}{30}$ feet of wrapping paper. She gave $6\frac{9}{10}$ feet to Jeremy. How much wrapping paper did she have left?

6. What number minus $\frac{3}{4}$ equals $4\frac{5}{8}$?

7. Jose needs $32\frac{3}{4}$ feet of fence board. He has $24\frac{1}{8}$ feet. How many more does he need?

8. Joan needs $13\frac{3}{4}$ yards to make a quilt. She has $9\frac{1}{2}$ yards. Her aunt gave her 2 more yards. How many more yards does she need to make the quilt?

9. Solve: $x - 18 = 12\frac{6}{9}$

10. Solve: $x - 10\frac{7}{16} = 8\frac{3}{16}$

11. Solve: $\frac{5}{7} + r = \frac{6}{7}$

12. Solve: $b - \frac{3}{16} = \frac{9}{16}$

13. Solve: $14\frac{11}{13} = 8\frac{7}{13} + f$

14. There were $3\frac{1}{6}$ chocolate cakes and $1\frac{1}{8}$ yellow cakes left at the party. How many cakes were left in all?

15. Solve: $n + \frac{11}{24} = \frac{23}{24}$

16. Solve: $25\frac{5}{8} = z + 22\frac{2}{5}$

17. Solve: $b - \frac{1}{7} = \frac{2}{7}$

18. Solve: $c - \frac{3}{10} = \frac{5}{10}$

19. Solve: $d + \frac{3}{11} = \frac{7}{11}$

20. Solve: $g + \frac{2}{13} = \frac{8}{13}$

MPA-039: Solving One-Step Equations with Fractions Using Multiplication and Division

By now, you know enough useful information to solve one-step algebra equations and multiply and divide fractions and mixed numbers. Here, you will put that information together. There are just a few rules that you must remember.

Multiplication Property of Equality	If you multiply one side of an equation by a number, then you must multiply the other side by that same number.
Division Property of Equality	If you divide one side of an equation by a number, then you must divide the other side by that same number.

Also, remember that when you divide a number by a fraction, it is the same as multiplying the number by the reciprocal of that fraction.

Here are the steps you must follow to solve one-step equations with fractions:

Step 1: Set up the equation if necessary.

Step 2: Use either the Multiplication or the Division Property of Equality to isolate the variable on one side of the equation.

Step 3: Multiply or divide.

Step 4: Reduce the answer to its simplest form.

Example 1 Solve $n \times \frac{1}{2} = \frac{3}{5}$

$n \times \frac{1}{2} \div \frac{1}{2} = \frac{3}{5} \div \frac{1}{2}$, so $n = \frac{3}{5} \div \frac{1}{2}$

Step 1: Use the Division Property of Equality to isolate the variable, n.

$n = \frac{3}{5} \div \frac{1}{2} = \frac{3}{5} \times \frac{2}{1}$

Step 1: Divide. You must divide both sides of the equation by $\frac{1}{2}$.

Remember to rewrite the problem to show the first fraction being multiplied with the reciprocal of the second fraction.

$n = \frac{3}{5} \times \frac{2}{1} = \frac{6}{5}$

Step 2: Multiply.

$n = \frac{6}{5} = 1\frac{1}{5}$

Step 3: Reduce your answer to simplest terms. In this case, you have to convert the improper fraction to a mixed number.

Answer: $n = 1\frac{1}{5}$

In this example, use the Multiplication Property of Equality.

Example 2 Solve $n \div 6 = 1\frac{2}{5}$

$n \div 6 \times 6 = 1\frac{2}{5} \times 6$, so $n = 1\frac{2}{5} \times 6$

Step 1: Use the Multiplication Property of Equality to isolate the variable. Multiply both sides of the equation by 6 to isolate the variable, n.

$n = 1\frac{2}{5} \times 6 = \frac{7}{5} \times \frac{6}{1}$	**Step 2:** Rewrite the numbers in a form that you can use to multiply. In this case, convert the mixed number to an improper fraction and convert the whole number to a fraction.
$n = \frac{7}{5} \times \frac{6}{1} = \frac{42}{5}$	**Step 3:** Multiply.
$n = \frac{42}{5} = 8\frac{2}{5}$	**Step 4:** Reduce the product to simplest terms.

Answer: $n = 8\frac{2}{5}$

Example 3 Mark and his family have been driving for $7\frac{2}{5}$ hours on their trip to Grandma's. If they are $\frac{1}{8}$ of the way there, then how many hours does it take to get to Grandma's house?

Note: The word ***of*** *in the problem is your clue that this is a multiplication problem.*

$n \times \frac{1}{8} = 7\frac{2}{5}$	**Step 1:** Set up the equation using the variable, *n,* to represent the hours it takes to get to Grandma's house.
$n \times \frac{1}{8} \div \frac{1}{8} = 7\frac{2}{5} \div \frac{1}{8}$, so $n = 7\frac{2}{5} \div \frac{1}{8}$	**Step 2:** Use the Division Property of Equality to isolate the variable. Divide both sides of the equation by $\frac{1}{8}$.
$n = 7\frac{2}{5} \div \frac{1}{8} = \frac{37}{5} \div \frac{1}{8}$	**Step 3:** Rewrite the problem by changing the mixed number to an improper fraction.
$n = \frac{37}{5} \div \frac{1}{8} = \frac{37}{5} \times \frac{8}{1}$	**Step 4:** Remember that when you divide a number by a fraction, it is the same as multiplying the number by the reciprocal of that fraction.
$n = \frac{37}{5} \times \frac{8}{1} = \frac{296}{5}$	**Step 5:** Multiply.
$n = \frac{296}{5} = 59\frac{1}{5}$	**Step 6:** Reduce to simplest terms.

Answer: $59\frac{1}{5}$ hours

Example 4 Jordan has a box of chocolate bars she bought at a school fund-raiser. She wants to divide the bars evenly among her four best friends. She gives each friend $3\frac{1}{2}$ chocolate bars. How many bars did she have in the box before she gave them away?

$n \div 4 = 3\frac{1}{2}$	**Step 1:** Set up the equation using the variable, *n,* to represent the number of bars Jordan had.
$n \div 4 \times 4 = 3\frac{1}{2} \times 4$, so $n = 3\frac{1}{2} \times 4$	**Step 2:** Use the Multiplication Property of Equality to isolate the variable. Multiply both sides of the equation by 4.

$n = 3\frac{1}{2} \times 4 = \frac{7}{2} \times \frac{4}{1}$	**Step 3:** Rewrite the problem by changing the mixed number to an improper fraction.
$n = \frac{7}{2} \times \frac{4}{1} = \frac{28}{2}$	**Step 4:** Multiply.
$n = \frac{28}{2} = 14$	**Step 5:** Reduce to simplest terms.
	Answer: 14 bars

Problem Set

Solve:

1. Solve: $\frac{8}{9}b = \frac{8}{9}$

2. Solve: $\frac{5}{3}h = \frac{7}{8}$

3. Solve: $\frac{3}{4}n = \frac{8}{11}$

4. Solve: $\frac{1}{11}n = \frac{1}{23}$

5. Hugh has used $\frac{2}{7}$ cup of flour in a recipe. He has only used $\frac{1}{2}$ of the flour he needs. How many cups does he need?

6. Genny is making cookies. The cookies have been baking for $\frac{1}{9}$ of an hour. They are $\frac{2}{5}$ of the way done. How many hours should the cookies bake?

7. Rena walked $\frac{2}{15}$ of a mile. She has completed $\frac{1}{7}$ of the total intended distance. How many miles does she want to walk?

8. Solve: $\frac{11}{12}m = 22$
Write the answer in simplest form.

9. Solve: $\frac{9}{13}p = 9$
Write the answer in simplest form.

10. Dominique paid \$10 for a poster. She paid $\frac{2}{3}$ of the original price. What was the original price of the poster? Write the answer in simplest form.

11. Solve: $2\frac{3}{5}b = 1\frac{1}{5}$
Write the answer in simplest form.

12. Solve: $3\frac{2}{3}c = 7\frac{1}{3}$
Write the answer in simplest form.

13. Solve: $\frac{1}{3}x = 7$

14. Solve: $\frac{1}{4}y = 9$

15. Solve: $\frac{1}{3}r = \frac{1}{5}$

16. Solve: $\frac{1}{5}t = \frac{1}{8}$

17. Solve: $\frac{1}{6}w = \frac{1}{7}$

18. Solve: $\frac{1}{7}x = \frac{1}{9}$

19. Solve: $\frac{1}{8}y = \frac{1}{11}$

20. Solve: $\frac{1}{9}x = \frac{1}{11}$

MPA-040: Solving One-Step Equations with Decimals Using All Four Operations

This lesson will require you to combine all the skills you have learned to solve one-step equations that include the four basic mathematical operations. Let's review which operations are opposites, or inverses, of each other. This information is essential to solving for the unknown variable.

Addition Property of Equality	The same number may be *added* to both sides of an equation without changing the equality. If $a = b$, then $a + c = b + c$.
Subtraction Property of Equality	The same number may be *subtracted* from both sides of an equation without changing the equality. If $a = b$, then $a - c = b - c$.
Multiplication Property of Equality	Both sides of an equation may be *multiplied* by the same number (except 0) without changing the equality. If $a = b$, then $a \cdot c = b \cdot c$.
Division Property of Equality	Both sides of an equation may be *divided* by the same number (except 0) without changing the equality. If $a = b$ and $c \neq 0$, then $a \div c = b \div c$.

In simple story problems involving equations, you must first choose a letter, or variable, to represent the unknown quantity you are solving for. Next, set up an equation using the information given. Then, you can solve for the variable using inverse operations. When solving an equation, we need to isolate the variable on one side and the solution on the other side of the equation. Let's work through a few examples of setting up equations and using inverse operations to solve for a variable.

Example 1 Kari bought two shirts for a total of \$37.18. If one shirt cost \$21.49, how much did the other shirt cost?

s = other shirt	**Step 1:** Choose a letter to represent the other shirt.
$\$21.49 + s = \$37.18$	**Step 2:** Set up an equation using the variable and the known information.
$\$21.49 + s = \$37.18$ $\underline{-21.49} \quad \underline{-21.49}$	**Step 3:** You want to isolate s, the variable. The opposite of adding \$21.49 is subtracting \$21.49. Subtract \$21.49 from both sides of the equation.
$s = \$15.69$	**Step 4:** Solve for the variable.
Check: 21.49 one shirt + 15.69 other shirt 37.18 total cost	**Answer:** Kari's other shirt cost \$15.69.

Example 2 Mrs. Keller says she is on a new diet. So far she has lost 52.5 pounds and now weighs 162 pounds. How much did she weigh when she started this new diet?

x = weight before diet	**Step 1:** Choose a variable to represent her weight when she started her diet.
$x - 52.5 = 162$	**Step 2:** Set up an equation using this variable and the known information.
$x - 52.5 = 162.0$ $\underline{+52.5} \quad \underline{+52.5}$	**Step 3:** You want to isolate the variable (x). The opposite of subtracting 52.5 is adding 52.5. Add 52.5 to both sides of the equation.

$x = 214.5$	**Step 4:** Solve for the variable.
Check: 214.5 -52.5 162.0	**Answer:** Mrs. Keller weighed 214.5 pounds when she started her diet.

Example 3 If 15 cherry colas cost Jim \$33.75, how much did each one cost?

c = 1 cherry cola	**Step 1:** Choose a variable to represent one cherry cola.
$15c = \$33.75$	**Step 2:** Set up an equation using the variable and known information.
$\frac{15c}{15} = \frac{\$33.75}{15}$	**Step 3:** You want to isolate *c,* the variable. The opposite of multiplying by 15 is to divide by 15. Divide both sides by 15.
$c = \$2.25$	**Step 4:** Solve for the variable.
Check: 2.25 × 15 1125 + 2250 33.75	**Answer:** Each cherry cola costs \$2.25.

Example 4 Brad wants to buy each of his three friends a gift and spend \$18.95 on each one. How much money does he need?

m = total money needed	**Step 1:** Choose a variable to represent the money Brad needs.
$\frac{m}{3} = \$18.95$	**Step 2:** Set up an equation using the variable and other known information.
$3\left(\frac{m}{3}\right) = (18.95)(3)$	**Step 3:** You want to isolate the variable (m). The opposite of dividing by 3 is to multiply by 3. Multiply both sides of the equation by 3.
$m = \$56.85$	**Step 4:** Solve for the variable.
Check: \$18.95 3)\$56.85 -3 26 -24 28 -27 15 -15 0	**Answer:** Brad needs \$56.85 to buy his gifts.

Problem Set

Solve:

1. Solve: $4n = 2.4$

2. Solve: $4n = 2.8$

3. Solve: $5n = 4.5$

4. Solve: $\frac{n}{8} = 7.2$

5. Solve: $\frac{n}{6} = 1.2$

6. Solve: $x + 2.45 = 3$

7. Solve: $x + 5.18 = 8.1$

8. Cherie buys one carton of eggs for $3.65, and gives the cashier a five dollar bill. How much change will Cherie get back?

9. Gavin buys 3 video tapes for $15.98. He gives the cashier a twenty dollar bill. How much change will Gavin get back?

10. Steve purchased 2 pairs of cowboy boots for $65.76. He gave the cashier eighty dollars. How much money will Steve get back?

11. Lexi purchased a can of tennis balls for $2.33. She gave the cashier five dollars. How much money will Lexi get back?

12. Kurt bought 10 cans of cola for $5.00. How much money did Kurt pay for each can of cola?

13. Joe ran a 26.2-mile marathon in 222.7 minutes. On average, how long did he take to run each mile?

14. Dwayne bought 20 bottles of juice for $11.40. How much did Dwayne pay for each bottle of juice?

15. Sandy bought 12 bananas for $3.12. How much did Sandy pay for each banana?

16. Fox bought 6 pairs of socks for $5.94. How much did Fox pay for each pair?

17. Anthony is training for the Olympics. He wants to reduce his 200 yard time by 0.4 second. If his desired time is 22.6 seconds, then what is his current time?

18. Solve: $3 + e = 42.5$

19. Solve: $y - (3.76 + 4.2) = 0.214$

20. Tranika bought 2 pairs of socks for $8.78. If she gave the cashier $10.00, how much change did she receive?

MPA-041: Writing Simple Algebraic Expressions from Phrases

Your boss offers to pay you double your wages this week. You think this is great, because you owe your friend \$100 and will take this money out of the overtime pay.

How can you write this situation using numbers and symbols? First, you must identify the information you don't know. In this case, you do not know what your wages will be for the week. Call this x. Remember that you use variables in place of numbers that you do not know. Since you are doubling this amount, you will multiply x by 2 and get $2x$. Since you will be paying your friend \$100, take this amount away from your total pay. Now you have $2x - 100$. In this situation, you have translated words into an algebraic expression. Remember that an algebraic expression is a combination of variables, numbers, and at least one operation.

You can look for keywords or phrases to help you decide which operation or operations will be used in your translation. These are NOT the only words or phrases, but are the most common.

Addition	Indicated by words or phrases such as "more than;" "plus;" "the sum of;" "greater than;" "increased by;" or "added to."
Subtraction	Indicated by words or phrases such as "less than;" "minus;" "difference;" "decreased by;" or "subtracted from."
Multiplication	Indicated by words or phrases such as "times;" "the product of;" "multiplied by;" or "twice" (indicates multiplication by 2).
Division	Indicated by words or phrases such as "divided by;" "the quotient of;" or "half" (indicates division by 2).

When you translate these words or phrases, it is important to remember a few rules or properties about the four operations.

1. **Addition is commutative.** This means that you can add numbers in any order. For example, $x + 5$ is the same as $5 + x$. Reversing the order won't change the sum. Replace x with a value and see for yourself.
2. **Multiplication is also commutative.** For example, 3×4 gives you the same answer as 4×3. Using variables, $5a$ is the same as $a5$, but you always write the number followed by the letter, so you use $5a$. Once you find the value of a, you can multiply in any order.
3. **Subtraction is NOT commutative.** The order in which you subtract numbers *does* matter. Three less than 4 is different than 4 less than 3. Three less than 4 translates into $4 - 3$, which equals 1. Four less than 3 translates into $3 - 4$, which equals -1. The answers are not the same.
4. **Division is NOT commutative.** Eight cookies shared between 4 people ($8 \div 4$) means each person gets 2 cookies. Four cookies shared between 8 people ($4 \div 8$) means each person gets half a cookie. The answers are not the same.

Example 1 Write "seven increased by some number" as an algebraic expression.

$+x$	**Step 1:** "Increased by" means add, and "some number" means you do not know the number, so call it x.
$7 + x$ $x + 7$	**Step 2:** Put the 7 into the expression. You can now translate this phrase into $7 + x$. You could also write $x + 7$, since addition is commutative.

Answer: $x + 7$ or $7 + x$

Example 2 Write "23 decreased by a number" as an algebraic expression.

$-x$	**Step 1:** "Decreased by" means subtract, and "a number" means you do not know the number, so call it x.
$23 - x$	**Step 2:** Put the 23 into the expression. You can now translate this phrase into $23 - x$. You cannot write this as $x - 23$, since subtraction is not commutative. *Note: You may sometimes need help deciding the order in which you should subtract. In those cases, try substituting real numbers. For example, 23 decreased by some number might be 23 decreased by 4. What does this mean? It means you would take 4 away from 23 or 23 – 4* not *4 – 23. So, from this example, you can see that 23 –* x *is correct.*
	Answer: $23 - x$

Example 3 Express "twice a number" algebraically.

$2 \cdot x$	**Step 1:** "Twice" means multiply by 2, and "a number" means that you do not know the number. Let x represent the number.
$2x$	**Step 2:** When you multiply a number and a variable, you do not need to write a multiplication symbol. Even though multiplication is commutative, remember to write the number first.
	Answer: $2x$

Example 4 Write "a number divided by 7" as an algebraic expression.

$x \div$	**Step 1:** "A number" means that you do not know the number, so call it x. Of course, "divided by" means division.
$x \div 7$	**Step 2:** Now put the 7 into the expression. Order does matter, since division is not commutative. Should the expression be $x \div 7$ or $7 \div x$? Let's pretend that the number you do not know is 14. "A number divided by 7" becomes $14 \div 7$. Now when you put the variable back in, the expression is $x \div 7$.
	Answer: $x \div 7$

Example 5 Express "2 more than 3 times a number" algebraically.

$+ 2$	**Step 1:** Break this expression down into smaller pieces; then translate. "Two more than" means that you will add 2.
$3 \times ?$	**Step 2:** "Three times" means that you will be multiplying something by 3.
$3x + 2$ $2 + 3x$	**Step 3:** "A number" means that you do not know this value, so let's call it x. When you put this together, it becomes $2 + 3x$. Since addition is commutative, you can also write this as $3x + 2$.
	Answer: $3x + 2$ or $2 + 3x$

Problem Set

1. Write as an algebraic expression: p increased by 9.

2. Write as an algebraic expression: Twelve plus w.

3. Write as an algebraic expression: A number b divided into four parts.

4. Write as an algebraic expression: 4 pies divided into y slices.

5. Write as an algebraic expression: Seventeen decreased by m.

6. Write as an algebraic expression: The sum of twice a number, k, and 8.

7. Write as an algebraic expression: 14 more than some number, a.

8. Write as an algebraic expression: 23 less than n.

9. Write as an algebraic expression: x less than 52.

10. Write as an algebraic expression: Twice the number, n.

11. Write as an algebraic expression: 16 subtracted from m.

12. Write as an algebraic expression: 4 decreased by c.

13. Write as an algebraic expression: The quotient of b and 4.

14. Write as an algebraic expression: The difference of p and 32.

15. Write as an algebraic expression: The product of b and 49.

16. Write as an algebraic expression: c less than 25.

17. Write as an algebraic expression: 8 less than the product of 20 and a number, b

18. If Cierra has $18 and she spends d dollars, how much will she have left? Write an algebraic expression that represents how much money Cierra will have left.

19. Write as an algebraic expression: The sum of 17 and the quotient of f and 3.

20. Eddie and Ernest were bowling. Eddie scored 12 points more than twice what Ernest scored. Let r represent Ernest's score. Write an expression for Eddie's score.

MPA-042: Solving Problems Using an Equation

In a previous lesson, you learned to create equations from the given information. Now, you will learn to solve equations to answer questions.

To create and solve an equation from a word problem:

1. Assign a letter variable to the unknown quantity.
2. Translate the word problem into an equation with numbers, symbols, and the variable.
3. Isolate the variable.
4. Solve for the variable.

The third step, isolating the variable, involves inverse operations. Addition and subtraction are inverse operations, and multiplication and division are inverse operations. For example, if the equation you write involves adding 3 to x, then with inverse operations, you subtract 3 from both sides of the equation to get x by itself.

Example 1 Solve $x + 3 = 4$

$x + 3 - 3 = 4 - 3$	**Step 1:** Subtract three from both sides of the equation to negate positive three on the left.
$x + 0 = 1$	**Step 2:** On the left side of the equation, $+ 3 - 3$ is the same as adding zero.
$x = 1$	**Step 3:** Remove the zero, and you are left with $x = 1$.
	Answer: $x = 1$

The same inverse operation relationship works for both multiplication and division.

Example 2 Solve $\frac{x}{3} = 4$

$\frac{x}{3} \cdot 3 = 4 \cdot 3$	**Step 1:** Since $\frac{x}{3}$ is the same as $\frac{1x}{3}$, multiply both sides of the equation by three to isolate x.
$x \cdot 1 = 12$	**Step 2:** Multiplying $\frac{x}{3} \cdot \frac{3}{1}$ equals 1. Since multiplying by one doesn't change the result, you are left with $x = 12$.
	Answer: $x = 12$

Before you can perform inverse operations to isolate a variable, you'll need to learn how to write an equation with a variable from a word problem. It's simply a matter of translation. Read the problem and substitute key words. For example, "more" indicates addition, while "less" indicates subtraction. "Times" or "of" usually indicates multiplication, while "among" or "between" usually indicates division.

Here's a simple example. Write an equation from the word problem and solve the equation:

Example 3 The supermarket has apples in a basket. If the stock clerk puts 27 more apples in the basket so that it contains 92 apples, how many apples did the basket originally have?

a = original number of apples	**Step 1:** Select a variable for the unknown—the original number of apples.
$a + 27 = 92$	**Step 2:** "More" means add, so add 27 to a.
$a + 27 - 27 = 92 - 27$	**Step 3:** The inverse operation of addition is subtraction, so subtract 27 from both sides of the equation to isolate a.
$a = 65$	**Step 4:** Simplify if necessary.
a = 65 apples	**Step 5:** Since a represents apples, $a = 65$ means 65 apples.
	Answer: The basket originally contained 65 apples.

The next problem demonstrates multiplication and division as inverse operations.

Example 4 Melinda has 12 bags of chocolate candies. If she has 240 candies in all, how many candies does each bag contain?

c = the number of chocolate candies in one bag	**Step 1:** Select a variable for the unknown—the number of chocolate candies in one bag.
$c \cdot 12 = 240$	**Step 2:** "Of" means multiply, so multiply 12 by c.
$c \cdot 12 \div 12 = 240 \div 12$	**Step 3:** The inverse of multiplication is division, so divide both sides of the equation by 12 to isolate c.
$c = 20$	**Step 4:** Simplify if necessary.
	Step 5: Since c represents chocolate candies, $c = 20$ means that each bag contains 20 candies.
	Answer: Each bag contains 20 chocolate candies.

It may be a little confusing to write equations from word problems, but it only gets easier with practice, so here are a few for you to try.

Example 5 Evan adds 14 marbles to Dade's collection. Now Dade has 216 marbles. How many did Dade have before Evan added to his collection?

m = original number of marbles	**Step 1:** Select a variable for the unknown—the original number of marbles in Dade's collection.
$m + 14 = 216$	**Step 2:** The question uses the word "adds," so add 14 to m.
$m + 14 - 14 = 216 - 14$	**Step 3:** The inverse operation of addition is subtraction, so subtract 14 from both sides of the equation to isolate m.
$m = 202$	**Step 4:** Simplify if necessary.
m = 202 marbles	**Step 5:** Since m represents marbles, $m = 202$ means 202 marbles.
	Answer: Dade's collection originally contained 202 marbles.

Example 6 Andrea has 36 episodes of a popular cartoon on tape. If she has 18 hours of tape, how long is each episode?

h = length of episode in hours	**Step 1:** Select a variable for the unknown—the length (in hours) of an episode.
$h \cdot 36 = 18$	**Step 2:** "Of" means multiply, so multiply 36 by h.
$h \cdot 36 \div 36 = 18 \div 36$	**Step 3:** The inverse of multiplication is division, so divide both sides of the equation by 36 to isolate h.
$h = \frac{1}{2}$	**Step 4:** Simplify if necessary.
$h = \frac{1}{2}$ hour	**Step 5:** Since h represents hours in an episode, $h = \frac{1}{2}$ means that each episode lasts $\frac{1}{2}$ hour.
	Answer: $\frac{1}{2}$ hour

Example 7 A group of players divided into 4 teams of 12 players each. How many players were in the whole group?

p = number of players in the whole group	**Step 1:** Select a variable for the unknown—the number of players in the whole group.
$p \div 4 = 12$	**Step 2:** The question uses the word "divided," so divide p by 4.
$p \div 4 \times 4 = 12 \times 4$	**Step 3:** The inverse of division is multiplication, so multiply both sides of the equation by 4 to isolate p.
$p = 48$	**Step 4:** Simplify if necessary.
$p = 48$ players	**Step 5:** Since p represents players, $p = 48$ means that the whole group had 48 players.
	Answer: 48 players

Problem Set

1. A number decreased by 21 is 13. Find the number.

2. A number multiplied by 12 is 60. Find the number.

3. Donald has 365 baseball cards. If he has 5 times as many baseball cards as Davey, how many baseball cards does Davey have?

4. James spent 3 times as much money today as he did yesterday. If James spent $10 yesterday, how much money did James spend today?

5. After Pinky loaned his friend Al $5.18, Pinky had $1.47 remaining in his wallet. Find the original amount in Pinky's wallet.

6. Find the equation which fits the statement: The difference of a number and 5 is 24.

7. Find the equation which best fits the statement: A number f less 12 is 16.

8. Find the equation that best fits the statement: 22 added to x is 33.

9. Find the equation that best fits the statement: The quotient of a number and 5 is 11.

10. Find the equation that best fits the statement: 12 less than a number n is 7.

11. Find the equation that best fits the following: A number multiplied by 3 is 24.

12. Find the equation that best fits the following: Five times a number is 35.

13. Find the equation that best fits the statement: The quotient of a number and 7 is 2.

14. Find the equation that best fits the statement: 3 times a number is 72.

15. Find the equation that best fits the statement: The difference of a number and 34 is 7.

16. Ms. Adams had 10 more students in her class than Mrs. Miller. If Ms. Adams had 35 students in her class, write an equation to show how many students Mrs. Miller had. Let m = Mrs. Miller's students.

17. Angela had \$10 less than Damien. If Angela had \$36, then write an equation to find how much money Damien had. Let d = amount of money Damien had.

18. A local high school has "m" minutes of class time in their school day. If there are 50 minutes in each class period and there are 7 class periods, write an equation to find out how many total minutes of class time there are in a day.

19. The difference of Tyrell's weight and Kendall's weight is 150 pounds. If Tyrell weighs 230 pounds, write an equation to show Kendall's weight. Let k = Kendall's weight.

20. Deishon tripled her hourly wage at work when she graduated from college. If her new hourly wage is \$18 per hour, write an equation to show how much her hourly wage was before she graduated from college. Let w = her wage before graduation.

MPA-043: Reading and Writing Integers

Have you ever heard someone say, "I lost 10 pounds on this new diet"? On the other hand, maybe a friend has confided that when she went on that cruise she gained 10 pounds. These situations are the *opposites* of one another. In one case there is a *loss* of 10 pounds and in the other there is a *gain* of 10 pounds. Think of other opposites. If you keep a checkbook, there are deposits (increases) and withdrawals (decreases). In football, a player can have a 3-yard gain or a 3-yard loss. There is a set of numbers in our number system that can help us numerically explain situations that involve opposites. It is called the set of **integers**.

Integers	The set of numbers consisting of whole numbers and their opposites.

Remember that the set of whole numbers begins with zero and keeps going $\{1, 2, 3, \ldots\}$. These are also called **positive numbers**. The opposites of these numbers are –1, –2, –3, –4, and so on. These are also called **negative numbers**. Zero is an integer that is neither positive nor negative. A number line may help you visualize this better.

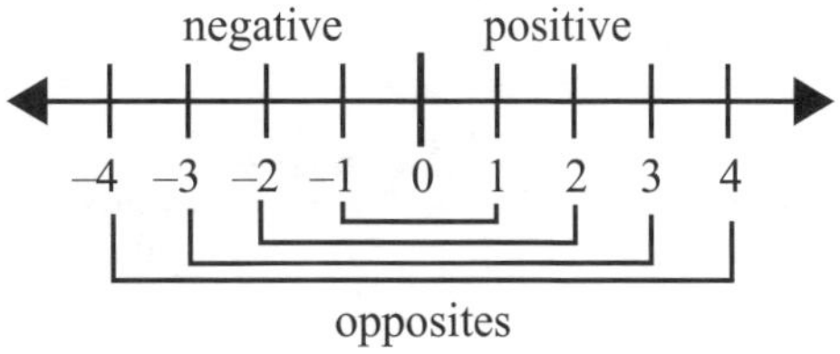

Positive numbers	Numbers that are greater than zero.
Negative numbers	Numbers that are less than zero.

So we could describe a deposit of $25 as +25, and a withdrawal of the same amount would be –25. A gain of 3 yards would be +3 and a loss of 3 yards would be –3. Similarly, a gain of 10 pounds is written +10 and a loss of 10 pounds is written –10.

In the following examples, we will be writing integers in standard form and in word form. **Standard form** means we will take words and write them using numbers and symbols. **Word form** means we will take numbers and symbols and translate them into words.

Example 1 Express "729" in word form.

Note: Notice that there is no positive or negative sign before the number. This means the number is automatically positive.

Answer: Positive seven hundred twenty-nine OR seven hundred twenty-nine. (Remember that compound numbers between 21 and 99 are usually hyphenated.)

Example 2 Express "–416" in word form.

Answer: Negative four hundred sixteen

Example 3 Express "two hundred eighty-four" as an integer in standard form.

Answer: +284 or 284

Example 4 Express "negative one thousand" as an integer in standard form.

Answer: –1,000 (Remember to use commas between periods of numbers: 10, 324, 421.)

Note: There are two ways to write positive integers: with or without the positive sign. There is only one way to write negative integers: with a negative sign before the number.

Problem Set

1. Write negative nine hundred seventy-seven in standard form.

2. A 20-ounce soda has about five hundred ninety-one milliliters of liquid. Express the number of milliliters as an integer.

3. The Dead Sea, which is located in the Dead Sea Rift, is at one thousand, three hundred seventy-one feet below sea level. Express this as an integer.

4. Write 548,322 in words.

5. Write negative eighty-five million, four hundred fifty-two thousand in standard form.

6. Write negative seven hundred sixty-three thousand, four hundred five in standard form.

7. Write –106 in words.

8. Write negative eight hundred thirty-seven in standard form.

9. The elevator went down two floors. Express this descent as an integer.

10. Write –1,125 in words.

11. Write negative one thousand, eight hundred one in standard form.

12. Write negative three thousand, one hundred three in standard form.

13. Write negative five thousand, five hundred twenty-six in standard form.

14. Write six thousand, four hundred thirty-two in standard form.

15. Write negative eight thousand, one hundred forty-four in standard form.

16. Write negative nine thousand, eight hundred one in standard form.

17. There are one thousand, seven hundred sixty yards in one mile. Express the number of yards as an integer.

18. The cavern was eight hundred, eighty-two feet below sea level. Express this depth as an integer.

19. Write –7,463,285 in words.

20. Write 97,335 in words.

MPA-044: Finding Opposite and Absolute Values of Integers

Think about the word "opposite." Can you think of examples? For instance, you can fast forward a videotape; you can also rewind it. These are opposite directions. You can drive north or you can drive south. Again, these are opposite directions. You can lose weight or gain it. You can make withdrawals or deposits into a savings account. You can turn a light on or off. Can you think of more?

Integers also have opposites. Let's look at a number line.

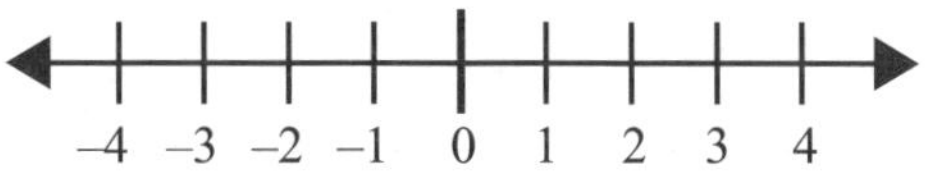

Numbers that are the same distance from zero are called **opposite numbers**. If you begin at zero and travel 3 spaces to the right, you end up at 3 or +3. If you travel the same distance from zero but to the left, you will end up at –3. This means that 3 and –3 are opposites.

Property of Opposites	For every number *a*, there is a number $-a$ such that $a + (-a) = 0$.

For example, if you gain 10 pounds (+10) and then lose ten pounds (–10), how many pounds have you really gained or lost overall? Zero! So, $10 + (-10) = 0$. If you deposit \$50 into your savings account, and then withdraw \$50 tomorrow, how much did you really save overall in those two transactions? Zero dollars, because $50 + (-50) = 0$.

Distance itself is never negative. If you decide to call traveling north positive and traveling south negative (remember these are opposites), then if you travel 6 miles south you can call this –6 just to show *direction*. In reality, though, you have traveled those 6 miles so the *distance* is positive. This distance is a number's **absolute value**. Absolute value is displayed as two straight lines on either side of the number. For example, the absolute value of 4 is written $|4|$, and the absolute value of –4 is written $|-4|$.

Absolute value	The number of units, or distance, from zero. Absolute value is always positive.

Let's look at the number line again.

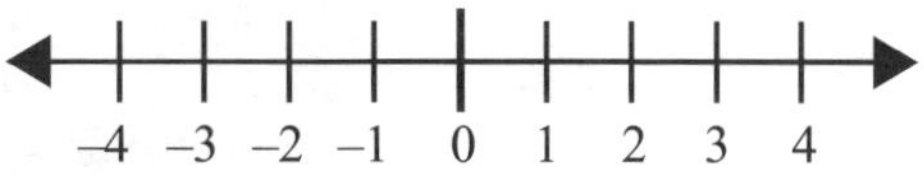

What is the absolute value of +4? The question you want to ask yourself is, "What is the distance of the number from 0?" In this case the distance is 4, so the absolute value is 4.

What is the absolute value of –1? What is the distance of –1 from 0? The *distance* is 1, so the absolute value is +1.

Example 1 What is the opposite of –9?

Answer: Since –9 is 9 spaces to the *left* of 0, its opposite will be 9 spaces to the *right* of 0. The answer is +9 or 9.

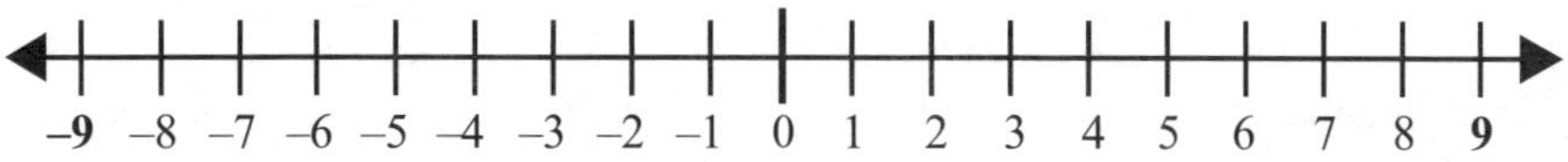

Example 2 What is the absolute value of 8?

Answer: How far away is 8 from 0? Answer: +8 or 8

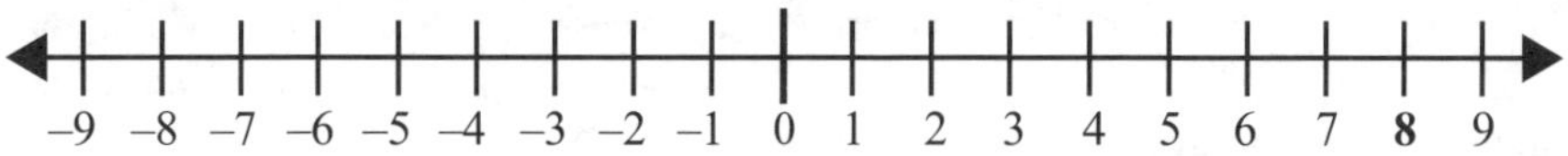

Example 3 What is the opposite of 60?

Answer: Since 60 is 60 spaces to the *right* of 0, its opposite will be 60 spaces to the *left* of 0. Answer: –60

Example 4 What is the absolute value of – 40?

Answer: How far away is –40 from 0? Answer: 40 or +40

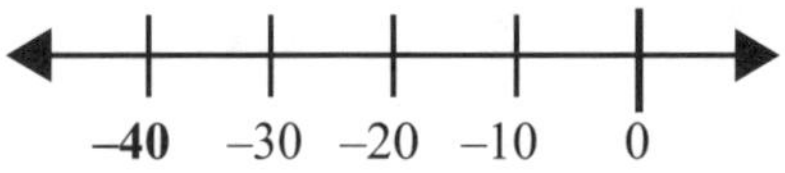

Problem Set

1. Find |150|

2. Find |–79|

3. Find |42|

4. Find the opposite of 211.

5. Find the opposite of –15.

6. Find the opposite of 9,500.

7. Find the opposite of 890.

8. Find the opposite of –45.

9. Find the opposite of –7,578.

10. Find the opposite of 384.

11. Find the opposite of –15.

12. The absolute value of 71 is _____.

13. The opposite of –89 is _____.

14. The opposite of 124 is _____.

15. The absolute value of 84 is –84.
True False

16. The Aardvarks lose 2 yards on a screen pass. What number of yards would they need on the next play to recover the yardage?

17. At noon the tide of a bay is six feet above sea level. How many feet must the tide drop to reach sea level?

18. The height of the water in a certain bay is 130 cm below sea level. How many centimeters must the water level of the bay rise to reach sea level?

19. Vishnu adds $600 to the $200 already in his checking account. How much money would he have to withdraw to empty his checking account?

20. Mandy has –$433 in her checking account. How much money must she put in the account by the end of the month to have $0 in the account?

MPA-045: Comparing and Ordering Integers

If you have \$5 and your friend has \$7, who has more money? Obviously your friend does. Using symbols and numbers, you can write this two ways: \$5 < \$7 or \$7 > \$5. The first example is read "\$5 is less than \$7", and the second example is read "\$7 is greater than \$5."

> *Note: Notice two things here. 1) You always read from left to right. 2) The "mouth," or the open end of the symbol is always open toward the* ***larger*** *number.*

If the temperature in Albany, New York is –10°, and the temperature in Albany, Georgia is 80°, which place is warmer? Georgia, of course! You can write this as –10 < 80, which reads as "–10 is less than 80". You can also write this as 80 > –10, which reads as "80 is greater than –10".

> *Note: If you think about it, a negative number will always be smaller than a positive number.*

Now let's say the temperature in Albany, New York is –10° and the temperature in Bar Harbor, Maine is –20°. Both temperatures are pretty cold, but which place is warmer? Let's look at a thermometer, which is really just a vertical number line.

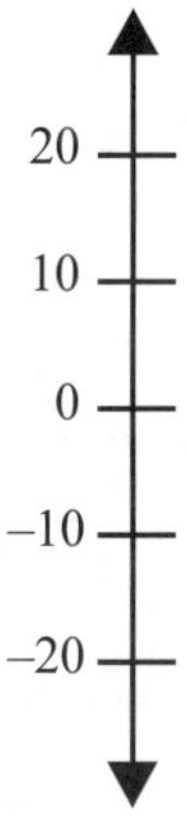

See how –10 is higher than –20? This means that –10° is warmer than –20°. So –10° is greater in temperature than –20°. You can write this as –10 > –20 or –20 < –10. Think of this general rule…

- When you compare or put numbers in order, think of where they are on a number line. The value of numbers increase as you move from left to right on the number line.

Example 1 The temperatures in 5 cities were 72°, –5°, 55°, 5°, and –22°. Put these temperatures in ascending order, from least to greatest.

Answer: Think of a vertical number line. The lowest temperature will be –22°. Moving up the number line, this will be followed by –5°, 5°, 55°, and 72°.

Answer: –22°, –5°, 5°, 55°, and 72°

Example 2 Insert the proper symbol: >, <, or =. –95 _ –91

Answer: –95 is to the left of –91 on a number line. Therefore, –95 is less than –91, or –95 < –91.

Example 3 In one week, Tammy earns \$231, Stacey earns \$395, and Holly earns \$367. Put these amounts in descending order, from greatest to least.

Answer: On a number line, the number furthest to the left (which is the least amount) is 231, followed by 367, then 395. On a number line, $231 < 367 < 395$, which reads as "231 is less than 367, which is less than 395." The problem asked us to put these amounts in order from greatest to least, so you need to write these numbers as $395 > 367 > 231$. The answer reads as "395 is greater than 367, which is greater than 231."

Answer: \$395, \$367, \$231

Example 4 The temperature in Bozeman, MT is –15 (degrees)F and the temperature in Duluth, MN is –23 (degrees) F, which city has the lowest temperature?

Answer: Think of a vertical number line. The lowest temperature will be –23 (degrees) therefore Duluth, MN has the lowest temperature.

Problem Set

1. Compare using <, >, or =

–45 ___ 1

2. Compare using <, >, or =

–9 ___ 23

3. Compare using <, >, or =

0 ___ –10

4. Which list of numbers is in order from least to greatest?

–15, 7, 5	7, 5, –15
5, –15, 7	–15, 5, 7

5. Which list of numbers is in order from least to greatest?

–5, 12, 2	–5, 2, 12
2, 12, –5	12, 2 – 5

6. Which list of numbers is in order from greatest to least?

5, –15, 7	7, –15, 5
7, 5, –15	5, 7, –15

7. The low temperature yesterday in Lamar, Colorado was –25 degrees. The low temperature today was 6 degrees. When was it warmer, yesterday or today?

8. The balance in Kaley's checking account was \$2 on Tuesday. The balance on Thursday was –\$20. When did Kaley have less money in her account?

Thursday   Tuesday

9. Three friends are playing a game with a die that is numbered with these numbers: –8, –6, –4, 0, 2, 3. Each friend rolled the die one time. The one with the highest roll wins. Danny rolled a –8, Jamie rolled a 0, Ethel rolled a 3. Which list shows the friends in order from highest roll to lowest roll?

Danny, Jamie, Ethel　　Jamie, Danny, Ethel
Ethel, Jamie, Danny　　Ethel, Danny, Jamie

10. Compare using <, >, or =
–6 ___ –9

11. Compare using <, >, or =
–9 ___ –14

12. Compare using <, >, or =
–12 ___ –22

13. Which list of numbers is in ascending order?

–4, 11, 0, 14, –13　　0, –4, 11, –13, 14
11, –13, –4, 0, 14　　–13, –4, 0, 11, 14

14. Which list of numbers is in ascending order?

0, 2, 3, –12, 14　　14, –12, 2, 3, 0
–12, 0, 2, 3, 14　　0, –12, 2, 3, 14

15. Which list of numbers is in ascending order?

–14, –11, 5, 9, –15　　–15, –14, –11, 5, 9
5, 9, –11, –14, –15　　–11, 5, 9, –15, –14

16. Which list of numbers is in descending order?

–5, 12, –1, 11, 8　　12, 11, 8, –1, –5
12, –1, 11, 8, –5　　12, 11, 8, –5, –1

17. Compare using <, >, or =
|–24| ___ –11

18. Compare using <, >, or =
–18 ___ |–8|

19. Compare using <, >, or =
|–8| ___ 8

20. Compare using <, >, or =
–23 ___ |21|

MPA-046: Graphing Points on a Coordinate Plane

Knowledge about the Cartesian coordinate plane is essential for many different tasks. For example, the coordinate plane is used to find directions, to track hurricanes, and in navigation systems, among other things. In this lesson, we will learn about the different parts of the coordinate plane and how to graph points on it as well.

Coordinate Plane	A plane formed by two perpendicular intersecting number lines.
Axes	The two perpendicular intersecting number lines on a coordinate plane.
Origin	The point at which the two axes of a coordinate plane intersect.

Each axis contains a positive and negative side. The ***x*-axis**, also known as the horizontal axis, has a negative left side and a positive right side. The ***y*-axis**, also known as the vertical axis, has a positive side and a negative side.

***x* - axis**	Horizontal axis on the coordinate plane.
***y* - axis**	Vertical axis on the coordinate plane.

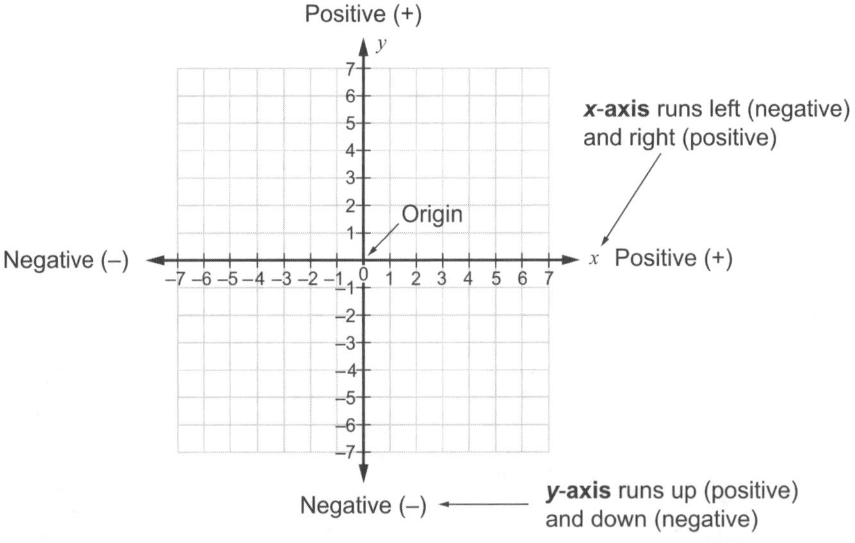

The location of points on the coordinate plane can be denoted as an ordered pair in the form (x, y), where x is the location of the point horizontally from the origin, and y is the location of the point vertically from the origin.

Ordered Pair (*x, y*)	A pair of numbers identifying the location of a point on a coordinate plane.

(x, y)

x-coordinate shows movement left or right from the origin

y-coordinate shows movement up or down from the origin

How to Locate / Graph a Point on a Coordinate Plane:

1. Always start at the origin.
2. Find the *x*-coordinate: Count the spaces horizontally from the origin to the given point.
   - Left of origin → negative *x*-coordinate
   - Right of origin → positive *x*-coordinate
3. Find the *y*-coordinate: Count the spaces vertically to the given point.
   - Up from origin → positive *y*-coordinate
   - Down from origin → negative *y*-coordinate

The axis of the coordinate plane also splits the coordinate plane into four equal parts called **quadrants**. The quadrants are numbered counter-clockwise starting with the upper right quadrant. Each quadrant has its own characteristics.

Quadrants	The four equal regions into which a coordinate plane is split by its axes.

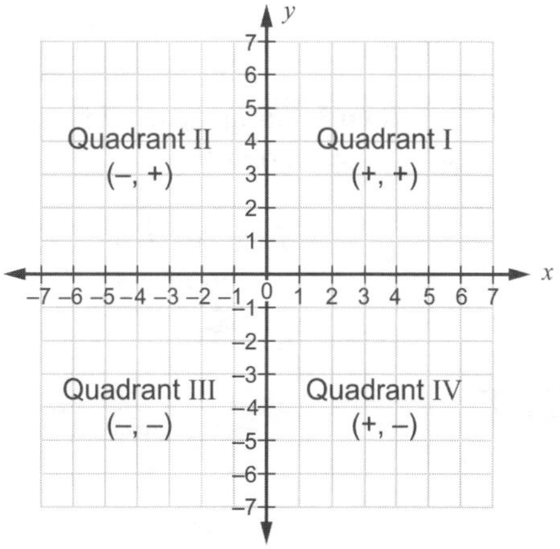

Every point in Quadrant I has a positive *x*-coordinate and a positive *y*-coordinate. The points in Quadrant II have a negative *x*-coordinate and a positive *y*-coordinate. The points in Quadrant III have a negative *x*-coordinate and a negative *y*-coordinate. The points in Quadrant IV have a positive *x*-coordinate and a negative *y*-coordinate.

Example 1 Find the coordinates of point H.

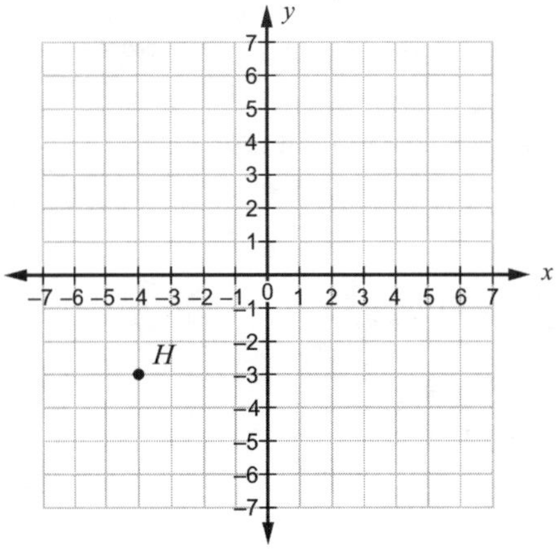

Step 1: Start at the origin.

Step 2: Count horizontal spaces to get closer to point H.

To get closer to H from the origin, move 4 spaces to the left. This means the x-coordinate is –4.

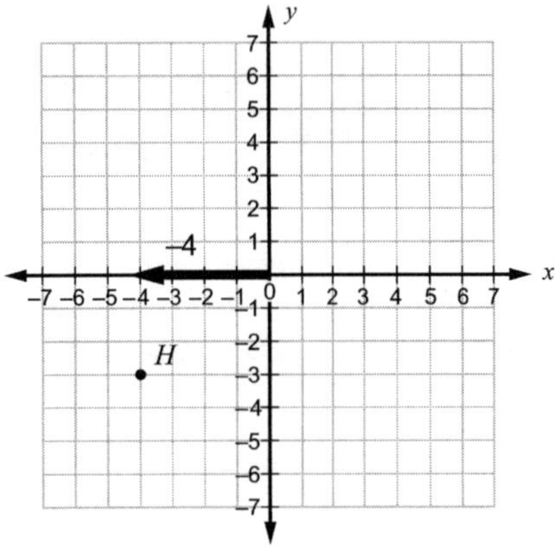

Step 3: From the previous location, count the vertical spaces to get to point H.

To get to H, move 3 spaces down. This means the y-coordinate is –3.

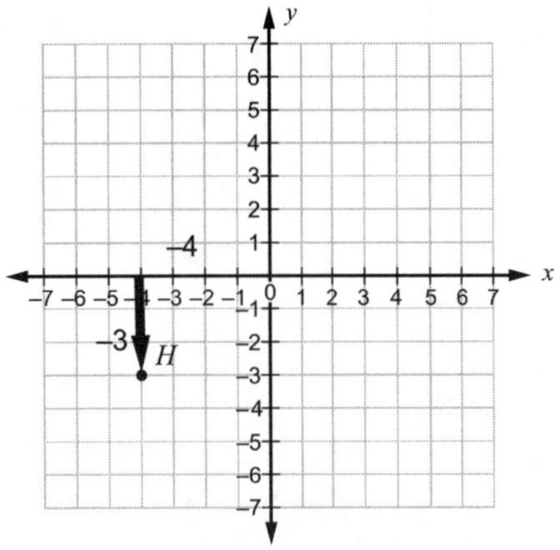

Answer: The coordinates for point H are (–4, –3).

Example 2

Point J has a negative x-coordinate and a positive y-coordinate. Determine the quadrant in which point J is located.

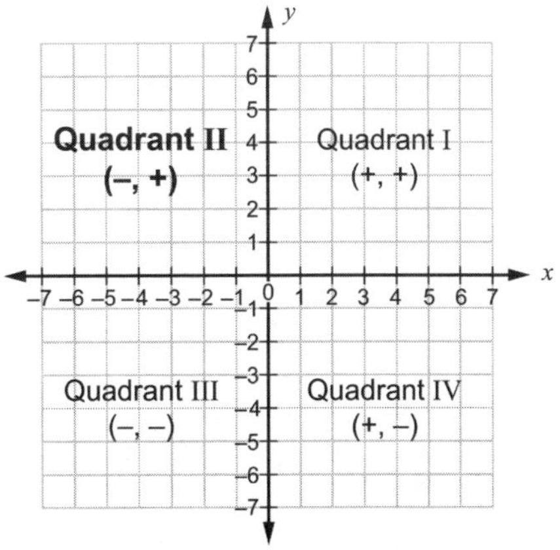

Step 1: A point with a negative x-coordinate and a positive y-coordinate is located in Quadrant II.

Answer: Point J is located in Quadrant II.

Example 3

Find the coordinates of point K.

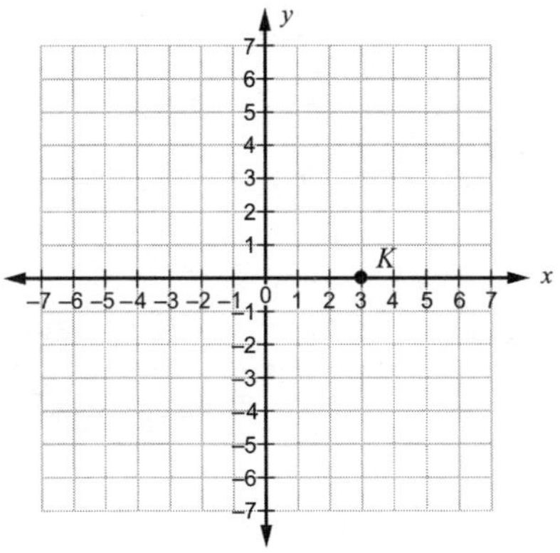

Step 1: Start at the origin.

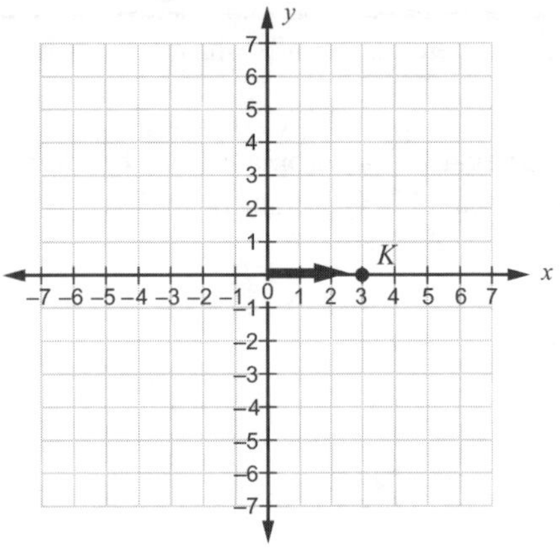

Step 2: To get to K from the origin, move 3 units to the right.

So, the x-coordinate is 3.

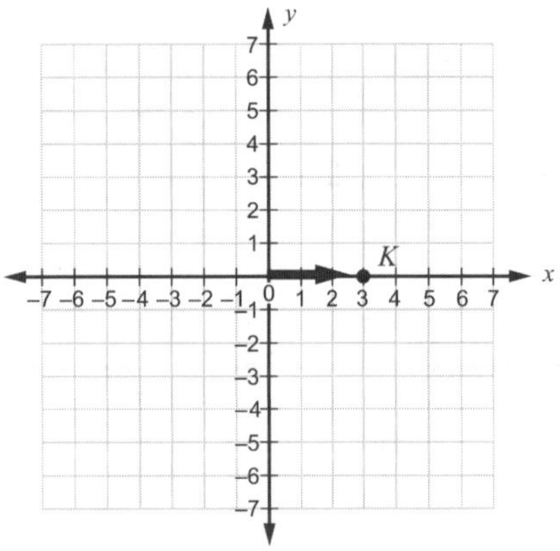

Step 3: Count the spaces vertically to get to point K.

Since it is not necessary to move up or down from this point to get to point K, the y-coordinate is 0.

Answer: The coordinates for point K are (3, 0).

Example 4 Plot the ordered pair (5, –2).

Step 1: Start at the origin.

Step 2: Since the x-coordinate is 5, move 5 spaces to the right of the origin.

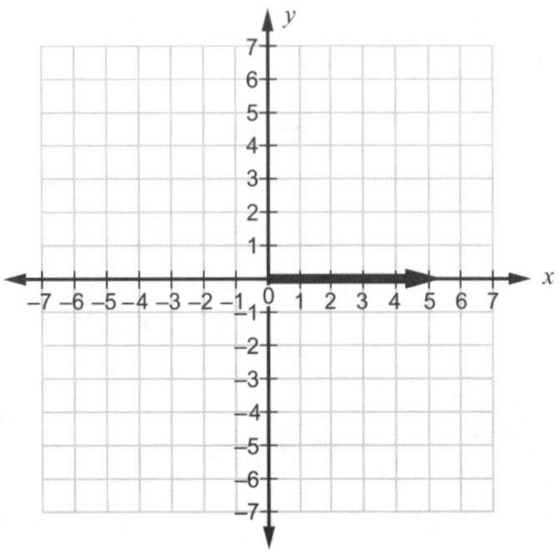

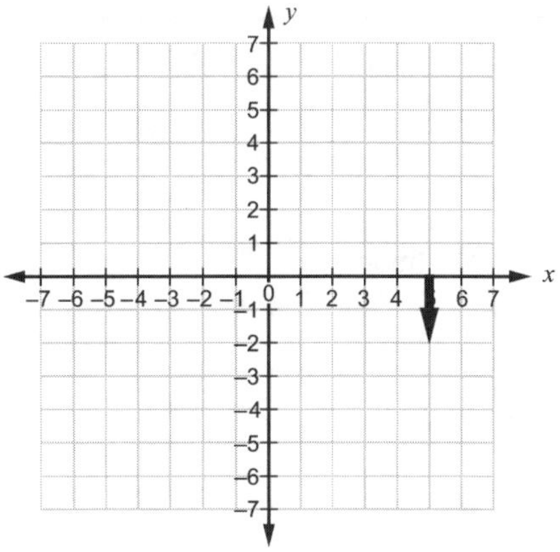

Step 3: Since the *y*-coordinate is –2, move down two spaces.

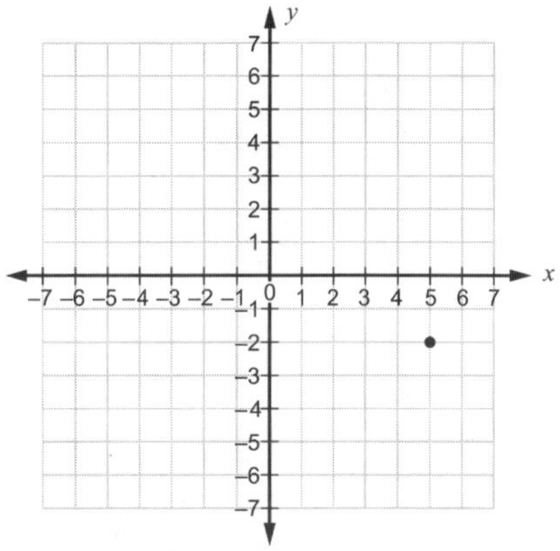

Step 4: Plot the point.

Answer:

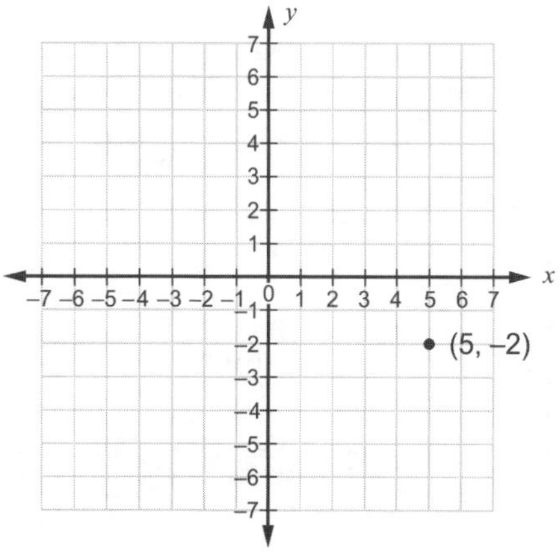

Example 5 Plot the ordered pair $\left(\frac{5}{2}, \frac{10}{3}\right)$.

Step 1: Start at the origin.

Step 2: Change the improper fraction to a mixed fraction, so it will be easier to read.

$$\frac{5}{2} = 2\frac{1}{2} \qquad \frac{10}{3} = 3\frac{1}{3}$$

$$\left(\frac{5}{2}, \frac{10}{3}\right) \rightarrow \left(2\frac{1}{2}, 3\frac{1}{3}\right)$$

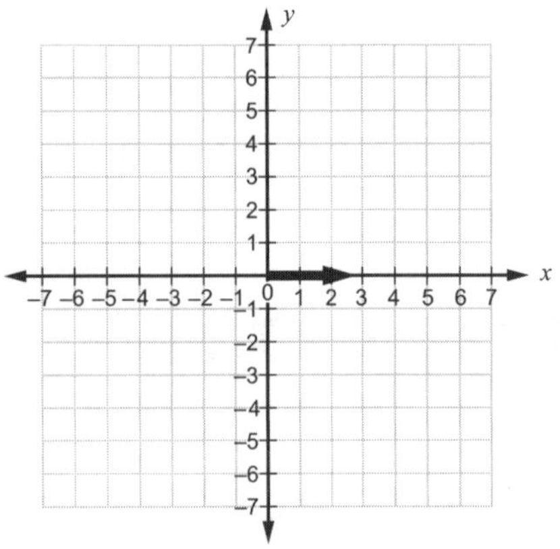

Step 3: Since the x-coordinate is $2\frac{1}{2}$, move two and one-half spaces to the right of the origin.

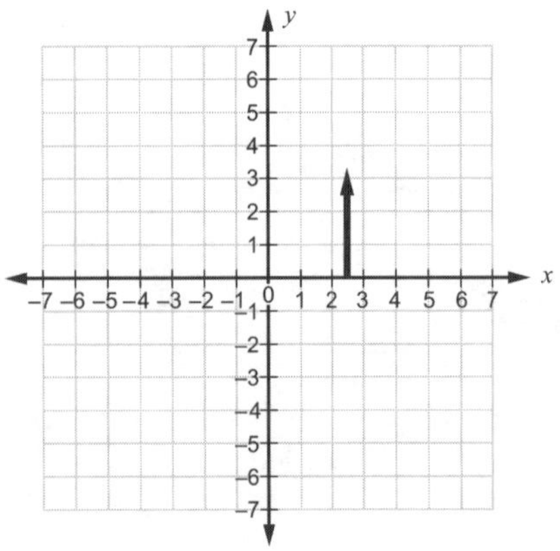

Step 4: Since the y-coordinate is $3\frac{1}{3}$, move up three and one-third spaces.

Step 5: Plot the point.

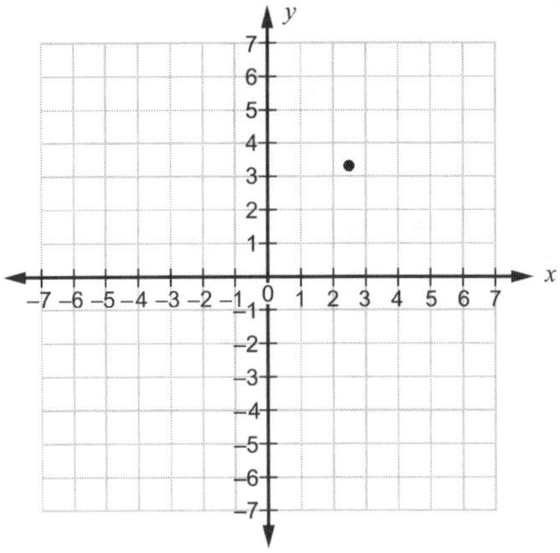

Answer:

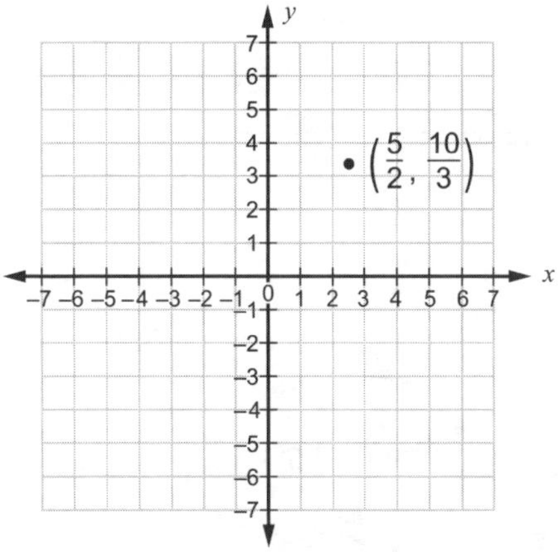

Problem Set

1. Find the coordinates of point A.

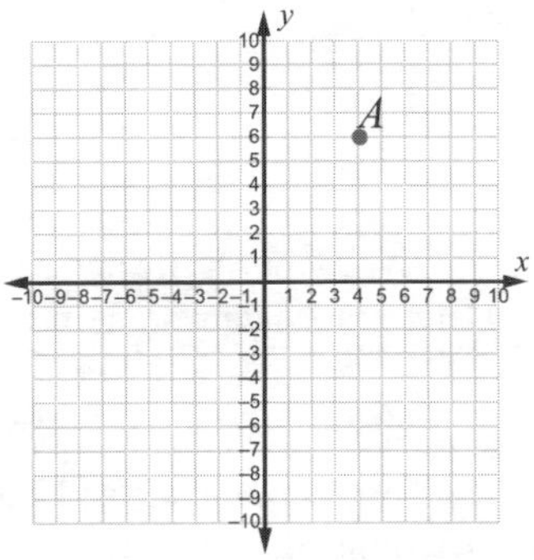

2. Find the coordinates of point C.

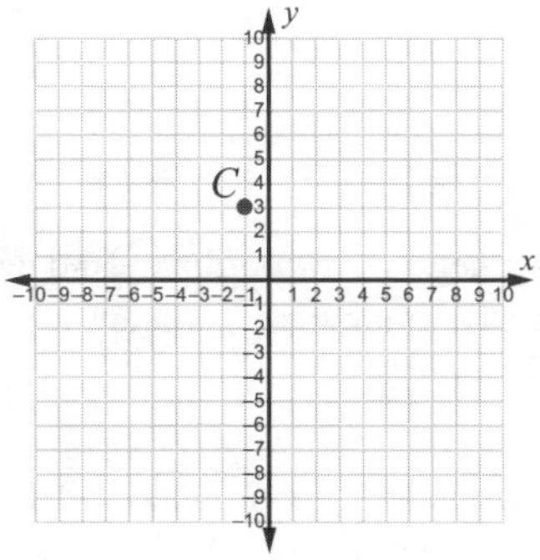

3. Find the coordinates of point B.

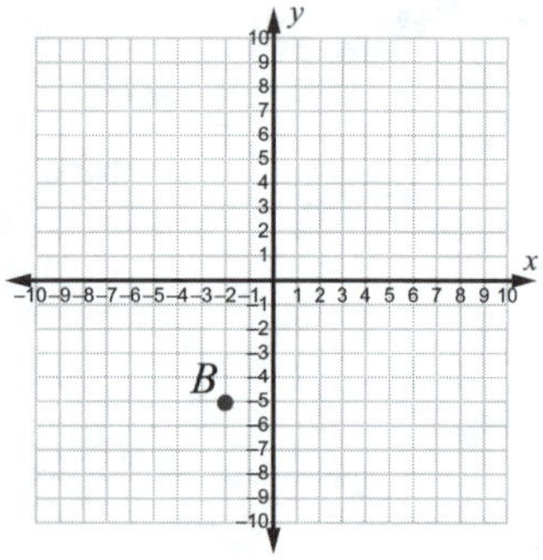

4. Find the coordinates of point D.

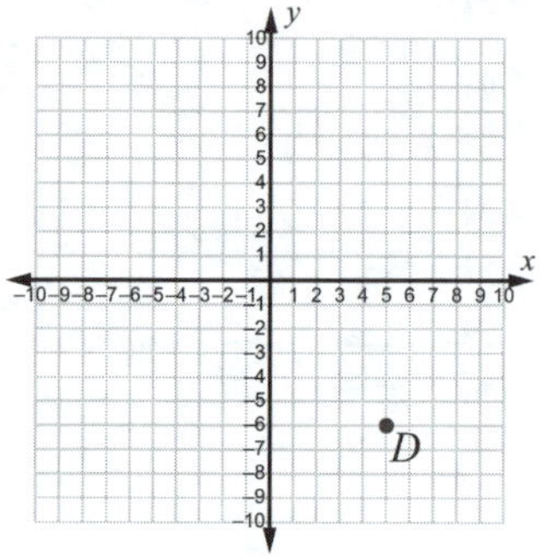

5. Find the coordinates of point F.

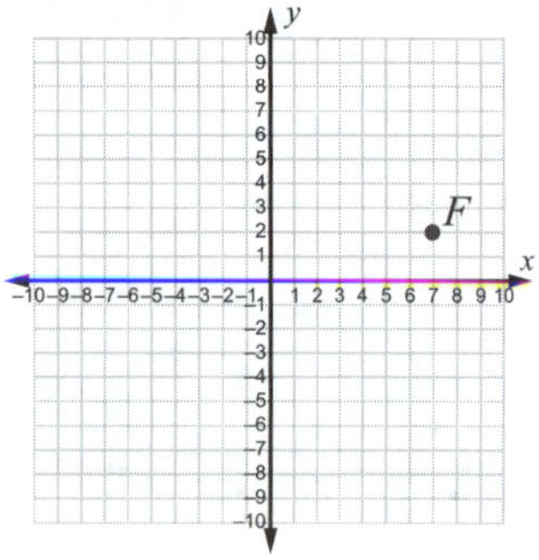

6. Find the coordinates of point M.

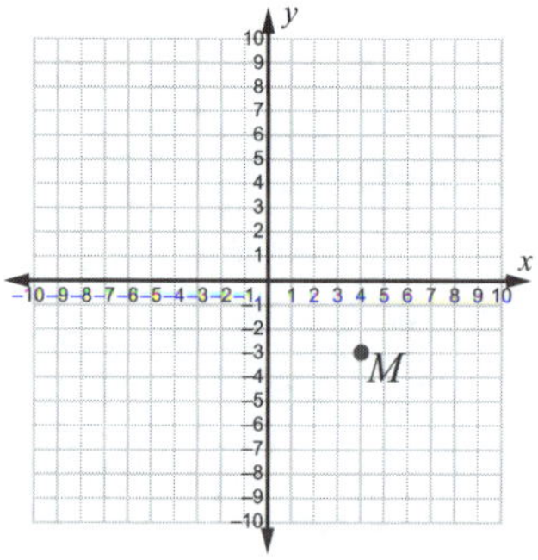

7. Find the coordinates of point T.

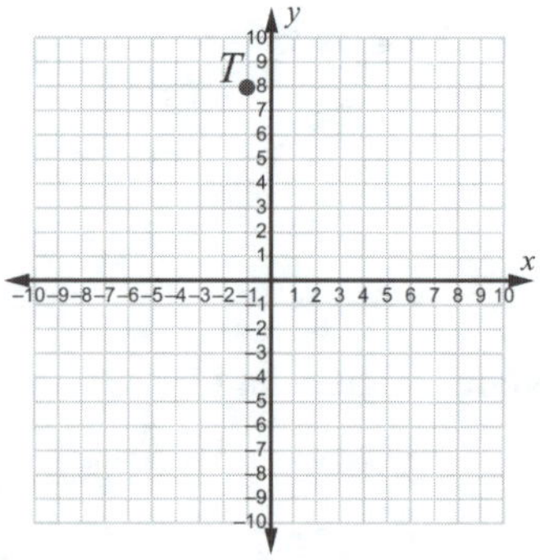

8. Find the coordinates of point S.

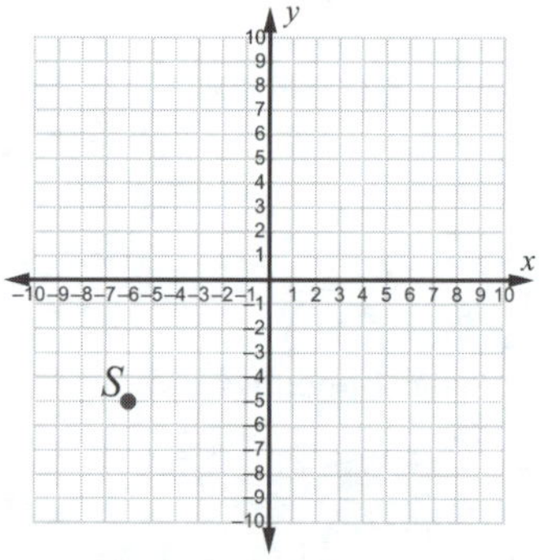

9. Point A has a negative x-coordinate and a negative y-coordinate. Determine the quadrant in which point A is located.

10. Point C has a positive x-coordinate and a negative y-coordinate. Determine the quadrant in which point C is located.

11. Point F has a positive x-coordinate and a positive y-coordinate. Determine the quadrant in which point F is located.

12. Point B has a negative x-coordinate and a positive y-coordinate. Determine the quadrant in which point B is located.

13. Point M has a negative x-coordinate and a negative y-coordinate. Determine the quadrant in which point M is located.

14. Point R has a positive x-coordinate and a negative y-coordinate. Determine the quadrant in which point R is located.

15. Point T has a negative x-coordinate and a positive y-coordinate. Determine the quadrant in which point T is located.

16. Find the coordinates of point C.

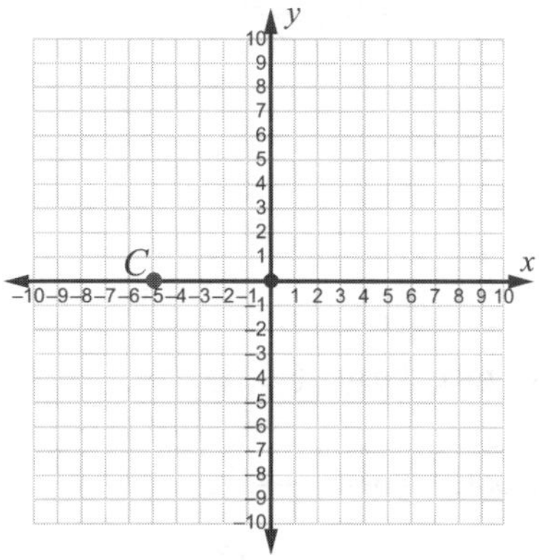

17. Find the coordinates of point D.

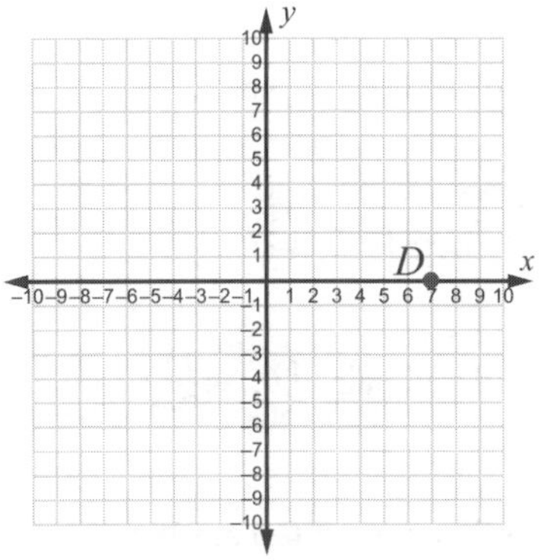

18. Find the coordinates of point F.

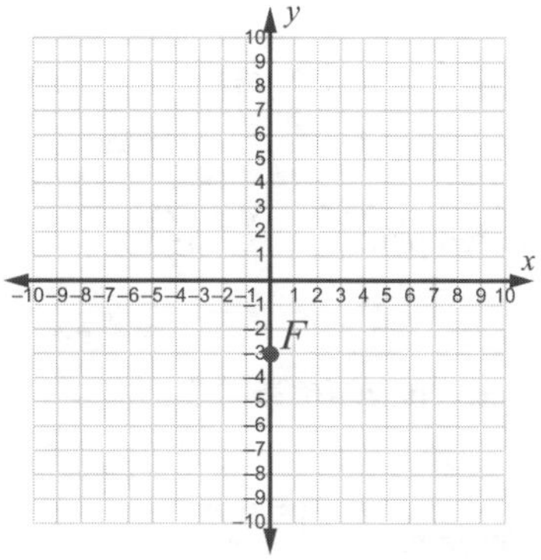

19. Find the coordinates of point G.

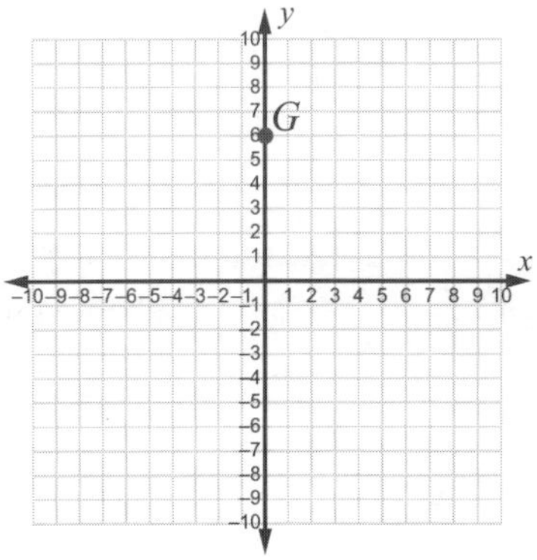

20. Find the coordinates of point S.

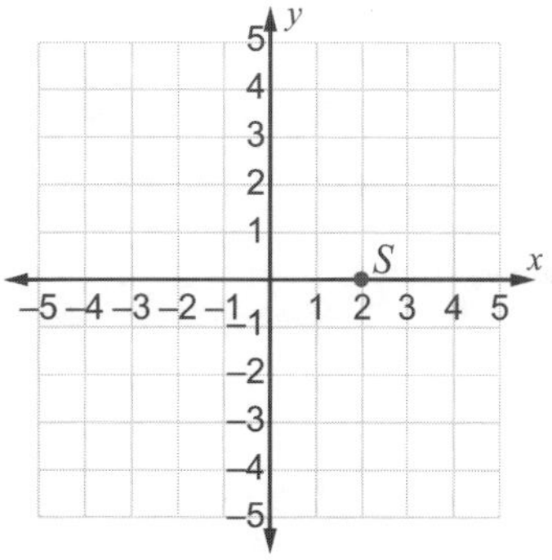

21. Find the coordinates of point M.

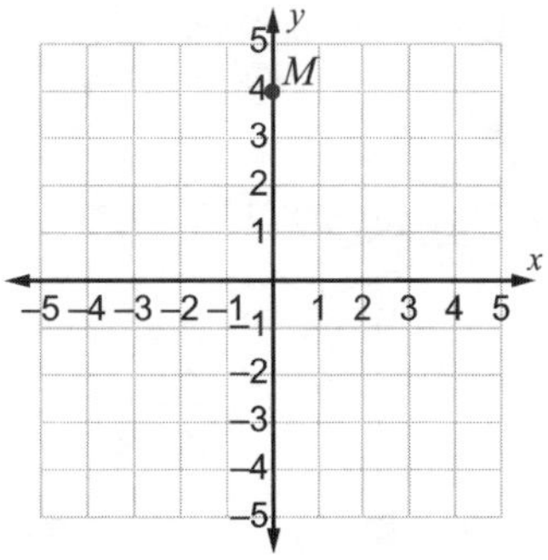

22. Find the coordinates of point B.

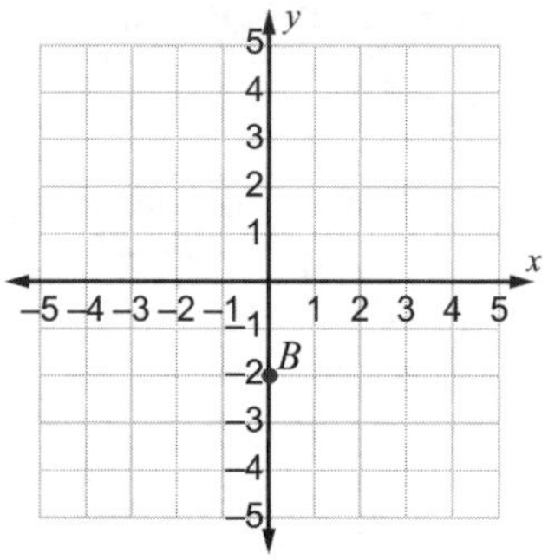

23. Plot the ordered pair: (–6, 1)

24. Plot the ordered pair: (7, 2)

25. Plot the ordered pair: (–4, –3)

26. Plot the ordered pair: (–6, 0)

27. Plot the ordered pair: (1, –4)

28. Plot the ordered pair: (0, 5)

29. Plot the ordered pair: (–7, –1)

30. Plot the ordered pair: (–5, 4)

31. Plot the ordered pair: $\left(\frac{5}{2}, \frac{3}{2}\right)$

32. Plot the ordered pair: (0.5, –2.5)

33. Plot the ordered pair: $\left(-3\frac{3}{4}, -6\frac{2}{3}\right)$

34. Plot the ordered pair: $\left(\frac{5}{2}, -6\right)$

35. Plot the ordered pair: (6.25, –3.5)

36. Plot the ordered pair: $\left(0, -4\frac{1}{2}\right)$

37. Plot the ordered pair: $\left(\frac{15}{4}, \frac{18}{4}\right)$

38. Plot the ordered pair: (–5.5, 0)

39. Plot the ordered pair: $\left(-\frac{11}{3}, \frac{17}{3}\right)$

40. Plot the ordered pair: $\left(-\frac{16}{3}, 0\right)$

MPA-047: Adding Integers with Like Signs

Number Line Method

If you borrow $5,000 from one bank and $3,000 from another bank, how much will you be in debt? You could write this equation as: –5,000 + (–3,000) = ? The negative signs mean you *owe* money. Let's look at the **number line method**.

You begin at –5000 and move *left* 3000, since this number is negative too. The answer is – $8000, or you *owe* $8000.

Let's look at another example. You deposited $500 into your savings account this week and deposit $100 into the same account next week. How much have you saved?

You could write this equation as: 500 + 100 = ? Both are positive numbers, because the money is being *deposited*.

You begin at +500 and move *right* 100, since this number is positive too. The answer is +$600, or $600.

Review the rules:
When you add two positive integers, you move to the right on the number line.
When you add two negative integers, you move to the left on the number line.

Counter Method

You can also use the **counter method** instead of a number line to understand how to add integers. Objects called **counters** are used to physically represent the number with which we are working. In this lesson,

 = +1 and  = –1.

For example, (–5) + (–2) = ?

The –5 can be shown as 

The –2 can be shown as 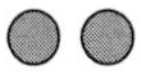

When you put these together (add), you have 

So (–5) + (–2) = –7

Absolute Value Method

You can also use the **absolute value** method to solve these problems.

Adding Integers with Like Signs	When adding integers with like signs, add the absolute value of each number and keep the same sign.

In the problem above, the absolute value of –5 is 5 and the absolute value of –2 is 2. Remember that absolute value is the distance the number is from zero, and that distance is always positive. When you add these, the sum is 7 and the sign is negative, because both integers are negative. The answer is –7

Example 1 A submarine is submerged 300 feet in the ocean. It then travels another 250 feet down. How far below sea level is the submarine?

Step 1: The equation you can use is –300 + (–250) = ? The negative signs mean you are below sea level.

Step 2: You can solve this using a number line or absolute value. Counters would be difficult because the numbers are large.

Step 3: **Number Line:** Begin at –300, because that is where the submarine begins in this problem. It then goes down an additional 250 feet, so you move to the left 250 units.

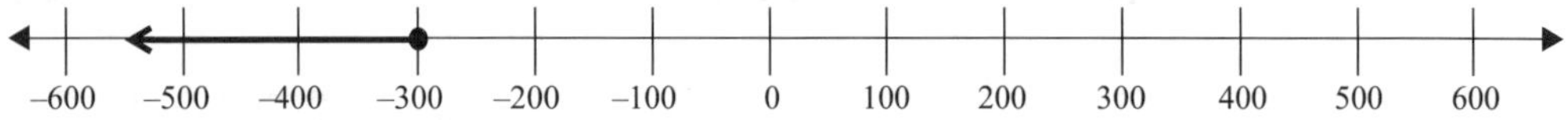

Absolute value: The absolute value of –300 is 300. The absolute value of –250 is 250. The sum of these is 550 and because both integers are negative, the answer is –550. You can write the problem as:

$$-300 + (-250) =$$
$$|-300| + |-250| =$$
$$300 + 250 = 550$$

Therefore, –300 + (–250) = –550.

Answer: The answer is –550, or the submarine is 550 feet below sea level.

Example 2 (–3) + (–7) = ?

Answer: **Number Line:** Begin at –3 and move to the left 7 spaces since you are adding a negative number. The answer is –10.

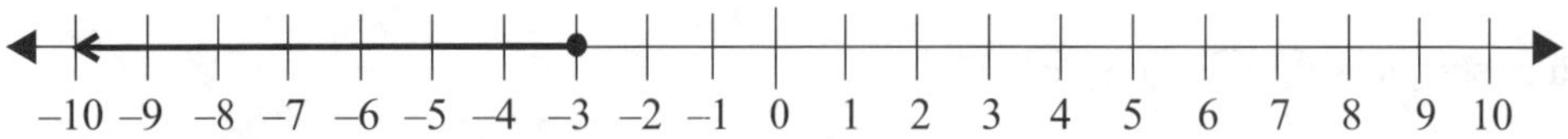

Counters:

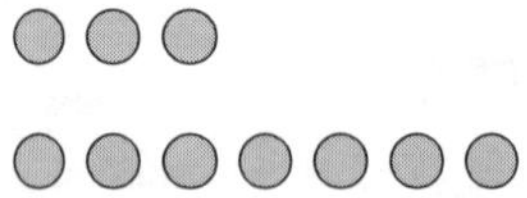

Absolute value: The absolute value of –3 is 3. The absolute value of –7 is 7. The sum of 3 and 7 is 10. Because both integers are negative, the answer is negative. The problem can be written as:

$$(-3) + (-7) =$$
$$|-3| + |-7| =$$
$$3 + 7 = 10$$

Therefore, (–3) + (–7) = –10.

Example 3 $1 + 4 = ?$

Answer: **Number Line:** Begin at +1 and move to the right 4 spaces. The answer is +5.

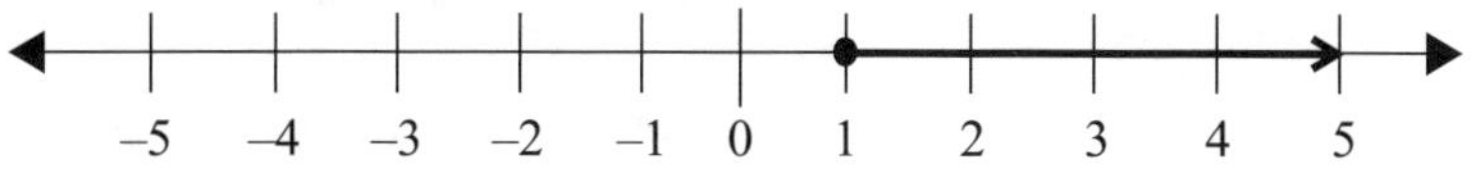

Counters:

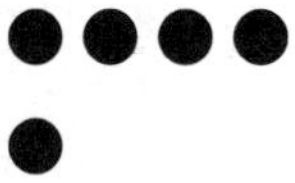

Absolute value: The absolute value of 1 is 1. The absolute value of 4 is 4. The sum is 5. Both integers are positive, so the answer is +5. The problem can be written as:

$$\begin{aligned} 1 + 4 &= \\ |1| + |4| &= \\ 1 + 4 &= 5 \end{aligned}$$

Therefore, $1 + 4 = 5$.

Problem Set

1. Simplify: $3 + 4$

2. Simplify: $-5 + (-1)$

3. The visiting football team had a total of –6 yards on the first play and –4 yards on the second play. What was the yardage for both plays?

4. Simplify: $-17 + (-12)$

5. Simplify: $26 + 8$

6. Simplify: $-63 + (-20)$

7. Simplify: $-100 + (-100)$

8. A quarterback was sacked and lost 12 yards on the first play. On the second play, he was sacked and lost 18 yards. What was the total lost yardage after the second play?

9. Simplify: $|-23| + |-47|$

10. Simplify: $-|-752| + (-|-836|)$

11. Using the number line, solve the following:
$-2 + (-5) =$

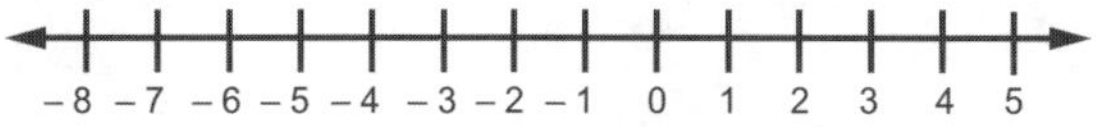

12. Simplify: $-|-5| + (-|-10|)$

13. Red counters are negative, yellow counters are positive. Add 3 yellow counters to 5 yellow counters.

14. Simplify: $-12 + (-3)$

15. What is the sum of the opposite of negative ten and the opposite of negative four? Give your answer as a number.

16. Simplify: $-52+(-35)$

17. The absolute value of the sum of –25 and –82 is ______

18. When adding two negative integers, are you moving closer to or farther from 0?

not moving	not enough information
closer to	farther from

19. When you add two positive integers, will their sum be greater or less than each addend?

equal to	greater than
not enough information	less than

20. The scores for the first two games played by the Jaguars golf team were $-15+-26$. Which of the following 4 teams had the same total score?

Fighters: $10+31$	Bears: $15+29$
Cubs: $-10+-31$	Razors: $-15+-25$

MPA-048: Adding Integers with Unlike Signs

Number Line Method

How would you write the following problem using numbers and symbols?

> You owe your friend \$10 and you have \$45 in your pocket.
> How much money will you have left after you pay back your friend?

If you *owe* money, you can represent this by a negative number, in this case –\$10.

If you *have* money, you can represent this by a positive number, in this case +\$45.

So, you could write the equation as: $-10 + 45 = ?$

Let's look at the **number line method**.

For this equation you would start at –10. Because you are *adding a positive* \$45, you will move to the *right* on the number line 45 units or spaces. Where do you end up? +\$35. This means you *have* \$35.

Suppose that in the stock market, a particular stock gained 2 points on Monday, lost 4 points on Tuesday, and gained 3 points on Wednesday. How much has this stock gained or lost over these three days? The equation would look like this: $+2 + (-4) + 3 = ?$ Use a number line:

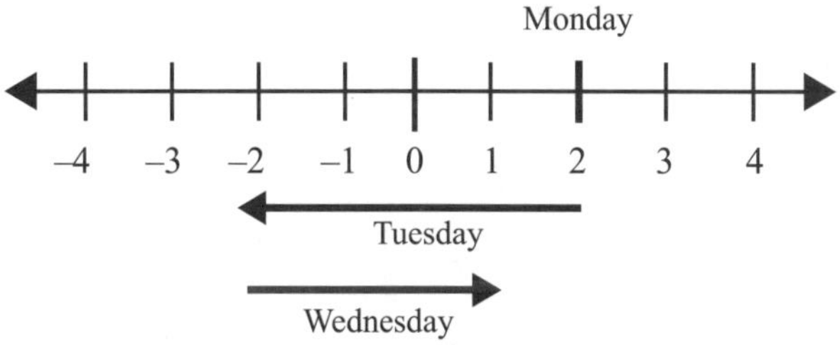

The stock gained 2 points on Monday so you begin at +2. Notice that you then add a negative number, –4. You will move to the left 4 spaces to show that loss on Tuesday; then it gained again on Wednesday so you will move to the right. The overall gain was 1 point.

Review the rules:
When you add a positive number, move to the right.
When you add a negative number, move to the left.

Counter Method

You can also use the **counter method** instead of a number line to understand adding integers. In this lesson, = +1 and  = –1. For example: $9 + (-4) = ?$

The 9 can be shown as . The –4 can be shown as 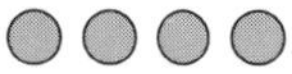. When you put these all together (add), you must remember that +1 and –1 are **additive inverses** and their sum is zero. You have four pairs that, when placed together, equal zero.

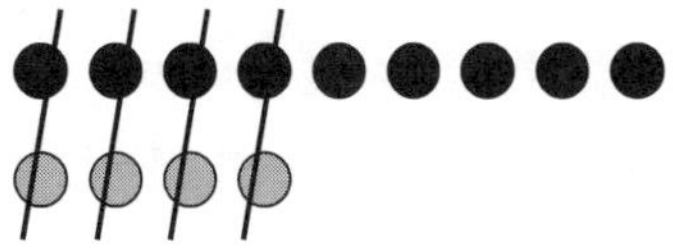

Five positive counters are left. Therefore, 9 + (–4) = +5.

Did you notice that adding integers with different signs is really like subtracting? In this problem, you actually took away counters until there were no more pairs that were equal to zero.

Absolute Value Method

Adding integers with unlike signs	When adding integers with unlike signs, subtract the absolute values of each of the integers. The answer will take the sign of the integer with the greater absolute value.

In the problem above, the absolute value of 9 is 9 and the absolute value of –4 is 4 (remember that absolute value is the distance the number is from zero and that distance is always positive). When you subtract these, the difference is 5 and because +9 had the greater absolute value, the answer is +5.

Example 1 8 + (–5) = ?

There are three ways to solve this: using a number line, counters, or absolute value.

Number line

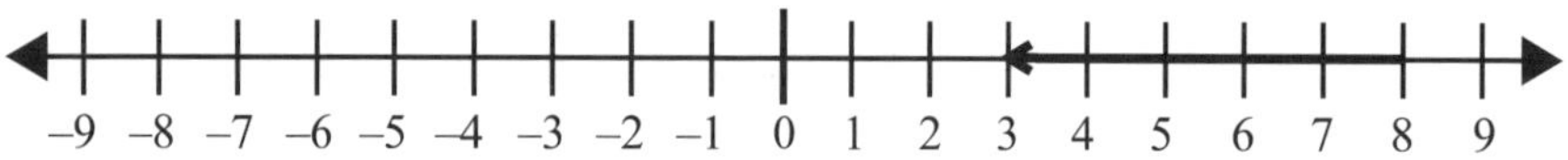

Begin at +8, and since you are adding a negative number, you will move to the *left* 5 spaces. The answer is +3.

Counters

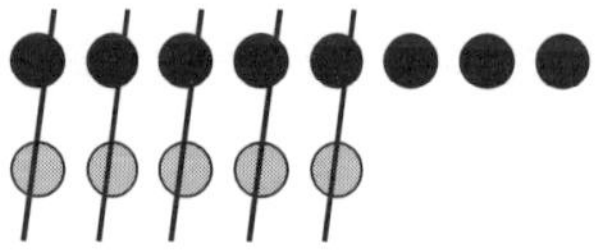

The answer is +3.

Absolute value

The absolute value of 8 is 8 ($|8| = 8$), and the absolute value of –5 is 5 ($|-5| = 5$). The difference is 3 ($|8| - |-5| = 3$). The integer with the greater absolute value is +8, so the answer will be positive. The answer is +3.

Example 2 (–6) + 2 = ?

Number line

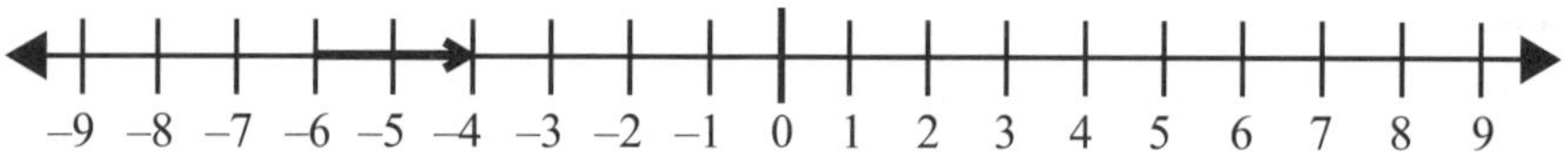

Begin at –6, and since you are adding a positive number, you will move to the *right* 2 spaces. The answer is –4.

Counters

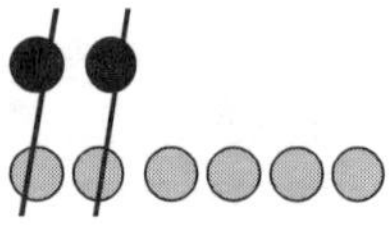

The answer is –4.

Absolute value

The absolute value of –6 is 6 ($|-6| = 6$), and the absolute value of 2 is 2 ($|2| = 2$). The difference is 4 ($|-6| - |2| = 4$). The integer with the greater absolute value is –6, so the answer will be negative. The answer is –4.

Problem Set

1. Determine whether the sum of the expression will be positive, negative, or zero: $7 + (-2)$

2. Determine whether the sum of the expression will be positive, negative, or zero: $-8 + 10$

3. Determine whether the sum of the expression will be positive, negative, or zero: $-19 + (-16)$

4. Determine whether the sum of the expression will be positive, negative, or zero: $-32 + 32$

5. Determine whether the sum of the expression will be positive, negative, or zero: $8 + (-3) + 30$

6. Determine whether the sum of the expression will be positive, negative, or zero: $11 + (-3) + (-8)$

7. Determine whether the sum of the expression will be positive, negative, or zero: $26 + (-12) + (-14)$

8. Determine whether the sum of the expression will be positive, negative, or zero: $100 + (-75) + (-25)$

9. Determine whether the sum of the expression will be positive, negative, or zero: $5 + (-8) + (-23)$

10. Determine whether the sum of the expression will be positive, negative, or zero: $5 + (-6) + (-13)$

11. Find the sum of the expression if $a = -20$, $b = 25$, $c = -13$, and $d = 22$.

    $a + a + d$

12. Find the sum of the expression if $a = -20$, $b = 25$, $c = -13$, and $d = 22$.

    $d + a + c$

13. Find the sum of the expression if $a = -20$, $b = 25$, $c = -13$, and $d = 22$.

    $d + c + d$

14. Find the sum: $-46 + 48$

15. Find the sum: $-12 + 12$

16. The foundation of a building is 120 feet below sea level and the total height of the building, including foundation, is 350 feet. What is the height of the building above sea level?

17. The temperature was recorded at 17° C. It rose 12°, fell 18°, then fell 6° before it rose again 3°. What is the current temperature?

18. Choose the correct answer: $-13 + 8$

    –21   21
    –5   5

19. Choose the correct answer: $19 + (-16)$

    –35   35
    3   25

20. Choose the correct sum: $-417 + (-956) + 754$

    –1,373   –619
    1,164   617

MPA-050: Subtracting Integers with Unlike Signs

Starting at the same point, Amanda runs east 100 yards and Lisa runs east 150 yards. When they stop, what is the *difference* in the distance between them?

Let's use a number line to look at this problem.

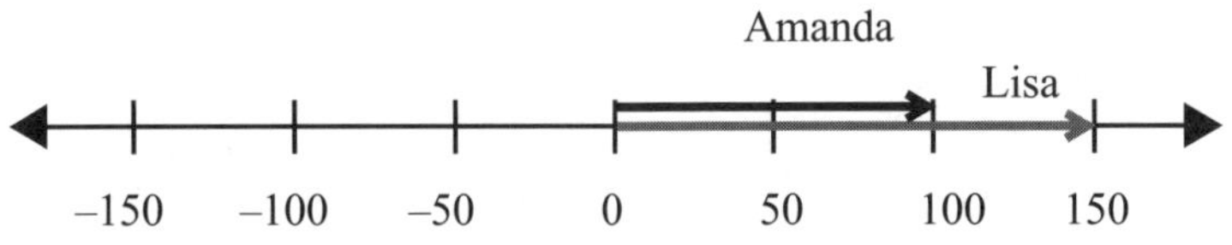

For this problem, "east" means traveling to the right on the number line. Therefore, Amanda will be at the 100 mark and Lisa will be at the 150 mark. The difference between them is 50 yards. Because you are finding a *difference*, you are actually *subtracting*. You can write this problem using numbers and symbols as $150 - 100 = ?$ This verifies the answer of 50.

Now, suppose that they start again at the same point. This time Amanda runs *east* 50 yards and Lisa runs *west* 150 yards. Look at this on a number line. Notice that a negative number is used to show that Lisa ran in the opposite direction.

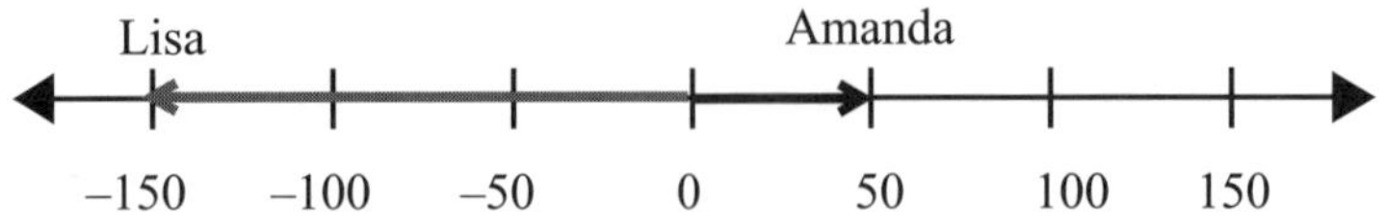

The question, again, is: What is the difference in the distance between them? This time, the difference is 200 yards. The equation would be: $50 - (-150) = ?$

You probably learned that subtraction is used when you "take away" objects from a group. You can also use subtraction when a problem asks for a difference, or how many things will be left over. When discussing integers, subtraction takes on a new meaning. It means to "add the opposite." Using the numbers from the example above, the problem can be written: $50 - (-150) = ?$

Using the idea of "adding the opposite," the equation becomes: $50 + 150 = ?$

The first integer stayed the same; the subtraction sign was changed to addition and then the number you are actually subtracting is its opposite (also known as the additive inverse). Now simply use addition rules to complete the problem. As you can tell from looking at the number line, the answer is 200.

Subtracting Integers	To subtract two integers, add the first number to the opposite (additive inverse) of the second number. For example, $3-4 = 3+(-4)$ or $-5-(-3) = -5+3$.

Let's look at a few more examples.

Example 1	Find the difference: $-2 - (-5) = ?$
$-2 - (-5) = ?$	**Step 1:** Since this is an equation involving integers, you will "add the opposite."
$-2 + (+5) = ?$	**Step 2:** The first integer stays the same. Turn the subtraction sign into addition, and the opposite of -5 is $+5$.
$-2 + 5 = +3$	**Step 3:** Now perform addition using the addition rules for integers with unlike signs. You can use the number line, counters, or absolute value. If you use absolute value, the absolute value of -2 is 2 and the absolute value of $+5$ is 5. The difference between them is 3 and the integer with greater absolute value is $+5$. This means that the answer will be positive.
	Answer: $-2 - (-5) = +3$

Example 2	Find the difference: $11 - 4 = ?$
$+11 - (+4) = ?$	**Step 1:** You probably already know the answer, but let's try "adding the opposite" and see if you get the same answer. $11 - 4$ really means $+11 - (+4) = ?$ *Note: Notice that 4 is positive. If it were negative, there would be two negative signs; one to show that it is a subtraction problem and another to show that 4 is negative.*
$+11 + (-4) = ?$	**Step 2:** Now using the "add the opposite" idea, $+11 - (+4)$ becomes $+11 + (-4) = ?$
$11 - 4 = 7$	**Step 3:** Using the addition rules for unlike signs, the absolute value of $+11$ is 11, and the absolute value of -4 is 4. The difference between the two numbers is 7, and the number with the greater absolute value is positive.
	Answer: $11 - 4 = 7$

Example 3	Find the difference: $29 - (-2) = ?$
$29 + (+2) = ?$	**Step 1:** Since this is an equation involving integers, you will "add the opposite." The first integer stays the same. Then, you turn the subtraction sign into addition, and change -2 to its opposite, which is $+2$.
$29 + 2 = ?$	**Step 2:** When you add the opposite, the equation becomes $29 + 2 = ?$
$29 + 2 = 31$	**Step 3:** Now perform the addition.
	Answer: $29 - (-2) = 31$

Example 4	Find the difference: $-5 - 4 = ?$
$-5 - (+4) = ?$	**Step 1:** Remember that 4 is positive in this example. The first negative sign means that 5 is negative, and the second negative is the subtraction sign. So, you can rewrite this problem as $-5 - (+4) = ?$
$-5 + (-4) = ?$	**Step 2:** When you add the opposite, the equation becomes $-5 + (-4) = ?$

$-5 + (-4) = -9$	**Step 3:** Now use the addition rule for integers with like signs. The absolute value of -5 is 5, the absolute value of -4 is 4. Their sum is 9 and because both integers are negative, the answer is negative.

Answer: $-5 - 4 = -9$

Problem Set

1. Subtract: $-12 - 4$

2. Subtract: $-441 - 89$

3. At 3 a.m. the water level of King Island was 13 feet below sea level. By 6 a.m. the water level had fallen an additional 7 feet. What was the water level at 6 a.m.?

4. Subtract: $12 - (-8)$

5. Subtract: $242 - (-78)$

6. Subtract: $11 - 7$

7. A football team gained 7 yards on one play and lost 16 yards on the next play. What was the change in yardage for both plays?

8. Subtract negative 8 from negative 5.

9. Subtract: $-66 - (-82)$

10. Subtract: $0 - 23$

11. Subtract: $-522 - (-418)$

12. Subtract: $44 - 7$

13. Subtract: $39 - 11$

14. Subtract: $103 - 89$

15. Subtract: $44 - 12$

16. Simplify: $15 + (-2)$

17. Simplify: $10 + (-4)$

18. Simplify: $-9 + 12$

19. Simplify: $-11 + 5$

20. Simplify: $-4 + 7$

MPA-051: Multiplying Integers with Like and Unlike Signs

Suppose that on a test, each problem is worth 2 points, and you miss three problems. Using integers, you can subtract 2 for each incorrect answer, because you experience a *loss* of points. For the first incorrect answer, you lose 2 points. For the second incorrect answer, your score would be another two points lower, and for the third, another loss of 2 points. When you put these numbers together (by addition), the result is: $(-2)+(-2)+(-2) = -6$, which means that you lost a total of 6 points on the test.

Another way to express the same idea is $3 \cdot (-2) = -6$ because you lost 2 points three times, making the result the same, a loss of 6 points. Notice that multiplication can be a shortcut for repeated addition.

Suppose that your teacher allows you to correct your mistakes. You try to solve the three problems again, and this time you answer them correctly. You can express this result as $-3 \cdot (-2)$, which means "the *opposite* of losing 2 points 3 times." Of course, the opposite of losing 2 points 3 times is *gaining* 2 points 3 times: $-3 \cdot (-2) = +6$. You have corrected the problems that caused a 6-point loss, which now created a 6-point gain.

The following rules will help you multiply integers:

Multiplying Integers	If the signs of the integers are alike, then the product will be positive. For example, the product of $3 \cdot 2$ or $-3 \cdot -2$ will be positive. If the signs of the integers are different, then the product will be negative. For example, the product of $-3 \cdot 2$ or $3 \cdot -2$ will be negative. *Note: When you multiply any integer by zero, the product will be zero.*

Example 1 $7 \cdot 9 = ?$

+63 **Answer:** The signs of the two integers are positive. Since they are alike, the product will be positive.

Example 2 $-8 \cdot 3 = ?$

– 24 **Answer:** The signs of the two integers are different. The product will be negative.

Example 3 $-5 \cdot -7 = ?$

+35 **Answer:** The signs of the two integers are alike. The product will be positive.

Example 4 Patty makes a monthly payment of $35 for her phone bill. Over the course of a year, what effect does this have on her checking account?

$12(-35) = x$

Step 1: Since Patty is *paying* $35 per month, this will have a negative effect on her account. Use –35 to represent the amount that she pays per month, and multiply it by 12 (12 months=1 year) to determine how much money she will pay over one year.

$12(-35) = x$
$-420 = x$

Step 2: Multiply to get your answer.

Answer: Since her monthly phone bill is $35, Patty will pay $420 for one year of service.

Example 5 Ed began an exercise program and is steadily losing weight at 2 pounds per week. If this steady loss continues, what is the effect on Ed's weight over 17 weeks?

$17(-2) = x$

Step 1: Since Ed is *losing* weight, illustrate this loss using negative numbers. The 2 pounds per week he is losing can be represented as –2. Then, multiply –2 by 17, representing the 17 weeks.

$17(-2) = x$
$-34 = x$

Step 2: Multiply to get your answer.

Answer: If Ed continues to lose 2 pounds per week, in 17 weeks, he will have lost 34 pounds.

Example 6 In her job as a salesperson, Theresa has been consistently earning $2,000 in commission each month. After 6 months, what effect will this have on her overall salary?

$6(\$2{,}000) = x$

Step 1: Since Theresa is *earning* money, the commission can be represented as +$2,000 or simply $2,000. Multiply this amount by 6 to determine how much money she will have made after 6 months.

$6(\$2{,}000) = x$
$\$12{,}000 = x$

Step 2: Multiply to get your answer.

Answer: Theresa's monthly commission of $2,000 will increase her salary by $12,000 in six months.

Problem Set

1. Multiply: $(-7) \times (-6)$

2. Multiply: $6 \times (-4)$

3. Multiply: $(-9) \times 6$

4. Multiply: $(-8) \times 5$

5. Multiply: $(-6) \times (-9)$

6. Multiply: $(-7) \times 0$

7. Multiply: $0 \times (-8)$

8. Which calculation yields the greater product, $(6) \times (-2)$ or $(-6) \times (-2)$?

9. Multiply: $(-31) \times 28$

10. Multiply: $(-54) \times (-21)$

11. A parachutist falls at a rate of 30 feet per second. How far will he fall after 8 seconds?

12. Multiply: $(-26) \times 19$

13. Multiply: $5 \times (-9) \times (-14)$

14. Multiply: $-12 \times 7 \times (-8)$

15. Multiply: $(-18) \times (-5) \times (-6)$

16. An oil well has been drilled to 4,200 feet below sea level. If it is drilled at a rate of 25 feet per hour, what is the well's depth after 18 hours?

17. Evaluate the term for the given value of the variable. Find $4x$ if $x = -2$.

18. Evaluate the term for the given value of the variable. Find $2c$ if $c = -5$.

19. Evaluate $-6ab$ if $a = -2$ and $b = -1$.

20. What is the sign of the product $(-1)(+2)(-1)(+4)$?

MPA-052: Dividing Integers with Like and Unlike Signs

Holly earns \$48,000 a year. How much is this per month? Because you are looking for the amount *per month*, you will use division to solve this problem. You can write a division problem in three ways:

- Using the long division symbol: $12\overline{)48,000}$
- In fractional form: $\frac{48,000}{12}$
- Or, horizontally: $48,000 \div 12$

In this example, Holly earns \$4,000 per month. This quotient is positive because she is *earning* this amount.

Now suppose that Holly paid \$6,000 in car payments last year. How much did she pay per month? You could write $-6,000 \div 12 = ?$ (You could also express this in fractional or long division form). Notice the 6,000 is *negative,* because this amount is being *paid*. The quotient is –500, which means that Holly is paying \$500 per month.

The rules for dividing integers are basically the same as when we multiply them.

Dividing integers	If the signs of the integers are alike, then the quotient will be positive. For example, $10 \div 2 = 5$ or $-10 \div -2 = 5$. If the signs of the integers are different, then the quotient will be negative. For example, $-10 \div 2 = -5$ or $10 \div -2 = -5$. *Note: Remember that you cannot divide by 0. For instance, you cannot have an expression such as $7 \div 0$.*

Example 1 Find the quotient for the following division problem: $56 \div 8$

$56 \div 8 = +$

Step 1: Because both integers are positive and therefore alike, the quotient will be positive.

$56 \div 8 = +7$

Step 2: When you perform the division, the answer is 7. Combine it with the positive sign to obtain your answer.

Answer: +7

Example 2 Find the quotient for the following division problem: $8\overline{)-64}$

Step 1: The signs of the two integers are different. The quotient will be negative.

$\begin{array}{r} -8 \\ 8\overline{)-64} \end{array}$

Step 2: When you perform the division, the answer is 8. Combine it with the negative sign to obtain your answer.

Answer: –8

Example 3 Find the quotient for the following division problem: $\frac{99}{-11}$

Step 1: The signs of the two integers are different. The quotient will be negative.

$\frac{99}{-11} = -9$	**Step 2:**	When you perform the division, the answer is 9. Combine it with the negative sign to obtain your answer.
	Answer:	–9

Example 4 A football team lost 8 yards in 2 plays. What was the average loss per play?

$-8 \div 2 = ?$	**Step 1:**	Set up an equation.
	Step 2:	Because the signs are different, the quotient will be negative.
$-8 \div 2 = -4$	**Step 3:**	When you perform the division, the answer is 4. Combine it with the negative sign to obtain your answer.
	Answer:	–4. This means that the team *lost* 4 yards per play on average.

Problem Set

1. Divide: $-60 \div (-12)$
2. Divide: $-35 \div 7$
3. Divide: $-30 \div 10$
4. Divide: $-12 \div (-3)$
5. Divide: $46 \div 2$
6. Divide: $\frac{-121}{-11}$
7. Divide: $\frac{126}{-18}$
8. Divide: $\frac{-142}{-71}$
9. Divide: $\frac{-460}{-5}$
10. Evaluate the expression if $x = -96$, $y = -8$, and $z = 2$.
    $\frac{-4x}{y}$
11. Evaluate the expression if $x = -96$, $y = -8$, and $z = 2$.
    $\frac{x}{-48}$
12. Evaluate the expression if $x = -96$, $y = -8$, and $z = 2$.
    $\frac{x}{yz}$
13. Stuart is doing math homework. He is dividing –270 by –90. What is his answer?
14. Determine which calculation yields the smaller quotient. Choose "same" if the quotients are equal.
    $\frac{12}{-2}$ or $\frac{-12}{2}$

    same $\frac{-12}{2}$

    none of these $\frac{12}{-2}$

15. Evaluate the term for the given value of the variable.

$\frac{-20}{x}$ when $x = -2$.

16. Evaluate the term for the given value of the variable.

$\frac{2x}{-3}$ when $x = -6$.

17. Find the value of $\frac{20}{-x}$ when $x = -4$.

18. Evaluate the expression $\frac{(-2x)}{(-4)}$ when $x = +8$.

19. Divide –64 by 8.

20. If $\frac{-20}{b}$ is negative, then b has which sign?

negative	positive
none of these	cannot be determined

MPA-053: Adding, Subtracting, Multiplying, and Dividing Integers

In this lesson, you will review the rules you have learned to add, subtract, multiply, and divide integers.

Addition		**Examples**
If the signs are the same...	add the absolute value of the numbers and keep the sign.	$-4+-3=-7$ $6+5=11$
If the signs are different...	subtract the absolute values of the numbers and take the sign of the integer with the greater absolute value.	$-7+10=3$ $4+(-5)=-1$

Subtraction		**Examples**
Turn the equation into an addition problem using "add the opposite"...	then use the addition rules above.	$-5-(-2)=?$ $-5+2=-3$ $-4-7=?$ $-4+(-7)=-11$

Multiplication and Division		**Examples**
If the signs are the same...	the product or quotient will be positive.	$-3\times-4=12$ $5\times2=10$ $-9\div-3=3$ $8\div2=4$
If the signs are different...	the product or quotient will be negative.	$-7\times3=-21$ $2\times-6=-12$ $18\div-3=-6$ $-15\div3=-5$

Now look at the following problems.

Example 1 Add 452 + 812 = ?

452 + 812 = 1,264	**Step 1:** You are adding two numbers with like signs (both are positive), so you will add the numbers.
+1,264	**Step 2:** Since both numbers have the same sign, the sum keeps the common sign.

Answer: +1,264

Example 2 Subtract $-56-(-43) = ?$

$-56+43 = ?$	**Step 1:** You are subtracting, so you must "add the opposite."
$-56+43 = -13$	**Step 2:** Now use the addition rules. Since you have integers with different signs, subtract the absolute values of the numbers. The absolute value of –56 is 56 and the absolute value of +43 is 43. The difference of these two numbers is 13. The integer with the greater absolute value is –56. Because this number is negative, the answer will be negative.
	Answer: –13

Example 3 Multiply $-32 \times 7 = ?$

$-32 \times 7 = -224$	**Step 1:** Set up the problem and perform the multiplication.
–224	**Step 2:** Since the signs are different, the product will be negative.
	Answer: –224

Example 4 Divide $-4\overline{)-324}$

$$\begin{array}{r} 81 \\ -4\overline{)-324} \\ \underline{-32} \\ 4 \\ \underline{-4} \\ 0 \end{array}$$

Step 1: Set up the problem and perform the division.

+81	**Step 2:** You are dividing integers with like signs, so the quotient will be positive.
	Answer: +81

Problem Set

1. Simplify: $49+(-13)$

2. Simplify: $93-(-16)$

3. Simplify: $62-137$

4. Simplify: $81-236$

5. Simplify: $(18)(-7)(13)$

6. Simplify: $-258 \div 3$

7. Simplify: $-78 \div (-13)$

8. Simplify: $-364+381$

9. Simplify: $236 \times (-18)$

10. Amy borrowed \$30 from each of her 3 friends. How much money did Amy borrow in all?

11. Hieu was selected to be on a TV game show. He played four rounds and had a total of –1,600 points. How many points did he average per round?

12. Johnny and Charlie were playing a video game. Johnny had –462 points. Charlie had 318 points. How many points did Johnny and Charlie have combined?

13. Mr. Graham has $300 in his savings account. If he gave each of his grandchildren $50 for Christmas, and was left with a balance of 0, how many grandchildren does he have?

14. Jade has $35 for the week to spend on lunch (a week is seven days). How much money can Jade spend on lunch per day?

15. Tony's math problem is to divide –760 by –10. Is Tony's answer positive or negative?

16. Write an expression for "the quotient of 11 and two times a number n".

17. Write an expression for "the sum of –2 times a number n and 45".

18. Write an expression for "the difference of six and twice n".

19. Write an expression for "twice a number n divided by negative 7".

20. Write an expression for "the quotient of a number n and the sum of 4 and twice the number n".

MPA-054: Solving One-Step Equations with Integers Using all Four Operations

Bill made a withdrawal of \$217 yesterday from his checking account. The balance today is \$912. How much money was in the account before the withdrawal?

Let's write this problem using numbers and symbols:

$$x - \$217 = \$912$$

x	$\$217$	$\$912$
amount originally in the account	amount Bill took out	amount left in the account

Think of this equation as a balance.

$x - 217$ ⎯⎯△⎯⎯ 912

The left side of the equation must equal (or balance) the right side. What value of x would make $x - 217$ balance 912?

You could just guess and check to find x. Using this method, you would keep putting different values in for x until one of them worked. This is a good method, but it is very time consuming. There is a better way!

You can keep this equation balanced as long as you perform the same operation on both sides. If you add 2 to the left side, then you must add 2 to the right side to keep the equation balanced.

Since you are solving for x (in other words, you want x isolated, by itself) what can you do with 217? Since it is being subtracted, you could use the inverse operation of subtraction: addition. In this case, add 217 to both sides. Why will this help? Remember that a positive and a negative of the same number equal zero. Some examples:

$-3 + 3 = 0$ $\quad$ $212 + (-212) = 0$ $\quad$ $-1{,}234 + 1{,}234 = 0$ $\quad$ $99{,}999 + (-99{,}999) = 0$

Example 1 In the example at the beginning of this lesson, you were asked to find how much money was in Bill's account before he withdrew \$217.

$x - \cancel{217} = 912$ $\cancel{+217} \quad +217$ $x = 1{,}129$	**Step 1:** To isolate the variable x, add 217 to both sides of the equation.
$x = 1{,}129$	**Step 2:** By adding 217 to both sides, you are left with $x = 1{,}129$. This means that Bill had \$1,129 in his checking account before he withdrew the money.
$x - 217 = 912$ $1{,}129 - 217 = 912$ $912 = 912$	**Step 3:** Let's check to be sure that this is correct. Substitute 1,129 for x. The answer is correct.

Answer: \$1,129

Look at another example.

Example 2 Sue made 3 equal payments for a bracelet that she bought at the jewelry store. What is the total price of the bracelet if each payment was $32?

$\frac{x}{3} = 32$

$3 \cdot \frac{x}{3} = 32 \cdot 3$

Step 1: One way to think about this problem is to remember that the total price divided into 3 payments equals 32. You can translate this into $\frac{x}{3} = 32$. How do you solve for x? The inverse operation of division is multiplication, so multiply each side of the equation by 3 to keep it in balance and to get x by itself.

$x = 96$

Step 2: By multiplying both sides by 3, you are left with $x = 96$, which means that the total price was $96.

Answer: $96

Example 3 $-20x = 120$

$\frac{-20x}{-20} = \frac{120}{-20}$

Step 1: The variable, x, is being multiplied by –20. The inverse operation is division. To make x positive, divide both sides by –20 because a negative number divided by a negative number is positive. The answer will be negative because positive 120 divided by negative 20 is negative.

$x = -6$

Step 2: By dividing both sides by –20, you are left with $x = -6$.

Answer: $x = -6$

Example 4 $-26 + x = -52$

$$\begin{array}{rcr} -26 + x & = & -52 \\ +26 & & +26 \\ \hline x & = & -26 \end{array}$$

Step 1: The variable, x, is being added to –26. To isolate x, add 26 to both sides of the equation. Remember the rule for adding integers with different signs.

Step 2: When you add –52 and +26, you are left with $x = -26$.

Answer: $x = -26$

Example 5 $y + 16 = -45$

$$\begin{array}{rcr} y + 16 & = & -45 \\ -16 & & -16 \\ \hline y & = & -61 \end{array}$$

Step 1: The variable, y, is being added to 16. The inverse operation is subtraction. You must subtract 16 from both sides to get y by itself.

$y = -61$

Step 2: When you subtract +16 from –45, you are really adding –16. (Recall the rule for subtracting integers). You then must remember how to add integers with the same signs.

Answer: $y = -61$

Example 6 $\frac{z}{-5} = 30$

$-5 \cdot \frac{z}{-5} = 30 \cdot -5$ $z = -150$	**Step 1:** The variable, z, is being divided by –5. The inverse operation is multiplication. Multiply both sides of the equation by –5.
$z = -150$	**Step 2:** Recall that a positive number multiplied by a negative number is negative.

Answer: $z = -150$

Problem Set

Solve:

1. An unknown number divided by 7 is –3. Find the number.

2. An unknown number divided by –9 is –3. Find the number.

3. Solve: $-7n = 84$

4. Solve: $6n = -12$

5. Solve: $x + (-240) = 300$

6. Solve: $x - (-13) = -7$

7. Solve: $\frac{n}{-4} = 3$

8. Solve: $-315 + x = -115$

9. Solve: $\frac{n}{4} = 20$

10. An unknown number divided by 8 is 4. Find the number.

11. An unknown number divided by 7 is 2. Find the number.

12. Solve: $\frac{n}{-9} = -10$

13. Solve: $\frac{n}{-4} = -2$

14. Solve: $\frac{n}{-13} = -2$

15. Gwen makes two withdrawals of $75 each from her savings account. She now has $800 in the account. Write and solve an equation to determine the amount in the account before the withdrawals were made.

16. B.J. spent $150 at the clothes shop, $200 at Sam's Sports Authority, and $50 at Chuck's Car Parts. After the shopping spree, he has $70 in his account. Use an equation to determine the balance before B.J. went shopping.

17. A withdrawal of $12 is added to Rico's account. The new balance of the account is – $10. What was the amount before the withdrawal?

18. A diver is located at a certain place in the ocean. He rises 15 feet and is now 16 feet below sea level. What was his depth before the rise?

19. The difference between the elevation of points a and b is 13 feet. The elevation of point b is –2 feet. What is the elevation of point a?

20. A number divided by –8 equals –9. What is the number?

MPA-055: Finding the Perimeter of a Figure

A basic property of any geometric figure is the distance around the figure, or the **perimeter**. If you know the dimensions of a figure, finding the perimeter is not difficult, you simply add the measurements of all the sides.

Perimeter	The distance around a figure.

The basic formula for the perimeter of a geometric figure is P = the sum of all the sides. Here is an example.

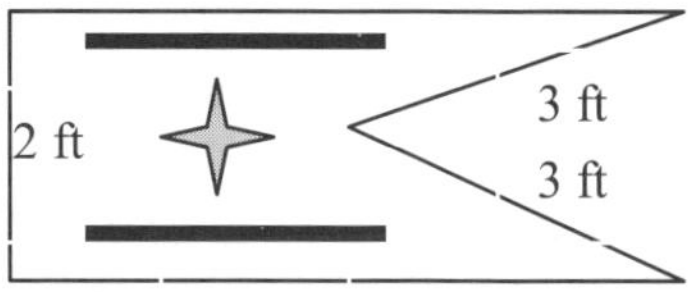

Example 1 Find the perimeter of the flag shown above.

P = Sum of all the sides	**Step 1:** Use the definition of perimeter to start solving the problem.
$P = 2+6+3+3+6$	**Step 2:** Insert all the measurements into the formula.
$P = 20$ feet	**Answer:** Using the basic definition, we just add up all the sides to get the answer.

Here's another example.

Example 2 A stop sign is in the shape of a regular octagon, meaning it has 8 sides that all have the same measurement. Find the perimeter of a stop sign, if each side measures 10 inches.

P = Sum of all the sides	**Step 1:** Use the definition of perimeter to start solving the problem.
$P = 10+10+10+10+10+10+10+10$	**Step 2:** Insert all the measurements into the formula.
$P = 80$ inches	**Answer:** Using the basic definition, we just add up all the sides to get the answer.

Notice in this example there is an easier way to find the perimeter: since you are adding up 8 numbers that are all the same, you can just multiply 8 by 10. In this example of the stop sign, $P = 8 \cdot 10$ or $P = 80$ inches. If you noticed this right away, you could have saved yourself some time and effort. Let's begin by looking at some simple figures that might have easier formulas.

For example, look at a rectangle. In rectangles, opposite sides always have the same measurement, so knowing the length and width of the rectangle means you know the measurements of all the sides.

You *could* say that the perimeter of a rectangle is $P = w+w+l+l$, where l = length and w = width. Let's try to make this formula a little easier. First, $w+w$ is the same as $2w$, and $l+l$ is the same as $2l$. Now we know that the perimeter of a rectangle could also be written as $P = 2(l+w)$. Notice that $2l+2w$ is the same as $2(l+w)$ by the distributive property.

So, we now have an even simpler formula for the perimeter of a rectangle: $P = 2(l+w)$, or 2 times the sum of the length and the width.

Perimeter of Rectangles	The perimeter of a rectangle is equal to two times the sum of the length and width. The formula is: $P = 2(l+w)$

Let's look at how to find the perimeter of a square. Since a square is a special type of rectangle, one in which all the sides have the same measurement, we can use the same idea to find an easy formula.

First, you can use the formula $P = w + w + l + l$, but the width and the length are the same. Since all the sides have the same length, an easier way to represent the measurement of a side in the formula is by using s. So our formula now becomes $P = s + s + s + s$. But, we can make it even simpler, since adding the same number 4 times is the same as multiplying it by 4. By doing this, we have a very easy formula: $P = 4(s)$, or the perimeter of a square equals four times the length of a side.

Perimeter of Squares	The perimeter of a square is equal to 4 times the length of one side of the square. $P = 4(s)$

Let's consider more examples.

Example 3 How do you find the perimeter of a rectangle with a length of 6 ft and a width of 2 ft?

$P = 2(w + l)$	**Step 1:** Use the formula for perimeter of a rectangle.
$P = 2(2 + 6)$	**Step 2:** The width is 2 ft and the length is 6 ft.
$P = 2(8)$	**Step 3:** Add inside the parentheses.
$P = 16$ ft	**Answer:** Multiplying gives the answer.

Example 4 Find the perimeter of a square field that has a length of 20 meters.

$P = 4(s)$	**Step 1:** Use the formula for perimeter of a square.
$P = 4(20)$	**Step 2:** The length of a side is 20 meters.
$P = 80$m	**Answer:** Multiplying gives the answer.

Problem Set

Solve:

1. Find the perimeter of the rectangle with the following dimensions:
width = 37, length = 67.

2. Find the perimeter of the rectangle with the following dimensions:
width = 53 m, length = 87 m.

3. Marco has a rectangular pool with sides of 30 ft and 45 ft. He wants to build a fence 3 feet away from the pool on all sides. What is the perimeter of the fence?

4. Kristina wants to buy a wallpaper border for her room. If the sides are 20 ft, 35 ft, 18 ft, and 12 ft, approximately how many rolls will she need to buy if the rolls cover 10 ft each? Round your answer to the nearest whole number.

5. A file folder has sides of 17 inches and 18 inches. A square piece of paper has sides of 16 inches. How much larger is the file folder's perimeter?

6. Marvin's yard has sides of 90 ft, 120 ft, 90 ft, and 120 ft. How much fencing will he need?

7. Find the perimeter of the rectangle.

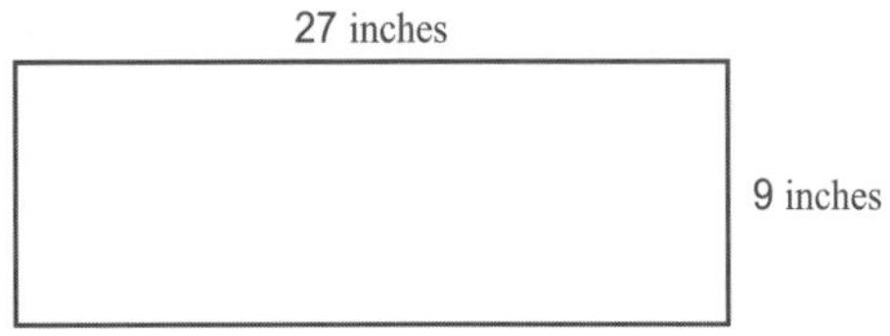

8. Find the perimeter of the rectangle.

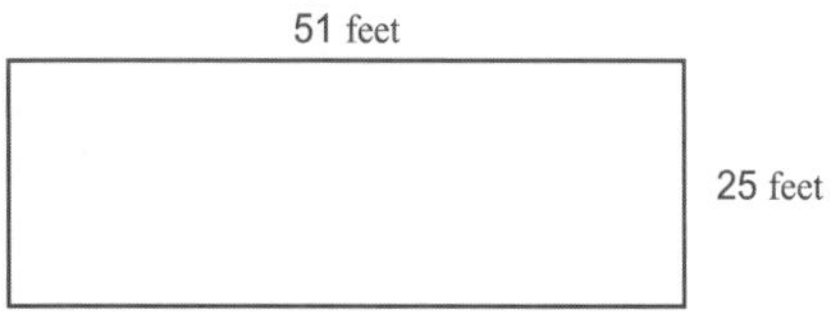

9. Find the perimeter.

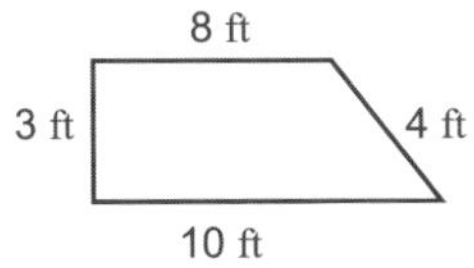

10. Find the perimeter.

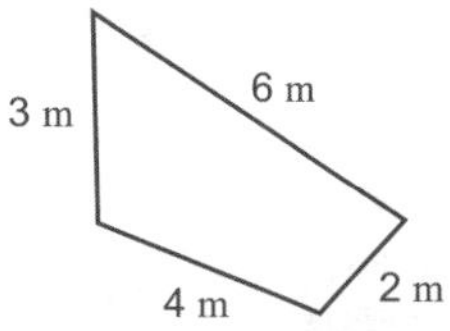

11. Find the perimeter.

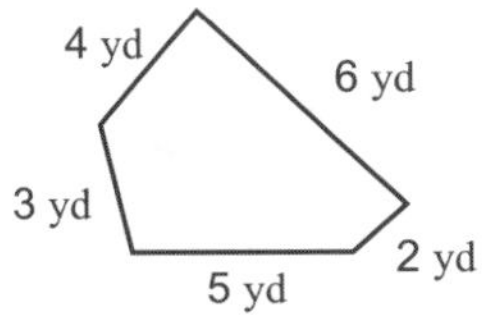

12. Find the perimeter.

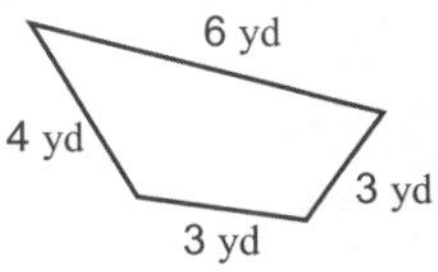

13. Find the perimeter.

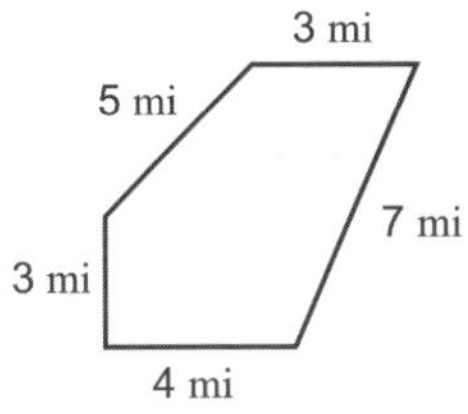

14. Find the perimeter.

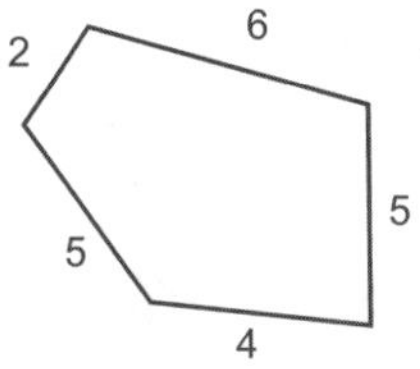

15. Find the perimeter.

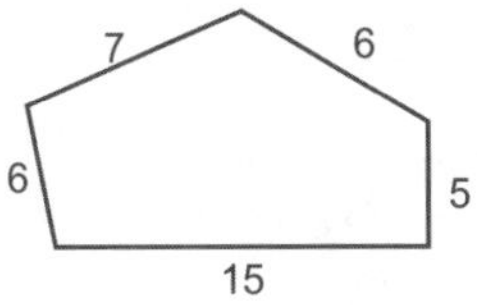

16. Super Corp's building has sides of 50 ft, 100 ft, 50 ft, and 100 ft. A sidewalk around the building is 15 ft away on all sides. What is the perimeter of the sidewalk?

17. Nathanial's swimming pool has sides of 15 ft, 25 ft, 15 ft, and 25 ft. Joe's swimming pool has sides of 10 ft, 27 ft, 10 ft, and 27 ft. How much larger is the perimeter of Nathanial's pool?

18. A computer company has an office building with sides of 720 ft, 900 ft, 720 ft, and 900 ft. The building has a security fence 15 feet away on all sides. What is the perimeter of the fence?

19. Find the perimeter.

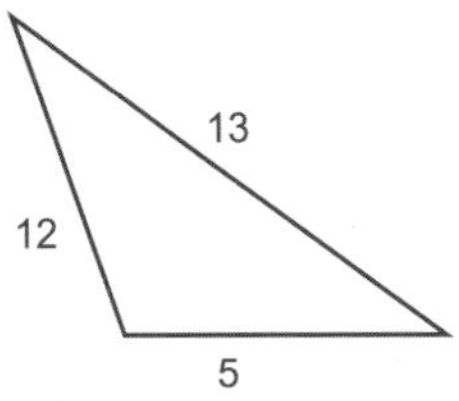

20. Find the perimeter.

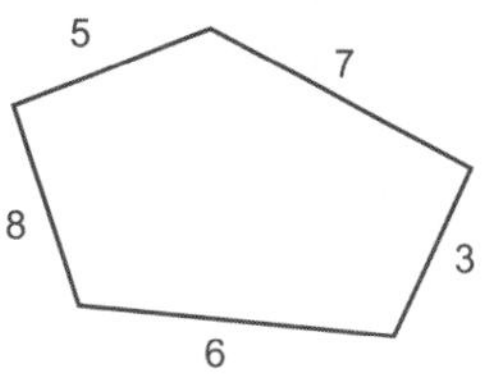

MPA-056: Classifying Angles

If you are going to build a house, how do you know what angle to use to build the roof? If you are standing on the side of a tall building, can you determine how tall the building is? If you are playing a game of pool, what angle do you hit the cue ball in order to sink the next ball? You can use your knowledge of angles to answer all these questions.

There are several different types of angles, and in this lesson, you will learn to identify and classify angles.

Ray

A line that has one endpoint and extends endlessly in the other direction.

Angle

Two rays that have a common endpoint.

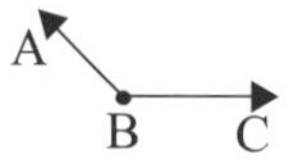

Vertex

The common endpoint where two rays meet.

The vertex is C. The vertex is F.

To name angles, start with one ray, name the vertex, then name the second ray. The vertex must always be in the middle. You can also name the angle using just the letter of the vertex if there is only one angle with that vertex. The symbol for an angle is similar to the less than sign ($\angle$).

The next thing you need to know is how to measure an angle. To measure an angle, you will use the units of degrees. The symbol for degrees is a small circle (for example, 55 degrees is written as 55°, and 45 degrees is written 45°).

Now look at different types of angles. A right angle is an angle that measures exactly 90°.

Right angle: An angle that measures exactly 90°.

You see right angles very frequently in everyday items, such as books, square picture frames, and CD cases. A right angle has a special symbol to identify it; a small box is placed in the interior of the angle by the vertex.

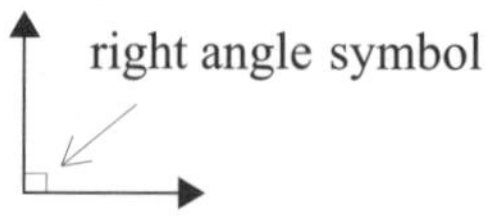

The next type of angle is an acute angle.

Acute angle: An angle that measures between 0° and 90°.

The following are all acute angles:

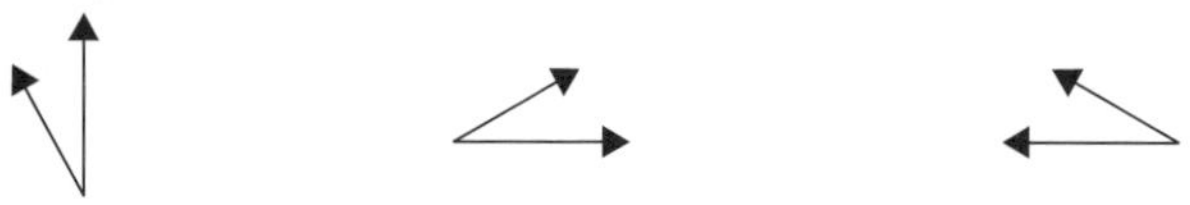

Another type of angle is the obtuse angle.

Obtuse angle	An angle that measures between 90° and 180°.

The following are all obtuse angles:

The last angle you will learn is the straight angle.

Straight angle	An angle that measures exactly 180°.

A straight angle looks like a line:

The final part of this lesson will demonstrate how to determine if two angles are adjacent angles.

Adjacent angles	Angles that share a common vertex and a common side, but do not overlap and have no common interior points.

The following are adjacent angles:

Look at the following examples.

Example 1 Name this angle:

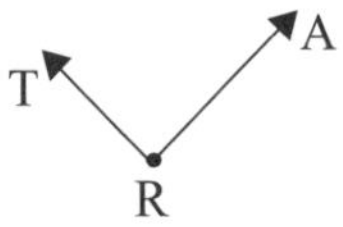

vertex: R rays: A, T	**Step 1:** To name angles, start with one ray, name the vertex, then name the second ray.
ART TRA R	**Step 2:** The vertex must always be in the middle. You can also name the angle using just the letter of the vertex if there is only one angle with that vertex.
$\angle ART$ $\angle TRA$ $\angle R$	**Step 3:** Now add the symbol for angle ($\angle$) to the angle names you just found.

Answer: $\angle ART$ or $\angle TRA$ or $\angle R$

Example 2 Name this angle:

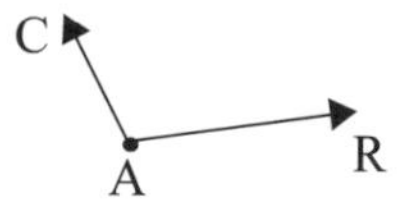

vertex: A rays: C, R	**Step 1:** To name angles, start with one ray, name the vertex, then name the second ray.
CAR RAC A	**Step 2:** The vertex must always be in the middle. You can also name the angle using just the letter of the vertex if there is only one angle with that vertex.
$\angle CAR$ $\angle RAC$ $\angle A$	**Step 3:** Now add the symbol for angle ($\angle$) to the angle names you just found.
	Answer: $\angle CAR$ or $\angle RAC$ or $\angle A$

Example 3 Use the following diagram to name three different angles.

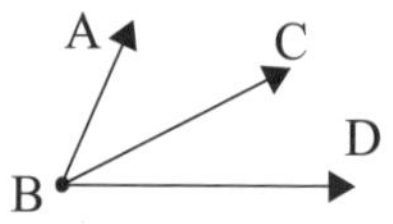

ray: A vertex: B ray: C	**Step 1:** Name the first angle. Start with the first ray, A. Then add the vertex, B. Finally, add ray C. You now have one angle named, $\angle ABC$.
ray: A vertex: B ray: D	**Step 2:** Name the second angle. Start with ray A. Then add the vertex, B. Finally, add ray D. You now have a second angle, $\angle ABD$.
ray: C vertex: B ray: D	**Step 3:** Name the third angle. Since you have already named two angles starting with ray A, start with ray C. Then add the vertex, B. Finally, add ray D. You now have a third angle, $\angle CBD$.

Answer: $\angle ABC$, $\angle ABD$, $\angle CBD$

Note: You cannot name angle B ($\angle B$) as one of your angles, because two different angles have the same vertex.

Example 4 Identify the type of angle:

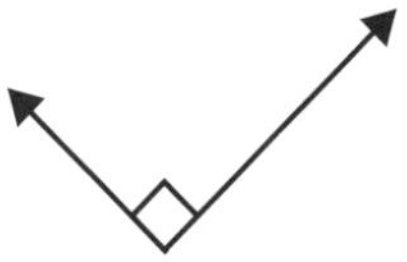

Step 1: Notice the symbol in the interior of the angle by the vertex. This symbol indicates that this is a right angle.

Answer: Right angle

Example 5 Identify the type of angle:

Step 1: This angle does not have the right angle symbol, so you know it is not a right angle. Its measurement is obviously larger than 90°, which makes it obtuse.

Answer: Obtuse angle

Example 6 Identify the type of angle:

Step 1: This angle does not have the right angle symbol, so you know it is not a right angle. It looks just like a line, which makes it a straight angle.

Answer: Straight angle

Example 7 Name the adjacent angles in the following diagram.

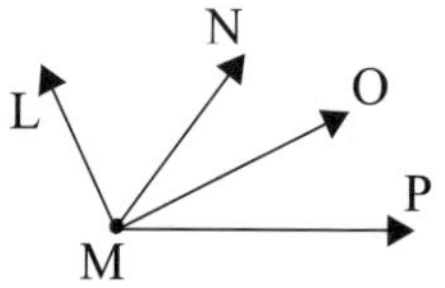

Not adjacent: $\angle LMN$ and $\angle LMO$

Not adjacent: $\angle LMN$ and $\angle LMP$

Adjacent: $\angle LMN$ and $\angle NMP$

Adjacent: $\angle LMN$ and $\angle NMO$

Step 1: See if you can find a set of adjacent angles using $\angle LMN$. The next angle after $\angle LMN$ is $\angle LMO$. The interior of $\angle LMN$ is also part of the interior of $\angle LMO$; therefore, these two angles would be considered overlapping angles, not adjacent angles. The next angle is $\angle LMP$, but it also overlaps $\angle LMN$; therefore, $\angle LMN$ and $\angle LMP$ are not adjacent. The next pair of angles is $\angle LMN$ and $\angle NMP$. They share a common vertex and a common side, but do not overlap. The same can be said of $\angle LMN$ and $\angle NMO$.

Adjacent: $\angle OMP$ and $\angle NMO$

Adjacent: $\angle OMP$ and $\angle LMO$

Step 2: Continue finding adjacent angles.

Answer: $\angle LMN$ and $\angle NMO$

$\angle LMN$ and $\angle NMP$

$\angle OMP$ and $\angle NMO$

$\angle OMP$ and $\angle LMO$

Problem Set

1. Name the vertex of this angle.

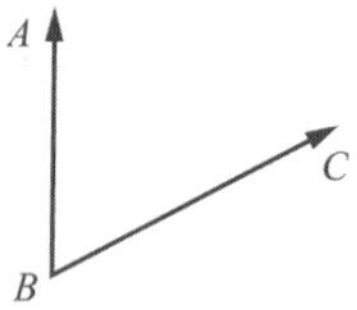

2. Name the vertex of this angle.

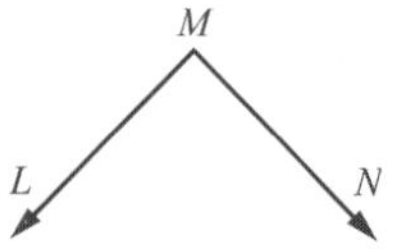

3. An angle that measures $90°$ is an obtuse angle. True or False?

4. An angle that measures $112°$ is an acute angle. True or False?

5. An angle that measures $86°$ is an acute angle. True or False?

6. An angle that measures $79°$ is an obtuse angle. True or False?

7. An angle that measures $100°$ is a straight angle. True or False?

8. Classify the angle.

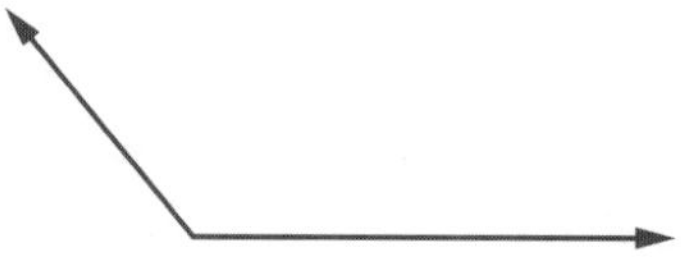

Acute angle
Straight angle
Right angle
Obtuse angle

9. Classify the angle.

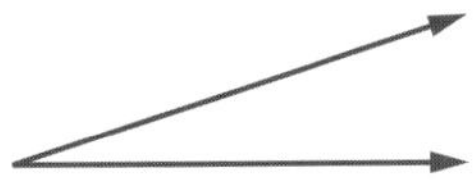

Right angle | Obtuse angle
Straight angle | Acute angle

10. Name this angle using a single letter.

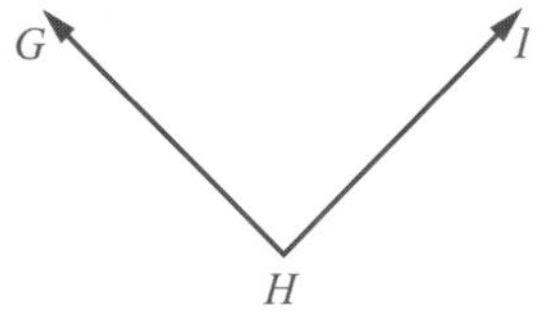

11. Name this angle using a single letter.

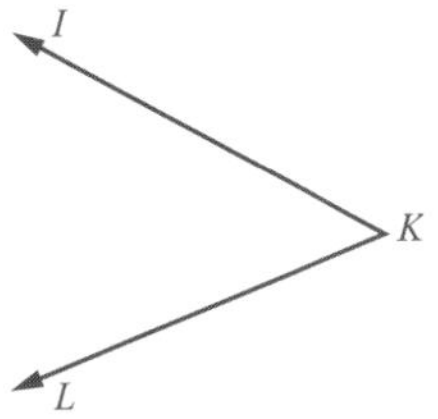

12. Which is the incorrect name for this angle?

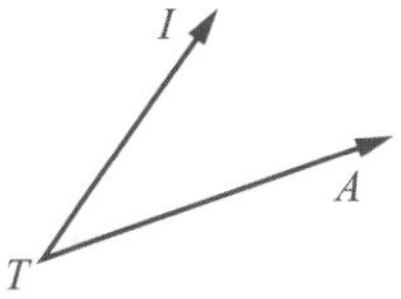

$\angle IAT$ | $\angle ITA$
$\angle ATI$ | $\angle T$

13. Which is the incorrect name for this angle?

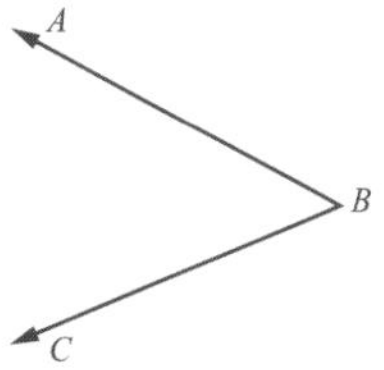

$\angle CBA$ | $\angle ABC$
$\angle B$ | $\angle A$

14. Which is the incorrect name for this angle?

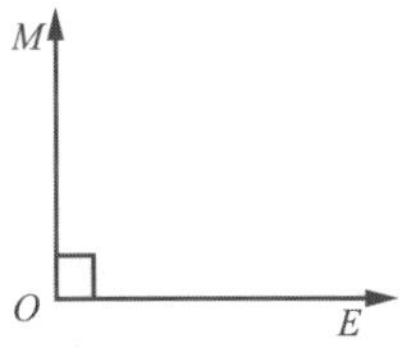

$\angle O$ | $\angle MOE$
$\angle EOM$ | $\angle E$

15. Choose the correct name for the angle.

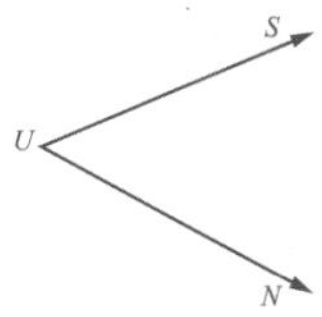

$\angle S$ | $\angle N$
$\angle U$ | $\angle SNU$

16. Choose the correct name for the angle.

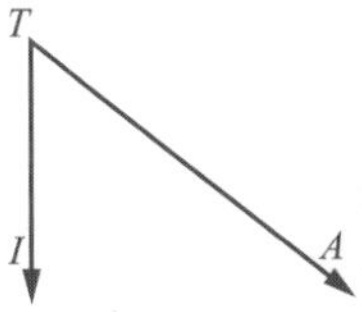

$\angle I$ | $\angle IAT$
$\angle T$ | $\angle A$

17. Choose the correct name for the angle.

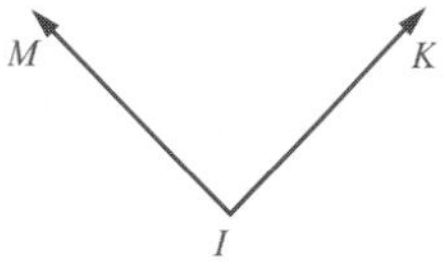

$\angle M$ | $\angle KMI$
$\angle I$ | $\angle K$

18. The lid on an ice chest is open at a $98°$ angle. Classify this angle.

Acute | Right
Obtuse | Straight

19. The tower makes a $90°$ angle with the ground. Classify this angle.

Obtuse | Right
Straight | Acute

20. The light pole made a $90°$ angle with the ground. Classify this angle.

Right | Acute
Straight | Obtuse

MPA-057: Identifying and Applying Supplementary and Complementary Angles

Complementary Angles

Two angles with measures that add up to 90° are called **complementary angles**. Since an angle measuring 90° is also called a **right angle**, complementary angles added together form a right angle. For example, an angle measuring 20° and an angle measuring 70° are complementary angles, since 20° + 70° = 90°.

The complementary angles measuring 20° and 70° are called a **complementary pair**. You can also say that an angle measuring 70° is complementary to an angle measuring 20°, or vice versa.

Complementary Angles	Two angles whose sum is 90° are called complementary angles.
Complementary Pair	Two complementary angles form a complementary pair. If two angles are complementary, each is said to be a complement of the other.

Supplementary Angles

Two angles with measures that add up to 180° are called **supplementary angles**. Since an angle measuring 180° is also called a **straight angle,** supplementary angles added together form a straight angle. For example, an angle measuring 50° and an angle measuring 130° are supplementary angles, since 50° + 130° = 180°.

The supplementary angles measuring 50° and 130° are called a **supplementary pair.** A supplementary pair is also called a **linear pair**, since a straight angle resembles a line. You can also say that an angle measuring 130° is **supplementary** to an angle measuring 50°, or vice versa.

Supplementary Angles	Two angles with a sum of 180° are called supplementary angles.
Supplementary Pair	Two supplementary angles form a supplementary pair. If two angles are supplementary, each is said to be a supplement of the other.
Vertical Angles	Opposite and congruent angles created by an intersecting line are called vertical angles.

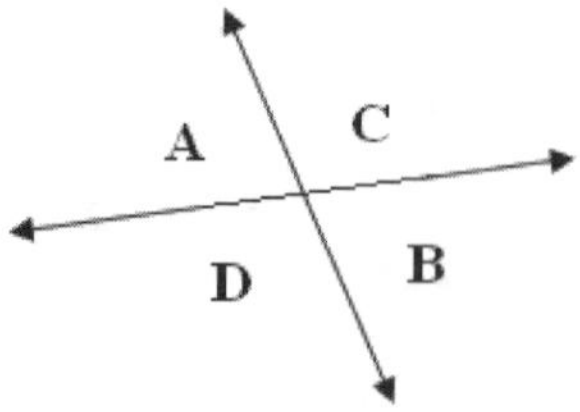

$\angle A$ and $\angle B$, $\angle C$ and $\angle D$ are vertical angles.

Example 1 Name the angle in the figure that is complementary to $\angle CBA$.

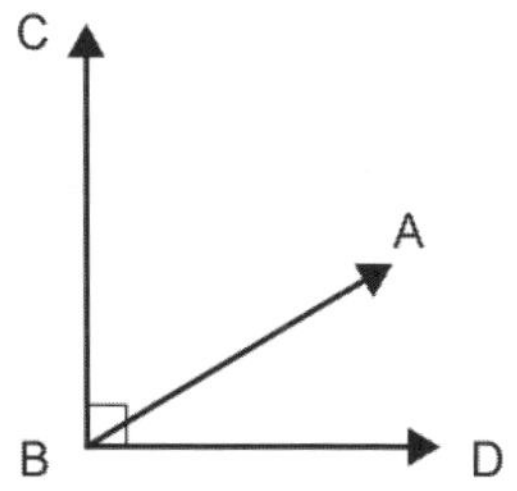

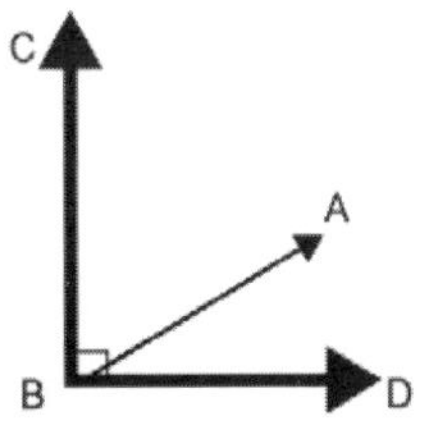

Step 1: Notice that $\angle CBD$ is a right angle.

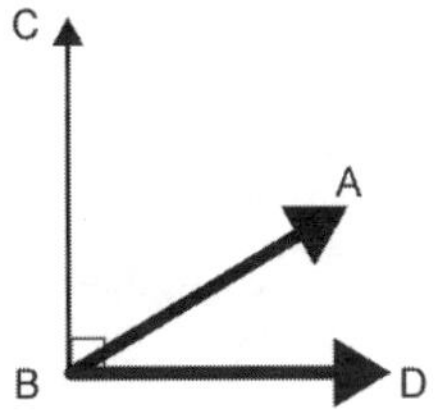

Step 2: Notice that $\angle ABD$ is adjacent to $\angle CBA$. They share the ray BA.

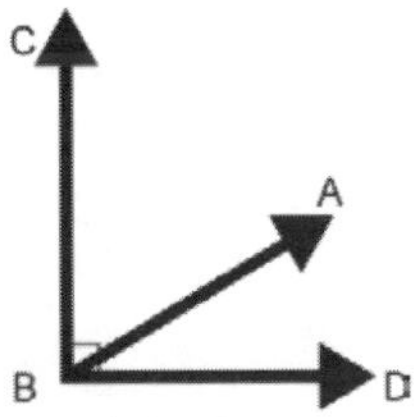

Step 3: Notice that $\angle ABD$ and $\angle CBA$ combined form the right angle $\angle CBD$. This means that $\angle ABD$ is complementary to $\angle CBA$.

Answer: $\angle ABD$ is complementary to $\angle CBA$.

Example 2 Name the angle in the figure that is supplementary to $\angle RST$.

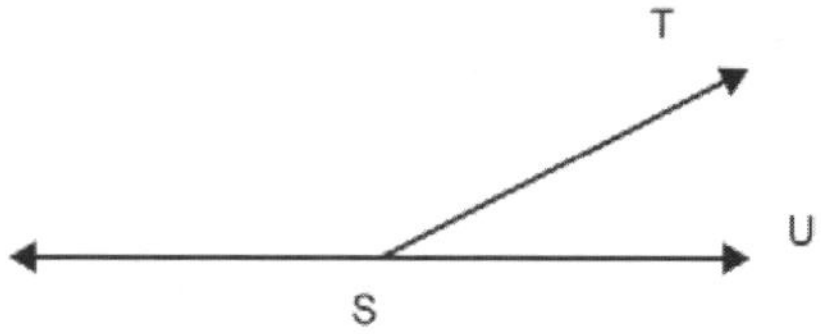

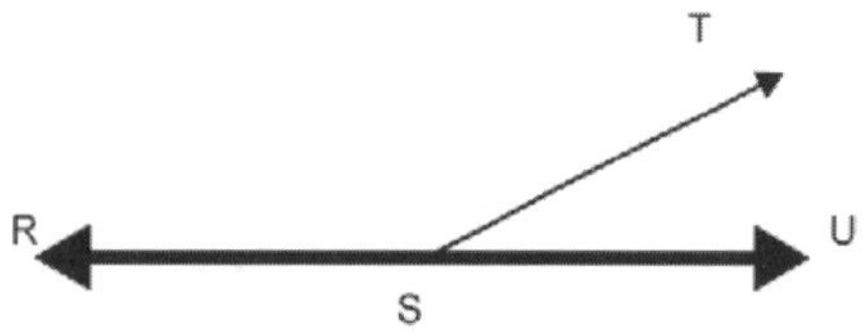

Step 1: Notice that $\angle RSU$ is a straight angle.

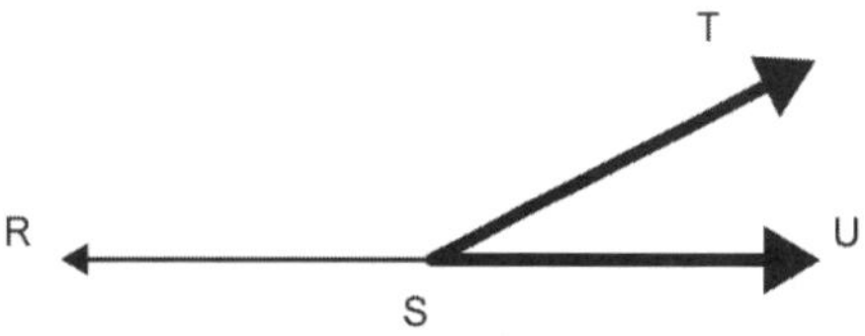

Step 2: Notice that $\angle TSU$ is adjacent to $\angle RST$. They share the ray ST.

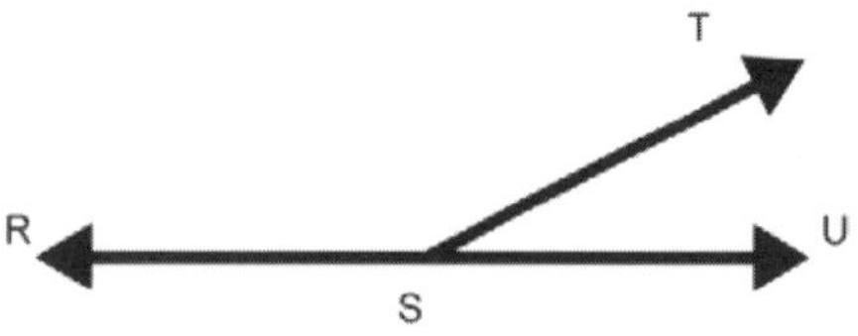

Step 3: Notice that $\angle TSU$ and $\angle RST$ combined form the straight angle $\angle RSU$. This means that $\angle TSU$ is supplementary to $\angle RST$.

Answer: $\angle TSU$ is supplementary to $\angle RST$.

Example 3 What angle is complementary to an angle measuring 36°?

$$36^\circ + x = 90^\circ$$

Step 1: Let x represent the measure of the unknown angle. This angle must sum with 36° to give 90°.

$$\begin{aligned} 36^\circ - 36^\circ + x &= 90^\circ - 36^\circ \\ x &= 54^\circ \end{aligned}$$

Step 2: Subtract 36° from both sides of the equation to isolate x.

Answer: An angle measuring 54° is complementary to an angle measuring 36°.

Example 4 What angle forms a supplementary pair with an angle measuring 95°?

$95° + x = 180°$

Step 1: Let x represent the measure of the unknown angle. This angle must sum with 95° to give 180°.

$95° - 95° + x = 180° - 95°$

$x = 85°$

Step 2: Subtract 95° from both sides of the equation to isolate x.

Answer: An angle measuring 85° is supplementary to an angle measuring 95°.

Example 5 Given that the measure of $\angle CBD$ is 55°, find the measure of $\angle ABC$ in the figure shown below.

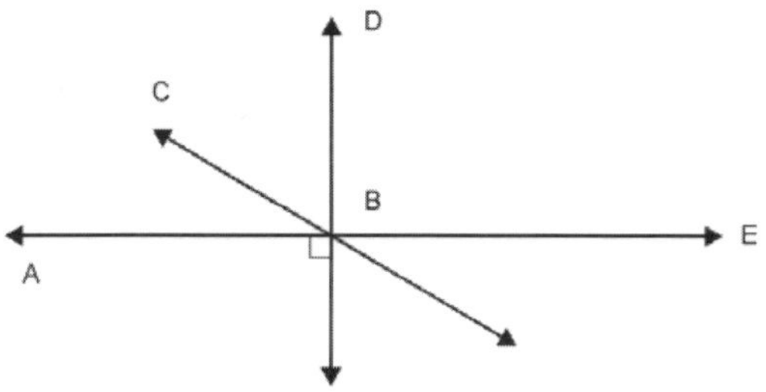

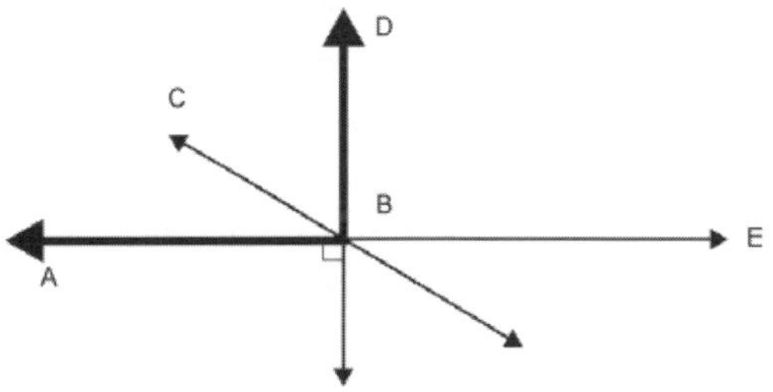

Step 1: Notice that $\angle ABD$ is a right angle.

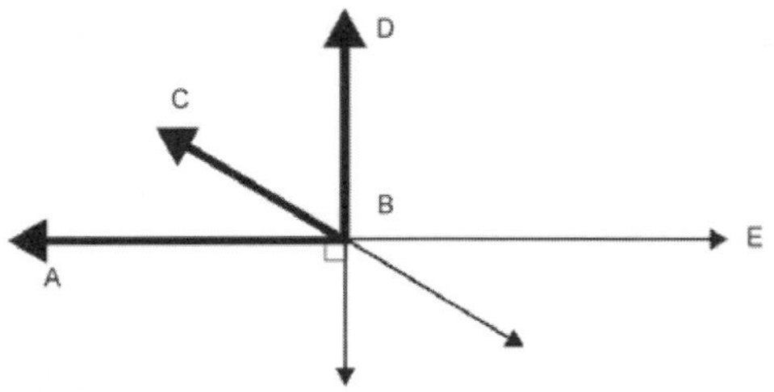

Step 2: Notice that $\angle ABC$ and $\angle CBD$ are adjacent angles that combine to form the right angle $\angle ABD$. Therefore, the two angles are complementary angles.

$55° + x = 90°$

Step 3: Let x represent the measure of $\angle ABC$. We see that the sum of x and 55° must be 90°.

$55° - 55° + x = 90° - 55°$

$x = 35°$

Step 4: Subtract 55° from each side of the equation.

Answer: $\angle ABC$ has a measure of 35°.

Problem Set

1. What is the angle complementary to $\angle QRS$?

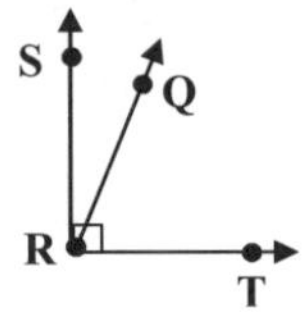

2. What is the angle complementary to $\angle VUW$?

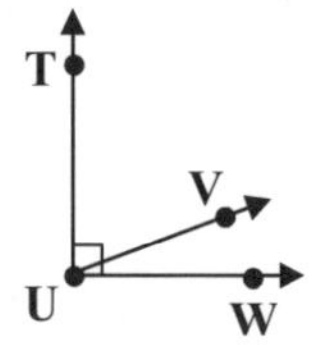

3. What is the angle complementary to $\angle WXY$?

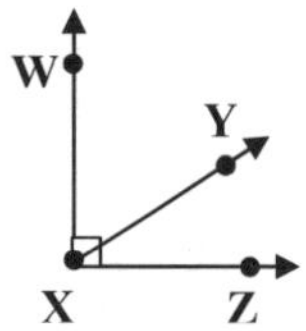

4. What is the angle complementary to $\angle ABC$?

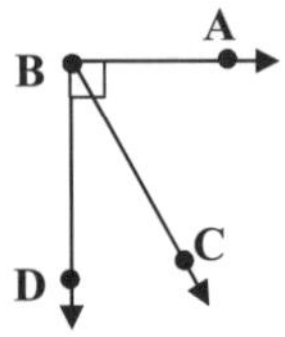

5. What is the angle complementary to $\angle LMN$?

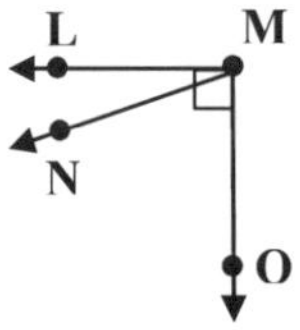

6. What is the angle complementary to $\angle GEF$?

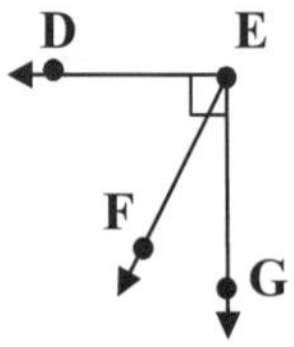

7. What is the angle complementary to $\angle HIJ$?

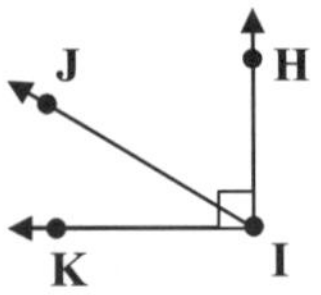

8. What is the angle supplementary to $\angle ABC$?

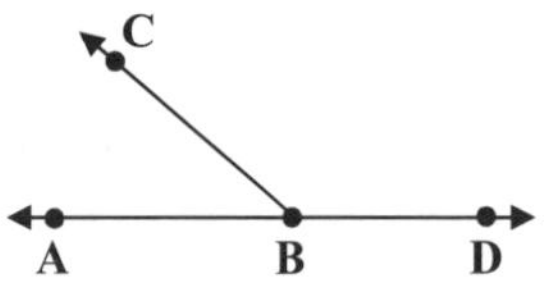

9. What is the angle supplementary to $\angle RQS$?

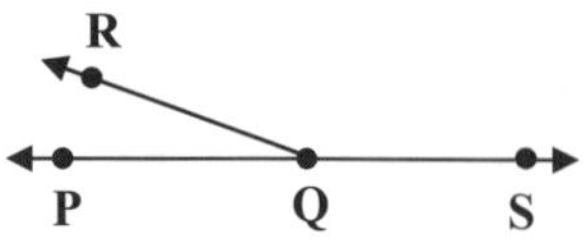

10. What is the angle supplementary to $\angle JLK$?

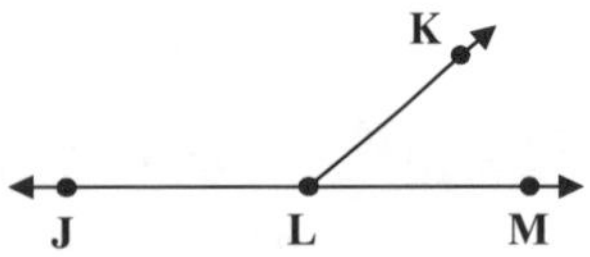

11. What is the angle supplementary to $\angle MLK$?

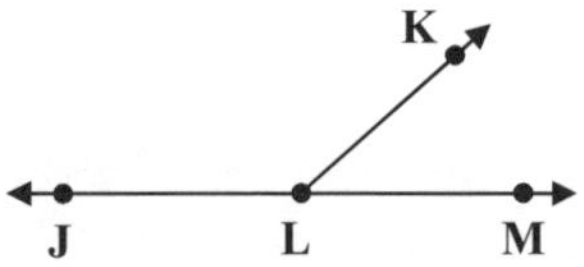

12. What is the angle supplementary to $\angle MNO$?

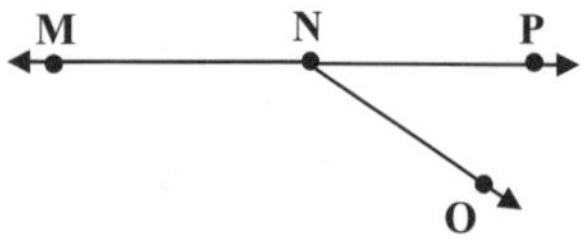

13. What is the angle supplementary to $\angle PNO$?

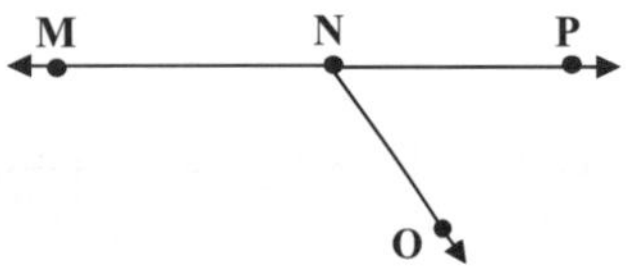

14. What is the angle supplementary to $\angle XYZ$?

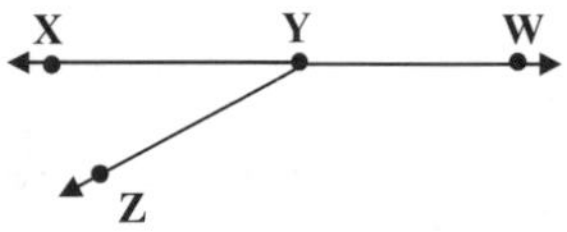

15. What is the angle supplementary to $\angle SRT$?

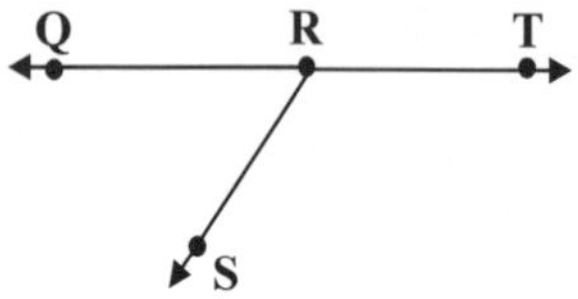

16. If the measure of one angle in a supplementary pair is double the other angle, find the measure of the larger angle.

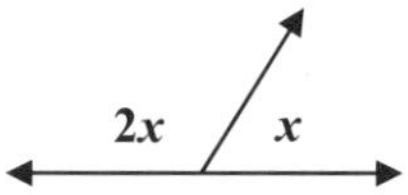

17. If the measure of one angle in a supplementary pair is three times the other angle, find the measure of the larger angle.

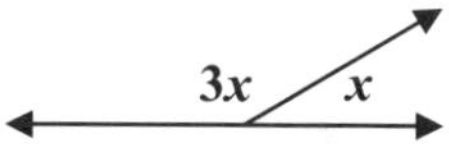

18. If the measure of one angle in a supplementary pair is four times the other angle, find the measure of the larger angle.

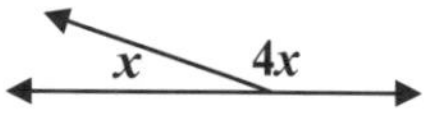

19. If the measure of one angle in a supplementary pair is 50° more than the other angle, find the measure of the smaller angle.

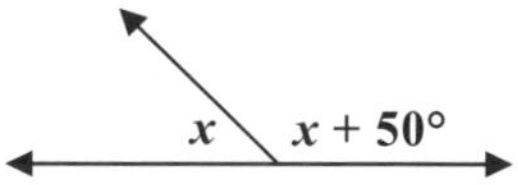

20. If the measure of one angle in a supplementary pair is 80° more than the other angle, find the measure of the larger angle.

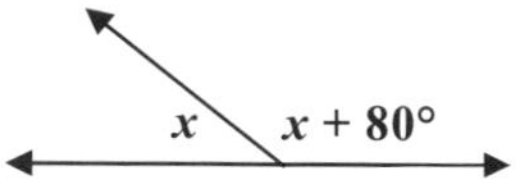

21. If the measure of one angle in a supplementary pair is 60° less than the other angle, find the measure of the smaller angle.

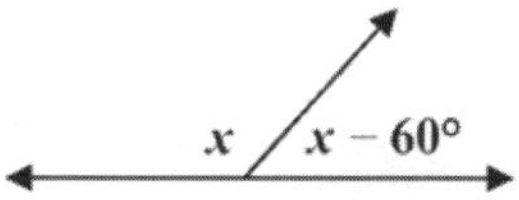

22. If the measure of one angle in a supplementary pair is 30° less than the other angle, find the measure of the larger angle.

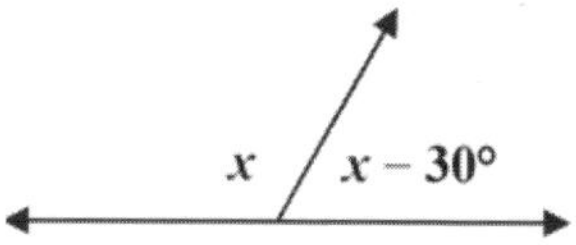

23. Find the measure of $\angle A$ from the given diagram.

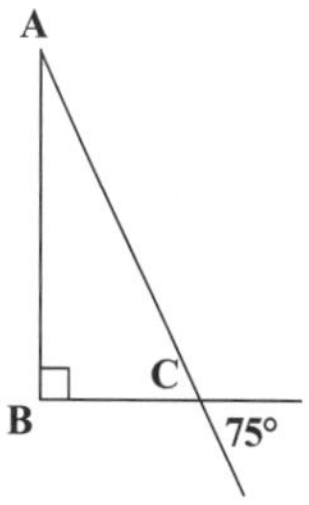

24. Find the measure of $\angle A$ from the given diagram.

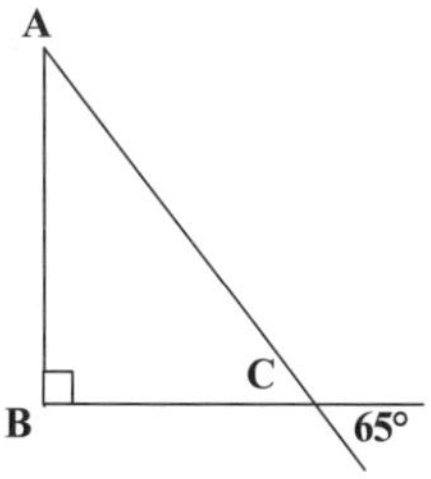

25. Find the measure of $\angle E$ from the given diagram.

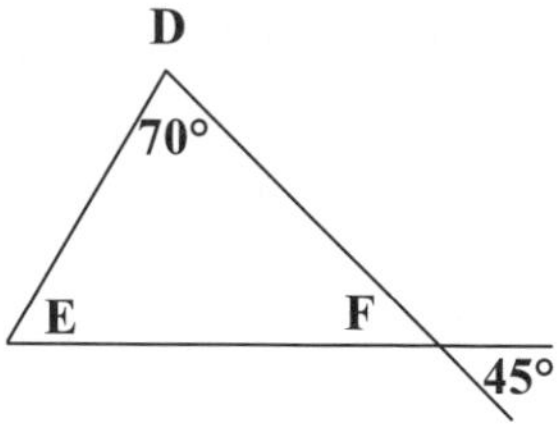

26. Find the measure of $\angle D$ from the given diagram.

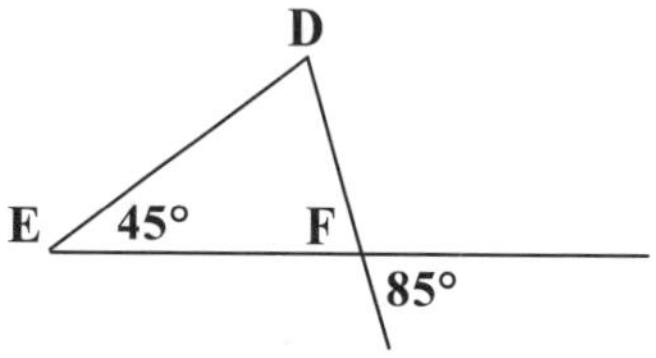

27. Find the measure of $\angle S$ from the given diagram.

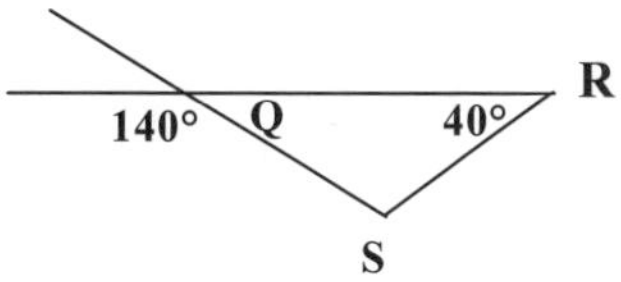

28. Find the measure of $\angle X$ from the given diagram.

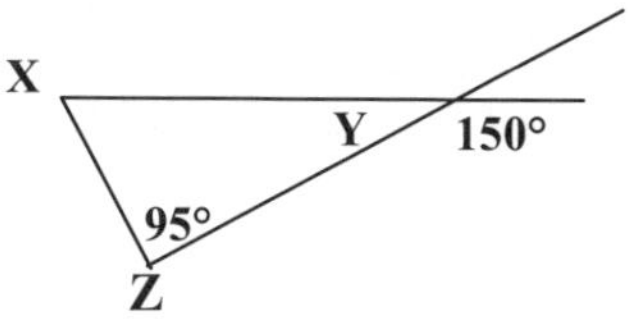

29. Find the measure of $\angle X$ from the given diagram.

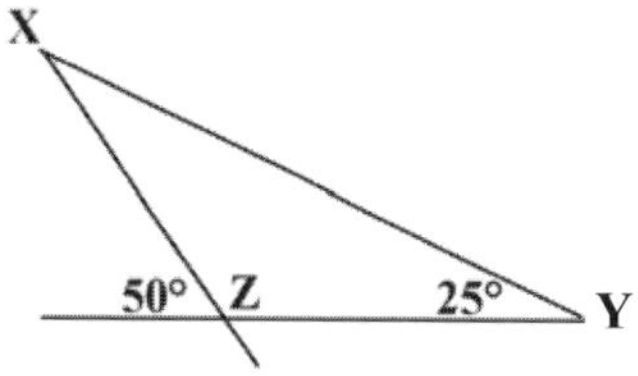

30. Find the measure of $\angle A$ from the given diagram.

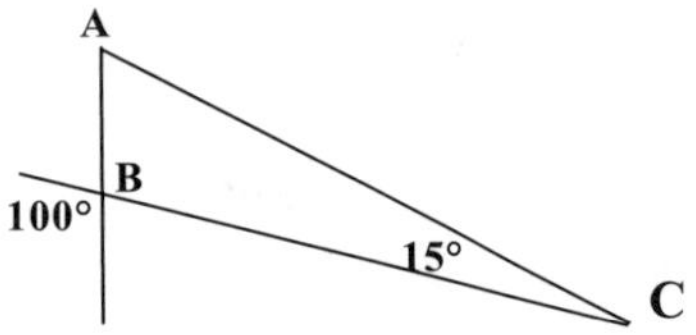

31. Find the measure of $\angle CGD$ if $m\angle 1 = 60°$ and $m\angle 2 = 80°$.

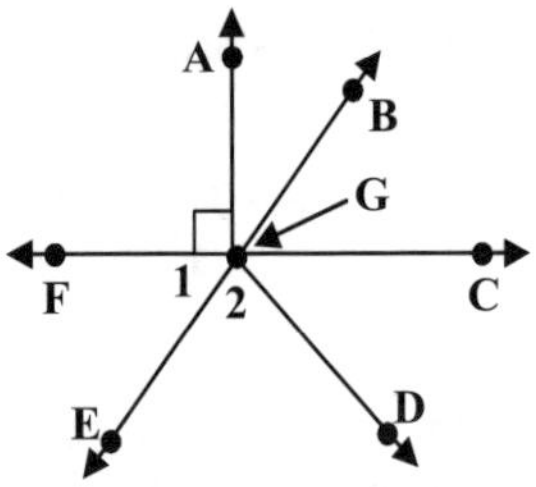

32. Find the measure of $\angle 5$ if $m\angle 4 = 65°$ and $m\angle 6 = 34°$.

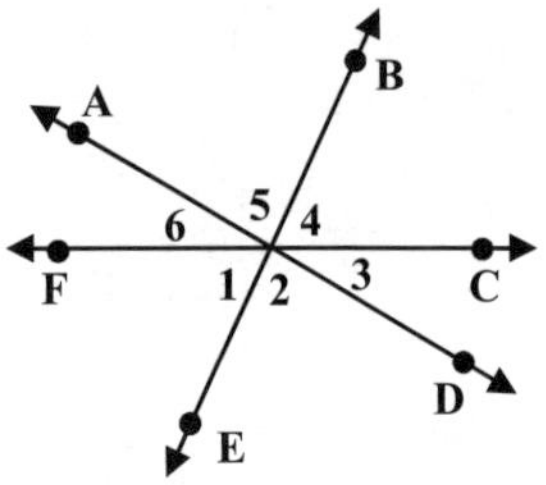

33. Find the measure of $\angle 3$ if $m\angle 4 = 53°$ and $m\angle 2 = 81°$.

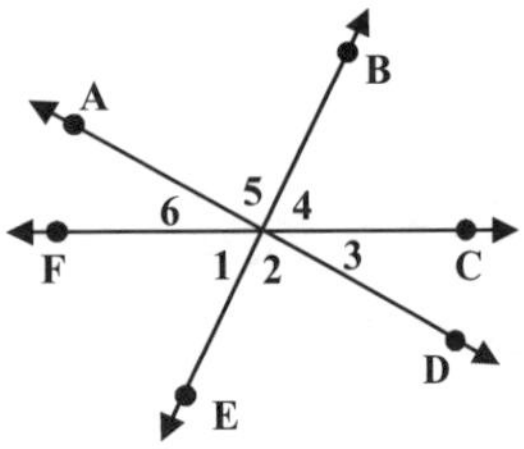

34. Find the measure of $\angle 3$ if $m\angle 1 = 45°$ and $m\angle 2 = 102°$.

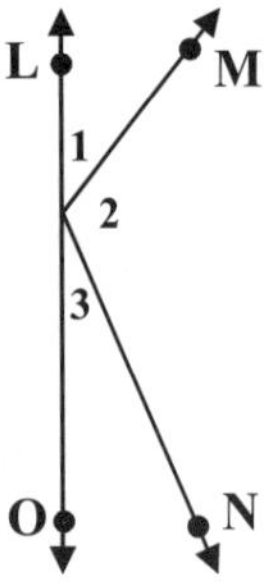

35. Find the measure of $\angle 3$ if $m\angle 1 = 45°$ and $m\angle 2$ is double $m\angle 1$.

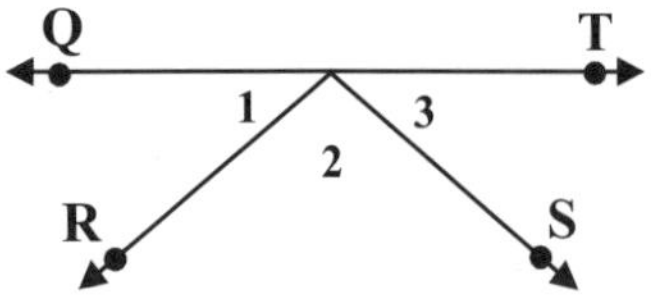

36. Find the measure of $\angle 3$ if $m\angle 1 = 36°$ and $m\angle 2$ is three times $m\angle 1$.

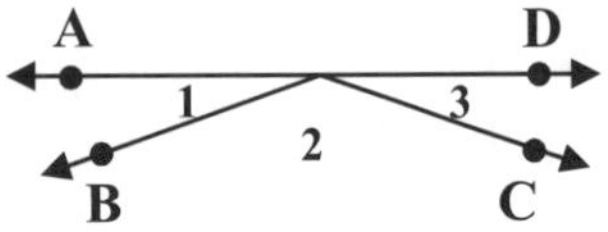

37. Find the measure of $\angle 3$ if $m\angle 2 = 90°$ and $m\angle 4 = 62°$.

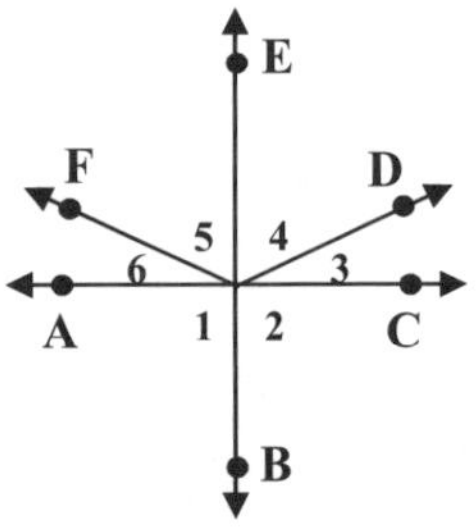

38. Find the measure of $\angle 3$ if $m\angle 4 + m\angle 5 = 90°$ and $m\angle 6 = 47°$.

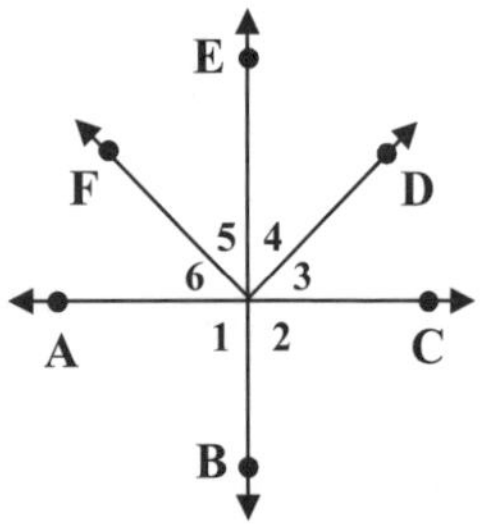

39. Find the measure of $\angle 3$ if $m\angle 2 = 114°$ and $m\angle 1$ is one-third of $m\angle 2$.

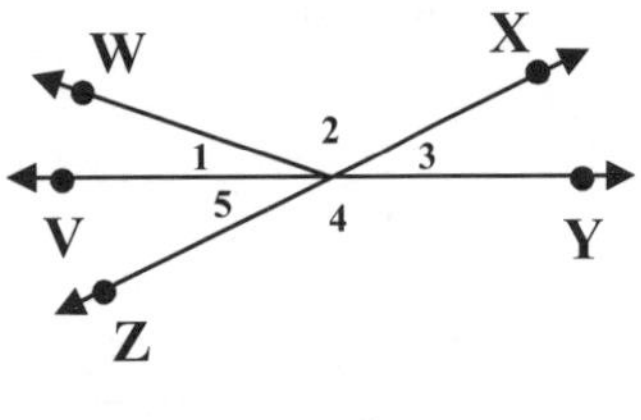

40. Find the measure of $\angle 2$ if $m\angle 5 = 46°$ and $m\angle 1$ is one-half of $m\angle 5$.

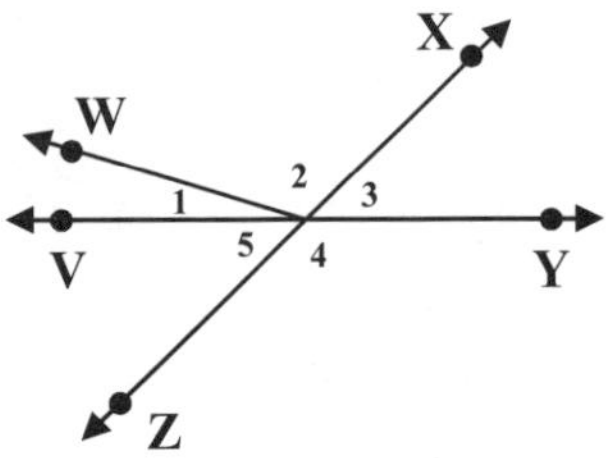

MPA-058: Identifying Polygons

When you are driving down the road, you see many road signs. What is the shape of a stop sign, a yield sign, or an information sign? All of these signs are polygons.

Polygon	A simple, closed figure made up of angles and line segments that do not cross each other.

A **closed** figure means that all the sides touch, or meet. Compare the following figures:

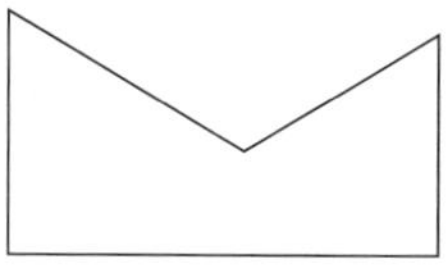
closed figure

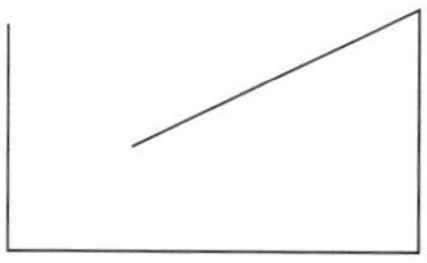
open figure

In every polygon, the number of sides is the same as the number of angles.

What is the fewest number of sides a polygon can have? The answer is three sides, which form a triangle. There is no way you can make two line segments close at both ends to form a two-sided polygon, and one line segment cannot form a polygon. Polygons are named for the number of sides they contain.

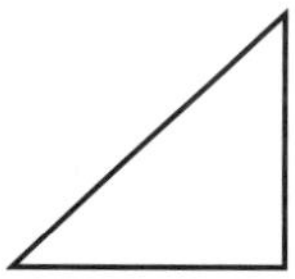
triangle
3 sides

quadrilateral
4 sides

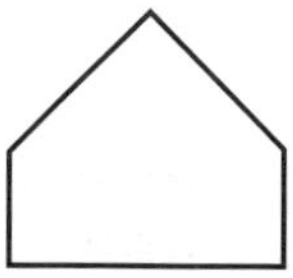
pentagon
5 sides

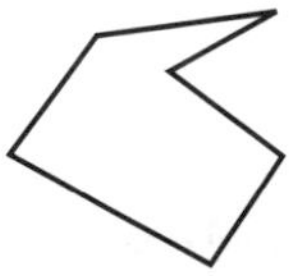
hexagon
6 sides

heptagon
7 sides

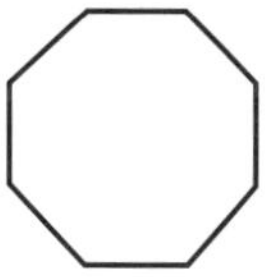
octagon
8 sides

nonagon
9 sides

decagon
10 sides

dodecagon
12 sides

There are two types of polygons: regular and irregular. **Regular polygons** are polygons that have congruent sides and angles. **Irregular polygons** are polygons that do not have congruent sides and angles. Look at the examples below.

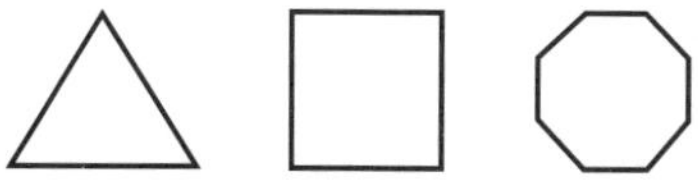
regular polygons

irregular polygons

Example 1 Is the figure below a polygon? Why or why not?

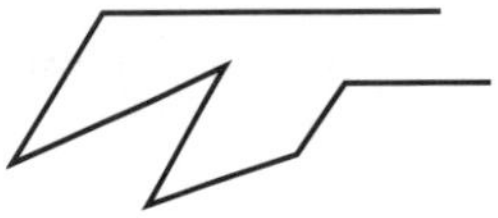

Answer: No, this is not a polygon. Even though all sides are line segments, the line segments do not meet. This figure is not closed, so it is not a polygon.

Example 2 Name and classify the following polygon.

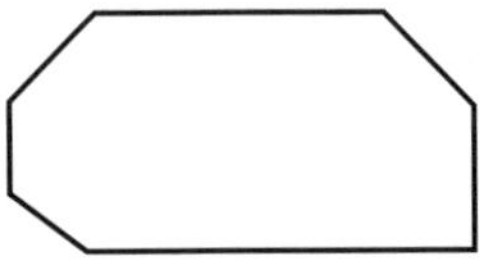

Answer: Since this figure has seven sides, it is a heptagon. Since the sides are not all congruent, and the angles are not all congruent, it is an irregular heptagon.

Example 3 Name and classify the following polygon.

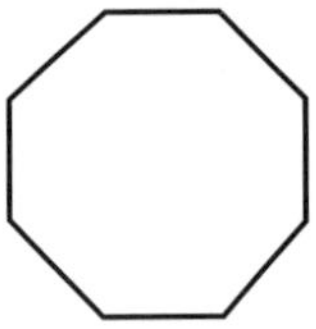

Answer: This is a closed figure that has eight equal sides. It is a regular octagon.

Problem Set

1. Classify the following as a polygon or not a polygon.

polygon not a polygon

2. Classify the following:

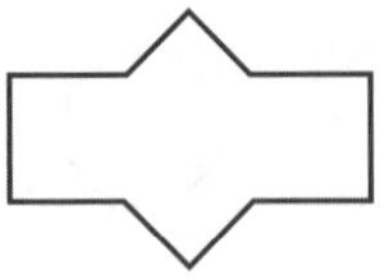

square hexagon
heptagon decagon

3. Classify the following polygon:

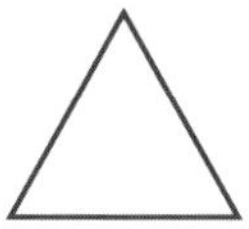

regular rectangle
regular triangle
quadrilateral
hexagon

4. This figure is equilateral. True or false?

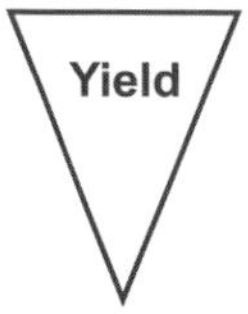

5. A rectangle is a quadrilateral. True or False?

6. Classify the following:

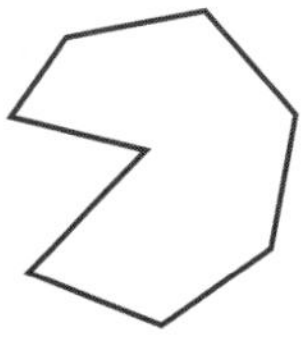

none of these
regular octagon
regular nonagon
regular hexagon

7. Classify a 7-sided figure.

hexagon
heptagon
heptahedron

8. This figure is not a polygon. True or false?

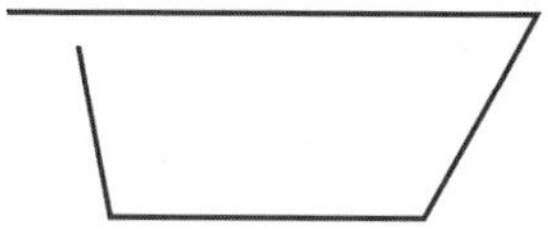

9. Identify the following:

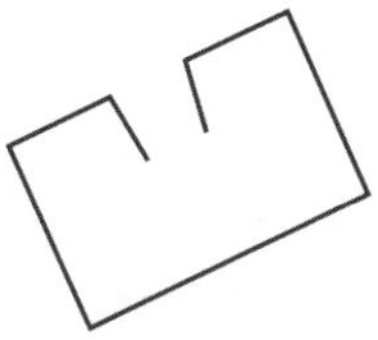

Not a polygon
Polygon

10. Classify the following:

Not a polygon
Polygon

11. The following figure is a polygon. True or false?

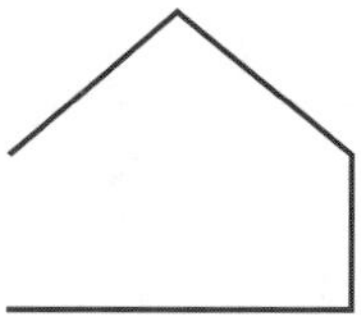

12. Classify the figure:

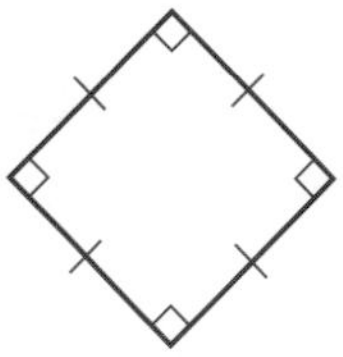

Irregular polygon
Regular polygon

13. Identify the following:

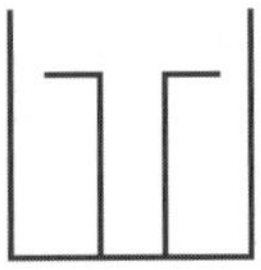

Polygon   Not a polygon

14. Identify the following:

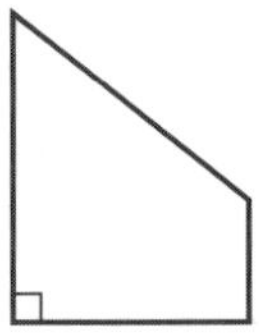

Polygon   Not a polygon

15. Identify the following:

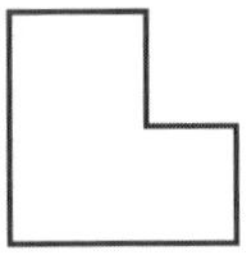

Polygon   Rectangle
Not a polygon

16. Identify the following polygon:

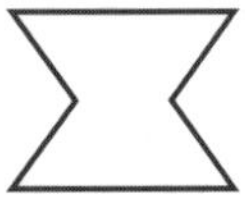

Regular   Irregular

17. Identify the following polygon:

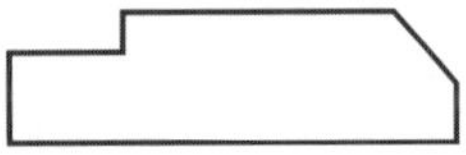

Decagon   Triangle
Heptagon

18. A decagon has 12 sides. True or False?

False   True

19. Classify the figure:

Decagon   Heptagon
Hexagon

20. Classify the figure:

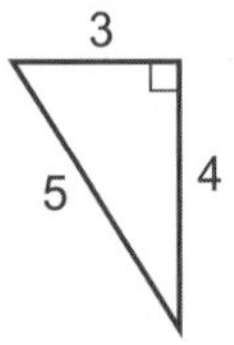

Not equilateral   Equilateral

MPA-059: Classifying Triangles and Quadrilaterals

Tonight you are going to order pizza for dinner. If you get a small pizza and a large pizza, will they be the same size? Or, will they be the same shape? When you use a copy machine, can you make copies of a picture? Can you make a copy of a picture but make it smaller or larger?

When you learned about polygons, you learned that you could draw an endless number of different polygons. You could draw two triangles that looked exactly alike. You could draw two triangles where one was just a smaller version of the other. Figures that are exactly alike are called congruent figures. Figures that are smaller or larger versions of each other are called similar figures.

Congruent figures	Figures that have the same shape and size.
Similar figures	Figures that have the same shape and corresponding angle measures but are different sizes.

A small pizza and a large pizza are similar figures. They are the same shape, but just different sizes. Using a copy machine to copy a picture means the pictures are the same shape and size (unless you specify "reduce" or "enlarge"), and are therefore congruent.

In this lesson, you will also learn how to classify triangles and quadrilaterals by their sides and angles. First, look at the six types of triangles:

Equilateral triangle	A triangle that has three congruent (equal) sides.
Isosceles triangle	A triangle that has two equal sides.
Scalene triangle	A triangle that has no equal sides.
Acute triangle	A triangle that has three acute angles (angles that measure between 0° and 90°).
Right triangle	A triangle that has one right angle (angle that measures exactly 90°).
Obtuse triangle	A triangle that has one obtuse angle (angle that measures between 90° and 180°).

Before discussing quadrilaterals, let's review the definition of parallel:

Parallel Lines	Lines that lie in the same plane but always remain the same distance apart.

There are several types of quadrilaterals: trapezium, trapezoid, parallelogram, rhombus, and rectangle.

Trapezium	A quadrilateral that has no parallel sides.
Trapezoid	A quadrilateral with one pair of parallel sides.
Parallelogram	A quadrilateral with two pairs of parallel sides.
Rhombus	A parallelogram with four congruent (equal) sides.
Rectangle	A parallelogram with four congruent angles.

Square	A parallelogram that has four congruent sides and four congruent angles. Thus, it is both a rhombus and a rectangle.

Since the rectangle has four congruent angles, and looks like a square, can you guess what each angle measures? The angles in a rectangle are all right angles; therefore, they have a measure of 90°.

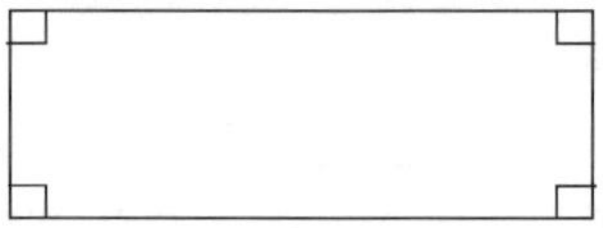

There is another situation in which there will always be right angles: when lines are perpendicular.

Perpendicular lines	Lines that intersect to form right angles.

Now, look at some examples of how to classify triangles and quadrilaterals.

Example 1 Classify the triangle below.

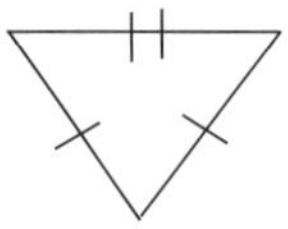

Step 1: The "tick marks" on the sides of the triangle indicate if the sides are congruent. The sides with the same number of marks are congruent. In this triangle, there are two sides that have the same number of tick marks.

Step 2: Two sides are congruent, so this triangle is an isosceles triangle.

Answer: Isosceles triangle

Example 2 Are these two quadrilaterals congruent or similar?

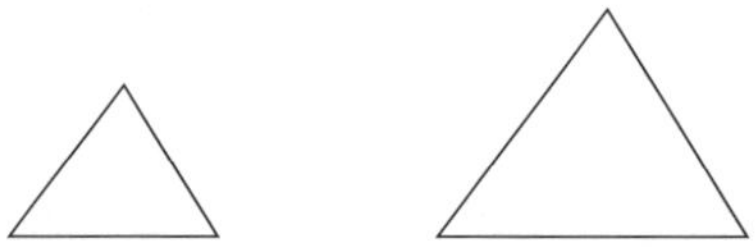

Step 1: The two quadrilaterals have the same shape.

Step 2: They are not the same size; therefore, they must be similar.

Answer: Similar

Example 3 Classify the quadrilateral below.

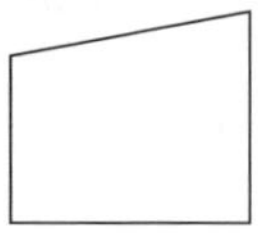

Step 1: Look at the sides of the quadrilateral. The sides are not congruent, so this cannot be a rhombus.

Step 2: See if this is a parallelogram. The quadrilateral does not have two pairs of parallel sides, so this is not a parallelogram.

Step 3: See if this is a trapezoid. The quadrilateral does have one pair of parallel sides, so this is a trapezoid.

Answer: Trapezoid

Note that the quadrilateral in the previous example is not a rectangle. Since a rectangle is a parallelogram with four congruent angles, and the quadrilateral was not a parallelogram, it could not be a rectangle.

Example 4 Classify the quadrilateral below.

Step 1: Since this quadrilateral has three sides and three angles, you know it is a triangle.

Step 2: The triangle has one obtuse angle; therefore, it is an obtuse triangle.

Answer: Obtuse triangle

Problem Set

1. A triangle can be classified by:

its sides | none of these
its angles | its angles and its sides

2. Choose the best classification:

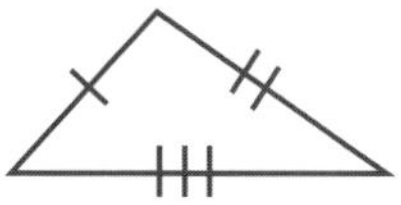

acute scalene | scalene
isosceles | equilateral

3. A trapezium is a quadrilateral. True or false?

4. Identify the rectangle.

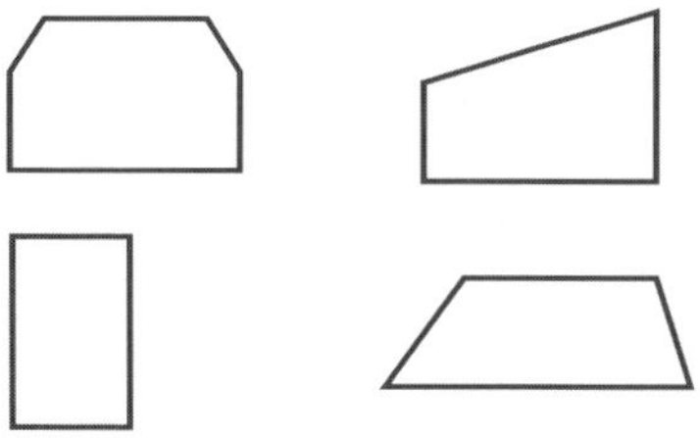

5. A parallelogram is always a quadrilateral. True or False?

6. Classify shape #5 in the picture.

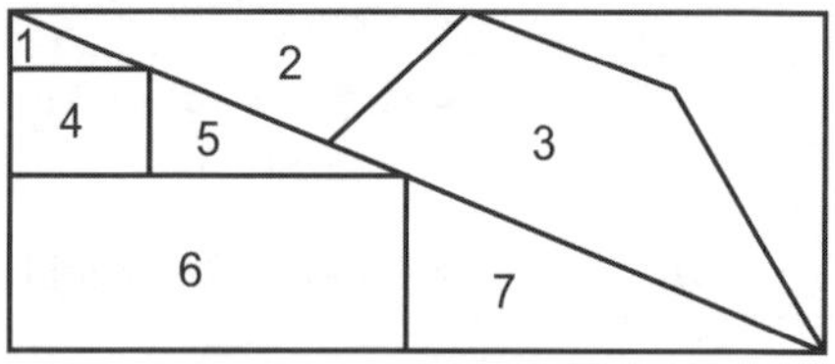

trapezium
acute triangle
right triangle
obtuse triangle

7. Classify this group of figures. Be as specific as possible.

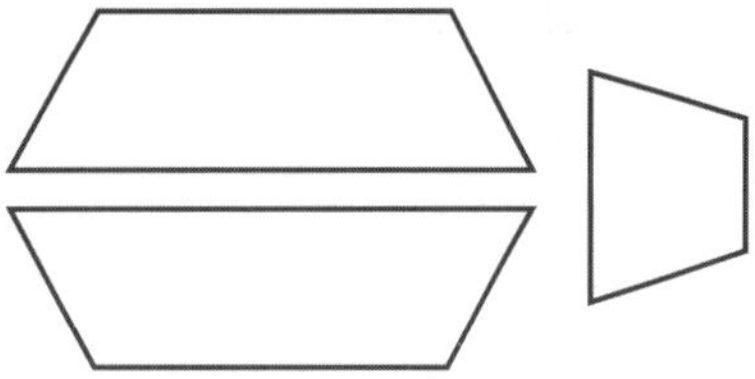

trapeziums
trapezoids
parallelograms
quadriaterals

8. Choose the best classification for this triangle.

right
acute
equilateral
obtuse

9. Classify this triangle by its sides and angles. Choose the best classification for this triangle.

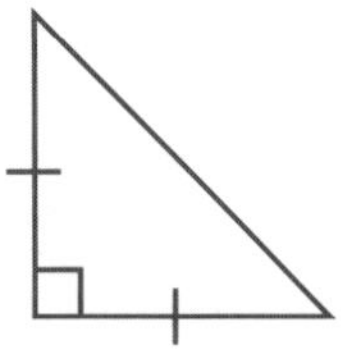

scalene, right
equilateral, right
isosceles, right
none of the above

10. Classify this triangle by its sides and angles.

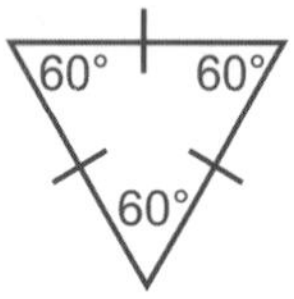

obtuse equilateral
acute scalene
acute isosceles
acute equilateral

11. Classify this triangle by its sides and angles.

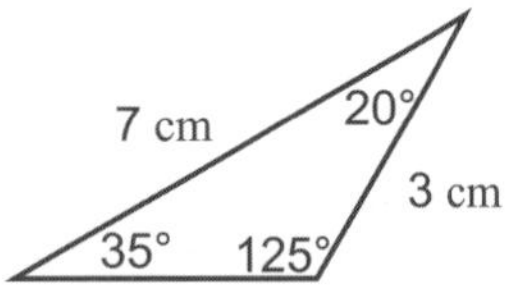

obtuse isosceles
obtuse scalene
obtuse equilateral
obtuse right

12. Classify this triangle by its sides and angles.

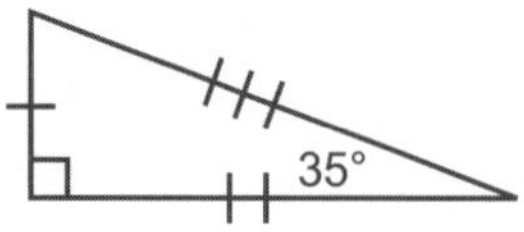

acute scalene
right scalene
right isosceles
right acute

13. Classify the following quadrilateral using the name that best describes it.

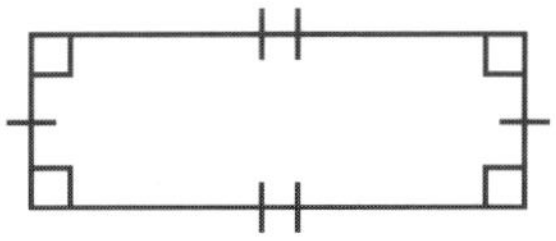

quadrilateral square
rectangle parallelogram

14. Classify the following quadrilateral using the name that best describes it.

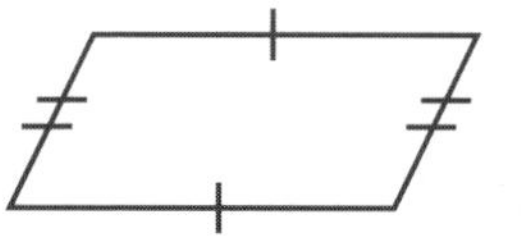

rectangle quadrilateral
parallelogram rhombus

15. Classify the following quadrilateral using the name that best describes it.

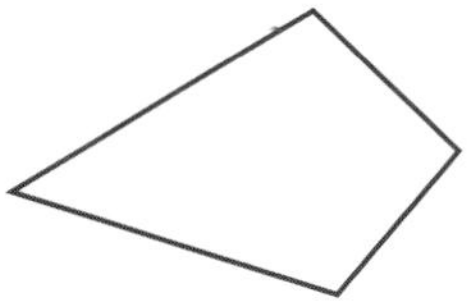

quadrilateral trapezoid
parallelogram rhombus

16. Buildings often have several shapes to give them a nice appearance. What shape is the front view of the church's roof?

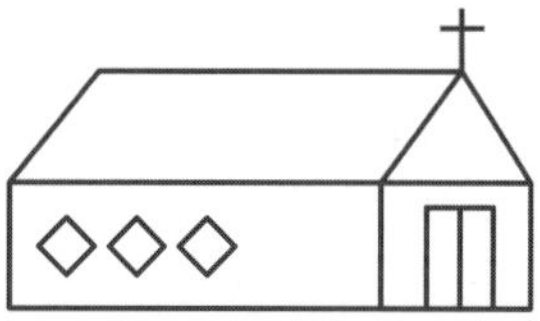

parallelogram triangle
rhombus trapezium

17. Classify the angle.

right isosceles
obtuse acute

18. Give the most specific name for this group of quadrilaterals:

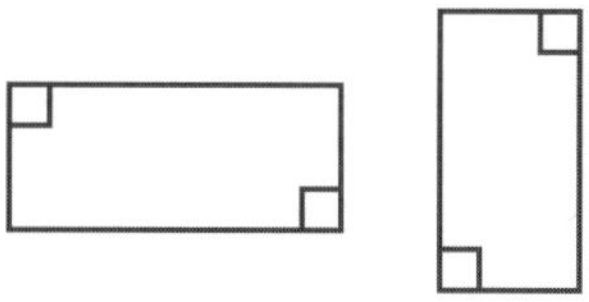

19. Give the most specific name for this group of triangles, according to sides:

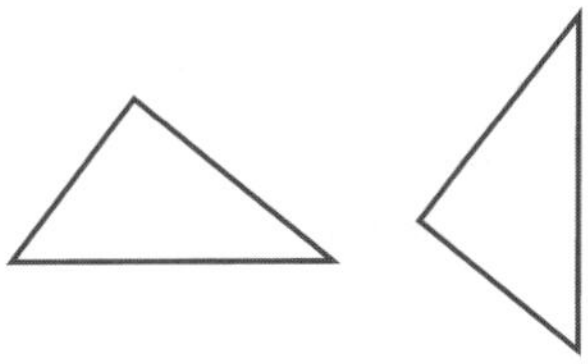

20. Look at the figures below: Are they similar or congruent?

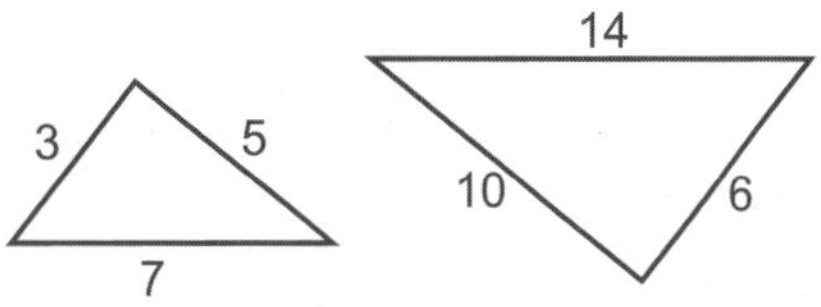

MPA-060: Determining Which Figures Tessellate

You've probably seen tiled floors or counters. When a surface is tiled, you may notice that there are patterns in the tile. These patterns always **tessellate**.

Tessellation	Repeating pattern of interlocking shapes that cover a plane.
Plane	Any flat surface.
Tiling	A special type of tessellation that uses polygons.

Kitchen floor tiles tessellate because they fit together without any gaps. Most tile floors are a simple pattern of squares, but several regular polygons tessellate.

triangle:

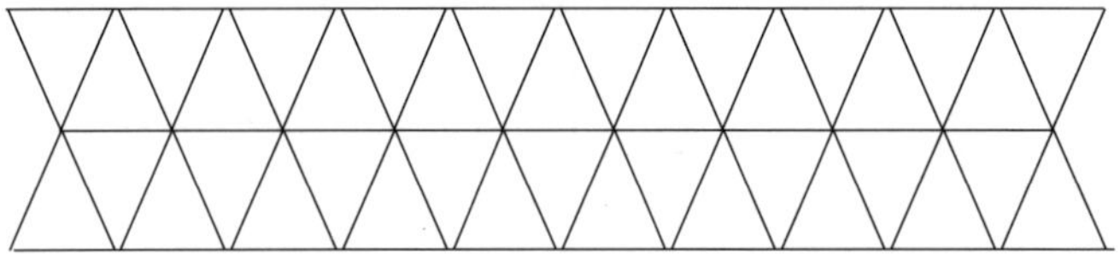

square:

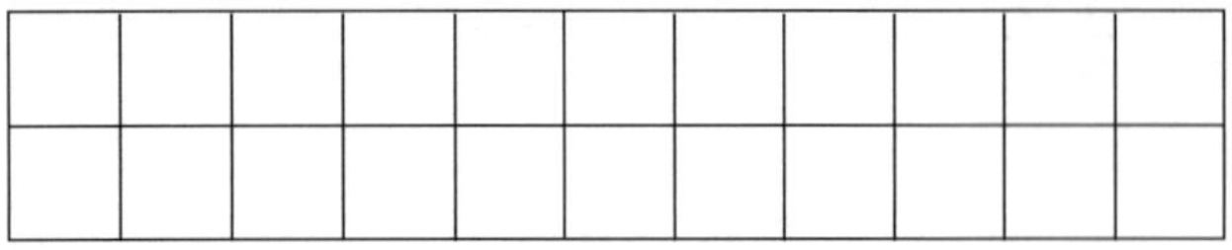

hexagon:

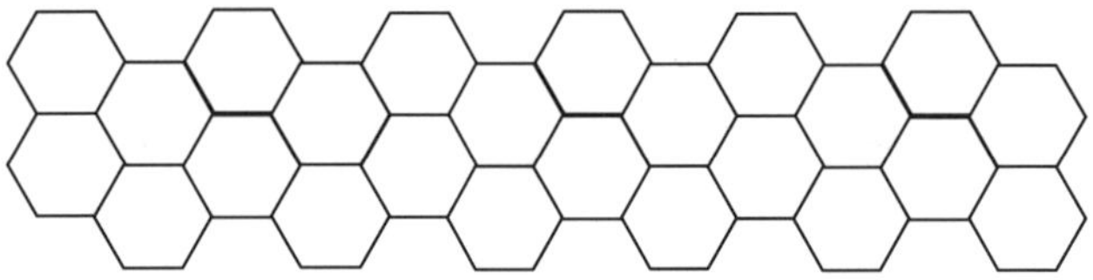

These three polygons are the only regular polygons that tessellate. (A regular polygon is a polygon with sides of equal length and angles of equal measure). Why is this? To find out, we must examine the polygons.

No matter what kind of triangle you draw, the sum of its interior angles will always be 180°.

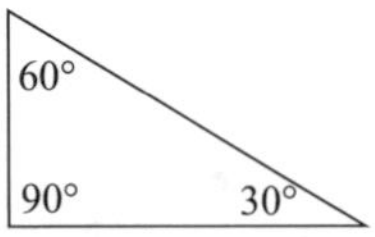

$90° + 60° + 30° = 180°$

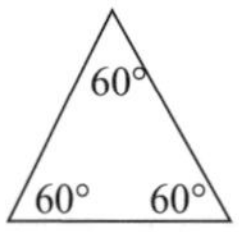

$60° + 60° + 60° = 180°$

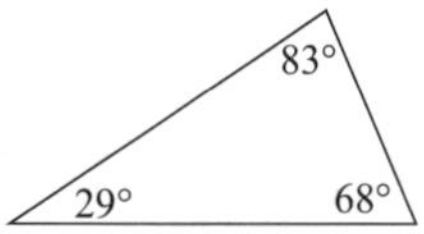

$83° + 68° + 29° = 180°$

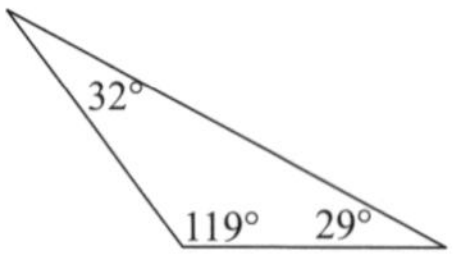

$119° + 32° + 29° = 180°$

Now, how can you apply this knowledge to find the sum of the interior angles of a regular polygon?

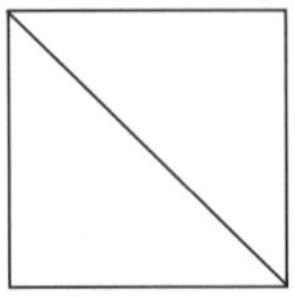

$2 \times 180° = 360°$

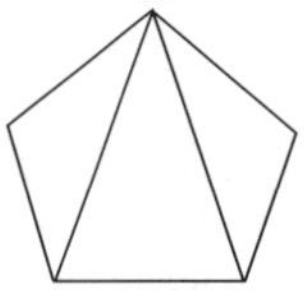

$3 \times 180° = 540°$

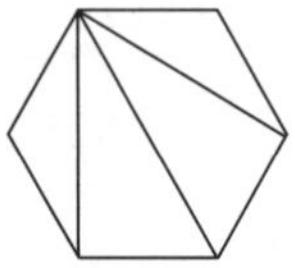

$4 \times 180° = 720°$

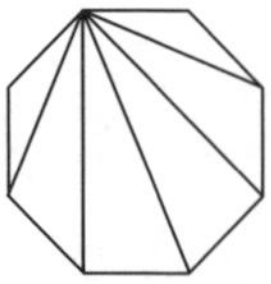

$6 \times 180° = 1080°$

Because a square has two triangles inside of it, the sum of a square's interior angles is twice the sum of a triangle's interior angles, or twice 180°. The sum of a square's interior angles is 360°. If you apply that procedure to other regular polygons, you can find the sum of their interior angles.

A pattern exists here. The number of triangles that fit inside a regular polygon seems to be two less than the number of sides of the polygon. If you multiply that number by 180°, you'll get the sum of the interior angles. So, if n is the number of sides of the polygon, the formula for the sum of the interior angles is $(n - 2) \times 180°$. Now it's easy to figure out the measure of each angle. Because the number of angles in a polygon is the same as the number of sides, you can just divide the sum of the interior angles by the number of sides to get the measure of each angle in a regular polygon:

$$\frac{(n-2) \times 180°}{n}$$

Apply these formulas to a chart for regular polygons:

Number of sides	Sum of interior angles	Measure of each angle
3	**180°**	**60°**
4	**360°**	**90°**
5	540°	108°
6	**720°**	**120°**
7	900°	128.571°
8	1080°	135°
9	1260°	140°
10	1440°	144°

The polygons in bold type are polygons that tessellate. Do you notice anything about them? The measurements of their angles (60°, 90°, and 120°) are the only angles that divide evenly into 360°. No other regular polygon has an interior angle that divides evenly into 360°. Therefore, equilateral triangles, squares, and regular hexagons tessellate because their angles can make a full circle. Here's why:

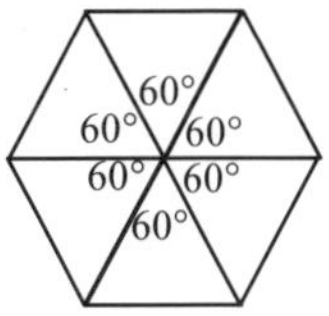

$6 \times 60° = 360°$

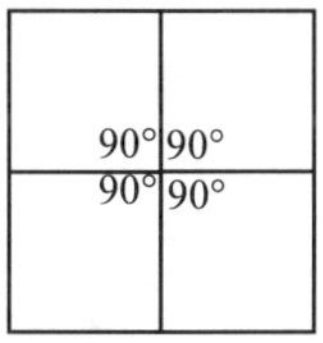

$4 \times 90° = 360°$

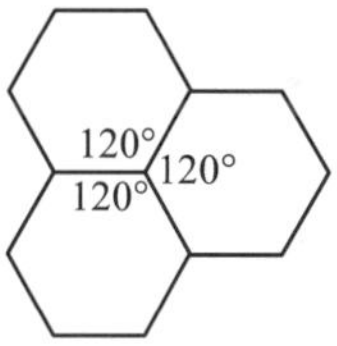

$3 \times 120 = 360°$

The polygons with angles that divide into 360° can be arranged around a circle. Other polygons don't quite make it all the way around.

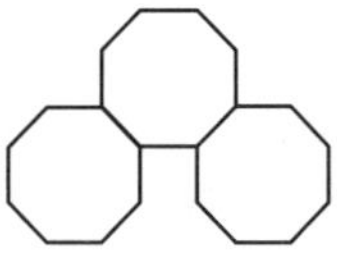

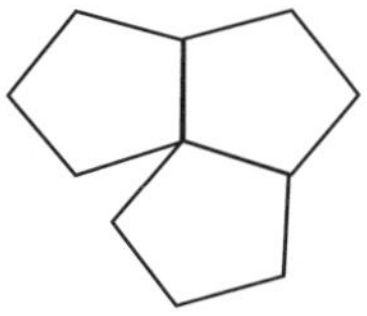

But that's not quite the whole story. Regular polygons aren't the only shapes that tessellate.

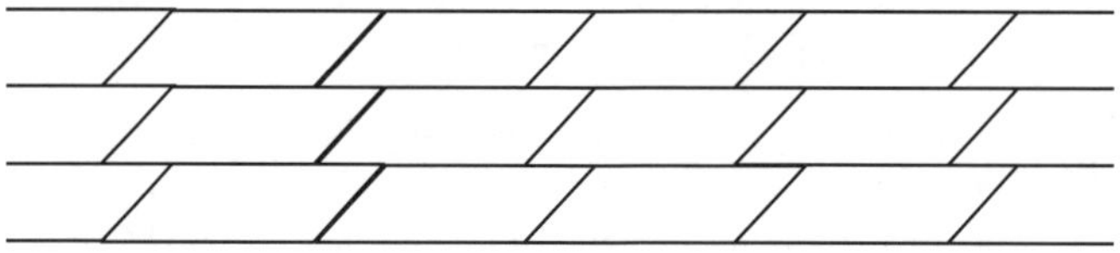

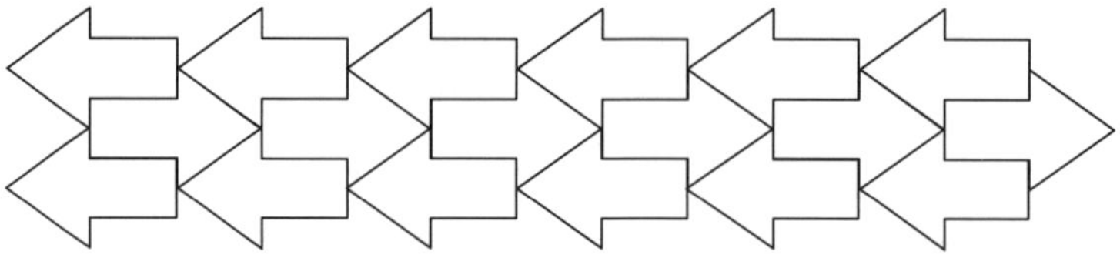

Let's examine the interior angles of these figures.

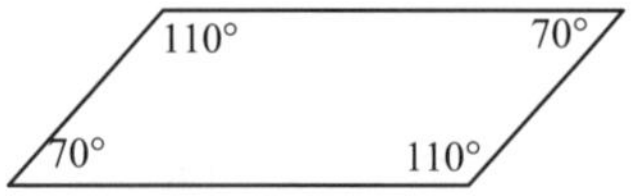

$$110° + 70° = 180°$$
$$110° + 70° = 180°$$

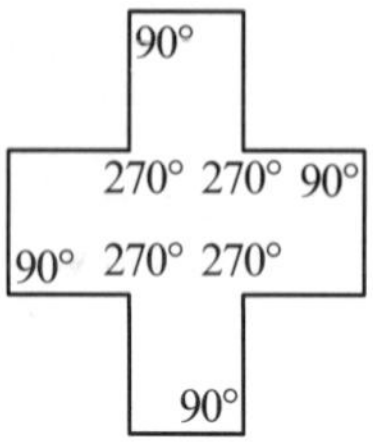

$$270° + 90° = 360°$$
$$270° + 90° = 360°$$
$$270° + 90° = 360°$$
$$270° + 90° = 360°$$
$$90° + 90° = 180$$
$$90° + 90° = 180$$
$$90° + 90° = 180$$
$$90° + 90° = 180$$

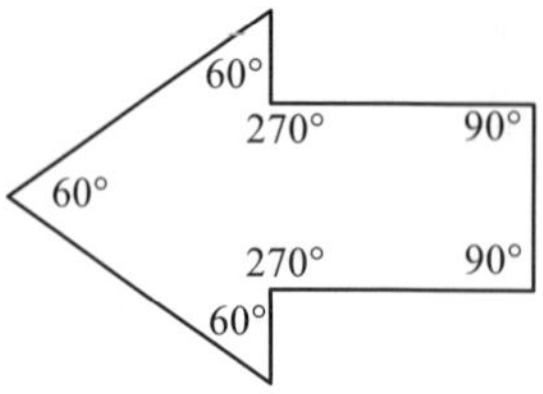

$$270° + 90° = 360°$$
$$270° + 90° = 360°$$
$$60° + 60° + 60° = 180°$$

Group these angles so that each group adds up to either 180° or 360°. If you can't make even groups of angles, your shape won't tessellate. The angle measurements must group evenly for the shape to tessellate, but an even grouping won't necessarily mean a shape tessellates. You just have to put them together to find out.

Finally, you can also tessellate with more than one shape, as long as you can group interior angles in sets of 360°.

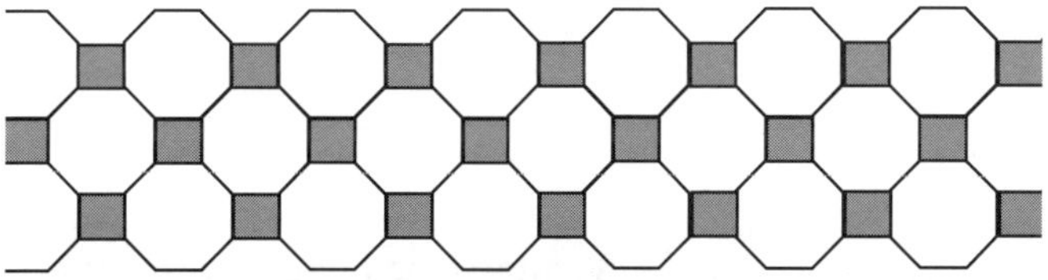

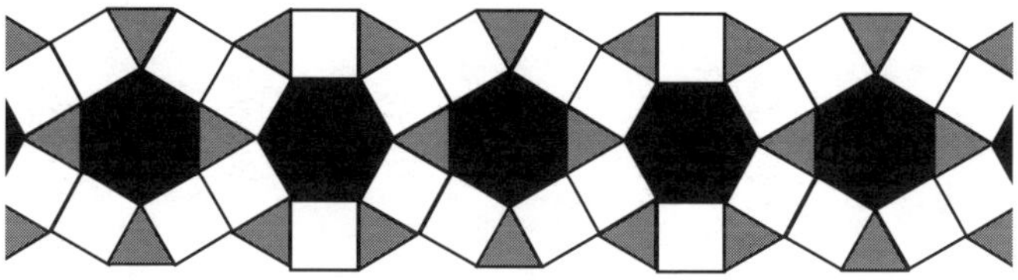

Look at the examples below and see if you can find the answer.

Example 1 Will this triangle tessellate?

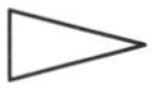

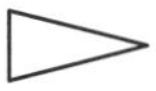

Step 1: Measure the angles and group them. The sum of the interior angles of any triangle is 180º, so it may tessellate.

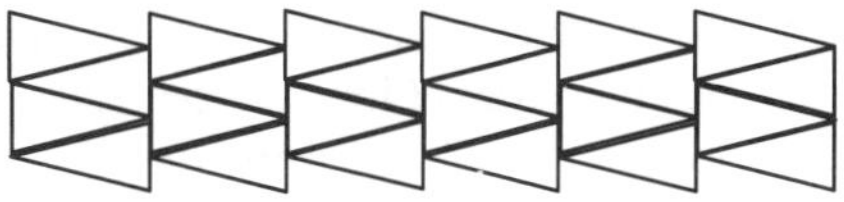

Step 2: Put them together to see if they tessellate.

Answer: The shapes fit together, so the triangle tessellates.

Example 2 Does this irregular figure tessellate?

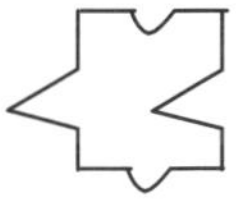

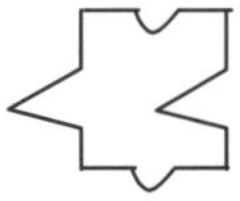

Step 1: The shape is basically a rectangle (rectangles tessellate), but pieces have been translated across the length and width.

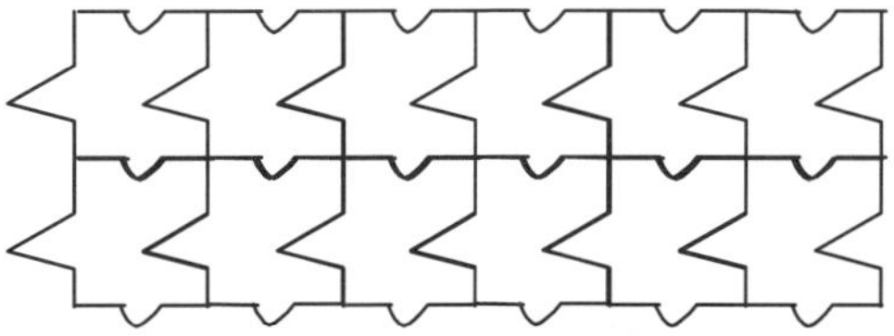

Answer: The shapes fit together, so it tessellates.

Example 3 Does this irregular figure tessellate?

Step 1: Try to fit the pieces together.

Answer: Because the pieces do not fit, this figure does not tessellate.

Example 4 Can a pentagon tessellate?

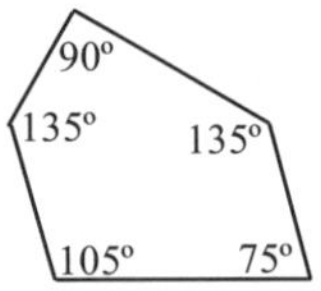

$$135° + 135° + 90° = 360°$$
$$105° + 75° = 180°$$

Step 1: Find a pentagon with the right combination of angles. It's important that there is exactly one pair of adjacent supplementary angles. This pentagon contains a pair of adjacent supplementary angles, which means you can arrange the angles in a straight line. By necessity, the other three angles total 360°. Therefore, they will fit together around a circle.

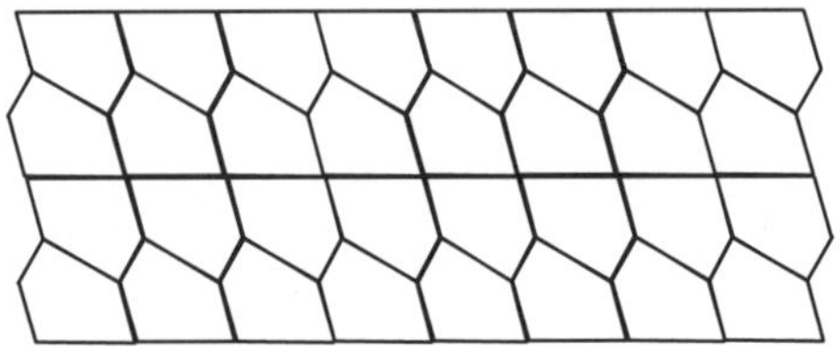

Solution: A pentagon can tessellate.

Problem Set

1. Will the shape of a heart tessellate? Answer yes or no.

2. Will the following figure tessellate? Answer yes or no.

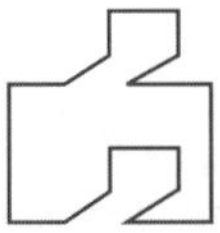

3. Will the following figure tessellate? Answer yes or no.

4. Kristian is making a mosaic for her mother for Mother's Day. Kristian is using glass for her mosaic. Can she make her pieces fit together like a puzzle? Answer yes or no.

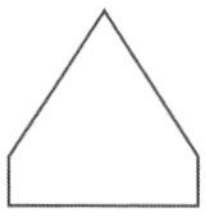

5. Jean-Luc is creating a mosaic for his art class. His tiles are shaped irregularly. Jean-Luc wants you to help him decide if he can use tiles shaped like the following to cover the surface. Answer yes or no.

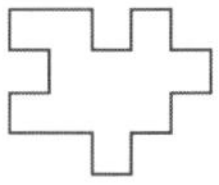

6. Can a regular pentagon tessellate? Answer yes or no.

7. Will the following figure tessellate? Answer yes or no.

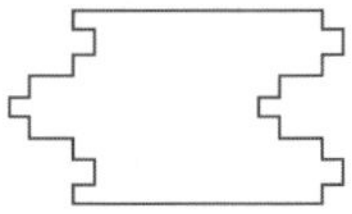

8. Will the following figure tessellate? Answer yes or no.

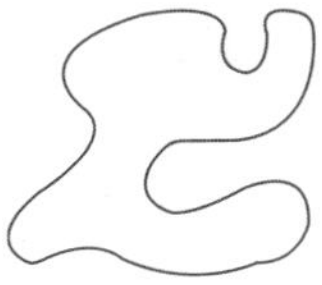

9. Will the following figure tessellate? Answer yes or no.

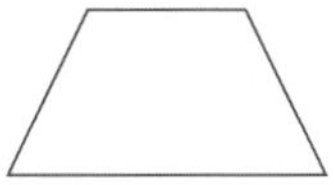

10. Will the following figure tessellate? Answer yes or no.

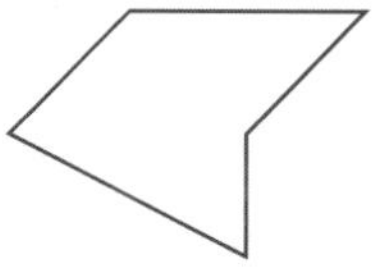

11. Will all squares tessellate? Answer yes or no.

12. Will all acute triangles tessellate? Answer yes or no.

13. Will all equilateral triangles tessellate? Answer yes or no.

14. Will all regular hexagons tessellate? Answer yes or no.

15. Will all regular pentagons tessellate? Answer yes or no.

16. Will a regular octagon tessellate? Answer yes or no.

17. Will a circle tessellate? Answer yes or no.

18. Does a regular triangle or a regular octagon tessellate?

19. Does a regular hexagon or a regular octagon tessellate?

20. Does a right triangle or a regular pentagon tessellate?

MPA-061: Converting Metric Units of Length, Capacity, and Mass

Have you seen the other scale on a car speedometer, the one in orange inside the white miles per hour scale? That's the kph scale; "kph" stands for "kilometers per hour." A kilometer isn't quite as long as a mile, so the numbers on the kph scale are larger than the numbers on the mph scale. You can convert the numbers on one scale on the speedometer to the other scale, but can you convert kph to m/s (meters per second) or even hmph (hectometers per hour)? You can if you use a chart of metric prefixes.

prefix	kilo (k)	hecto (h)	deca (da)	units (no prefix)	deci (d)	centi (c)	milli (m)
value	1,000	100	10	1	$\frac{1}{10}$	$\frac{1}{100}$	$\frac{1}{1,000}$

Unfortunately, prefixes won't help you without units. You'll also need a chart of the three metric units.

unit	meter (m)	liter (L)	gram (g)
measure	length	volume/liquid measure	mass

These units by themselves are whole in meaning. A liter is a definite measure of volume. However, it may be combined with a metric prefix to mean something larger or smaller than just a liter. For example, a milliliter (mL) is the same as 0.001 liters (L), and a hectoliter (hL) is the same as 100 liters.

To **convert metric units**, you only need to move the decimal point to the right or the left.

- To convert from a larger unit to a smaller unit—from left to right on the prefix chart—move the decimal to the right as many places as the units moved on the chart.
- To convert from a smaller unit to a larger unit—from right to left on the prefix chart—move the decimal to the left as many places as the units moved on the chart.

Why does this process work? If a hectometer is 100 meters, then converting from hectometers to meters involves moving the decimal two places to the right. 5 hectometers equals 500 meters. Moving the decimal two places to the right is equivalent to multiplying by 100.

Converting from meters to hectometers works similarly. Moving the decimal two places to the left is the same as dividing by 100. 500 meters equals 5 hectometers.

The following is an example of converting larger units to smaller units.

Example 1 How many millimeters is 1 decimeter?

1 dm = ?	**Step 1:** Find millimeters in relation to decimeters on the metric prefix chart. Millimeters lie two places to the right of decimeters on the prefix chart.
1 dm = 100 mm	**Step 2:** Move the decimal two places to the right.
	Answer: 100 mm

The following is an example of converting smaller units to larger units.

Example 2 How many kilometers is 50 centimeters?

50 cm = ?	**Step 1:** Find kilometers in relation to centimeters on the metric prefix chart. Kilometers lie five places to the left of centimeters on the prefix chart.
50 cm = 0.0005 km	**Step 2:** Move the decimal five places to the left.
	Answer: 0.0005 km

Converting units within the metric system is as easy as multiplying and dividing by multiples of 10.

Example 3 How many decagrams is 15.6 grams?

15.6 g = ?

Step 1: Find decagrams in relation to grams on the metric prefix chart. Decagrams lie one place to the left of grams on the prefix chart.

15.6 g = 1.56 dag

Step 2: Move the decimal one place to the left.

Answer: 1.56 dag

Example 4 How many centiliters is 0.144 liters?

0.144 L = ?

Step 1: Find centiliters in relation to liters on the metric prefix chart. Centiliters lie two places to the right of liters on the prefix chart.

0.144 L = 14.4 cL

Step 2: Move the decimal two places to the right.

Answer: 14.4 cL

Example 5 How many liters is 0.0332 kiloliter?

0.0332 kL = ?

Step 1: Find liters in relation to kiloliters on the metric prefix chart. Liters lie three places to the right of kiloliters on the prefix chart.

0.0332 kL = 33.2 L

Step 2: Move the decimal three places to the right.

Answer: 33.2 L

Example 6 How many kilometers is 2,120.2 millimeters?

2,120.2 mm = ?

Step 1: Find kilometers in relation to millimeters on the metric prefix chart. Kilometers lie six places to the left of millimeters on the prefix chart.

2,120.2 mm = 0.0021202 km

Step 2: Move the decimal six places to the left.

Answer: 0.0021202 km

Problem Set

Solve:

1. 83.4 hectometers = ______ centimeters

2. 1,000 grams = ______ decagrams

3. 102.3 kiloliters = ______ hectoliters

4. 100 milligrams = ______ centigrams

5. Determine which method of measurement would be best for the situation. Cynthia is measuring the length of her new dress for her senior prom.

meters   kilometers

grams   millimeters

6. Jonathon needs an oil change. He needs to determine how much oil to pour into the engine of his new sports car. Determine which method of measurement would be best for the situation.

milliliters   liters

meters   grams

7. William is measuring the amount of copper sulfate, a solid, to mix in his Chemistry experiment. Determine which method of measurement would be best for the situation.

kilograms   grams

hectograms   millimeters

8. 634 centigrams ____ 10 hectograms

=   none of these

<   >

9. Place in order from least to greatest:
100 milliliters; 99 deciliters; 99.9 centiliters, 1,000 liters

10. Place in order from least to greatest:
45 liters; 435 milliliters; 9 kiloliters; 140 centiliters

11. Tess is making agar for her bacterial plates. The instructions call for 230 grams of agar and 65 grams of lactose. Her weighing equipment measures everything in centigrams. How many centigrams of agar and lactose does Tess need for her microbial plates?

12. 250 kilometers = ___ meters

13. 93 decaliters = ____ milliliters

14. 59 decagrams = ____ kilograms

15. 12 grams ____ 210 decigrams

<   =

none of these   >

16. Chrissy drives a big rig truck cross country. Her truck takes 168 deciliters of gas. The gas pump measures volume in liters. How many liters of gas should Chrissy fill up?

17. Carol is storing oil for the winter. She stores 65 kiloliters of oil in her basement. On average, she will use 58,000 liters of oil to heat the house during the entire winter season. How many kiloliters of oil does Carol use each winter? Did she store enough oil?

18. How many millimeters is 25 centimeters?

19. How many meters is 17 decimeters?

20. How many milligrams is 574 centigrams?

MPA-062: Converting Units in Customary System

Have you ever thought about running in a five-mile race and wondered how that distance was measured in feet? Have you ever looked at milk products and thought about how many quarts equaled one gallon? Units of length, weight, and volume are all around you every day!

You already know, for example, that a yardstick is three feet long; thus, you know that one yard is equivalent to three feet. You also know that a standard ruler is one foot, or twelve inches, long. Therefore, twelve inches is equivalent to one foot. You see, then, that you are already familiar with several common conversions. In this lesson, you will learn more about these conversions and many others, such as those listed below.

Length	12 inches (in.) = 1 foot (ft)
	3 feet = 1 yard (yd)
	5,280 feet = 1 mile (mi)
	1,760 yards = 1 mile
Weight	16 ounces (oz) = 1 pound (lb)
Volume	2 cups (c) = 1 pint (pt)
	2 pints = 1 quart (qt)
	4 quarts = 1 gallon (gal)

To convert from one unit to another, set up a proportion to use as a multiplier. Set the units you are converting to over the units you are converting from:

$$\frac{\text{units converting to}}{\text{units converting from}}$$

The units you are converting from cancel each other out, and you are left with the units you are converting to. Take a look at the examples below.

Example 1 Convert 7 feet to inches.

$$12 \text{ in.} = 1 \text{ ft}$$
$$7 \text{ ft}\left(\frac{12 \text{ in.}}{1 \text{ ft}}\right) = ? \text{ in.}$$

Step 1: Set up the equation as a proportion. Keep in mind that 12 inches equals 1 foot.

$$7 \cancel{\text{ft}}\left(\frac{12 \text{ in.}}{1 \cancel{\text{ft}}}\right) = ? \text{ in.}$$

Step 2: The common units cancel each other out, leaving just inches behind.

$$7(12 \text{ in.}) = ? \text{ in.}$$
$$7(12 \text{ in.}) = 84 \text{ in.}$$

Step 3: Finish solving the problem.

Answer: 84 inches

Example 2 While shopping at the market, you see a block of cheese that weighs 112 ounces. What is the equivalent weight in pounds?

$$16 \text{ oz} = 1 \text{ lb}$$
$$112 \text{ oz}\left(\frac{1 \text{ lb}}{16 \text{ oz}}\right) = ? \text{ lb}$$

Step 1: Set up a proportion, keeping in mind that 16 ounces equals 1 pound.

$$112 \cancel{\text{oz}}\left(\frac{1 \text{ lb}}{16 \cancel{\text{oz}}}\right) = ? \text{ lb}$$

Step 2: The common units cancel each other out, leaving pounds behind.

$$112\left(\frac{1 \text{ lb}}{16}\right) = ? \text{ lb}$$
$$\frac{112 \text{ lb}}{16} = 7 \text{ lb}$$

Step 3: Finish solving the problem.

Answer: 7 pounds

Example 3 Sasha rode 12 miles on his bike to the comic book shop. How far is that in yards?

$$1 \text{ mi} = 1{,}760 \text{ yd}$$
$$12 \text{ mi}\left(\frac{1{,}760 \text{ yd}}{1 \text{ mi}}\right) = ? \text{ yd}$$

Step 1: Set up the equation as a proportion. Recall that 1,760 yards is equivalent to one mile.

$$12 \cancel{\text{mi}}\left(\frac{1{,}760 \text{ yd}}{1 \cancel{\text{mi}}}\right) = ? \text{ yd}$$

Step 2: Cancel the common units, leaving yards behind.

$$12(1{,}760 \text{ yd}) = 21{,}120 \text{ yd}$$

Step 3: Solve the equation.

Answer: 21,120 yards

Example 4 The supermarket has a sale on 64 pints of tropical punch. If your refrigerator can only hold 10 gallons of liquid, would it be able to hold 64 pints of tropical punch?

$$2 \text{ pt} = 1 \text{ qt}$$
$$64 \text{ pt}\left(\frac{1 \text{ qt}}{2 \text{ pt}}\right) = ? \text{ qt}$$

Step 1: To solve this problem, you will need to convert pints to quarts and then quarts to gallons. Set up the equation as a proportion.

$$64 \cancel{\text{pt}}\left(\frac{1 \text{ qt}}{2 \cancel{\text{pt}}}\right) = ? \text{ qt}$$
$$64\left(\frac{1 \text{ qt}}{2}\right) = 32 \text{ qt}$$

Step 2: The pints cancel each other out, leaving quarts behind. 64 pints is equivalent to 32 quarts.

$4 \text{ qt} = 1 \text{ gal}$ $32 \text{ qt}\left(\frac{1 \text{ gal}}{4 \text{ qt}}\right) = ? \text{ gal}$	**Step 3:** Now, convert quarts to gallons. To do so, set up the equation as a proportion. Keep in mind that 4 quarts equals 1 gallon.
$32 \cancel{\text{qt}}\left(\frac{1 \text{ gal}}{4 \cancel{\text{qt}}}\right) = ? \text{ gal}$	**Step 4:** Cancel out the common units, leaving gallons behind.
$32\left(\frac{1 \text{ gal}}{4}\right) = ?$ $\frac{32}{4} \text{gal} = 8 \text{ gal}$	**Step 5:** Solve the equation. Thus, 64 pints is equivalent to 8 gallons.

Answer: Since your refrigerator has room for 10 gallons, and 64 pints = 8 gallons, you will be able to fit all of the punch.

Problem Set

Solve:

1. 144 ounces = ___ pounds

2. 40 quarts = ___ gallons

3. 20 quarts = ___ gallons

4. 11 gallons = ___ quarts

5. 18 gallons = ___ quarts

6. 168 inches = ___ feet

7. 9 feet = ___ inches

8. 27 feet = ___ yards

9. 6 yards = ___ feet

10. 5 yards = ___ feet

11. Chef Juanita needs 95 quarts of cream for her pastries. She can only find cream sold in gallons. How many gallons should she buy to get exactly 95 quarts? Use decimals if needed, not fractions.

12. Harold told his teacher that he is 2.1 yards tall. Help his teacher by converting Harold's height to feet. How tall is Harold in feet?

13. After many day of hard work, Willie Worm wiggled 7,920 feet. How many miles did Willie go?

14. Convert 258 ounces to pounds. Use the labels 'pounds' and 'ounces' in your answer.

15. Express 81 quarts as gallons. Use the labels 'gallons' and 'quarts' in your answer.

16. Convert 302 inches to feet. Use the labels 'feet' and 'inches' in your answer.

17. Express 302 feet as yards. Use the labels 'yards' and 'feet' in your answer.

18. 20.37 miles = ___ yards

19. A mail package weighs 12 pounds and 3 ounces. A mail carrier needs to know if it is more than 200 ounces so she can determine if it has enough postage. Help the mail carrier by converting 12 pounds and 3 ounces to just ounces.

20. A person must be at least 54 inches tall to ride Mad Insanity, a roller coaster at Wacky Wally's Wonderland. Jeannie is 6 feet 2 inches tall. She needs to change her height to inches to see if she can ride the ride. Help her by converting 6 feet 2 inches to just inches.

MPA-063: Converting Units Between Metric and Customary System

You may not use metric units, but that's probably because they're not part of the customary system. It'll be easier to understand how long a kilometer is if you convert it to miles. You'll have an idea of how much a kilogram weighs if you convert it to pounds. To switch back and forth between the systems, you'll need unit conversion factors, equivalent measures of customary and metric units.

Customary		Metric
1 oz	=	28 g
1 gal	=	3.79 L
1 fluid oz	=	29.6 mL
1 in.	=	2.54 cm
1 yd	=	0.91 m
1 mi	=	1.61 km

Now that you have the metric equivalents for customary units for weight, you can convert any unit to any other unit (even if they're not in the same system). With the proper conversion factors, this is simple.

Use the following charts to assist you when converting:

prefix	kilo (k)	hecto (h)	deca (da)	units (no prefix)	deci (d)	centi (c)	milli (m)
value	1,000	100	10	1	$\frac{1}{10}$	$\frac{1}{100}$	$\frac{1}{1,000}$

unit	meter (m)	liter (L)	gram (g)
measure	length	volume/liquid measure	mass

Conversion factors are similar to the conversion ratios you learned in a previous lesson. The process for conversion is the same:

1. Choose the correct unit conversion factor.
2. Set up an equation using the conversion equivalents.
3. Substitute the factor of the given units into the equation.
4. Balance the equation by applying the same factor to the other side of the equation.
5. Solve the equation.

Let's look at some examples.

Example 1 How many liters is 22.4 gallons?

1 gal = 3.79 L	**Step 1:** Set up the equation using the conversion equivalents.
22.4 gal =	**Step 2:** Substitute 22.4 as the factor for gallons.

$22.4 \text{ gal} = 22.4(3.79)\text{L}$	**Step 3:** Because you multiplied the left side of the equation by 22.4, you must also multiply the right side by 22.4.
$22.4 \text{ gal} = 84.896 \text{ L}$	**Step 4:** Perform all operations.
	Answer: 84.896 L

Example 2 How many inches is 3.44 centimeters?

$1 \text{ in.} = 2.54 \text{ cm}$ $\frac{1}{2.54} \text{ in.} = 1 \text{ cm}$	**Step 1:** Set up the equation using the conversion equivalents. Because you are converting from centimeters to inches, solve the equation for centimeters.
$\frac{1}{2.54}(3.44) \text{ in.} = 3.44(1 \text{ cm})$	**Step 2:** Apply the given factor, 3.44, to both sides of the equation.
$\frac{1}{2.54}(3.44) \text{ in.} = 3.44(1 \text{ cm})$ $1.354 \text{ in.} = 3.44(1 \text{ cm})$ $1.354 \text{ in.} \approx 3.44 \text{ cm}$	**Step 3:** Perform all operations.
	Answer: 1.354 in.

Example 3 How many meters is 0.8 yards?

$1 \text{ yd} = 0.91 \text{ m}$	**Step 1:** Set up the equation using the conversion equivalents.
$1(0.8) \text{ yd} = 0.91(0.8) \text{ m}$	**Step 2:** Apply the given factor, 0.8, to both sides of the equation.
$1(0.8) \text{ yd} = 0.91(0.8) \text{ m}$ $0.8 \text{ yd} = 0.91(0.8) \text{ m}$ $0.8 \text{ yd} = 0.728 \text{ m}$	**Step 3:** Perform all operations.
	Answer: 0.728 m

Example 4 How many meters is 0.54 miles?

$1 \text{ mi} = 1.61 \text{ km}$ $1 \text{ mi} = 1.61(1000) \text{ m}$ $1 \text{ mi} = 1{,}610 \text{ m}$	**Step 1:** Set up the equation using the conversion equivalents. Because you were not given a conversion for miles to meters, you must create one. As 1 km = 1,000 meters, multiply 1.61 km by 1,000 meters.
$0.54 \text{ mi} = 1{,}610(0.54) \text{ m}$	**Step 2:** Multiply both sides of the equation by the given factor, 0.54.
$0.54 \text{ mi} = 869.4 \text{ m}$	**Step 3:** Perform all operations.
	Answer: 869.4 m

Example 5 How many centimeters is 14 feet?

1 in. = 2.54 cm 12 in. = 2.54(12) cm 1 ft = 30.48 cm	**Step 1:** Set up the equation using the conversion equivalents. For feet to centimeters, use the conversion for inches to centimeters, then multiply both sides by 12 to convert from feet.
14 ft = 14(30.48) cm	**Step 2:** Multiply both sides of the equation by the given factor, 14.
14 ft = 426.72 cm	**Step 3:** Perform all operations.
	Answer: 426.72 cm

Problem Set

Solve:

1. Convert 4 ounces to milliliters.

2. Convert 7 inches to centimeters.

3. Convert 11 yards to meters.

4. How many meters are equivalent to 5 yards?

5. Two ounces is the same as ___ milliliters.

6. Convert 8 yards to meters using the conversion factor 0.91 meter = 1 yard.

7. Are 22.54 kilometers the same as 1 mile?

No Yes

8. True or False? 13,500 ounces = 378,000 grams

True False

9. True or False? 20 ounces = 490 milliliters

True False

10. Convert 50 inches to centimeters.

11. Convert 86 grams to ounces. Round to the nearest thousandth.

12. Convert 16 grams to ounces. Round to the nearest thousandth.

13. Convert 225 grams to ounces.

14. Convert 355 grams to ounces.

15. Convert 105 ounces to grams.

16. Convert 9 ounces to milliliters.

17. In the U.S., candy is sold by the pound. Chocolates are $3.50 per pound. A man from Germany wants to buy chocolate, but he is only familiar with weight in the metric system. Help him convert pounds to grams.

18. On a residential street in the U.S., the average speed is 35 miles per hour. What is the equivalent speed in kilometers?

19. Running 35 meters is equivalent to running how many yards? Round your answer to the nearest hundredth.

20. Steven was taking a test on centimeters and inches. He was having some trouble on the conversions. Can you help him out? 24 inches = ___ cm

MPA-064: Finding Square Roots of Perfect Squares

A chessboard has 64 squares. Depending on the size of the board, each square is about an inch on each side. You could say that they are perfect squares.

It just so happens that the number of squares on the chessboard, 64, is the product of 8 and 8. Because there are exactly two identical factors, 64 is a **perfect square.**

Perfect Square	The product of exactly two identical factors.

A number squared is expressed as a superscript two: $8^2 = 64$. In this case, the 2 is the **exponent** and the 8 is the **base**.

base $\longrightarrow x^y \longleftarrow$ exponent

A **square root** is one of those identical factors. The square root of a number, multiplied by itself, is equal to that number's perfect square.

Square Root	One of two identical factors of a number.

Because $64 = 8 \times 8$, the square root of 64 is 8. Square roots have a special symbol called a radical. Here is a translation from words to symbols:

Words	Symbols
The square root of 64 is 8	$\sqrt{64} = 8$

The following rules will help you find the square root of a number:

- Use the radical sign when looking for the positive square root of a number.
- *Because* $-8 \times -8 = 64$, *technically* $\sqrt{64} = \pm 8$. *However, for simplicity, use only the positive root when solving square root problems in this textbook.*
- For all positive y, the square root of N is y if $y \times y = N$.
- If $\sqrt{N} = y$, then $y^2 = N$. Conversely, if $y^2 = N$, then $\sqrt{N} = y$.
- A radical sign cancels with a two in the exponent under the radical: $\sqrt{y^2} = y$.

Example 1 Find $\sqrt{4}$

$4 = 2 \times 2 = 2^2$

Step 1: To find the square root, you must find the identical factors that equal 4 when multiplied.

$\sqrt{2^2} = 2$ or $\sqrt{4} = 2$

Step 2: The radical and the square cancel, leaving 2. Therefore, the square root of 4 is 2.

Answer: 2

Perfect squares and square roots will help you with the formula for the area of a square. The area of a square is the length of one side squared.

area of square = (length of side)$^2 \rightarrow A = s^2$

Consequently, according to the rules above, $s = \sqrt{A}$.

Example 2 What is the length of a side of a square with an area of 81 square units?

$s = \sqrt{A}$	**Step 1:** Write the formula for the length of the side of a square.
$s = \sqrt{81}$	**Step 2:** Replace A with 81.
$81 = 9 \times 9 = 9^2$	**Step 3:** Find the identical factors that equal 81 when multiplied. The radical and the square cancel, leaving 9. Therefore, the square root of 81 is 9.
$\sqrt{81} = \sqrt{9^2} = 9$	**Step 4:** You can factor 81 as 9×9, or 9^2.
	Answer: 9 units

Example 3 Find $\sqrt{64}$.

$64 = 8 \times 8 = 8^2$	**Step 1:** Find the identical factors that equal 64 when multiplied. You can factor 64 as 8×8, or 8^2.
$\sqrt{64} = \sqrt{8^2} = 8$	**Step 2:** The radical and the square cancel, leaving 8. The square root of 64 is 8.
	Answer: 8

Example 4 Find $\sqrt{121}$.

$121 = 11 \times 11 = 11^2$	**Step 1:** Find the identical factors that equal 121 when multiplied. You can factor 121 as 11×11, or 11^2.
$\sqrt{121} = \sqrt{11^2} = 11$	**Step 2:** The radical and the square cancel, leaving 11. The square root of 121 is 11.
	Answer: 11

Example 5 What is the length of a side of a square with an area of 49 square units?

$s = \sqrt{A}$	**Step 1:** Write the formula for the length of the side of a square.
$s = \sqrt{49}$	**Step 2:** Replace A with 49.
$49 = 7 \times 7 = 7^2$	**Step 3:** Find the identical factors that equal 49 when multiplied. You can factor 49 as 7×7, or 7^2.
$\sqrt{49} = \sqrt{7^2} = 7$	**Step 4:** The radical and the square cancel, leaving 7. The square root of 49 is 7.
	Answer: 7 units

Problem Set

1. Simplify: $\sqrt{100}$

2. Simplify: $\sqrt{121}$

3. Simplify: $\sqrt{144}$

4. Simplify: $\sqrt{225}$

5. Simplify: $\sqrt{400}$

6. Simplify: $\sqrt{289}$

7. A square has an area of 144 square inches. Which of the following equals its perimeter?

72 in. 48 in.

12 in. 36 in.

8. Josie has a square backyard that measures 196 square feet. What is the perimeter of the yard?

9. What is the square root of 9?

10. What is the square root of 16?

11. What is the square root of 36?

12. What is the square root of 576?

13. If a square has an area of 81 square inches, what is the length of each side? _______ inches

14. Sandra has a square garden with an area of 16 square feet. What is the length of each side of the garden? _______ feet

15. Courtney cut her brownies into squares with an area of 9 square centimeters. What was the length of 1 side of a brownie? _______ cm

16. What is the length of one side of a square with an area of 1 square unit? _______ unit

17. The square tiles on the floor have an area of 1 square foot. If an ant marches around the edge of one tile with food in his mouth, how far did he march? (Include units in your answer).

18. If a square has an area of 25 square km, what is its perimeter? (Do not include units in your answer).

19. Darryl ran around a square room with an area of 256 square meters. How far did Darryl run? (Do not include units in your answer).

20. A square has an area of 9 square miles. What is its perimeter? (Do not include units in your answer).

MPA-065: Estimating Square Roots

When learning about square roots, you have primarily worked with perfect squares such as 4, 9, or 25. What about numbers that are *not* perfect squares, numbers like 20, 12, or 17? The square root of a number like 15 or 8 extends to several decimal places, which is why it is often easier to *estimate* the square roots of some numbers.

You can estimate the square root of a number by using one of four tools:

- A calculator.
- A number line.
- Trial-and-error.
- Factoring.

Estimating Square Roots Using a Calculator

To find an estimated square root using a calculator, enter the number, then press the key with the radical symbol.

Example 1 Estimate $\sqrt{17}$ to the nearest tenth using a calculator.

17	**Step 1:** Enter 17 into the calculator.
$\sqrt{\ }$	**Step 2:** Press the key with the radical symbol on it.
$4.1231056 \approx 4.1$ $\sqrt{17} \approx 4.1$	**Step 3:** Estimate the results to the nearest tenth. To do so, look at the number in the hundredths place, which is 2. Because 2 is less than 5, the number in the tenths place, 1, stays the same.
4.1^2 x^2 $4.1^2 \approx 16.81$	**Step 4:** Check your answer by entering 4.1 into your calculator and pressing the key with x^2 on it. The result is 16.81, which is close to 17. You can also multiply 4.1 by 4.1 to check your answer.

Answer: The estimated square root of 17 is 4.1.

Estimating Square Roots Using a Number Line

When using a number line to find an estimated square root, first decide between which two perfect squares the number lies. You can then determine that the estimated square root is greater than that of one perfect square but less than that of another.

Example 2 Estimate $\sqrt{13}$ to find the two numbers between which the answer will fall.

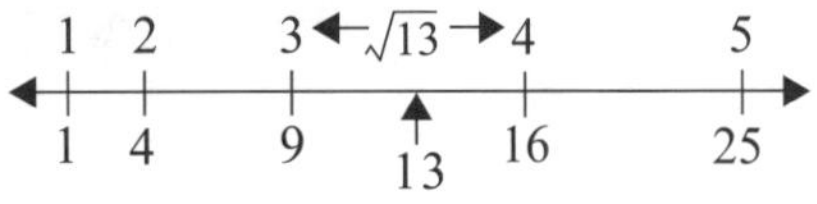

Step 1: Before setting up the number line, think about which perfect squares 13 lies between. 3^2 is 9, and 4^2 is 16. 13 is greater than 9 but less than 16, so you know that $\sqrt{13}$ lies between 3 and 4. Set up a number line with the consecutive numbers on top and the same numbers squared along the bottom. Mark where 13 would lie along the bottom line, and mark where $\sqrt{13}$ would lie along the top line.

$3 < \sqrt{13} < 4$

Step 2: Check your answer using a calculator. Enter 13 and then press the key with the radical sign on it. $\sqrt{13}$ is approximately 3.61, which is more than 3 but less than 4.

Answer: Thus, you can see that $\sqrt{13}$ lies somewhere between 3 and 4.

Estimating Square Roots Using Trial-and-Error

If you are familiar with the square roots of some numbers, then you can estimate the square roots of other numbers using trial-and-error.

Example 3 Find $\sqrt{11}$.

3×3	9
3.1×3.1	9.61
3.2×3.2	10.24
3.3×3.3	10.89
3.4×3.4	11.56

Step 1: To estimate $\sqrt{11}$, first think about what number 11 is close to. 11 is only a little greater than 9, whose square root is 3. You know, then, to set up a chart starting at 3.

$\sqrt{11} \approx 3.3$

Step 2: Looking at the chart, you can see that 10.89 is closest to 11.

Answer: Thus, $\sqrt{11}$ is approximately 3.3.

Estimating Square Roots Using Factoring

You can also use factoring to estimate the square root of a number.

Example 4 Find $\sqrt{28}$.

$\sqrt{28} = \sqrt{4} \times \sqrt{7}$

Step 1: To estimate the square root of 28, break 28 down to its factors.

$$\sqrt{28} = \sqrt{4} \times \sqrt{7}$$
$$= 2 \times \sqrt{7}$$
$$= 2\sqrt{7}$$

Step 2: Since 4 is a perfect square, you can break down $\sqrt{4}$ to 2.

Answer: Thus, $\sqrt{28} = 2\sqrt{7}$.

Problem Set

1. 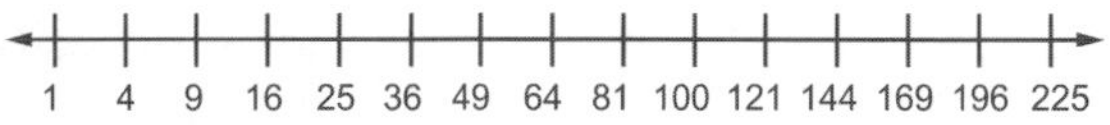

Select the two consecutive perfect squares that make this inequality true.

$\square < 50 < \square$

2. 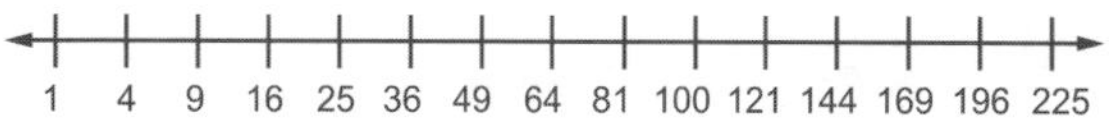

Select the two consecutive perfect squares that make this inequality true.

$\square < 3 < \square$

3. Fill in the blanks with two positive integers that best complete the inequality.

□ < $\sqrt{48}$ < □

4. Use a calculator to find the value of $\sqrt{39}$ to the nearest tenth.

5. Use a calculator to find the value of $\sqrt{99}$ to the nearest tenth.

6. Jack has a poster that he knows is a square. The length of one side of the poster is between 11 inches and 12 inches. He knows that the area is one of the numbers below. Which of the following numbers could be the area of the poster? Use the number line to help you.

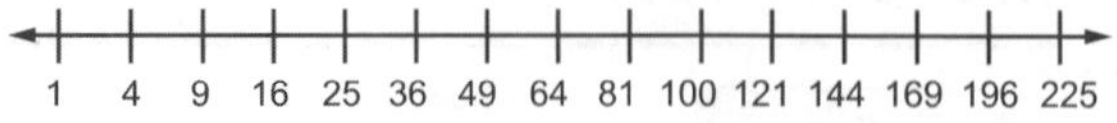

130 square inches 11.7 square inches

145 square inches 100 square inches

7. Use a calculator to find the square root of 33. Round your answer to the nearest tenth.

8. Use a calculator to find the square root of 28. Round your answer to the nearest tenth.

9. Enter the two consecutive perfect squares that make this inequality true.

□ < 142 < □

10. Fill in the blanks with two positive integers that best complete the inequality.

□ < $\sqrt{170}$ < □

11. You are playing a guessing game with your friend Libby. She tells you that the square root of the number she has written down is about 3.5. What number do you think she has written? Round your answer to the nearest whole number.

12. Which one of the following is closest to the square root of 101?

10.05 10

10.04 10.11

13. Which one of the following is closest to the square root of 230?

15.20 15.12

15 15.17

14. Use the factoring method to find the missing number.

$\sqrt{12}$ = _____ $\sqrt{3}$

15. Give an estimate for the following square root by filling in the blank to complete the equation.

$\sqrt{44} = 2\sqrt{\_\_}$

16. Give an estimate for the following square root by filling in the blank to complete the equation.

$\sqrt{99} = 3\sqrt{\_\_}$

17. Estimate the square root of 21 rounded to the nearest tenth.

18. Estimate to the nearest tenth, the square root of 35.

19. Estimate the length of one side of a square with an area of 375 km squared. (Round your answer to the nearest tenth.) _____ km

20. James painted a large square on the cement. Its area was 146 ft squared. What was the length of one side of the square? (Round your answer to the nearest tenth.) _____ ft

MPA-066: Solving Problems Using the Pythagorean Theorem

You have already been introduced to certain types of triangles such as isosceles triangles, equilateral triangles, and right triangles. The sum of the interior angles of a triangle is 180°, and a right triangle is a triangle that contains exactly one right (90°) angle. For a right triangle, the two sides adjacent to the 90° angle are called the legs, while the side opposite the 90° angle is called the hypotenuse.

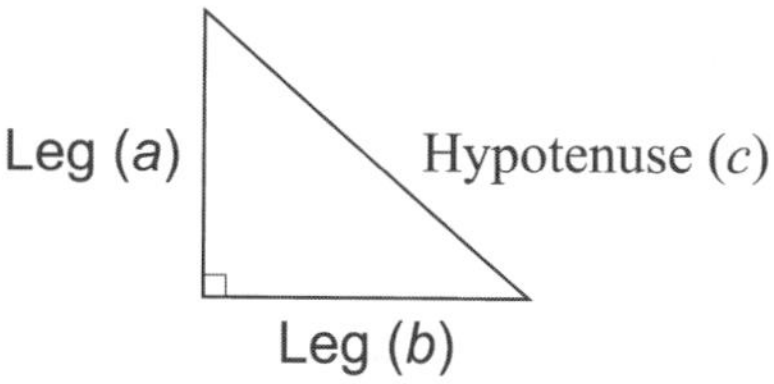

Pythagoras of Samos was a Greek mathematician who was responsible for numerous discoveries in music and astronomy. In approximately 532 BC, Pythagoras founded a school of philosophy that eventually made many notable contributions to the field of mathematics.

Pythagoras is perhaps most famous for proving the Pythagorean Theorem. Together with the students of his school, Pythagoras demonstrated that the sum of squares of the legs of a right triangle is equivalent to the length of the hypotenuse squared, $\text{Leg}^2 + \text{Leg}^2 = \text{Hypotenuse}^2$.

Pythagorean Theorem	The sum of the squares of the legs of a right triangle is equal to the square of the hypotenuse.

The Pythagorean Theorem can be written as a formula, $a^2 + b^2 = c^2$, where the variables a and b each represent one leg of the right triangle, and the variable c always represents the length of the hypotenuse.

You can use the Pythagorean Theorem to determine the length of a missing side of a right triangle.

Example 1 The length of one leg of a right triangle is 6 cm, and the length of the hypotenuse is 10 cm. How long is the other leg?

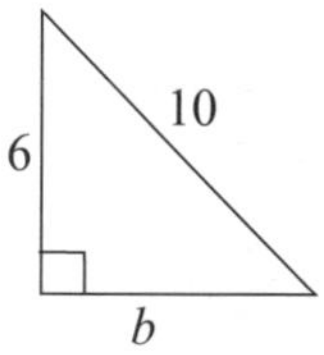

$a^2 + b^2 = c^2$ $6^2 + b^2 = 10^2$	**Step 1:** Set up the equation, substituting 6 for a and 10 for c.
$36 + b^2 = 100$	**Step 2:** Solve for b. First, square 6 and 10.
$36 - 36 + b^2 = 100 - 36$ $b^2 = 64$	**Step 3:** Then, subtract 36 from both sides of the equation.
$\sqrt{b^2} = \sqrt{64}$	**Step 4:** Next, take the square root of both sides of the equation.
$b = 8$	**Step 5:** The length of the hypotenuse is 8 cm.

$6^2 + 8^2 = 10^2$ $36 + 64 = 100$ $100 = 100$	**Step 6:** Check your answer by substituting 8 for *b* in the original equation.
	Answer: The leg is 8 cm.

Example 2 One side of the city park is 24 meters long, and the other side is 32 meters long. How far is it from one corner to the other?

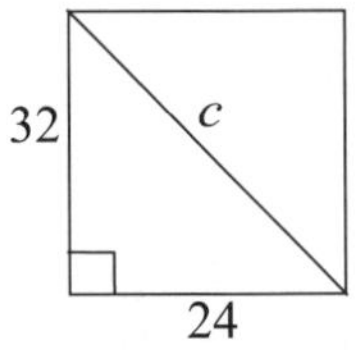

$a^2 + b^2 = c^2$ $32^2 + 24^2 = c^2$	**Step 1:** Set up the equation, substituting 32 for *a* and 24 for *b*.
$1{,}024 + 576 = c^2$ $1{,}600 = c^2$	**Step 2:** Next, solve for *c* by first squaring 32 and 24 and then adding the results.
$\sqrt{1{,}600} = \sqrt{c^2}$	**Step 3:** Find the square root of each side of the equation.
$c = 40$	**Step 4:** Thus, the distance from one corner of the city park to the other is 40 meters.
$32^2 + 24^2 = 40^2$ $1{,}024 + 576 = 1{,}600$ $1{,}600 = 1{,}600$	**Step 5:** Check the answer by substituting 40 for *c* in the original equation.
	Answer: 40 meters

Example 3 One leg of a right triangle measures 4 inches. The hypotenuse measures 5 inches. How long is the other leg?

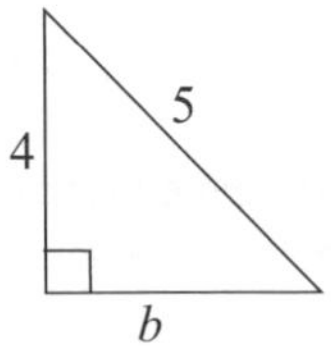

$a^2 + b^2 = c^2$ $4^2 + b^2 = 5^2$	**Step 1:** Set up the equation by substituting 4 for *a* and 5 for *c*.
$16 + b^2 = 25$	**Step 2:** Solve for *b* by first squaring 4 and 5.
$16 - 16 + b^2 = 25 - 16$ $b^2 = 9$	**Step 3:** Subtract 16 from both sides of the equation to isolate the variable.
$\sqrt{b^2} = \sqrt{9}$	**Step 4:** Take the square root of both sides.

$b = 3$	**Step 5:** The length of the leg is 3 inches.
$4^2 + 3^2 = 5^2$ $16 + 9 = 25$ $25 = 25$	**Step 6:** Check your answer by substituting 3 for *b* in the original equation.
	Answer: 3 inches

Example 4 One leg of a right triangle measures 12 inches. The hypotenuse measures 20 inches. How long is the other leg?

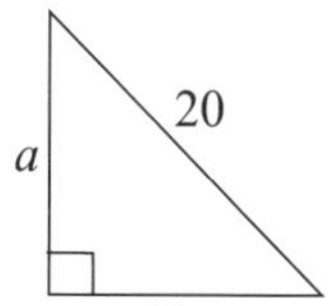

$a^2 + b^2 = c^2$ $a^2 + 12^2 = 20^2$	**Step 1:** Set up the equation. Substitute 12 for *b* and 20 for *c*.
$a^2 + 144 = 400$	**Step 2:** Square 12 and 20.
$a^2 + 144 - 144 = 400 - 144$ $a^2 = 256$	**Step 3:** Subtract 144 from both sides of the equation to isolate the variable.
$\sqrt{a^2} = \sqrt{256}$	**Step 4:** Take the square root of both sides.
$a = 16$	**Step 5:** The length of the other leg is 16 inches.
$16^2 + 12^2 = 20^2$ $256 + 144 = 400$ $400 = 400$	**Step 6:** Check your answer by substituting 16 for *a* in the original equation.
	Answer: 16 inches

Example 5 A baseball diamond is actually composed of two right triangles. Suppose that your neighborhood team is constructing a makeshift baseball diamond in a playground. The distance from second base to home plate is 15 feet. The distance from third base to home plate is 9 feet. How far is second base from third base?

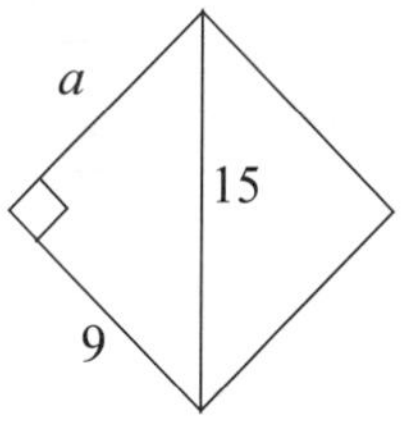

$a^2 + b^2 = c^2$ $a^2 + 9^2 = 15^2$	**Step 1:** Set up the equation. Substitute 9 for *b* and 15 for *c*.

$a^2 + 81 = 225$	**Step 2:** Solve for a by first squaring 9 and 15.
$a^2 + 81 - 81 = 225 - 81$ $a^2 = 144$	**Step 3:** Isolate the variable by subtracting 81 from both sides of the equation.
$\sqrt{a^2} = \sqrt{144}$	**Step 4:** Take the square root of both sides.
$a = 12$	**Step 5:** The distance from second base to third base is 12 feet.
$12^2 + 9^2 = 15^2$ $144 + 81 = 225$ $225 = 225$	**Step 6:** Check your answer by substituting 12 for a in the original equation.

Answer: 12 feet

Problem Set

Solve:

1. Find the missing side. Round your answer to the nearest tenth.

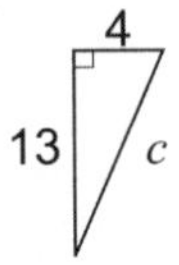

2. Find the missing side. Round your answer to the nearest tenth.

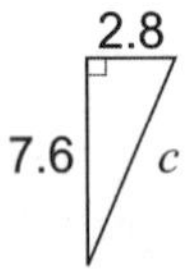

3. Find the missing side. Round your answer to the nearest tenth.

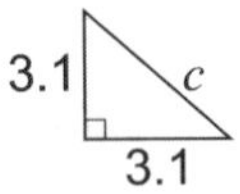

4. The square sign on the wall has sides that equal 3 feet. If someone drew a line diagonally from one corner to the other, how long would the line be?

2.4 ft   none of these

42 ft   4.2 ft

5. Find the missing side. Round your answer to the nearest tenth.

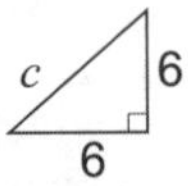

6. Find the missing side. Round your answer to the nearest tenth.

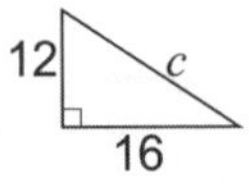

7. Find the missing side. Round your answer to the nearest tenth.

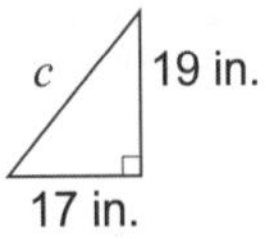

8. Solve using the Pythagorean Theorem:
$a = 24, b = 32, c = ?$
Round to the nearest tenth.

9. Solve using the Pythagorean Theorem:
$a = 27, b = 36, c = ?$
Round to the nearest tenth.

10. Solve using the Pythagorean Theorem:
$a = ?, b = 12, c = 15$
Round to the nearest tenth.

11. Solve using the Pythagorean Theorem:
$a = 7, b = ?, c = 21$
Round to the nearest tenth.

12. Solve using the Pythagorean Theorem:
$a = 38, b = 48, c = ?$
Round to the nearest tenth.

13. Solve using the Pythagorean Theorem:
$a = 61, b = 71, c = ?$
Round to the nearest tenth.

14. Find the missing side. Round your answer to the nearest tenth.

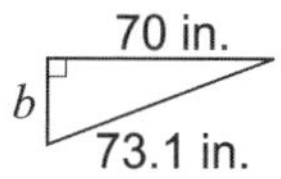

15. Find the missing side. Round your answer to the nearest tenth.

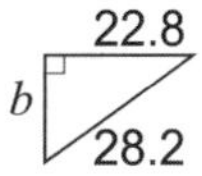

16. Find the missing side. Round your answer to the nearest tenth.

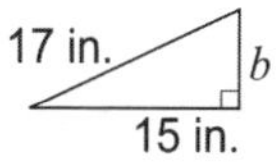

17. Jenny put two pencils on a table, eraser to eraser, forming a 90° angle. Each pencil is 8 inches long. How many inches is it from one tip of the pencil to the tip of the other pencil? Round your answer to the nearest tenth.

16,384 in.   113 in.
11.3 in.   none of these

18. Josh's window is shaped like a rectangle. Two sides equal 3 feet each, and the other two sides equal 8 feet each. He watches a bug crawl diagonally from one corner to the next. How many feet did it crawl? Round to the nearest tenth.

85 ft   none of these
5,329 ft   8.5 ft

19. Determine whether or not the triangle formed by the given sides is a right triangle. Answer yes or no.
$a = 9, b = 39, c = 41$

No   Yes

20. The lengths of the sides of a right triangle are 9 m, 12 m, and 15 m. Is the triangle a right triangle?

No   Yes

MPA-067: Finding the Area of Rectangles and Parallelograms

The ability to calculate the area of geometric figures is a very practical skill. Area is used whenever you want to cover a two-dimensional space, such as floors and walls. For example, if you want to repaint your walls or install new flooring, you must first find the area of the walls or floors, so you'll know how much paint or flooring you will need. Of course, there are many more uses for area. In this lesson, you will learn how to find the area of rectangles, squares, and parallelograms.

Area of Rectangles and Squares

Begin thinking about what area means by using a simple example, a tic-tac-toe board:

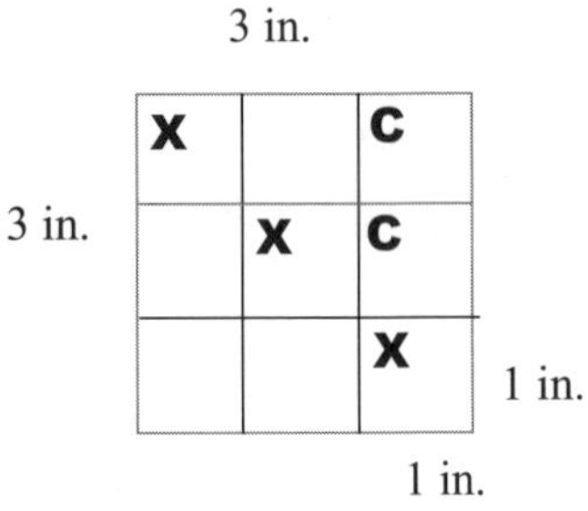

A square that measures 1 inch long and 1 inch wide is called a **square inch**. Area is measured in square inches, square feet, square meters, and so forth. You can easily count the square inches in the tic-tac-toe board above to get 9 square inches or 9 in^2. Note that a square like the one above, which is 3 inches by 3 inches, does **not** have an area of 3 in^2; it actually has an area of $A = (3 \cdot 3)in^2 = 9in^2$. As you can probably see by now, to find the area of any square, you simply multiply two sides, or square the measurement of a side.

Area of a Square	The area of a square is two times the length of one side. The formula for area of a square is $A = s \cdot s$ or $A = s^2$.

Squares are the simplest type of rectangles. Because the length and width are the same in a square, you only need the measurement of a side. However, when dealing with rectangles, there are two different measurements. Let's find the area of the following rectangle:

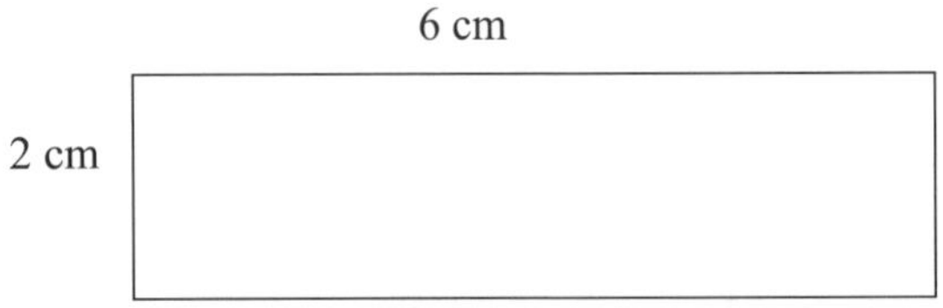

Without a formula, you must count the number of square centimeters that would fit inside the rectangle. Divide the rectangle and see how many squares there are:

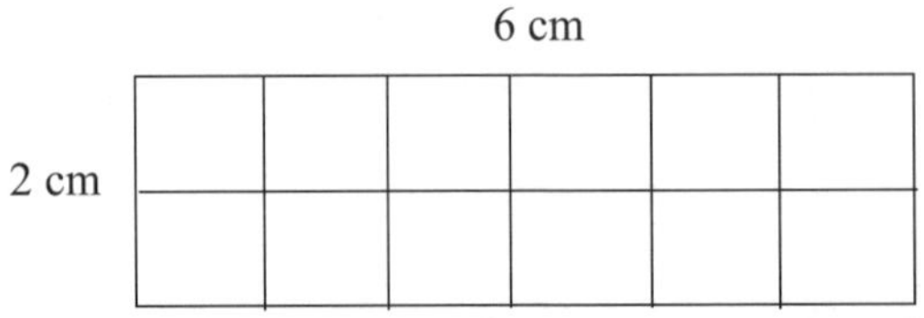

Instead of counting each one, you can see that there are 2 rows of 6 square centimeters, or $A = (2 \cdot 6)\text{cm}^2$, which makes the area 12 square centimeters or 12 cm^2. As you have probably guessed by now, the formula for the area of a rectangle is just the product of the length and the width:

Area of a Rectangle	The area of a rectangle is the product of the length times the width. The formula for area of a rectangle is: $A = l \cdot w$ or $A = lw$.

With these two examples, you can actually count the number of square inches or square centimeters. You may not always be able to count the number of square inches or centimeters, but the two formulas will always work. Let's look at the following example.

Example 1 Find the area of a rectangle when $l = 7.2\text{m}$ and $w = 2.3\text{m}$.

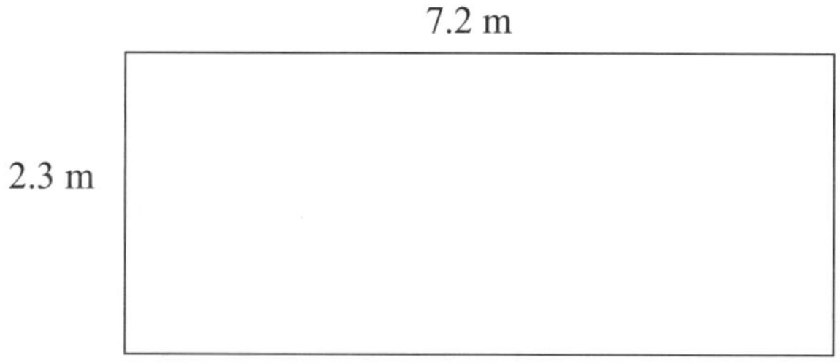

$A = lw = (7.2\text{m})(2.3\text{m})$	**Step 1:** Insert the length and width into the formula for the area of a rectangle.
$A = (7.2)(2.3) \cdot (\text{m})(\text{m})$ $A = 16.56\text{m}^2$	**Step 2:** Multiply length by width to find the answer.

Answer: The area equals $16.56\ \text{m}^2$.

Area of Parallelograms

Parallelograms basically look like rectangles that are leaning. However, you can't fit as many squares into a parallelogram as you can in rectangles. If you try to cover a parallelogram with squares, you will have some leftover space that isn't covered.

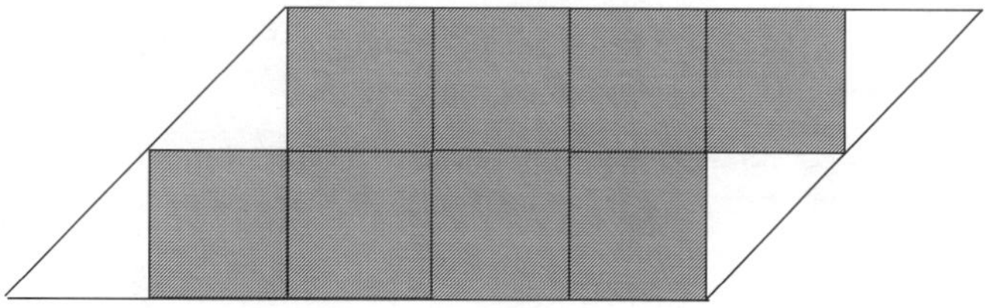

Therefore, to find the area of a parallelogram, you must have a different formula. Unlike squares and rectangles, the area of a parallelogram is not found by multiplying the length by the width.

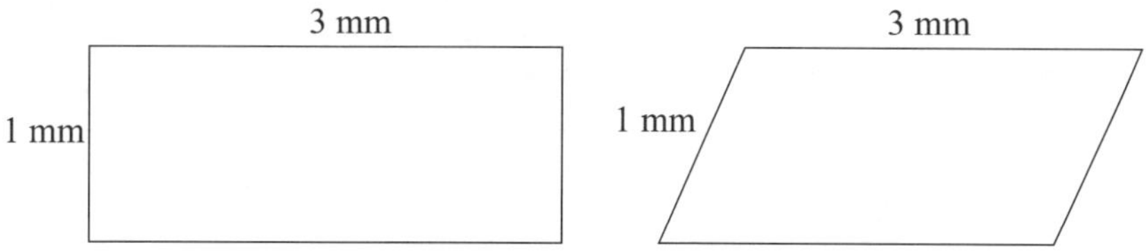

If you draw a rectangle on top of the parallelogram, you'll notice that the parallelogram is simply a rectangle with a right triangle on both sides.

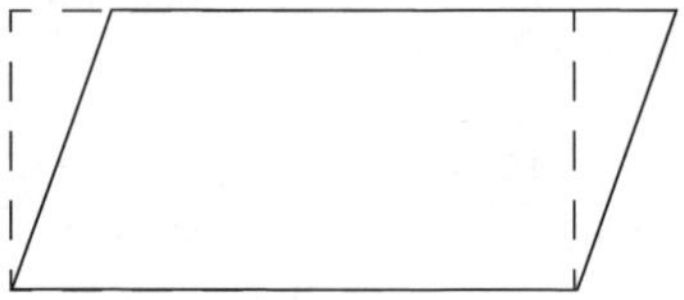

Notice that the two right triangles are exactly the same (see the shaded areas in the figure below).

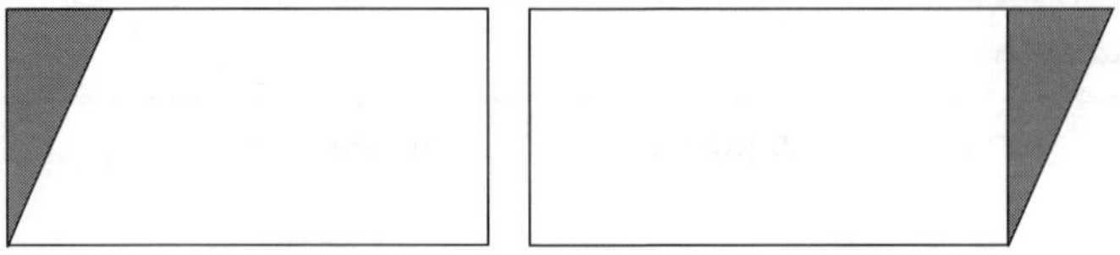

The area of the parallelogram is *exactly* the same as the area of the rectangle. Therefore, to find the formula for the area of a parallelogram, you need two measurements: the length and width of the rectangle you can put on top of the parallelogram. To avoid confusion, however, when measuring the area of a parallelogram, use the words base (width) and height (length).

Area of a Parallelogram	The formula for finding the area of a parallelogram is $A = b \cdot h$ or $A = bh$, where A is area, b is base, and h is height.

Try these examples.

Example 2 Find the area of the parallelogram below.

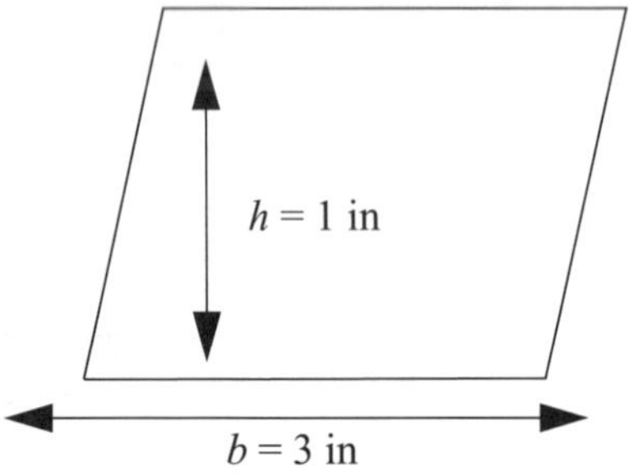

$A = bh = (3)(1)$

Step 1: Insert the base and height into the formula for the area of a parallelogram.

$A = (3)(1)$
$A = 3 \text{ in}^2$

Step 2: Multiply base by height to find the answer.

Answer: The area equals 3 in^2.

Example 3 Find the area of the parallelogram below.

*Note: 3 m is **not** one of the measurements in the formula.*

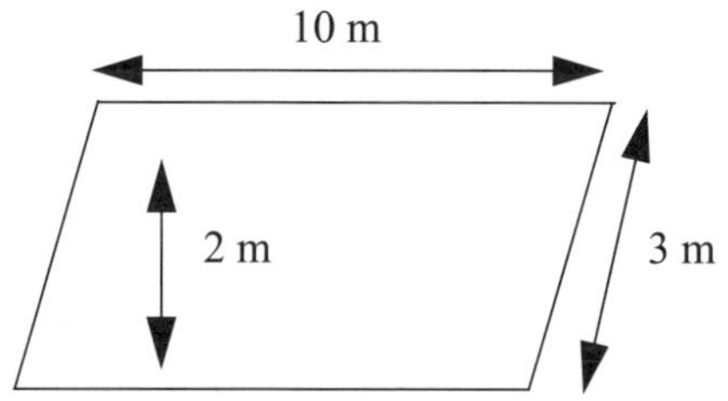

$A = bh = (10)(2)$	**Step 1:** Insert the base and height into the formula for the area of a parallelogram.
$A = (10)(2)\cdot(\text{m})(\text{m})$ $A = 20\text{m}^2$	**Step 2:** Multiply base by height to find the answer.

Answer: The area equals $20\ \text{m}^2$.

Problem Set

Solve:

1. Find the area, in square inches, of a rectangle with a length of 16 inches and a width of 14 inches.

2. Find the area, in square meters, of a rectangle with a length of 34 meters and a width of 8 meters.

3. Find the length of a rectangle with a width of 12 inches and an area of 144 square inches.

4. Find the area, in square inches, of a rectangle with a length of $12\frac{1}{2}$ inches and a width of $3\frac{1}{5}$ inches.

5. Find the area, in square feet, of a rectangle with a length of 22.4 feet and a width of 12.6 feet.

6. Find the area, in square inches, of a rectangle with a length of $22\frac{1}{2}$ inches and a width of $16\frac{1}{2}$ inches.

7. Find the width of a rectangle with a length of $3\frac{1}{4}$ feet and an area of $35\frac{1}{4}$ square feet.

8. Find the area of a square with one side measuring 21.1 meters.

9. Ling has just bought her first house. The bathroom of the house has a square floor that has a length of 7 ft. What is the area, in square feet, of Ling's bathroom floor?

10. LaToya is an engineer. Her square office floor has a length of 44.5 feet. What is the area of LaToya's office floor?

11. Find the width, in feet, of a rectangle with an area of 42 square feet and a length of 7 feet.

12. Find the area, in square meters, of a rectangle with a length of 10.5 meters and a width of 6.75 meters.

13. Find the length, in inches, of a rectangle with an area of $100\frac{3}{4}$ square inches and a width of $10\frac{3}{4}$ inches.

14. Long is buying carpet for his square bedroom. The carpet he buys has a length of 20 feet. The price of the carpet is $3.00 per square foot. What did Long pay for carpeting his bedroom?

15. What is the area of the figure below?

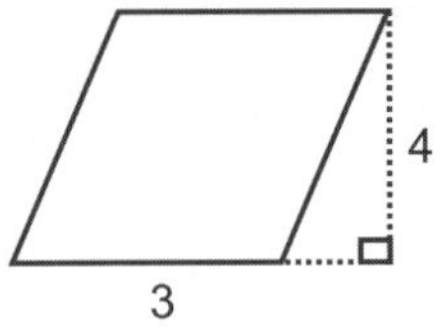

16. What is the area of the figure below?

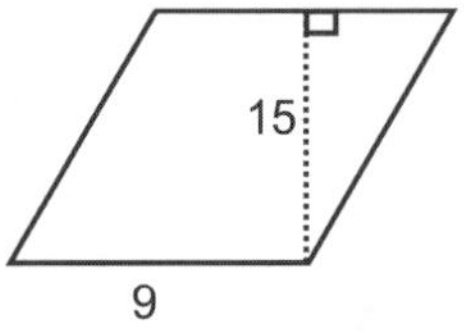

17. What is the area of the figure below?

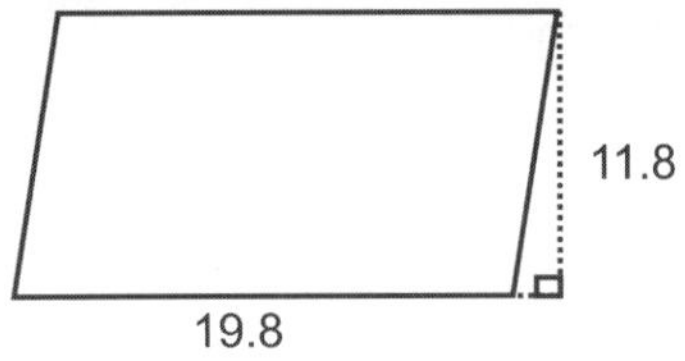

18. A parallelogram-shaped piece of fabric has a height of 32 inches and a base length of 14 inches. What is the area of the piece of fabric in square inches?

19. The cover of a magazine is 11 inches high and 8.5 inches wide. What is the area of the page in square inches?

20. A parallelogram has an area 134.784 square centimeters. The length of the base is 8.64 centimeters. What is the height?

MPA-068: Finding the Area of Irregular Figures

Occasionally in real-world situations, you might need to calculate the area of a simple geometric figure. However, most of the time, you will need to find the area of an irregular, or unusually shaped, figure. When you're faced with one of these problems, the easiest way is to divide the figure into smaller, simpler shapes and find their areas. Then you will use the basic idea that the area of the whole figure is equal to the sum of the areas of its parts.

Look at the following example. Notice that this figure is simply two different rectangles joined together.

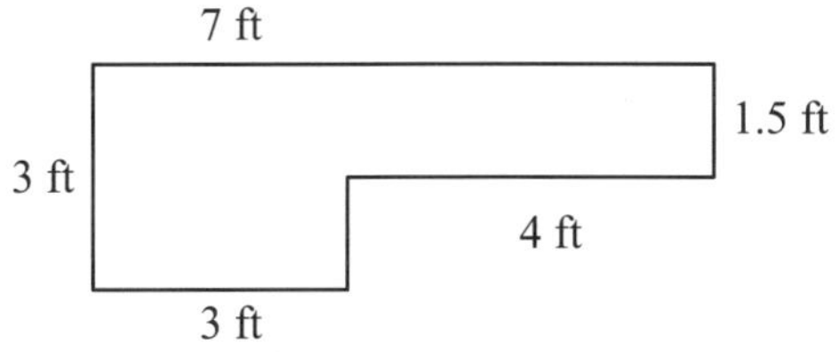

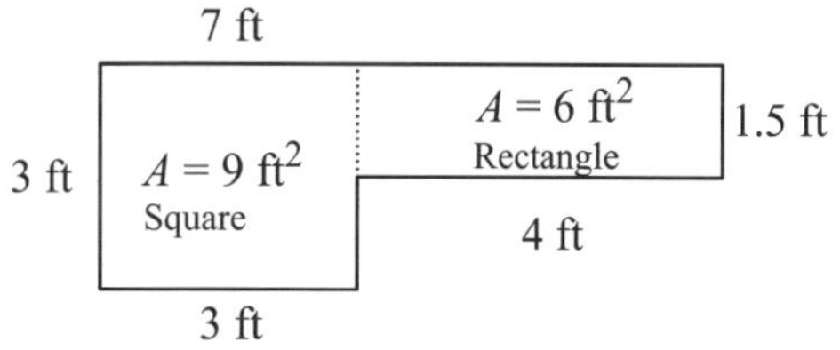

You already know how to calculate the area of both of these rectangles: you simply multiply length times width. The area of the square is $A = s^2 = 3^2 = 9 \text{ ft}^2$, and the area of the other rectangle is $A = l \times w = (4)(1.5) = 6 \text{ ft}^2$. It makes sense then that the total area of the figure is just the sum of these two areas: $A = 9 \text{ ft}^2 + 6 \text{ ft}^2 = 15 \text{ ft}^2$.

You have just seen the basic steps for finding the area of an irregular geometric figure.

Step 1: Divide the figure into smaller, regular shapes.

Step 2: Find the area of the smaller shapes.

Step 3: Add all of the smaller areas.

The basic underlying idea here is a very important geometric concept:

Area of a Geometric Figure	The area of a geometric figure is equal to the sum of the areas of its parts.

The example above showed you how to find the area of a basic geometric figure. Now, let's look at more complicated examples.

Example 1 See if you can determine how to divide this figure and find its area. Pay close attention to the dimensions of the smaller figures.

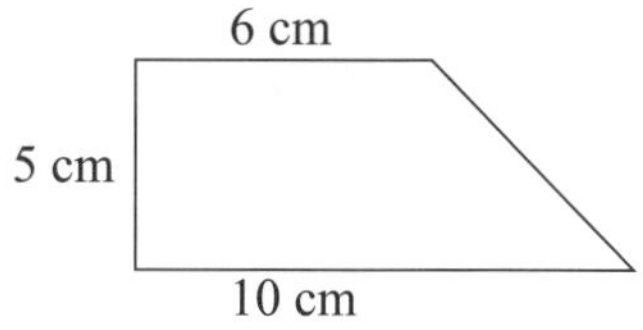

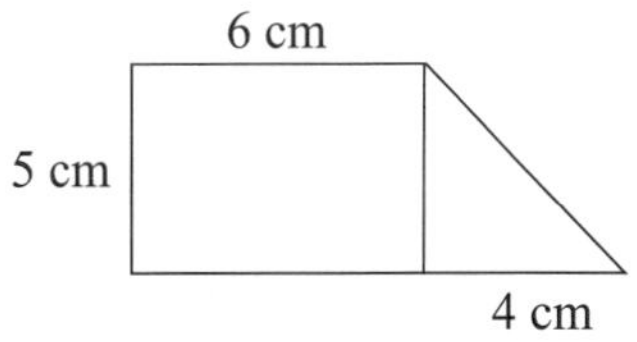

Step 1: You can divide this shape into two figures: a rectangle that has measurements 6 cm (length) by 5 cm (width), and a right triangle that has measurements 4 cm (base) by 5 cm (height).

Area of rectangle $= lw$

Area of triangle $= \frac{1}{2}bh$

Step 2: Find the area of the rectangle and the triangle.

Area of rectangle $= (6\text{ cm})(5\text{ cm})$ Area of triangle $= \frac{1}{2}(4\text{ cm})(5\text{cm})$	**Step 3:** Insert the measurements of the rectangle and the triangle into the formulas.
Area of rectangle $= 30\text{ cm}^2$ Area of triangle $= 10\text{ cm}^2$	**Step 4:** Multiply across from left to right to get the areas of the two figures. Remember, centimeters times centimeters equals centimeters squared ($\text{cm} \times \text{cm} = \text{cm}^2$).
Area of figure $= 30\text{ cm}^2 + 10\text{ cm}^2$	**Step 5:** The area of the figure is equal to the sum of the areas of its parts.
Area of figure $= 40\text{ cm}^2$	**Step 6:** Add the two areas to get the area of the irregular figure.
	Answer: 40 cm^2

Now try this one.

Example 2 Find the area of this figure:

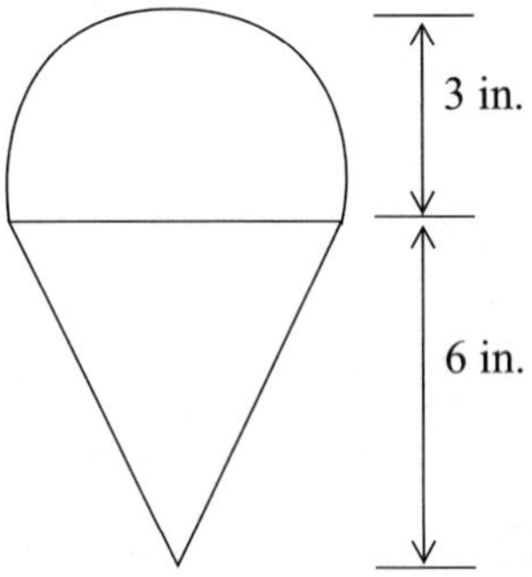

This ice- cream cone shape is basically just a half-circle on top of an isosceles triangle. For the half-circle, you can see that the radius is 3 inches. That's the only dimension you need to calculate its area. For the triangle, you need the base and height. We already know the height; it's 6 inches. Now, find the measurement of the base. Notice that the base of the triangle (at the top) is actually the diameter of the circle, or twice the radius: 6 inches. Now you have all the information needed to calculate the area of the figure. Once you find the area of both shapes, add them to find the area of the entire figure: Area of figure = Area of half-circle + Area of triangle.

Area of half-circle $= \frac{1}{2}\pi r^2$ Area of triangle $= \frac{1}{2}bh$	**Step 1:** Start with the formulas for area of each figure.
Area of half-circle $= \frac{1}{2}(3.14)(3\text{ in.})^2$ Area of triangle $= \frac{1}{2}(6\text{ in.})(6\text{ in.})$	**Step 2:** Insert the measurements of the half-circle and the triangle into the formulas.We will use 3.14 as our approximation for π.

Area of half-circle $= \left(\frac{3.14}{2}\right)(3 \text{ in.})^2$ $= (1.57)(9 \text{ in}^2)$ $= 14.13 \text{ in}^2$ Area of triangle $= \frac{1}{2}(36 \text{ in.}) = 18 \text{ in}^2$	**Step 3:** Multiply across from left to right to get the areas of the two figures.
Area of figure $= 14.13 \text{ in}^2 + 18 \text{ in}^2$	**Step 4:** Insert the areas into your original formula.
Area of figure $= 32.13 \text{ in}^2$	**Step 5:** Add the two areas to get the area of the irregular figure.
	Answer: 32.13 in^2

Now let's try one with a twist. It uses the same geometric property: all geometric figures have the same area as the sum of their parts.

Example 3 Find the area of the shaded part of this figure:

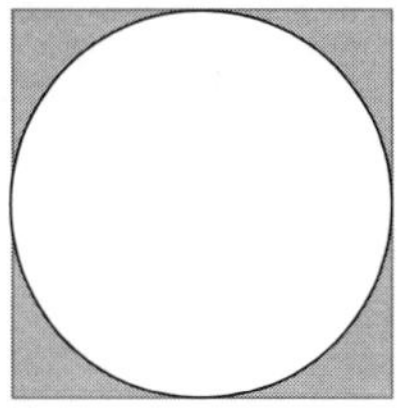

You need to find the area of the shaded part only. Using the basic geometric property mentioned at the beginning, the area of the shaded part plus the area of the unshaded circle equals the area of the whole square. To find the area of the shaded part, subtract the area of the circle from the area of the square:

$$\text{Area of figure} = \text{Area of square} - \text{Area of circle}$$

The remaining measurement will be the area of the shaded region.

Area of square $= s^2$ Area of circle $= \pi r^2$	**Step 1:** Start with the area formulas for each shape.
Area of square $= (4 \text{ mm})^2$ Area of circle $= (3.14)(2)^2$	**Step 2:** Insert the measurements of the circle and the square into the formulas. Use 3.14 to approximate π.
Area of square $= (4 \text{ mm})^2 = 16 \text{ mm}^2$ Area of circle $= (3.14)(2 \text{ mm})^2 = 12.56 \text{ mm}^2$	**Step 3:** Multiply across from left to right to get the areas of the two figures.
Area of figure $= 16 \text{ mm}^2 - 12.56 \text{ mm}^2$	**Step 4:** Insert the areas into your original formula.
Area of figure $= 3.44 \text{ mm}^2$	**Step 5:** Subtract the two areas to get the area of the irregular figure.
	Answer: 3.44 mm^2

Problem Set

Solve and use 3.14 to approximate π where appropriate:

1. What is the area of the figure below?

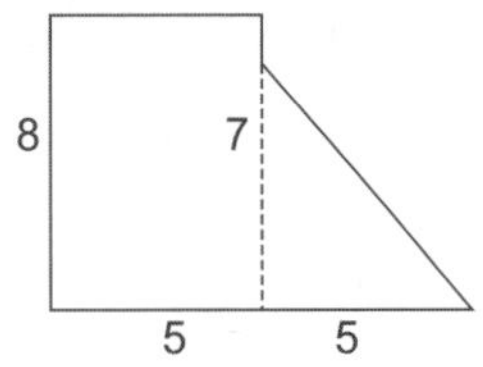

2. Farmer Minnie needs to know the area of her pig pen so she can determine if the pen is overcrowded. Below is the plan for the pig pen measured in yards. Help Farmer Minnie by finding the area of the pen. Include units in your answer.

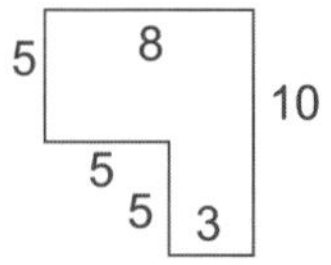

3. Farmer Kelly needs to know the area of her pig pen so she can determine if the pen is overcrowded. Below is the plan for the pig pen measured in inches. Help Farmer Kelly by finding the area of the pen. Include units in your answer.

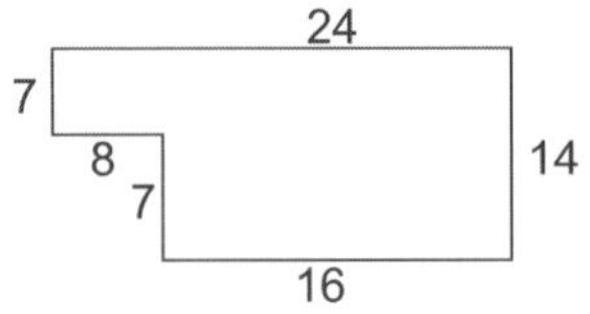

4. Find the area of the following figure. Include units in your answer.

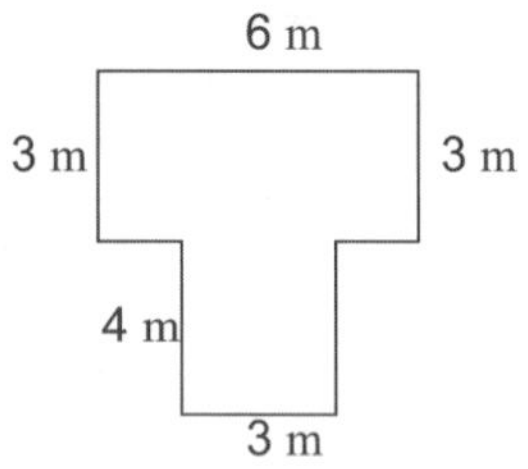

5. The given figure is a stage measured in inches for Kiki's dolls. Kiki wants to make the stage for her shop project at school. Find the area of the stage so Kiki will know if it is large enough to use for the project. Include units in your answer.

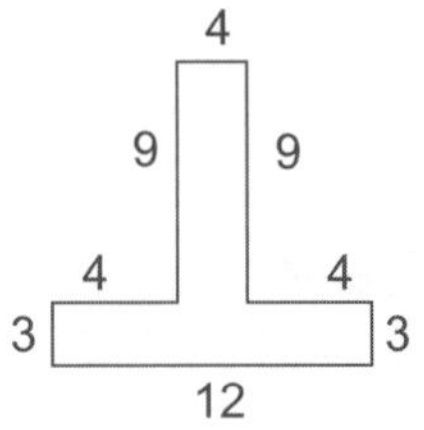

6. The Tigard Tigards want to make a giant "T" to win a biggest sign contest sponsored by the local radio station. Below is a plan for their sign measured in yards. It will be made of green fabric. How much fabric will they need? Include units in your answer.

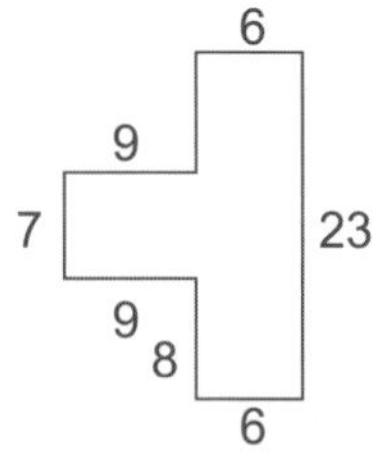

7. Perfect Prints (a framing and photography store) is designing a display for framing. Below is a picture of the display measured in inches. Help the owner of the store figure out how much material to use by finding the area of the shaded part of the figure. Include units in your answer.

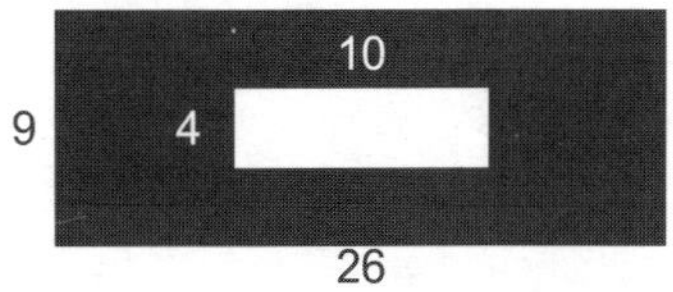

8. What is the area of shaded part of the figure? The measurements are given in miles. Include units in your answer.

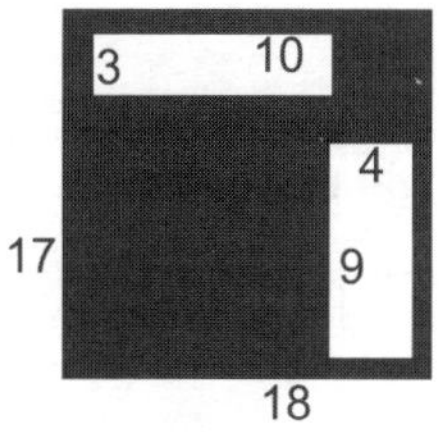

9. Find the area of the following figure. The measurements are given in centimeters. Include units in your answer.

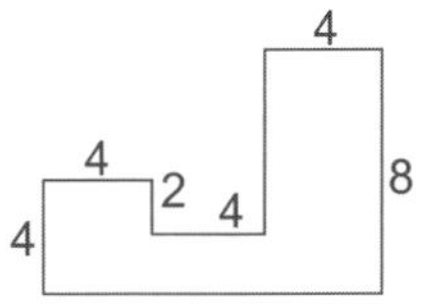

10. Find the area of the following figure. The measurements are given in inches. Include units in your answer.

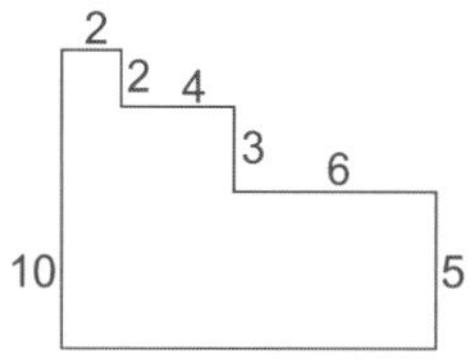

MPA-069: Finding the Area of Triangles and Trapezoids

Let's continue the discussion of area and take a look at more basic geometric figures: triangles and trapezoids.

Area of Triangles

Given a triangle such as the right triangle below, it is obvious that the triangle is exactly one-half of a rectangle.

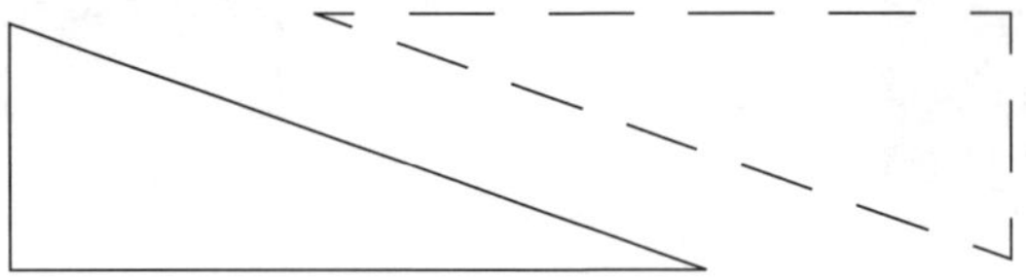

As the diagram shows, you could just turn the triangle around and put the two together in order to get a complete rectangle. But what about other triangles, such as the isosceles and scalene triangles shown below?

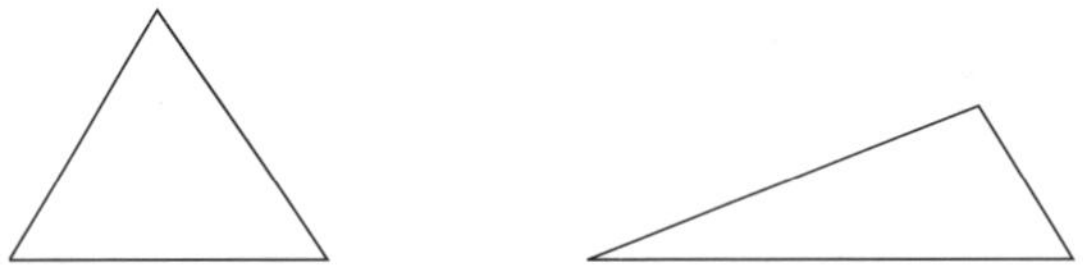

If you look closely, you can see that these are also halves of rectangles:

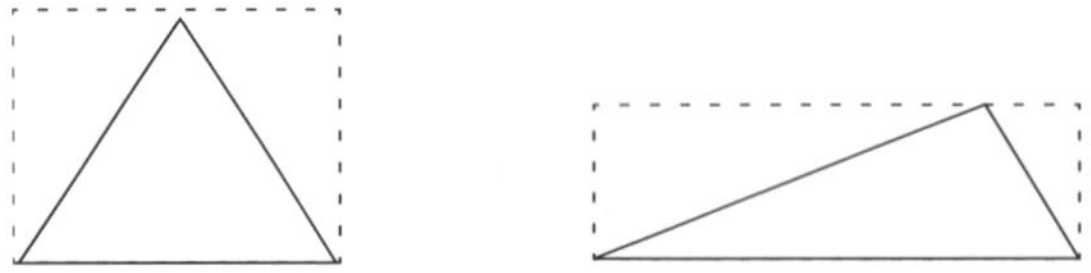

As you know, the area of a rectangle is given by the formula $A = lw$. However, since a triangle is exactly half of a rectangle, it stands to reason that the formula for the area of a triangle is exactly half of the area of the rectangle, or $A = \frac{1}{2}lw$. However, to distinguish between the dimensions of rectangles and the dimensions of triangles, we use base and height to describe these measurements.

Area of Triangles	The area of a triangle is one-half the base times the height. The formula for the area of a triangle is: $A = \frac{1}{2}bh$.

Example 1 Find the area of the triangle shown below.

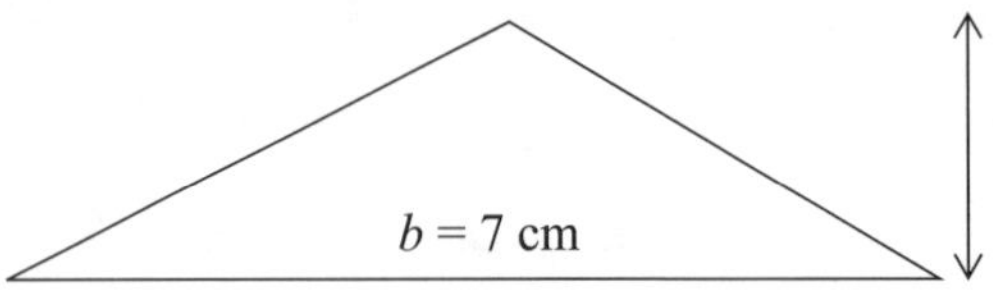

Step 1: Substitute the values into the formula for the area of a triangle.

$$A = \frac{1}{2}bh$$

$$A = \frac{1}{2}(7 \times 3)$$

$$A = \frac{1}{2}(21)$$

Step 2: Multiply inside the parentheses, then multiply by $\frac{1}{2}$.

$$A = 10.5$$

Answer: 10.5 cm^2

Here's another example.

Example 2 Find the area of the right triangle below:

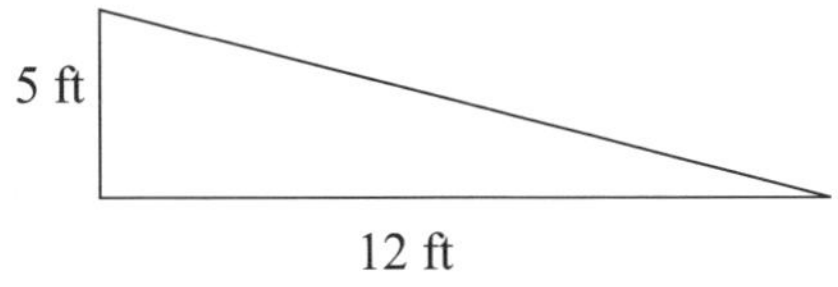

Step 1: Substitute the values into the formula for the area of a triangle.

$$A = \frac{1}{2}bh$$

$$A = \frac{1}{2}(12 \times 5)$$

Step 2: Multiply inside the parentheses, then multiply by $\frac{1}{2}$.

$$A = \frac{1}{2}(60)$$

$$A = 30$$

Answer: 30 ft^2

Area of Trapezoids

The first thing you might notice about a trapezoid is that it looks sort of like a rectangle with one pair of opposite sides ***(called bases)*** that are parallel, but not the same length. So, if it were a rectangle, or rather if a rectangle did have the same area as the trapezoid below, how would it look?

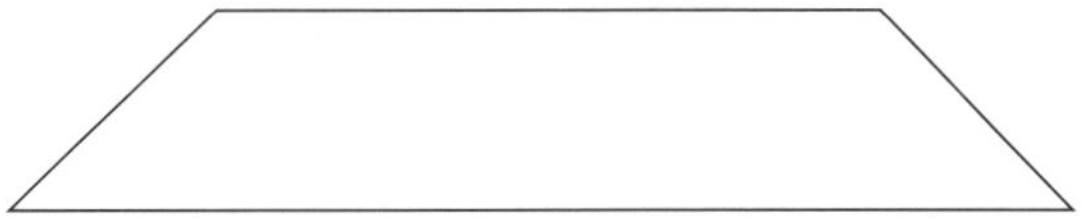

It would look something like this:

The *area* of the missing parts in the rectangle is equal to the *area* of the leftover parts of the trapezoid.

It makes sense that in order to get the two areas to match, you must cut the two sides exactly in half, forming two identical triangles (as in the figure shown above). So, how do you do that?

All you must do is come up with the dimensions of a rectangle that will have the same area. You already have the width of the rectangle: the height of the trapezoid. So, all you need is its length in order to calculate its area (and hence, the area of the trapezoid). The length of the rectangle is just the mean or average of the measurements of the two bases. The formula for the average of the two bases (denoted b_1 and b_1) is simply $\frac{1}{2}(b_1 + b_2)$. Now that we have both dimensions of the rectangle, the formula is just length times width.

Area of a Trapezoid	The area of a trapezoid is the average of the bases times the height, or one-half the sum of the bases multiplied by the height. The formula for the area of a trapezoid is: $A = \frac{1}{2}(b_1 + b_2)h$.

Here are some examples of how to find the area of a trapezoid.

Example 3 Find the area of the trapezoid below:

$$A = \frac{1}{2}(b_1 + b_2)h$$

$$A = \frac{1}{2}(10 + 6)5$$

Step 1: Place all the known values into the formula for the area of a trapezoid.

$$A = \frac{1}{2}(16)(5)$$

Step 2: Add within the parentheses.

$$A = 8(5)$$

$$A = 40$$

Step 3: Multiply from left to right.

Answer: 40 cm^2

Example 4 Find the area of the right trapezoid shown below:

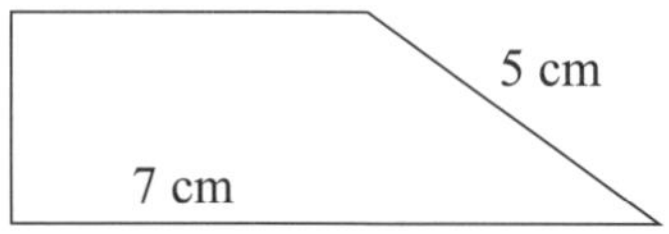

*Note: 3 and 5 are **not** bases since they are not parallel.*

$$A = \frac{1}{2}(b_1 + b_2)h$$

$$A = \frac{1}{2}(4 + 7)3$$

Step 1: Place all the known values into the formula for the area of a trapezoid.

$$A = \frac{1}{2}(4 + 7)3$$

$$A = \frac{1}{2}(11)(3)$$

Step 2: Add within the parentheses.

$A = 5.5(3)$	**Step 3:** Multiply from left to right.
$A = 16.5$	

Answer: 16.5 cm^2

Problem Set

Solve:

1. Find the area of the figure.
Area = ______ square units

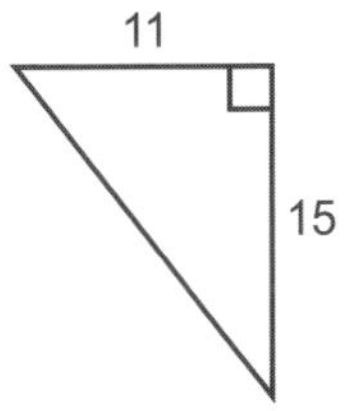

2. Find the area of the triangle.
Area = ______ square units

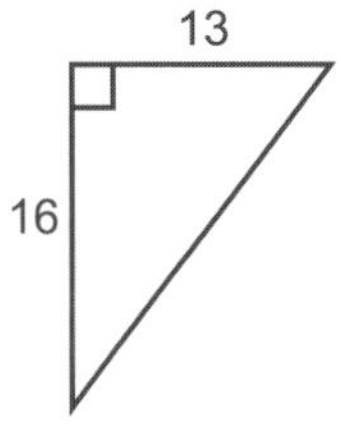

3. Find the area of the triangle.
Area = ______ square units

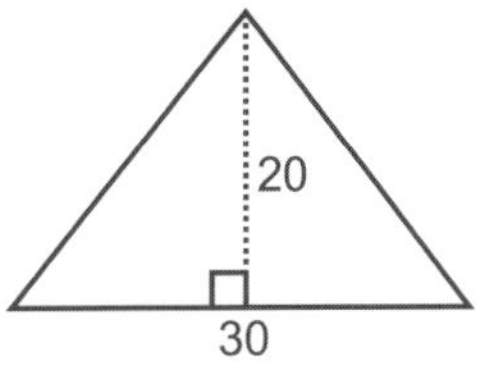

4. Find the area of the triangle.
Area = ______ square units

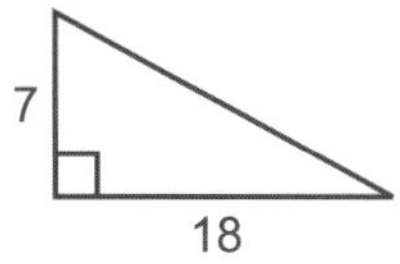

5. Find the area of the trapezoid.
Area = ______ square units

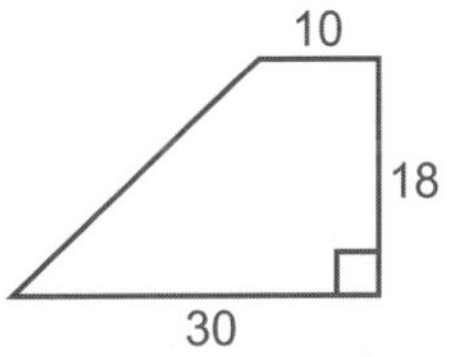

6. Find the area of the figure.
Area = ______ square units

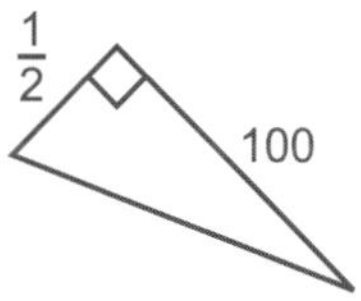

7. Find the area of the triangle.
Area = ______ square units

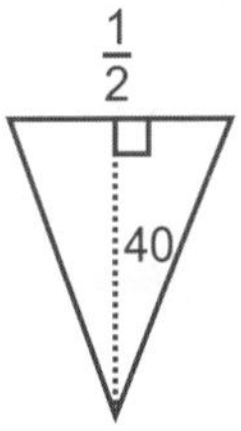

8. A triangular plot of land has a height of $\frac{3}{8}$ and a base of 24. What is the area of the land?

9. What is the area of a trapezoid with bases of length $\frac{2}{3}$ and $\frac{4}{3}$ and a height of 3?

10. Find the area of the trapezoid.
Area = ______ square units

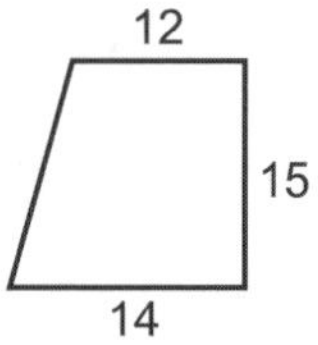

11. Find the area of the trapezoid.
Area = ______ square units

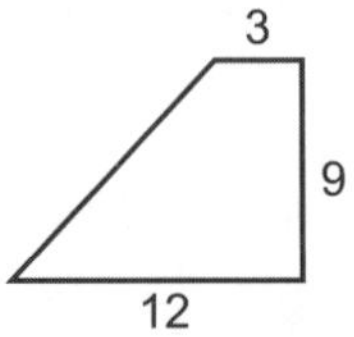

12. Find the area of the trapezoid.
Area = ___ square units

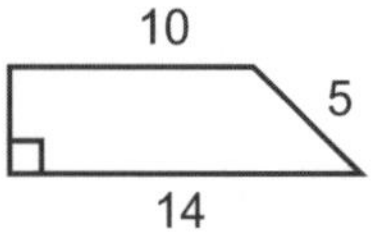

13. Find the area of the trapezoid.
Area = ______ square units

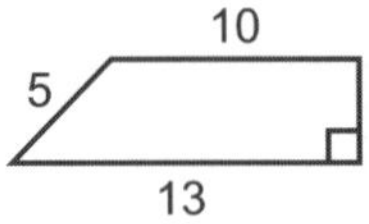

14. Find the area of the trapezoid.
Area = ______ square units

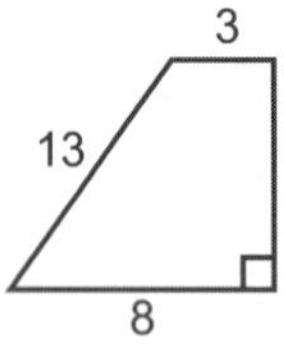

15. Find the area of the trapezoid.
Area = ______ square units

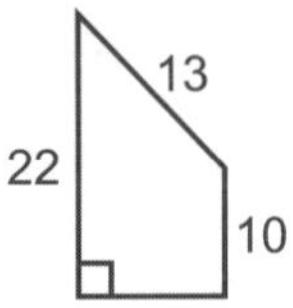

16. Find the area of the trapezoid.
Area = ______ square units

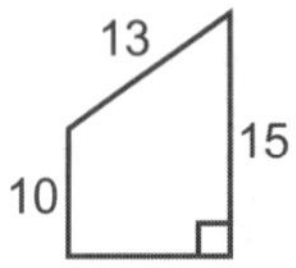

17. A triangle has an area of 60 in^2 and a base of 10 in. Find the height of the triangle. Include units in your answer.

18. A triangle has an area of 212.5 m^2 and a base of 25 m. Find the height of the triangle. Include units in your answer.

19. A triangle has an area of 3 ft^2 and a base of 3 ft. Find the height of the triangle. Include units in your answer.

20. A trapezoid has bases of 12 cm and 3 cm. It has an area of 67.5 cm^2. Find the height of the trapezoid. Include units in your answer.

MPA-070: Finding the Circumference of a Circle

You previously learned how to find the perimeter of polygons (geometric shapes formed from line segments). Now, consider circles. If a circle had sides, then you could just use the basic formula for perimeter: add the measurements of all the sides. However, you can't really add the sides of a circle, because there aren't any. For this reason, the word **circumference** is used instead of perimeter to describe the distance around a circle.

Before you begin finding the circumference, you must know some basic terms that involve circles.

Think about what exactly a circle is. A circle is a set of points that are all the same distance (equidistant) from one point located in the center of the circle. As circles are flat, or two-dimensional figures, it's also important to include in the definition that a circle must lie in a plane.

Circle	A set of points in a plane that are equidistant from a given point in the plane.

The "given point" mentioned in the definition above is a central point in the circle called the **center**.

Center	The given point in the plane from which all the points on the circle are equidistant.

Since all the points of a circle are the same distance from the center, a line segment from the center to any one of those points would have the same length as any other such line segment. Also, the length of one of these line segments is exactly the distance from the center to any other point on the circle. For this reason, this line segment has a special name, the **radius** (the plural form is radii).

Radius	Any line segment from the center of the circle to any point on the circle.

Another special line segment on a circle runs directly across the circle, cutting it in half. This is known as the **diameter**. It's easy to see that a diameter is simply two radii joined together to form a line segment. However, you'll need a more precise definition. The key element of the definition is that a line segment passes through the center and connects two points on the circle.

Diameter	Any line segment from one point on the circle to another point on the circle that passes through the center.

The final definition you will need is the subject of this lesson, circumference. It is one of the most interesting aspects of circles. The circumference of a circle is the distance around a circle. Essentially, the circumference is exactly the same as the perimeter of a polygon. If you were to cut the circle at some point, then uncurl it into a straight line segment, the circumference would be the length of that line segment.

Circumference	The distance around a circle.

Note that it is standard practice to use the terms radius and diameter to refer to the lengths of the radius and diameter (since the lengths of these line segments are more important than the line segments themselves). So, when asked to find the radius of a circle, usually this means to find the length of a radius (likewise for diameter).

The following circles will help you visualize the terms just defined.

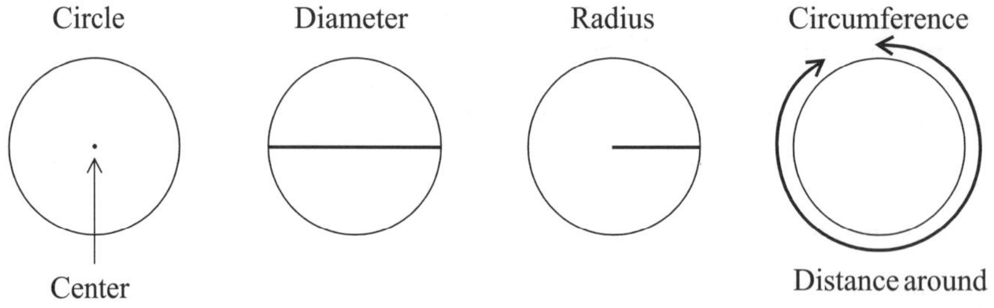

Now that you have a basic understanding of the vocabulary for circles, look at the relationships between the diameter, radius, and circumference of a circle.

The first and most obvious relationship is between the radius and the diameter. Because a diameter is basically just two radii joined together, it's clear that the length of the diameter of a circle is twice the length of the radius. Similarly, it's clear that the length of the radius is half the length of the diameter.

Diameter	The diameter of a circle is twice the length of the radius. The formula for diameter is $d = 2r$.

Radius	The radius of a circle is half the length of the diameter. The formula for radius is $r = \frac{1}{2}d$.

Now look at how the radius and diameter are related to the circumference. To find such a relationship, you must use a simple formula developed by ancient Greek mathematicians. The formula involves a mathematically interesting number, pi (π). Pi is a non-ending, non-repeating decimal. You have probably already heard of pi, and know that its value is 3.141592654 . . .

To make calculations easier, use 3.14 for the value of pi. Occasionally, the fraction $\frac{22}{7}$ is used instead.

The ancient Greeks found pi when they were trying to find a relationship between the diameter and circumference of a circle. After much experimenting, they discovered that the ratio of the circumference to the diameter of **any** circle is equal to pi.

Pi	Circumference divided by diameter is equal to pi: $\pi = \frac{\text{circumference}}{\text{diameter}} \approx 3.14 \approx \frac{22}{7}$

So, if circumference divided by diameter is equal to pi, you can use algebra to solve for the equation for circumference. Simply use the inverse operation of division: multiplication. If you multiply both sides of the equation by diameter, you will get an equation for circumference. This equation will be $C = \pi d$. Also, since $d = 2r$, you can rewrite the formula for circumference as $C = 2\pi r$.

Circumference	The circumference of a circle is equal to pi times the diameter. Since the diameter of a circle is twice the radius, circumference can also be calculated by multiplying pi by twice the radius: $C = \pi d = 2\pi r$.

Now you have all the information you need in order to calculate the circumference of a circle given either the radius or diameter. Here are some examples:

Example 1 A pizzeria sells pizzas that are 12 inches in diameter. Find the circumference of one of their pizzas.

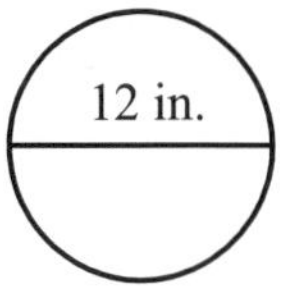

$d = 12$ in. $C = \pi d$	**Step 1:** Choose the formula for circumference that matches the information you are given. In this case, you know the diameter.
$C = \pi d$ $C = (3.14)12$	**Step 2:** Insert the measurement of the diameter and the value of pi into the formula.
$C = (3.14)12$ $C = 37.68$ in.	**Step 3:** Multiply to get the answer. *Note: The units of circumference are units of length since circumference is a distance.*

Answer: C = 37.68 in.

Example 2 A bottle of water is a cylinder that has a radius of 4 centimeters. The label on the bottle is rectangular and wraps around to where the ends just meet. How long is the label?

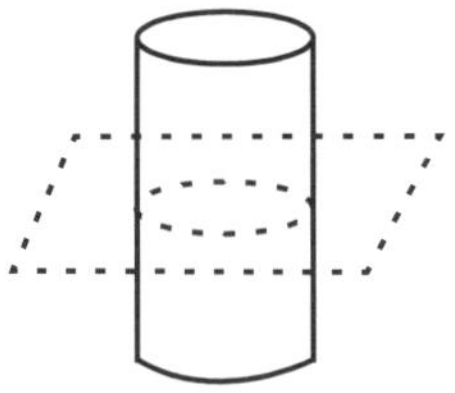

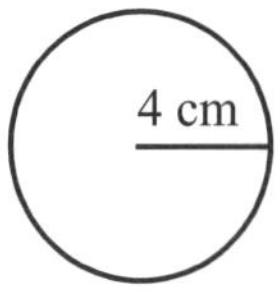

To answer this question, you must know the distance around the cylinder. A cross-section of the cylinder is actually just a circle, and the distance around the cylinder is the same as the distance around this circle. To find the length of the label, you only need to calculate the circumference of the circle.

$r = 4$ cm $C = 2\pi r$	**Step 1:** Choose the formula for circumference that matches the information you are given.
$C = 2\pi r$ $C = 2(3.14)(4)$	**Step 2:** Insert the measurement of the radius and the value of pi into the formula.
$C = 2(3.14)(4)$ $C = 25.12$ cm	**Step 3:** Multiply to get the answer.

Answer: C = 25.12 cm

Example 3 Nancy wants to tie a ribbon around a tree. The diameter of the tree is 2 feet 4 inches. She needs an extra 2 feet of ribbon to make the bow. Rounding up to the nearest inch, how much ribbon will she need to go around the tree and tie the bow?

$d = 2$ ft 4 in. $d = 2(12) + 4 = 28$ in.	**Step 1:** You must first convert the measurement into one type of unit, inches. Since there are 12 inches in a foot, multiply 2 by 12 and add the other 4 inches.
$C = \pi d$	**Step 2:** Choose the formula for circumference that matches the information you are given.
$C = \pi d$ $C = (3.14)(28)$	**Step 3:** Insert the measurement of the diameter and the value of pi into the formula.
$C = (3.14)(28)$ $C = 87.92$ in. ≈ 88 in.	**Step 4:** Multiply to get the circumference. Round to the nearest inch.
$l = C + 2$ ft $l = 88$ in. $+ 24$ in.	**Step 5:** The length of the ribbon needed is equal to the circumference of the tree plus 2 feet, or 24 inches.
$l = 88$ in. $+ 24$ in. $l = 112$ in.	**Step 6:** Add to get the answer. Since the problem doesn't specify whether the answer should be in feet or inches, leave it in inches.

Answer: 112 in.

Problem Set

Solve:

1. What is the circumference of a circle with a diameter of 7 in.? Use $\pi \approx 3.14$.

2. What is the diameter of a circle with a circumference of 26.69 in.? Use $\pi \approx 3.14$.

3. What is the diameter of a circle with a circumference of 307.72 in.? Use $\pi \approx 3.14$.

4. What is the radius of a circle with a circumference of 84.78 in.? Use $\pi \approx 3.14$.

5. What is the radius of a circle with a circumference of 263.76 inches? Use $\pi \approx 3.14$.

6. The radius of a circle is 91 cm. What is the circumference of this circle? Use $\pi \approx 3.14$.

7. What is the circumference of a circle with a radius of 46 in.? Use $\pi \approx 3.14$.

8. What is the diameter of a circle with a radius of 77 cm?

9. What is the radius of a circle with a diameter of 70 in.?

10. A circle with a circumference of 12.56 in. has a radius of 4 in. True or False? Use $\pi \approx 3.14$.

11. A circle with a circumference of 72.22 in. has a radius of 23 in. True or False? Use $\pi \approx 3.14$.

12. A circle with a circumference of 21.98 in. has a radius of 7.5 in. True or False? Use $\pi \approx 3.14$.

13. Mario's Pizza Shop sells three different sizes of pizza. The small pizza is 8 inches in diameter. What is the circumference of the small pizza? Use $\pi \approx 3.14$.

14. Mario's Pizza Shop sells three different sizes of pizza. The small pizza is 6.5 inches in diameter. What is the circumference of the small pizza? Use $\pi \approx 3.14$.

15. Javier owns a bike shop in North Carolina. He has a wide variety of bikes for children and adults. One of the tires on a bike has a circumference of 48.67 inches. What is the radius of this tire? Use $\pi \approx 3.14$.

16. Chantelle owns a pastry shop in New York City. Her store is famous for the cookies it sells. Her oatmeal cookie has a 3 inch radius. What is the circumference of the cookie? Use $\pi \approx 3.14$.

17. Harry works in a department store. Part of his job is to sell plates. If Harry has a plate with a circumference of 28.26 inches, what is the radius of the plate? Use $\pi \approx 3.14$.

18. A dime has a diameter of 14 mm. What is its circumference? Use $\pi \approx 3.14$.

19. Find the circumference of a circle if $d = 8$ in. Use $\pi \approx 3.14$.

20. A Pizza Shop sells a large pizza that has a diameter of 16 inches. What is the circumference of the circle? Use $\pi \approx 3.14$.

MPA-071: Finding the Area of a Circle

Finding the area of a geometric figure is not always an easy task. In many of the examples that you've seen so far, you could have used squares to "tile" the figure to find the area. However, in most cases, this is not possible or not practical. In these cases, you need a formula. Ancient Greek mathematicians found the formula for area and for circumference, and in so doing, discovered the value of pi (π).

Pi is a non-repeating, non-ending decimal. To simplify the process of finding the area of a circle, use 3.14 for the value of pi.

The formula that the Greeks developed for the area of a circle is pi times the measurement of the radius squared.

Area of a Circle	The measure of the region inside a circle's edge. The formula for the area of a circle is πr^2.

Recall that the radius of a circle is the distance from the center of the circle to any point on the circle.

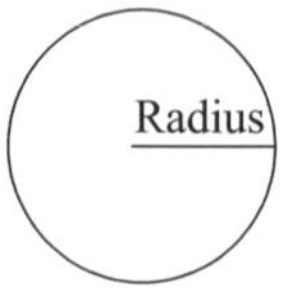

To find the area of a circle, simply substitute 3.14 for pi, and substitute the radius measurement of the circle into the area formula.

Example 1 You have a lawn sprinkler that sprays in a circle. If it sprays 12 feet, how large is the area covered by the sprinkler?

Note: Think of the sprinkler as the center of a circle, and the distance it sprays as its radius.

$A = \pi r^2$ $A = (3.14)(12)^2$	**Step 1:** Substitute the value of the radius and pi into the formula.
$A = (3.14)(144)$	**Step 2:** Square the radius.
$A = 452.16$	**Step 3:** Multiply.

Answer: 452.16 ft^2

Example 2 The local recreational center has a pool with a radius of 15 feet. They would like to protect the pool during the winter with a pool cover. If the cover must include one extra foot of coverage on all sides of the pool, what is the area of the cover?

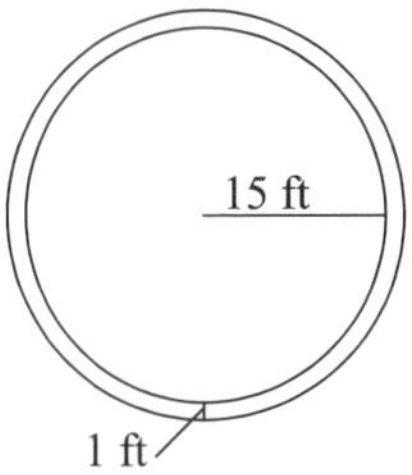

Since the radius of the pool is 15 feet, and you need an extra foot on all sides of the circle, the radius of the cover will be 16 feet.

$A = \pi r^2$ $A = (3.14)(16)^2$	**Step 1:** Substitute the value of the radius and pi into the formula.
$A = (3.14)(256)$	**Step 2:** Square the radius.
$A = 803.84$	**Step 3:** Multiply.

Answer: 803.84 ft^2

Example 3 Suppose you know that the circumference of a circle is 62.8 inches. Find the area.

To find the area, you must find the radius first. Although the measurement of the radius isn't directly given, you can find it using the formula for circumference. Recall that the formula for the circumference of a circle is $C = d\pi$, or the measurement of the diameter times pi. Also recall that the diameter is twice the length of the radius. With that information, you can find the value of the radius.

$C = d\pi$ $62.8 = d(3.14)$	**Step 1:** Set up an equation using the formula for circumference, substituting all known values.
$\frac{62.8}{3.14} = \frac{d(3.14)}{3.14}$	**Step 2:** To solve for d (diameter), you need to undo the multiplication by 3.14 using the inverse operation of division. Divide both sides of the equation by 3.14.
$20 = d$	**Step 3:** On the right side of the equation, only d remains after canceling. On the left side, $62.8 \div 3.14 = 20$.
$d = 2r$ $20 = 2r$	**Step 4:** Set up an equation using the relationship between the diameter and the radius of a circle.
$\frac{20}{2} = \frac{2r}{2}$	**Step 5:** Again, you must undo the multiplication using division. Divide both sides by 2.
$10 = r$	**Step 6:** The right side of the equation now contains only r, and r is equal to 10.
$A = \pi r^2$ $A = (3.14)(10)^2$	**Step 7:** Substitute values for r and pi in the equation for the area of a circle.

$A = (3.14)(100)$	**Step 8:** Square the radius.
$A = 314$	**Step 9:** Multiply.
	Answer: 314 in^2

Problem Set

Solve:

1. The radius of a circle is 3 m long. What is the area of the circle? Round the final answer to the nearest hundredth.

2. The radius of a circle is 20 yd long. What is the area of the circle? Round the final answer to the nearest hundredth.

3. The radius of a circle is 50 in. long. What is the area of the circle? Round the final answer to the nearest hundredth.

4. The radius of a circle is 60 m long. What is the area of the circle? Round the final answer to the nearest hundredth.

5. The radius of a circle is 25 yd long. What is the area of the circle? Round the final answer to the nearest hundredth.

6. The radius of a circular pizza is 23 cm long. What is the area of the pizza? Round the final answer to the nearest hundredth.

7. The radius of a wheel is 34 cm long. What is the area of the wheel? Round the final answer to the nearest hundredth.

8. The radius of a circular piece of cloth is 16 in. long. What is the area of the piece of cloth? Round the final answer to the nearest hundredth.

9. The radius of a circular game board is 39 cm long. What is the area of the game board? Round the final answer to the nearest hundredth.

10. The diameter of a circle is 10 yd long. What is the area of the circle? Round the final answer to the nearest hundredth.

11. The diameter of a circular pizza is 200 mm long. What is the area of the pizza? Round the final answer to the nearest hundredth.

12. The diameter of a circular table is 300 mm long. What is the area of the table? Round the final answer to the nearest hundredth.

13. The diameter of a circular game board is 5 in. long. What is the area of the game board? Round the final answer to the nearest hundredth.

14. The circumference of a circle is 6.28 cm long. What is the area of the circle? Round the final answer to the nearest hundredth.

15. Find the area of the circle shown. Use 3.14 for π and round all calculations to the nearest hundredth.

16. Find the area of the circle shown. Use 3.14 for π and round all calculations to the nearest hundredth.

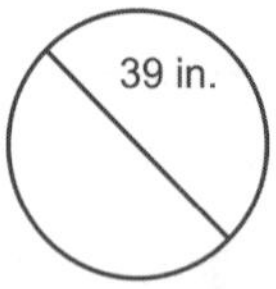

17. The area of a circle is 153.86 feet squared. What is the diameter?

18. A round sticker has an area of 200.96 mm squared. What is its diameter?

19. A circle has an area of 1,256 miles squared. What is its diameter?

20. When a dairy company orders their milk jugs, the opening at the top of the jug is 28 mm in diameter. For the caps to fit the jugs, the cap must be 2 mm larger than the opening on all sides. What is the area of the milk jug tops?

MPA-072: Identifying 3-D Figures

You've heard of 3-D movies, holograms and 3-D comic books with the red and blue glasses, but how is a 3-D image technically different from a flat image? You may know that 3-D means three-dimensional. But do you know what a dimension is? If a contractor measures the dimensions of your living room, what is he doing?

There are three spatial dimensions: length, width, and height (also called depth).

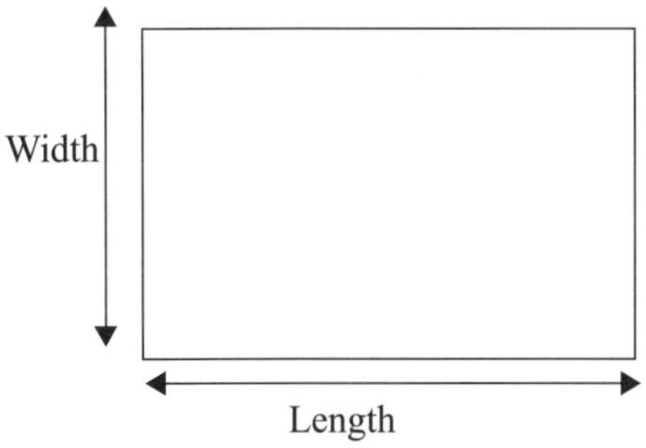

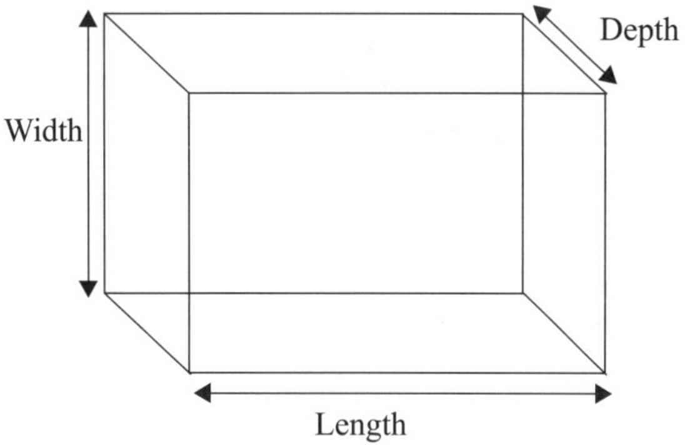

Therefore, the difference between 2-D figures and 3-D figures is the additional dimension of depth (or height). The following are terms used to identify 3-D figures.

Polyhedron	Three-dimensional solids composed of faces, or flat sides. The base of a polyhedron is a polygon.
Edges	Line segments that join the faces of polyhedrons.
Vertex	Point where the line segments of polyhedrons intersect. (The plural of "vertex" is "vertices.")
Base	The face of a polyhedron in the shape around which the figure is constructed.
Lateral face	Polyhedron faces that are not bases.

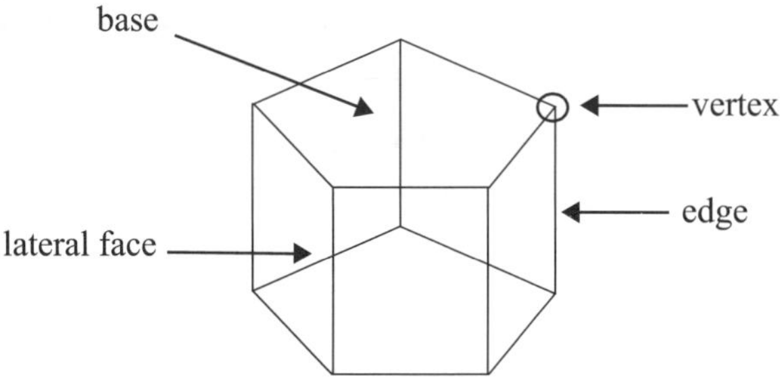

Different 3-D figures have different names. Here are a few:

Prism	A polyhedron with two parallel and congruent bases. The lateral faces of a prism are rectangles.
Pyramid	A polyhedron with a polygonal base and lateral faces that meet at a vertex. The lateral faces of a pyramid are triangles.

Platonic solid	Polyhedron with regular polygons for all the faces.
Cylinder	A three-dimensional solid that has two parallel, circular bases. The radii of the bases have the same measure.
Cone	A polyhedron with a single circular base and a single vertex at the opposite end.
Sphere	A solid whose points are all equidistant from a center point. A sphere has no base.
Tetrahedron	A platonic solid with four sides.
Hexahedron	A platonic solid with six sides; a cube.

You identify a three-dimensional figure by the shape of its base. For example, a pyramid with a triangle as its base is a triangular pyramid. The prism shown below with a pentagon as its base is a pentagonal prism.

Prisms

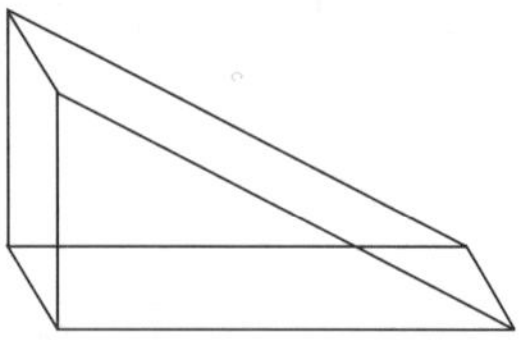

Triangular prism

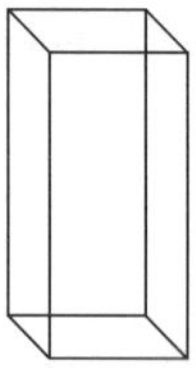

Rectangular prism

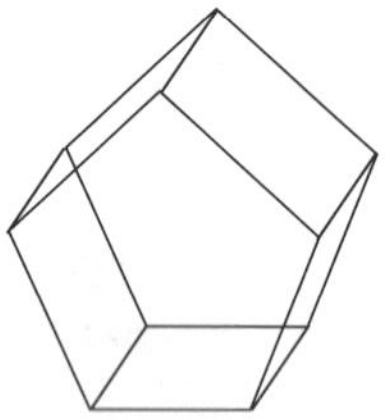

Pentagonal prism

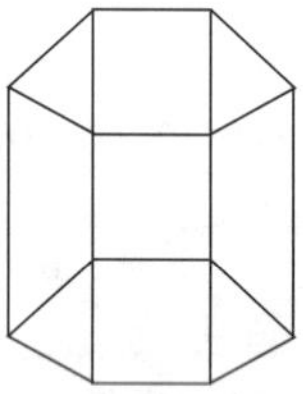

Hexagonal prism

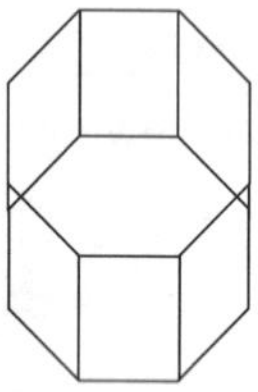

Octagonal prism

Pyramids

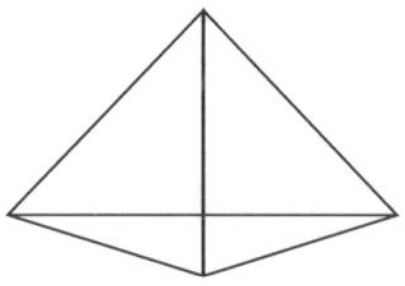

Triangular pyramid

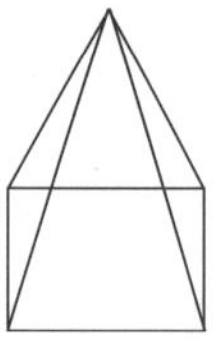

Rectangular pyramid

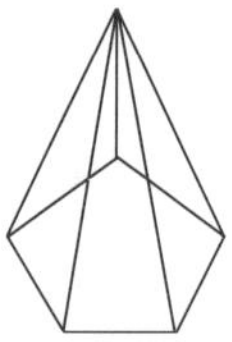

Pentagonal pyramid

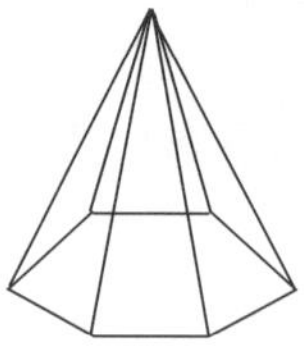

Hexagonal pyramid

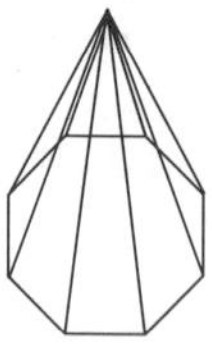

Octagonal pyramid

Figures based on circles

Cylinder

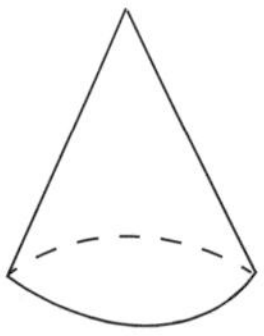

Cone

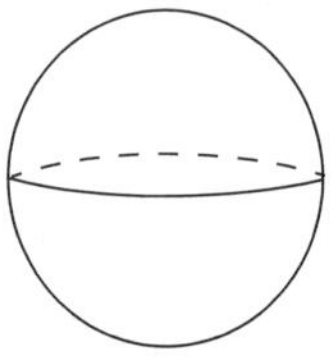

Sphere

Platonic solids

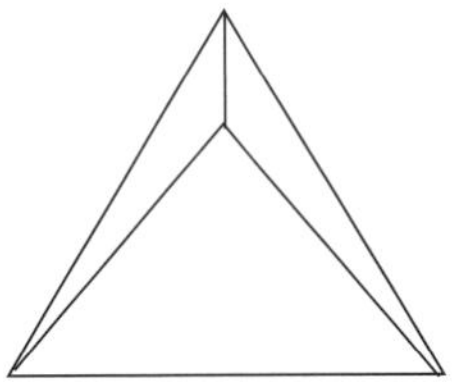

Tetrahedron

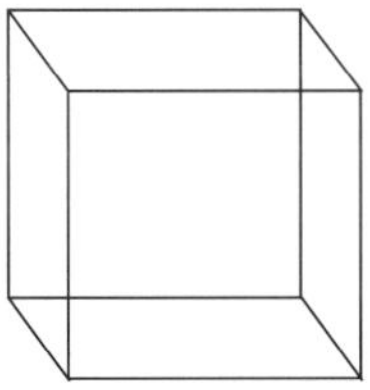

Hexahedron (cube)

If you've ever played games that require multiple dice, any die (four-sided, six-sided, twelve-sided, and so on) is a platonic solid.

Now that you know about 3-D figures, try solving these examples.

Example 1 How many edges does an octagonal pyramid have?

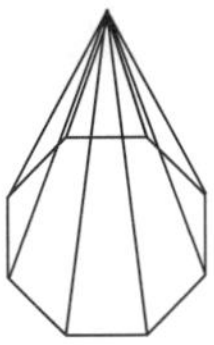

Step 1: Draw the figure for visual reference. Draw an octagonal pyramid so that you can count its edges. Edges are the boundaries between faces.

Step 2: There are 16 edges on the figure.

Answer: 16 edges

Example 2 Which prism has 15 edges?

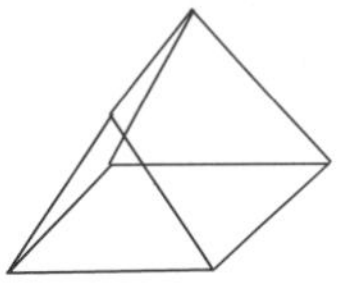

9 edges

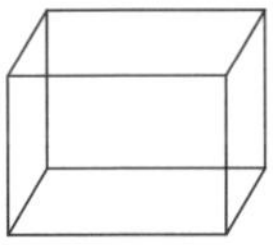

12 edges

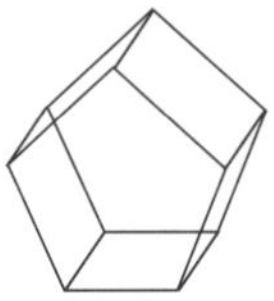

15 edges

Step 1: Draw prisms with successively larger bases until you find one with 15 edges.

pentagon

Step 2: The prism with fifteen edges has a pentagon as its base, so it is a pentagonal prism.

Answer: Pentagonal prism

Example 3 What is the most specific name for this figure?

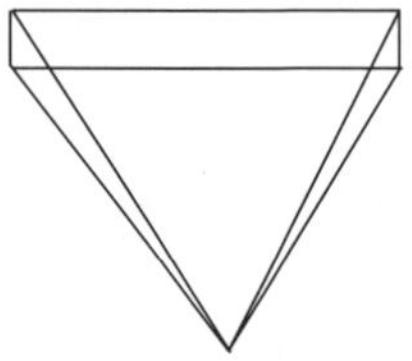

pyramid	**Step 1:** Identify the type of polyhedron. The figure is a pyramid because its lateral faces are triangles. Also, it has a single base and lateral faces that meet at a single vertex.
rectangle	**Step 2:** Identify the base.
rectangular pyramid	**Step 3:** Put the two names together.

Answer: Rectangular pyramid

Example 4 What is the most specific name for this figure?

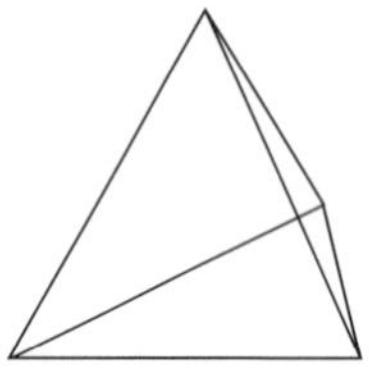

pyramid	**Step 1:** Identify the type of polyhedron. The figure is a pyramid because its lateral faces are triangles. Also, it has a single base and lateral faces that meet at a single vertex.
triangle	**Step 2:** Identify the base. The base is a triangle.
triangular pyramid	**Step 3:** Put the two names together. *Note: This is not a tetrahedron. The triangles are not regular polygons.*

Answer: Triangular pyramid

Problem Set

1. Which of the following is the general name for the shape below?

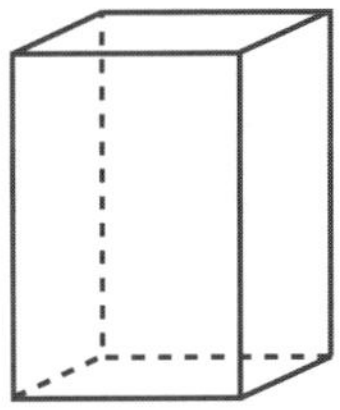

sphere pyramid
prism cone

2. Which of the following is the general name for the shape below?

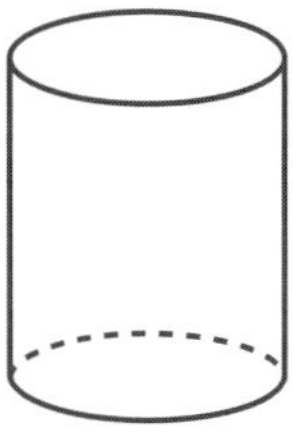

cone cylinder
prism pyramid

3. Which of the following is the general name for the shape below?

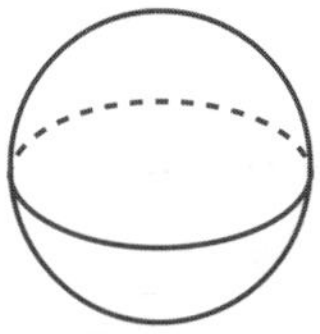

prism cone
pyramid sphere

4. Which of the following shapes is closest to a prism?

basketball waffle ice cream cone
box of crackers doughnut

5. Which of the following shapes is closest to a pyramid?

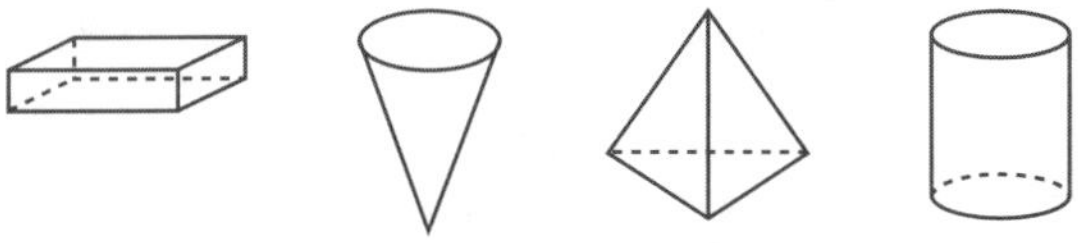

6. Which of the following shapes is closest to a sphere?

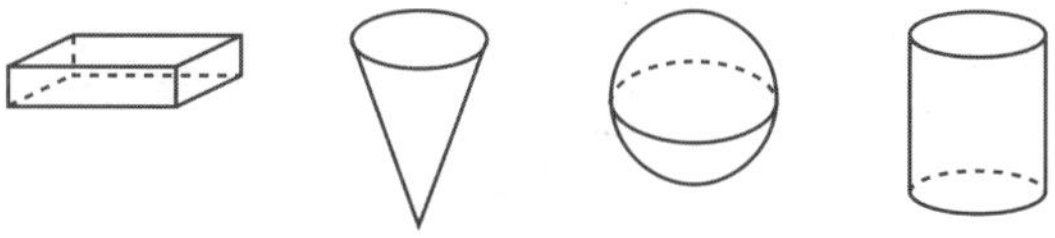

7. What is the specific name for the figure below?

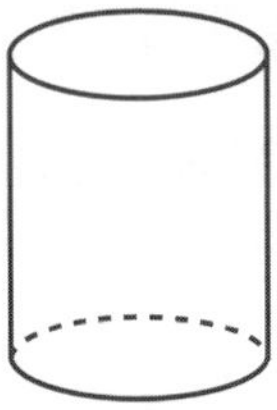

8. What is the specific name for the figure below?

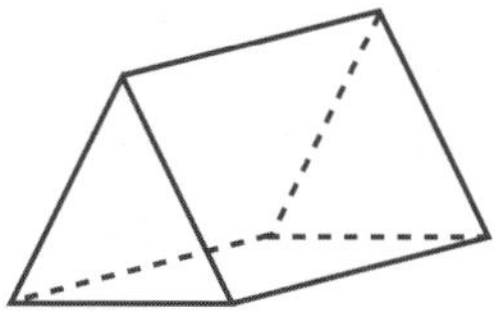

9. Which of the figures below is a square pyramid?

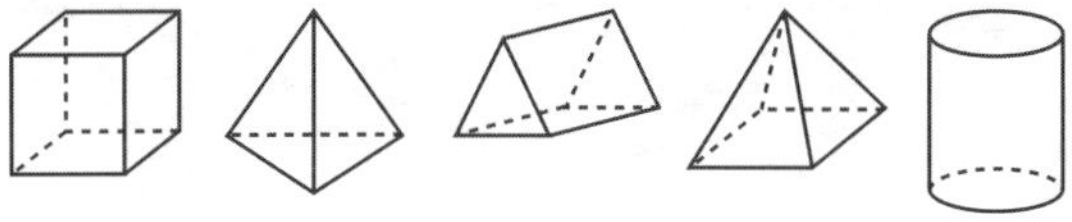

10. How many faces does a cube have?

11. How many edges does a cube have?

12. How many vertices does a rectangular prism have?

13. How many faces does a square pyramid have?

14. How many bases does a triangular prism have?

15. I am a 3-dimensional figure. I have 4 faces and a triangular base. Which of the following best describes me?

tetrahedron	cube
sphere	triangular prism

16. How many edges are on the figure below?

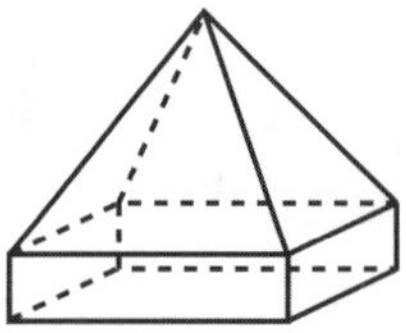

17. How many bases does the figure below have?

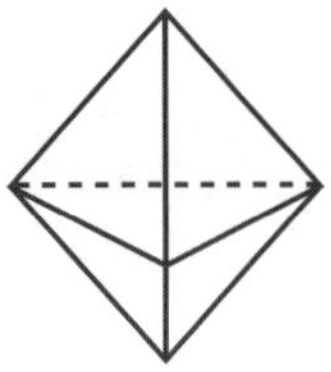

18. How many faces are on the figure below?

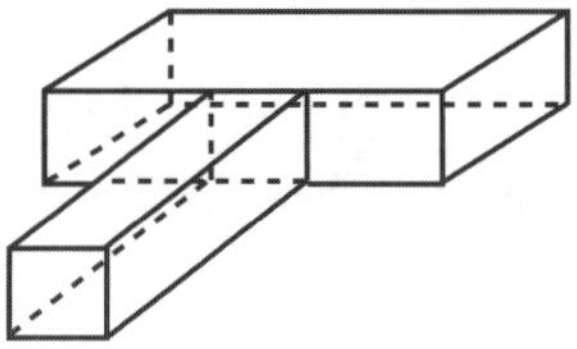

19. How many bases are on the figure below?

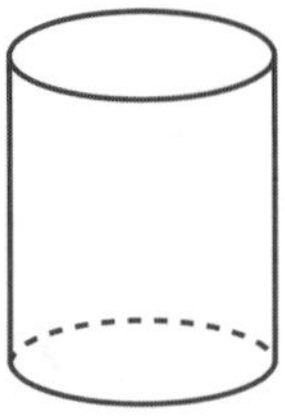

20. How many edges are on the figure below?

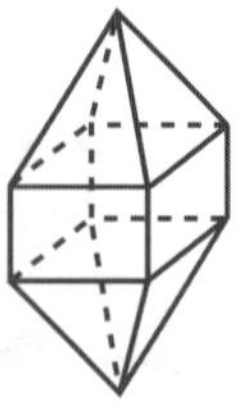

MPA-073: Finding the Surface Area of Rectangular Prisms

Now that you know how to find the area of geometric figures, let's look at an application of this concept. In the visible world, there aren't any two-dimensional objects (since all tangible objects have some thickness); there are only three-dimensional (3-D) objects. Occasionally, you may need to find the area of all the sides of a three-dimensional figure, such as a rectangular prism.

Surface Area	The total area of the faces of a 3-D object.

Each face of a rectangular prism is just a rectangle, and you already know how to find the area of rectangles. To find the total area of all the faces of a rectangular prism, you need to find the area of each face and then add them together. Since you are finding area, the unit of measurement will still be square units (for example, in^2, ft^2, m^2, and so forth).

Surface Area of Rectangular Prisms		
Surface area =	Area of top	$l \times w$
	Area of bottom	$l \times w$
	Area of right	$w \times h$
	Area of left	$w \times h$
	Area of front	$l \times h$
	+ Area of back	$+\ l \times h$

Notice that because you're working with a rectangular prism, opposite sides will always have exactly the same area.

Area of top = Area of bottom

Area of right = Area of left

Area of front = Area of back

Since a rectangular prism only has three dimensions, you can multiply the areas of the top, front, and right faces by two and add.

The area of top face is $l \times w$.

The area of front face is $w \times h$.

The area of the right face is $l \times h$.

Surface area $= 2(l \times w) + 2(w \times h) + 2(l \times h)$

Now that you have the basic formula, try a few examples.

Example 1 John needs to wrap a gift in a box measuring 10 in. × 10 in. × 20 in. How much wrapping paper does he need?

$$\begin{aligned} \text{Area of front} &= l \times w = 20 \times 10 \\ \text{Area of back} &= l \times w = 20 \times 10 \\ \text{Area of left} &= w \times h = 10 \times 10 \\ \text{Area of right} &= w \times h = 10 \times 10 \\ \text{Area of top} &= l \times h = 20 \times 10 \\ +\ \text{Area of bottom} &= l \times h = 20 \times 10 \end{aligned}$$

Step 1: First, find the area of all the faces of the box. Use the formula to set up the area equation.

Area of front $= 20 \times 10 = 200$ Area of back $= 20 \times 10 = 200$ Area of left $= 10 \times 10 = 100$ Area of right $= 10 \times 10 = 100$ Area of top $= 20 \times 10 = 200$ + Area of bottom $= 20 \times 10 = 200$	**Step 2:** Calculate the areas of all the sides.
$200 + 200 + 100 + 100 + 200 + 200 = 1{,}000$	**Step 3:** Add all the areas together to get the surface area.

Answer: 1,000 in^2 (not counting any overlap)

Example 2 A painter needs to paint some toy blocks that are in the shape of a cube. She has a bucket of paint that will cover 10,000 square centimeters. How many cubes can she paint if a side of each cube measures 9 cm? (First, you must calculate the surface area of a cube, then you can determine how many cubes she can paint.)

Surface Area $= 6(\text{Area of each face})$	**Step 1:** Since each face of the cube has the same area, multiply the area of a face by six to find the surface area.
$SA = 6(s)^2$ $SA = 6(9)^2$ $SA = 6(81)$	**Step 2:** Square the side to find the area of each square.
$SA = 486$	**Step 3:** Multiply to find the surface area of one cube.
$SA = \dfrac{10{,}000 \text{ cm}^2}{486 \text{ cm}^2} = 20.576$ cubes	**Step 4:** Divide the total paint coverage by the surface of a cube. Round down since you want to know how many cubes can be painted completely.

Answer: 20 cubes

Example 3 Juan needs to cover his math textbook with a book cover. The book cover will only cover the front and back of the book and the spine. Find the surface area of the book cover if the book's dimensions are 12 in. × 9 in. × 1 in.

In this case, you only need to calculate part of the surface area, the part consisting of the spine, back, and front.

Area of front $= l \times w$ Area of back $= l \times w$ + Area of spine $= l \times h$	**Step 1:** Use the formula for the area of a rectangle to find the area of each side.
$SA = (12 \times 9) + (12 \times 9) + (12 \times 1)$	**Step 2:** Substitute all the values into the formula.
$SA = (108) + (108) + (12)$	**Step 3:** Multiply to find the area of each side.
$SA = 228$	**Step 4:** Add to find the total surface area covered by the book cover.

Answer: 228 in^2

Problem Set

Solve:

1. Find the surface area of the rectangular prism.

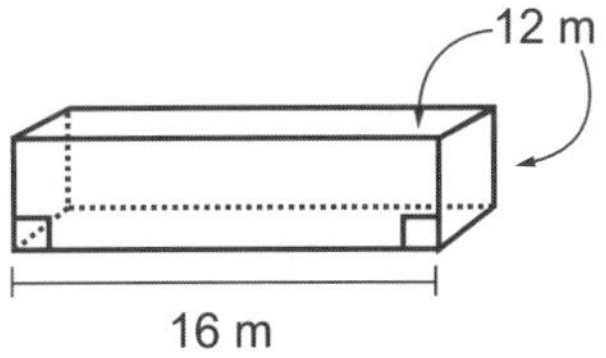

2. Find the surface area of the rectangular prism.

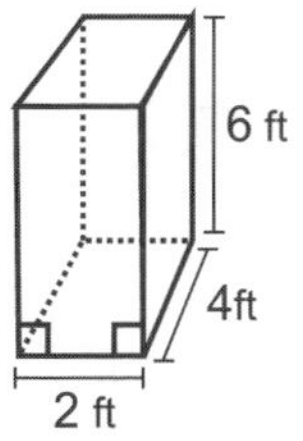

3. Find the surface area of the rectangular prism.

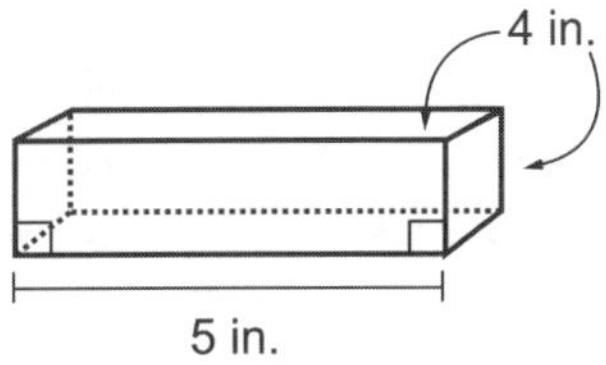

4. Find the surface area of the cube.

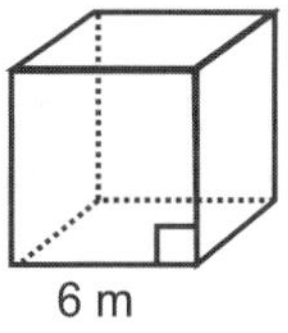

5. Find the surface area of the rectangular prism.

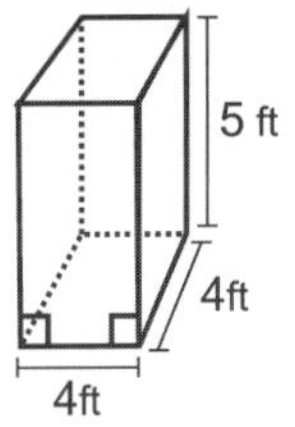

6. Find the surface area of the rectangular prism.

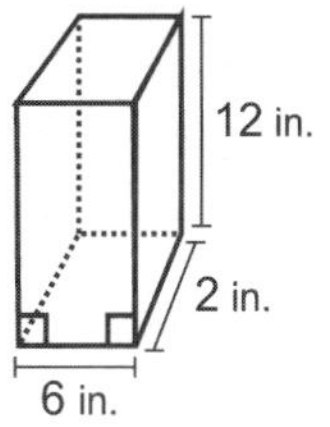

7. Find the surface area of the rectangular prism.

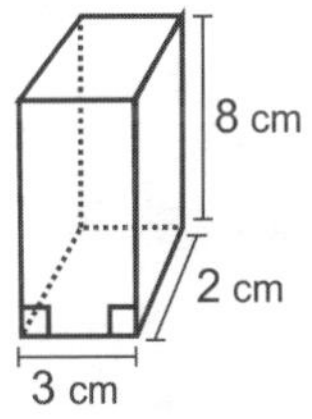

8. Find the surface area of the rectangular prism.

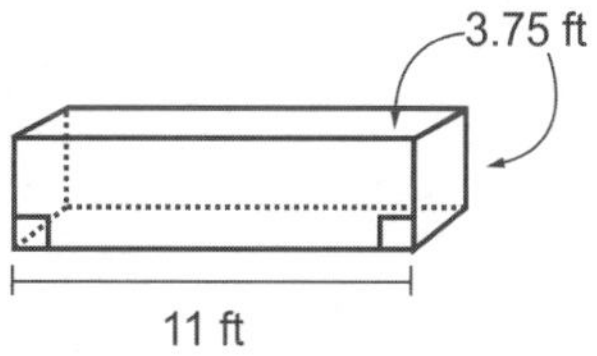

9. Find the surface area of the rectangular prism.

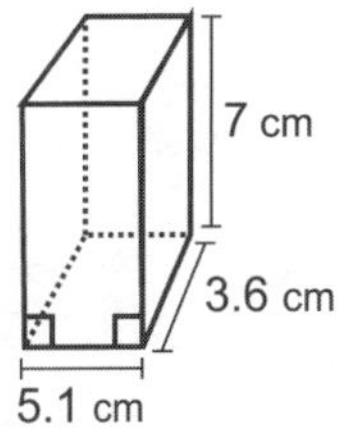

10. Find the surface area of a rectangular prism, given length = 25 cm, width = 19 cm, and height = 13 cm.

11. Find the surface area of the rectangular prism.

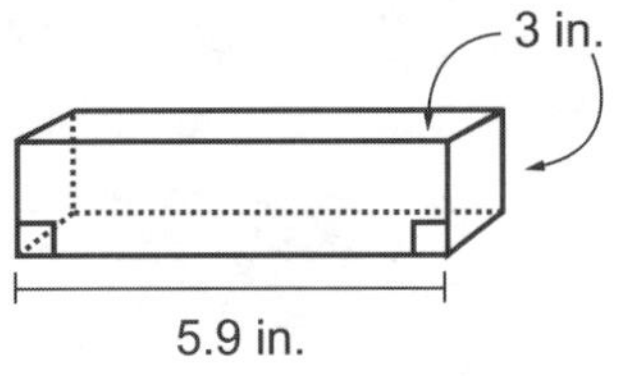

12. Find the surface area of the rectangular prism.

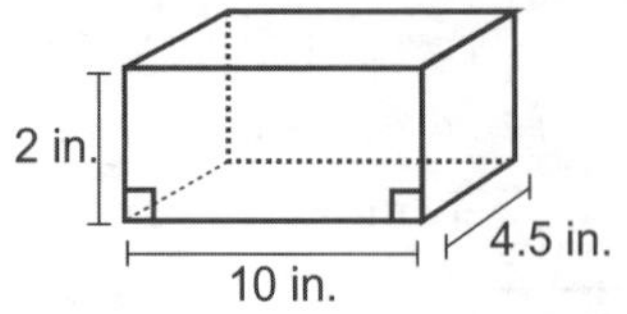

13. Find the surface area of the cube.

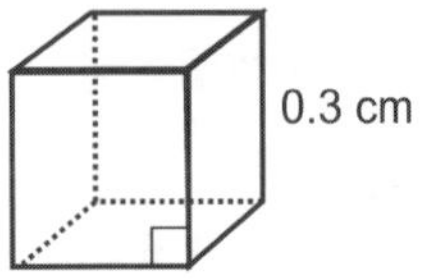

14. Jim wants to wrap a package with the following dimensions: $l = 8$ in., $w = 8$ in., and $h = 10$ in. Determine the surface area of the box.

15. If the surface area of a cube is 36 square inches, find the length of an edge of the cube.

16. A room is 15 feet long, 12 feet wide, and 8 feet high. Find the number of square feet of wallpaper needed to cover the four walls.

17. Jerry wrapped a package that has a surface area of 726 in^2. He wants to trim two edges with a decorative ribbon. If the package is a cube, how much ribbon is needed to trim two edges?

18. Find the surface area of a cube, given that each edge is equal to 25 ft.

19. What is the surface area of a rectangular solid that is 4 inches long, 4 inches wide and 4 inches tall?

20. If the total area of a cube is 54 square inches, find the length of an edge of the cube.

MPA-074: Finding the Surface Area of Cylinders

In a previous lesson, you learned how to find the surface area of a rectangular prism. Now let's look at cylinders. Essentially, the basic process is the same. To find the surface area of any figure, you just add up the areas of all the sides. However, with cylinders you have two bases (the circles) and the side of the cylinder. So, you'll need to figure out how to find the area of the side of the cylinder.

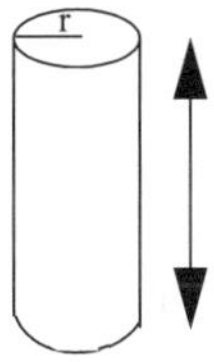

The first definition you must recall is that of surface area:

Surface Area	The sum of the areas of all of the faces of a three-dimensional figure.

The first part of the formula for the surface area of a cylinder is the area of the two bases, the circles. Recall that the formula for the area of a circle is as follows:

Area of a Circle	The area of a circle is equal to pi multiplied by the square of the radius. The formula is $A = \pi r^2$.

The area of each circle is πr^2, and both circles will have the same radius, which is the radius of the cylinder. Now you have the first part of the formula for the surface area of a cylinder: Surface area $= 2\pi r^2 +$ Area of side .

Now, you must determine how to calculate the area of the side of the cylinder. To do this, imagine that the cylinder is actually a tube, like the tube in the center of a roll of paper towels. If you cut the tube straight down from top to bottom and then uncurl it, it will look like a rectangle.

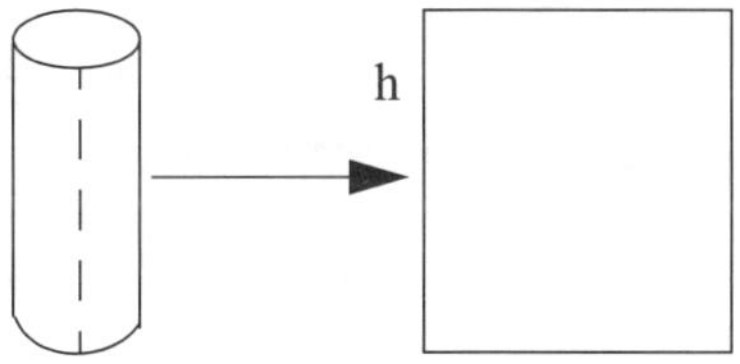

Notice that the height (or width) of the rectangle is the height of the cylinder, since you didn't alter it. You only need to find the length in order to calculate its area. Remember that the length of a circle is simply the distance around the circle at the top (or bottom). This is called the circumference; the formula for circumference is $C = 2r\pi$.

This was all the information you needed to complete the formula for surface area of a cylinder. Now you can calculate the area of the side: Area of side $= C \times h = 2r\pi \times h$ or Area of side $= 2r\pi h$ (since the order of multiplication doesn't really matter).

Now you have a complete formula for the surface area of a cylinder:

Surface Area of a Cylinder	The formula for surface area of a cylinder is: Surface Area $= 2\pi r^2 + 2\pi rh$.

Let's look at some examples.

Example 1 Luc's father carved a hockey puck out of wood. He used a 4 cm long string tacked to a 2 cm thick board to draw a circle. Then he cut along the circle to form the puck. Find the surface area of the homemade puck.

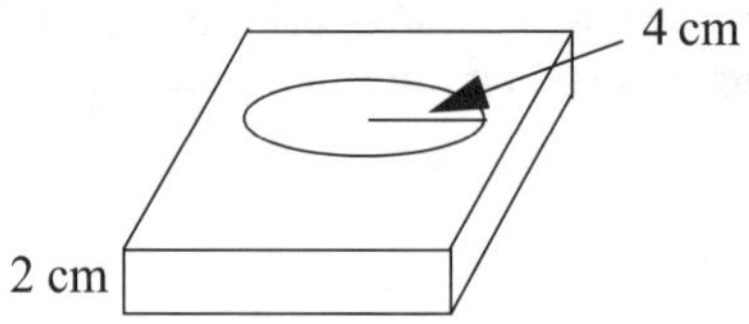

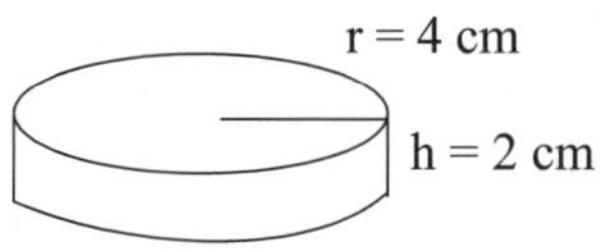

$r = 4$ cm
$h = 2$ cm
$SA = 2\pi r^2 + 2\pi rh$
$SA = 2(3.14)(4)^2 + 2(3.14)(4)(2)$

Step 1: Substitute the known values for radius and height into the formula for the surface area of a cylinder. Use 3.14 to approximate π.

$SA = 2(3.14)(4)^2 + 2(3.14)(4)(2)$
$SA = 2(3.14)(16) + 2(3.14)(4)(2)$

Step 2: Using the order of operations, calculate the exponent first.

$SA = 2(3.14)(16) + 2(3.14)(4)(2)$
$SA = 100.48 + 50.24$

Step 3: Multiply from left to right, and round to the nearest hundredth.

$SA = 100.48 + 50.24$
$SA = 150.72 \text{ cm}^2$

Step 4: Add the two numbers to get the total surface area.
Note: The units for area are always squared.

$SA = 150.72 \text{ cm}^2$

Answer: The surface area measures 150.72 square centimeters.

Example 2 The diameter of a normal can of soda is 2.5 inches and the height is 4.8 inches. Find the surface area of the can.

$d = 2.5$ in.
$h = 4.8$ in.
$SA = 2\pi r^2 + 2\pi rh$

Step 1: Start with the formula for surface area. Before you can work this problem using the formula, you must make sure you have the correct information for the formula.

$d = 2.5$ in.
$h = 4.8$ in.
$r = \frac{1}{2}d = \frac{1}{2}(2.5) = 1.25$ in.

Step 2: The formula requires the height and the radius of the cylinder. However, you only know the height and the diameter. So, you must find the radius using its relationship with the diameter: $d = 2r$ or $r = \frac{1}{2}d$.

$SA = 2\pi r^2 + 2\pi rh$
$SA = 2(3.14)(1.25)^2 + 2(3.14)(1.25)(4.8)$

Step 3: Substitute the known values for radius and height into the formula for the surface area of a cylinder. Use 3.14 to approximate π.

$SA = 2(3.14)(1.25)^2 + 2(3.14)(1.25)(4.8)$
$SA = 2(3.14)(1.5625) + 2(3.14)(1.25)(4.8)$

Step 4: Using the order of operations, calculate exponents, multiply, then add.

$SA = 2(3.14)(1.5625) + 2(3.14)(1.25)(4.8)$
$SA = 9.8125 + 37.68$

Step 5: Multiply from left to right, rounding to the nearest hundredth.

$SA = 9.8125 + 37.68$ $SA = 47.49 \text{ in}^2$	**Step 6:** Finally, add the two numbers to get the total surface area. *Note: Round all decimals to the hundredths' position.*
$SA = 47.49 \text{ in}^2$	**Answer:** The surface area measures 47.49 square inches.

Problem Set

Solve:

1. Find the surface area of a cylinder with a radius of 6 feet and a height of 7 feet. Use $\pi \approx 3.14$.

2. Find the surface area of a cylinder with a radius of 1 meter and a height of 4 meters. Use $\pi \approx 3.14$.

3. Find the surface area of a cylinder with a radius of 3 inches and a height of 12 inches. Use $\pi \approx 3.14$.

4. Find the surface area of a cylinder with a radius of 6 centimeters and a height of 11 centimeters. Use $\pi \approx 3.14$.

5. If a can of green beans has a radius of 1 inch and a height of 5 inches, what is its surface area? Use $\pi \approx 3.14$.

6. If a can of soup has a radius of 1.5 inches and a height of 7 inches, what is its surface area? Use $\pi \approx 3.14$.

7. John has an above-ground pool. If the pool has a radius of 5 ft., and a height of 4 ft., what is its surface area? (Consider the water's surface as the circular "top" of the pool.) Use $\pi \approx 3.14$.

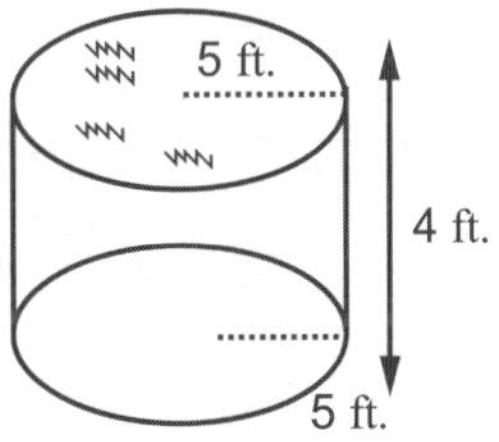

8. Bill works in a paint store. His favorite color of paint is "cherry red". If a can of cherry red paint has a radius of 6 centimeters and a height of 12 centimeters, what is its surface area? Use $\pi \approx 3.14$.

9. Find the surface area of a cylinder with a radius of 4.4 centimeters and a height of 10 centimeters. Round your answer to the nearest hundredth. Use $\pi \approx 3.14$.

10. Find the surface area of a cylinder with a radius of 2.5 meters and a height of 10 meters. Use $\pi \approx 3.14$.

11. Find the surface area of a cylinder with a diameter of 4.6 inches and a height of 9 inches. Round your answer to the nearest hundredth. Use $\pi \approx 3.14$.

12. Find the surface area of a cylinder with a diameter of 60 centimeters and a height of 65 centimeters. Use $\pi \approx 3.14$.

13. Maisie wishes to cover a cylindrical pillow with a striped fabric. To help Maisie estimate the amount of fabric she needs, calculate the surface area of the pillow. Its dimensions are a height of 13 inches and a diameter of 20 inches. Use $\pi \approx 3.14$.

14. If a pole has a diameter of 2 feet and a height of 20 feet, what is its surface area? Use $\pi \approx 3.14$.

15. If a cylindrical carton of ice cream has a diameter of 12 centimeters and a height of 12 centimeters, what is its surface area? Use $\pi \approx 3.14$.

16. What is the surface area of a cylinder with a base circumference of 2π inches and a height of 1 inch? Use $\pi \approx 3.14$. (Round your answer to the nearest tenth of a square inch.)

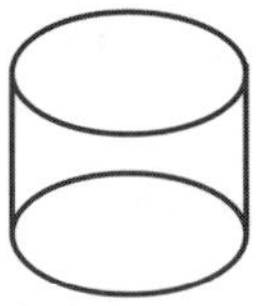

17. What is the surface area of a cylinder with a base circumference of 1π inches and a height of 11.5 inches? Use $\pi \approx 3.14$. (Round your answer to the nearest tenth of a square inch.)

18. What is the surface area of a cylinder with a base circumference of 4π inches and a height of 1 inch? Use $\pi \approx 3.14$. (Round your answer to the nearest tenth of a square inch.)

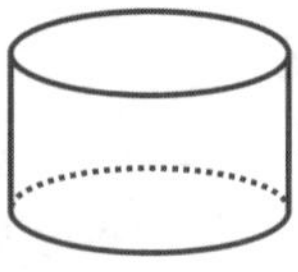

19. What is the surface area of a cylinder with a base circumference of 6π inches and a height of 30 inches? Use $\pi \approx 3.14$. (Round your answer to the nearest tenth of a square inch.)

20. What is the surface area of a cylinder with a base circumference of 5π inches and a height of 1 inch? Use $\pi \approx 3.14$. (Round your answer to the nearest tenth of a square inch.)

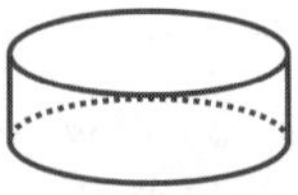

MPA-075: Finding the Volume of Rectangular Prisms

Until now, everything you have learned to calculate has been a one or two-dimensional property of geometric figures. For example, length and circumference are one-dimensional properties, since they use basic units, such as meters or feet. Area and surface area are two-dimensional, since they use square units, such as m^2 or ft^2.

The last basic property of three-dimensional geometric objects is the amount of space they take up. This property will have three-dimensional units, cubic units such as m^3 (cubic meters) or ft^3 (cubic feet). This final property is called the **volume** of the object.

Volume	Number of cubic units it takes to fill a space.

A cubic unit is just as you would imagine it: a cube with every side having a length of one unit.

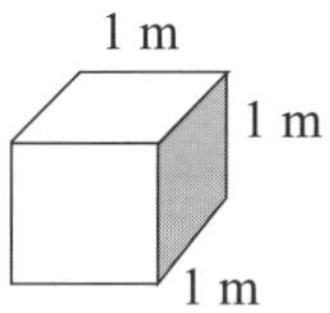

It's not difficult to find the volume of rectangular prisms. Just try to stack the cubes together inside the prism. The first step in finding a formula for the volume of a rectangular prism, or any other three-dimensional object, is to determine how many cubes you can use to completely cover the bottom, or *base,* of the figure.

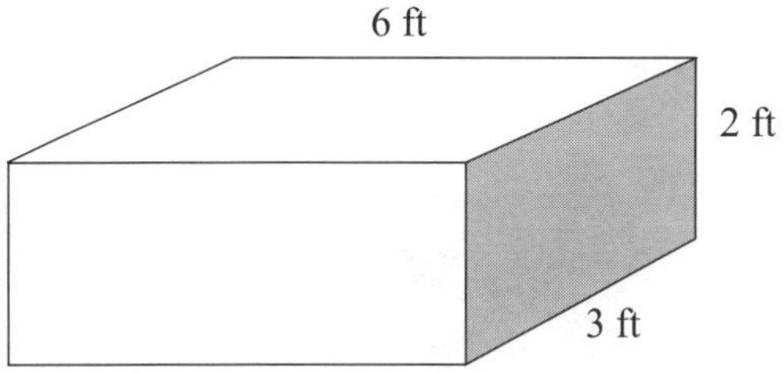

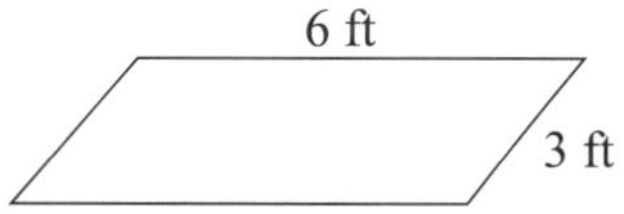

base of the prism

As you can see, you have a rectangular prism with a height of two feet, a width of three feet, and a length of six feet. The base is a rectangle that has dimensions six feet by three feet. So, to cover the base, you can use six cubic feet in a row, like this:

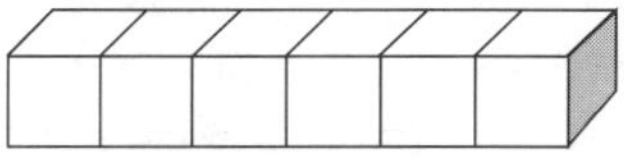

However, you will need three of these rows to completely cover the base, since its dimensions are 6×3 . So, you will need 6×3 cubes to completely cover the base. Notice that 6×3 is the formula for the area of the base (Area of rectangle $= l \times w$). It only makes sense that the base of any figure can be completely covered by a number of cubic units equal to its area.

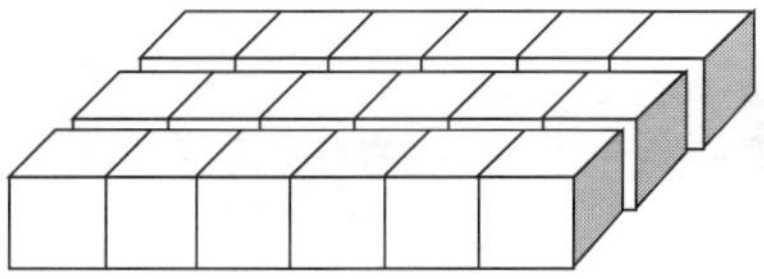

However, you are not finished filling in the entire space yet. You still need to stack other sets of cubes on top of this one to fill the space completely. Since you must go only two units high, you will need one more level of cubes:

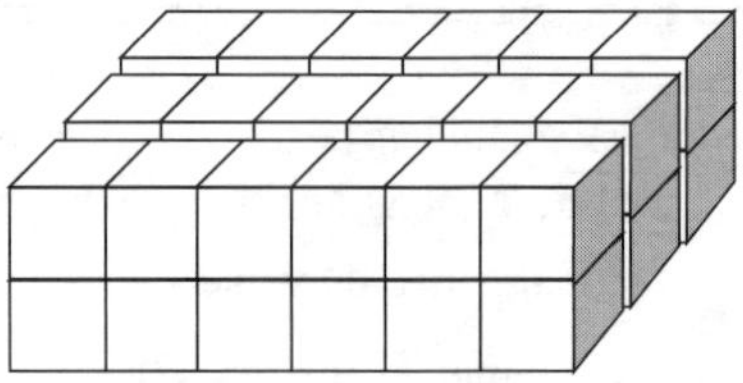

You need two levels of 18 cubic feet, or 36 cubic feet to fill the space completely. To find this, multiply the number of cubes it took to cover the base by the number of levels of cubes you'll need to fill in the space.

Recall now that the number of cubes it took to cover the base is the same number as the area of the base. Then, you multiplied this number by the height of the prism. Again, if the height is not an integer, you don't really have a problem—it's alright to have partial cubes; the formula will still work.

In general, a prism is just a tube formed by using a two-dimensional polygon (in this case a rectangle) as a base. The tube is formed from the base to another base (identical to the first one). The angle of the intersection between a base and the sides of the tube must be 90°. For example, a rectangular prism is just a tube from one rectangle to an identical rectangle.

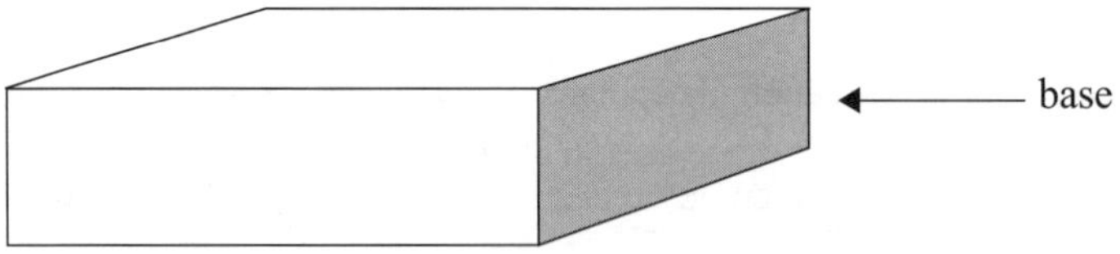

You can form a prism from a regular or irregular shape, like these:

The basic formula for finding the volume of any prism is just to find the area of the base and multiply it by the height of the prism (you will always refer to it as height, even if the prism is laying on its side like the ones shown above).

Volume of a prism	The volume of a prism is the area of the base multiplied by the height of the prism. The formula is $V = \text{Area of base} \times h$.

Or, in the case of rectangular prisms:

Volume of a rectangular prism	$V = (lw) \cdot h$ or $V = lwh$

Volume of a cube	For side s, $V = s^3$

Now that you have an easy formula for the volume of a rectangular prism, apply it to some examples.

Example 1 Julian is an abstract artist who has just finished sculpting his latest masterpiece: a cube standing up on its corner in the middle of a blue circle. The cube is a solid piece of white marble. He knows that one cubic centimeter of marble is 4 grams or 0.004 kilograms, but his cube is 50 centimeters long on a side. How many kilograms of marble did he use?

$$l = 50 \text{ cm}$$
$$weight = 0.004 \text{ kilograms, or } 4 \text{ grams}$$

Step 1: Identify the known values.

$$V = lwh$$
$$V = (50)(50)(50)$$

Step 2: Insert the known values into the formula for volume of a rectangular prism.

$$V = (50)(50)(50) = 125{,}500 \text{ cm}^3$$

Step 3: Multiply from left to right to find the volume.

$$\text{Weight} = \text{Volume of cube} \times \text{weight of } 1 \text{ cm}^3$$
$$\text{Weight} = (125,000)(0.004) = 500 \text{ kg}$$

Step 4: To find the weight, multiply the number of cubic centimeters in the cube by the weight of one cubic centimeter.

Answer: 500 kg

Example 2 You have a set of blocks in the shape of cubes that are all $2 \text{ in.} \times 2 \text{ in.} \times 2 \text{ in.}$, and a box that will hold them exactly. The dimensions of the box are $2 \text{ ft} \times 1 \text{ ft} \times 6 \text{ in.}$ How many cubes do you have?

The basic strategy for solving this problem will be to divide the volume of the box by the volume of a cube, to see how many cubes will fit inside the box. First you must find the volume of a cube and the volume of the box.

$$2 \text{ ft} \times 1 \text{ ft} \times 6 \text{ in.} = 24 \text{ in.} \times 12 \text{ in.} \times 6 \text{ in.}$$

Step 1: Always make sure that all units of measurement are the same. In this case, you must convert all units to inches. The new dimensions of the box:
$24 \text{ in.} \times 12 \text{ in.} \times 6 \text{ in.}$

$$\text{Cube: } l = 2 \text{ in.}, w = 2 \text{ in.}, h = 2 \text{ in.}$$
$$\text{Prism: } l = 24 \text{ in.}, w = 12 \text{ in.}, h = 6 \text{ in.}$$

Step 2: Identify the known values.

$$\text{Volume of cube} = (2)(2)(2)$$
$$\text{Volume of prism} = (24)(6)(12)$$

Step 3: Insert the known values into the formula for the volume of a rectangular prism.

$$\text{Volume of cube} = (2)(2)(2) = 8 \text{ in}^3$$
$$\text{Volume of prism} = (24)(6)(12) = 1728 \text{ in}^3$$

Step 4: Multiply from left to right to find the volumes.

$$\text{Number of cubes} = \frac{\text{Volume of box}}{\text{Volume of cube}} = \frac{1728}{8}$$
$$\frac{1728}{8} = 216$$

Step 5: To find the number of cubes it will take to fill the box, divide the volume of the box by the volume of a cube.

Answer: 216 cubes

Problem Set

Solve:

1. Find the volume of the cube.

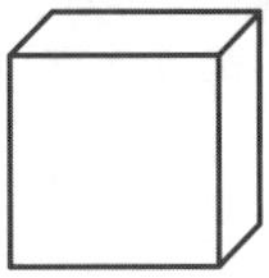

3 dm

Volume = ______ cubic decimeters

2. Find the volume of the cube.

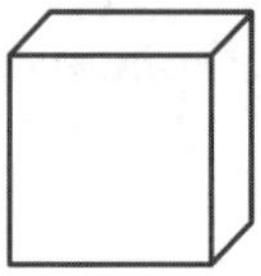

5 in.

Volume = ______ cubic inches

3. Find the volume of the cube.

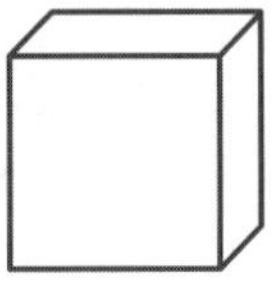

18 cm

Volume = ______ cubic centimeters

4. Find the volume of the cube.

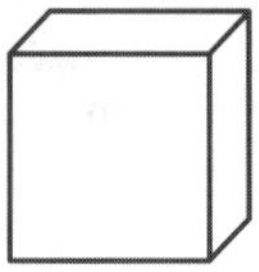

19 cm

Volume = ______ cubic centimeters

5. Find the volume of the rectangular prism.

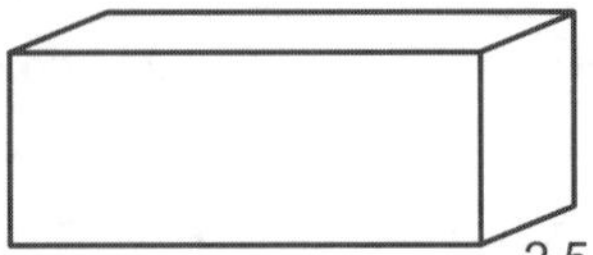

3.8 in.

2.5 in.

4.6 in.

Volume = ______ cubic inches

6. Find the volume of the rectangular prism.

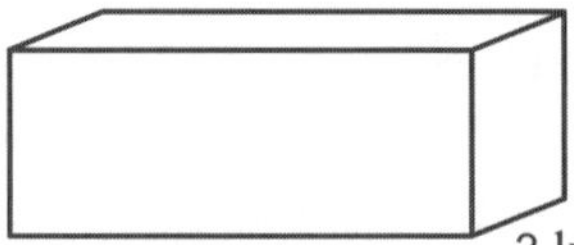

4.3 km

3 km

5.5 km

Volume = ______ cubic kilometers

7. Find the volume of the cube.

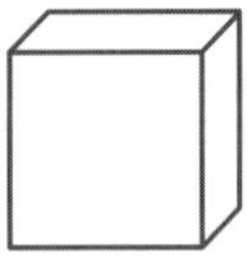

1.4 km

Volume = ______ cubic kilometers

8. Find the volume of the cube.

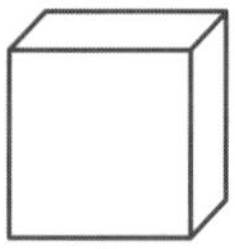

$\frac{1}{4}$ ft

Volume = ______ cubic feet

9. Find the volume of the rectangular prism.

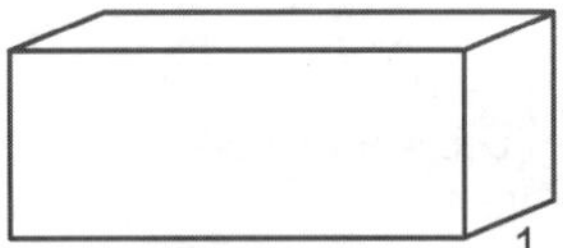

$\frac{10}{11}$ in.

$\frac{2}{5}$ in.

$\frac{1}{8}$ in.

Volume = ______ cubic inches

10.

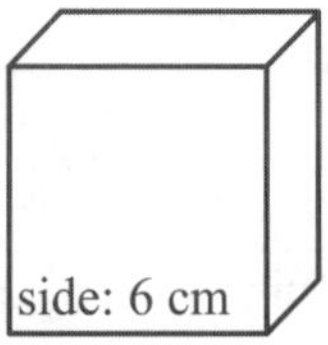

The volume of the cube where each side is 6 centimeters long is _____ cubic centimeters.

11.

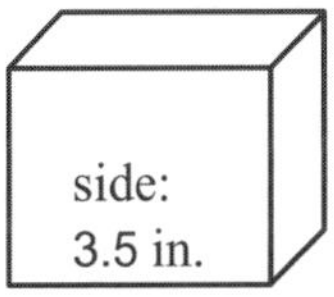

The volume of the cube where each side is 3.5 inches long is _____ cubic inches.

Do not round the answer.

12.

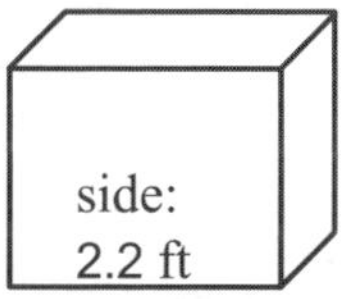

The volume of the cube where each side is 2.2 feet long is _____ cubic feet.

Do not round the answer.

13.

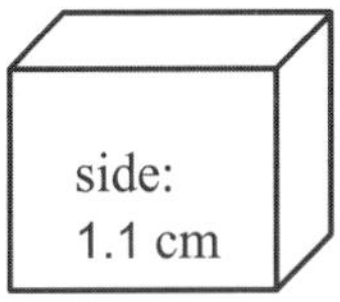

The volume of the cube where each side is 1.1 cm long is _____ cubic centimeters.

Do not round the answer.

14.

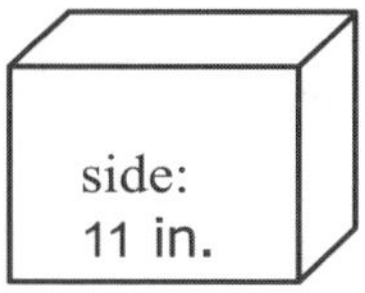

The volume of the cube where each side is 11 inches long is ____ cubic inches.

15.

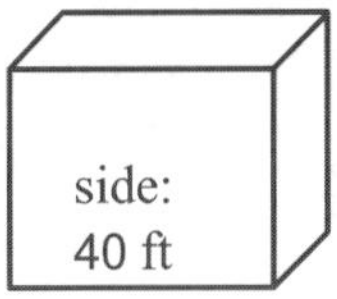

The volume of the cube where each side is 40 feet long is _____ cubic feet.

Do not round the answer.

16. A rectangular prism has a base area of 18 square centimeters and a height of 3 centimeters. It has a volume of _____ cubic centimeters.

17. A rectangular prism has a base area of 84 square centimeters and a height of 9 centimeters. It has a volume of _____ cubic centimeters.

18. A rectangular prism has a base area of 40 square centimeters and a height of 6 centimeters. It has a volume of _____ cubic centimeters.

19. A rectangular prism has a base area of 72 square centimeters and a height of 6.5 centimeters. It has a volume of _____ cubic centimeters.

20. A rectangular prism has a base area of 84 square inches and a height of 8.5 inches. It has a volume of _____ cubic inches.

MPA-076: Finding the Volume of Cylinders

Let's look at how to find the volume of a cylinder. Calculating the volume of cylinders is more difficult than calculating the volume of rectangular prisms. The main reason is that you can't actually cover the base of the figure with cubic units, which is the first step in finding the formula for the volume of rectangular prisms. So, you need to use another method for finding the volume of a cylinder.

First, recall the definition of volume: Volume is the number of cubic units it takes to fill a space.

When you're faced with a new problem in math, usually it helps to look at an easier case of the problem, or to look at something similar to which you know the answer. With this in mind, it might help to look at the formula for finding the volume of prisms, because cylinders are closely related to prisms.

Recall from the lesson on finding the volume of rectangular prisms, the general formula for finding the volume of any prism. The volume of a prism is the area of the base multiplied by the height of the prism. The formula is $V = \text{Area of base} \times h$.

Circles aren't polygons, and cylinders aren't exactly prisms, but if you know the area of the base, then you know how many square units will fit exactly into it. Once you know that, you can use that number of cubic units to cover the base, then multiply that number by the height of the cylinder.

Remember that the area of a circle is given by the formula πr^2. The formula for the volume of a cylinder is similar.

Volume of a cylinder	$V = \text{Area of base} \times h = \pi r^2 \times h$ or $V = \pi r^2 h$

Now let's look at some examples.

Example 1 Find the volume of a circular swimming pool that has a radius of 10 meters, and is 3 meters deep.

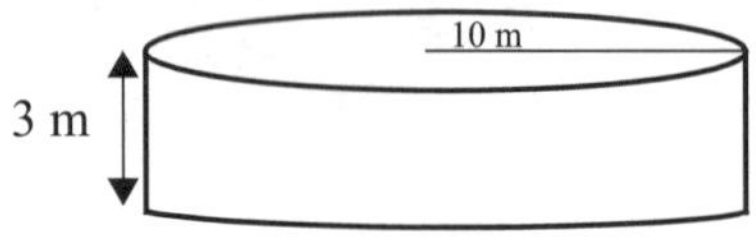

$r = 10$ m $h = 3$ m	**Step 1:** Identify the known variables.
$V = \pi r^2 h$ $V = (3.14)(10)^2(3)$	**Step 2:** Insert the known values into the formula. Use $\pi \approx 3.14$.
$V = (3.14)(10)^2(3)$ $V = (3.14)(100)(3)$	**Step 3:** Using the order of operations, calculate 10^2 first.
$V = (3.14)(100)(3)$ $V = 942 \text{ m}^3$	**Step 4:** Then multiply from left to right to find the answer.
$V = 942 \text{ m}^3$	**Answer:** 942 cubic meters or 942 m^3

Example 2 The diameter of a normal can of soda is 2.5 inches and the height is 4.8 inches. Find the volume of the can.

$d = 2.5$ in., $h = 4.8$ in., $r = 1.25$ in.	**Step 1:** First, you must make sure you have all the information needed for the formula. The formula uses radius and height, but you have diameter and height. You must divide the diameter by two to find the radius of the cylinder. Remember, $r = \frac{1}{2}d$.
$r = 1.25$ in $h = 4.8$ in $V = \pi r^2 h$ $V = (3.14)(1.25)^2(4.8)$	**Step 2:** Insert the known values into the formula. Use $\pi \approx 3.14$.
$V = (3.14)(1.25)^2(4.8)$ $V = (3.14)(1.5625)(4.8)$	**Step 3:** Using the order of operations, calculate the exponent first. Round to the nearest hundredth.
$V = (3.14)(1.5625)(4.8)$ $V = 23.55 \text{ in}^3$	**Step 4:** Multiply from left to right to find the volume.
$V = 23.55 \text{ in}^3$	**Answer:** 23.55 cubic inches or 23.55 in^3

Problem Set

Solve:

1. Find the volume of the cylinder whose radius is 4 feet and height is 3.5 feet. Use $\pi \approx 3.14$.

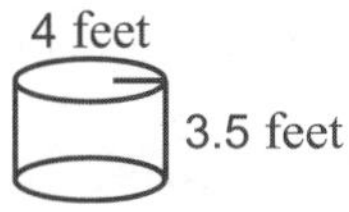

2. A bottle of correction fluid, shaped like a cylinder, has a diameter of 8 centimeters and a height of 6 centimeters. Find the volume of the bottle. Use $\pi \approx 3.14$.

3. A bucket of ice cream has a radius of 10 inches and a height of 9 inches. Find the volume of the bucket. Use $\pi \approx 3.14$.

4. A coffee pot has a radius of 15 centimeters and a height of 8 centimeters. Find the volume of the pot. Use $\pi \approx 3.14$.

5. Gordon's Gas and Go buys gas for $13 per barrel. If the barrel has a radius of 2 feet and a height of 4 feet, what is its volume? Use $\pi \approx 3.14$.

6. Mr. Bill discovered a small reserve of oil on his property. He is selling the oil for $9 per barrel. If the barrel has a radius of 3 feet and a height of 4.5 feet, what is its volume? Use $\pi \approx 3.14$.

7. If a bottle of perfume has a radius of 0.5 inches and a height of 3 inches, what is its volume? Use $\pi \approx 3.14$.

8. Ryan is preparing a pitcher of juice for breakfast. The pitcher, shaped like a cylinder, has a radius of 2.7 inches and a height of 13 inches. What is the volume of the pitcher? Use $\pi \approx 3.14$.

9. Find the volume of the cylinder whose radius is 18 centimeters and whose height is 14 centimeters. Use $\pi \approx 3.14$.

10. Find the volume of the cylinder whose radius is 4.5 feet and whose height is 10 feet. Use $\pi \approx 3.14$.

11. Find the volume. Use $\pi \approx 3.14$.

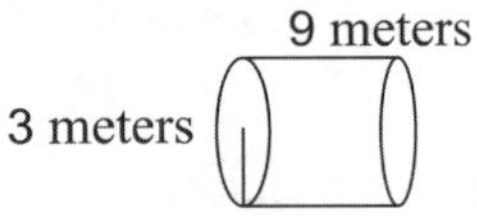

12. Find the volume. Use $\pi \approx 3.14$.

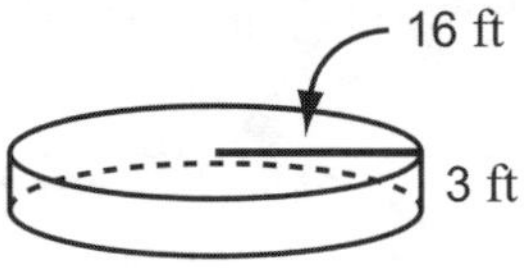

13. Find the volume. Use $\pi \approx 3.14$.

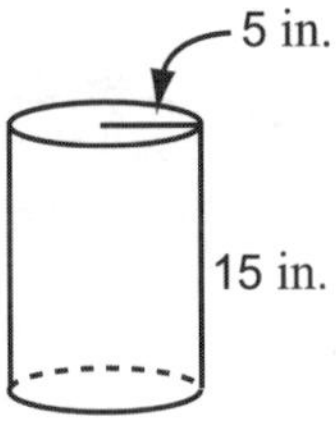

14. Find the volume of the cylinder whose radius is 13 feet and whose height is 8 feet. Use $\pi \approx 3.14$.

15. Find the volume of the cylinder whose radius is 14 meters and whose height is 12 meters. Use $\pi \approx 3.14$.

16. Find the height of a cylinder with a radius of 1.1 feet and a volume of 56 cubic feet.
Use $\pi \approx 3.14$
Round the number to the nearest tenth of a foot.

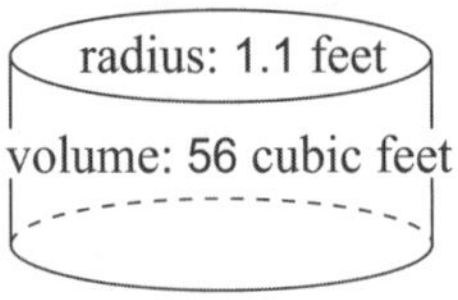

17. Find the height of a cylinder with a radius of 2 feet and a volume of 40 cubic feet.
Use $\pi \approx 3.14$
Round the number to the nearest foot.

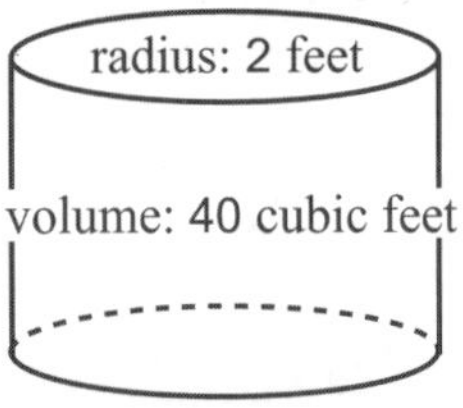

18. Find the height of a cylinder with a radius of 8 feet and a volume of 64 cubic feet.
Use $\pi \approx 3.14$
Round the number to the nearest hundredth of a foot.

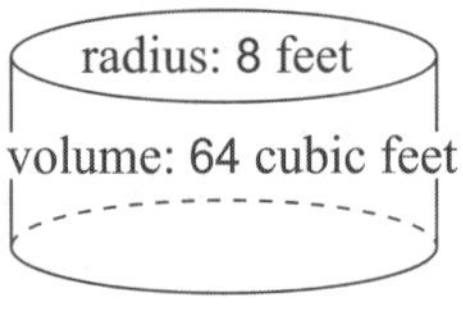

19. Find the height of a cylinder with a radius of 7 feet and a volume of 343 cubic feet.
Use $\pi \approx 3.14$
Round the number to the nearest tenth of a foot.

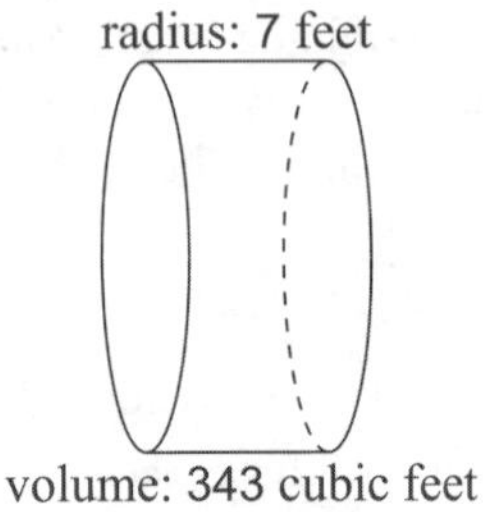

20. Find the height of a cylinder with a radius of 4 feet and a volume of 848 cubic feet.
Use $\pi \approx 3.14$
Round the number to the nearest tenth of a foot.

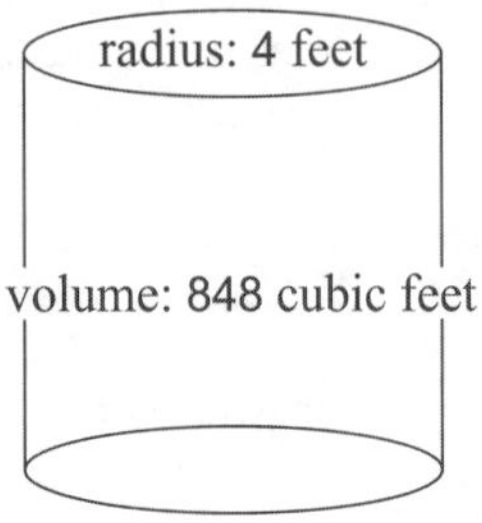

MPA-077: Solving Problems Using a Formula

You've heard of Einstein's famous equation, $E = mc^2$, but do you know how to use it? It may sound like a complex equation that's hard to understand, but it's actually pretty easy to use. It's a **formula**, just like the formulas for the area of a square, or the perimeter of a circle.

Formula	An equation that states a rule for the relationship that exists between certain quantities.

You may use formulas to derive information from pieces of data. For instance, if you know how fast you're going in your car, and if you know how long you have been traveling, then you can figure out how far you have gone. The relationship obeys the distance formula: $d = rt$. You can see that letters represent the pieces of data. $d = rt$ is simply a shorter form of writing *distance* = *rate* × *time*.

The same formula may be written several ways. However, you only need to memorize one form of the formula. Simple arithmetic will give you the other forms of the formula. Here is the distance formula as a demonstration:

Example 1 How would you solve for time instead of distance using the formula $d = rt$?

$\frac{d}{r} = \frac{\cancel{r}t}{\cancel{r}}$ **Step 1:** Divide both sides by r to isolate t.

$\frac{d}{r} = t$ or $t = \frac{d}{r}$ **Step 2:** Simplify. Now t is by itself on one side of the equation.

Answer: $t = \frac{d}{r}$ is an equivalent form of $d = rt$.

See if you can solve for rate. If you got $r = \frac{d}{t}$, you're correct. Check your work by looking at the next example.

Example 2 How would you solve for rate instead of distance using the formula $d = rt$?

$\frac{d}{t} = \frac{r\cancel{t}}{\cancel{t}}$ **Step 1:** Divide both sides by t to isolate r.

$\frac{d}{t} = r$ **Step 2:** Simplify. Now r is by itself on one side of the equation.

Answer: $\frac{d}{t} = r$

The important point about this formula is that you don't need to memorize units. Because distance is divided by time, your unit for rate will be a unit of distance over a unit of time. So, if your distance is given in miles and your time is given in hours, the unit for rate will be miles over hours, or miles per hour (mph). Try the next problem using that formula, and remember to use the proper units.

Example 3 Phil started traveling down the interstate at 2:00 p.m. If, at 5:00 p.m., he had traveled 180 miles, what was his speed?

5:00 – 2:00 = 3 hours **Step 1:** Calculate how long Phil traveled.

$d = 180$ miles
$t = 3$ hours **Step 2:** Determine the information you have and the category to which it belongs. In this problem, you have a distance of 180 miles and a time of 3 hours.

$r = \frac{d}{t}$	**Step 3:** Pick the appropriate formula. If you know distance (*d*) and time (*t*), pick the equation that allows you to solve for rate (*r*).
$r = \frac{180 \text{ miles}}{3 \text{ hours}}$	**Step 4:** Substitute the values for the letters. Remember the units!
$r = 60$ mph	**Step 5:** Simplify. Be sure to use the proper units. Miles over hours is the same as miles per hour, abbreviated mph.
	Answer: 60 mph

Scientists and mathematicians use so many formulas that no one can memorize them all. Here's a reference guide for some formulas:

Formula	Explanation	Name of formula
$d = rt$	distance = rate × time	distance formula
$E = mc^2$	energy = mass × (speed of light)^2	mass-energy conversion formula
$F = \frac{9}{5}C + 32$	degrees Fahrenheit = $\frac{9}{5}$ (degrees Celsius) + 32	Fahrenheit-Celsius conversion formula
$I = PRT$	interest = principal × rate × time Principal is how much money you initially invest.	simple interest formula

Remember that you can make more formulas from these. For instance, $m = \frac{E}{c^2}$ is an equivalent form of $E = mc^2$. But you don't need to alter the formula to solve for *m* or for *c*. As an example, you can work with the Fahrenheit-Celsius conversion formula two ways: (1) Formula manipulation, where you isolate part of the formula, then solve; or, (2) Immediate substitution, where you substitute a given temperature into the formula, then solve.

The following example shows how to convert a temperature using formula manipulation:

Example 4 Convert 68° F to Celsius degrees.

$F = \frac{9}{5}C + 32$	**Step 1:** Write the conversion formula.
$F - 32 = \frac{9}{5}C + 32 - 32$	**Step 2:** Start isolating *C* on one side of the equation. Subtract 32 from both sides.
$\frac{5}{9}(F - 32) = \frac{5}{9}\left(\frac{9}{5}C\right)$	**Step 3:** Multiply both sides by $\frac{5}{9}$, the reciprocal of $\frac{9}{5}$.
$\frac{5}{9}(F - 32) = C$	**Step 4:** Now *C* is by itself.
$C = \frac{5}{9}(68 - 32)$	**Step 5:** Substitute 68 for *F*. Subtract.
$C = \frac{5}{9}(36)$	**Step 6:** Multiply.
$C = 20°$	**Answer:** 68°F = 20°C

The following example shows how to convert a temperature using immediate substitution:

Example 5 Convert 68° F to Celsius degrees.

$F = \frac{9}{5}C + 32$	**Step 1:** Write the conversion formula.
$68 = \frac{9}{5}C + 32$	**Step 2:** Substitute 68 for *F*.
$68 - 32 = \frac{9}{5}C + 32 - 32$	**Step 3:** Start isolating *C* on one side of the equation. Subtract 32 from both sides.
$\frac{5}{9}(36) = \frac{5}{9}\left(\frac{9}{5}C\right)$	**Step 4:** Multiply both sides by $\frac{5}{9}$, the reciprocal of $\frac{9}{5}$. Simplify.
$20° = C$	**Answer:** 68°F = 20°C

You may choose whichever method you prefer. Both are equally effective.

The following examples will use only the formulas given above. Remember that you can use these processes for solving any formula.

Example 6 Alice traveled at 56 mph for 84 miles. How long did the trip take?

$t = \frac{d}{r}$	**Step 1:** Use the distance formula to solve for time.
$t = \frac{84 \text{ miles}}{56 \text{ miles per hour}}$	**Step 2:** Substitute the given values for the letters. Include units.
$t = 1\frac{1}{2}$ hours	**Step 3:** Simplify. Be sure to use the proper units.
	Answer: $1\frac{1}{2}$ hours

Example 7 How much mass energy does a 1-kg telephone book have? (Use the mass-energy conversion formula.) *Note: The speed of light is* 300,000 m/s, *and the unit of energy is one joule* (J), *which is the same as* $1\frac{kg \cdot m^2}{s^2}$.

$E = mc^2$	**Step 1:** Write the mass-energy conversion formula.
$E = (1 \text{ kg})\left(\frac{300,000 \text{ m}}{\text{s}}\right)^2$	**Step 2:** Substitute the given values for the letters.
$E = (1 \text{ kg})\left(\frac{90,000,000,000 \text{ m}^2}{\text{s}^2}\right)$	**Step 3:** Square the speed of light.

$E = \frac{90,000,000,000(\text{kg} \cdot \text{m}^2)}{\text{s}^2}$	**Step 4:** Multiply by 1 kg.
$E = 90{,}000{,}000{,}000$ J	**Answer:** 90,000,000,000 J

Problem Set

Solve:

1. Find the missing value in the equation: $d = r \times t$
$d = 12, t = 4$

2. Find the missing value in the equation: $d = r \times t$
$d = 8, t = 16$

3. Find the missing value in the equation: $d = r \times t$
$d = 12, t = 2$

4. Find the missing value in the equation: $d = r \times t$
$d = 18, t = 9$

5. Find the missing value in the equation:
$F_1 \times D_1 = F_2 \times D_2$

$F_1 = 11$
$D_1 = 4$
$F_2 = ?$
$D_2 = 2$

6. Find the missing value in the equation:
$F_1 \times D_1 = F_2 \times D_2$

$F_1 = ?$
$D_1 = 21$
$F_2 = 7$
$D_2 = 6$

7. Find the missing value in the equation:
$D = 4.9t^2$
$t = 6$

8. Find the missing value in the equation:
$D = 4.9t^2$
$D = 313.6$

9. Identify the formula that is true for the given variables:
$C = 25.12$
$D = 8$

$C = 3.14 \times D^2$ $\quad C = D + 9$

$C = 3.14 \times D$ $\quad C = \frac{1}{2}D^2$

10. Find the height in meters of an airplane if an object dropped from the plane falls for 45 seconds before hitting the ground.

Use the formula $D = \frac{1}{2} \times 9.8 \times t^2$, where t is time in seconds.

11. Mike drove his car for 2 hours at 55 mph. How many miles did he travel?
Use the formula $d = r \times t$.

12. Jeff drove his truck for four hours at a rate of 17 mph. How many miles did he travel?
Use the formula $d = r \times t$.

13. Jose walked twelve miles in four hours. What was Jose's rate in miles per hour?
Use the formula $d = r \times t$.

14. Glenn rode his bike eighteen miles in two hours. What was his rate in miles per hour?
Use the formula $d = r \times t$.

15. Jamie walks at a rate of 3 miles per hour. She traveled a total of 15 miles. For how many hours did she walk?
Use the formula $d = r \times t$.

16. What is the force in newtons of a 28.2 kg object with an acceleration of 41 m/s^2?
The formula is Force = mass × acceleration

17. What is the force in newtons of a 16.8 kg object with an acceleration of 6 m/s^2?
The formula is Force = mass × acceleration

18. A force of 892.9 newtons is created by an object with a mass of 62 kg. What was the acceleration of the object?
The formula is Force = mass × acceleration

19. A force of 252 newtons is created by an object with a mass of 12 kg. What was the acceleration of the object?
The formula is Force = mass × acceleration

20. A force of 1,678.89 newtons is created by an object with an acceleration of 29.3 m/s^2. What was the mass of the object?
The formula is Force = mass × acceleration

MPA-078: Expressing Ratios as Fractions and Determining Equivalency

Ratios are used to make comparisons between two quantities. A ratio can be expressed three ways:

- Using a colon (:), as in 2:1.
- Using the word "to," as in "2 to 1."
- Using a fraction bar, as in $\frac{2}{1}$.

Ratio	A comparison of two quantities using a colon, the word "to," or a fraction bar.

Look at the following example that demonstrates the primary focus of this lesson: ratios expressed as fractions.

Example 1 Suppose that you ride in a carpool to school every Monday through Friday. Last week, you were late for the carpool on Tuesday and Wednesday. What is the ratio of days you were late to the days you were on time? Express the answer as a fraction.

5 days in the week
−2 days late
3 days on time

Step 1: First, remember that there are 5 days in the school week. Out of those 5 days, you were late for 2 of them. Subtract 2 from 5 to determine the number of days that you were on time.

$\frac{2}{3}$

Step 2: The problem asked for the ratio of days tardy to days on time expressed as a fraction. This means that you should set 2 over 3. The ratio is read as "2 to 3."

Answer: $\frac{2}{3}$

In this lesson, you will work with ratios expressed as fractions to determine whether the ratios are *equivalent*. To do so, you will use **cross-multiplication**. If the cross-products of two ratios are equal, then the ratios are equivalent.

Equivalent ratios	Two ratios are considered equivalent if their cross-products are equal.

When working with equivalent ratios, keep in mind that you must maintain *proportionality*. In other words, if one quantity increases, then the other quantity must also increase to keep the ratios equivalent.

Proportionality	Proportionality is when two quantities maintain an equivalent relationship.

The examples below demonstrate the process of determining if two ratios are equivalent.

Example 2 Are $\frac{15}{21}$ and $\frac{5}{7}$ equivalent?

$\frac{15}{21} \times \frac{5}{7}$

$15 \times 7 = 105$

Step 1: To determine whether the two ratios are equivalent, use cross-multiplication. Multiply the first numerator, 15, by the second denominator, 7.

$21 \times 5 = 105$
$105 = 105$

Step 2: Then, multiply the first denominator, 21, by the second numerator, 5. The cross-products are equal.

Answer: Yes. Since both cross-products are equal, you know that the two ratios are equivalent.

Example 3 Are $\frac{16}{19}$ and $\frac{27}{22}$ equivalent?

$\frac{16}{19}$ ⤫ $\frac{27}{22}$

$16 \times 22 = 352$

Step 1: To determine whether the two ratios are equivalent, use cross-multiplication. Multiply the first numerator, 16, by the second denominator, 22.

$19 \times 27 = 513$

Step 2: Then, multiply the first denominator, 19, by the second numerator, 27.

$352 \neq 513$

Step 3: The cross-products are *not* equal, which means that the two ratios cannot be considered equivalent.

Answer: No. Since the cross-products are not equal, the ratios are not equivalent.

Example 4 One vending machine sells 12-ounce sodas for 45 cents. Another vending machine sells 20-ounce sodas for 75 cents. Which is the better deal?

$\frac{12}{45}$ $\frac{20}{75}$

Step 1: To determine equivalency, first set the numbers up as proportions. The first proportion can be read as "12 ounces for 45 cents," and the second proportion can be read as "20 ounces for 75 cents."

$\frac{12}{45}$ ⤫ $\frac{20}{75}$

$12 \times 75 = 900$

$45 \times 20 = 900$

$900 = 900$

Step 2: Now, cross-multiply. Multiply the first numerator, 12, by the second denominator, 75. Then, multiply the first denominator, 45, by the second numerator, 20. The cross-products are equal.

Answer: The cross-products are the same. Therefore, the ratios are equivalent, which means that you are paying the same amount per ounce of soda.

Example 5 At the discount store, you can buy 1 pair of socks for 2 dollars, or you can buy an economy pack of 7 pairs of socks for 12 dollars. Which pack gives you more socks for your money?

$\frac{1}{2}$ $\frac{7}{12}$

Step 1: To determine equivalency, first set the numbers up as proportions. Read the first proportion as "1 pair for 2 dollars," and read the second proportion as "7 pairs for 12 dollars."

$\frac{1}{2}$ ⤫ $\frac{7}{12}$

$2 \times 7 = 14$

$1 \times 12 = 12$

$14 \neq 12$

Step 2: Cross-multiply the ratios to determine whether they are equivalent. Multiply 2 by 7 and 1 by 12. The cross-products are not the same.

Answer: The cross-products are not equivalent. The second product, 12, is less than the first product, 14. Therefore, the second ratio represents the better bargain.

Problem Set

1. What's the ratio of vowels to consonants in "Boston"?

$\frac{1}{2}$ $\frac{1}{6}$

$\frac{4}{3}$ $\frac{3}{3}$

2. What's the ratio of odd to even digits in "789,941"?

$\frac{2}{1}$ $\frac{5}{8}$

$\frac{3}{5}$ $\frac{5}{3}$

3. What's the ratio of odd to even digits in "7,921"?

$\frac{8}{11}$ $\frac{3}{2}$

$\frac{3}{1}$ $\frac{8}{3}$

4. What's the ratio of odd to even digits in "789,113"?

$\frac{2}{3}$ $\frac{5}{1}$

$\frac{3}{9}$ $\frac{3}{2}$

5. Is a ratio of six cats to two dogs equivalent to eighteen cats to four dogs? Answer Yes or No.

6. Is a ratio of one mouse to two traps equivalent to two mice to six traps? Answer Yes or No.

7. 

What is the ratio of ovals to triangles?

$\frac{5}{3}$ $\frac{5}{6}$

$\frac{3}{5}$ $\frac{5}{7}$

8. 

What is the ratio of triangles to squares?

$\frac{6}{7}$ $\frac{3}{7}$

$\frac{5}{6}$ $\frac{5}{7}$

9. 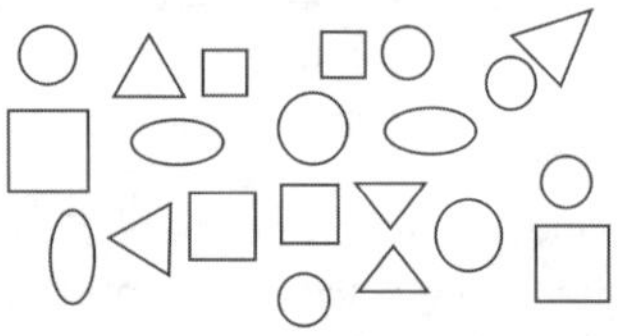

What is the ratio of triangles to ovals?

$\frac{5}{3}$ $\frac{3}{6}$

$\frac{5}{7}$ $\frac{3}{5}$

10. 

What is the ratio of triangles to circles?

$\frac{5}{3}$ $\frac{3}{7}$

$\frac{5}{7}$ $\frac{6}{3}$

11. 

What is the ratio of pens to books?

$\frac{4}{9}$ $\frac{7}{3}$

$\frac{7}{9}$ $\frac{7}{4}$

12. Simplify the ratio: $\frac{10}{2}$

13. Simplify the ratio: $\frac{8}{4}$

14. Simplify the ratio: $\frac{39}{3}$

15. Simplify the ratio: $\frac{40}{42}$

16. Simplify the ratio: $\frac{8}{14}$

17. Tell whether these ratios are equivalent:

$\frac{2}{5}$ and $\frac{100}{250}$

Answer Yes or No.

18. Choose the ratio equivalent to $\frac{39}{3}$.

$\frac{11}{1}$ $\frac{13}{3}$

$\frac{13}{1}$ $\frac{13}{2}$

19. Choose the ratio equivalent to $\frac{42}{7}$.

$\frac{1}{17}$ $\frac{1}{21}$

$\frac{126}{21}$ $\frac{124}{23}$

20. Choose the ratio equivalent to $\frac{2}{11}$.

$\frac{8}{22}$ $\frac{16}{84}$

$\frac{4}{11}$ $\frac{16}{88}$

MPA-079: Unit Rates

Now that you have a little experience with ratios, let's look at a special type of ratio: **rates**. You're already familiar with the concept of rate, mainly dealing with speeds. For example, when driving a car, you measure the speed or rate of the car in miles per hour. You can also use rates to measure other things, such as how fast water is dripping from a faucet. For example, you might have a very leaky faucet that drips three times every five seconds. Rates can also be used to describe prices, such as buying oranges at the grocery store at 3 pounds for $2.

Learning how to calculate rates is an important skill that you will use many times in your everyday life. First, let's define the concept of rate:

Rate	A ratio comparing two measurements using different units of measure.

This definition makes perfect sense if you think about it. Look at the examples given above. In the first example, you are comparing the miles you drive in your car to the number of hours it takes to drive that far. In the second one, you're comparing drips to the amount of time it takes for three drops to fall. In the last example, you're comparing the number of pounds to the number of dollars it takes to buy them.

In each of these examples, you can write the rate as a fraction:

a) $\frac{35 \text{ miles}}{1 \text{ hour}}$ b) $\frac{3 \text{ drips}}{5 \text{ seconds}}$ c) $\frac{3 \text{ pounds}}{2 \text{ dollars}}$

Notice that the first rate is a special type of rate, since the denominator is one unit. We could convert these ratios to single units in the denominators by dividing both the numerator and denominator by the denominator:

a) $\frac{35 \text{ miles}}{1 \text{ hour}}$ b) $\frac{\frac{3}{5} \text{ drips}}{1 \text{ second}}$ c) $\frac{\frac{3}{2} \text{ pounds}}{1 \text{ dollar}}$

In these cases, use decimals to represent the fractions in the numerator. To obtain the decimals in the numerator, divide the numerator by the denominator.

a) $\frac{35 \text{ miles}}{1 \text{ hour}}$ b) $\frac{0.6 \text{ drips}}{1 \text{ second}}$ c) $\frac{1.5 \text{ pounds}}{1 \text{ dollar}}$

These three rates are often said in the same way: 35 miles per hour, 0.6 drips per second, or 1.5 pounds per dollar. These special types of rates are called **unit rates.** Unit rates are especially useful when you want to compare two rates to see which is higher. Most people will use them to compare prices to find the better bargain.

Unit Rate	A rate with a denominator of one.

If you're given a rate and you want to find the unit rate, just divide the numerator by the denominator. It's important to remember to include the units in your answers.

Let's look at some examples.

Example 1 John and his family went on vacation from New Orleans to Carson National Forest in New Mexico. If the trip was 1,850 miles one-way, and it took them 25 hours to get there and 30 hours to get back, what was their speed each way?

$\frac{1,850 \text{ miles}}{25 \text{ hours}}$ $\frac{1,850 \text{ miles}}{30 \text{ hours}}$

Step 1: Set up ratios using the information given.

$\frac{1,850}{25} = \frac{74 \text{ mi}}{1 \text{ hr}}$ $\frac{1,850}{30} = \frac{61.\overline{6} \text{ mi}}{1 \text{ hr}}$

Step 2: To reduce to unit rates, simply divide the numerator by the denominator. Round the decimal to the nearest hundredth.

Note: The line over the 6 indicates that this is a repeating decimal.

Answer: 74 mph going to New Mexico
61.67 mph returning to New Orleans

Now let's consider an example where you must compare two prices using unit rates.

Example 2 Natalie is in the grocery store deciding which can of peas to buy. Brand A peas are on sale at three cans for two dollars, but Brand B peas cost $1.25 for two cans. Which brand is the better bargain?

Brand A Brand B

Step 1: Set up ratios using the information given.

$\frac{\$2.00}{3 \text{ cans}}$ $\frac{\$1.25}{2 \text{ cans}}$

$\frac{2.00}{3}$ $\frac{1.25}{2}$

$\frac{2.00}{3} = 0.6\overline{6}$ $\frac{1.25}{2} = 0.625$

Step 2: To reduce to unit rates, simply divide the numerator by the denominator. Round the decimal to the nearest hundredth.

$0.67 per can $0.63 per can

Answer: Brand B is the better bargain.

Problem Set

Solve:

1. Billy ran in a 32-mile race. He completed the race in 8 hours. What was his average speed?

2.

Caramel Sticks	Chocolate Bars
5.2 grams: 260 calories	40 grams: 208 calories
price: $1.82	price: $1.60

What is the price per gram of chocolate bars?

3.

Clam Chowder Soup
65 grams: 195 calories
price: $1.82

Vegetable Soup
17 grams: 68 calories
price: $2.04

Chicken Soup
75 grams: 135 calories
price: $1.80

What is the price per gram of vegetable soup?

4.

Clam Chowder Soup
65 grams: 195 calories
price: $1.82

Vegetable Soup
65 grams: 195 calories
price: $1.82

Chicken Soup
75 grams: 135 calories
price: $1.80

Calculate the unit rate that expresses the number of calories per gram of vegetable soup.

5. Mary drove 286 miles in 2 hours. What was her average speed?

6.

Crab Delight	Shrimp Surprise
62 grams: 314 calories	47 grams: 287 calories
price: $3.41	price: $2.87

What is the price per gram of Shrimp Surprise? Round to the nearest hundredth.

7. Michael drove 459 miles in 6 hours. What was his average speed?

8. A 16-ounce soda costs $1.00. Express as a unit rate the price per ounce.

9. John's grandmother says that 4 cups of salt are needed to make 36 batches of cookies. Express as a unit rate the amount of salt per batch of cookies. Round to the nearest hundredth.

10. Mike borrowed $160 from Joe and he spent all the money in 16 days. How much money per day on average did he spend?

11. The price of American cheese at the downtown grocery store is $8.20 for 5 pounds. What is the price of the cheese per pound?

12. Find the unit price for a 32-ounce box of cereal that costs $4.16.

13. Teddy complains about his new luxury car because it burns a lot of gasoline. On the last trip, for example, Teddy's car consumed 5 gallons of gasoline for a 45-mile distance. How many miles per gallon on average did the car drive?

14. Express 44 miles in 4 gallons as a unit rate.

15. Twenty-two meters per second is an example of a unit rate. True or false?

16. Twenty-four rotations in 6 minutes is equal to saying 4 rotations per minute. True or false?

17. Greg is considering two jobs. Job 1 pays $200 for 8 hours of work. Job 2 pays $297 for 11 hours of work. If Greg wants to take the job that pays more per hour, which job should he take?

18. The Rockmore pattern requires 18 yards of material for 8 dresses, and the Sageway pattern requires 36 yards of material for 15 dresses. Which pattern uses the least material per dress?

19. Which product is better to buy?

Hardy: 136 grams; price: $4.25
Zesty: 162 grams; price: $4.50

20. Which vehicle gets the best gas mileage?

Luxury car: 300 miles; 12 gallons

Compact truck: 286 miles; 11 gallons

MPA-080: Solving Proportions

The next step in learning how to use ratios is to use them to calculate equivalent ratios. First, you must solve for the unknown ratio using equations and algebra. An equation stating that two ratios are equal is called a **proportion**.

Remember that a **ratio** is a comparison of two quantities, and is often expressed as a fraction.

Proportion	An equation stating that two ratios are equal.

Sometimes you can say that two ratios are proportional to indicate that they are equal, or you can set up an equation stating that they are equal.

An example of a proportion is:

$$\frac{\$1.00}{4 \text{ pieces of gum}} = \frac{\$0.25}{1 \text{ piece of gum}}$$

When setting up a proportion, you usually omit the units of measurement, so that the equation above looks like this:

$$\frac{1.00}{4} = \frac{0.25}{1}$$

Let's look at how to solve a proportion when one of the quantities is unknown.

For example, you know that 5 out of 7 students in your class prefer to have pizza for dinner instead of hamburgers. If there are 49 students in your class, how many prefer pizza?

The proportion would be set up using the ratio of students who prefer pizza to the total number of students:

$$\text{ratio: } \frac{\text{number of students who prefer pizza}}{\text{total number of students}} \qquad \text{proportion: } \frac{5}{7} = \frac{p}{49}$$

Here, we're using the variable p to represent the number of students who prefer pizza.

To solve this equation, use a technique called cross-multiplication, or finding the **cross-product.** The cross-product is simply multiplying the numerator of each ratio by the denominator of the other. If the ratios are proportional (equal) then the products will be the same.

Cross-product	The result of multiplying the numerator of one ratio by the denominator of the other. $\frac{\text{numerator 1}}{\text{denominator 1}} \times \frac{\text{numerator 2}}{\text{denominator 2}}$

When comparing the cross-products of two ratios, you can determine whether or not the ratios form a proportion by making sure that the two products are equal.

For example, in the first proportion, $\frac{1.00}{4} = \frac{0.25}{1}$, the cross-products would be:

$(1.00)(1) = (0.25)(4)$ or $1.00 = 1.00$, so the ratios are proportional.

Let's look at an example of how to determine if two ratios are in proportion.

Example 1 Juan did a survey to determine how many of his classmates prefer tea instead of milk with their lunch. He found that 32 out of the 40 students in his class would rather have tea. Juan then decided to survey three other classes to see what they said. The total number of students he asked was 160, and of that 160, 140 said they would prefer tea. Are these two ratios proportional?

First, set up the ratios, and then set up a proportion between them:

In Juan's class: $\frac{32 \text{ prefer tea}}{40 \text{ total students}}$

In all 4 classes: $\frac{140 \text{ prefer tea}}{160 \text{ total students}}$

$$\frac{32}{40} \stackrel{?}{=} \frac{140}{160}$$

Step 1: Set up a proportion between the two ratios. Use a question mark to indicate that you aren't sure if they're equal.

$$\frac{32}{40} \quad \frac{140}{160}$$

$$(32)(160) \stackrel{?}{=} (140)(40)$$

Step 2: Cross-multiply to determine if the ratios are proportional.

$$\begin{aligned}(32)(160) &\stackrel{?}{=} (140)(40)\\ 5,120 &\stackrel{?}{=} 5,600\end{aligned}$$

Step 3: Multiply from left to right.

$$5,120 \neq 5,600$$

Answer: Since the two cross-products are not equal, the ratios are not proportional.

Let's look at an example of how to solve for an unknown ratio if you know that two ratios are proportional.

Example 2 In a bag of 240 marbles, if you know that 2 out of 3 of them are blue, how many blue marbles are there in the bag?

$$\frac{2 \text{ blue marbles}}{3 \text{ marbles}} = \frac{\text{total \# blue marbles}}{240 \text{ marbles}}$$

Step 1: Set up a proportion between the two ratios. Use the variable b to represent the total number of blue marbles in the bag:

$$\begin{aligned}\frac{2}{3} &= \frac{b}{240}\\ (2)(240) &= (b)(3)\end{aligned}$$

Step 2: Cross-multiply the equation.

$$\begin{aligned}(2)(240) &= (b)(3)\\ 480 &= 3b\end{aligned}$$

Step 3: Multiply from left to right.

Note: b times 3 is the same as 3b.

$$\begin{aligned}480 &= 3b\\ \frac{480}{3} &= \frac{3b}{3}\end{aligned}$$

Step 4: Divide both sides of the equation by 3 to isolate the variable, b, by itself on the right side of the equation.

$$\begin{aligned}\frac{480}{3} &= \frac{\cancel{3}b}{\cancel{3}}\\ 160 &= b\\ b &= 160\end{aligned}$$

Step 5: Dividing gives the answer. There are 160 blue marbles in the bag.

Answer: 160 blue marbles

Here's one last example that uses similar figures. Similar figures have congruent, corresponding angles, and proportional, corresponding sides.

Example 3 You know that the ratios of the sides (width to length) of two similar rectangles are proportional. The measurements of one rectangle are 14 cm (length) by 6 cm (width), and the width of the second rectangle is 9 cm. Find the length of the second rectangle.

$$\frac{6\text{ cm}}{14\text{ cm}} = \frac{9\text{ cm}}{l}$$

Step 1: Set up a proportion between the two ratios. Use the variable *l* to represent the length of the second rectangle.

$$\frac{6}{14} = \frac{9}{l}$$
$$(6)(l) = (9)(14)$$
$$6l = 126$$

Step 2: Cross-multiply to get an equation that you know how to solve. Multiply from left to right.

$$6l = 126$$
$$\frac{6l}{6} = \frac{126}{6}$$

Step 3: Divide both sides of the equation by 6 to get the variable *l* by itself on the left.

$$\frac{\cancel{6}l}{\cancel{6}} = \frac{126}{6}$$
$$l = 21\text{ cm}$$

Step 4: Dividing gives the answer.

Answer: 21 cm

Problem Set

Solve:

1. Determine if the ratios are equal (=) or not equal (≠).

$\frac{5}{40} \bigcirc \frac{9}{72}$

2. Determine if the ratios are equal (=) or not equal (≠).

$\frac{4}{16} \bigcirc \frac{12}{48}$

3. Determine if the ratios are equal (=) or not equal (≠).

$\frac{6}{12} \bigcirc \frac{16}{32}$

4. Determine if the ratios are equal (=) or not equal (≠).

$\frac{20}{50} \bigcirc \frac{42}{84}$

5. Determine if the ratios are equal (=) or not equal (≠).

$\frac{24}{40} \bigcirc \frac{42}{80}$

6. Determine if the ratios are equal (=) or not equal (≠).

$\frac{21}{60} \bigcirc \frac{36}{54}$

7. Determine if the ratios are equal (=) or not equal (≠).

$\frac{42}{60} \bigcirc \frac{26}{34}$

8. Solve the proportion:

$\frac{30}{50} = \frac{15}{n}$

9. Solve the proportion:

$$\frac{16}{64} = \frac{s}{44}$$

10. Use a proportion to determine if the two triangles are similar. Answer Yes or No.

11. Use a proportion to determine if the two triangles are similar. Answer Yes or No.

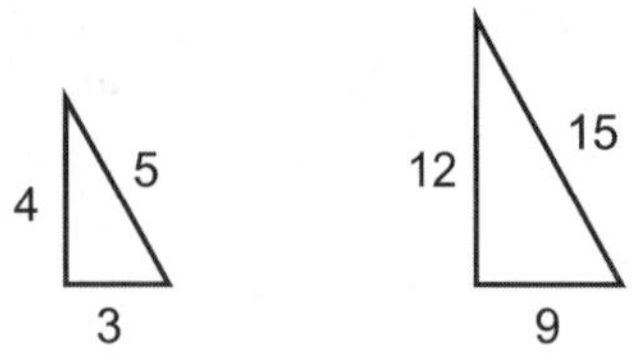

12. Use a proportion to determine if the two rectangles are similar. Answer Yes or No.

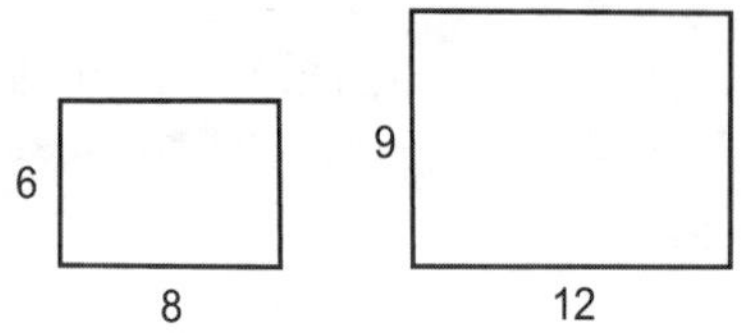

13. The triangles below are similar. Write a proportion and solve to find the missing measure.

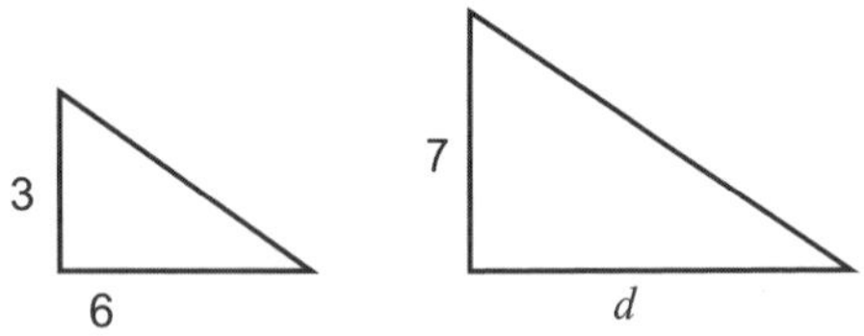

14. The triangles below are similar. Write a proportion and solve to find the missing measure.

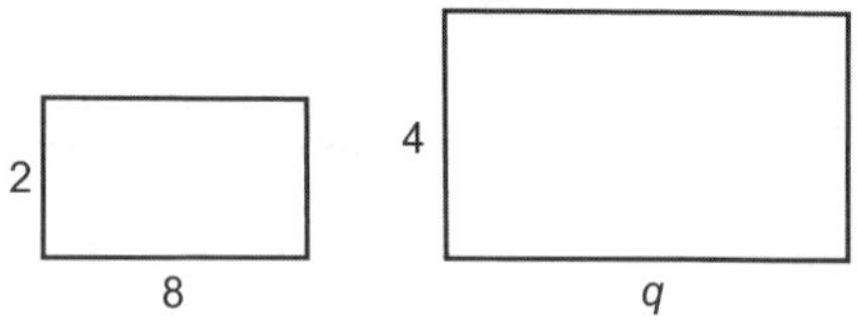

15. Out of every 50 girls in the world, 20 of them will have blonde hair. If there are 10,000 girls, how many will have blonde hair?

16. Out of every 160 fans that the company produces, 6 are defective. If the company produces 6400 fans, how many of them will be defective?

17. Tammy can jog 30 miles in 5 hours. How many hours will it take her to jog 270 miles?

18. Carrie has to shred 63 documents for her lawyer. If she can shred 9 documents every 2 hours, how many hours will it take her to finish this task?

19. Taki has to sharpen 24 knives for her restaurant. If she can sharpen 3 knives every 5 hours, how many hours will it take her to sharpen the knives?

20. Hope is planting a garden. She wants every 3 red flowers to have 7 white flowers to complement them. If she plants a garden with 72 red flowers, how many white flowers will she need?

MPA-081: Converting Fractions, Decimals, and Percents I

You go into a clothing store and your purchases total \$49.99. If you live in a state with no tax on clothing, this will be your total cost. What if you live in a state with a 4% tax on clothing? What will be your total cost then?

What does "percent" mean? This word is used all the time. A percentage of your paycheck goes to the federal government. You go into a store and everything is 50% off. A college will accept a percentage of its applicants. "Percent" actually means "per one hundred" or "for every one hundred." When you pay 4% sales tax, it means that you will pay an extra 4¢ for every \$1.00 (100¢) that you spend.

Sometimes it is useful and necessary to convert a percent to a decimal. For instance, when you calculate the tax for the example above, you will need to change 4% to a decimal. How do you do this?

Convert a percent to a decimal

1. Move the decimal point two places to the left. If there is no decimal point in the number, it is understood to be at the end.
2. Add zeroes to any blank place values.

Example 1 Convert 4% to a decimal.

4. 4. (moved 1 2 ←)	**Step 1:** Move the decimal point two places to the left. If there is no decimal point in the number, it is understood to be at the end. In this case, put a decimal point to the right of the number 4. Now, you will move it two places to the left.
.04	**Step 2:** Add zeroes to any blank place values. Because you have a blank place value, you must fill it with a zero.
	Answer: 4% = .04 or 0.04

As you can see from the example above, 4% sales tax is equal to 4 cents, or 0.04. At other times it might be useful to know how to change a decimal to a percent (the opposite of the example above). For instance, on your receipt, you might notice that your total purchases are multiplied by .08. What percentage is this?

Converting from a decimal to a percent

1. Move the decimal point two places to the right.
2. Add zeroes to any blank place values.

Example 2 Convert .08 to a percent.

.08 .08 (moved 1 2 →)	**Step 1:** Move the decimal point two places to the right.
8.	**Step 2:** Add zeroes to any blank place values. There are no blank place values in this example.
	Answer: 0.08 = 8%

You often hear the claim "nine out of 10 dentists surveyed prefer" a particular type of toothpaste, or "two out of three doctors surveyed prefer" a certain type of aspirin. Suppose you want to compare these claims, to determine which company can make the better claim. Where would you start?

The first thing you must do is write these numbers in fractional form. Nine out of 10 can be written $\frac{9}{10}$. Two out of three can be written $\frac{2}{3}$.

These are difficult to compare, because one of the fractions is *out of 10* and the other is *out of 3*. If you convert these to percents, they will be easier to compare, since both will then be *out of 100* (remember that percent means out of 100). Let's examine how to convert a fraction to a percent.

Convert a fraction to a percent

1. Change the fraction to a decimal by dividing the numerator by the denominator.
2. Round if the decimal does not terminate.
3. Change the decimal to a percent.

Example 3 Using the information above, convert $\frac{9}{10}$ and $\frac{2}{3}$ to percent, and then determine which company can make the better claim.

$10\overline{)9.0}$ with quotient $.9$	**Step 1:** Convert $\frac{9}{10}$ to a percent. Divide the numerator by the denominator.
.9	**Step 2:** Round if the decimal does not terminate. The decimal terminates (ends), so do not round.
$0.9 = 90\%$	**Step 3:** Change the decimal to a percent using the steps above. Change the decimal, 0.9, to a percent by moving the decimal point two places to the right. Add zeroes to the blank place values.
$\frac{9}{10} = 90\%$	**Step 4:** You now have the first part of your answer: 90% of the dentists surveyed prefer a certain type of toothpaste.
$3\overline{)2.000}$ with quotient $.666$	**Step 5:** Now convert $\frac{2}{3}$. Divide the numerator by the denominator.
$0.66\overline{6}$	**Step 6:** Round if the decimal does not terminate. This decimal does not end. Stop after three decimal places and round this number to 0.667.
$0.66\overline{6} = 66.7\%$	**Step 7:** Change the decimal to a percent using the steps above. Change the decimal 0.667 to a percent by moving the decimal point two places to the right.
$\frac{2}{3} = 67\%$	**Step 8:** You now have the second part of your answer: approximately 67% of the doctors surveyed prefer a certain type of aspirin.

Answer: Because both fractions were converted into percents, you can compare them more easily now. The toothpaste company can make the better claim, since 90% is larger than 67%.

Now, suppose that a family budgets 20% of their income for their mortgage payment. What fraction of their income is this? In this case, you need to convert a percent to a fraction. How do you do this?

Convert a percent to a fraction

1. Convert the percent to a decimal. Move the decimal two places to the left.
2. Convert the decimal to a fraction.
3. Simplify the fraction.

Example 4 Convert 20% to a fraction.

20. .20	**Step 1:** Convert the percent to a decimal. Move the decimal two places to the left.
$0.20 = \frac{20}{100}$	**Step 2:** Convert the decimal to a fraction. Now you must convert 0.20 to a fraction. You can read this decimal as "twenty hundredths." As a fraction, this looks like $\frac{20}{100}$.
$\frac{20 \div 20}{100 \div 20} = \frac{1}{5}$	**Step 3:** Simplify. The fraction $\frac{20}{100}$ can be simplified. You can divide both the numerator and denominator by 20.
	Answer: $20\% = \frac{1}{5}$. So the family budgets $\frac{1}{5}$ of their income for the mortgage.

Problem Set

1. Convert 86% to its decimal equivalent.

2. Convert 23% to its decimal equivalent.

3. Convert 77% to its decimal equivalent.

4. Convert 88% to its decimal equivalent.

5. Convert 0.29 to its percent equivalent.

6. Convert 0.17 to its percent equivalent.

7. Convert 0.93 to its percent equivalent.

8. Convert 0.1 to its percent equivalent.

9. Convert 0.6 to its percent equivalent.

10. Convert 0.2 to its percent equivalent.

11. Convert 6% to its decimal equivalent.

12. Convert 2% to its decimal equivalent.

13. Convert $\frac{4}{5}$ to its percent equivalent.

14. Convert 19% to a fraction in simplest form.

15. Convert 31% to a fraction in simplest form.

16. Convert 63% to a fraction in simplest form.

17. Convert $\frac{7}{20}$ to its equivalent percent. If necessary, round to the nearest percent.

18. Convert $\frac{11}{20}$ to its equivalent percent. If necessary, round to the nearest percent.

19. Convert $\frac{9}{25}$ to its equivalent percent. If necessary, round to the nearest percent.

20. Convert $\frac{17}{40}$ to its equivalent percent. If necessary, round to the nearest percent.

MPA-082: Converting Fractions, Decimals, and Percents II

Let's review the rules you learned in a previous lesson about converting numbers.

To convert a percent to a decimal	1. Move the decimal point two places to the left. 2. Add zeroes to any blank place values.
To convert a decimal to a percent	1. Move the decimal point two places to the right. 2. Add zeroes to any blank place values.
To convert a fraction to a percent	1. Divide the numerator by the denominator. 2. Round if the decimal does not terminate. 3. Change the decimal to a percent.
To convert a percent to a fraction	1. Convert the percent to a decimal. 2. Convert the decimal to a fraction. 3. Simplify.

Now you'll apply these same rules to percents greater than 100 and less than 1. Percentages over 100 do not always make sense in every situation. For example, your total budget should not exceed 100% of your income, but you could make 120% of the salary you made last year. What does this mean? If you break it down, 100% of the 120% means you made at least the same salary as last year. Because the percentage is higher than 100%, it means you made more than you did last year, in this case 20% more. To change this percent to a fraction, follow the same rules as before.

Example 1 Convert 120% to a fraction.

1 2 0. $1\underset{\smile}{2}\underset{\smile}{0}.$ 1 2 ← 1 .2 0	**Step 1:** Change the percent to a decimal by moving the decimal point two places to the left. Remember that if you do not see a decimal, it is assumed to be at the end of the number.
$1.20 = 1\frac{20}{100}$	**Step 2:** Convert the decimal to a fraction. The decimal part of this number is 0.20, and this can be read as "twenty hundredths." you can rewrite this decimal number as $1\frac{20}{100}$.
$1\frac{20 \div 20}{100 \div 20} = 1\frac{1}{5}$	**Step 3:** Simplify the fraction if possible. In this case, both the numerator and denominator can be divided by 20.

Answer: $120\% = 1\frac{1}{5}$

Now look at very small percentages, like $\frac{1}{4}$%. This is less than one percent. How do you change this to a decimal?

Example 2 Convert $\frac{1}{4}$% to a decimal.

$$\begin{array}{r} 0.25 \\ 4\overline{)1.00} \\ \underline{-\ 8} \\ 20 \\ \underline{-20} \\ 0 \end{array}$$

Step 1: In this example, you need to turn the fraction into decimal form first. Do this by dividing the numerator by the denominator (not the other way around—remember that division is not commutative). Now you have 0.25%.

.25

$\smile\smile$.25
1 2
←

Step 2: Move the decimal point two places to the left.

.0025

Step 3: Add zeroes to any blank place values.

Answer: 0.0025

Example 3 Convert $2\frac{3}{8}$ to a percent.

$$\begin{array}{r} 0.375 \\ 8\overline{)3.000} \\ \underline{-24} \\ 60 \\ \underline{-56} \\ 40 \\ \underline{-40} \\ 0 \end{array}$$

Step 1: Follow the rules for changing a fraction into a percent. First change the mixed fraction to a decimal number. Divide the numerator by the denominator.

2.375

2.375
1 2
→

237.5

Step 2: You can write $2\frac{3}{8}$ as a decimal, 2.375. Change this to a percent by moving the decimal two places to the right.

Answer: 237.5%

Example 4 Change twenty-three ten thousandths from a decimal to a percent.

0.0023

Step 1: Write the words as a number first.

0.0023
1 2
→

0.23

Step 2: Then follow the rules for changing a decimal into a percent. Move the decimal point two places to the right.

Answer: 0.23%

Example 5 Convert 0.3% to a fraction.

0.003 .3 (decimal point moved 1 2 places left)	**Step 1:** Follow the rules for changing a percent to a fraction. First convert the percent to a decimal. Move the decimal point two places to the left. This produces the number 0.003.
$0.003 = \frac{3}{1000}$	**Step 2:** To change this to a fraction, read the number. This number is read "three thousandths" and can be written as $\frac{3}{1000}$.
	Step 3: Check to see if the fraction can be simplified; in this case, it cannot.
	Answer: $\frac{3}{1000}$

Problem Set

1. Convert 1.20 to a percent.

2. Change 0.0013 to a percent.

3. Convert 5.80 to a percent.

4. Change 225% to a decimal.

5. Express 0.45% as a decimal.

6. Convert 864% to a decimal.

7. Convert to a decimal: 0.1%

8. Convert $\frac{9}{5}$ to a percent. Round answer to the nearest whole percent.

9. Express the following fraction as a percent: $\frac{8}{3}$. Round answer to the nearest whole percent.

10. Change $\frac{7}{5}$ to a percent. Round answer to the nearest whole percent.

11. Convert 125% to a mixed number.

12. Change this percent to a fraction: 0.5%

13. Meredith, a realtor, informs her clients that their home has been rated 185% above the current market value of their neighborhood. Convert this percent to a decimal.

14. Lanagin's Department Store always marks up its merchandise 2.35 times. The store manager of Lanagin's is preparing a report and must express this decimal as a percent. What percent will he report?

15. Lin Toa was extremely upset when she learned that she had paid almost 310% over the value price for her car. Express the percent as a decimal.

16. Camelia's scores for a national, standardized test she took was 0.09 below the national average. Express the decimal as a percentage.

17. Kelger's Department Store had a sales increase of 225% last month. Express this percent as a decimal.

18. Convert $12\frac{2}{5}$% to a decimal. Continue to divide until there is no remainder.

19. Ellen collected angels. She has $2\frac{3}{4}$ times as many angels this year than she had last year. What percentage of angels has Ellen collected since last year?

20. Dr. Bruner realizes that he has a $\frac{1}{335}$ chance of finding a cure for a rare disease he is researching. What is this fraction expressed as a percent? Round your answer to the nearest tenth of a percent.

MPA-083: Finding Number Given Percent and Total

Probably the most useful mathematical tool in finance is the percent. Tax, capital gains, dividends, store sales, interest, mortgage, and many other money matters depend on percents.

Percent	Per 100, of 100, or out of 100.

If a percent means "out of 100," it is identical to division by 100; for example, 70% is "70 out of 100," or "70 divided by 100."

$$70\% = \frac{70}{100} = 0.70$$

An easy shortcut for converting percents to decimals is to move the decimal point in the percent two places to the left. Conversely, to convert a decimal to a percent, move the decimal point two places to the right, and add the percent symbol.

$62\% = 0.62$	$81.2\% = 0.812$	$0.75\% = 0.0075$
$112\% = 1.12$	$317.3\% = 3.173$	$0.42 = 42\%$
$0.004 = 0.4\%$	$1.2 = 120\%$	$1.0072 = 100.72\%$

When you answer questions involving the calculation of a percent, "a percent of" indicates multiplication. Multiply the specified percent by the total to get the answer.

Example 1 What is 20% of 50?

$20\% \times 50 =$	**Step 1:** Rewrite the question as a multiplication problem.
$0.20 \times 50 = 10$	**Step 2:** Convert the percent to a decimal and solve.
	Answer: 10

Example 2 What is 50% of 152?

$50\% \times 152 =$	**Step 1:** Rewrite the question as a multiplication problem.
$0.50 \times 152 = 76$	**Step 2:** Convert the percent to a decimal and solve.
	Answer: 76

Example 3 What is 32% of 96?

$32\% \times 96 =$	**Step 1:** Rewrite the question as a multiplication problem.
$0.32 \times 96 = 30.72$	**Step 2:** Convert the percent to a decimal and solve.
	Answer: 30.72

Example 4 What is 175% of 200?

$175\% \times 200 =$	**Step 1:** Rewrite the question as a multiplication problem.
$1.75 \times 200 = 350$	**Step 2:** Convert the percent to a decimal and solve.
	Answer: 350

Example 5 What is 219.6% of 132.9?

$219.6\% \times 132.9 =$	**Step 1:** Rewrite the question as a multiplication problem.
$2.196 \times 132.9 = 291.8484$	**Step 2:** Convert the percent to a decimal and solve.
	Answer: 291.8484

Problem Set

Solve:

1. There are thirty-six houses in Michelle's neighborhood. 25% are two-story homes. How many are two-story homes?

2. There are 15 street lights on Main Street. 80% are working. How many street lights are working?

3. If 68% of the schools in the city of Silvertown are elementary schools and there are 150 schools total, how many elementary schools are there?

4. 27% of the rides at Zodiac Amusement park are slow-moving rides. If there are 81 rides in all, how many slow rides are there? Round your answer.

5. Tammy spent $19.75 on 15 yo-yos and the tax rate is 9.5%. How much tax did she have to pay? Round your answer to the nearest cent.

6. Find 18% of 854

7. Find 78% of 8,123

8. Find 92% of 750

9. Find 16% of 35,000

10. Find 82% of 500

11. The election returns for the senate race are listed in the chart. 46,000 people voted. How many votes did Mackenzie receive?

Candidate	Votes
Mackenzie	37%
Lucky	32%
Pine	20%
Felton	6%
Cragie	5%

12. The junior football team made 40 touchdowns this season. The chart shows the percentage of touchdowns scored by each player. How many touchdowns did Smith score?

Player	Touchdowns
Mason	20%
Anderson	15%
Jones	10%
Smith	30%
Dunn	25%

13. 1,648 people immigrate to the United States daily. 41% of them will become naturalized American citizens. How many of the daily immigrants to the U.S. will become naturalized? Round your answer to the nearest whole person.

14. Be Beautiful cosmetics is the leading brand of cosmetics in the U.S. Last year it had 20% of the total sales in the industry. The total sales were $26 million. What was Be Beautiful's sales? Round your answer to the nearest tenth.

15. What is 40% of 80?

16. What is 80% of 362?

17. What is 70% of 170?

18. 25% of 69 is what number?

19. 10% of 28 is what number?

20. Japan has approximately 14,000,000 people. Males make up 64% of the population and females make up 36% of the population. Approximately how many females live in Japan?

MPA-084: Finding Percent Given Number and Total

Now that you have learned how to calculate the percent of a number, you should be ready to solve for the percent when you are given just the number and the total. This skill will help you in your everyday life when you are faced with figuring out, for instance, the percent of problems answered correctly on a test.

When given the number and the total, you can find the percent by following several steps:

1. Read the question, and identify key information.
2. Write the question, "What percent of the total is a number?"
3. Substitute the given values into this question.
4. Use this question to write an equation.
5. Solve the equation.
6. Convert the decimal to a percent.
7. Check your work.

Look at the following examples.

Example 1 What percent of 2,500 is 72?

$n \times 2{,}500 = 72$
$2{,}500\,n = 72$

Step 1: Identify the key information. The word "what" will become the variable *n*, and you will multiply *n* by 2,500 because "of" indicates multiplication. You will set *n* times 2,500 equal to 72, as indicated by "is." Use this information to set up an equation. Since multiplication is commutative, you can change "*n* × 2,500" to simply "2,500*n*."

$$\frac{n(2{,}500)}{2{,}500} = \frac{72}{2{,}500}$$
$$n = 0.0288$$

Step 2: Solve for *n* by dividing both sides of the equation by 2,500. The result is 0.0288.

$n = 0.0288 = 2.88\%$

Step 3: Change 0.0288 to a percent by moving the decimal two places to the right and adding a percent sign.

$2{,}500 \times 0.0288 = 72$

Step 4: Check your answer by multiplying 2,500 by 0.0288. The product is 72, which means that your answer is correct.

Answer: 2.88%

Example 2 You see a sweater on sale for $7.50 off of the original price of $25.00. The discount is what percent of the total?

$n \times \$25.00 = \$7.50$
$\$25.00n = \$7.50$

Step 1: Set up an equation by identifying key information. You must answer the question, "What percent of $25.00 is $7.50?" Let *n* represent "what percent." "Of" indicates that you will use multiplication, and "is" indicates that you will set the product of *n* times $25.00 equal to $7.50.

$$\frac{n(\$25.00)}{\$25.00} = \frac{\$7.50}{\$25.00}$$
$$n = 0.3$$

Step 2: Solve the equation for *n* by dividing both sides by $25.00. Thus, *n* equals 0.3.

$n = 0.3 = 30\%$

Step 3: Change 0.3 to a percent by moving the decimal point two places to the right.

$0.3 \times \$25.00 = \$7.50$	**Step 4:** Check your answer by multiplying 0.3 by \$25.00. The product is \$7.50, which means that your answer is correct.
	Answer: 30%

Example 3 65 is what percent of 175? Round your answer to the nearest hundredth.

$65 = n \times 175$ $65 = 175n$	**Step 1:** Examine the wording in the question, and then set up the equation. Set 65 equal to ("is") the variable *n* ("what percent") multiplied by ("of ") 175. Remember that since multiplication is commutative, you can write "*n* x 175" as "175*n*."
$\frac{65}{175} = \frac{\cancel{175}n}{\cancel{175}}$ $n \approx 0.37142$	**Step 2:** Solve the equation for *n* by dividing both sides by 175. Since you will round your percent answer to the nearest hundredth, divide to the hundred thousandths place.
$n \approx 0.37142$ $n \approx 37.14\%$	**Step 3:** Change the result to a percent by moving the decimal point two places to the right. The 4 remains unchanged, because the number following it is a 2.
$0.3714 \times 175 = 64.995$ $64.995 \approx 65$	**Step 4:** Check your answer by multiplying 0.3714 by 175. The product is 64.995, which is approximately 65. Your answer is correct.
	Answer: 37.14%

Example 4 On your last math test, you answered 82 questions correctly out of 125 questions. What percentage of questions did you answer correctly?

$82 = n \times 125$ $82 = 125n$	**Step 1:** Set up the equation. Set 82 equal to the product of *n* times 125.
$\frac{82}{125} = \frac{\cancel{125}n}{\cancel{125}}$ $0.656 = n$	**Step 2:** Solve for *n*. Divide both sides by 125. The result is that *n* equals 0.656.
$n = 0.656$ $n = 65.6\%$	**Step 3:** Change *n* to a percent by moving the decimal two places to the right.
$0.656 \times 125 = 82$	**Step 4:** Check your answer by multiplying 0.656 by 125. The result is 82, which means that your answer is correct.
	Answer: 65.6%

Example 5 Looking back at the same test in which you answered 82 out of 125 questions correctly, what percentage of questions did you answer incorrectly?

$125 - 82 = 43$	**Step 1:** First, subtract the number of correctly answered questions, 82, from the total number of questions, 125, to get the number of incorrectly answered questions, 43.
$43 = n \times 125$ $43 = 125n$	**Step 2:** Set up the equation. Set 43 equal to n times 125.
$\frac{43}{125} = \frac{\cancel{125}n}{\cancel{125}}$ $0.344 = n$	**Step 3:** Solve for n by dividing both sides by 125. The result is 0.344.
$0.344 = n$ $34.4\% = n$	**Step 4:** Change 0.344 to a percent by moving the decimal two places to the right.
$0.344 \times 125 = 43$	**Step 5:** Check your answer by multiplying 0.344 by 125. The product is 43, which means that your answer is correct.

Answer: 34.4%

Problem Set

Solve:

1. What percent of 200 is 24?
2. What percent of 150 is 9?
3. What percent of 100 is 69?
4. What percent of 80 is 8?
5. What percent of 300 is 93?
6. What percent of 614 is 135.08?
7. What percent of 382 is 267.4?
8. What percent of 2,337 is 584.25?
9. What percent of 1,834 is 917?
10. What percent of 2,039 is 856.38?
11. 30 is what percent of 75?
12. 20 is what percent of 100?
13. Hoai had $345. He used $310.50 to buy a new entertainment system. What percent of his money was used to buy the entertainment system?
14. Donny had 90 marbles. He gave his little brother 27 marbles. What percent of Donny's marbles were given to his little brother?
15. Thu has $400. She wants to use $100 on a new jacket. What percent of Thu's money will be used to buy the jacket?
16. The floor in Tuyen's house is made of 3,000 tiles. 900 tiles were used for the restroom. What percent of the tiles were used for the restroom?
17. What percent of 24 is 6?
18. What percent of 8 is 3?
19. What percent of 32 is 24?
20. Andrew had $38. He used $17.10 to buy a video game. What percent of Andrew's money was used to buy the video game?

MPA-085: Finding Total Given Number and Percent

In a previous lesson, you learned how to find an unknown percent when you are given only a number and the total. In this lesson, you will use the same principles to find the total when you are given only the number and the percent. Finding the total involves three steps:

1. Translate the sentence into a numerical equation.
2. Convert the given percent to a decimal.
3. Solve for the unknown variable.

Look at the following examples.

Example 1 48 is 12% of what number?

$48 = 12\% \times t$ $48 = 12\%t$	**Step 1:** First, translate the question into an equation. Set 48 equal to ("is") 12% times ("of") *t* ("what number").
$12\% = 0.12$ $48 = 0.12t$	**Step 2:** Convert 12% to a decimal by moving the decimal two places to the left. 12% becomes 0.12.
$\frac{48}{0.12} = \frac{\cancel{0.12}t}{\cancel{0.12}}$ $400 = t$	**Step 3:** Solve for *t* by dividing both sides of the equation by 0.12.
$48 = 0.12 \times 400$ $48 = 48$	**Step 4:** Check your answer by substituting 400 for *t* in the original equation.
$400 = t$	**Answer:** Thus, *t* equals 400.

Example 2 Suppose that after you've eaten 4 cookies, your mother informs you that you have eaten 25% of the cookies in the jar. How many cookies were in the jar?

$4 = 25\% \times t$ $4 = 25\%t$	**Step 1:** Interpret the important information in the question in order to set up an equation. Set 4 equal to ("is") 25% times ("of") *t* ("cookies in the jar").
$25\% = 0.25$ $4 = 0.25t$	**Step 2:** Convert 25% to a decimal by moving the decimal two places to the left. 25% becomes 0.25.
$\frac{4}{0.25} = \frac{\cancel{0.25}t}{\cancel{0.25}}$ $16 = t$	**Step 3:** Solve the equation by dividing both sides by 0.25.
$4 = 0.25 \times 16$ $4 = 4$	**Step 4:** Check your answer by substituting 16 in place of *t* in the original equation. Your answer is correct.
$t = 16$	**Answer:** 16 cookies

Example 3 189 is 12% of what number?

$189 = 12\% \times t$ $189 = 12\%t$	**Step 1:** Use the information in the question to form an equation. Set 189 equal to ("is") 12% times ("of") t ("what number").
$12\% = 0.12$ $189 = 0.12t$	**Step 2:** Convert 12% to a decimal by moving the decimal point two places to the left. 12% becomes 0.12.
$\frac{189}{0.12} = \frac{\cancel{0.12}t}{\cancel{0.12}}$ $1{,}575 = t$	**Step 3:** Solve the equation by dividing both sides by 0.12.
$189 = 0.12 \times 1{,}575$ $189 = 189$	**Step 4:** Check your answer by substituting 1,575 for t in the original equation.
$t = 1{,}575$	**Answer:** 1,575

Example 4 Antonio would like to buy a new SuperZoom bicycle. He told the clerk that he had \$42, which the clerk informed him is only 21% of the total price. How much does the bicycle cost?

$\$42 = 21\% \times t$ $\$42 = 21\%t$	**Step 1:** Set up the equation. Set \$42 equal to 21% of t, which represents the total price.
$21\% = 0.21$ $\$42 = 0.21t$	**Step 2:** Convert 21% to a decimal by moving the decimal point two places to the left. 21% becomes 0.21
$\frac{\$42}{0.21} = \frac{\cancel{0.21}t}{\cancel{0.21}}$ $\$200 = t$	**Step 3:** Solve the equation by dividing both sides by 0.21.
$\$42 = 0.21 \times \$200$ $\$42 = \$42$	**Step 4:** Check your answer by substituting \$200 for t in the original equation.
$t = \$200$	**Answer:** \$200

Example 5 Your friend Mia has 11 chickens in her backyard. She tells you, "That's nothing! Those 11 chickens are only 25% of all that we have!" How many chickens does Mia's family own?

$11 = 25\% \times t$ $11 = 25\%t$	**Step 1:** Set up the equation. Set 11 equal to 25% times t.
$25\% = 0.25$ $11 = 0.25t$	**Step 2:** Convert 25% to a decimal by moving the decimal point two places to the left.
$\frac{11}{0.25} = \frac{\cancel{0.25}t}{\cancel{0.25}}$ $44 = t$	**Step 3:** Solve the equation by dividing both sides by 0.25.
$11 = 0.25 \times 44$ $11 = 11$	**Step 4:** Check your answer by substituting 44 for t in the original equation.
$t = 44$	**Answer:** 44 chickens

Problem Set

Solve:

1. A package of brand A candy claims that the candy has only 60% of the calories of brand B. Brand A candy has 180 calories. How many calories are in brand B?

2. Auto shop A claims that an oil change will cost 70% of the cost of an oil change at auto shop B. Auto shop A charges $14 for an oil change. What does auto shop B charge?

3. The price of a blue shirt is 50% of the price of a green shirt. The price of a blue shirt is $18. What is the price of the green shirt?

4. A book is discounted 10%. The amount of the discount is $1. What was the original price of the book?

5. A CD is discounted 20%. The amount of the discount is $4. What was the original price of the CD?

6. A computer is discounted 30%. The amount of the discount is $240. What was the original price of the computer?

7. A necklace is discounted 40%. The amount of the discount is $100. What was the original price of the necklace?

8. 63 is 75% of what number?

9. 7 is 35% of what number?

10. 36 is 45% of what number?

11. 22 is 55% of what number?

12. 39 is 65% of what number?

13. 40.2 is 33.5% of what number?

14. 291 is 48.5% of what number?

15. 76.3 is 54.5% of what number?

16. 129 is 64.5% of what number?

17. 7.55 is 75.5% of what number?

18. Karleene's weight is equal to 31.5% of Simon's weight. Karleene weighs 75.6 pounds. How much does Simon weigh?

19. Reg's weight is equal to 64.5% of his father's weight. Reg weighs 77.4 pounds. How much does his dad weigh?

20. Bibi's weight is equal to 25.5% of her mother's weight. Bibi weighs 10.2 kilograms. How much does her mother weigh?

MPA-086: Solving Problems Using Percent

Previous lessons have demonstrated how to solve problems involving percent. This lesson combines all three types of percent problems. However, you may find that only one basic type of percent problem exists. There are simply three variations of this problem.

The basic phrase for a percent problem is, "A percent of a total is a number." To construct the questions about percent, replace exactly *one* of the articles, "a," with the question word, "what." That yields three different questions:

- "A percent of a total is what number?"
- "What percent of a total is a number?"
- "A percent of what total is a number?" Or "A number is a percent of what total?"

Which question you choose to solve a problem depends on the unknown.

Unknown	Question to ask
number	A percent of a total is what number?
percent	What percent of a total is a number?
total	A number is a percent of what total?

This example requires the first type of percent question.

Example 1 Dan has $2,000 invested in the bank at a 5% annual interest rate. How much interest does his investment accrue after one year?

A percent of a total is what number?

Step 1: Choose the question based on the unknown. The only unknown value is the number; therefore, choose the "what number?" question.

5% of $2,000 is what number?

Step 2: Replace "a percent" and "a total" in the question with known values.

$5\% \times \$2,000 = x$

Step 3: Write an equation based on the question.

$5\% = 0.05$

$0.05 \times \$2,000 = x$

Step 4: Convert 5% to a decimal.

$0.05 \times \$2,000 = x$

$\$100 = x$

Step 5: Solve the equation.

Answer: $100

This next example requires the second type of percent question.

Example 2 Tamika paid $30 for a knapsack with a list price of $37.50. What percent of the list price did she pay?

What percent of a total is a number?

Step 1: Choose the question based on the unknown. The only unknown value is the percent, so choose the "what percent?" question.

What percent of $37.50 is $30?

Step 2: Replace "a total" and "a number" in the question with known values.

$p \times \$37.50 = \$30$	**Step 3:** Write an equation based on the question.
$p \times \frac{\$37.50}{\$37.50} = \frac{\$30}{\$37.50}$ $p = 0.8 = 80\%$	**Step 4:** Solve the equation by dividing both sides of the equation by \$37.50, which will isolate the unknown variable, *p*. Convert 0.8 to a percent.
	Answer: 80%

Finally, this example requires the third type of percent question:

Example 3 Jay's cat ate 3.3 ounces, or 60%, of the cat food out of a can. How much food was in the can when it was full?

A number is a percent of what total?	**Step 1:** Choose the question based on the unknown. The only unknown value is the total, so choose the "what total?" question.
3.3 ounces is 60% of what total?	**Step 2:** Replace "a number" and "a percent" in the question with known values.
$3.3 = 60\% \times t$	**Step 3:** Write an equation based on the question.
$60\% = 0.6$ $3.3 = 0.6 \times t$	**Step 4:** Convert 60% to a decimal.
$3.3 = 0.6 \times t$ $\frac{3.3}{0.6} = \frac{0.6 \times t}{0.6}$ $5.5 = t$	**Step 5:** Solve the equation by dividing both sides by 0.6, which will isolate the unknown variable, *t*.
	Answer: 5.5 ounces

Do you see that the three different questions are really variations of the same basic phrase? This means that really only one type of percent problem exists. As long as you remember that "a percent of a total is a number," you'll have no trouble solving any percent problem.

Example 4 If Jamal's \$4,000 investment accrued \$300 in one year, what was the annual interest rate?

What percent of a total is a number?	**Step 1:** Choose the question based on the unknown. The only unknown value is the percent, so choose the question that begins with "what percent?"
What percent of \$4,000 is \$300?	**Step 2:** Replace "a total" and "a number" in the question with the known values.
$p \times \$4{,}000 = \$300$	**Step 3:** Write an equation based on the question.
$\frac{p \times \$4{,}000}{\$4{,}000} = \frac{\$300}{\$4{,}000}$ $p = 0.075$ $p = 7.5\%$	**Step 4:** Solve the equation by dividing both sides of the equation by \$4,000, which will isolate the variable *p*. Convert 0.075 to a percent.
	Answer: 7.5%

Example 5 Anton bought a shirt at a 25%-off sale for $15.00. What is the full price of the shirt?

A number is a percent of what total?	**Step 1:** Choose the question based on the unknown. The only unknown value is the total, so choose the question that begins with "what total."
$15.00 is 75% of what total?	**Step 2:** Replace "a number" and "a percent" in the question with known values. Since $100\% - 25\% = 75\%$, 25% off means 75% of the total.
$\$15.00 = 75\% \times t$	**Step 3:** Write an equation based on the question.
$75\% = 0.75$ $\$15.00 = 0.75 \times t$	**Step 4:** Convert 75% to a decimal.
$\$15.00 = 0.75 \times t$ $\frac{\$15.00}{0.75} = \frac{\cancel{0.75} \times t}{\cancel{0.75}}$ $\$20.00 = t$	**Step 5:** Solve the equation by dividing both sides by $0.75, which will isolate the variable *t*.
	Answer: $20.00

Example 6 After taxes, Belinda was left with 63% of her $1,005,000 lottery winnings. How much money was she left with?

A percent of a total is what number?	**Step 1:** Choose the question based on the unknown. The only unknown value is the number, so choose the "what number?" question.
63% of $1,005,000 is what number?	**Step 2:** Replace "a percent" and "a total" in the question with known values.
$63\% \times \$1{,}005{,}000 = x$	**Step 3:** Write an equation based on the question.
$63\% = 0.63$ $0.63 \times \$1{,}005{,}000 = x$	**Step 4:** Convert 63% to a decimal.
$0.63 \times \$1{,}005{,}000 = x$ $\$633{,}150 = x$	**Step 5:** Solve the equation.
	Answer: $633,150

Example 7 Julian donated 12% of the money in his wallet to a charity collecting money outside a discount mart. If he donated \$3.00, how much was in his wallet before he donated?

A number is a percent of what total?	**Step 1:** Choose the question based on the unknown. The only unknown value is the total, so choose the "what total?" question.
\$3.00 is 12% of what total?	**Step 2:** Replace "a number" and "a percent" in the question with known values.
$\$3.00 = 12\% \times t$	**Step 3:** Write an equation based on the question.
$12\% = 0.12$ $\$3.00 = 0.12 \times t$	**Step 4:** Convert 12% to a decimal.
$\$3.00 = 0.12 \times t$ $\frac{\$3.00}{0.12} = \frac{0.12 \times t}{0.12}$ $\$25.00 = t$	**Step 5:** Solve the equation by dividing both sides of the equation by 0.12, which will isolate the variable t.

Answer: \$25.00

Problem Set

Solve:

1. 60% of 80 is what number?
2. 380% of 30 is what number?
3. What percent of 128 is 24? Round your answer to the nearest tenth of a percent.
4. What percent of 250 is 25?
5. 40% of what number is 32?
6. What number is 150% of 80?
7. What number is 25% of 560?
8. What number is 18% of 450?
9. Jack has 14 video games, which is 200% of the amount that Robbie has. How many video games does Robbie have?
10. Gina has 9 pairs of shoes, which is 180% of the pairs of shoes Pam has. How many pairs of shoes does Pam have?
11. Berry's light bill was \$120, which was 80% of the cost of his brother's bill. How much was his brother's bill?
12. Crystal worked 30 hours this week. That is 150% of the hours that Tiffany worked. How many hours did Tiffany work?
13. Marcy ran 14 miles this week, which was 70% of her goal. What was her goal?
14. Little Joey caught 60% of the balls that his big brother threw to him. If he caught 24 balls, how many did his brother throw?

15. In a bag of 50 candies, 12 are green. What percent of the candies are green?

16. The Moving Truck Company has 48 trucks. Twelve of them are new. What percent of the company's trucks are new?

17. The Mahoneys' spent $250 on groceries for the month. They spent $95 on meat. What percent of the money was spent on meat?

18. The Stingers soccer team spends 14 minutes of each practice on drills. This activity makes up 35% of their total practice time. How long is each practice?

19. Johan spends 25% of each workweek writing for the local newspaper. Last week he spent 12 hours writing for the paper. What is the total amount of hours he worked last week?

20. Pewter, Michigan is a small town that spends about $75 on street signs each year. This amount represents 15% of the town's total yearly budget. How much is the town's total yearly budget?

MPA-087: Finding Percent Increase and Decrease

The world population increased by 1.26% from 1999 to 2000, but it increased by 137.96% from 1950 to 2000. Do these statistics indicate that the world is in danger of overpopulation?

Percent change is a mathematical tool used in statistics and in prediction.

$$\text{percent change} = \frac{\text{change}}{\text{original}} \text{ where change} = \text{new} - \text{original}$$

Percent change will sometimes be positive and sometimes be negative. If the new value is more than the original value, then the change (and therefore the percent change) will be positive. In such cases, percent change is percent increase. If the new value is less than the original value, then the change (and the percent change) will be negative. In those cases, percent change is percent decrease. Be sure to express percent decrease as a positive number.

Here's an example involving percent increase.

Example 1 Linda had $2,500 in her savings account. A year later, she had $2,750 because interest accrued on her savings. Find the percent increase.

change = new – original = $2,750 – $2,500 = $250	**Step 1:** Find the change in value.
$\text{percent change} = \frac{\text{change}}{\text{original}} = \frac{\$250}{\$2{,}500} = 0.1 = 10\%$	**Step 2:** Use the formula for percent change.
	Answer: 10%

Here's an example involving percent decrease.

Example 2 Fernando monitors crime at the police station. If the number of burglaries was 20 last year and 16 this year, find the percent decrease.

$16 - 20 = -4$	**Step 1:** Find the change in value.
$\frac{\text{change}}{\text{original}} = \frac{-4}{20} = -0.2 = -20\%$	**Step 2:** Use the formula for percent change. The result contains a negative sign, which indicates a decrease. Because the problem asks for a decrease, the negative sign is not needed in the solution.
	Answer: The crime decreased 20%.

Now, follow along with these examples. These problems involve calculating percent increase or decrease.

Example 3 Otis participated in the United States Census. He counted 3,625 residents of the Apalain reservation in Utah in 1990, and 4,060 residents in 2000. What was the ten-year percent change?

$4{,}060 - 3{,}625 = 435$	**Step 1:** Find the change in value.
$\frac{\text{change}}{\text{original}} = \frac{435}{3{,}625} = 0.12 = 12\%$	**Step 2:** Use the formula for percent change.
	Answer: The population increased (positive percent change) by 12%.

Example 4 Pablo bought diapers for his baby girl. Last week the diapers cost \$7.75, but today they cost \$5.89. What is the percent change in the price of the diapers?

$\$5.89 - \$7.75 = -\$1.86$

Step 1: Find the change in value.

$\frac{\text{change}}{\text{original}} = \frac{-\$1.86}{\$7.75} = -0.24 = -24\%$

Step 2: Use the formula for percent change. Because the problem asks for the percent change, indicate the word "decrease" by using a negative sign or by writing the word "decrease."

Answer: –24%, or 24% decrease

Example 5 Zeke was 5 feet tall in January, and by October he was 5.4 feet tall. What was the percent change in his height?

$5.4 - 5 = 0.4$

Step 1: Find the change in value.

$\frac{\text{change}}{\text{original}} = \frac{0.4}{5} = 0.08 = 8\%$

Step 2: Use the formula for percent change.

Answer: +8%, or 8% increase

Example 6 Wallace invested \$12,525 into one stock. Unfortunately, his capital at the end of the fiscal year was only \$6,513. What was the percent change of his capital?

$\$6{,}513 - \$12{,}525 = -\$6{,}012$

Step 1: Find the change in value.

$\frac{\text{change}}{\text{original}} = \frac{-\$6{,}012}{\$12{,}525} = -0.48 = -48\%$

Step 2: Use the formula for percent change. Because the problem asks for the percent change, indicate the word "decrease" by using a negative sign or by writing the word "decrease."

Answer: –48%, or 48% decrease

Problem Set

Solve:

1. Find the percent of increase of a jump from 2.9 to 3.5.
2. Find the percent of increase of a rise from 18.23 to 24.81. Round your answer to the nearest whole percent. (Use % in your answer.)
3. Find the percent of increase of an advance from 0.49 to 0.65. Round your answer to the nearest whole percent. (Use % in your answer.)
4. Find the percent of increase of a rise from 6 to 6.25. Round your answer to the nearest whole percent. (Use % in your answer.)
5. Find the percent of increase of a rise from 13,500 to 15,000. Round your answer to the nearest whole percent. (Use % in your answer.)
6. Find the percent of increase of a change from 221 to 331. Round your answer to the nearest whole percent. (Use % in your answer.)

7. When the price of oil rose from $22.50 to $34.00 a barrel, by what percent was the price increased? Round your answer to the nearest whole percent. (Use % in your answer.)

8. The toll for the San Marie Turnpike increased from $2.00 to $3.50. Express this increase as a percent. Round your answer to the nearest whole percent. (Use % in your answer.)

9. The population of Summerville rose from 20,282 in 1995 to 23,615 in 1996. Find the percent of increase. Round your answer to the nearest whole percent. (Use % in your answer.)

10. The latest report on new orders for manufactured goods in August showed an increase from 6 billion to 8 billion. Find the percent of increase. Round your answer to the nearest whole percent. (Use % in your answer.)

11. North-South Air noticed an increase in the number of delayed flights from 235 in June to 284 in July. Find the percent of increase in delayed flights.

12. In 1994 the average American ate 55.7 pounds of beef. This was 2.8 pounds more than the average in 1993. Find the percent of increase. Round your answer to the nearest whole percent. (Use % in your answer.)

13. In the new city budget, the amount of money to be spent on recreation facilities has decreased from 823,000 to 250,000. Find the percent of decrease.

14. The enrollment at Green Elementary School dropped from 1,295 last year to 1,023 this year. Find the percent of decrease in enrollment. Round your answer to the nearest whole percent. (Use % in your answer.)

15. Does a change from 419 to 457 represent an increase or decrease?

16. Does a change from 5.9 to 4.9 represent an increase or decrease?

17. Does a change from 741 to 742 represent an increase or decrease?

18. Does a change from 9.83 to 9.8 represent an increase or decrease?

19. Does a change from 0.74 to 0.91 represent an increase or decrease?

20. Does a change from 7.64 to 7.55 represent an increase or decrease?

MPA-088: Solving Real-World Problems Involving Percent

Did you ever stop to think about percents? You may not realize it, but you use percents all the time in your everyday life. For example, when you shop at the local mall or at the grocery market, you must add a percentage of sales tax to your total. Do you sometimes have trouble figuring out sales tax? It's easy! You can figure out the total price of any purchase, including sales tax, in three steps.

Calculate Sales Tax

1. Convert the percent sales tax to a decimal.
2. Multiply the price by the decimal, rounding if necessary. The product is the dollar amount of the sales tax.
3. Add the product (the sales tax) to the original price. This yields the total price.

The examples below demonstrate how to figure the total price including sales tax.

Example 1 In New Haverbrook, the sales tax is 6.7%. If Morris wants to buy a new lunch box priced at \$6.57, what would the total price be, including sales tax?

$6.7\% = 0.067$

Step 1: First, convert the percent sales tax to a decimal by moving the decimal point two places to the left.

$\$6.57 \times 0.067 = \$0.44019$

$\$0.44019 \approx 0.44$

Step 2: Multiply the price by the decimal. Since money is expressed to the hundredths place, round the product, 0.044019, to the hundredths place. The sales tax is 44 cents.

$\$0.44 + \$6.57 = \$7.01$

Step 3: Add the sales tax to the original price of \$6.57.

Answer: Thus, the total price including tax is \$7.01.

Example 2 Luanne has \$10 and would like to buy both a pair of socks and some batteries for her radio. The socks cost \$3.17, and the batteries cost \$4.99. The sales tax in her town is 8.2%. Does she have enough money to buy both the socks and the batteries? (Remember to consider the sales tax when finding the answer).

$\$3.17 + \$4.99 = \$8.16$

Step 1: First, add the prices of the socks and the batteries together.

$8.2\% = 0.082$

Step 2: Then, change the percent sales tax to a decimal by moving the decimal point two places to the left.

$\$8.16 \times 0.082 = \$0.66912$

$\$0.66912 \approx \$0.67$

Step 3: Multiply the sales tax by the subtotal. Since money is expressed to the hundredths place, round 0.66912 to the hundredths place. Since the 9 in the thousandths place is higher than 5, round the 6 in the hundredths place to 7. The sales tax is 67 cents.

$\$8.16 + \$0.67 = \$8.83$

Step 4: Add 67 cents to the subtotal to get the total price.

Answer: Yes, Luanne has enough money. The total price of the socks and batteries, including sales tax, is \$8.83. The total is less than \$10, which means that Luanne has enough money for the purchase.

Find the Total Price Including Discounts and Sales Tax

If you are a frugal shopper, then you might enjoy buying merchandise that is on sale. You can calculate a sale price by subtracting a percentage from the regular price and adding sales tax. This process may sound complicated, but you can do it using these four steps:

1. Convert the percent to a decimal.
2. Multiply the decimal times the original price, rounding if necessary.
3. Subtract the product from the original price.
4. Add the sales tax, using the steps you already have learned.

The following examples involve calculating the total price, including sales tax and discounts.

Example 3 Paulo has been saving his money for a shirt that costs $35.50. He notices one day that the shirt is on sale at 10% off. The sales tax in his city is 9%. What is the total cost of the shirt?

$10\% = 0.1$	**Step 1:** First, convert the sale percentage to a decimal by moving the decimal point two places to the left.
$\$35.50 \times 0.1 = \$3.55$	**Step 2:** Multiply the shirt price, $35.50, by 0.1.
$\$35.50 - \$3.55 = \$31.95$	**Step 3:** Subtract the product from the original price of the shirt.
$9\% = 0.09$	**Step 4:** Convert the sales tax, 9%, to a decimal by moving the decimal point two places to the left.
$\$31.95 \times 0.09 = \$2.8755$ $\$2.8755 \approx \$2.88$	**Step 5:** Multiply the product, 0.09, by the price of the shirt minus the discount, $31.95. Since money is only expressed to the hundredths place, round $2.8755 to the hundredths place. Since the digit in the thousandths place is 5, round the 7 in the hundredths place to 8.
$\$31.95 + \$2.88 = \$34.83$	**Step 6:** Add the rounded product to the price of the shirt.
	Answer: $34.83

Example 4 Lin-Mai wants to buy a skirt and a jacket. The skirt costs $24.00, and is on sale at 15% off. The jacket costs $67.00, and is on sale at 35% off. In Lin-Mai's town, the sales tax is 7.75%. What is the total cost of her purchase?

$15\% = 0.15$	**Step 1:** First, calculate the sale price of the skirt. Change the sale percentage to a decimal by moving the decimal point two places to the left.
$\$24.00 \times 0.15 = \$3.60$	**Step 2:** Multiply the original cost of the skirt, $24.00, by the decimal 0.15.
$\$24.00 - \$3.60 = \$20.40$	**Step 3:** Subtract the product from the original price to get the sale price.
$35\% = 0.35$	**Step 4:** Change the sale percentage of the jacket to a decimal by moving the decimal point two places to the left.
$\$67.00 \times 0.35 = \$23.45$	**Step 5:** Multiply the price of the jacket by 0.35.
$\$67.00 - \$23.45 = \$43.55$	**Step 6:** Subtract the product from the original price.
$\$20.40 + \$43.55 = \$63.95$	**Step 7:** Now, add the sale price of the skirt ($20.40) to the sale price of the jacket ($43.55) to get a subtotal.

$7.75\% = 0.0775$	**Step 8:** Calculate the amount of sales tax by moving the decimal point two places to the left.
$\$63.95 \times 0.0775 = \$4.956125$ $\$4.956125 \approx \$4.96$	**Step 9:** Multiply the sales tax in decimal form by the subtotal. Since money is only expressed to the hundredths place, round the decimal to the hundredths place. The number in the thousandths place, 6, is greater than 5, so round the 5 in the hundredths place to 6.
$\$63.95 + \$4.96 = \$68.91$	**Step 10:** Add the amount of sales tax to the subtotal.
	Answer: $68.91

Calculate Interest Rate

Do you have a savings account at the bank? If so, then you may be familiar with the concept of *interest*. Interest is money that is added to the *principal*, which is the money you originally deposited.

Principal	The initial amount of money deposited in a bank.
Simple Interest	The amount the bank pays to the borrower on the principal for a stated period of time.

What determines how much interest is paid is the *interest rate*, the percent of money the bank adds to the account.

Interest Rate	The percentage of interest that the bank adds to the principal.

For example, the principal in your account might be $300, and the bank may have offered you an interest rate of 2% per year. This means that for each year that you keep your money in the bank, the bank adds 2% of the principal to your account.

To calculate the amount of interest earned, you can use one simple formula:

$$\text{Simple Interest} = \text{Principal} \times \text{Interest Rate} \times \text{Time}$$

When you use this formula, represent all time spans less than one year as a decimal. Also, remember that if you are given the interest amount but not the principal, interest rate, or amount of time, then you can use the formula to solve for any of the unknowns.

The examples below illustrate the concepts of interest, principal, and interest rate.

Example 5 One year ago, Juan-Carlos deposited $175.00 into a savings account that has an annual interest rate of 3.4%. How much interest has accrued thus far?

$m = \$175.00 \times 3.4\% \times 1$	**Step 1:** First, set up an equation using the formula for calculating interest. Let m represent the interest.
$3.4\% = 0.034$	**Step 2:** Convert the interest rate to a decimal by moving the decimal point two places to the left.
$m = \$175.00 \times 0.034 \times 1$ $m = \$5.95$	**Step 3:** Solve the equation for m by multiplying.
	Answer: $5.95

Example 6 Bonnie deposited \$4,000 into a savings account. Her bank statement after one year indicates that her account has accrued \$250.44 in interest. What is the annual interest rate of the account?

$$\$250.44 = \$4,000 \times m \times 1$$
$$\$250.44 = \$4,000\,m$$

Step 1: Set up the equation using the formula for calculating interest. Let *m* represent the unknown interest rate.

$$\$250.44 = \$4,000\,m$$
$$\frac{\$250.44}{\$4,000} = \frac{\$4,\cancel{000}\,m}{\cancel{\$4},000}$$

Step 2: Solve for *m* by isolating the variable.

$$m = 0.06261 = 6.261\%$$

Step 3: Convert the decimal to a percent by moving the decimal point two places to the right.

Answer: The interest rate on Bonnie's account is 6.261%.

Problem Set

Solve:

1. Find the simple interest on the unchanged balance of \$2,122.00 at 7.5% for a quarter of a year. Round all answers to the nearest cent.
2. Find the simple interest on the unchanged balance of \$614.13 at 9.5% for three quarters of a year. Round all answers to the nearest cent.
3. Add 6.5% sales tax to \$6.99. Round all answers to the nearest cent.
4. Add 9.75% sales tax to \$99.95. Round all answers to the nearest cent.
5. Find the discounted price of an item costing \$162.00 if there is a 60% discount. Round all answers to the nearest cent.
6. Find the discounted price of an item costing \$242.00 if there is a 75% discount. Round all answers to the nearest cent.
7. Ann wants to purchase a mahogany canopy bed at the antique store. Its price is \$692.25. Adding the 7% sales tax, how much will it cost? Round all answers to the nearest cent.
8. Jamie's favorite brand of jeans are on sale for 30% off. If the regular price is \$28.99, what is the sale price? Round all answers to the nearest cent.
9. Santana wants to buy a purse that regularly costs \$76.75 but is on sale for 10% off. What is the sale price? Round all answers to the nearest cent.
10. Dewitt wants to buy a stereo, but he decides to wait for it to go on sale. The regular retail price is \$93.88. It has just been discounted 15%. How much will he have saved by waiting? Round all answers to the nearest cent.
11. Shantel has been wanting a computer. A particular floor model which retails for \$1,150.00 is priced at 60% off. Shantel has \$400.00. How much more will she need? Round all answers to the nearest cent.
12. Maurie wants to buy his mother a CD of her favorite music. The regular price is \$16.50, but it has just been discounted 55%. How much will it cost him? Round all answers to the nearest cent.

13. Raoul wants to buy a pair of sneakers that cost $61.92. He is pleased when the store puts them on sale at 25% off. How much will he save? Round all answers to the nearest cent.

14. Chris wants to buy a computer but doesn't want to spend too much money. He found a floor model that was discounted 75%. The original retail price was $1,012.50. How much will Chris save? Round all answers to the nearest cent.

15. Samantha bought her prom dress for 15% off the retail price of $89.86. Add 6% sales tax to find the total price. Round all answers to the nearest cent.

16. Alexis bought a high-powered telescope at a 35% discount. The regular price was $309.15. Sales tax was 8.75%. What was her total cost? Round all answers to the nearest cent.

17. Herbie bought a computer for $1,002.99 plus 8.75% sales tax. His mother paid for 60% of the total. How much did Herbie pay? Round all answers to the nearest cent.

18. Kyle bought a belt for 35% off the regular price of $19.15. He paid 6.5% sales tax. How much change did he get from his $20.00 bill? Round all answers to the nearest cent.

19. Terez bought an ankle bracelet for 55% off the regular price of $9.98. Sales tax was 9.75%. How much change did she get from her $10.00 bill? Round all answers to the nearest cent.

20. A new pair of jeans costs $26.00. The sales tax for the jeans is 9%. What will the price of the jeans be with tax? Round all answers to the nearest cent.

21. A box of popsicles costs $4.57 at the grocery store. The sales tax is 3.2%. What is the price of the popsicles with tax? Round your answers to the nearest cent.

22. A hamburger costs $5.60. The sales tax on the hamburger is 4%. What is the total price of the hamburger with sales tax? Round your answers to the nearest cent.

MPA-089: Using Tree Diagrams

Have you ever heard of a fractal? With the right equation, you can make a picture of an infinitely repeating pattern. A good example of a fractal is a tree. A trunk splits into branches, which split into stems. If the pattern continued infinitely, the tree itself would be a fractal.

A **tree diagram** exhibits the same fractal-like nature of a tree. Each of a tree diagram's branches split into similar "stems." Use a tree diagram to find out how many outcomes or possibilities exist for a situation.

Tree Diagram	A diagram used to determine the number of outcomes or possibilities that exist for a multistage experiment.

To create a tree diagram:

- Write the first items and draw branches from them.
- Write the second items at the end of the first branches and draw branches from them.
- Continue writing the choices and branches until you have written all the choices.

The number of possibilities or outcomes for the situation is the same as the number of terminating branches. Collectively, the outcomes are the **sample space**.

Sample Space	The set of all possible outcomes of an experiment.

However, you don't always need to create a tree diagram to find the number of outcomes. The **counting principle** states that the number of outcomes is the product of the number of choices for each item. If the first item has three choices, the second item has four choices and the third item has two choices, the number of outcomes is $3 \times 4 \times 2$, or 24.

Counting Principal	The total number of outcomes for a multistage experiment is the product of the number of choices at each stage.

The best way to learn how to construct a tree diagram is by example.

Example 1 Manuel wants to buy a television. The screen may be 12, 14, or 15 inches. The TV may come with or without a remote control. And, the TV may come with or without a built-in VCR. What is the number of possible TV combinations Manuel can buy?

Step 1: Write the possibilities for the first choice: screen size.

12"

14"

15"

Step 2: Draw branches and write the possibilities for the second choice: remote control inclusion.

12" — remote / no remote

14" — remote / no remote

15" — remote / no remote

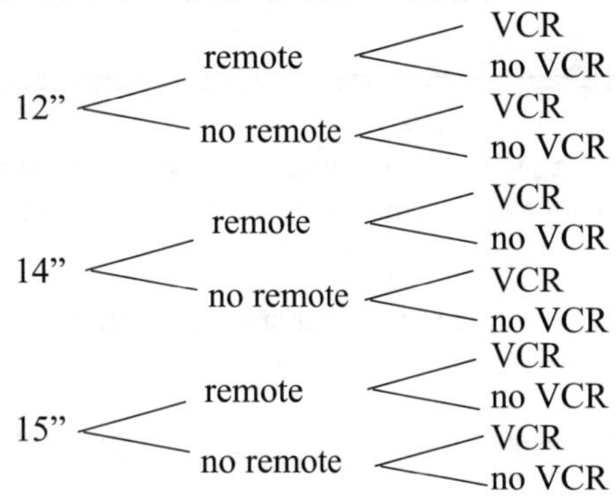

Step 3: Draw branches and write the possibilities for the third choice: VCR inclusion.

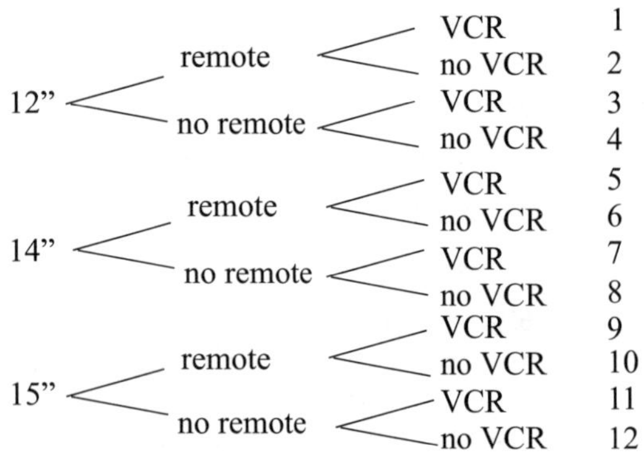

Step 4: Count the terminating branches. The number of terminating branches, 12, is the number of possible TV combinations.

Answer: 12

Now answer the same question using the counting principle.

Example 2 Manuel wants to buy a television. The screen may be 12, 14, or 15 inches. The TV may have a remote control. The TV may also have a built-in VCR. What is the number of possible TV combinations Manuel can buy?

screen size: 3 choices
remote control: 2 choices
VCR: 2 choices

Step 1: Write the number of choices for each feature of the TV.

$3 \times 2 \times 2 = 12$

Step 2: Multiply the numbers. Their product, 12, is the number of possible TV combinations.

Answer: 12

The tree diagram and the counting principle give the same, correct answer. Both methods are equally valid to determine the possibilities or outcomes for a situation. Follow these examples using tree diagrams.

Example 3 Henri wants to see *Lizards from Mars* or *The Proton Strikes Again* at either 5:15 p.m. or 7:45 p.m. The only theaters playing the shows at those times are the Superplex 20 or the Superduperplex 45. How many choices does Henri have?

Lizards

Proton

Step 1: Write the possibilities for the first choice: movie.

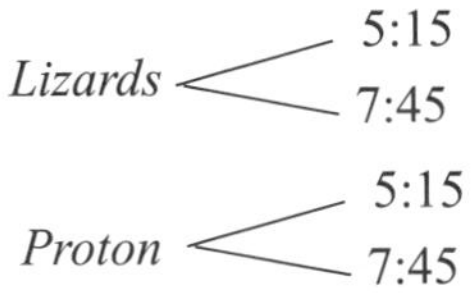

Step 2: Draw branches and write the possibilities for the second choice: time.

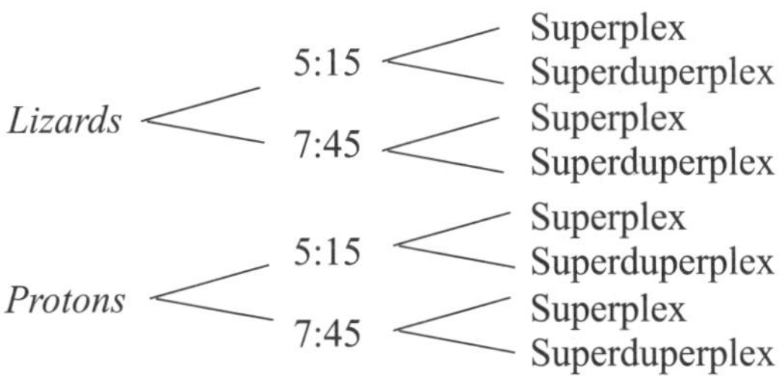

Step 3: Draw branches and write the possibilities for the third choice: theater.

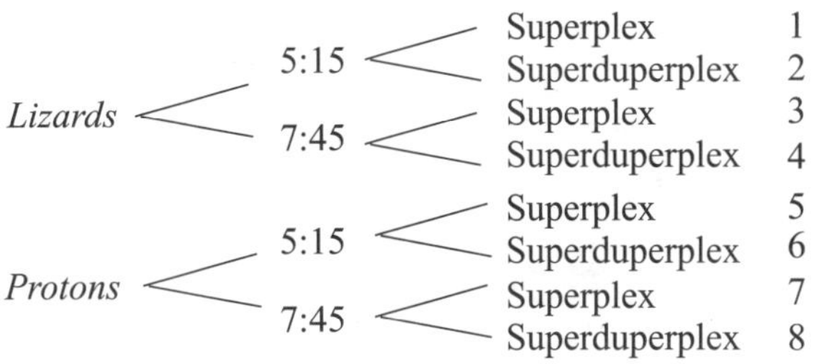

Step 4: Count the terminating branches. The number of terminating branches, 8, is the number of possible movie combinations.

Answer: 8

Example 4 Lauren wants a pet. She can buy a cat, a dog, or a horse. The pet can be white, black, or brown. Finally, the pet can be male or female. How many choices does she have?

Cat

Dog

Horse

Step 1: Write the possibilities for the first choice: pet.

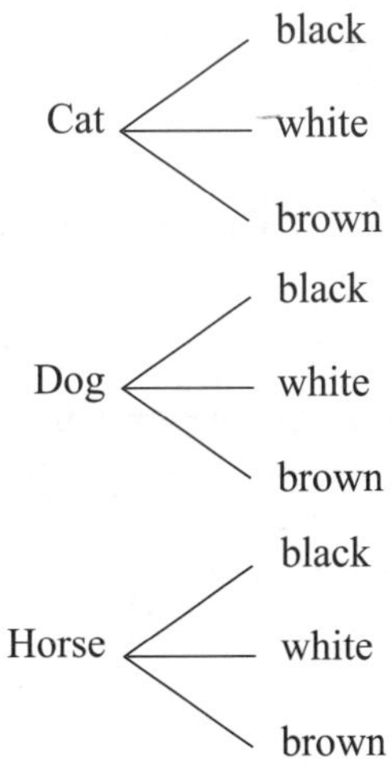

Step 2: Draw branches and write the possibilities for the second choice: color.

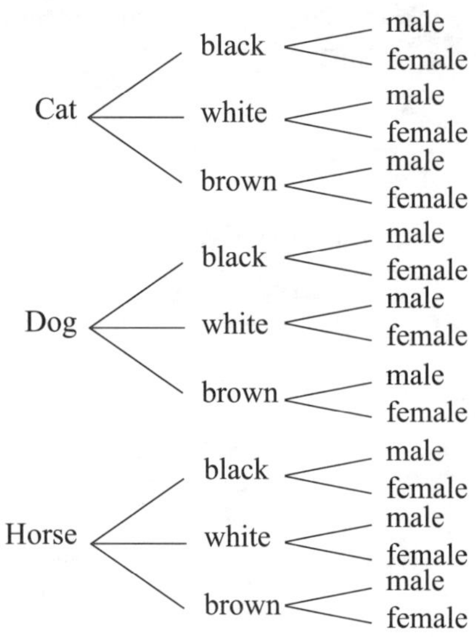

Step 3: Draw branches and write the possibilities for the third choice: gender.

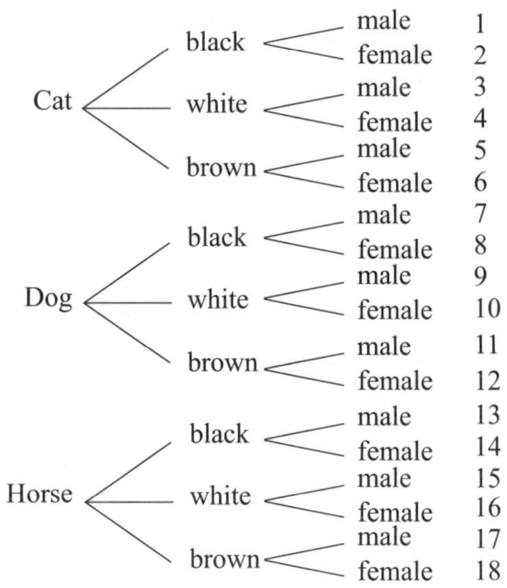

Step 4: Count the terminating branches. The number of terminating branches, 18, is the number of possible pet combinations.

Answer: 18

Problem Set

1. Jane bought 2 pairs of jeans and 3 shirts. How many different outfits can be made with these choices?

2. Sara has 3 pairs of shoes and 3 purses. How many different combinations can be made with these choices?

3. Lucy has 2 pairs of earrings and 2 hats. How many different combinations can be made with these choices?

4. Karen has 3 pairs of shoes (High Jumpers, Fast Runners, and Skywalkers) and 4 pairs of laces (white, black, red, and blue). How many different combinations can be made with these choices?

5. Lacey wants a snack. She can choose from cookies, cake, or candy. She can have root beer, cola, or tea to drink. How many different combinations can be made with these choices?

6. Nina is going to a ball tonight. She can wear one of three dresses. She can wear her hair up or down. She can bring a purse or choose not to. How many different combinations can be made from these choices?

7. There are two beds and two sets of sheets (red and blue). How many different combinations can be made with these choices?

8. Jennifer has 4 pairs of shorts and 4 shirts. The shirts and shorts are white, black, pink and green. She doesn't want to wear the same color shorts and shirt. How many different combinations can be made from these 8 items?

9. How many possible uniform combinations are there if a coach can choose from 2 pants, 3 shirts and 3 hats?

10. James wants a pet. He can choose a cat, a dog, or a bird. He can get a male or female. If he chooses a male, he will name it Sam or Jake; if he chooses a female, he will call it Rita or Pam. How many different combinations can be made from these choices?

11. Use the Counting Principle to find the number of different outcomes:

Select 1 sandwich and 1 drink:
Sandwiches: cheese, ham, chicken
Drink: iced tea, cola, milk

12. Use the Counting Principle to find the number of different outcomes:

Select 1 cat and 1 color of fur:
Cats: long-haired, short-haired
Color of fur: white, black, ginger

13. Use the Counting Principle to find the number of different outcomes:

Select 1 pizza and 1 topping:
Pizzas: small, medium, large
Topping: cheese, pepperoni, sausage, onions

14. Use the Counting Principle to find the number of different outcomes:

Select 1 exterior car color and 1 interior car color:
Exterior car color: white, red, blue, tan
Interior car color: black, beige, light blue, burgundy

15. Use the Counting Principle to find the number of different outcomes:

Select 1 sandwich and 1 bag of chips:
Sandwiches: ham, roast beef, cheese, turkey
Chips: barbecue, sour cream and onion, nacho cheese

16. Use the Counting Principle to find the number of different outcomes:

Select 1 bicycle and 1 color:
Bicycle: racing, mountain, exercise
Color: metallic blue, red, silver, black

17. Use the Counting Principle to find the number of different outcomes:

Select 1 shirt and 1 tie:
Shirt: white, blue, cream, gray, black
Tie: striped, solid

18. Use the Counting Principle to find the number of different outcomes:

Select 1 t-shirt and 1 type of jeans:
T-shirt: v-neck, tank top, long-sleeved, scoop neck
Type of jeans: boot-cut, straight-legged, low-rise, baggy

19. Use the Counting Principle to find the number of different outcomes:

Select 1 sandwich, 1 drink, and 1 fruit:
Sandwich: vegetarian, bologna, cheese
Drinks: juice, lemonade, chocolate milk
Fruit: orange, banana, apple, grapes

20. Use the Counting Principle to find the number of different outcomes:

Select 1 type of flower and 1 color:
Flower: rose, carnation, daisy
Color: red, pink, yellow, white, peach

MPA-090: Finding the Probability of Simple Real-Life Events

In an **experiment**, the result is not known until after the experiment has been performed. The result of the experiment is called an **outcome**. A set of outcomes of an experiment is called the **sample space**. An **event** is a subset of the sample space and can be made up of one or more outcomes. A **simple event** is an event which consists of a single outcome in the sample space.

Experiment	A procedure in which the result is unknown until after the experiment has been performed.
Outcome	A single possible result of performing an experiment.
Sample Space	The set of all possible outcomes.
Event	A subset of the sample space that can be made up of one or more outcomes.
Simple Event	An event which consists of a single outcome in the sample space.

The **probability** of an event from a random experiment is a measurement of the likelihood that an event will occur. For example, when tossing a quarter into the air and letting it fall to the ground, two possible events can occur. Either the coin will land on the head side (the side with the bust of President Washington) or the coin will land on the tail side (the opposite of the head side).

Probability	A number between zero and one that is used to show how likely it is that something will happen.
Equally Likely	If each member of a sample space has the same chance of happening as each of the other members of the sample space.

The probability of getting a head is $\frac{1}{2}$, since each of the results (heads or tails) is equally likely. We can expect that about 50% of the times that we toss a coin, the result will be a head.

A slightly more complicated random experiment involves rolling a six-sided number cube. In this experiment, there are six equally likely outcomes numbered 1, 2, 3, 4, 5, and 6. The probability of getting a five as an outcome is $\frac{1}{6}$.

Number cube showing the outcome "5"

A **compound event** is an event that consists of more than one outcome in the sample space. For example, consider the event that rolling a number cube will result in a number greater than or equal to 3. The possible outcomes in this event are 3, 4, 5, and 6. There are four equally likely outcomes out of the six total possible outcomes. Therefore, the probability of this event occurring is $\frac{4}{6}$ or $\frac{2}{3}$.

Compound Event	An event which consists of more than one outcome in the sample space.

The general formula for finding the probability of an event for a given experiment is:

$$\text{Probability (Event)} = \frac{\text{number of outcomes in event}}{\text{total number of possible outcomes}}$$

Example 1 What is the probability of tossing a fair coin three times and getting all heads or all tails?

Total number of outcomes = 8

Head, Head, Head
Head, Head, Tail
Head, Tail, Head
Head, Tail, Tail
Tail, Head, Head
Tail, Head, Tail
Tail, Tail, Head
Tail, Tail, Tail

Step 1: We must know how many outcomes there are for this experiment. The first toss of the coin has two possible outcomes, for each of these there are two possible outcomes for the second toss. This means that for two tosses there are four possible outcomes. For each of these outcomes, the third toss has two possibilities, for a total of eight possible outcomes for the experiment.

Outcomes in Event = 2

Head, Head, Head
Tail, Tail, Tail

Step 2: Since the event we are interested in is the one where all the tosses are the same, we can see that there are two outcomes that are either all heads or all tails.

Probability (all heads or all tails) = $\frac{2}{8} = \frac{1}{4}$	**Step 3:** The probability of getting all heads or all tails is computed by dividing the number of outcomes in the event by the total number of outcomes.
	Answer: The probability of getting all heads or all tails in three tosses of a coin is $\frac{1}{4}$.

Example 2 If a drawer contains 6 blue socks, 4 green socks, and 8 black socks, what is the probability that if you reach into the drawer you will pull out a black sock?

Total number of outcomes = 6 + 4 + 8 = 18	**Step 1:** We must know how many possible outcomes there are for this experiment. Since there are a total of 18 socks in the drawer, there are 18 equally likely outcomes for this experiment.
Outcomes in Event = 8	**Step 2:** Next, we must know how many ways our event can occur. Since there are 8 black socks, there are 8 ways for the event to occur.
Probability (black) = $\frac{8}{18} = \frac{4}{9}$	**Step 3:** The probability of getting a black sock is the number of ways to get the black sock divided by the total number of outcomes, which is $\frac{8}{18}$ or $\frac{4}{9}$.
	Answer: The probability of pulling out a black sock is $\frac{4}{9}$.

Example 3 Suppose that you chose a letter from the word CAPITOLS at random. What is the probability that you will choose a vowel?

Total number of outcomes = 8	**Step 1:** First, we compute the number of outcomes of this experiment. We are choosing a letter from the word CAPITOLS. Since there are 8 letters, there are 8 equally likely outcomes to this experiment.
Outcomes in event = 3	**Step 2:** Next we find the number of outcomes in our event "choose a vowel." Since A, I, and O are vowels, there are 3 possible, equally likely outcomes in this event.
Probability (vowel) = $\frac{3}{8}$	**Step 3:** Finally, we compute the probability of getting a vowel by dividing the number of ways to get a vowel by the total number of possible outcomes, which is $\frac{3}{8}$.
	Answer: The probability of choosing a vowel from the letters of the word CAPITOLS is $\frac{3}{8}$.

To compute the probability of an event, sometimes it is easier to determine when an event does not occur, the complement.

Complement	The complement of an event includes all the outcomes in a sample space that are not included in the outcomes of that event.

In this instance, use the following formulas to find the probability of the event.

Probability (Event) + Probability (complement of Event) = 1

or

Probability (complement of Event) = 1 – Probability (Event)

Example 4 What is the probability of getting at least one head when tossing a coin three times?

Step 1: Getting at least one head is a complicated event. You could get either one head, two heads, or three heads. By considering the event that will not occur means that you will have no heads, or all tails.

Total number of outcomes = 8

Step 2: Looking back to Example 1, we see that there are eight possible outcomes for this experiment.

Head, Head, Head
Head, Head, Tail
Head, Tail, Head
Head, Tail, Tail
Tail, Head, Head
Tail, Head, Tail
Tail, Tail, Head
Tail, Tail, Tail

(All tails) $= \frac{1}{8}$

Step 3: The probability of getting all tails is just $\frac{1}{8}$.

P (at least 1 head) $+ \frac{1}{8} = 1$

Step 4: Now use the formula to determine the probability of getting at least one head.

$$\text{P (at least 1 head)} + \frac{1}{8} - \frac{1}{8} = 1 - \frac{1}{8}$$
$$= \frac{8}{8} - \frac{1}{8}$$
$$= \frac{7}{8}$$

Step 5: Subtracting $\frac{1}{8}$ from both sides gives us that the probability of tossing at least one head is $\frac{7}{8}$.

Answer: The probability of getting at least one head in three coin tosses is $\frac{7}{8}$.

Example 5 After rolling a number cube two times, what is the probability that the sum of the numbers is seven?

Total possible outcomes = $6 \times 6 = 36$

Step 1: First, compute the number of possible outcomes in this experiment. For each of the six possible outcomes of the first roll of the cube, there are six possible outcomes of the second roll, for a total of 36 possible outcomes.

First roll	Second roll
1	6
2	5
3	4
4	3
5	2
6	1

Total number of outcomes in event = 6

Step 2: Next, determine how many of the outcomes lead to the event that the sum of the cubes is seven. The sum of seven occurs when the cubes are 1 and 6, 2 and 5, or 3 and 4, as summarized in the chart. There are a total of 6 outcomes in this event.

Probability (sum is 7) = $\frac{6}{36} = \frac{1}{6}$

Step 3: The probability of getting a sum of two number cubes that is seven can be computed by dividing the number of outcomes that lead to the event by the total number of outcomes which is $\frac{6}{36}$ or $\frac{1}{6}$.

Answer: The probability of getting a sum of seven when adding the result of two rolls of a number cube is $\frac{1}{6}$.

Problem Set

1. A box has 7 blue, 5 orange, and 3 yellow tiles. If a tile is chosen at random from the box, what is the probability that it is yellow?

2. In Matthew's backpack, he has 5 pencils, 3 pens, and 1 marker. If he reaches into the backpack and randomly selects a writing tool, what is the probability that he will select a pencil?

3. Mrs. Jackson baked cupcakes for her class. She has 10 chocolate, 11 vanilla, and 4 strawberry cupcakes. If the cupcakes are passed out randomly, what is the probability of receiving a vanilla cupcake?

4. There are 15 pints of blueberries, 12 pints of raspberries, 16 pints of blackberries, and 30 pints of strawberries arranged randomly in the refrigerator at a berry farm. If Scott reaches into the refrigerator to randomly select a pint of berries, what is the probability that he will pick a pint of raspberries?

5. A cookie jar has 12 chocolate chip, 8 oatmeal raisin, and 10 peanut butter cookies. If Sharon selects one cookie at random, what is the probability that it will be peanut butter?

6. A bike will be awarded to one student whose name will be randomly drawn from a bag that contains the names of honor roll students. There are 12 sixth grade, 9 seventh grade, and 13 eighth grade names in the bag. What is the probability that a 7th grader will win the bike?

7. A pack of construction paper has 17 blue, 14 green, and 15 red pieces of paper randomly arranged. If you open the pack, what is the probability a red paper is first in the pack?

8. A vase contains 12 daisies, 5 tulips, 6 roses, and 4 lilies randomly arranged. If you select a flower without looking, what is the probability you will pick a tulip?

9. In the school book room, there are 7 math, 7 spelling, 3 English, and 12 history books randomly arranged on a shelf. If you select a book off the shelf without looking, what is the probability you will pick a spelling book?

10. On a field trip, 10 ham, 14 turkey, and 13 roast beef sandwiches have been packed for lunch. If a student randomly selects a sandwich, what is the probability that the student will select a ham sandwich?

11. There are 5 mysteries, 3 biographies, 1 book of short stories, and 5 novels in the librarian's prize bag. If a book is chosen at random, what is the probability of selecting a biography?

12. A group of students were asked to vote for their favorite sport. Each vote was recorded on a slip of paper and dropped into a box. There were 15 votes for soccer, 17 for baseball, and 12 votes for football. If a slip is randomly selected, what is the probability it will be a vote for baseball?

13. In a cooler, there are 11 ginger ales, 15 colas, 8 bottles of water, and 4 grape sodas. If Paul reaches into the cooler and randomly selects a drink, what is the probability that he will select a ginger ale?

14. Erin has 11 cans of corn, 13 cans of peas, 17 cans of green beans, and 8 cans of beets. If she selects one can at random, what is the probability that she will select peas?

15. There are 2 principals, 7 teachers, 8 classroom aides, and 1 guidance counselor that will each be assigned a group of 10 students to chaperone on the next field trip. If students are randomly placed in groups, what is the probability of having a teacher as a chaperone?

16. The numbers 2 through 10 are written on flashcards and arranged randomly in a pile. What is the probability of not selecting a prime number?

17. There are 12 sixth graders, 7 seventh graders, and 10 eighth graders trying out to be the school mascot. If the mascot is randomly picked, what is the probability that the mascot will not be a seventh grader?

18. There are 6 triangles, 7 squares, 3 pentagons, and 8 hexagons in a bucket. If you reach into the bucket without looking, what is the probability that you will not select a square?

19. A basket contains 6 red, 1 green, and 5 blue eggs. If an egg is selected at random, what is the probability that it will not be a green egg?

20. John's coin bank contains 31 pennies, 17 nickels, 12 dimes, and 15 quarters. If the bank is turned upside down, what is the probability that a coin dropped out of the bank will not be a penny?

21. Josh reaches into a bag of candy that contains 22 lime, 18 lemon, and 23 orange flavored candy pieces. What is the probability of selecting a piece that is not orange?

22. There are 11 swimmers, 12 baseball players, 20 football players, and 10 golfers who are eligible to win the Athlete of the Year trophy. If the winner is randomly picked, what is the probability that a golfer will not be selected?

23. Jack is running for class president. If the president is randomly picked and the probability that Jack will be elected is $\frac{2}{3}$, what is the probability that he will not be elected?

24. Tim is reaching into a fruit bowl to randomly select a piece of fruit. The probability that he will select a peach is $\frac{4}{5}$. What is the probability that he will not select a peach?

25. Margaret is randomly giving slices of pizza to guests at her party. If the probability that she will give a slice with pepperoni is $\frac{4}{7}$, what is the probability that she will not give a slice with pepperoni?

26. Michael has a $\frac{3}{8}$ chance of taking a plane ride. What is the probability that he will not ride in a plane?

27. There is a $\frac{4}{9}$ probability that Angie will go to the movies this weekend. What is the probability that she will not go to the movies?

28. The probability that Jessica will be the next student selected to receive a candy bar from her teacher is $\frac{1}{18}$. What is the probability that she will not be selected to receive a candy bar?

29. The probability that the next television show will be a cartoon is $\frac{7}{12}$. What is the probability that the next show will not be a cartoon?

30. The probability that Jason will select a lime drink is $\frac{5}{11}$. What is the probability that he will not select a lime drink?

31. A number from 2 to 10 is chosen at random from a bowl. What is the probability that the number chosen is prime?

State whether or not the events (prime or not prime) are equally likely or not equally likely to occur.

32. Brendan tosses a number cube that has six sides. Each side is labeled with a number from 1 to 6. What is the probability that he will roll an odd number?

State whether or not the events (odd or not odd) are equally likely or not equally likely to occur.

33. The following numbers are on flashcards:

2, 4, 6, 8, 5, 15, 25, 35

If John pulls a card from a hat, what is the probability that he will pull a multiple of 5?

State whether or not the events (multiple of 5 or not multiple of 5) are equally likely or not equally likely to occur.

34. There are 5 red, 3 green, and 3 yellow pieces of candy in a bag. One piece is chosen at random. What is the probability of choosing a red piece?

State whether the events (red or not red) are equally likely or not equally likely to occur.

35. To win a game, Todd must draw an odd number from slips of paper numbered 2 through 15. What is the probability of choosing an odd number?

State whether the events (odd or not odd) are equally likely or not equally likely to occur.

36. The numbers from 1 to 15 are placed in a box and are randomly selected. If a multiple of 3 is selected, the red team is the winner. If a multiple of 4 is chosen, the blue team wins. What is the probability of the red team winning?

State whether the outcomes of the experiment are equally likely or not equally likely to occur.

37. Mrs. Tate will flip a coin to find out if her class will have homework. If the coin lands on heads, the class will have homework. If it lands on tails, the class will not have homework. What is the probability that Mrs. Tate's class will have homework?

State whether or not the events (heads or not heads) are equally likely or not equally likely to occur.

38. The factors of 36 are written on tiles that are face down in random order on Mrs. Smith's desk. If an odd factor is selected, the girls will line up for lunch first. If an even factor is selected, the boys will get to line up first. What is the probability that the girls will line up for lunch first?

State whether or not the events (odd or not odd) are equally likely or not equally likely to occur.

39. There are 5 purple, 5 yellow, and 5 blue folders in a stack. One folder is chosen at random. What is the probability of choosing a blue folder?

State whether the outcomes of the experiment are equally likely or not equally likely to occur.

40. There are 7 goldfish, 7 eels, and 7 starfish in an aquarium. One fish is chosen at random. What is the probability of choosing a goldfish?

State whether the outcomes of the experiment are equally likely or not equally likely to occur.

MPA-091: Finding the Number of Combinations of a Set of Objects

Previously, you learned that *probability* is the study of chance occurrences. Probability is used to attempt to predict outcomes of events with a certain degree of accuracy. Recall that if the outcomes for a certain experiment are equally likely, then thc probability of an event can be found as follows:

P (event) = $\dfrac{\text{the number of ways an event can occur}}{\text{the total number of equally likely outcomes}} = \dfrac{f}{N}$, where N is also known as the sample space. It is often the case that the sample space is unknown. For instance, suppose that a balanced coin is tossed and a fair die is rolled. What is the probability of the coin landing with heads facing up and the die landing with six dots facing up? First, you need to know the total number of possible outcomes for this experiment. One way to find the total number of all possible outcomes is to construct a *tree diagram* as shown below.

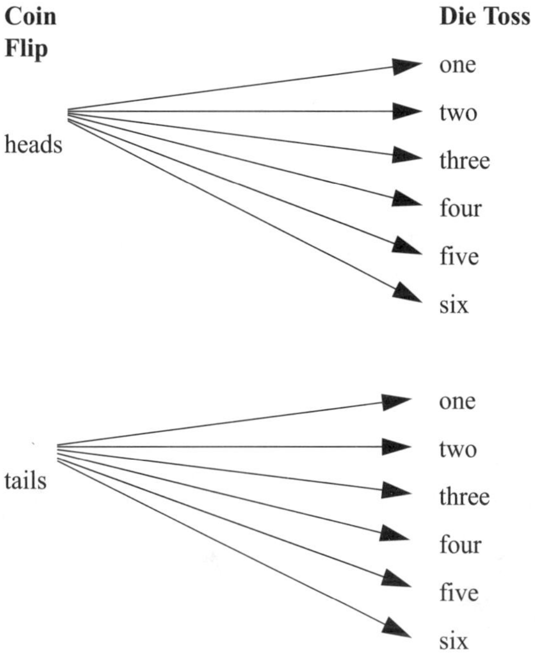

From this diagram, you can see that you have twelve different possibilities:

1. heads, one	2. heads, two	3. heads, three	4. heads, four	5. heads, five	6. heads, six
7. tails, one	8. tails, two	9. tails, three	10. tails, four	11. tails, five	12. tails, six

You can also see that the probability of the coin landing with heads facing up and the die landing with six dots facing up is $\dfrac{f}{N} = \dfrac{1}{12}$.

In general, you can see that the number of different outcomes above can be obtained by calculating the product of the total number of possible outcomes for the coin toss and the total number of possible outcomes for the die toss. What you have just illustrated is known as the fundamental counting principle.

Fundamental Counting Principle	If one task A can be performed in *m* different ways and a second task B can be performed in *n* different ways, then the task A followed by task B can be performed in $m \cdot n$ ways. This concept can be extended in a similar manner if there are additional tasks: $m \cdot n \cdot r \ldots \cdot t$

Example 1 If you have five shirts, four pairs of slacks, and three pairs of shoes, how many different ways can you create an outfit consisting of one shirt, one pair of slacks, and one pair of shoes?

$m \cdot n$	**Step 1:** Start with the fundamental counting principle.
$A = 5$	**Step 2:** Let task A = choosing a shirt; task B = choosing a pair of slacks; and task C = choosing a pair of shoes. Since you have five different shirts, task A can be performed in five different ways.
$B = 4$	**Step 3:** Similarly, task B can be performed in four different ways.
$C = 3$	**Step 4:** Task C can be performed in three different ways.
$5 \times 4 \times 3 = 60$	**Step 5:** From the fundamental counting principle, you can see that the total number of ways to create an outfit consisting of one shirt, one pair of slacks, and one pair of shoes is $5 \times 4 \times 3 = 60$.

Answer: 60

The last two concepts in this lesson involve *factorials*. The factorial of a positive integer is the product of all the integers from 1 up to the integer in question.

Factorial	The factorial of any positive integer k is defined by $k! = k(k-1)(k-2)\ldots 3 \times 2 \times 1$ A factorial is represented by the exclamation point (!) after the number. Also, zero factorial is defined as 0! = 1.

Example 2 Compute the result of 6!.

$6! = 6 \times 5 \times 4 \times 3 \times 2 \times 1$	**Step 1:** Using the definition above, set up the problem.
$6! = 6 \times 5 \times 4 \times 3 \times 2 \times 1$ $6! = 30 \times 4 \times 3 \times 2 \times 1$ $6! = 120 \times 3 \times 2 \times 1$ $6! = 360 \times 2 \times 1$ $6! = 720 \times 1$ $6! = 720$	**Step 2:** Multiply to find the answer.

Answer: 720

The next concept in this lesson is a counting formula known as a **permutation**. A permutation refers to an arrangement obtained by selecting r objects from a group of n objects where the order of selection is important. The number of permutations of n objects, taken r at a time, is summarized as follows:

Permutation	The number of permutations of n objects, taken r at a time is given by the formula ${}_nP_r = \frac{n!}{(n-r)!}$, where the order of selection is important.

Example 3 — A driver wants to have a personalized license plate that consists of only letters. If a certain state requires that five distinct letters be displayed on a license plate, how many different arrangements of letters does this driver have to choose from?

$${}_nP_r = \frac{n!}{(n-r)!}$$

Step 1: Note that the order of the letters is important. This problem consists of selecting five letters from a group of 26 letters. Start with the formula for the number of permutations given above.

$$\frac{26!}{(26-5)!} = \frac{26!}{21!}$$

Step 2: Insert the values into the formula.

$$\frac{26 \times 25 \ldots 2 \times 1}{21 \times 20 \ldots 2 \times 1} = 7{,}893{,}600$$

Step 3: Multiply to find the answer.

Answer: 7,893,600

The final concept you will study in this lesson is a counting formula known as a **combination**. A combination refers to any arrangement of r objects from a group of n objects where the order of selection is not important. The number of combinations of n objects, taken r at a time is summarized as follows:

Combination	The number of combinations of n objects, taken r at a time, is given by the formula ${}_nC_r = \binom{n}{r} = \frac{n!}{r!(n-r)!}$, where the order of selection is not important.

Example 4 — Determine the number of five-card poker hands that can be dealt from a standard 52-card deck.

$${}_nC_r = \binom{n}{r} = \frac{n!}{r!(n-r)!}$$

Step 1: In this case, the order in which you receive the cards does not matter, so you must find the number of combinations that can be made of 52 cards taken 5 at a time. Start with the formula for combinations.

$$\frac{52!}{5!(52-5)!} = \frac{52!}{5!(47!)}$$

Step 2: Insert the values into the formula.

$$\frac{52 \times 51 \ldots 2 \times 1}{5 \times 4 \times 2 \times 1(47 \times 46 \ldots 2 \times 1)} = 2{,}598{,}960$$

Step 3: Multiply to find the answer.

Answer: 2,598,960

Problem Set

1. Mrs. Stevens is a social studies teacher. Her class is working on a project about the diversity of heritages in America. Mrs. Stevens decides to select one student from each group for a class presentation. In her class she has 7 Hispanic-American students, 8 Asian-American students, 9 African-American students, and 6 white students. How many ways can she select one student from each group for her project?

2. Louise must buy diapers for her little cousin. There are 12 different brands. Each brand comes in five different sizes, and each size is made in two different kinds, for boys or girls. How many ways can Louise select diapers for her little cousin?

3. How many permutations can be made from the digits 1, 3, 5, 2, 4?

4. How many permutations can be made with the numbers 3, 4, 7, 5?

5. How many permutations can be made with the numbers 9, 8, 2?

6. How many permutations are possible for the numbers 3, 5, 2?

7. How many permutations are possible for the numbers 4, 2, 1, 3?

8. The neighborhood pizza parlor has a choice of ten toppings for the pizza. How many pizzas can be made if each pizza must have two different toppings?

9. The coach has to select three representatives from his team of seventy players to attend a conference. How many combinations of players can he select?

10.

Jose's Hamburger Shack

SANDWICHES	DESSERTS
Hamburger	Apple Pie
Cheese Burger	Strawberry Ice Cream
Grilled Chicken	Vanilla Ice Cream
Fish	Chocolate Ice Cream

FRIES	SODAS
Large Fries	Large Soda
Small Fries	Small Soda

How many combinations can be made with ice cream, soda and pie?

11. Find the value of 3!

12. Find the value of 12!

13. Evaluate 4!

14. Evaluate 6!

15. Evaluate 10!

16. Evaluate (4!)(4!)

17. Evaluate (3)(4!)

18. Evaluate $\frac{5!}{15}$

19. Evaluate $\frac{6!}{40}$

20. Evaluate $\frac{5!}{4!}$

MPA-092: Reading and Interpreting Bar, Line, and Circle Graphs

The next time that you read a newspaper or magazine, note the different types of graphs that you see. How do the writers and editors decide which type of graph or plot to use? Their decision is based on the kind of **data** that they are presenting.

Data	A set of facts or a collection of information.

We use graphs to **interpret** data.

Interpret	To understand and manipulate data in order to arrive at answers to questions about the data.

Some graphs, for example, display **quantitative data**.

Quantitative data	Data obtained by observing values of a quantitative variable.
Quantitative variable	A variable with a numeric value.

Consider the following data set:

Fasting Plasma Glucose Level for 20 Non-Diabetic Patients (mg/dL)			
90	95	85	85
70	85	85	90
80	90	95	85
85	95	80	90
85	90	90	80

This quantitative data can easily be grouped into classes, and each data value can be plotted on a number line. Unlike quantitative data, **qualitative data** is derived from the observation of a non-numeric value.

Qualitative data	Data obtained by observing values of a qualitative variable.
Qualitative variable	A variable with a non-numeric value.

Consider the following data set:

State of Birth	Number in Classroom
Louisiana	17
Pennsylvania	1
Georgia	5
California	2

Plotting this data on a number line would not make sense, because the data is qualitative. Qualitative data is more suitable for presentation in a bar graph or circle graph. A bar graph compares related pieces of data using vertical bars of equal width. The lengths of the bars are frequently used for comparison. The following is an example of a bar graph:

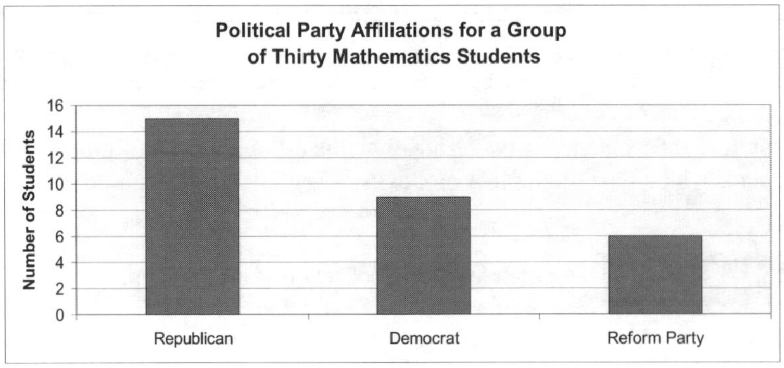

Let's look at general characteristics of a graph. First, the **title** is the first part of the graph you should read. The title of a graph tells you the subject being presented in the graph. From the title in the graph above, you can see that the political party affiliations of a group of 30 mathematics students are being presented. Another characteristic is the **category**. The category of a graph represent the class of data being presented. The categories for this graph are Republican, Democrat, and Reform Party. Lastly, the **units** of a graph tell you how the data is being measured. In this case, the units are "number of students." Use the bar graph to work through the following example.

Example 1 Determine the number of Republicans, Democrats, and Reform party affiliates in the group of mathematics students represented by the graph on the previous page.

Republican = 15	**Step 1:** Since the bar above the Republican category is halfway between 14 and 16, the number of Republican Party affiliates is 15.
Democrat = 9 Reform = 6	**Step 2:** The number of Democrat affiliates and Reform Party affiliates are 9 and 6, respectively.
$15 + 9 + 6 = 30$	**Step 3:** Notice that the title reveals that there is a total of 30 mathematics students represented in the graph, and the sum of the results for each respective party is $15 + 9 + 6 = 30$.

Answer: Republican: 15; Democrat: 9; Reform 6

Another method for displaying qualitative data graphically is the circle graph. The circle graph is a circle partitioned into pie-shaped pieces, and is particularly effective when you want to show how a certain quantity is sectioned into parts. A **legend** is used when the categories are represented by colors and symbols. A legend explains how the categories are displayed on a graph. In the circle graph below, the legend tells you which color represents each political party. The following is an example of a circle graph:

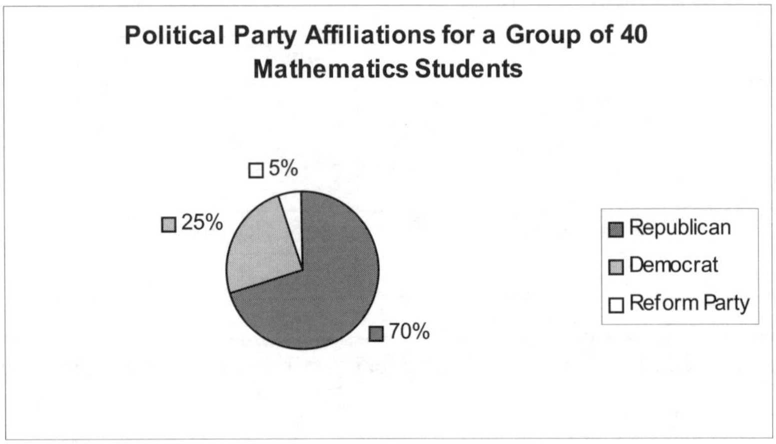

Example 2	Determine the number of Republicans, Democrats, and Reform party affiliates in the group of mathematics students from the preceding graph.
70% of 40 = $0.70 \times 40 = 28$	**Step 1:** From the graph, you can see that 70 percent of the 40 students are Republicans. Therefore, the number of Republican party affiliates can be determined as follows: 70% of 40 = $0.70 \times 40 = 28$ students.
$0.25 \times 40 = 10$	**Step 2:** Similarly, the number of Democratic party affiliates is 10 students.
$0.05 \times 40 = 2$	**Step 3:** The number of Reform party affiliates is 2 students.
	Answer: Republican: 28; Democrat: 10; Reform: 2

One of the most commonly used graphs is the line graph. A line graph associates values with all points over a continuous interval; thus, they are more appropriate for continuous data (data in which all intermediate values have meaning). For example, consider the previous example. A line graph would not be appropriate in this case, as the number of students must be depicted by a whole number. The following is an example of a line graph:

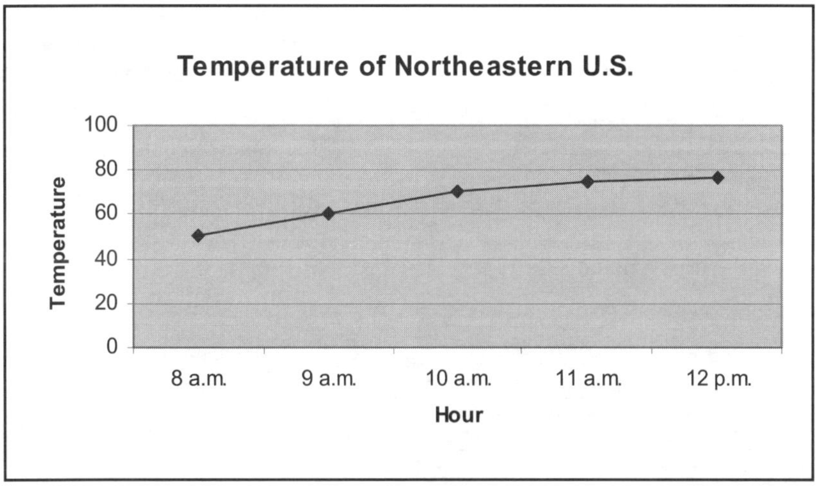

Example 3	Determine the approximate temperature at 9:30 a.m. from the graph above.
	Step 1: Move along the horizontal axis to the location halfway between 9 a.m. and 10 a.m.
	Step 2: Move vertically upward to the corresponding location on the graph. The location on the graph corresponds to a temperature of approximately 65°.
	Answer: 65°

Problem Set

1. Jessica never seems to have enough money. She made a pie chart of her monthly budget so she could improve her spending habits.

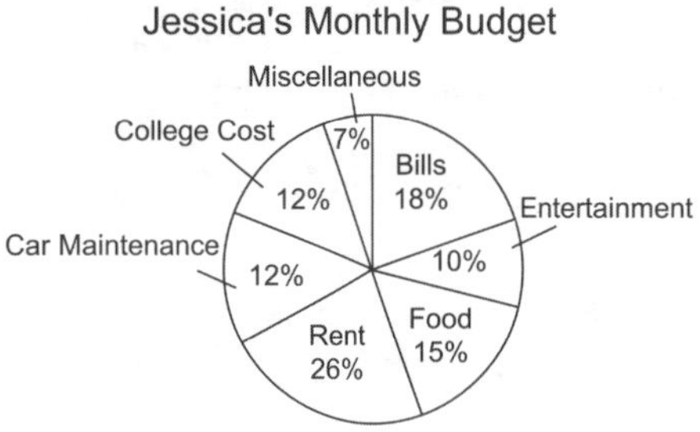

What is the least expensive item on Jessica's budget?

2. Lin made a bar graph of her classmates' favorite colors. Help Jamie answer the question about Lin's graph.

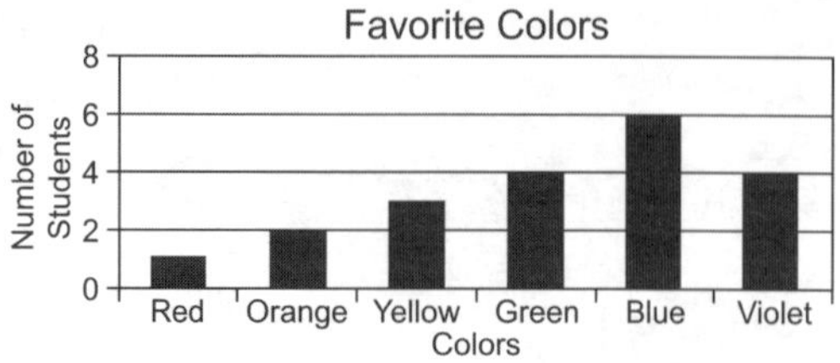

What is the most popular color among Lin's classmates?

3. For his science project, Santosh wanted to find the best method for growing grass. He grew grass for three weeks.

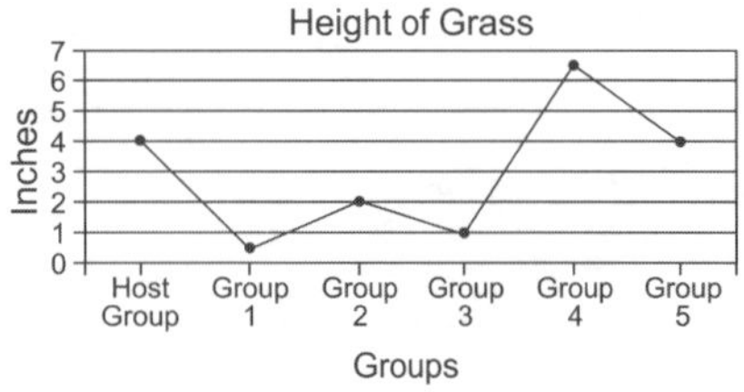

What group grew the least amount?

4.

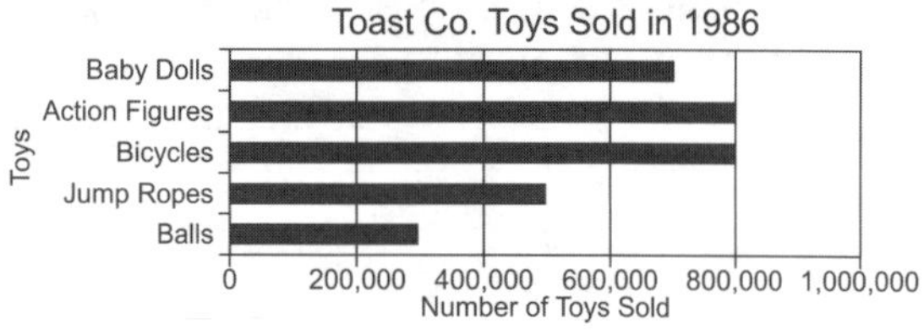

How many bicycles were sold in 1986?

5.

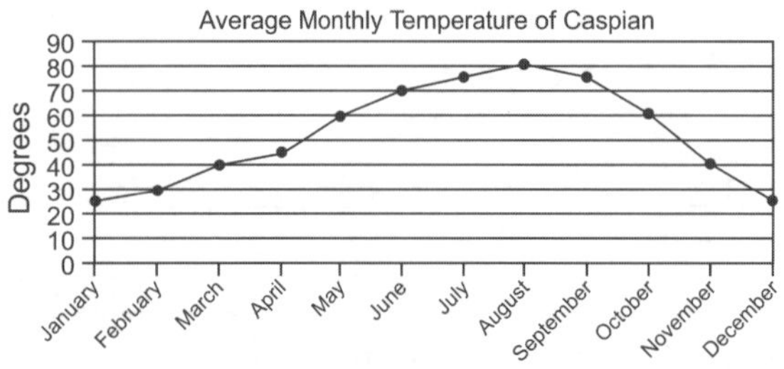

What was the hottest month?

6.

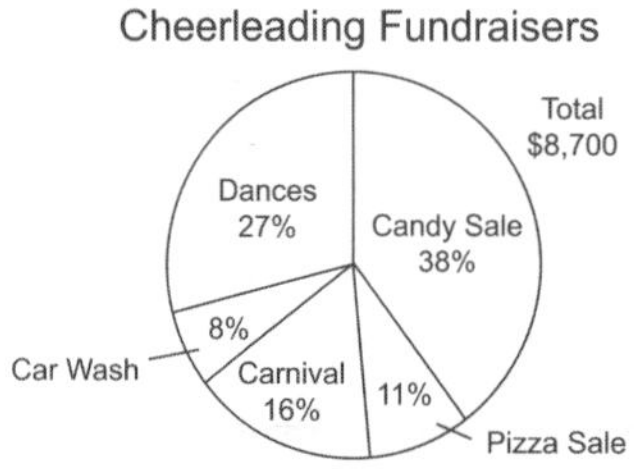

How much money was raised during the Fall Carnival?

7. Art Class/Club Budget

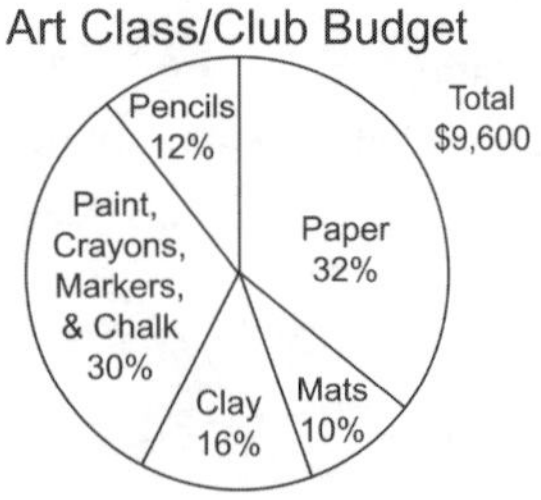

How much money was spent on mats and clay?

8. Art Class/Club Budget

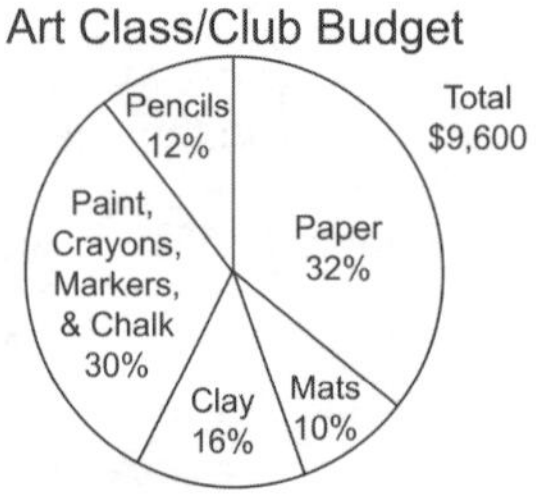

How much money was spent on paper and pencils?

9.

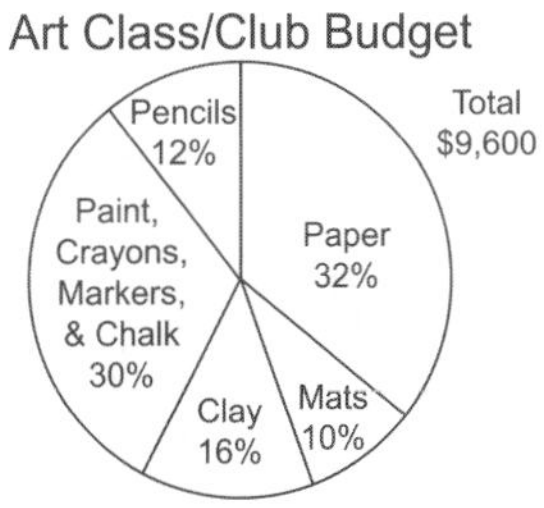

How much money was spent on paint, crayons, markers, and chalk?

10.

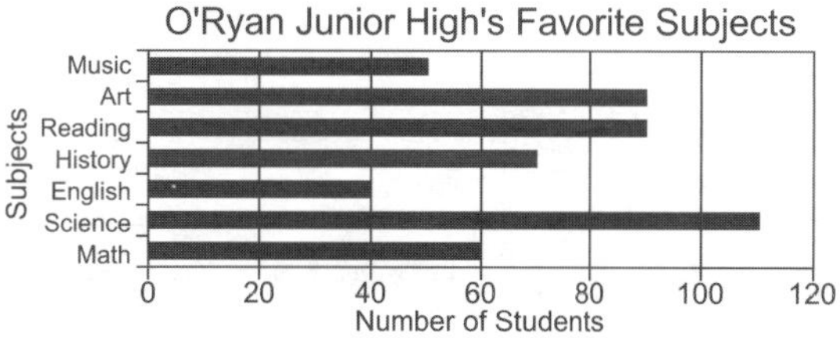

What is the favorite subject for students at O'Ryan?

11.

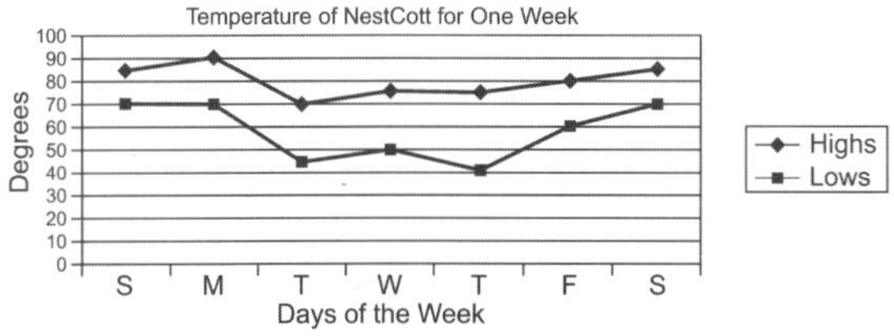

What was Saturday's low?

12.

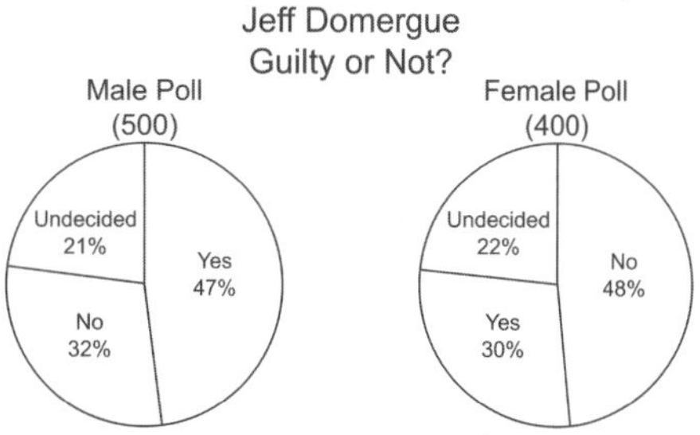

How many males voted "yes"?

13.

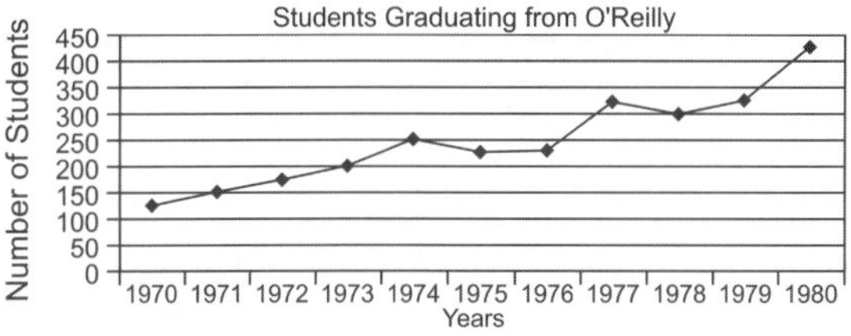

How many more students graduated in 1978 than in 1974?

14.

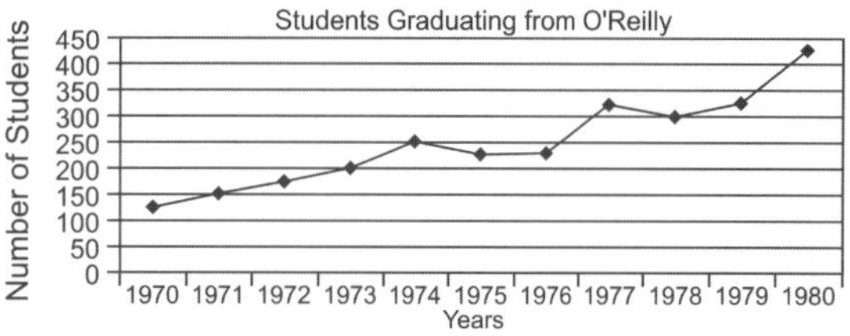

How many students graduated in 1973?

15.

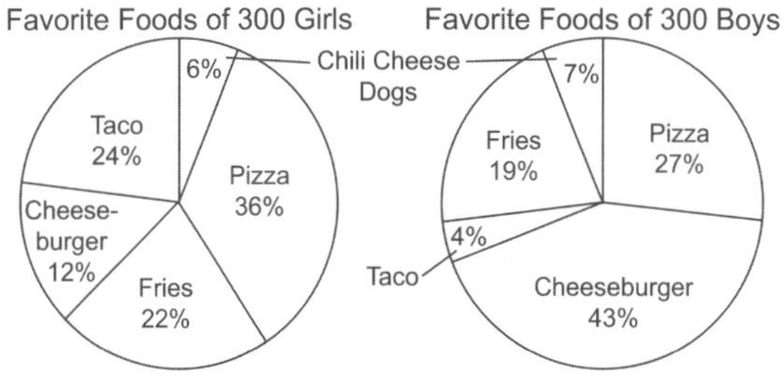

How many girls like tacos?

16.

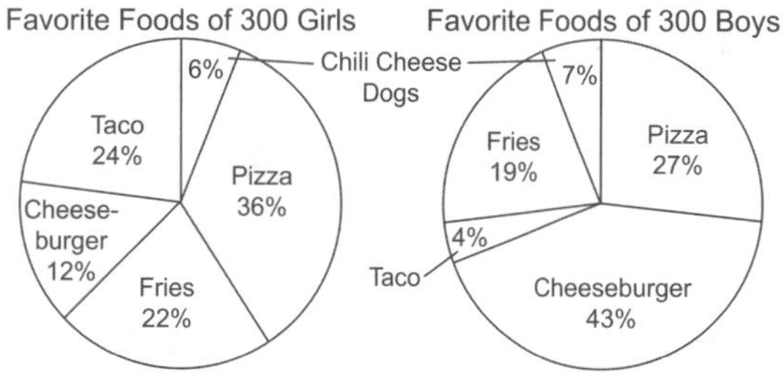

How many more girls like fries than cheeseburgers?

17.

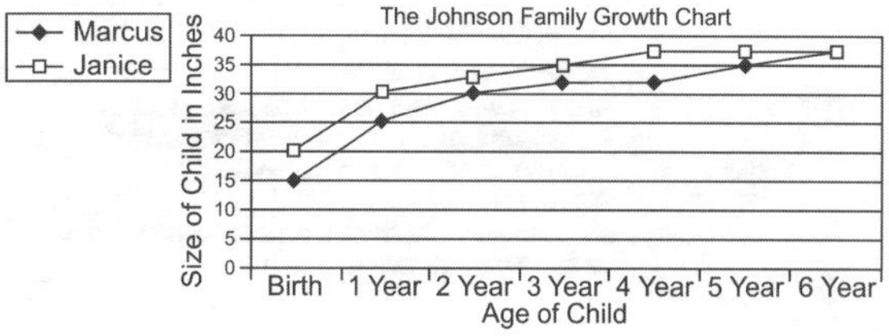

How many inches did Marcus grow between three and four years old?

18.

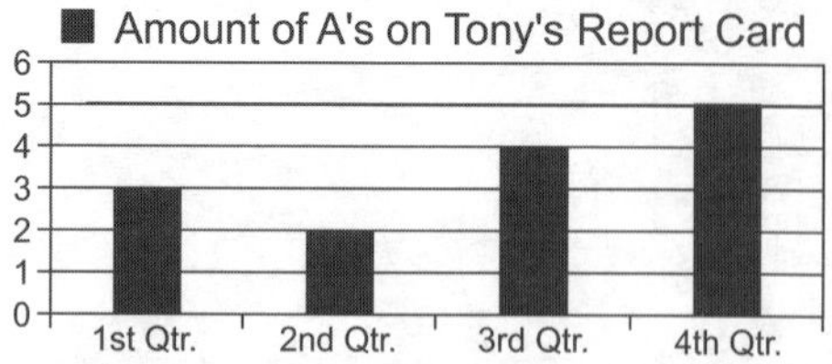

Which quarter did Tony have the most A's on his report card?

19. Students Who Go on To Post-Secondary Schools

Total: 800 students

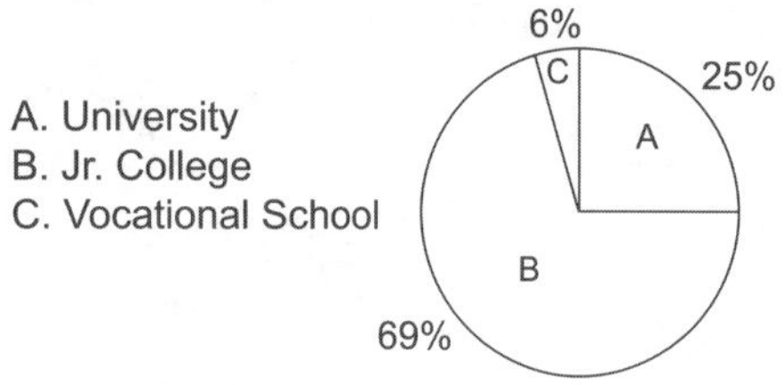

According to the graph, how many students went to Vocational School?

20. Teen's Favorite Soft Drinks

Total: 250 teens

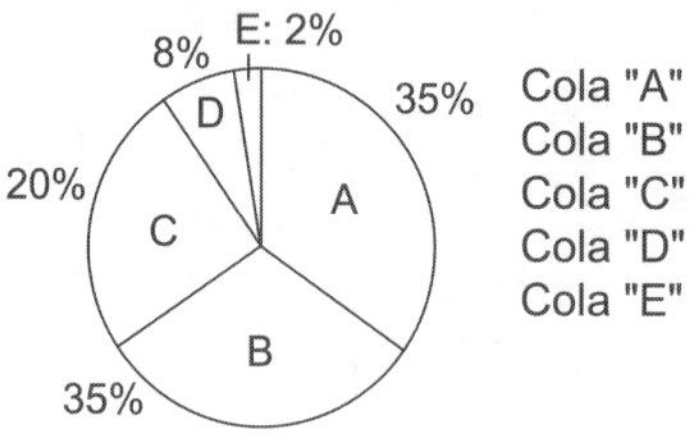

How many teens chose Cola "C" as their favorite soft drink?

MPA-093: Choosing Appropriate Scales and Intervals for Data

Every day we are presented with graphs via newspapers, magazines, television, Internet, radio, and various other media sources. The reason for this type of advertising is to induce readers (often consumers) to follow a certain course of action. However, many people accept these graphic representations as facts without a basic understanding of how to properly interpret such graphs.

Graphic representations can be conveyed in a misleading or convoluted manner. Sometimes this is intentional and sometimes it is not. This is why it is important to learn some fundamentals about statistics. In this lesson, you will learn several basic concepts about graphs.

The first concept you will learn is the **range** of a graph. The range of a given set of data is commonly referred to as a **measure of dispersion,** and is a descriptive measure that indicates the amount of variation (or spread) in a data set.

Range	The range of a set of data is the difference between the maximum and minimum values in the data set. Range = Maximum - Minimum = |Minimum -Maximum|

The following example demonstrates how to find the range.

Example 1 Calculate the range of the data on the x-axis and the y-axis of the following graph:

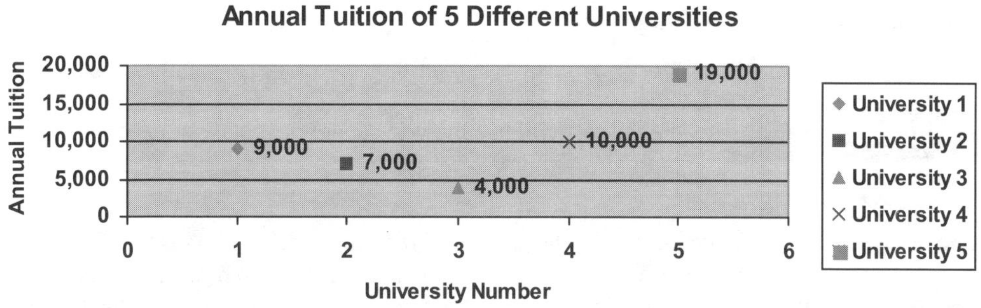

Maximum – Minimum = Range $5 - 1 = 4$	**Step 1:** The maximum x-value that corresponds to an observation is 5 and the minimum x-value that corresponds to an observation is 1. The range of the data on the x-axis is 4.
Maximum – Minimum = Range $19{,}000 - 4{,}000 = 15{,}000$	**Step 2:** Similarly, since the maximum y-value that corresponds to an observation is 19,000 and the minimum y-value that corresponds to an observation is 4,000, the range of the data on the y-axis is 15,000.

Answer: Range of data on the x-axis: 4
Range of data on the y-axis: 15,000

The next concept is the **scale** of a graph. The scale of a graph is the way in which the axes are numbered. This scale consists of a series of marks along the axes at graduated intervals that are used to measure the data. The numbering scale for each axis will always be given in equally-spaced intervals.

Scale	The scale of a graph is the number of equally-spaced units between each mark on a given axis.
Scale Selection	The length of the interval between the grid markings on a graph.

The following example demonstrates how to determine the scale.

Example 2 Determine the scale of the x-axis and the y-axis of the following graph:

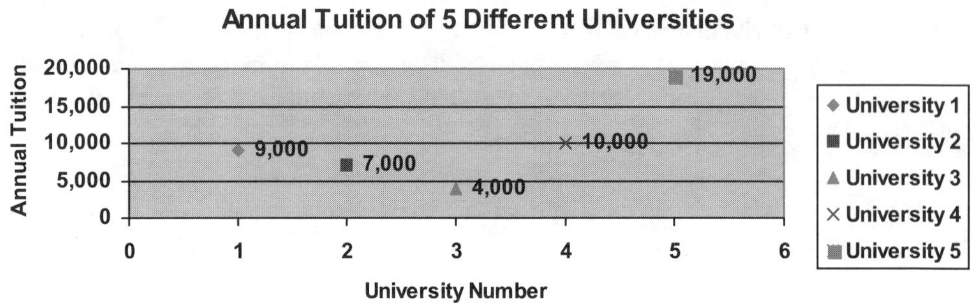

x-axis: 1	**Step 1:** Since there is one unit between each mark on the x-axis, the scale of the x-axis is one.
y-axis: 5,000	**Step 2:** Similarly, since there are 5,000 units between each mark on the y-axis, the scale of the y-axis is 5,000.

Answer: Scale of the x-axis: 1
Scale of the y-axis: 5,000

The final concept is the **interval** of a graph.

Interval	The interval of a graph is the vertical or horizontal ranges of the axes of a graph.

Note that the difference between the interval of the graph and the range of the graph is that the interval is the gap between the extreme values on the axes, where the range of the graph is the gap between extreme values of the actual observations. Since the observations of the graph are plotted *within* the horizontal and vertical intervals of the graph, the intervals of the graph will always be greater than the corresponding ranges of the data. Choose intervals that encompass all the observations without leaving large amounts of unused space on the graph.

The following example demonstrates how to determine intervals.

Example 3 Determine the horizontal and vertical intervals of the following graph:

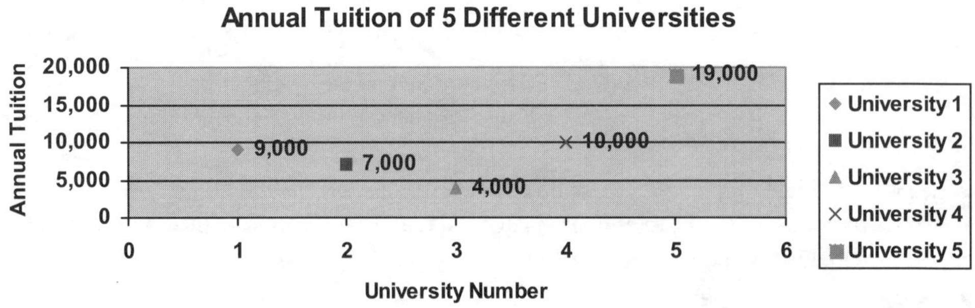

Maximum – Minimum = Range $6 - 0 = 6$	**Step 1:** Since the maximum x-value on the x-axis is 6 and the minimum x-value on the x-axis is 0, the horizontal interval is 6.

Maximum – Minimum = Range 20,000 – 0 = 20,000	**Step 2:** Similarly, since the maximum y-value on the y-axis is 20,000 and the minimum y-value on the y-axis is 0, the vertical interval is 20,000.

Answer: Horizontal interval: 6
Vertical interval: 20,000

Rules of Selecting Scale and Interval

1. The interval selected must be greater than the range of the data.
2. No matter what scale is selected, the numbering system must remain consistent along the axis.

Problem Set

1. What is the range of the data set? 90, 83, 81, 73, 68, 49, 37, 49, 62, 80, 43, 64, 23, 42, 40

2. What is the range of the data set? 111, 126, 126, 182, 199, 114, 151, 191, 120, 143, 104, 138, 100, 167, 166

3. What is the range of the data set? 9, 12, 19, 2, 18, 17, 3, 8, 13, 9

4. Would 3 – 14 be an appropriate interval for the data? 9, 12, 19, 2, 18, 17, 3, 8, 13, 9

Yes No

5. What is the range of the data set? 13, 55, 42, 67, 31, 37, 81, 22, 12, 29

6. Would 10 – 85 be an appropriate interval for the data? 13, 55, 42, 67, 31, 37, 81, 22, 12, 29

Yes No

7. What is the range of the data set? 65, 60, 14, 39, 26, 42, 64, 47, 27, 24

8. Would 0 – 10 be an appropriate interval for the data? 7, 7, 6, 6, 7, 7, 7, 4, 3, 5

Yes No

9. Would 0 – 60 be an appropriate interval for the data? 37, 1, 15, 43, 3, 56, 57, 1, 52, 22

Yes No

10. Would 0 – 50 be an appropriate interval for the data? 14, 3, 28, 28, 23, 25, 43, 38, 39, 47, 10, 13

Yes No

11. Are the intervals in the graph consistent?

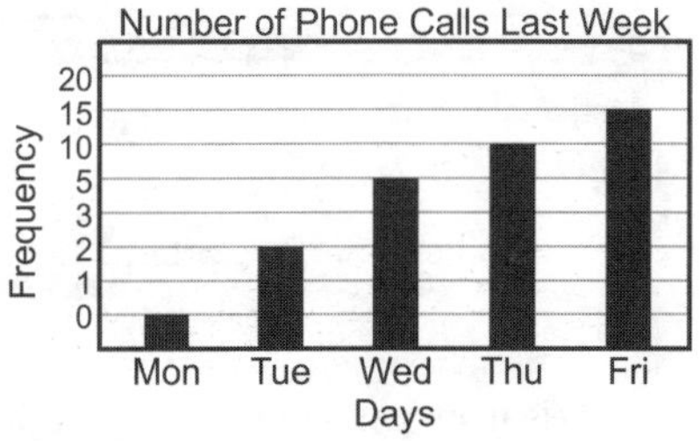

Yes No

12. Are the intervals in the graph consistent?

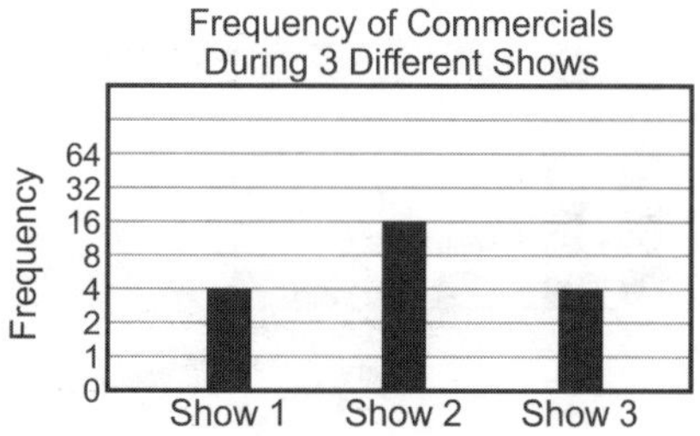

Yes No

13. Are the intervals in the graph consistent?

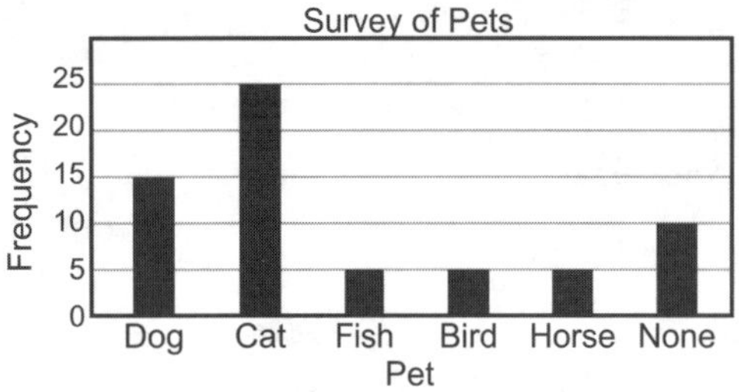

Yes   No

14. A data set has an interval of 57 points. How many tick marks will be needed to display it if the scale markings are in steps of 3 points?

15. A data set has an interval of 72 points. How many tick marks will be needed to display it if the scale markings are in steps of 8 points?

16. What, if anything, is wrong with the graph?

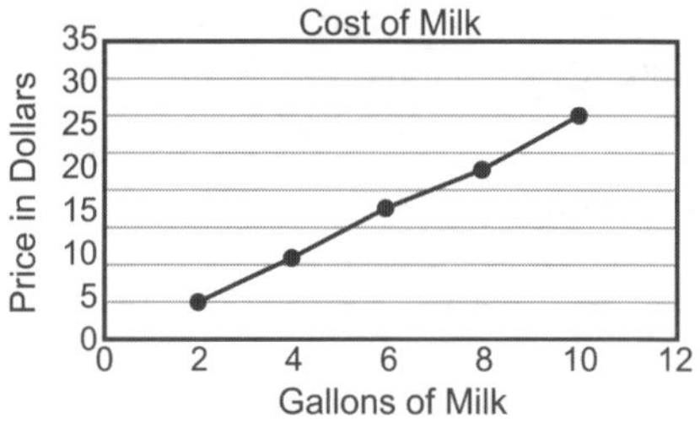

The vertical axis is inconsistent   Nothing
The horizontal axis is inconsistent

17. What, if anything, is wrong with the graph?

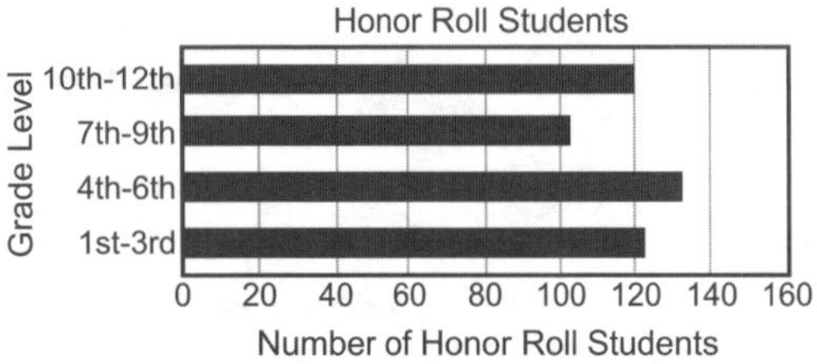

The vertical axis is inconsistent   Nothing
The horizontal axis is inconsistent

18. What, if anything, is wrong with the graph?

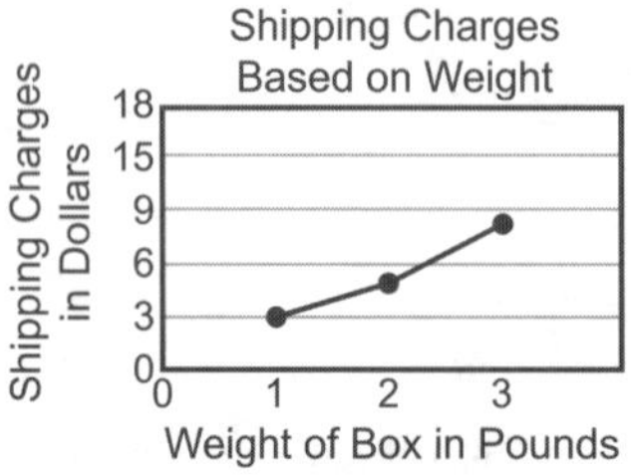

The vertical axis is inconsistent   Nothing
The horizontal axis is inconsistent

19. What, if anything, is wrong with the graph?

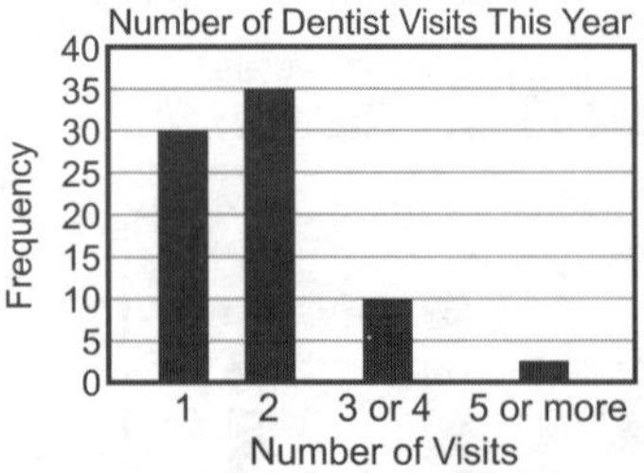

The vertical axis is inconsistent   Nothing
The horizontal axis is inconsistent

20. What, if anything, is wrong with the graph?

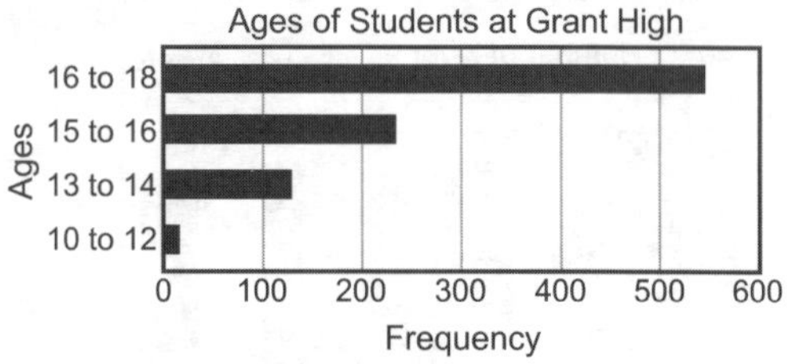

The vertical axis is inconsistent   Nothing
The horizontal axis is inconsistent

MPA-094: Interpreting and Constructing Line Plots

Previously, you learned that a *statistic* is an item of information expressed in numbers. Each day, whether you realize it or not, you use statistics to express and interpret your thoughts and ideas. When you are dealing with a small set of data, these statistics can be easily understood without using pictorial representations of the data. However, when you deal with large sets of data, such pictorial representations become essential.

One such type of pictorial representation is a line plot.

Line Plot	A line plot provides a visual display of information on a number line.

A line plot involves first putting data into groups, or classes. Next, you represent the observations with an "x" above the respective locations on the number line. The frequency of the data is the number of values within a group or class. Perhaps the most useful property of the line plot is that it reveals the frequency distribution of the data. The frequency distribution can be thought of as the relative positions of the data in the data set.

In this lesson you will investigate the principles of interpreting and constructing line plots. The basic procedure for constructing line plots is summarized below.

Constructing Line Plots	1. Determine the scale and the interval of the data. 2. Construct a number line. 3. Plot the data by placing an "x" above the numbers on the number line for each time the numbers are listed in the data set.

The following example demonstrates how to construct a line plot.

Example 1 Construct a line plot for the data given below.

Outcome of a fair die rolled 20 times			
6	4	2	1
6	2	2	2
4	4	4	5
3	1	5	4
5	3	6	3

Step 1: Begin by determining the scale and interval of the data. You have six distinct outcomes: 1, 2, 3, 4, 5, and 6; therefore, the scale of the data is one because there is a difference of one between one outcome and the next. Since the largest outcome is six and the smallest is one, the interval of the data is 6 – 1 = 5.

Step 2: Next, construct a number line for the data. Note that the numbers are arranged in ascending order.

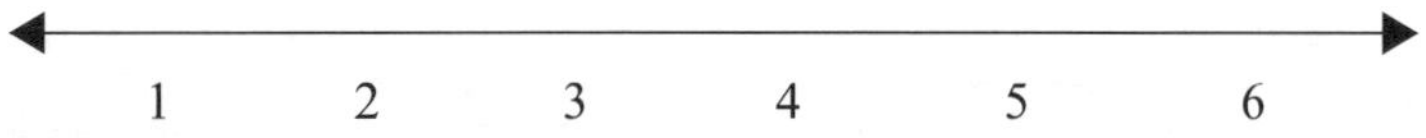

Step 3: Finally, plot the data by placing an "x" above the numbers on the number line for each time that the data is listed in the data set. It is often helpful to cross out the data as you plot it.

Answer:

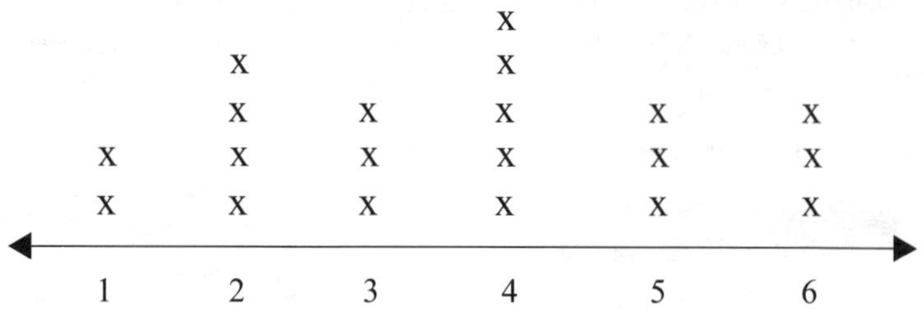

Let's look at another example.

Example 2 Construct a line plot for the data given below.

Fasting Plasma Glucose Level for 20 Non-Diabetic Patients (mg/dL)			
90	95	85	85
70	85	85	90
80	90	95	85
85	95	80	90
85	90	90	80

Step 1: Begin by determining the scale and interval of the data. Since you have five distinct outcomes (70, 80, 85, 90, and 95), a convenient scale is 5, because 5 is the smallest difference between numbers.

Step 2: Since the largest measurement is 95 and the smallest measurement is 70, the interval of the data is 95 – 70 = 25.

Step 3: Next, construct a number line for the data. Note that the numbers are placed in ascending order.

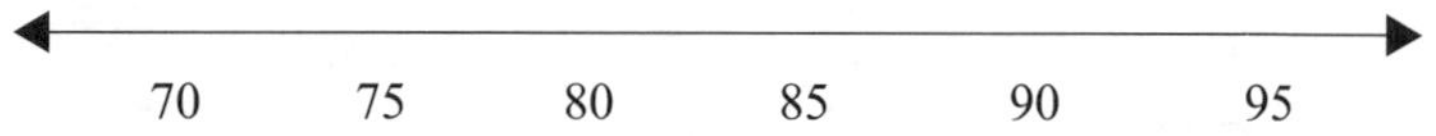

Step 4: Finally, plot the data by placing an "x" above the number line for each time that the data is listed in the data set. It is often helpful to cross out the data as you plot it.

Answer:

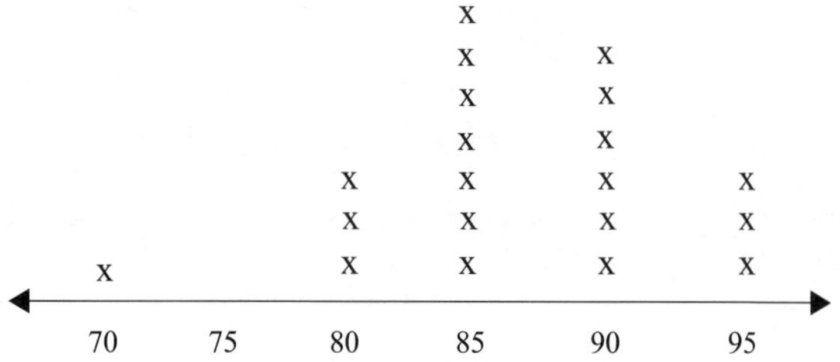

Since the measurement 85 has the most “x’s” above it, 85 occurs with the most frequency. An observation that lies far apart from the rest of the data is known as an **outlier**.

Outlier	An observation that lies far apart from the rest of the data.

In this plot, the measurement 70 is an outlier.

Look at the plot again. Notice that the largest amount of “x’s” occur between 80 and 95. This arrangement is called a **cluster**.

Cluster	Data that are grouped closely together on a line plot.

The data do not have to be consecutive to form a cluster but must lie in close proximity.

Problem Set

1.

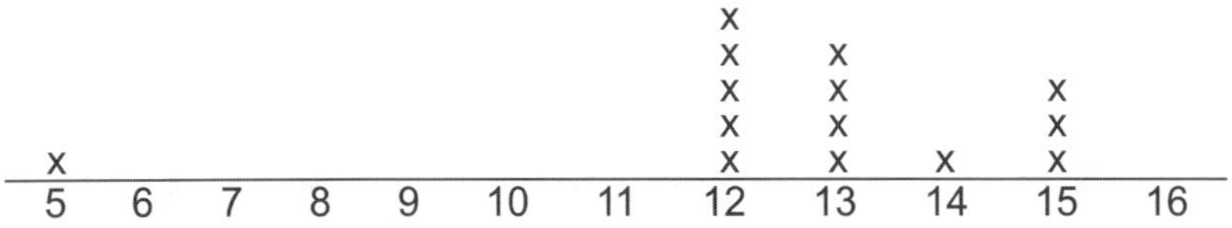

Identify the outlier or outliers in the given plot.

2.

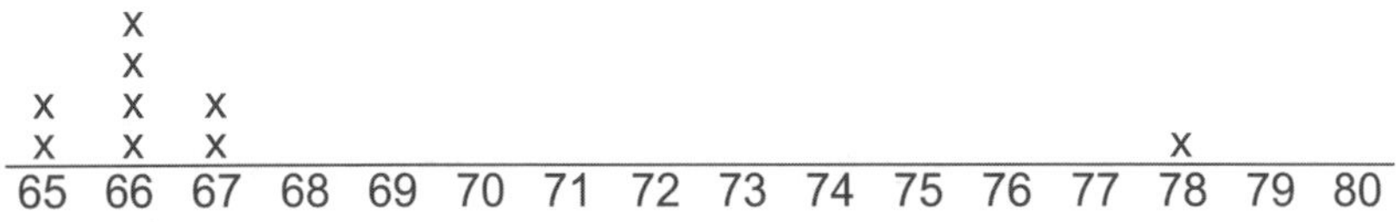

Identify the cluster in the given line plot by the range of the cluster. For example, if the range of the cluster is 0 – 1, answer "0 – 1". If there is no cluster, answer "no cluster".

3.

Which number or numbers appear with the most frequency in the data set used to construct the given line plot?

4.

Identify the cluster in the given line plot by the range of the cluster. For example, if the range of the cluster is 0 – 1 , answer "0 – 1". If there is no cluster, answer "no cluster".

5.

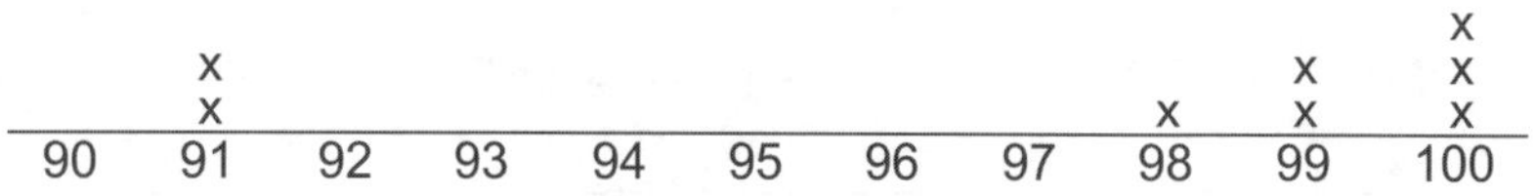

Which number or numbers appear with the most frequency in the data set used to construct the given line plot?

6.

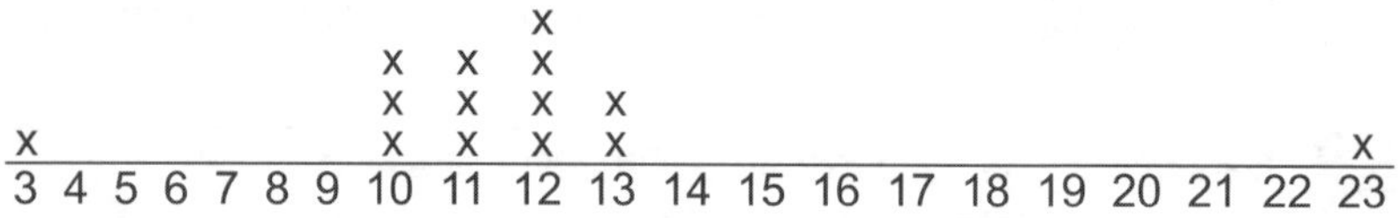

Identify the cluster in the given line plot by the range of the cluster. For example, if the range of the cluster is 0 – 1, answer "0 – 1". If there is no cluster, answer "no cluster".

7.

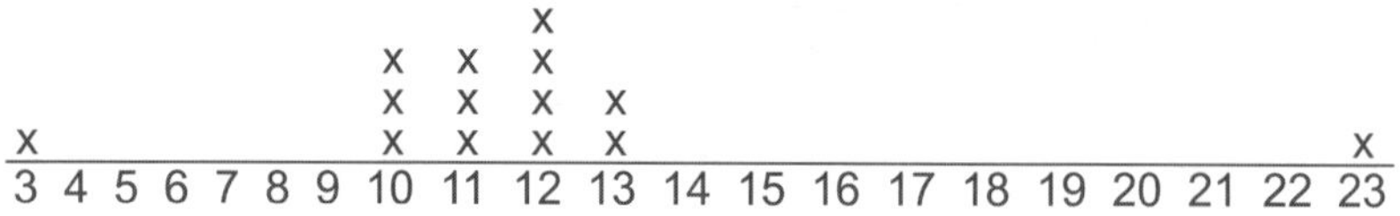

Which number or numbers appear with the most frequency in the data set used to construct the given line plot?

8.

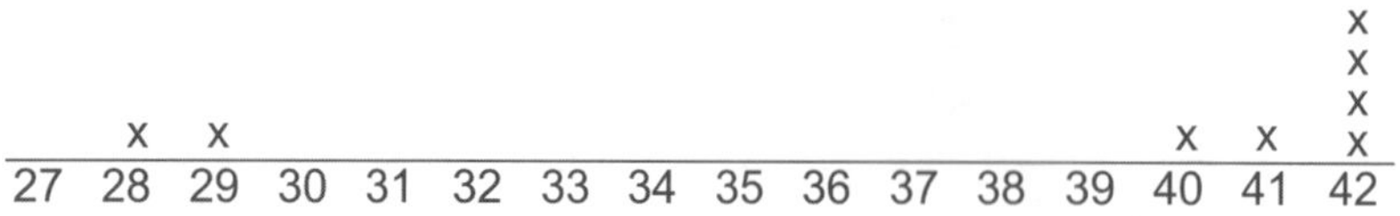

Identify the outlier or outliers in the given plot.

9. Is the following line plot correct for this set of data? Answer yes or no.
93, 74, 79, 84, 79, 96, 95, 83, 90, 99, 87, 100, 93, 81, 82, 93, 87, 95, 77

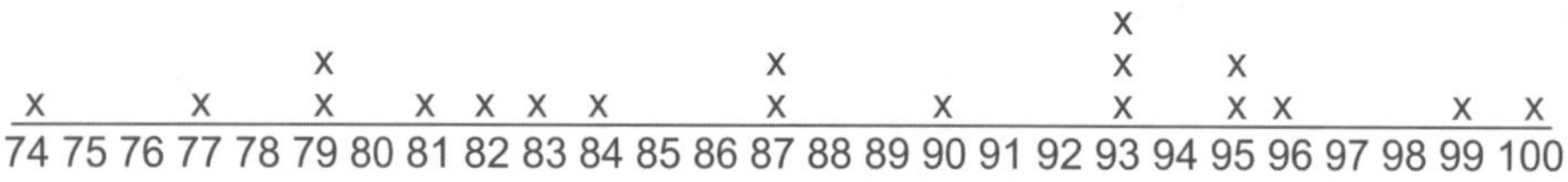

10. Is the following line plot correct for this set of data? Answer yes or no.
11, 10, 1, 6, 7, 4, 10, 8, 10, 10, 8, 13, 4, 8, 9, 9, 8, 13, 8, 10

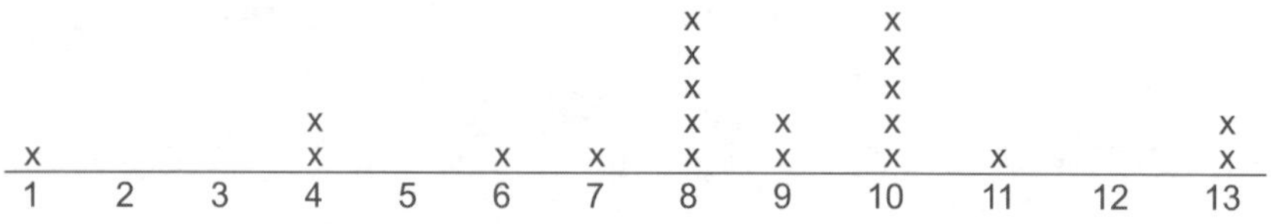

11. Is the following line plot correct for this set of data? Answer yes or no.
3, 2, 8, 5, 1, 0, 7, 2, 1, 0, 6, 0, 1, 6, 3, 5, 4, 1, 6

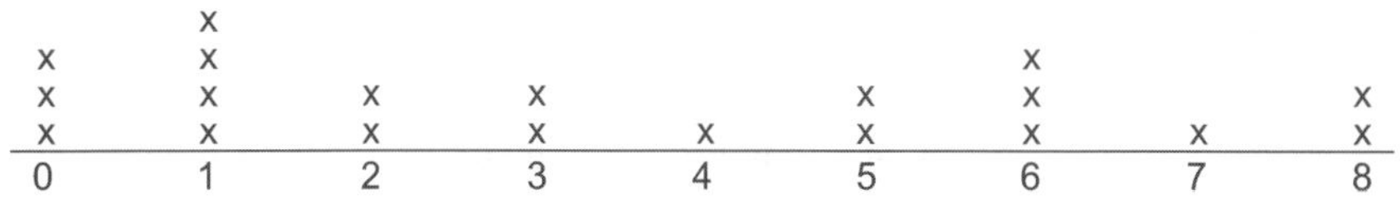

12. Is the following line plot correct for this set of data? Answer yes or no.
10, 0, 9, 0, 0, 5, 0, 8, 3, 2, 6, 2, 5, 9, 2, 5, 5, 0, 6, 1

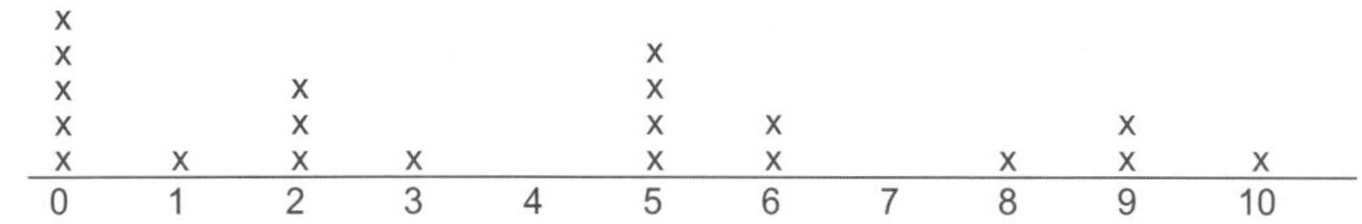

13. Is the following line plot correct for this set of data? Answer yes or no.
9, 0, 10, 7, 2, 9, 1, 1, 8, 6, 4, 9, 1, 1, 1, 5, 4, 0, 2, 6

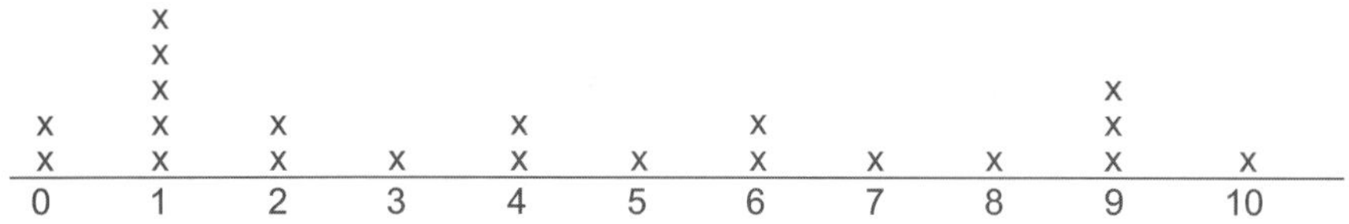

14.

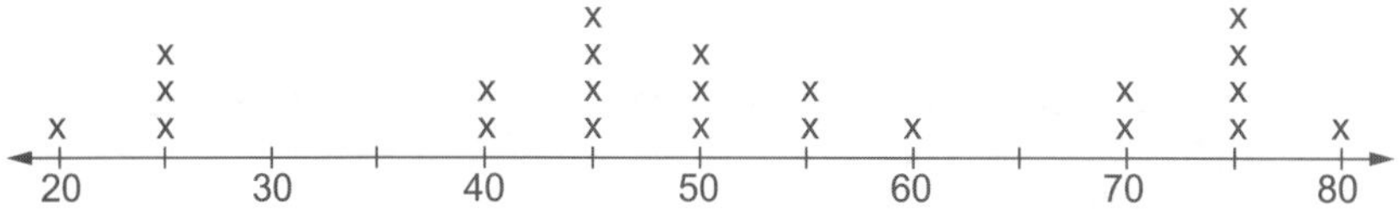

Which number or numbers appear with the most frequency in the data used to construct the given plot?

15.

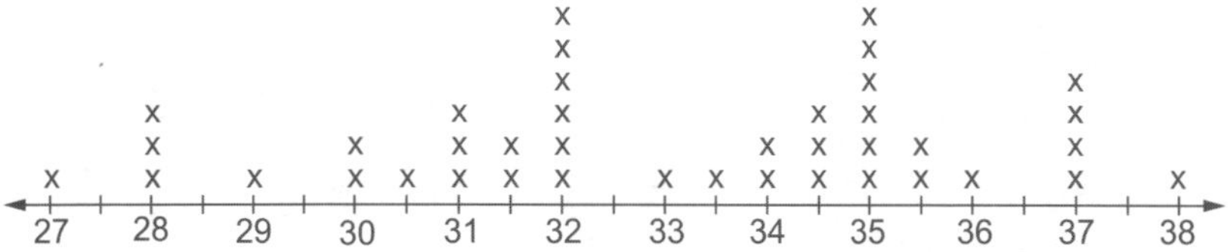

Which number or numbers appear with the most frequency in the data used to construct the given plot?

16.

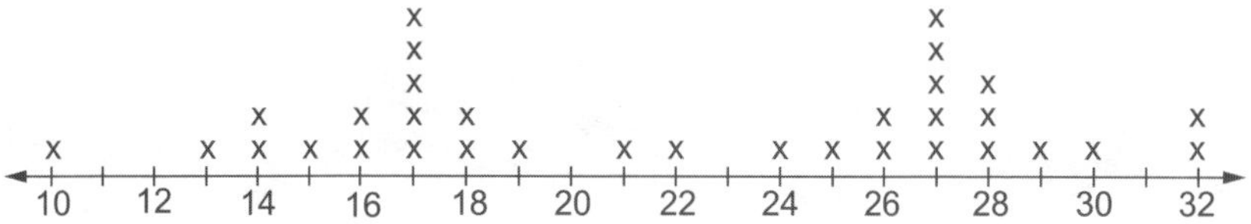

Which number or numbers appear with the most frequency in the data used to construct the given plot?

17.

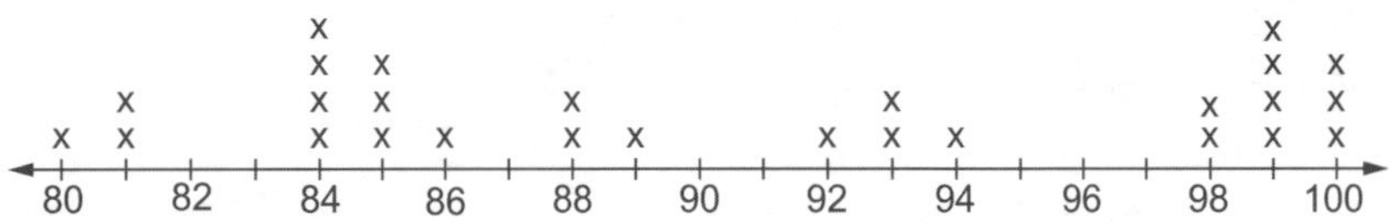

Which number or numbers appear with the most frequency in the data used to construct the given plot?

18.

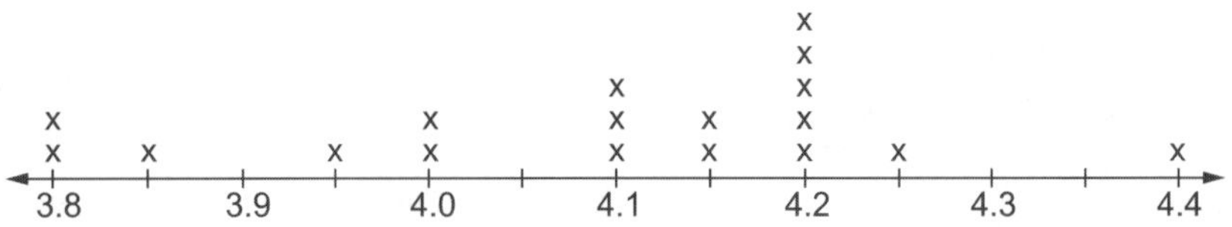

Which number or numbers appear with the most frequency in the data used to construct the given plot?

19.

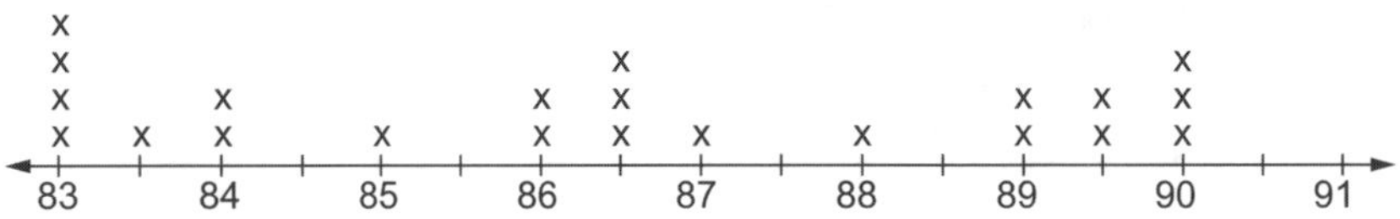

Which number or numbers appear with the most frequency in the data used to construct the given plot?

20.

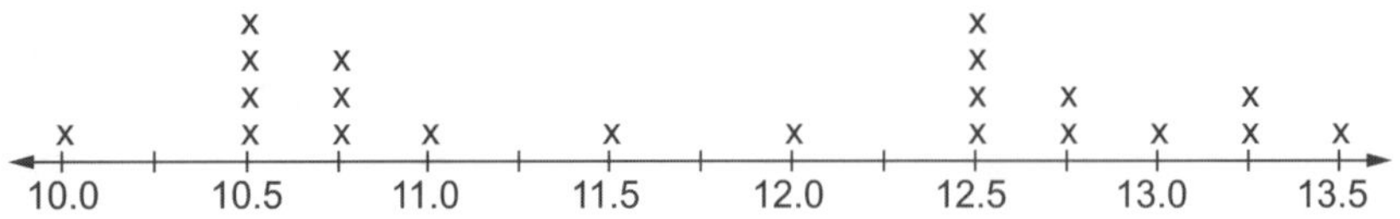

Which number or numbers appear with the most frequency in the data used to construct the given plot?

MPA-095: Find the Mean, Median, and Mode

Previously, you learned that pictorial representations, such as graphs, can be useful tools for interpreting large sets of data. In this lesson, you will investigate how to describe large data sets by using a single number. A number that is used to describe a data set is called a **descriptive measure.**

Descriptive measure	A number that is used to describe a data set.

To describe a large set of data with a single number, you must find a number that best represents the complete data set. Descriptive measures that indicate the most likely value of a set of data are known as **measures of central tendency.**

Measures of central tendency	Descriptive measures that indicate the most likely value of a set of data.

Measures of central tendency are commonly referred to as averages. The three measures of central tendency we will learn in this lesson are: mean, median, and mode. All three of these measures of central tendency involve different methods for computing the average for a set of data. The **arithmetic mean**, or simply the **mean**, is the most commonly used method of computing the average of a set of data. The procedure for computing the mean of a data set is summarized below.

Mean	The mean of a data set is the sum of the data divided by the total number of pieces of data in a given data set. $\text{Mean} = \dfrac{\text{the sum of the data}}{\text{the total number of pieces of data}}$

The following example demonstrates how to find the mean of a data set.

Example 1 Calculate the mean of the following data set: 5, 60, 77, 11, 7, 9, 89, 42, 1, –1

$$\text{Mean} = \frac{\text{the sum of the data}}{\text{the total number of pieces of data}}$$

Step 1: Start with the formula for mean.

$$\text{Mean} = \frac{5 + 60 + 77 + 11 + 7 + 9 + 89 + 42 + 1 + (-1)}{10}$$

Step 2: Insert the known values into the formula.

$$\text{Mean} = \frac{300}{10} = 30$$

Step 3: Reduce the answer to lowest terms.

Answer: 30

The second measure of central tendency that you will study in this lesson is the **median.**

Median	The number in the middle when data are arranged in ascending order.

The median is a method of computing the average that involves dividing the bottom half of the data from the top half of the data in a given data set. The procedure for calculating the median of a data set is summarized below.

Determining the Median of a Data Set	1. Arrange the data in ascending order. 2. If the data set has an odd number of data values, then the median is the middle number of the data set. 3. If the data set has an even number of data values, then the median is the mean of the two middle numbers in the data set.

Example 2 Calculate the median of the following data set: 0, 10, 9, 4, 5, 6, 1, 2, 11

0, 1, 2, 4, 5, 6, 9, 10, 11 — **Step 1:** Arrange the data in ascending order.

Step 2: Since this data set has an odd number of data values (9 numbers), the median is the middle number of the data set.

Answer: 5

Example 3 Calculate the median of the following data set: 0, 10, 9, 4, 5, 6, 1, 2, 11, 18.

0, 1, 2, 4, 5, 6, 9, 10, 11, 18 — **Step 1:** Arrange the data in ascending order.

$$\text{Median} = \frac{5+6}{2} = \frac{11}{2} = 5.5$$

Step 2: Since this data set has an even number of data values (10 numbers), the median is the mean of the two middle numbers in the data set.

Answer: 5.5

The last measure of central tendency is the **mode**.

Mode	The number or item that appears most often in the data set. The mode is best used to find the average when the data is in the form of categories.

The mode is a method of computing the average that involves finding the data value that occurs most frequently in a given data set. The procedure for finding the mode of a data set is summarized below.

Determining the Mode of a Data Set	The mode is the data value that occurs most frequently in a given data set. Some special cases are as follows: 1. If all data values in the data set occur only once, then there is no mode. 2. If more than one value occurs most frequently, then there is more than one mode. 3. A set that has two modes (two values that occur most frequently) is said to be *bimodal*.

The following examples demonstrate how to determine the mode.

Example 4 Find the mode of the following data set: 8, 2, 2, 1, 5, 21.

Step 1: The data value 2 occurs most frequently.

Answer: 2

Example 5 Find the mode of the following data set: 1, 2, 3, 4, 5, 8, 9, 100.

Step 1: Each of the data values in this data set occur only once.

Answer: This data set has no mode.

Example 6 Find the mode of the following data set: –1, 7, 7, 2, 1, 3, 5, 5.

Step 1: The data values 5 and 7 both occur twice.

Answer: The modes of this data set are 5 and 7. This data set is said to be bimodal.

Note that for larger data sets, it is often helpful to arrange the data values in ascending order before determining the mode.

Problem Set

1. What is the mode of the given data set? Answer "no" if the set does not have a mode. Answer "more" if the set has more than one mode.
14, 94, 5, 7, 88

2. What is the mean of the given data set? Round to the nearest tenth if necessary. 3, 3, 15, 3, 18, 23

3. What is the mode of the given data set?
24, 17, 30, 22, 1, 1

4. What is the mean of the given data set? Round to the nearest tenth if necessary. 24, 17, 30, 22, 1, 1

5. What is the mean of the given data set? Round to the nearest tenth if necessary. 14, 16, 26, 32, 35, 36, 45

6. What is the mean of the given data set? Round to the nearest tenth if necessary. 45, 11, 29, 33, 46, 10, 37

7. What is the median of the given data set? Round to the nearest tenth if necessary.
50, 52, 48, 44, 42, 45, 49

8. What is the median of the given data set? Round to the nearest tenth if necessary.
57, 40, 57, 43, 59, 52, 54, 43

9. What is the median of the given data set? Round to the nearest tenth if necessary.
67, 70, 34, 48, 41, 53, 64, 60

10. What is the median of the given data set? Round to the nearest tenth if necessary.
62, 63, 66, 34, 45, 48, 45, 54

11. What is the mode of the given data set? Answer "no" if the set does not have a mode. Answer "more" if the set has more than one mode.
19, 13, 1, 6, 32, 32, 34, 20

12. What is the median of the given data set? Round to the nearest tenth if necessary.
1, 10, 17, 39, 36, 48, 27, 29, 15

13. What is the median of the given data set? Round to the nearest tenth if necessary.
50, 42, 39, 29, 12, 13, 22, 18, 19

14. The sales price of six watches are listed in the table below. What is the median price?

Ladies' gold band	$89.99
Men's gold band	$99.99
Ladies' leather band	$59.99
Men's leather band	$49.99
Children's character	$49.99
Pocket watch	$59.00

15. The produce clerk was asked to find the median weight of 5 sacks of corn. The weights were $10\frac{1}{2}$ pounds, 12 pounds, $10\frac{1}{2}$ pounds, 11 pounds, and $11\frac{1}{4}$ pounds. What is the median weight?

16. What is the mode for this set of numbers? 120, 345, 237, 423, 237, 120. If there is more than one mode, separate each one with a single space.

17. Find the mode for this set of numbers: 26, 89, 77, 26, 56, 77, 89, 61, 77, 26. If there is more than one mode, separate each one with a single space.

18. Sarah is reading a novel. She records the amount of pages she reads each day until she finishes the book: 10, 6, 12, 7, 6, 3, 11, 8, 8, 4, 15, and 12. What is the average amount of pages Sarah read in a day?

19. Adam takes a survey of his classmates who had summer jobs. He asked how much money students made in a week. These were the amounts he found: \$103, \$120, \$48, \$52, \$120, \$110, and \$78. What is the mode of this set?

20. Shawnique times (in seconds) how long she can jump rope without stopping. What is the average amount of time she can jump rope?
25, 47, 68, 123, 178, 203, 392

MPA-096: Constructing Stem-and-Leaf Plots

In earlier lessons, you learned several techniques for organizing and summarizing large data sets in order to make them easier to understand and interpret. Such methods included measures of dispersion, measures of central tendency, and line plots. Recall that one of the most useful properties of the line plot is that it displays the frequency distribution of the data set.

In this lesson, you will learn how to construct stem-and-leaf plots, which also display the frequency distribution of a data set.

Stem-and-Leaf Plot	A plot that displays the frequency distribution of a data set.

The stem-and-leaf plot is useful because it displays the minimum and maximum data values for a given set of data. This type of plot allows you to discern between typical and atypical data values in a data set and also allows you to determine how many data values fall into a certain range within a data set. To construct a stem-and-leaf plot, all data must have the same number of digits. Use leading zeroes if necessary. The leading digit becomes the stem, and the remaining digits form the leaves. The basic procedure for constructing a stem-and-leaf plot is summarized below.

Constructing Stem-and-Leaf Plots

1. Determine the minimum and maximum data values for the given data.
2. Determine the distinct stems (the leading digits for the data values) and place them in a vertical list.
3. Determine the leaves (the final digits for the data values) and place them to the right of the corresponding stem.
4. Rearrange the leaves for each stem so that they are in ascending order.
5. Devise a key for the stem-and-leaf plot.

The following example demonstrates how to create a stem-and-leaf plot.

Example 1 Construct a stem-and-leaf plot for the data given below.

Fasting Plasma Glucose Levels for 20 Non-Diabetic Patients (mg/dL)			
51	62	83	88
78	85	87	92
86	99	93	84
85	94	80	65
83	92	49	81

Stems	Leaves
4	9
5	
6	
7	
8	
9	9

Step 1: The minimum data value is 49, and the maximum data value is 99. Thus, the extreme stems will be 4 and 9. Since you have leaves for 5, 6, 7, and 8, the stems consist of numbers 4 through 9. The minimum data of 49 is represented as stem 4 and leaf 9. The maximum data of 99 is represented as stem 9 and leaf 9.

Stems	Leaves
4	9
5	1
6	2 5
7	8
8	6 5 3 5 3 7 0 8 4 1
9	9 4 2 3 2

Step 2: Match the leaves of the data values with the corresponding stem. For example, 85 becomes 8 | 5.

Stems	Leaves
4	9
5	1
6	2 5
7	8
8	0 1 3 3 4 5 5 6 7 8
9	2 2 3 4 9

Step 3: Rearrange the leaves for each stem so that they are in ascending order.

Step 4: Finally, devise a key for the stem-and-leaf plot. The key consists of a stem, a horizontal line, a leaf, and the equivalent number: 4 | 9 = 49.

Answer:

Stems	Leaves
4	9
5	1
6	2 5
7	8
8	0 1 3 3 4 5 5 6 7 8
9	2 2 3 4 9

4 | 9 = 49

Another type of stem-and-leaf plot is called the back-to-back stem-and-leaf plot. This type of plot is similar to the single stem-and-leaf plot, but the back-to-back plot lets you construct stem-and-leaf plots for two different sets of data simultaneously by using the same stems for each data set. The back-to-back stem-and-leaf plot is useful in comparing the frequency distributions and extreme values for two sets of data. The procedure for constructing back-to-back stem-and-leaf plots is illustrated in the following example.

Example 2 Construct a back-to-back stem-and-leaf plot for the two sets of data given below.

Fasting Plasma Glucose Levels for 20 Non-Diabetic Patients (mg/dL)			
51	62	83	88
78	85	87	92
86	99	93	84
85	94	80	65
83	92	49	81

Ages for 20 Non-Diabetic Patients (years)			
69	40	77	23
48	35	85	92
76	20	54	41
51	24	44	38
49	56	60	89

Step 1: Refer to Example 1 for a detailed explanation of the stem-and-leaf plot for the fasting plasma glucose level data. The stem-and-leaf plot for the age data is constructed in a similar manner.

Step 2: For the back-to-back stem-and-leaf plot, list the distinct stems for both sets of data in the middle column. The first data set has a range of 49 to 99 and the second has a range of 20 to 92. Therefore, the stems consist of the numbers 2 through 9. The leaves are constructed the same way as in the single stem-and-leaf case for each data set and are tabulated on opposite sides of the stem column.

Leaves	Stems	Leaves
	2	0 3 4
	3	5 8
9	4	0 1 4 8 9
1	5	1 4 6
5 2	6	0 9
8	7	6 7
8 7 6 5 5 4 3 3 1 0	8	5 9
9 4 3 2 2	9	2

Step 3: Arrange the leaves in ascending order from the stem. Add the keys.

Leaves	Stems	Leaves
	2	0 3 4
	3	5 8
9	4	0 1 4 8 9
1	5	1 4 6
5 2	6	0 9
8	7	6 7
8 7 6 5 5 4 3 3 1 0	8	5 9
9 4 3 2 2	9	2
1 | 6 = 16		2 | 0 = 20

Problem Set

1. Heights of the Members of the West Ogden Junior High Student Council (in inches): 37, 33, 59, 43, 50, 47, 58, 52, 34, 65, 55, 38. How many leaves will be on the 40 stem on a stem-and-leaf plot for the data?

2. Height of the Members of the West Ogden Junior High School Council (in inches):

3	3 4 7 8
4	3 7
5	0 2 5 8 9
6	5

3|7 = 37 inches

How many data points are included in the plot?

3. Pencil prices at Various School Store (in cents): 16, 21, 40, 44, 45, 33, 49, 38, 49, 27. How many leaves would be on the 4 stem on the stem-and-leaf plot for this data?

4. Pencil prices at Various School Store (in cents): 16, 21, 40, 44, 45, 33, 49, 38, 49, 27. What will be the leaves for the 3 stem on the stem-and-leaf plot for this data?

5. Pencil prices at Various School Store (in cents):

1	6
2	1 7
3	8
4	0 3 4 5 9

1|6 = 16 cents

How many data points are on the plot?

6. Weights of Students' Baseball Card Collections (in pounds): 44, 87, 16, 60, 17, 55, 51, 58, 81, 71, 92, 35, 29, 34, 61. How many stems would be on the stem-and-leaf plot for this data?

7. Weights of Students' Baseball Card Collections (in pounds): 44, 87, 16, 60, 17, 55, 51, 58, 71, 92, 35, 29, 34, 61. Is this the correct stem-and-leaf plot for the data?

1	6
2	9
3	4 5
4	4
5	1 5 8
6	0 1
7	1
8	7
9	2

No Yes

8. Number of Students in Mrs. Starr's Music Classes: 25, 26, 30, 17, 15, 22, 37, 24. How many leaves would be for the 1 stem on the stem-and-leaf plot for this data?

9. Jored is buying a phone. The prices of the phones are 27, 33, 42, 49, 46, 36, 50, 40, 29, 15, and 12 dollars.

Is this the correct stem-and-leaf plot for the data?

1	2 5
2	7 9
3	3 6
4	0 2 6
5	0

2|6 = $26

Yes No

10. Miles Run in One Week by Members of a Cross Country Team: 17, 33, 42, 49, 46, 36, 50, 40, 19, 15, 12. How many stems would be on the stem-and-leaf plot for this data?

11. Height of marshmallow towers built in a contest:

0	8
1	
2	2
3	7
4	2 6

0|8 = 8 inches

What is the mean of the data represented by the stem-and-leaf plot?

12. Height of marshmallow towers built in a contest:

0	8
1	
2	2
3	7
4	2 6

0|8 = 8 inches

What is the median of the data represented by the stem-and-leaf plot?

13. Prices of Different Packs of Gum:

4	1 8
5	1
6	
7	2 3

4|1 = 41 cents

What is the largest number represented by the stem-and-leaf plot?

14. Prices of Different Packs of Gum:

4	1 8
5	1
6	
7	2 3

4|1 = 41 cents

What is the range of data represented by the stem-and-leaf plot?

15. Prices of Different Magazines:

1	5 9
2	0 5 5
3	6
4	
5	8
6	
7	
8	8

1|5 = 1.5 dollars

What is the mode of the data represented by this stem-and-leaf plot?

16. Weight of Candy Bags (in ounces):

1	6
2	4
3	
4	
5	1
6	
7	4
8	8 9

1|6 = 1.6 ounces

What is the smallest number represented by the stem-and-leaf plot?

17. Weight of Candy Bags (in ounces):

1	6
2	4
3	
4	
5	1
6	
7	4
8	8 9

1|6 = 1.6 ounces

What is the median of the data represented by the stem-and-leaf plot?

18. Length of My Favorite Movies:

9	3
10	3
11	3 3 4
12	1 8
13	1
14	0 6
15	
16	1

9|3 = 93 minutes

What is the smallest number represented by the stem-and-leaf plot?

19. Distance Traveled to Grandma's House:

11	4 6
12	4
13	0 5 9
14	
15	
16	1
17	5 6
18	
19	0

7|5 = 75 miles

What is the largest number represented by the stem-and-leaf plot?

20. Distance Traveled to Grandma's House:

11	4 6
12	4
13	0 5 9
14	
15	
16	1
17	5 6
18	
19	0

7|5 = 75 miles

What is the mean of the data represented by the stem-and-leaf plot?

MPA-097: Constructing Box-and-Whisker Plots

You learned in previous lessons how to compute the median of a data set. The median of a data set separates the data into two equal parts, namely, the upper 50 percent of the data and the lower 50 percent of the data. In this lesson, you will separate data sets into **percentiles**. Percentiles divide data into 100 equal parts. For instance, if a test score is at the 80th percentile, then 80 percent of all scores fall below this score, and 20 percent of all scores lie above this score.

Distinctive names are given to percentiles with certain characteristics. **Quartiles** are percentiles that partition data into four equal parts. The first quartile is the 25th percentile of a data set. The second quartile is the 50th percentile of the data set. Recall that a median partitions a data set into its upper 50 percent of data values and its lower 50 percent of data values, which implies that the median is the 50th percentile of a data set. Thus, the second quartile is simply the median of the data set. The third quartile is the 75th percentile of the data set. The quartiles are calculated as follows:

Steps for Calculating Quartiles of a Data Set	1. Arrange the data in ascending order. 2. Find the median of the entire data set. The median is the second quartile. 3. Find the median of the data lying at or below the median of the entire data set. This median is the first quartile. 4. Find the median of the data lying at or above the median of the entire data set. This median is the third quartile.

In this lesson, you will also learn how to construct a box-and-whisker plot.

Box and Whisker Plot	A graph that summarizes and illustrates the central tendencies of a data set.

Box-and-whisker plots summarize and illustrate the central tendencies of a data set. These plots are a graphical display of a five-number summary of a data set. A five-number summary consists of the following values of a data set in ascending order:

1. Minimum data value
2. First quartile (often denoted Q_1)
3. Second quartile, or the median (often denoted Q_2)
4. Third quartile (often denoted Q_3)
5. Maximum data value

You will now learn how to display this summary graphically by constructing a box-and-whiskers plot:

1. Put the data values in ascending order.
2. Draw a horizontal axis that is scaled to the data.
3. Determine and plot the first, second, and third quartiles above their respective locations on the horizontal axis.
4. Plot the minimum and maximum data values above their respective locations on the horizontal axis.
5. Draw a rectangular box above the horizontal axis, with the left and right sides drawn at the first and third quartiles.
6. Draw a vertical line segment inside the rectangular box at the second quartile, or the median.
7. Draw a horizontal line segment from the minimum data value to the midpoint of the left side of the box. This line is called the left whisker.

8. Draw a horizontal line segment from the midpoint of the right side of the box to the maximum data value. This line is called the right whisker.

9. Adjust for outliers by using the number closer to the overall pattern for the minimum or maximum data value. An outlier is a data value that lies well beyond the overall pattern of the data.

Example 1 Determine the five-number summary for the data given below.

Fasting Plasma Glucose Levels for 20 Non-Diabetic Patients (mg/dL)			
50	60	85	75
75	70	90	25
85	100	90	85
80	95	80	70
80	90	45	80

Minimum of the data set = 25

Maximum of the data set = 100

Step 1: Arrange the data in ascending order: 25, 45, 50, 60, 70, 70, 75, 75, 80, 80, 80, 80, 85, 85, 85, 90, 90, 90, 95, and 100. This illustrates the minimum (25) and maximum (100) numbers in the data set.

$$\frac{80+80}{2} = \frac{160}{2} = 80$$

Step 2: The second quartile, Q_2, is the median of the entire data set (80).

$$\frac{70+70}{2} = \frac{140}{2} = 70$$

Step 3: For the first quartile, Q_1, find the median of the data lying at or below the median of the entire data set. The median of the entire data set is 80. The data values lying at or below 80 are 25, 45, 50, 60, 70, 70, 75, 75, 80, and 80. The median of this data set is $\frac{70+70}{2} = \frac{140}{2} = 70$. Thus, the first quartile is 70.

$$\frac{85+90}{2} = \frac{175}{2} = 87.5$$

Step 4: For the third quartile, Q_3, you want to find the median of the data lying at or above the median of the entire data set. The median of the entire data set is 80. The data values lying at or above 80 include the following: 80, 80, 85, 85, 85, 90, 90, 90, 95, 100. The median of this data set is $\frac{85+90}{2} = \frac{175}{2} = 87.5$. Thus the third quartile is 87.5.

Answer: The five-number summary for this data set is 25, 70, 80, 87.5, and 100.

Example 2 Construct a box-and-whiskers plot for the data given below.

Fasting Plasma Glucose Levels for 20 Non-Diabetic Patients (mg/dL)			
50	60	85	75
75	70	90	25
85	100	90	85
80	95	80	70
80	90	45	80

Step 1: Begin by drawing a horizontal axis that is scaled to the data.

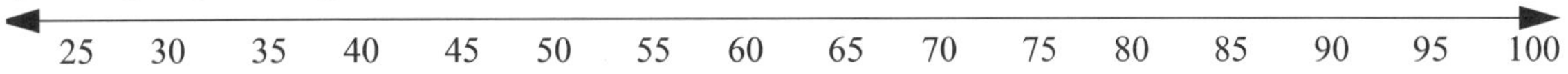

Step 2: From the preceding example, you found the following five-number summary:

Minimum data value = 25

Quartile one = 70

Quartile two = 80

Quartile three = 87.5

Maximum data value = 100

Step 3: Plot the values above their respective locations on the horizontal axis. Observe that the data value 25 appears to be well beyond the overall pattern of the data and is therefore an outlier. Thus, you will not include the data value 25 in the box-and-whiskers plot.

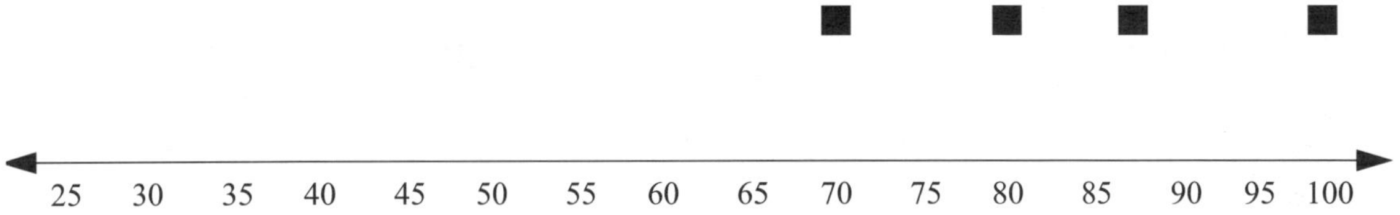

Step 4: Draw a rectangular box with the left and right sides drawn at the first and third quartiles. Also, draw a vertical line segment inside of the rectangular box at the second quartile, or the median.

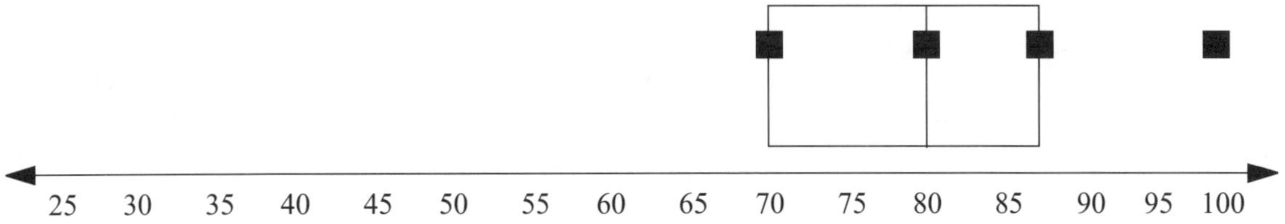

Step 5: Draw the whiskers of the box plot by connecting the midpoints of each side of the rectangular box with the minimum and maximum data values. Since the data value 25 is considered to be an outlier, use the next data value (45) as the minimum data value.

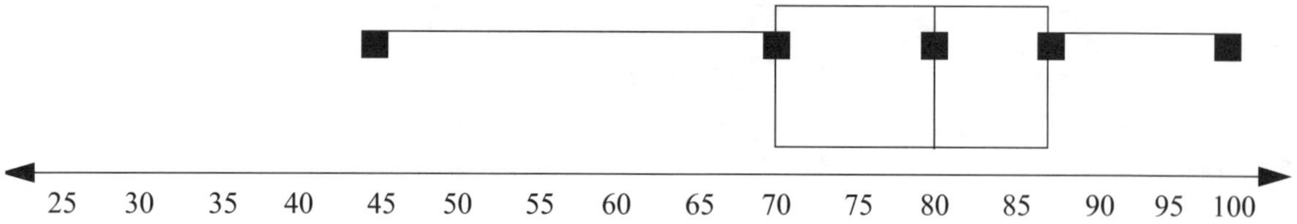

Answer:

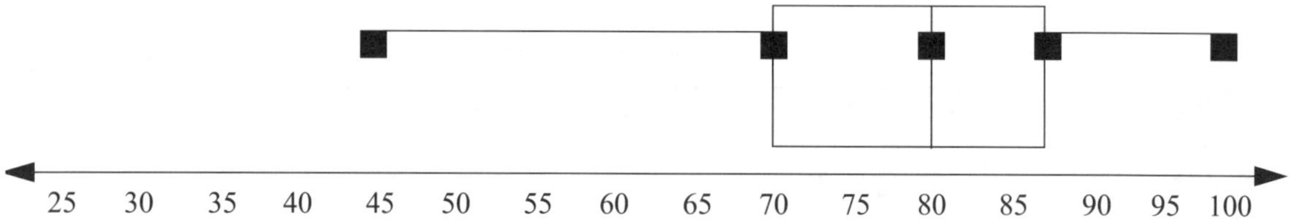

Problem Set

1. Using the following data set: 16, 25, 29, 42, 38, 9, 32, 16, 4. What is the median of the data set?

2. Using the following data set: 32, 13, 21, 9, 11, 15, 17, 32, 34, 15, 9. What is the 1st quartile of the data set?

3. Using the following data set: 32, 13, 21, 9, 11, 15, 17, 32, 34, 15, 9. What is the 3rd quartile of the data set?

4. Using the following data set: 9, 45, 37, 13, 46, 17, 4, 22, 39, 32, 35, 11, 50, 23. What is the lowest data point in the set?

5. Using the following data set: 20, 9, 49, 35, 27, 42, 11, 16, 23, 7, 34, 8, 7, 5. What is the lowest data point in the set?

6. Using the following data set: 20, 9, 49, 35, 27, 42, 11, 16, 23, 7, 34, 8, 7, 5. What is the highest data point in the set?

7. Using the following data set: 50, 11, 49, 23, 22, 1, 17, 4, 33, 4, 48, 32, 32, 39, 16. What is the lowest data point in the data set?

8. Using the following data set: 50, 11, 49, 23, 22, 1, 17, 4, 33, 4, 48, 32, 32, 39, 16. What is the highest data point in the set?

9.

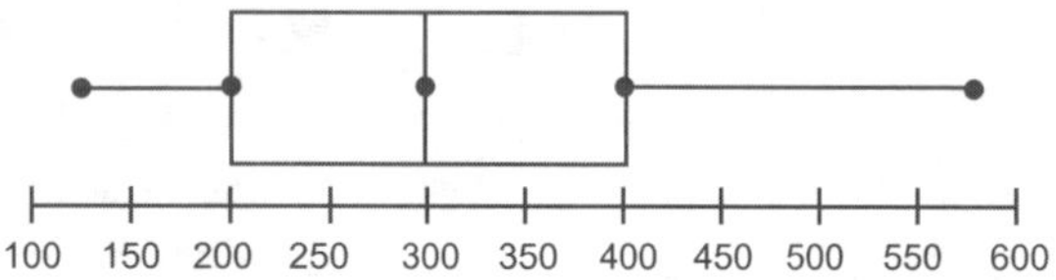

What is the median of the set of data used to make this box-and-whisker plot?

10.

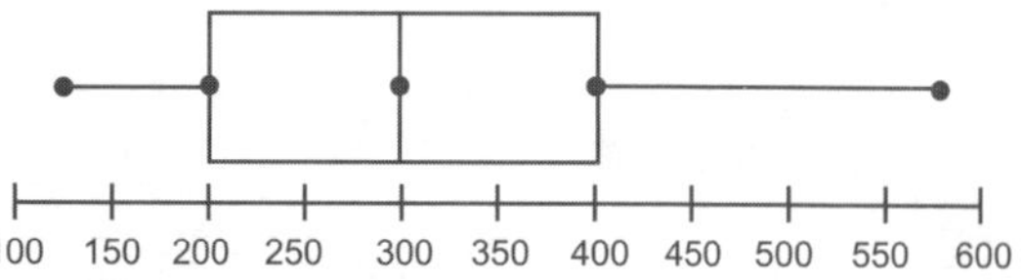

What is the first quartile of the set of data used to make this box-and-whisker plot?

11.

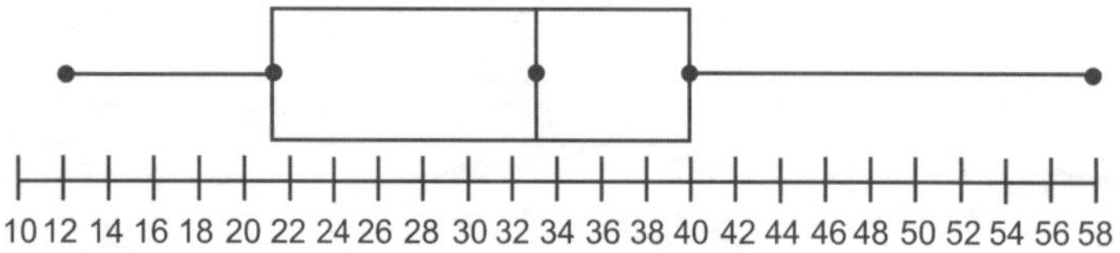

What is the lowest value of the set of data used to make this box-and-whisker plot?

12.

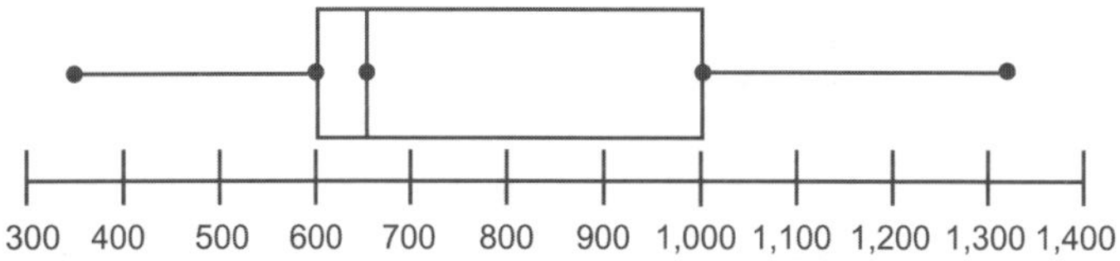

What is the lowest value of the data used to make this box-and-whisker plot?

13.

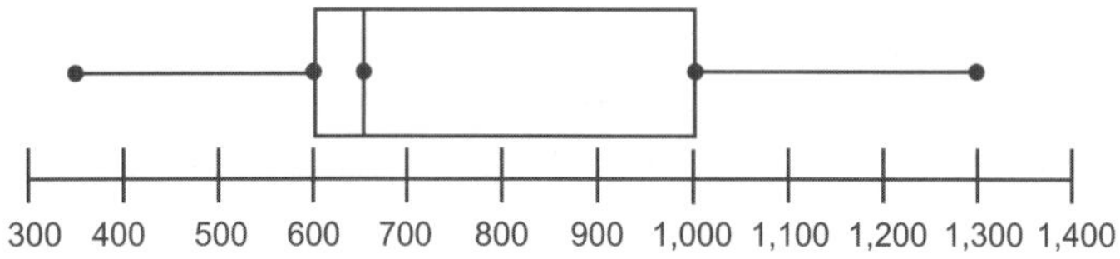

What is the highest value of the set of data used to make this box-and-whisker plot?

14. 10, 30, 20, 26, 42, 12, 26, 9, 4, 50
What is the highest data point in the set?

15. 90, 44, 41, 83, 18, 31, 6, 2, 24, 99, 5
What is the highest data point in the set?

16.

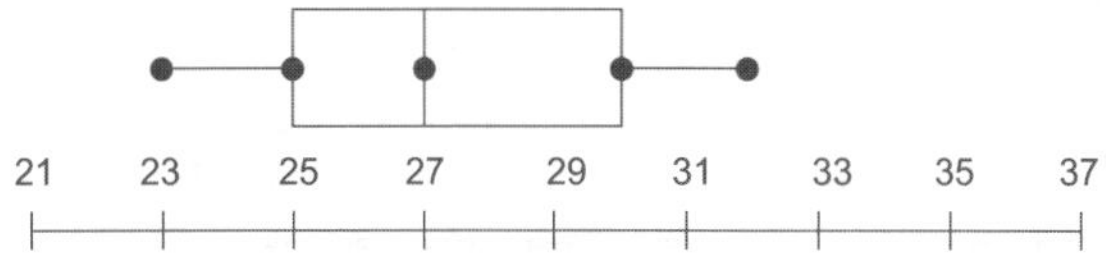

There are 1,000 numbers in the data set for the plot shown above.
What percent of the numbers are between 21 and 25? Use the percent sign in your answer.

17.

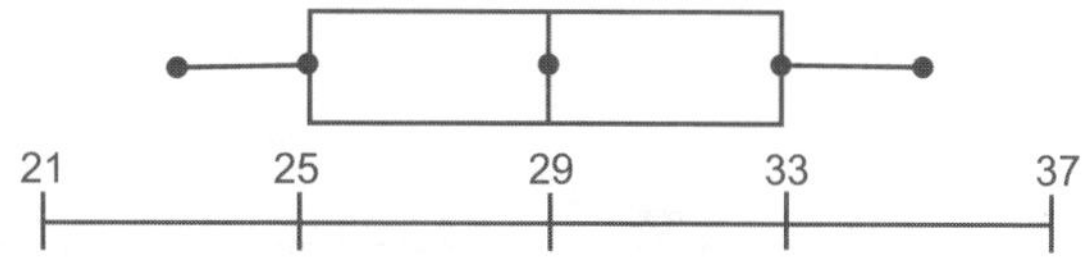

What percent of the numbers are greater than 37? Use the percent sign in your answer.

18.

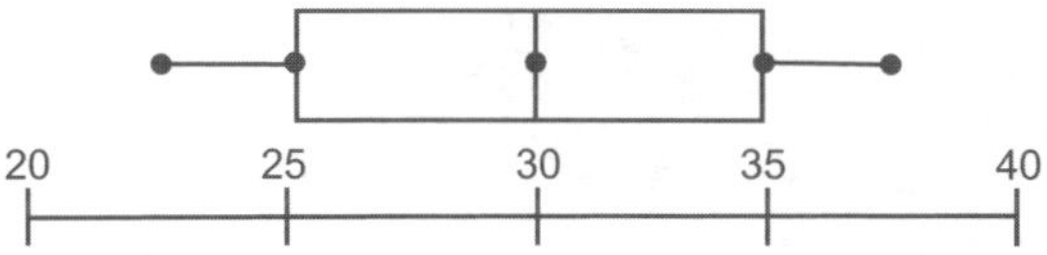

If there are 1,000 numbers in the data set for the above plot, what percent of the numbers are between 25 and 35? Use the percent sign in your answer.

19.

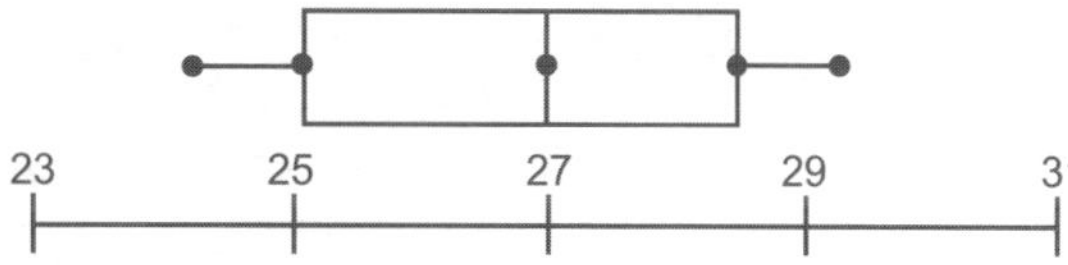

How many numbers are greater than 27 if there are 1,000 numbers in the data set?

20.

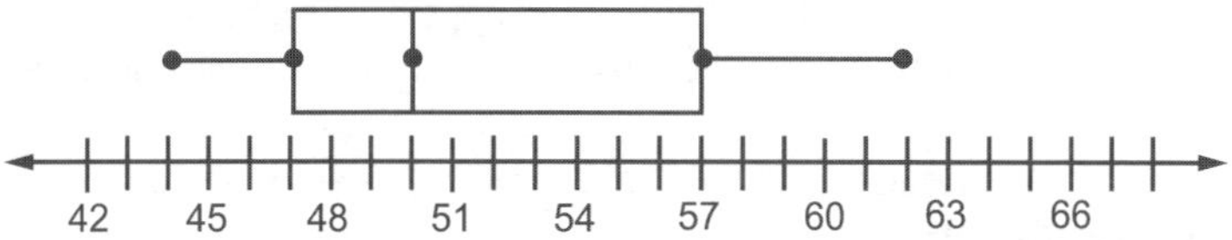

Using the box-and-whisker plot, what percent of the numbers are between 47 and 57?

MPA-098: Making Predictions from Graphs and Choosing the Correct Graph

Previously, you learned how to interpret bar, line, and circle graphs. You will find that you often want to make some kind of prediction based on the trend or comparison displayed by the graph. For instance, stock traders routinely make decisions about certain stocks based on the general tendency or trend of their performance. Stock traders are able to determine the trends of certain stocks by analyzing their performance data graphically. In this lesson, you will learn how to make predictions based on a graph's trend or general tendency.

When making predictions from graphs, there is always some degree of uncertainty involved. The goal of this lesson is to help you understand how to choose a graph that best represents the given data. To help you choose which graph is best, you can refer to the table below for the different types of graphs and their uses:

Bar Graph	Best for comparing values to each other.
Pie Graph	Compares the size of different parts to the whole.
Line Graph	Shows trends and predicts trends over time.
Scatter Plot	Shows relationships between two sets of data.

Example 1 Predict the population of the United States in the year 2010 using the graph below. Also, explain why a line graph is used for this population data.

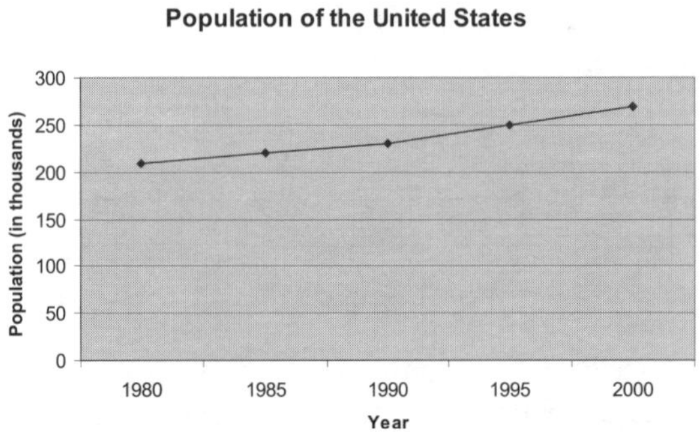

Step 1: Since this graph displays a linear, upward trend, you can conclude that the population of the United States will rise to approximately 300,000,000 by the year 2010. Note that a line graph is used for this data because the population of the United States varies continuously. A line graph associates values with all points over a continuous interval. Thus, line graphs are more appropriate for continuous data such as these.

Answer: Population: approximately 300,000,000. A line graph is used because the population of the United States varies continuously.

Example 2 A hospital patient with diabetes takes a glucose tolerance test, which involves the physician taking hourly plasma glucose level measurements each hour after the patient drinks a high-glucose beverage. At 8:00 a.m., this patient is given a high-glucose beverage. What effect does this beverage seem to have on this patient's plasma glucose level? Also, explain why a bar graph is used to represent this plasma glucose level data.

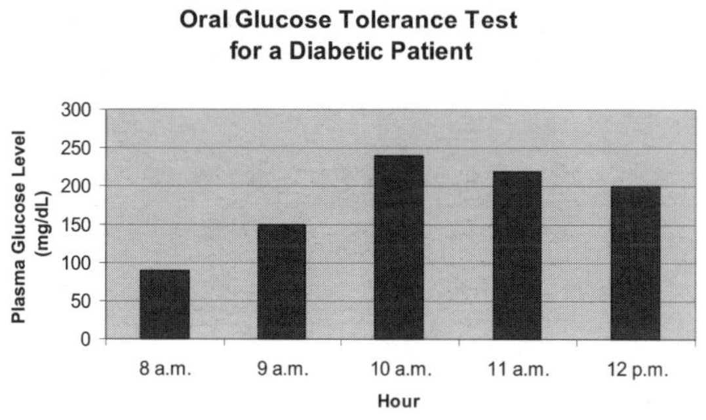

Step 1: Since you can see that this patient's plasma glucose level increases significantly after 8:00 a.m., you can conclude that the beverage caused the patient's glucose level to rise. A bar graph is used to illustrate this data, because the problem states that the plasma glucose level measurements are taken hourly; therefore the data is not continuous. You want to compare the plasma glucose levels at each hour; therefore, a bar graph is the appropriate choice, because it is used primarily for comparisons. A line graph would be inappropriate, because the data is not continuous. A circle graph is not appropriate, because you do not want to divide anything.

Answer: The beverage caused the patient's glucose level to rise. A bar graph is used because you are comparing the patient's glucose levels at hourly intervals.

Example 3 Using the graph below, what party do you think the newly elected president will represent? Why is a circle graph used to represent this data?

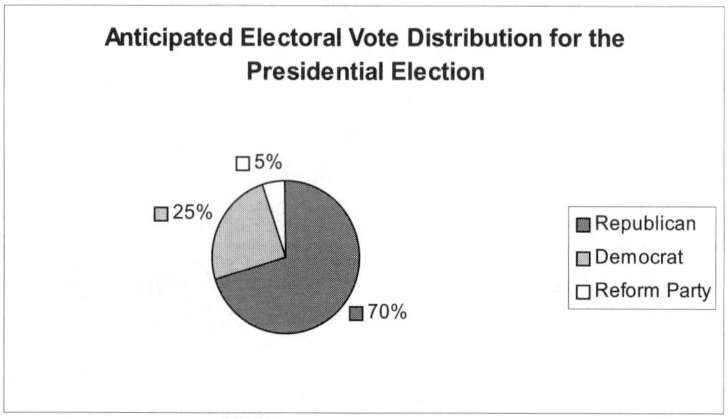

Step 1: Since it is anticipated that the majority of the electoral votes will be for the Republican party, you conclude that the newly elected president will probably represent the Republican party. A circle graph is appropriate in this case because you are dealing with qualitative data.

Answer: The president will probably represent the Republican party. A circle graph is used because you are dealing with qualitative data.

Example 4 Determine whether or not there is a relationship between weight and average daily exercise based on the graph below. Also, why is a scatter plot used to represent this data?

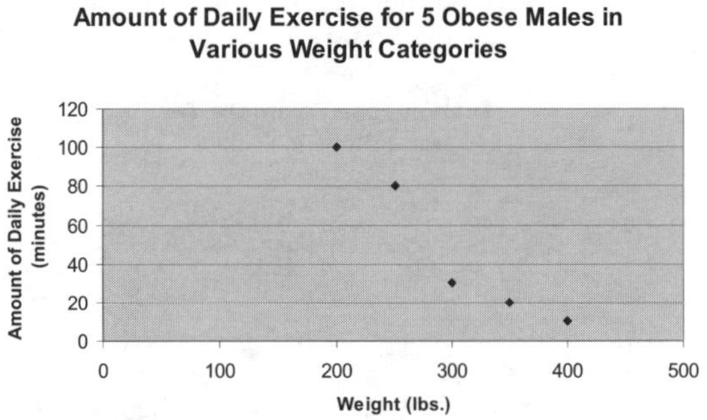

Step 1: There appears to be a relationship between weight and amount of daily exercise for this experiment. The heavier obese males seem to exercise much less than the lighter obese males. For example, the male who weighs 200 pounds exercises 100 minutes daily, and the male who weighs 400 pounds exercises about 10 minutes daily. A scatter plot is used because you want to demonstrate the relationship between two different sets of data taken from the same subjects.

Answer: The heavier obese males seem to exercise much less than the lighter obese males. A scatter plot is used to demonstrate the relationship between two different sets of data taken from the same subjects.

Problem Set

1. What kind of graph would you use to show which part of the body different percentages of people said they would change if they could?

none of these | pie and line graph
line graph only | pie graph only

2. What kind of graph would you use to compare the number of speeding tickets earned by men and women?

none of these | pie graph only
bar graph and pie graph | line and pie graph

3. Which type of graph is used for comparing parts to a whole amount?

line and bar graph | none of these
pie graph | line graph

4.

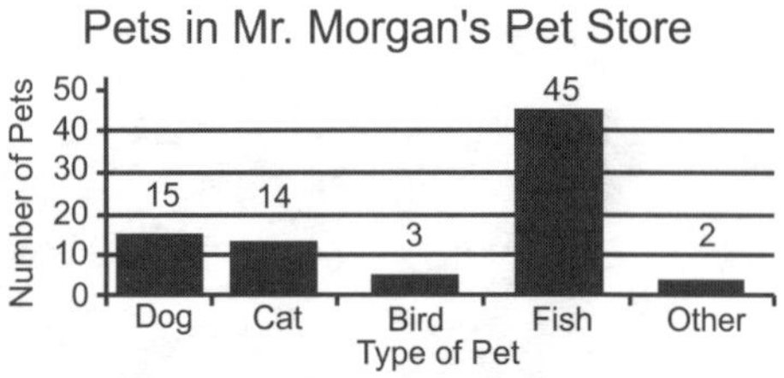

Next Wednesday Mr. Morgan will groom all of his cats and dogs. How many will be groomed?

5. 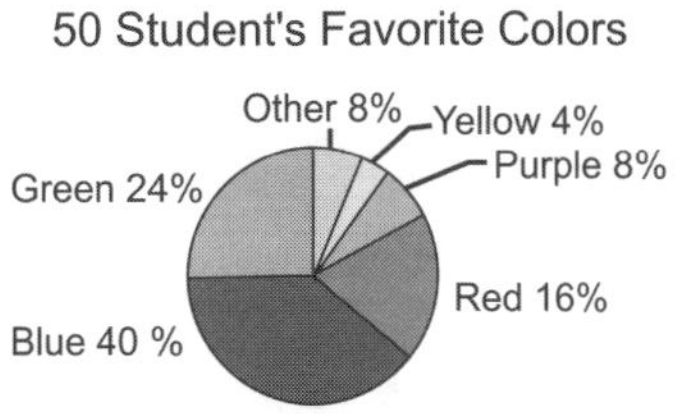

Mrs. Kinsha will use the graph to decide what color balloons to buy for her students for field day. What color balloon should she buy in the largest amount?

6. 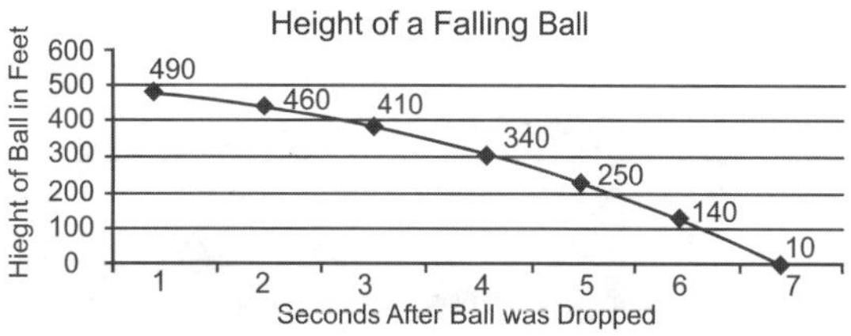

True or False: The ball is falling at the same rate throughout the fall.

7. 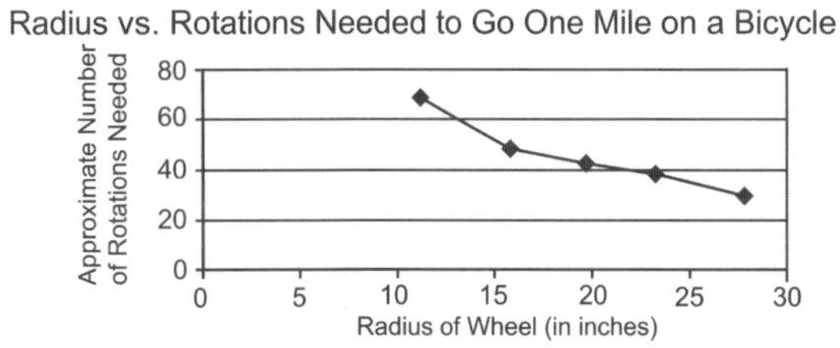

A certain bicycle needs about 30 rotations of the wheel to travel one mile. Which of the following is the best approximation of the radius of the wheel?

25 cm　　28 inches

none of these　　20 inches

8. Length of Segments in the 6 o'clock News (in minutes)

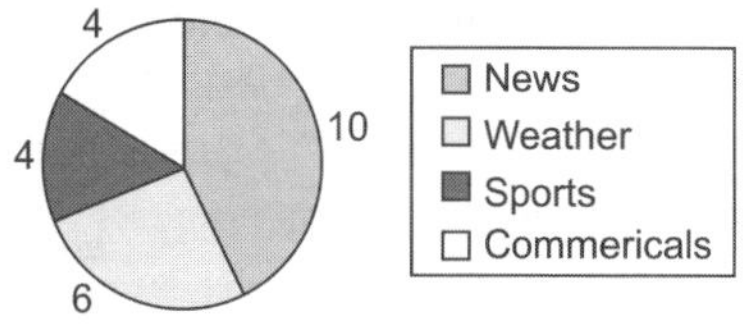

The program director for the station that shows this news program wants to increase the length of the show to 36 minutes. If she wants the proportions to stay the same, how many minutes of sports will be on the new show?

9. 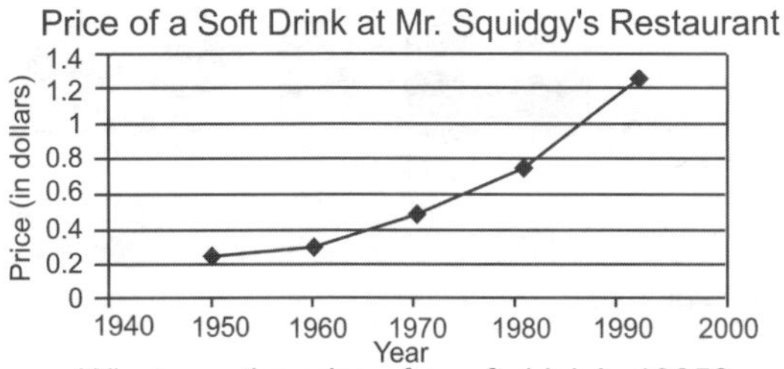

What was the price of a soft drink in 1985?

\$1.00　　none of these

\$.70　　\$1.80

10. 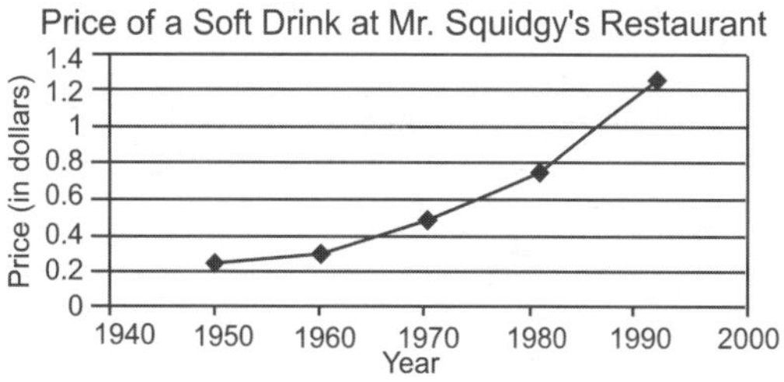

Can we conclude from this graph that Mr. Squidgy's was first opened in the year 1950?

11. 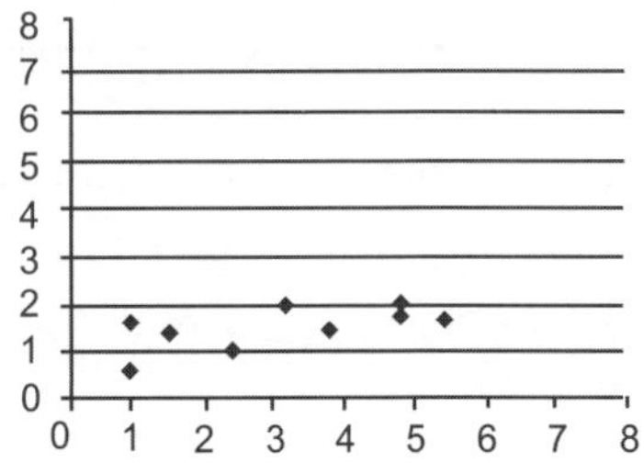

Based on the scatterplot above, predict which of the following values would complete this point. (7, ___)

none of these　　5

6　　3

12. 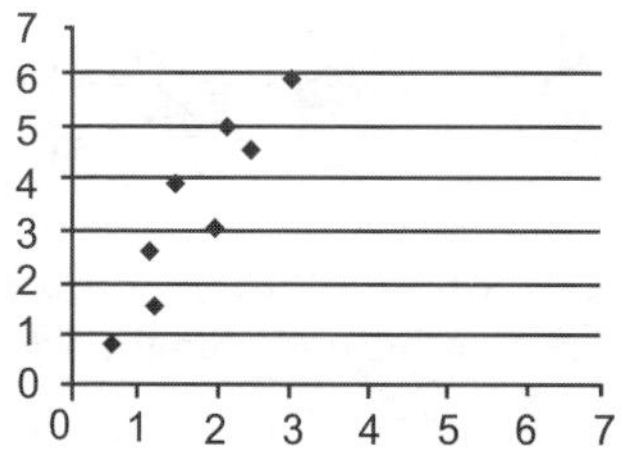

Based on the scatterplot above, predict which of the following values would complete this point. (9, ___)

5　　15

13.

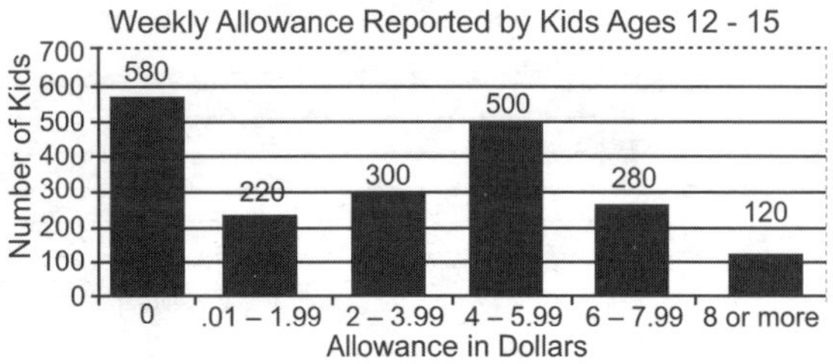

What percent of the kids do not get an allowance?

14.

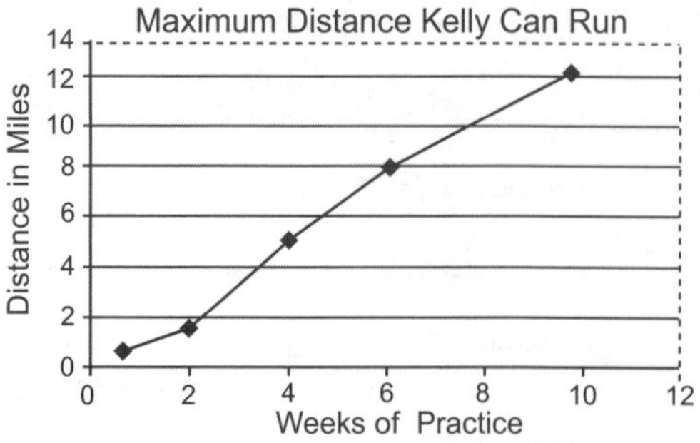

Kelly met her goal distance after 5 weeks of practice. What was her goal distance?

15. Identify the type of graph that should be used for the following set:

A data set that shows what percentage of all boys in the United States play baseball.

pie graph bar graph line graph

16. Identify the type of graph that should be used for the following set:

A data set that shows a decrease in cotton candy sales at the parish fair over the last 5 years.

pie graph bar graph line graph

17. Identify the type of graph that should be used for the following set:

A data set that compares the heart-rates of a grown man, a grown woman, a teenage girl.

pie chart bar graph line graph

18.

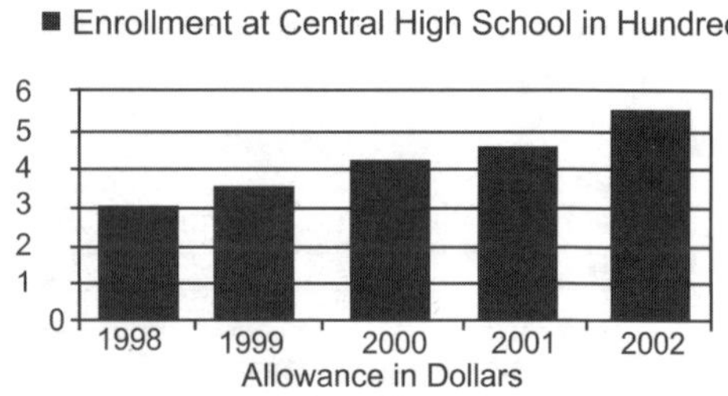

Is the trend of this graph increasing, decreasing, or neither?

19.

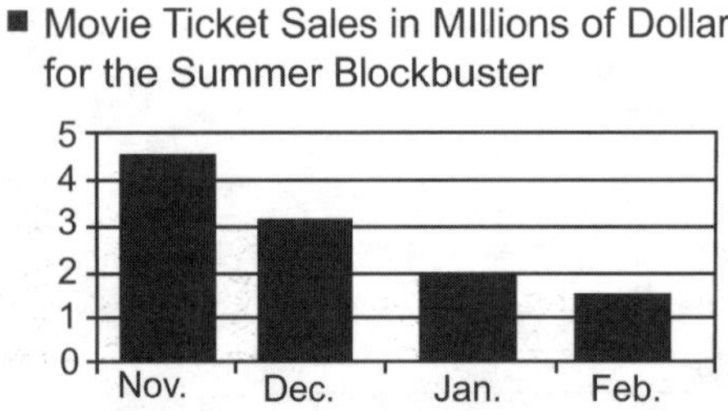

Are the movie ticket sales increasing, decreasing, or staying the same over a four-month period?

20.

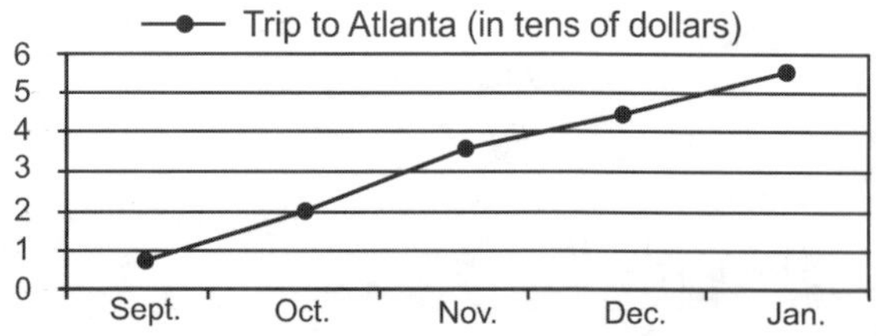

Is the trend in this graph increasing, decreasing, or neither?

MPA-099: Recognizing Misleading Statistics and Graphs

Previously, you learned how to interpret data sets represented graphically. However, such representations can be misleading. Because of this, you must understand potentially misleading graphs before accepting them as facts. This lesson points out several different types of such potentially misleading graphs.

Example 1 Describe the percentage of the population increase in the United States from 1980 to 2000.

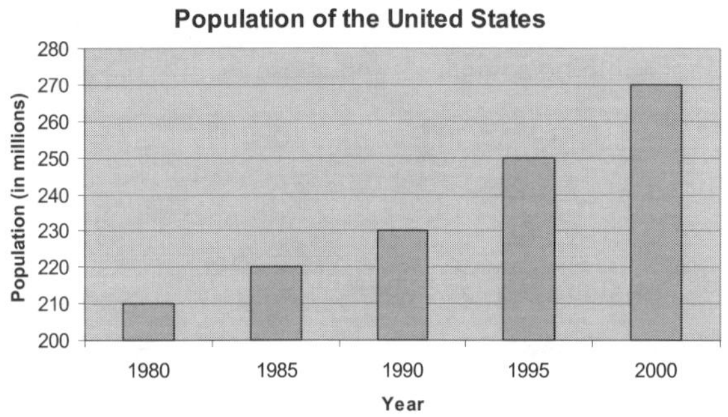

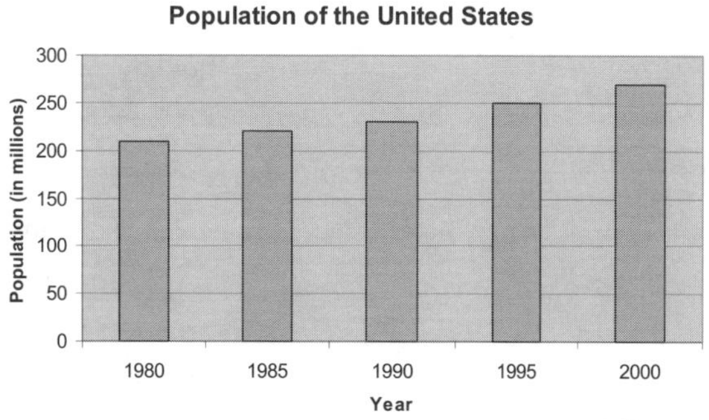

Step 1: From the first graph, it **appears** that the population of the United States has grown by more than 600 percent from 1980 to 2000. It only appears that way because the scale does not start at zero.

Step 2: Now, look at the second graph. These two graphs represent the same data set. The difference between the two graphs is that the scale of the first graph's vertical axis starts at 200,000,000, and the scale of the second graph's vertical axis starts at 0. The first graph is certainly the more deceptive of the two graphs. The first graph is an example of a *truncated graph*. A truncated graph is a graph in which part of the graph has been cut off. The part of the first graph from 0 to 200,000,000 has been cut off. The second graph is known as a non-truncated graph, because none of the graph has been cut off.

Step 3: From the non-truncated graph, you can see that the actual increase in the population of the United States is approximately 29 percent.

Answer: The population increased approximately 29 percent.

Statistics is the science of gathering, analyzing, and presenting data. One branch of statistics is known as **inferential statistics**. Inferential statistics involves making assumptions or predictions about a certain **population** based on information obtained from a **sample** of that population. A population consists of the set of all items of interest. Since it is not always possible to collect data for an entire population, you must select a sample, a group of items chosen from the population, to represent the entire population. However, choosing a sample of a population to represent the entire population often leads to misconstrued results.

Example 2 A survey is conducted over the telephone to determine the number of Independents in a certain city. A sample of 40 people is chosen to represent the entire population of the city. Give several reasons why this sample may not accurately represent the entire population of the city.

Step 1: First, a person may not be comfortable revealing such information, political affiliation, to someone he or she does not know. To maintain privacy, the person may give false information.

Step 2: Since the sample size is just 40, this false information would yield misleading results. Also, since the survey is conducted over the telephone, people who do not have phones are unable to be contacted, and consequently the entire population is not properly represented.

Answer: (1) The information may be too personal for some people to reveal.

(2) The sample size is too small.

(3) The entire population of the city is not represented, since this is a phone survey.

Sometimes the mean of a data set is not the best measure of central tendency. When one or more observations are far from others, it may overestimate or underestimate the average. Consider the following example:

Example 3 Jennifer wants to go shopping at a certain store. However, her mother will not take her to this store because she knows the mean price of a shirt at this store is \$130. If Jennifer's mother has only \$45 to spend on a shirt for Jennifer, should she go to this store? Explain.

1,000	41	38	20	38
30	40	31	25	37

Step 1: Consider the data set of shirt prices on the left.

Step 2: At this store, a cashmere sweater sells for \$1,000. This costly sweater greatly influences the mean of the entire data set. However, if this data set represents the shirt prices at this store, Jennifer's mother can afford most of the shirts, since most are under \$45.

Step 3: Yes, she should go to the store. If this data set represents the shirt prices at this store, Jennifer's mother can afford nine of the ten shirts.

Problem Set

1. Which of the following statistics would you use if you want a misleading statistic?

8,000 20,000 19,500 89,000 19,500

mode mean median

2. Which of the following statistics would you use if you did not want a misleading statistic?

109 182 1,450 3,800 3,800

mode mean median

3. These numbers represent the purchase prices of eight new computers and parts for a city's community center:

109 182 1,450 3,800 3,800

You want to use them to show the city council that the community center is spending a lot of money on technology. Which statistic would be the most effective?

mode mean median

4. Which of the following statistics would you use if you want a misleading statistic?

12,500 14,300 16,890 12,750 16,890 13,000
52,000 17,000

mode mean median

5. Which of the following statistics would you use if you want a misleading statistic?

100 110 112 105 104 106
111 105 107 113 109

none of these mode
mean median

6. The librarian at Rutherford Junior High believes that students in his school would enjoy having access to the Internet in the library. Which of the following methods should he use to support his claim?

Poll the computer club
Poll students at the school
None of these
Poll the library club.

7. To determine what day to hold the Neighborhood Homeowner's Meeting, President Moody calls her five best friends in the neighborhood to get their opinions. Does this sampling method produce results that are a good representation of the opinion of the entire neighborhood of 50 homes?

No, because the president only called her friends, whose schedules are similar to hers. She should choose to call a variety of different people in the neighborhood.
Yes, because the president took the time to get the opinions of others before setting the date.
Yes, because 5 is one-tenth of 50.
No, because the only way to make this decision is to call all 50 homes.

8. Mean Price of a Ticket to an Ice Hockey Game

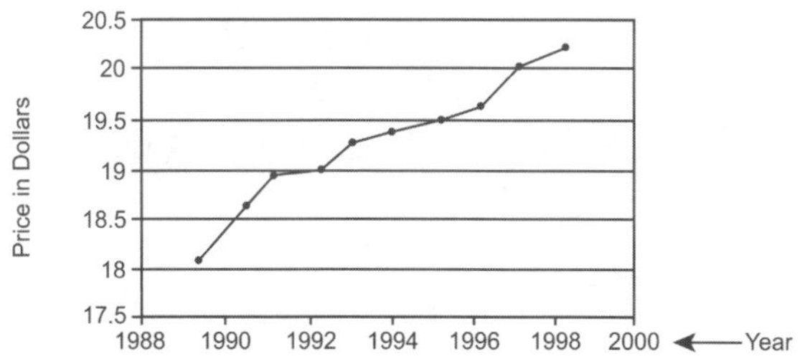

The price of a ticket doubled from 1990 to 1992.

True False

9. Years of College Required for Your Job

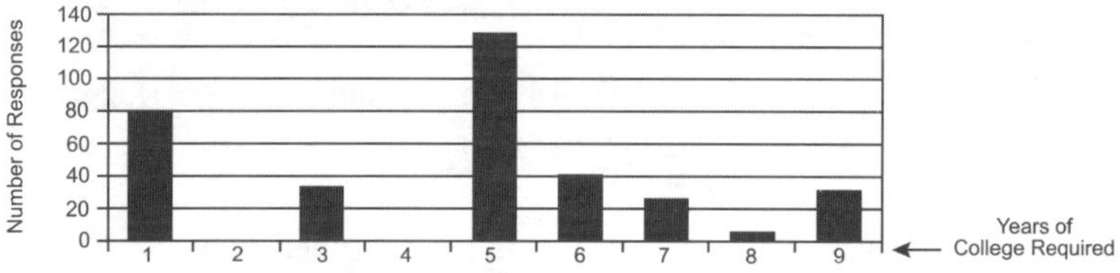

Does this graph imply that you can't get a job if you have completed exactly 2 years of college?

No Yes

10. Money Spent per Student for Education

The town spends at least twice as much as the state.

True False

11. Attendance at School Last Week

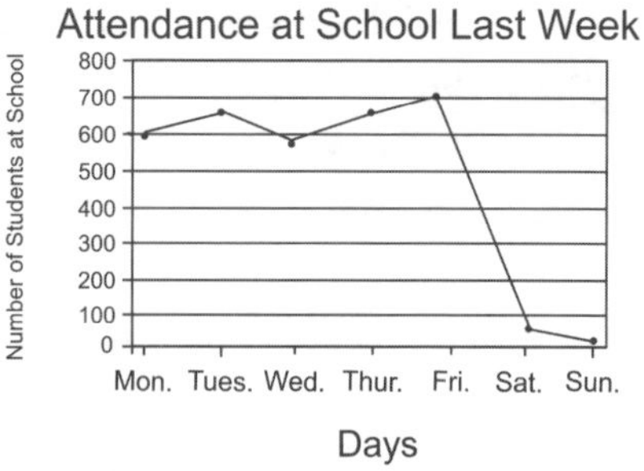

Which day had the highest attendance?

12. Attendance at School Last Week

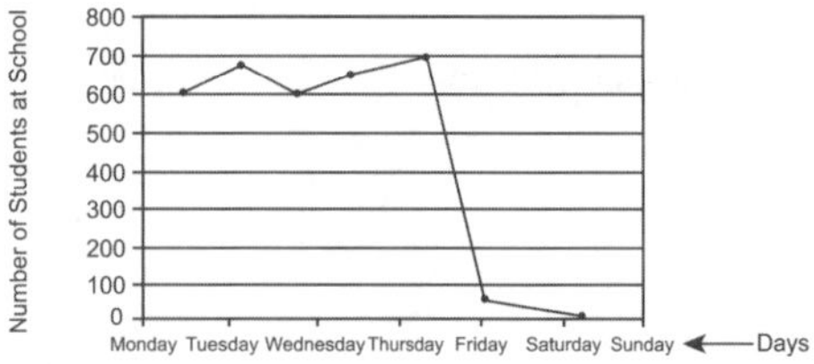

Which of the following statements best describes the graph?

The graph is not misleading
The graph is misleading because the vertical scale is too large.
The graph is misleading because the horizontal scale is too large.
The graph is misleading because the vertical scale is too small.

13. Students' Favorite Field Trips

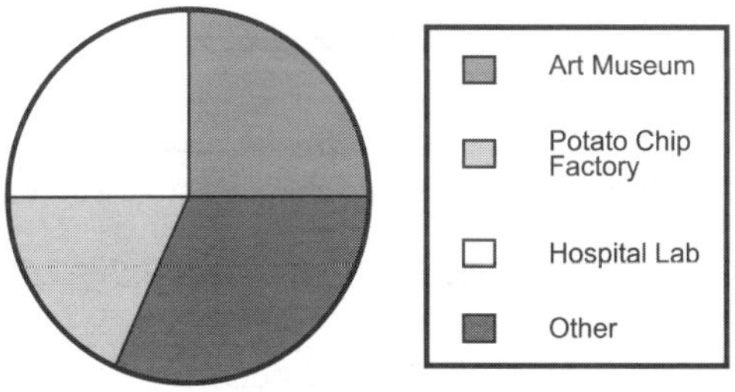

The students took exactly 4 field trips this year.

Not enough information True

False

14. Superhero Costumes Sold at Hal's Costume Shop

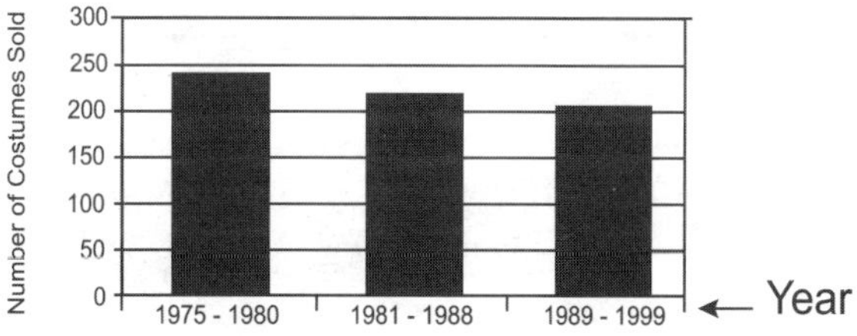

Should the company use this graph to decide how many costumes to buy for the year 2000?

No Yes

15. Mean Waiting Times for Elevators on Different Floors

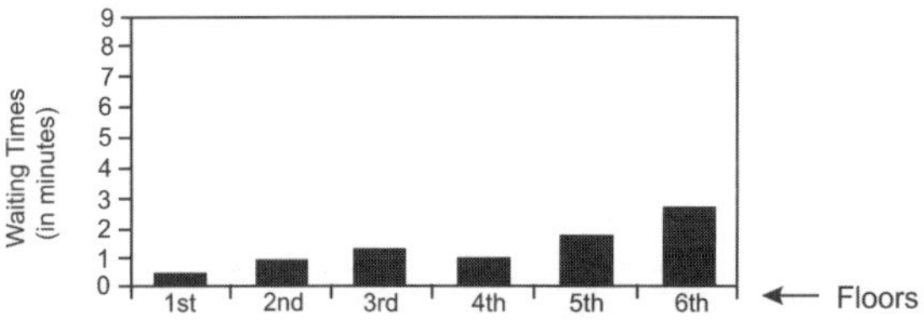

The mean waiting time on the 4th floor is longer than the 3rd floor.

True False

16. Consider the following test scores from six students: 50, 60, 70, 80, 90, & 100.
Is 75 a reasonable measure of central tendency for the given scores? Answer yes or no.

17. Consider the data 66, 73, 83, 77, 59. Is 61 a reasonable measurement of the central tendency of the given data? Answer yes or no.

18. Consider the sales to five sample shoppers: $42.75, $44.50, $57.00, $58.25, $47.50.
Is $50.00 a reasonable value for the central tendency? Answer yes or no.

19. Consider the sample shoe sizes: 8, 9, 7, 10, 9, 9.5, 9, 7, 10, 11, 13.5, 6.
Is 9 a reasonable value for the measure of central tendency? Answer yes or no.

20. Consider the sample number of pages in five sample textbooks: 912; 476; 890; 1,324; 908.
Is 900 a reasonable value for the central tendency? Answer yes or no.

MPA-100: Solving Two-Step Equations

By this point, you have learned about the order of operations, the steps you should follow when solving equations. To solve two-step equations, however, you will have to reverse the order of operations, which means that you will add or subtract first, multiply or divide second, and reduce exponents last.

How to solve a two-step equation:

1. Reverse the order of operations. To do so, first add or subtract; then, multiply or divide.
2. Use inverse operations to isolate the variable.
3. Check your answer.

Look at the following examples of two-step equations.

Example 1 Solve for y: $2y + 6 = 18$

$$\begin{aligned} 2y + 6 &= 18 \\ 2y + 6 - 6 &= 18 - 6 \\ 2y &= 12 \end{aligned}$$

Step 1: Reverse the order of operations. Subtract 6 from both sides of the equation.

$$\frac{2y}{2} = \frac{12}{2}$$
$$y = 6$$

Step 2: Divide both sides by 2.

$$\begin{aligned} 2(6) + 6 &= 18 \\ 12 + 6 &= 18 \\ 18 &= 18 \end{aligned}$$

Step 3: Check your answer by substituting 6 for y in the original equation.

Answer: $y = 6$

Example 2 Solve for n: $4n + 10 = 26$

$$\begin{aligned} 4n + 10 &= 26 \\ 4n + 10 - 10 &= 26 - 10 \\ 4n &= 16 \end{aligned}$$

Step 1: Reverse the order of operations. Subtract 10 from both sides of the equation.

$$\frac{4n}{4} = \frac{16}{4}$$
$$n = 4$$

Step 2: Divide both sides by 4.

$$\begin{aligned} 4(4) + 10 &= 26 \\ 16 + 10 &= 26 \\ 26 &= 26 \end{aligned}$$

Step 3: Check the answer by substituting 4 in place of n in the original equation.

Answer: $n = 4$

Example 3 Solve for d: $3d - 7 = 14$

$3d - 7 = 14$ $3d - 7 + 7 = 14 + 7$ $3d = 21$	**Step 1:** Reverse the order of operations by first adding 7 to both sides of the equation.
$\frac{3d}{3} = \frac{21}{3}$ $d = 7$	**Step 2:** Next, divide both sides by 3.
$3(7) - 7 = 14$ $21 - 7 = 14$ $14 = 14$	**Step 3:** Check your answer by substituting 7 for d in the original equation.
	Answer: $d = 7$

Problem Set

Solve:

1. Stu and Laura had pictures taken of their baby. The sitting fee was $10. They paid $15 per sheet for the pictures. Their total was $160. How many sheets did they buy?

2. Mr. Zing made his iced tea a little too strong, so he added water to dilute it. He had 3 cups of tea. He added 2 containers of bottled water to the tea to make a total of 7 cups of tea. How many cups of water were in each container?

3. Jefferson is making his own Halloween costume. He bought a pattern for $5.00 and material that cost $7.00 a yard. Jefferson spent a total of $68. How many yards of material did he buy?

4. The Parker family went to Thrill-O-Rama Amusement Park. Mrs. Parker was the only adult, and her ticket was $12. The kids' tickets were $8.00 each. The total cost to get into the park was $52. How many kids went to the amusement park?

5. Solve: $2f + 7 = 35$

6. Holly made a new dress for her baby sister. She bought a pattern for $4. She paid $6 per yard for the fabric. Her total for the pattern and material was $28. How many yards of fabric did she buy?

7. Sarah took her 3 younger siblings to the movies. Her adult ticket was $6.00. The total cost of admission for the movie was $18. All of her siblings bought tickets at the kids' price. What was the price of one kids' ticket?

8. Allison's age is 3 more than 3 times Bubba's age. Allison is 24. How old is Bubba?

9. Our most recent family trip took 3 hours of driving time. That is 9 hours less than 2 times the time we spent driving on our family trip four years ago. How many hours did we spend driving on our family trip four years ago?

10. Solve: $y + 3 = 7 - 2$

11. Solve: $5h + 7 = 32$

12. Solve: $9c - 4 = 41$

13. Solve: $3c + 6 = 48$

14. Solve: $5 + 6m = 35$

15. Solve: $3 + 2n = -19$

16. Solve: $7 - 9g = 61$

17. Solve: $-12 - 6m = 48$

18. Solve: $15 - 9g = 132$

19. Solve: $15a - 4 = 56$

20. Solve: $-14m + 12 = -184$

MPA-101: Solving Equations by Combining Like Terms

Try to solve the following equation for both x and y: $4x + y = 15$.

Don't try too hard. It's impossible to solve by itself. If you tried to combine $4x$ and y somehow, you probably wouldn't get too far. The terms $4x$ and y do not share any variables, so they are not **like terms**. This means that they cannot be combined at all. Only terms that share all their variables are like terms.

Like Terms	Expressions that contain the same variable or variables.

Here are a few examples of like terms:

$3x$ and $9x$ $\qquad$ y and $-2y$ $\qquad$ $\frac{1}{2}p$ and $\frac{3}{5}p$ $\qquad$ $0.5h$ and $2.6h$ $\qquad$ $7d$ and $-3d$

When you need to solve an equation with like terms, you may combine their coefficients. For example:

$$3x + 9x = (3+9)x = 12x$$
$$y - 2y = (1-2)y = -1y = -y$$
$$7d - 3d = (7-3)d = 4d$$
$$0.5h + 2.6h = (0.5+2.6)h = 3.1h$$

This process is called combining like terms. Once you have combined the like terms in an equation, then it becomes a simple equation, and you solve it like other equations you solved in previous lessons. The following equation is an easy example.

Example 1 Solve $8p + 3p = 88$

$11p = 88$	**Step 1:** Combine the like terms, $8p$ and $3p$, by adding their coefficients.
$\frac{\cancel{11}p}{\cancel{11}} = \frac{88}{11}$	**Step 2:** Divide both sides by 11, the coefficient of the variable p.
$p = \frac{88}{11} = 8$	**Step 3:** Simplify.
	Answer: $p = 8$

After you understand how like terms work, try the following examples.

Example 2 Solve: $19x - 4x = 100$

$15x = 100$	**Step 1:** Combine the like terms, $19x$ and $4x$, by adding their coefficients.
$\frac{\cancel{15}x}{\cancel{15}} = \frac{100}{15}$	**Step 2:** Divide both sides by 15, the coefficient of the variable x.
$x = \frac{100}{15} = \frac{20}{3}$	**Step 3:** Simplify.
	Answer: $x = \frac{20}{3}$

Example 3 Solve: $8y + 7 = 9y$

$8y + 7 = 9y$ $8y - 8y + 7 = 9y - 8y$	**Step 1:** Combine like terms by subtracting $8y$ from both sides. This isolates y on one side of the equation.
$7 = 9y - 8y$ $7 = y$	**Step 2:** Simplify.
	Answer: $y = 7$

Example 4 Jack has four packets of candy, and Jill has seven packets of candy. They both pour the contents of all their packets into a pile. Each packet contained the same amount of candy. If the pile has 1,100 candies, how many candies were in each packet?

$4c + 7c = 1{,}100$	**Step 1:** Write an equation using c as the number of candies in one packet.
$11c = 1{,}100$	**Step 2:** Combine the like terms, $4c$ and $7c$, by adding their coefficients.
$11c = 1{,}100$ $\frac{11c}{11} = \frac{1{,}100}{11}$	**Step 3:** Divide both sides by 11, the coefficient of the variable c.
$c = \frac{1{,}100}{11} = 100$	**Step 4:** Simplify.
	Answer: 100

Example 5 Tiniqua has 18 boxes of golf balls. On the course, she loses 9 balls. If she has 15 boxes left, how many golf balls are in a box?

$18g - 9 = 15g$	**Step 1:** Write an equation using g as the number of golf balls in one box.
$18g - 18g - 9 = 15g - 18g$ $-9 = 15g - 18g$	**Step 2:** Combine like terms by subtracting $18g$ from both sides. This isolates g on one side of the equation.
$-9 = -3g$	**Step 3:** Simplify.
$\frac{-9}{-3} = \frac{-3g}{-3}$	**Step 4:** Divide both sides by –3, the coefficient of the variable g.
$3 = \frac{-9}{-3} = g$	**Step 5:** Simplify.
	Answer: 3

Problem Set

Solve:

1. Solve for r: $5r + 4r = 54$

2. Solve for f: $6f - 5f = 7$

3. Solve for g: $9g + 2g = 88$

4. Solve for s: $10s + 2s - 40 = 44$

5. Solve for v: $14v - 4v = 37$
(Give your answer as a decimal.)

6. Solve for y: $2y + 2y = 25$
(Give your answer as a decimal.)

7. Solve for b: $10b - 5b = 48$
(Give your answer as a decimal)

8. A mother is three times older than her son. The boy's father is four times older than he is. The sum of the mother's age and the father's age is 70. How old is the son?

9. Solve for h: $\frac{2}{5}h + \frac{3}{10}h = 14$

10. Solve for s: $3s = 2s + 7$

11. Solve for u: $4u = 8u - 20$

12. Solve for x: $3 - 4x = 10x + 17$

13. Solve for y: $3y - 5 = 7y + 7$

14. Walker is making a salad for the cafeteria. He bought 10.8 pounds of lettuce and 6.2 pounds of tomatoes. Together, the ingredients cost $42.50. The ingredients were the same price per pound. What was the price per pound?

15. Solve for z: $5.15z - 29 = -2.1z$

16. Solve for x: $-\frac{2}{3}x + 21 = \frac{5}{3}x$

17. Solve: $-7a + 2a - 4 = -109$

18. Solve: $15a - 5a - 4 = 56$

19. Solve: $7g - 11g = 104$

20. Solve: $4m + 12m = -192$

MPA-102: Graphing Equations by Plotting Points

Previously, you learned how to plot points in a rectangular coordinate system. This rectangular coordinate system is a plane, often called the Cartesian plane in tribute to Rene Descartes (1596-1650). In this lesson, you will learn how to graph linear equations on the Cartesian plane.

A **linear equation** is of the form $ax + by + c = 0$, where a, b, and c are real numbers and a and b cannot both be zero. Real numbers make up the set of all numbers that can be represented by points on the number line, often denoted by R, and a and b cannot both be zero. To graph a straight line, you must determine at least two points on the line. It is often practical to choose the x-intercept and y-intercept, because calculations are not as difficult. Recall that an ***x*-intercept** is the point where the line crosses the x-axis, so the y-value in this case is 0. Similarly, the ***y*-intercept** is the point where the line crosses the y-axis, so the x-value in this case is 0. A **vertical line** does not need a y-intercept. Also, a **horizontal line** does not need an x-intercept. In this lesson, you will apply these concepts to graph linear equations.

Linear Equation	An equation of the form $ax + by + c = 0$, where a, b, and c are real numbers and a and b cannot both be zero.

***x*-intercept**	The point where a graph crosses the x-axis.

***y*-intercept**	The point where a graph crosses the y-axis.

Vertical Line	The graph of an equation of the form $y = h$, where h is a real number.

Horizontal Line	The graph of an equation of the form $y = k$, where k is a real number.

How to graph a line:

Remember that you need only two points to graph a linear equation, but graphing three points helps eliminate calculation errors.

1. Draw a table with columns labeled x-pick, solution, y-answer, and point.
2. Pick three different values for x. It is usually a good idea to pick zero (0) as one of the values. Then the y value will be the y-intercept.
3. Substitute the first value for x into the original equation, and solve for the matching y.
4. Write the ordered pair.
5. Repeat Steps 3 and 4 for the two remaining x-values.
6. Plot the three points on the graph.
7. Draw a line through the points.

Example 1 Graph the equation $y = 3x + 4$.

Step 1: Draw a table with columns labeled x-pick, solution, y-answer, and point.

***x*-pick**	**solution ($y = 3x + 4$)**	***y*-answer**	**point**

Step 2: Pick three different values for x.

***x*-pick**	**solution ($y = 3x + 4$)**	***y*-answer**	**point**
–2			
0			
1			

Step 3: Substitute the first value for x into the original equation, and solve for the matching y.

***x*-pick**	**solution ($y = 3x + 4$)**	***y*-answer**	**point**
–2	$y = 3(-2) + 4$	–2	

Step 4: Write the ordered pair.

***x*-pick**	**solution ($y = 3x + 4$)**	***y*-answer**	**point**
–2	$y = 3(-2) + 4$	–2	(–2, –2)

Step 5: Repeat Steps 3 and 4 for the two remaining x-values.

***x*-pick**	**solution ($y = 3x + 4$)**	***y*-answer**	**point**
–2	$y = 3(-2) + 4$	–2	(–2, –2)
0	$y = 3(0) + 4$	4	(0, 4)
1	$y = 3(1) + 4$	7	(1, 7)

Step 6: Plot the points on the graph.

Step 7: Draw a line through the three points. Since $y = 3x + 4$ is a linear equation, its graph should resemble a straight line.

Answer:

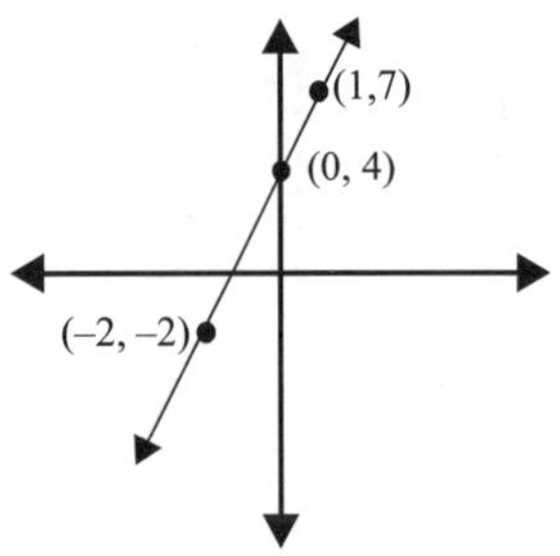

Example 2 Graph the equation $y = \frac{1}{2}x - 3$.

Step 1: Following Steps 1-5, construct the table. For convenience, choose x-values that will yield y-values that are whole numbers.

x-pick	solution ($y = \frac{1}{2}x - 3$)	y-answer	point
–2	$y = \frac{1}{2}(-2) - 3$	–4	(–2, –4)
0	$y = \frac{1}{2}(0) - 3$	–3	(0, –3)
2	$y = \frac{1}{2}(2) - 3$	–2	(2, –2)

Step 2: Complete the problem by plotting the points on the Cartesian plane and drawing a line through them. Since $y = \frac{1}{2}x - 3$ is a linear equation, its graph should resemble a straight line.

Answer:

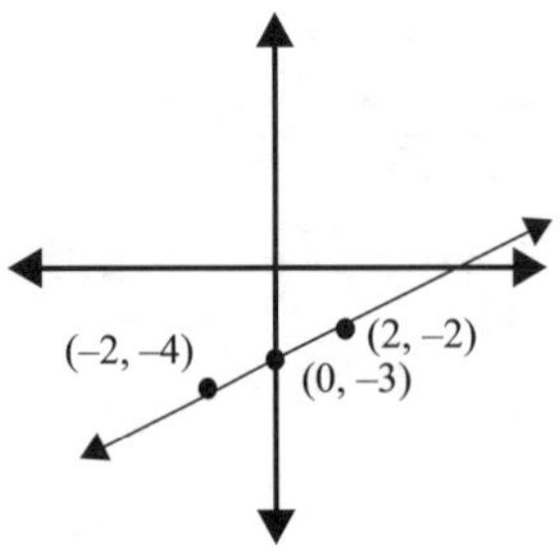

Example 3 Graph the equation $y = 2$.

Step 1: Following Steps 1-5, construct the table. Note that for every choice of x, the y-value is 2.

x-pick	**solution ($y = 2$)**	**y-answer**	**point**
–1	$y = 2$	2	(–1, 2)
0	$y = 2$	2	(0, 2)
+1	$y = 2$	2	(1, 2)

Step 2: Complete the problem by plotting the points on the Cartesian plane and drawing a line through them. Note that $y = 2$ has a horizontal line as its graph.

Answer:

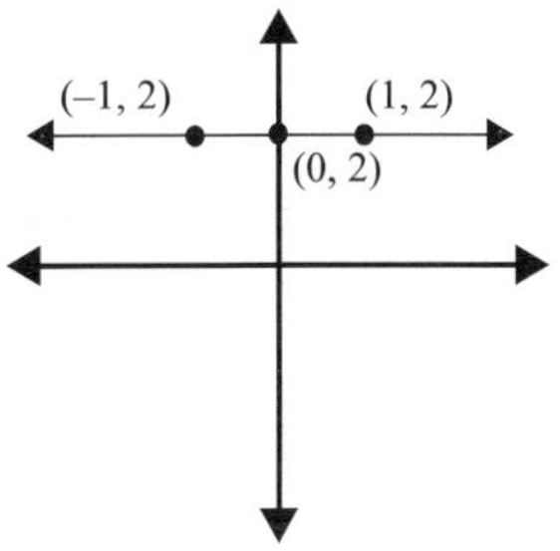

Example 4 Graph the equation $x = 2$.

This is a special instance of a linear equation where the x-values are all 2 and the y-values can be any number.

Step 1: We create a table with the value 2 for all x-values and different numbers for the y-values.

x	2	2	2
y	–1	0	1

Step 2: We plot the points on a Cartesian plane and draw a line through them. Note that $x = 2$ has a vertical line as its graph.

Answer:

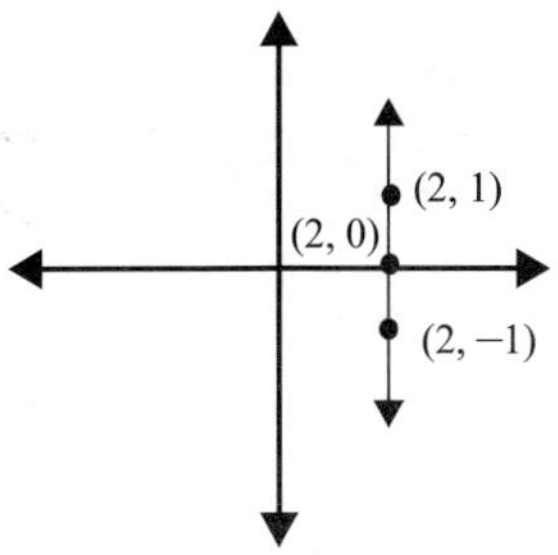

Problem Set

1. Determine whether the graph of this line is vertical, horizontal, or neither.
$x = 2$

2. Determine whether the graph of this line is vertical, horizontal, or neither.
$y = 8$

3. Determine whether the graph of this line is vertical, horizontal, or neither.
$y = 14$

4. Will the line for the graph of the equation $y = 12x$ be vertical, horizontal, or neither?

5. For the equation $y = 6x$, find y if $x = 3$.

6. For the equation $y = x - 11$, find y if $x = -10$.

7. In the equation $y = x$, if $x = 3$, what is the ordered pair?

8. In the equation $y = x + 5$, if $x = -9$, what is the ordered pair?

9. In the equation $y = x - 12$, if $x = 11$, what is the ordered pair?

10. Choose the graph that best represents the equation $y = x + 4$

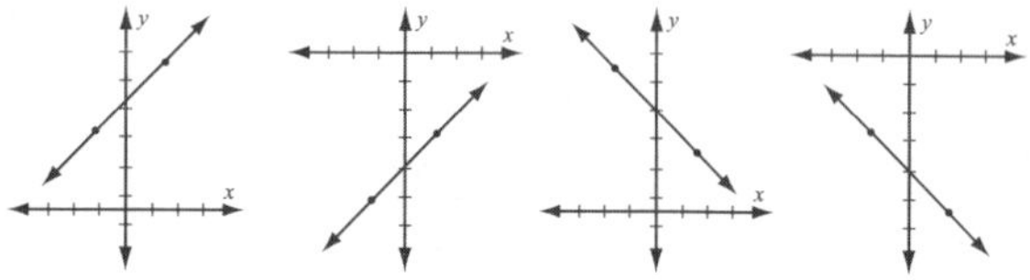

11. Choose the equation that best represents the equation: $y = 3x$

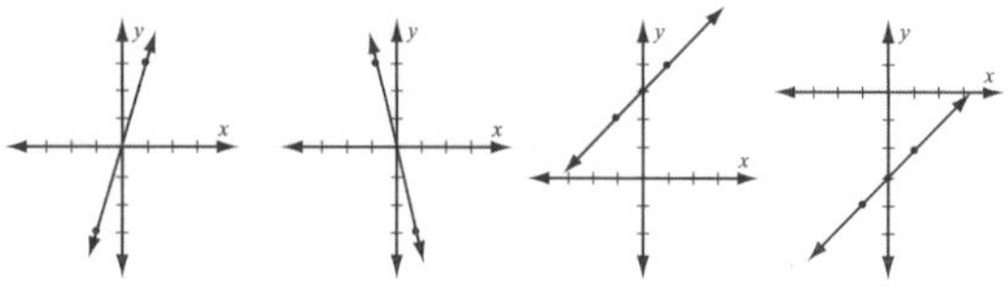

12. Choose the graph that best represents the equation: $y = -3x + 1$

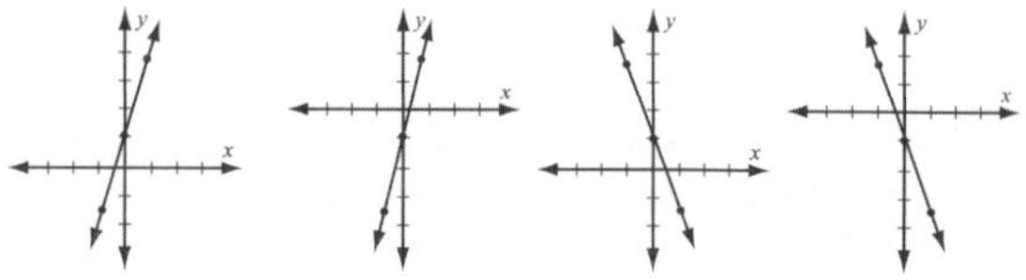

13. Choose the graph that best represents the equation: $x = -5$

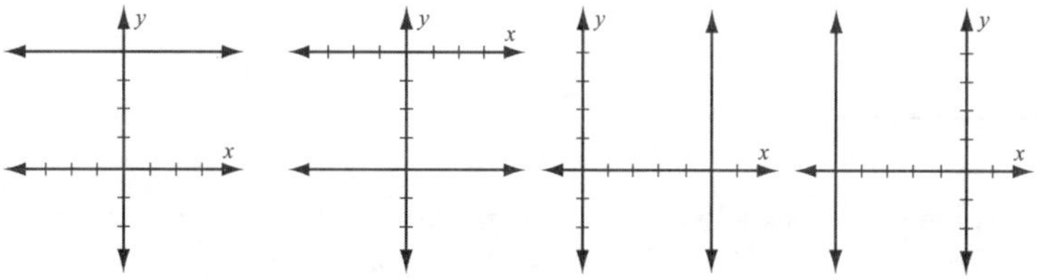

14. Choose the graph that best represents the equation: $y = 8$

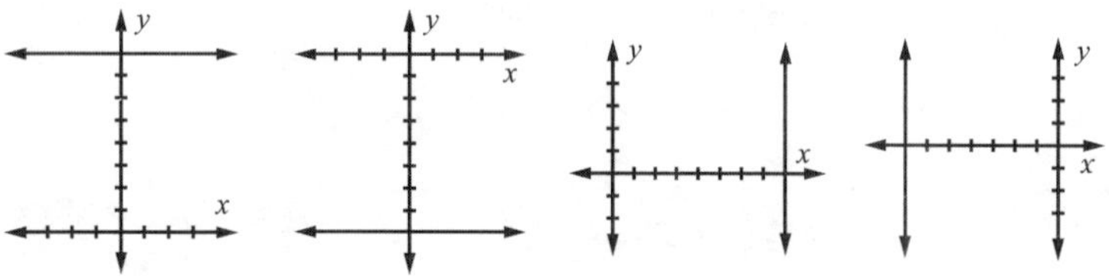

15. Determine whether the graph of this line is vertical, horizontal, or neither: $x = 7$

16. Determine whether the graph of this line is vertical, horizontal, or neither: $x = 12$

17. Determine whether the graph of this line is vertical, horizontal, or neither: $y = 7$

18. Determine whether the graph of this line is vertical, horizontal, or neither: $y = -8$

19. Determine whether the graph of this line is vertical, horizontal, or neither: $y = -6x$

20. Tom, Mickey and Damon are examining the equation $y = x - 3$. Tom says the graph of this line will be horizontal, Damon says the graph will be vertical, and Mickey says the graph will be neither. Who is correct?

MPA-103: Distinguishing Between Relations and Functions

Recall that an ordered pair is a set of two numbers where the order in which the numbers are written has an agreed upon meaning. In other words, a *relationship* exists between the x- and y-values of an ordered pair. In this lesson, you will build upon this concept by learning how to distinguish between relations and functions.

A **relation** is simply a set of ordered pairs. If a relation has the special property such that for each value of x, there exists a unique value of y, then this relation is said to be a **function**.

Relation	A set of ordered pairs.
Function	A set of ordered pairs that has only one unique value of y for each value of x.

Since a function is a set of ordered pairs, it is also a relation; however, a relation is not necessarily a function because it may have more than one y-value for a particular x-value. You can use one of the following three methods to determine whether a relation is a function:

- Perform a Vertical Line Test.
- Draw a chart.
- Draw arrows.

Perform a Vertical Line Test

To determine whether a relation is a function using the Vertical Line Test, draw a vertical line through the graph of the relation. If the line intersects the graph only once, then the relation is a function.

Example 1 Determine whether or not the following graph is a function.

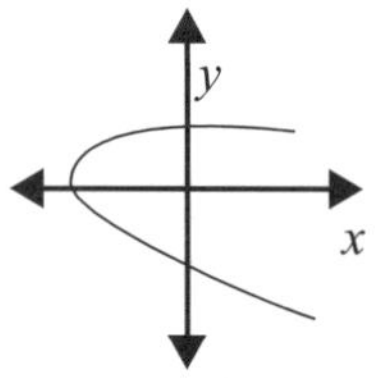

Step 1: When you draw a vertical line through the graph of the relation above, the vertical line intersects with the graph more than once. Thus, the relation is not a function.

Answer: This graph is not a function.

Use a Chart

You can also use a chart to determine whether a relation is a function. To do so, place all x-coordinates and y-coordinates of the relation in a vertical chart. Then, place the x-coordinates in the left column of the chart, and place the y-coordinates in the right column of the chart. Ensure that no repeat observations occur. If no x-coordinate is repeated, then the relation is a function. If a x-coordinate is repeated, then the relation is not a function.

Example 2 Determine whether or not the relation $\{(8, 2), (1, 0), (9, 20), (0, 1), (2, 8)\}$ is a function.

8	2
1	0
9	20
0	1
2	8

Step 1: Place all x-coordinates and y-coordinates of the relation in a vertical chart.

Step 2: Determine if any x-coordinate is repeated.

Answer: Since each coordinate has exactly one y-coordinate, this relation is a function.

Use Arrows

Drawing arrows may also help you determine whether a relation is also a function. You can list all distinct x-coordinates and map them to their corresponding y-coordinates.

If a distinct x-coordinate is mapped to more than one y-coordinate, then the relation is not a function. If each distinct x-coordinate gets mapped to exactly one y-coordinate, then the relation is a function.

To use the arrow method:

1. Place all distinct x-coordinates in a vertical list and all y-coordinates in a vertical list adjacent to the distinct x-coordinates.
2. Map each x-coordinate to its corresponding y-coordinate by drawing arrows.
3. If one or more x-coordinates get mapped to more than one y-coordinate, then the relation is not a function. If each x-coordinate gets mapped to exactly one y-coordinate, then the relation is also a function.

Example 3 Determine whether or not the function $\{(8, 2), (1, 0), (9, 20), (0, 1), (2, 8), (1, 20)\}$ is a function.

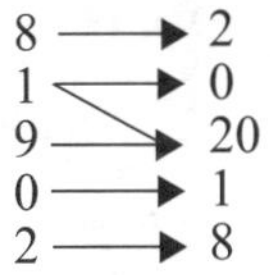

Step 1: Place all x-coordinates and y-coordinates of the relation in a vertical chart.

Step 2: Since the x-coordinate 1 is assigned to more than one y-coordinate, this relation is not a function.

Answer: This relation is not a function.

Problem Set

Identify if the following relations are functions. Answer yes or no.

1.

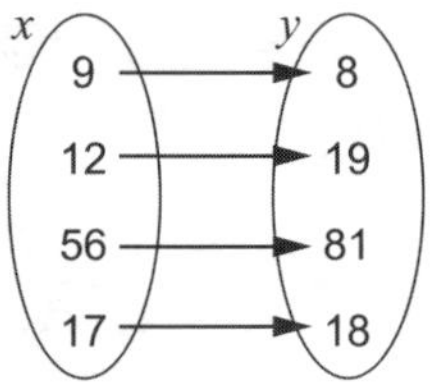

2.

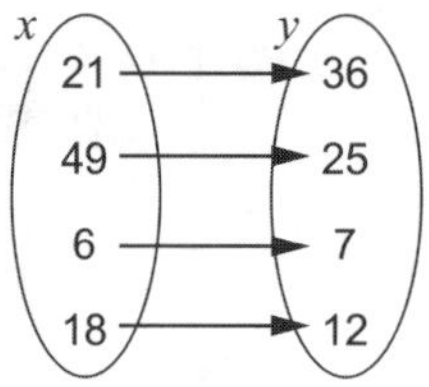

3.

x	56	27	14	31	43
y	27	18	3	26	3

4.

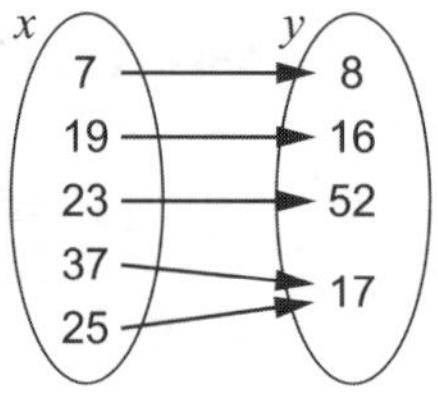

5.

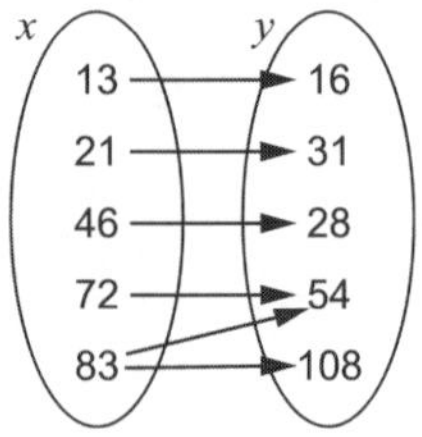

6.

x	y
4	3
6	8
12	9
6	7
28	19

7.

x	y
8	13
14	26
52	9
33	21
14	27

8.

x	y
9	15
6	8
21	70
52	65

9.

x	45	54	25	45
y	30	19	26	84

10.

x	35	68	186	490
y	568	392	27	236

11.

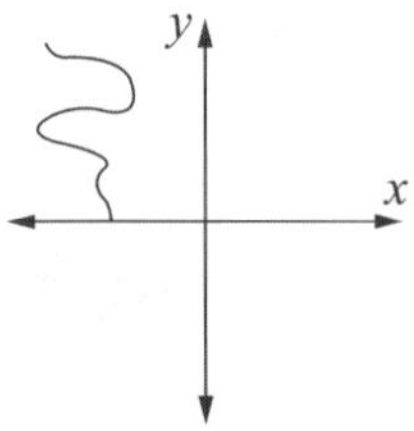

12.

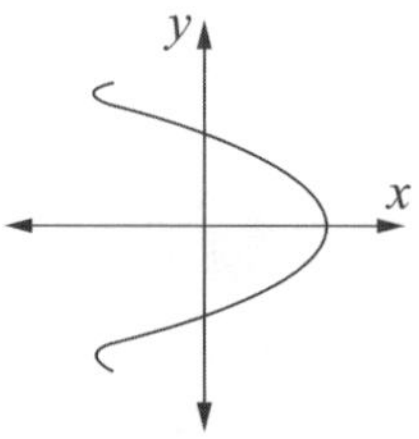

13.

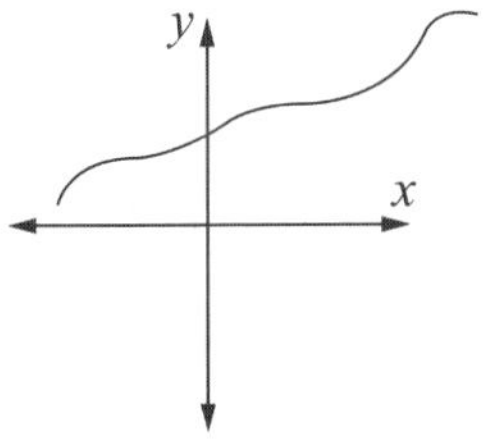

14.

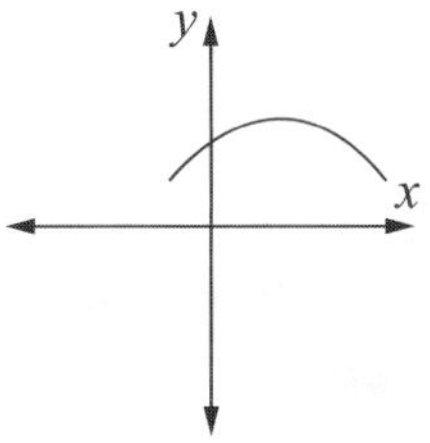

15.

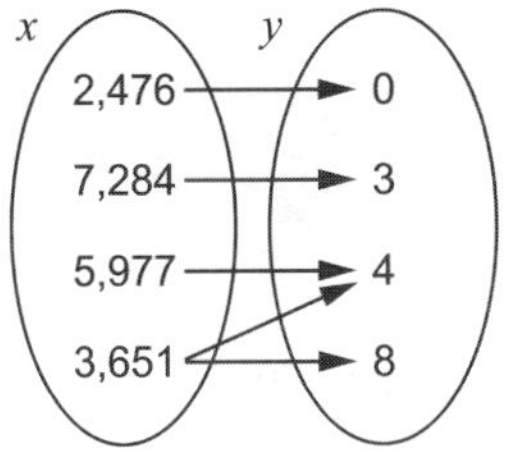

16.

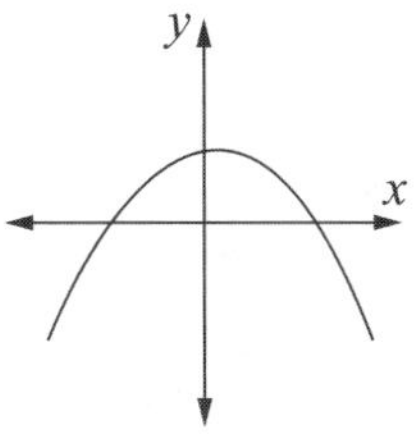

17.

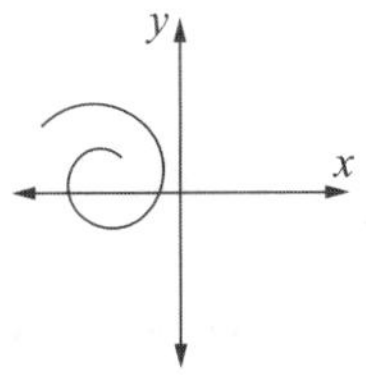

18.

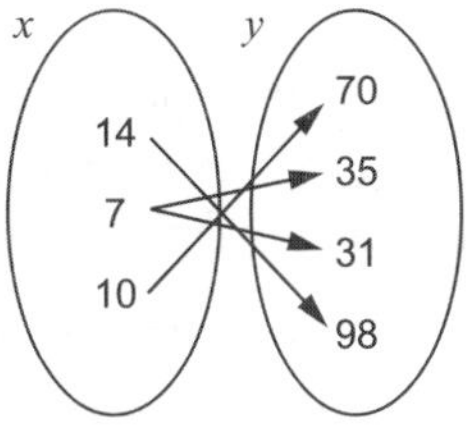

19.

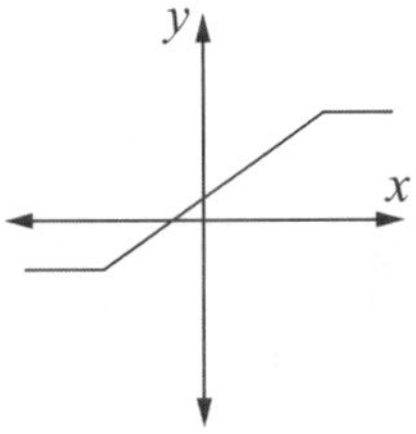

20.

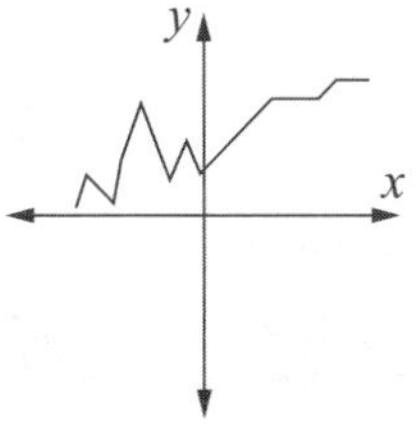

MPA-104: Recognizing Patterns

In this lesson, you will not have to establish equations or mull over units or apply any formulas. In fact, you'll get to make your own formulas, or rules. You see, this lesson is all about **number patterns**, which can be fun.

Number Pattern	A set of three or more numbers that obey a rule.

Can you determine the next number in the pattern?

1, 2, 4, 7, 11, 16, 22, …

The number is…29. Did you find the correct answer? The pattern obeys a specific rule: Add one more to each number than you added to the previous one. Add 1 to 1 to get 2. Add 2 to 2 to get 4. Add 3 to 4 to get 7. Add 4 to 7 to get 11. Eventually you reach 29, the next number in the pattern.

You may find it difficult to see a pattern immediately. If you have a hard time visualizing a pattern, then ask yourself these questions:

- Are the numbers getting larger or smaller?
- What is the difference between the terms?

If the numbers in a pattern increase, then the rule may involve addition or multiplication, because those operations tend to increase the values of terms in a pattern. If the numbers in a pattern decrease, then the rule may involve subtraction or division because those operations tend to decrease the values of terms in a pattern.

The difference between terms may be immensely helpful or not much help at all. Find the difference between the terms of the first example:

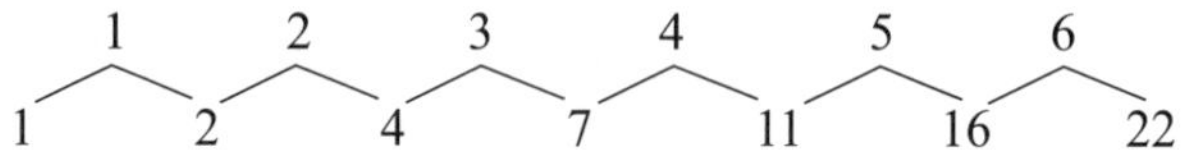

The difference increases by one for each element. The next step is clear: Add 7 to 22 to get 29.

Patterns do not have to be composed of numbers. **Figural patterns** pertain to pictures, or figures.

Figural Patterns	A set of three or more shapes that relate to each other according to a rule.

Find the next shape in this figural pattern:

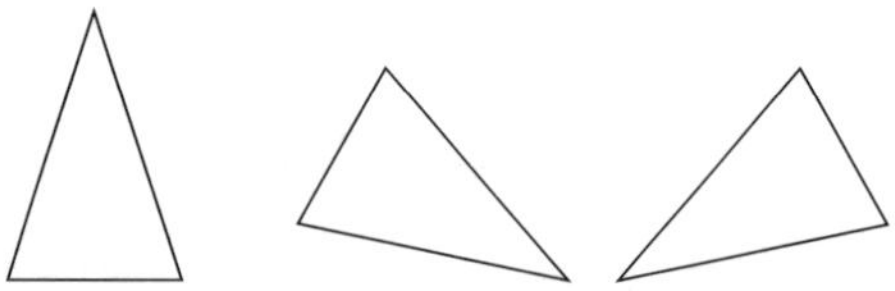

The rule is fairly simple here. You can think of it in two ways: The triangle points at the vertex located clockwise from the previous one, or the shape rotates 120° clockwise at each step. The next figure would be the last shape rotated 120° clockwise. The next figure is the same as the first.

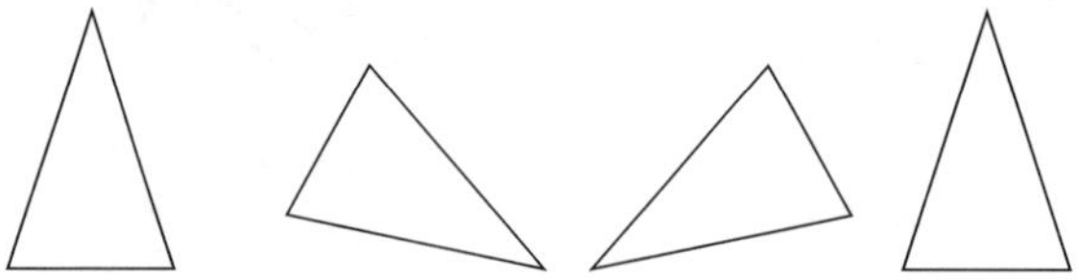

Rules do not have to be so simple. They can involve more than one step. Try to find the next element in the following number pattern:

3, 5, 9, 17, 33, 65…

The next number is 129. Ask yourself one of the basic pattern questions: Are the numbers increasing or decreasing? They are increasing, so the rule probably involves addition or multiplication. What is the difference between the terms?

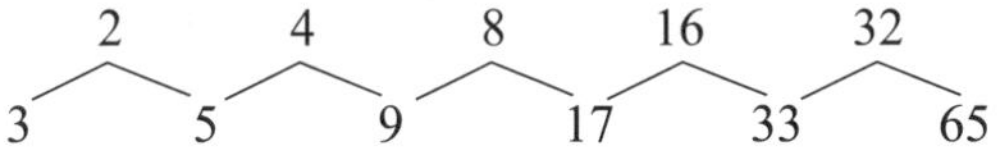

The differences display their own pattern. The difference between each term is doubled. So, the difference between 65 and the next term should be twice 32, or 64. 65 plus 64 is 129. The next term is 129.

Another rule would work just as well, and you don't need to find the difference between terms. If you double a term and subtract 1, then you'll get the next term. This dual-rule pattern shows that more than one rule can apply to the same pattern.

Here's a figural pattern with a complex rule. Try to find the next figure.

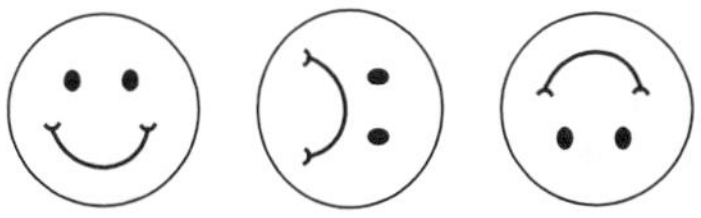

The rule has two steps. First, the faces rotate 90° clockwise for each step. Next, a smile turns into a frown, or a frown turns into a smile. So, the next face would be frowning and rotated clockwise 90° from the last face:

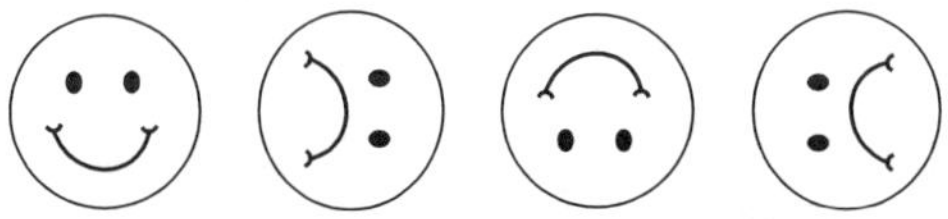

Now try some examples. After you find a rule, find the next element in the pattern.

Example 1 Find the next term in the pattern: 34, 18, 10, 6, 4, . . .

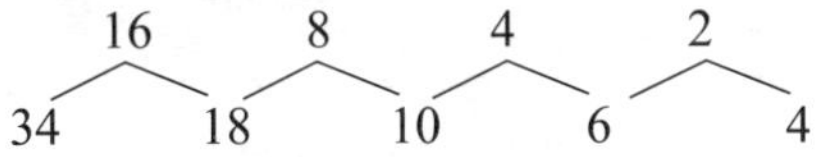

Step 1: Find the differences between the terms.

$\frac{2}{2} = 1$

$4 - 1 = 3$

Step 2: The differences obey a rule: The differences are halved for each term. Apply this rule to find the next term.

$\frac{4}{2} + 1 = 3$

Step 3: The terms obey another rule: Divide by 2; then add 1. You could also apply this rule to find the next term.

34, 18, 10, 6, 4, 3

Answer: The next term is 3.

Example 2 Find the next term in the pattern: 1, 4, 9, 16, 25, . . .

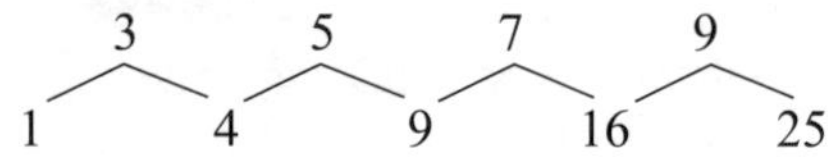

Step 1: Find the differences between the terms.

$25 + 11 = 36$

Step 2: The differences obey a rule: The differences are increasing odd numbers. The next odd number is 11. Apply this rule to find the next term.

The next term is $6^2 = 36$

Step 3: The terms obey another rule: The terms are their positions squared. You could also apply this rule to find the next term.

1, 4, 9, 16, 25, 36

Answer: Using either rule, you will find that the next term is 36.

Example 3 Find the next term in the pattern: 2, 4, 16, 256, . . .

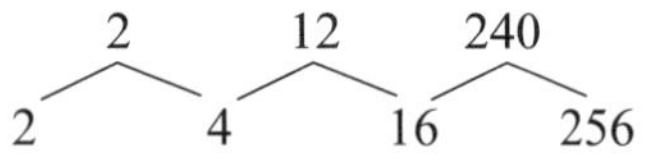

Step 1: Find the differences between the terms.

Step 2: The differences do not seem to have any rule.

$256^2 = 65,536$

Step 3: Find a rule for the terms themselves. The rule is to square the previous term to find the next term. Apply the rule to find the next term.

2; 4; 16; 256; 65,536

Answer: The next term is 65,536.

Example 4 Find the next term in the pattern:

Step 1: Find a rule for the pattern. Each figure has one more side. Also, every other figure is shaded. The next shape would be a shaded hexagon.

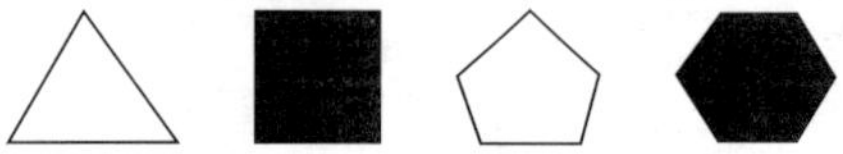

Answer: The next figure in the pattern is a shaded hexagon (6-sided figure).

Problem Set

1. The pattern below is what type of pattern?
40, 44, 48, 52, 56, ...

division multiplication
subtraction addition

2. The pattern below is what type of pattern?
33, 45, 57, 69, ...

division multiplication
subtraction addition

3. The pattern below is what type of pattern?
9, 18, 27, 36, 45,...

division multiplication
subtraction addition

4. The pattern below is what type of pattern?
37, 29, 21, 13, 5,...

division multiplication
subtraction addition

5. The pattern below is what type of pattern?
30, 15, 0, –15,...

division multiplication
subtraction addition

6. The pattern below is what type of pattern?
2, –4, 8, –16,...

division multiplication
subtraction addition

7. What is the next number in the pattern?
2.24, 4.48, 8.96, 17.92,...

8. What is the next number in the pattern?
8.1, 2.7, 0.9, 0.3,...

9. What is the next number in the pattern?
$3\frac{1}{2}, 4\frac{1}{4}, 5, 5\frac{3}{4},...$

10. What is the next number in the pattern below?
–10, –9, –19, –28, –47, ...

11. What is the 10th number in the pattern below?
6, 10, 14, 18,...

12. What is the 8th number in the pattern below?
–9, –10, –11, –12,...

13. What is the 9th number in the pattern below?
–8, –3, 2, 7,...

14. What is the 7th number in the pattern below?
1, 11, 21, 31,...

15. What is the 9th number in the pattern below?
7, 6, 5, 4,...

16. In which corner will you find the shaded block in the next picture of the pattern below?

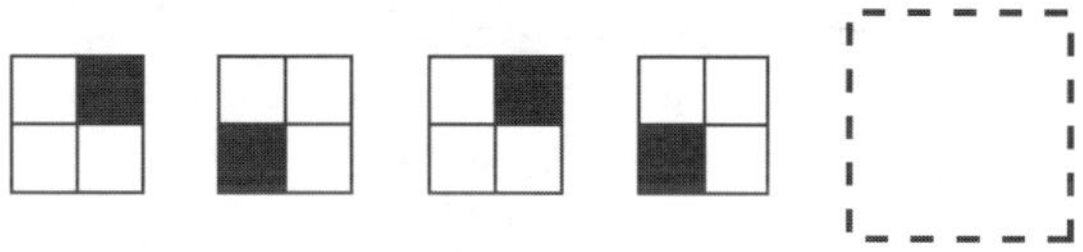

Upper right Upper left
Lower right Lower left

17. The pattern below is what type of pattern:
–32, –38, –44, –50,...

division multiplication
subtraction addition

18. What type of pattern is this? 7000, 700, 70, 7,...

division multiplication
subtraction addition

19. The pattern below is what type of pattern:
$\frac{2}{7}, \frac{1}{14}, \frac{1}{56}, \frac{1}{224},...$

division multiplication
subtraction addition

20. This pattern is what type of pattern? 108, 36, 12, 4,...

division multiplication
subtraction addition

MPA-105: Determining the Measure of Angles Made by Parallel Lines and a Transversal

Parallel lines are straight lines in the same plane that never intersect. Since parallel lines do not intersect, the distance between the two lines also stays the same or constant. Two parallel lines are displayed below.

Parallel Lines	Parallel lines are straight lines in the same plane that never intersect.

When two lines intersect, the distance between the lines does not stay constant. Consider the two non-parallel lines shown in the next figure.

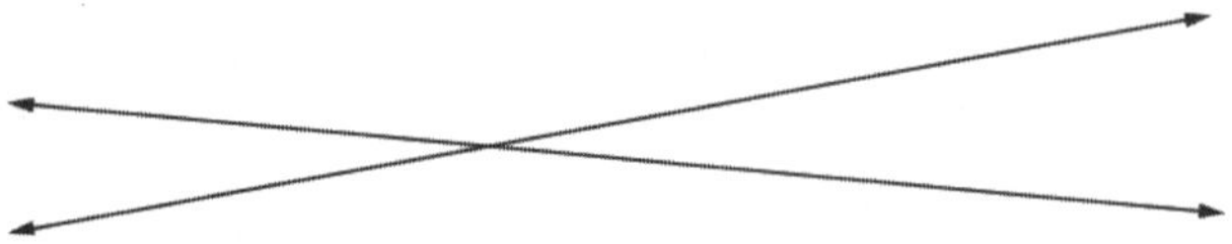

Suppose we have two parallel lines that are both intersected by a third line. The third line is called a **transversal**.

Transversal	A transversal is a line which intersects two or more lines.

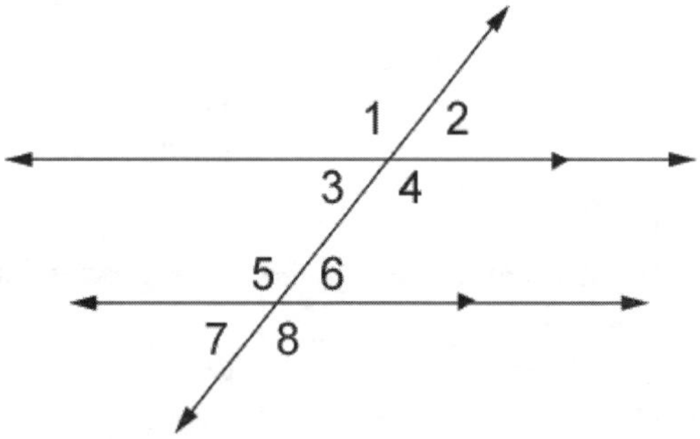

At the intersection of the lines, eight angles are formed. There are four sets of **corresponding angles**. In the figure above, the sets of corresponding angles include the following: angles 1 and 5, angles 2 and 6, angles 3 and 7, and angles 4 and 8.

Corresponding Angles	A pair of congruent angles which have the same position on two parallel lines that are cut by a transversal.

If the two lines intersected by the transversal are parallel lines, the corresponding angles formed at the intersection are congruent (equal).

Within the diagram on the previous page, there are other angles that are congruent. There are four sets of **vertical angles**. Recall that vertical angles are opposite angles formed by intersecting lines. Vertical angles are always congruent. In the diagram on the previous page, the sets of vertical angles include the following: angles 1 and 4, angles 2 and 3, angles 5 and 8, and angles 6 and 7.

Since the corresponding angles and the vertical angles are congruent, we can conclude that of the eight angles formed when a transversal intersects two parallel lines, there are two sets of congruent angles:

$$\angle 1 \cong \angle 4 \cong \angle 5 \cong \angle 8$$

and

$$\angle 2 \cong \angle 3 \cong \angle 6 \cong \angle 7$$

Additional terms are used to describe the congruent angles based on their location. **Alternate Interior Angles** are the angles inside the parallel lines and on opposite sides of the transversal. The angles ($\angle 3 \cong \angle 6$ and $\angle 4 \cong \angle 5$) are congruent, as illustrated in the diagram below.

Alternate Interior Angles	A pair of congruent angles formed when a transversal crosses a pair of parallel lines, and which lie between the parallel lines on opposite sides of the transversal.

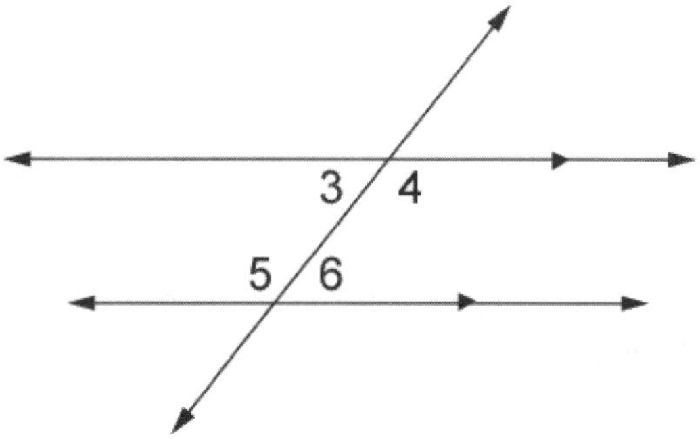

Alternate Exterior Angles are the angles outside the parallel lines and on opposite sides of the transversal. The angles ($\angle 1 \cong \angle 8$ and $\angle 2 \cong \angle 7$) are congruent, as illustrated in the diagram on the following page.

Alternate Exterior Angles	A pair of congruent angles formed when a transversal crosses a pair of parallel lines, and which lie outside the parallel lines but on opposite sides of the transversal.

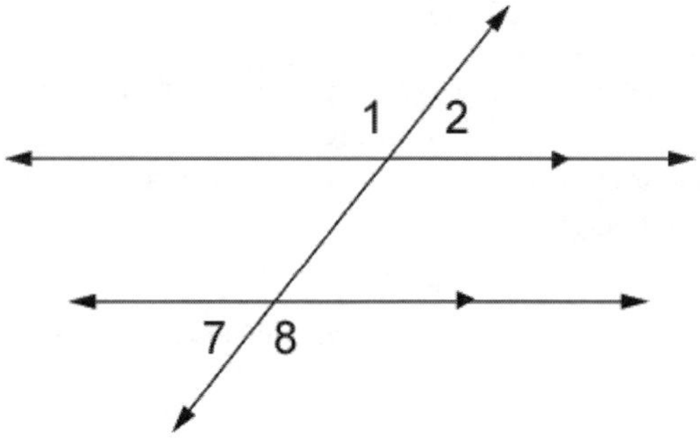

In addition to congruent angles, many **supplementary angles** are also formed when two parallel lines are cut by a transversal. Recall that two angles are supplementary if they form a straight angle or the sum of their measures is 180º.

Supplementary Angles	Two angles are supplementary if the sum of their measures is 180º.

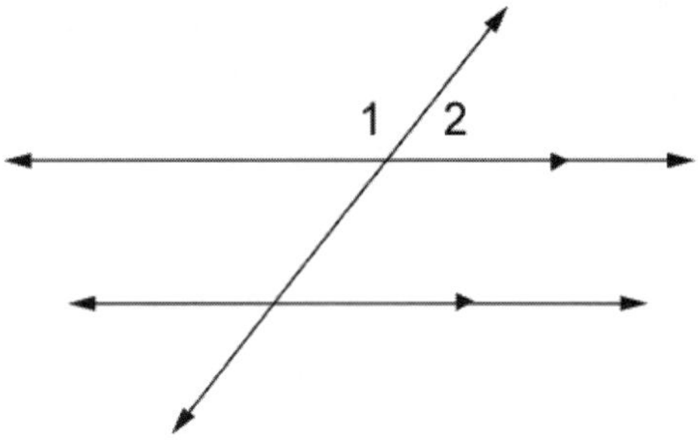

The most obvious pairs of supplementary angles are the ones that share a common ray. These angles are called **Adjacent Supplementary Angles**.

Adjacent Supplementary Angles	Two angles are adjacent supplementary angles if they share a common ray, and the sum of their measures is 180º.

$\angle 1, \angle 2$	$\angle 1, \angle 3$
$\angle 3, \angle 4$	$\angle 2, \angle 4$
$\angle 5, \angle 6$	$\angle 5, \angle 7$
$\angle 7, \angle 8$	$\angle 6, \angle 8$

From the table above we can conclude that $\angle 3$ and $\angle 5$ are supplementary and $\angle 4$ and $\angle 6$ are also supplementary. Therefore, when two parallel lines are intersected by a transversal, the interior angles on one side of the transversal are supplementary.

In addition, we can conclude that $\angle 1$ and $\angle 7$ are supplementary and $\angle 2$ and $\angle 8$ are also supplementary. Therefore, when two parallel lines are intersected by a transversal, the exterior angles on one side of the transversal are also supplementary.

Note: In each of the examples given below, consider the angles (1-8) created when two parallel lines are intersected by a transversal.

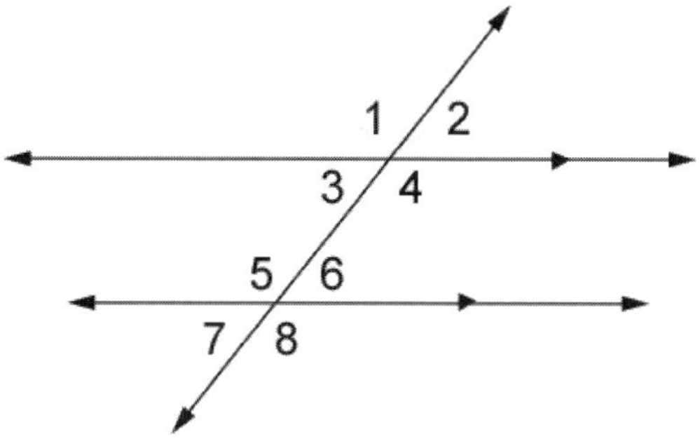

Example 1 If the measure of $\angle 1$ is $112°$, what is the measure of $\angle 2$?

$112° + x = 180°$

Step 1: Notice that $\angle 1$ and $\angle 2$ are adjacent supplementary angles. Let x represent the measure of $\angle 2$.

$112° - 112° + x = 180° - 112°$
$x = 68°$

Step 2: Subtract $112°$ from both sides of the equation.

Answer: The measure of $\angle 2$ is $68°$.

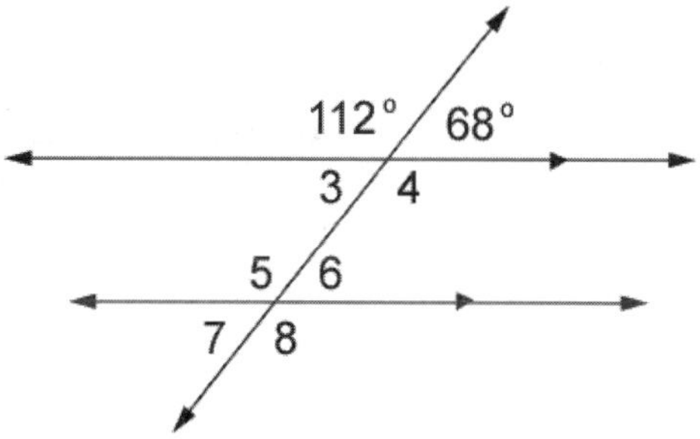

Example 2 If the measure of $\angle 1$ is $112°$, what is the measure of $\angle 4$?

Step 1: Notice that $\angle 1$ and $\angle 4$ are vertical angles.

Step 2: Vertical angles are congruent, so the measure of $\angle 4 = 112°$.

Answer: The measure of $\angle 4$ is $112°$.

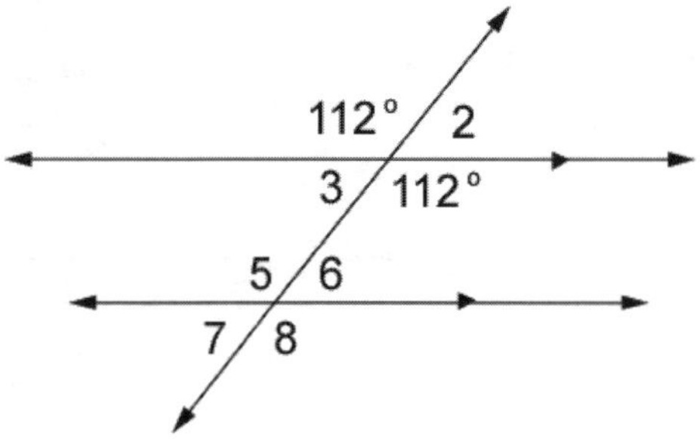

Example 3 If the measure of $\angle 1$ is $112°$, what is the measure of $\angle 6$?

Step 1: Since the measure of $\angle 1 = 112°$ and $\angle 1$ and $\angle 4$ are vertical angles, the measure of $\angle 4 = 112°$.

$112° + x = 180°$

Step 2: Note that $\angle 4$ and $\angle 6$ are supplementary angles. Let x represent the measure of $\angle 6$.

$112° - 112° + x = 180° - 112°$
$x = 68°$

Step 3: Subtract $112°$ from both sides of the equation.

Answer: The measure of $\angle 6$ is $68°$.

Example 4 If the measure of $\angle 6$ is $71°$, what are the measures of the other interior angles?

Step 1: Note that the interior angles are $\angle 3$, $\angle 4$, $\angle 5$, and $\angle 6$.

$x + 71° = 180°$

Step 2: Note also that $\angle 5$ is adjacent and supplementary to $\angle 6$. Let x represent the measure of $\angle 5$.

$x + 71° - 71° = 180° - 71°$ $x = 109°$	**Step 3:** Subtract $71°$ from both sides of the equation. So, the measure of $\angle 5$ is $109°$.
$m\angle 3 = m\angle 6 = 71°$ $m\angle 4 = m\angle 5 = 109°$	**Step 4:** Notice that $\angle 3$ and $\angle 6$ are alternate interior angles, and $\angle 4$ and $\angle 5$ are alternate interior angles. This means that each of these pairs of equal angles have congruent measures.

Answer: $m\angle 3 = m\angle 6 = 71°$ and $m\angle 4 = m\angle 5 = 109°$.

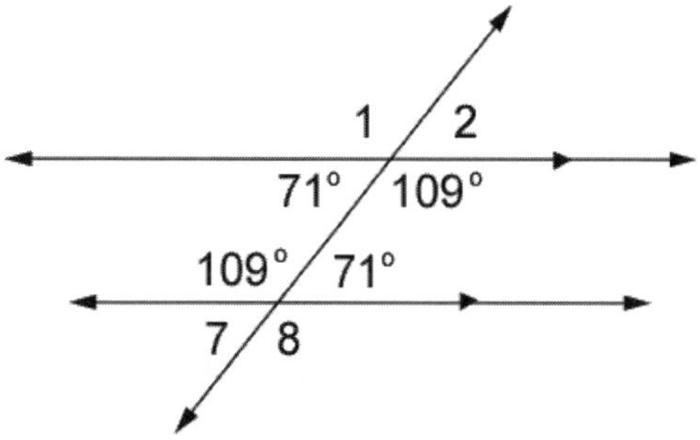

Example 5 If the measure of $\angle 1$ is three times as large as the measure of $\angle 7$, what is the measure of each angle?

$\angle 1 + \angle 7 = 180°$	**Step 1:** Note that $\angle 1$ and $\angle 7$ are exterior angles on the same side of the transversal. This means that they are supplementary.
$\angle 7 = x$ $\angle 1 = 3x$	**Step 2:** Let x represent the measure of $\angle 7$. So, the measure of $\angle 1$ is $3x$, since it is three times as large.
$3x + x = 180°$	**Step 3:** Since these angles are supplementary, their sum is $180°$.
$4x = 180°$	**Step 4:** Combine like terms.
$x = 45°$	**Step 5:** Divide both sides by 4. The measure of $\angle 7$ is $45°$.
$3x = 135°$	**Step 6:** Multiply both sides by 3 to find the measure of $\angle 1$.

Answer: $m\angle 1 = 135°$ and $m\angle 7 = 45°$

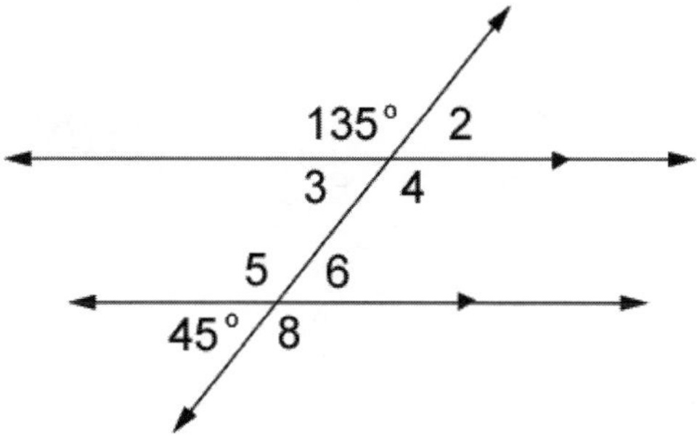

Problem Set

1. In the given figure, $\ell \parallel m$.
Identify a pair of corresponding angles from the given choices.

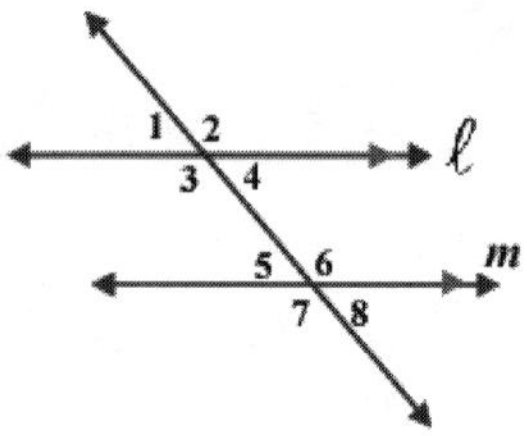

$\angle 1$ and $\angle 2$ $\quad$ $\angle 2$ and $\angle 3$

$\angle 2$ and $\angle 8$ $\quad$ $\angle 1$ and $\angle 5$

2. In the given figure, $\ell \parallel m$.
Identify a pair of corresponding angles from the given choices.

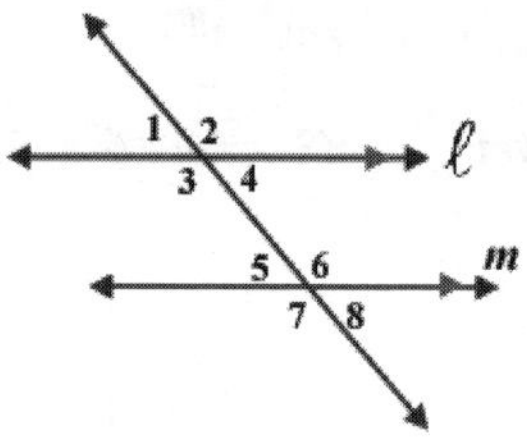

$\angle 1$ and $\angle 7$ $\quad$ $\angle 2$ and $\angle 6$

$\angle 2$ and $\angle 4$ $\quad$ $\angle 3$ and $\angle 5$

3. In the given figure, $a \parallel b$. Identify a pair of corresponding angles from the given choices.

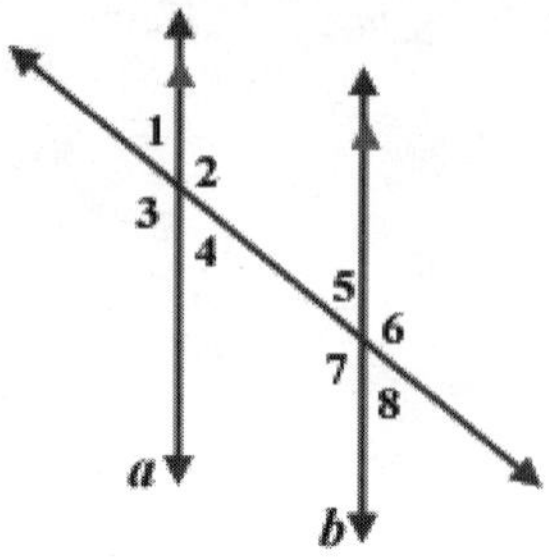

$\angle 4$ and $\angle 7$ $\quad$ $\angle 2$ and $\angle 8$

$\angle 3$ and $\angle 7$ $\quad$ $\angle 7$ and $\angle 5$

4. In the given figure, $a \parallel b$. Identify a pair of corresponding angles from the given choices.

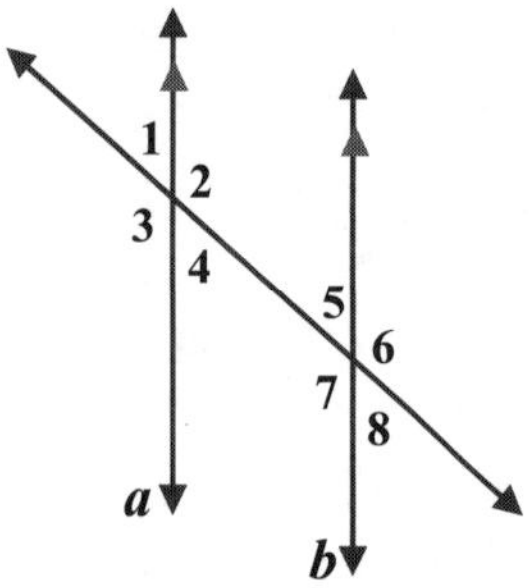

$\angle 4$ and $\angle 6$ $\quad$ $\angle 4$ and $\angle 8$

$\angle 7$ and $\angle 8$ $\quad$ $\angle 2$ and $\angle 5$

5. In the given figure, $c \parallel d$. Identify a pair of corresponding angles from the given choices.

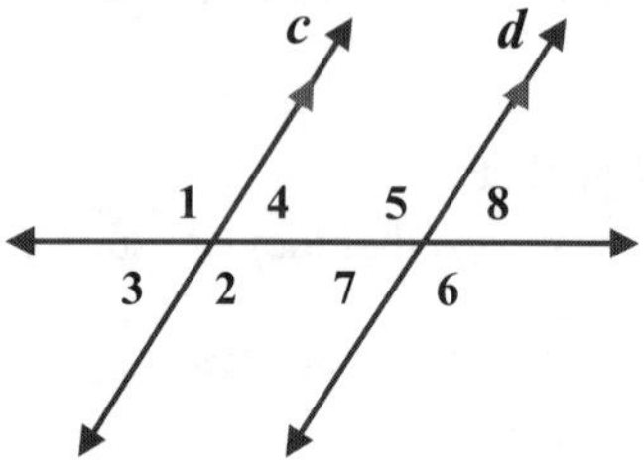

$\angle 6$ and $\angle 5$ $\quad$ $\angle 5$ and $\angle 1$

$\angle 5$ and $\angle 7$ $\quad$ $\angle 2$ and $\angle 1$

6. In the given figure, $c \parallel d$. Identify a pair of corresponding angles from the given choices.

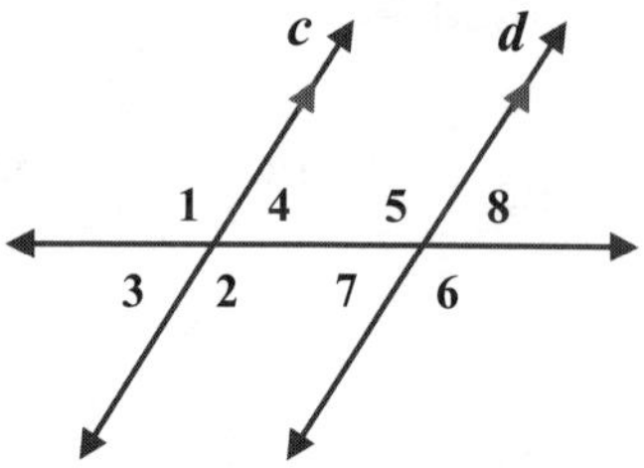

$\angle 3$ and $\angle 6$ $\quad$ $\angle 8$ and $\angle 1$

$\angle 5$ and $\angle 7$ $\quad$ $\angle 3$ and $\angle 7$

7. In the given figure, $g \parallel h$. Identify a pair of corresponding angles from the given choices.

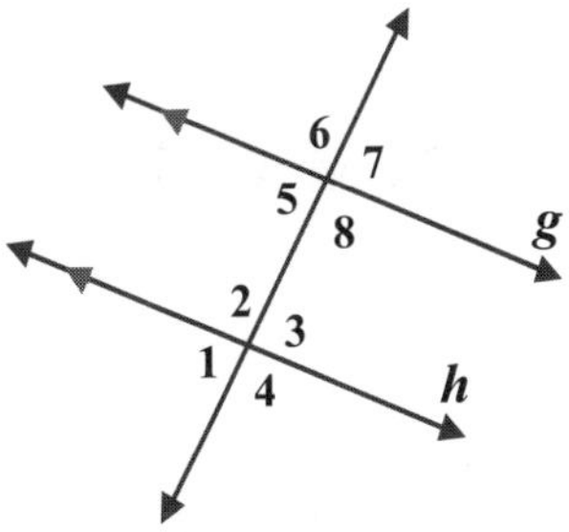

∠1 and ∠5   ∠1 and ∠3

∠1 and ∠7   ∠1 and ∠4

8. In the given figure, $g \parallel h$. Identify a pair of corresponding angles from the given choices.

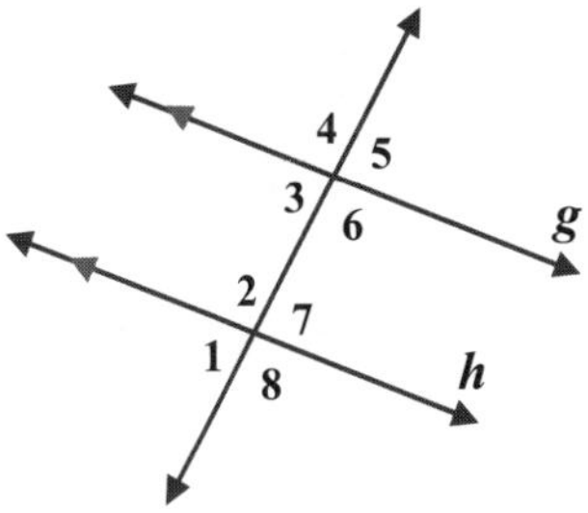

∠6 and ∠3   ∠6 and ∠4

∠6 and ∠7   ∠6 and ∠8

9. In the given figure, $g \parallel h$. Identify a pair of alternate exterior angles from the given choices.

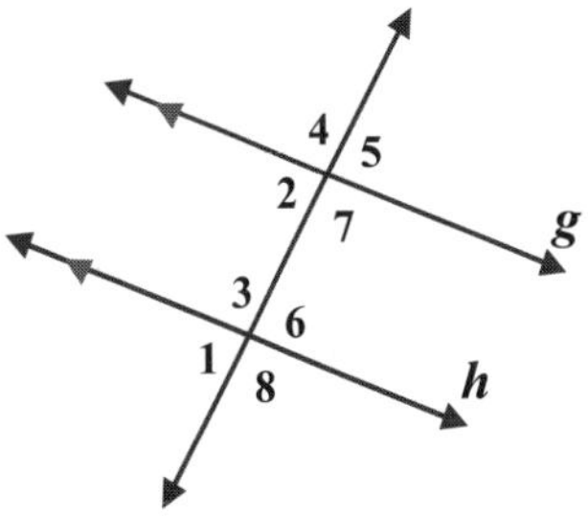

∠5 and ∠6   ∠4 and ∠8

∠1 and ∠7   ∠4 and ∠7

10. In the given figure, $g \parallel h$. Identify a pair of alternate exterior angles from the given choices.

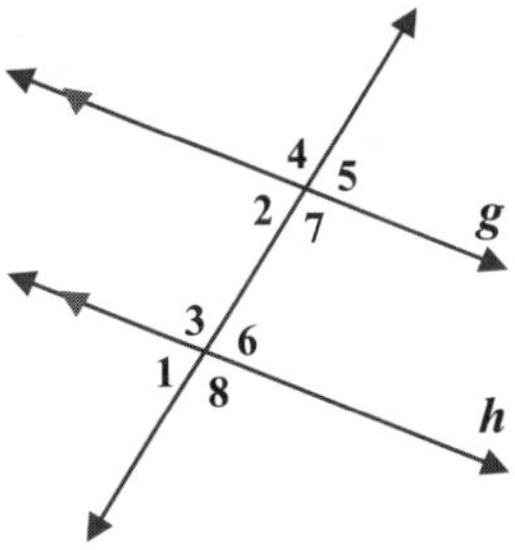

∠1 and ∠5   ∠1 and ∠4

∠6 and ∠7   ∠1 and ∠7

11. In the given figure, $m \parallel n$. Identify a pair of alternate exterior angles from the given choices.

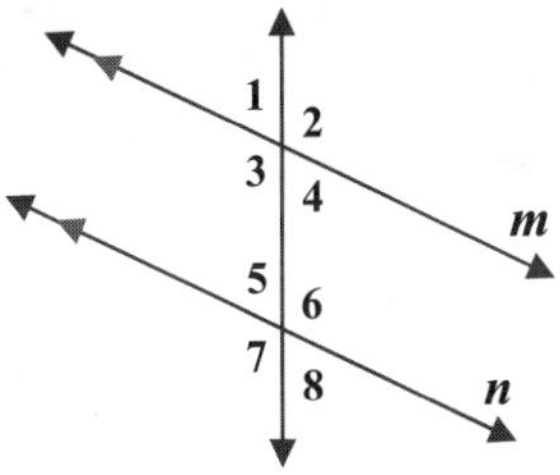

∠1 and ∠2   ∠2 and ∠8

∠1 and ∠7   ∠1 and ∠8

12. In the given figure, $m \parallel n$. Identify a pair of alternate exterior angles from the given choices.

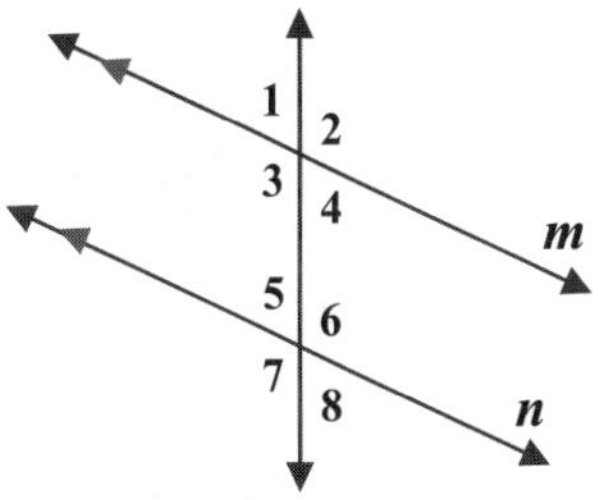

∠2 and ∠8   ∠1 and ∠7

∠1 and ∠2   ∠2 and ∠7

13. In the given figure, $r \parallel s$. Identify a pair of alternate exterior angles from the given choices.

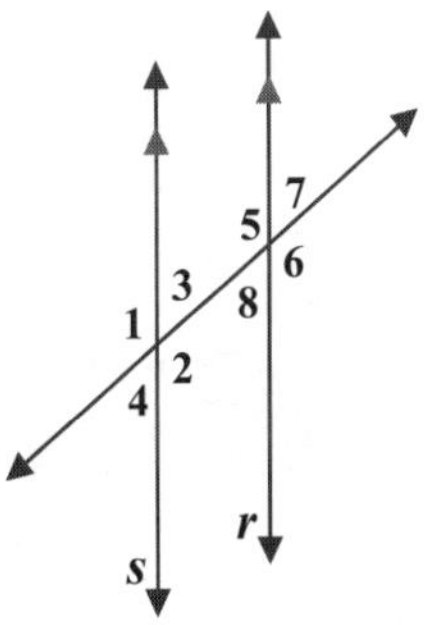

∠4 and ∠5   ∠4 and ∠7

∠1 and ∠7   ∠1 and ∠4

14. In the given figure, $r \parallel s$. Identify a pair of alternate exterior angles from the given choices.

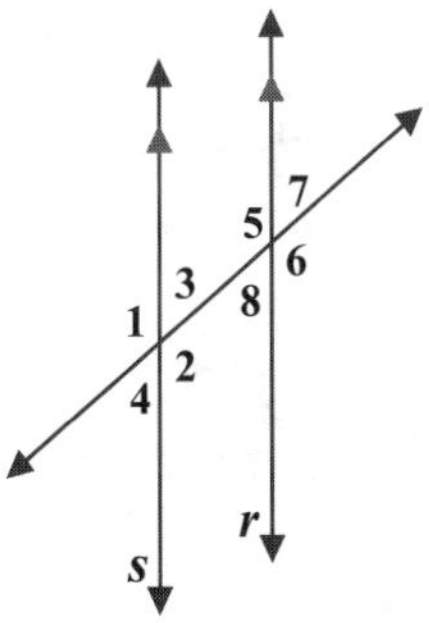

∠4 and ∠5   ∠1 and ∠7

∠4 and ∠6   ∠1 and ∠6

15. In the given figure, $q \parallel t$. Identify a pair of alternate exterior angles from the given choices.

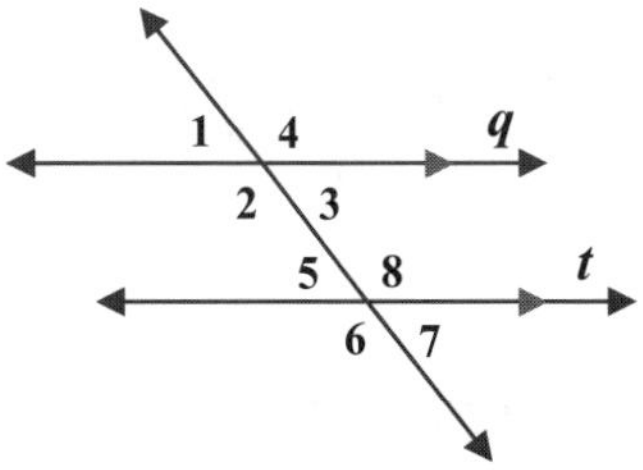

∠4 and ∠7   ∠1 and ∠7

∠1 and ∠4   ∠4 and ∠2

16. In the given figure, $\ell \parallel m$. The measure of $\angle 1$ is 45°. Find the measure of $\angle 8$.

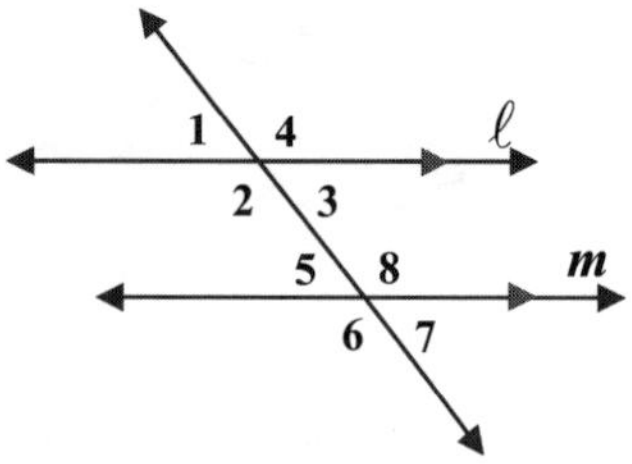

17. In the given figure, $\ell \parallel m$. The measure of $\angle 4$ is 37°. Find the measure of $\angle 5$.

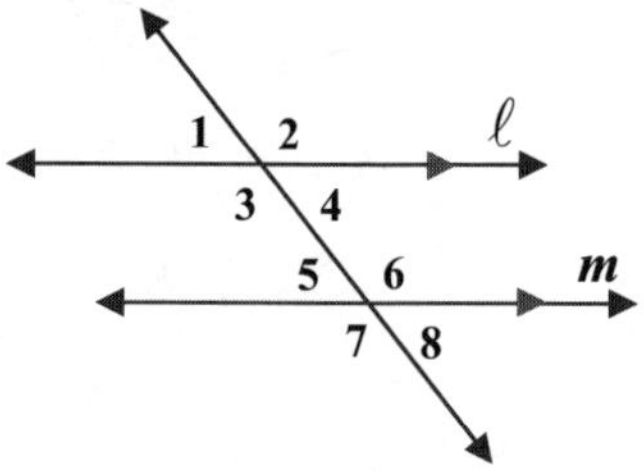

18. In the given figure, $a \parallel b$. The measure of $\angle 3$ is 115°. Find the measure of $\angle 7$.

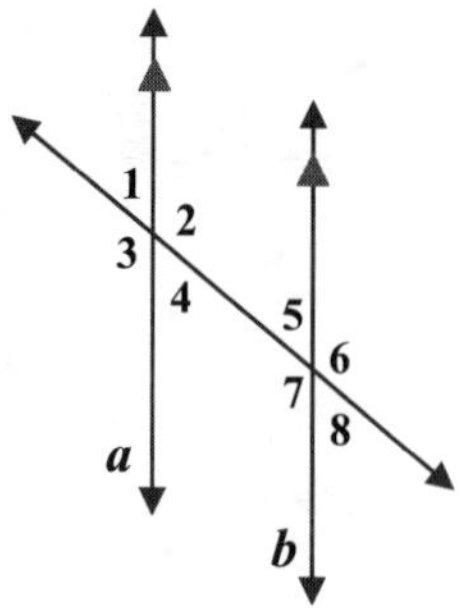

19. In the given figure, $a \parallel b$. The measure of $\angle 2$ is 130°. Find the measure of $\angle 7$.

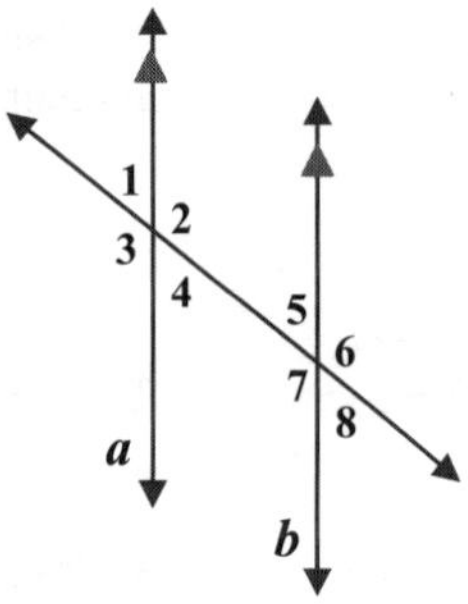

20. In the given figure, $c \parallel d$. The measure of $\angle 3$ is 50°. Find the measure of $\angle 8$.

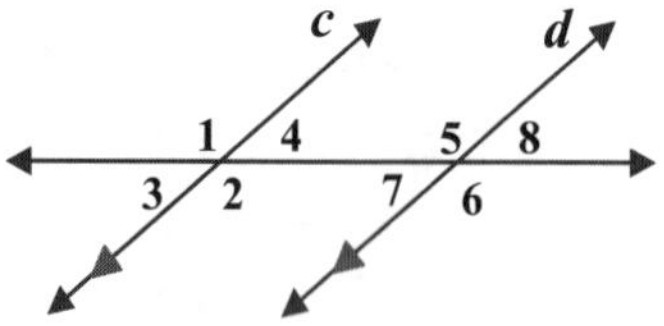

21. In the given figure, $c \parallel d$. The measure of $\angle 3$ is 50°. Find the measure of $\angle 6$.

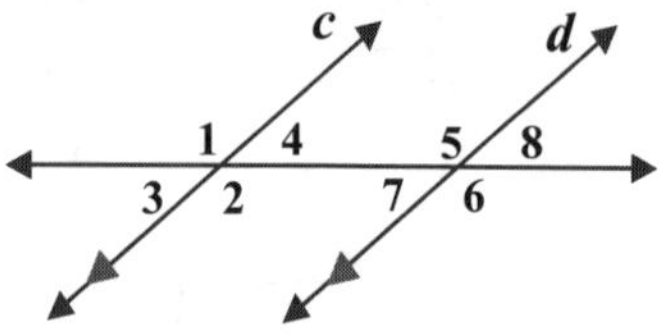

22. In the given figure, $g \parallel h$. The measure of $\angle 1$ is 82°. Find the measure of $\angle 3$.

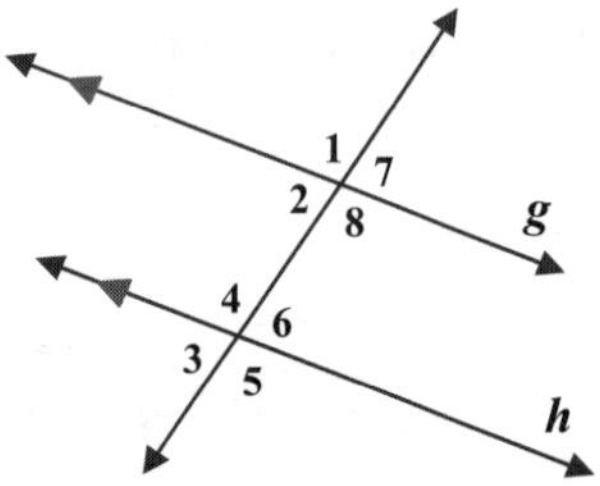

23. In the given figure, $g \parallel h$. The measure of $\angle 5$ is 79°. Find the measure of $\angle 8$.

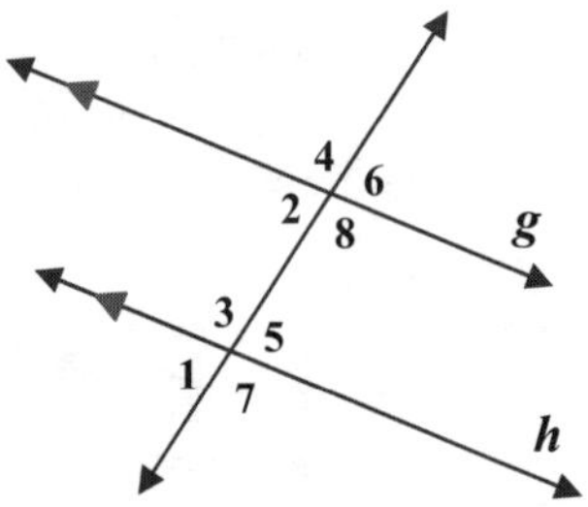

24. In the figure, $\ell \parallel m$. The measure of $\angle 1$ is 50°. The measure of $\angle 2$ is 80°. Find the measure of $\angle 5$ and $\angle 6$.

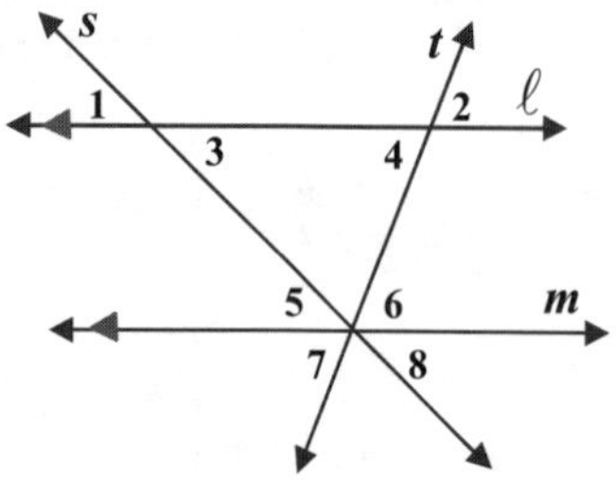

25. In the figure, $\ell \parallel m$. The measure of $\angle 3$ is 42°. The measure of $\angle 4$ is 77°. Find the measures of $\angle 5$ and $\angle 6$.

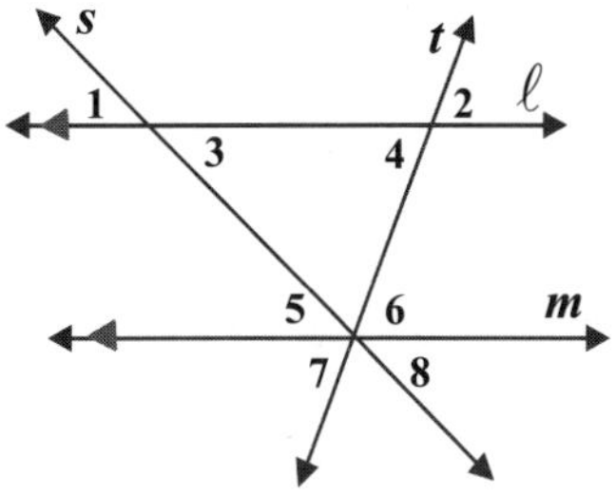

26. In the figure, $\ell \parallel m$. The measure of $\angle 1$ is 45°. The measure of $\angle 2$ is 85°. Find the measures of $\angle 7$ and $\angle 8$.

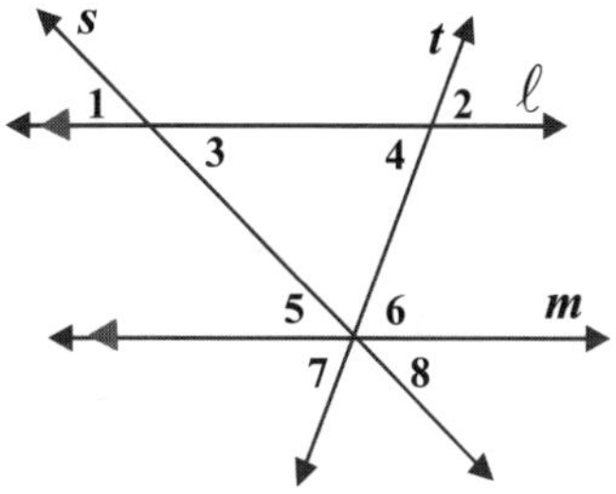

27. In the figure, $p \parallel q$. The measure of $\angle 3$ is 83°. The measure of $\angle 4$ is 80°. Find the measures of $\angle 7$ and $\angle 8$.

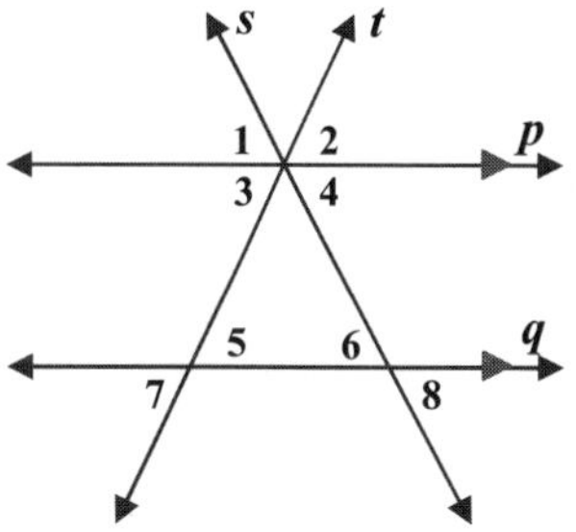

28. In the figure, $p \parallel q$. The measure of $\angle 3$ is 75°. The measure of $\angle 4$ is 87°. Find the measures of $\angle 5$ and $\angle 6$.

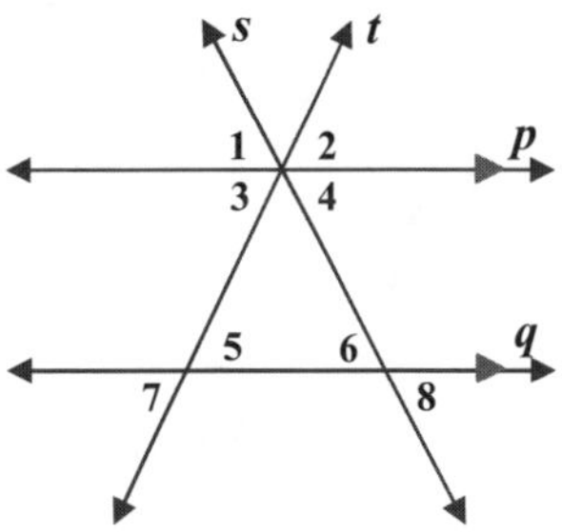

29. In the figure, $p \parallel q$. The measure of $\angle 1$ is 69°. The measure of $\angle 2$ is 71°. Find the measures of $\angle 7$ and $\angle 8$.

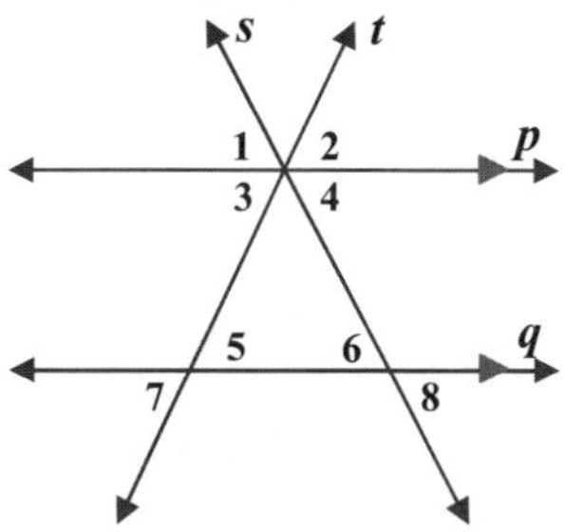

30. In the figure, $r \parallel s$. The measure of $\angle 1$ is 73°. The measure of $\angle 2$ is 56°. Find the measures of $\angle 7$ and $\angle 8$.

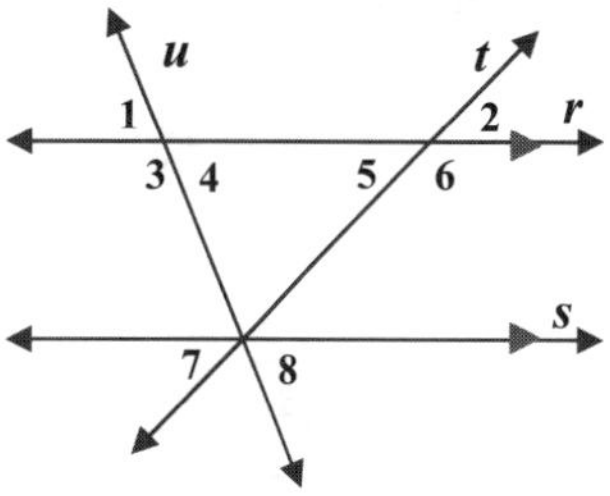

31. In the figure shown, $\overline{BC} \parallel \overline{DE}$. If the measures of $\angle B$ and $\angle AED$ are 75°, what is the measure of $\angle A$?

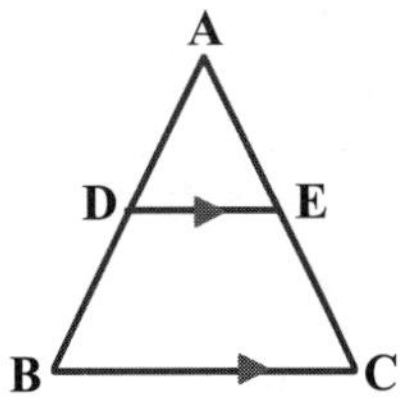

32. In the figure shown, $\overline{BC} \parallel \overline{DE}$. If the measure of $\angle A$ is 25° and the measure of $\angle AED$ is 80°, what is the measure of $\angle B$?

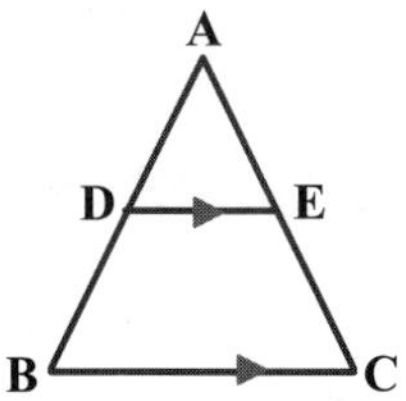

33. In the figure shown, $\overline{BC} \parallel \overline{DE}$. If the measure of $\angle A$ is 65° and the measure of $\angle AED$ is 70°, what is the measure of $\angle B$?

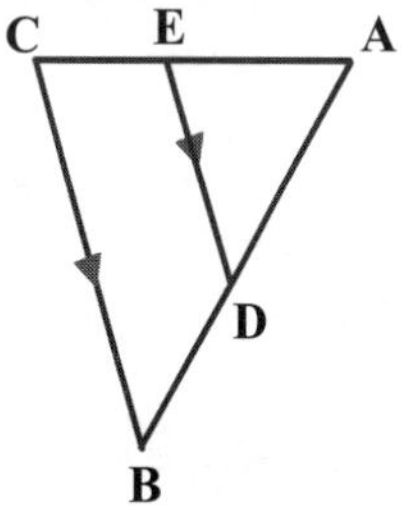

34. In the figure shown, $\overline{BC} \parallel \overline{DE}$. If the measure of $\angle B$ is 35° and the measure of $\angle AED$ is 77°, what is the measure of $\angle A$?

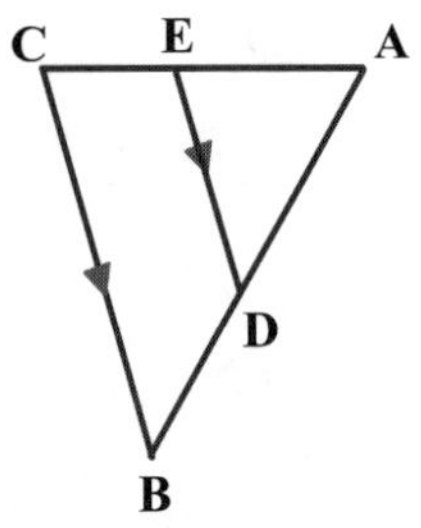

35. In the figure shown, $\overline{RS} \parallel \overline{TU}$. If the measure of $\angle R$ is 90° and the measure of $\angle QUT$ is 58°, what is the measure of $\angle Q$?

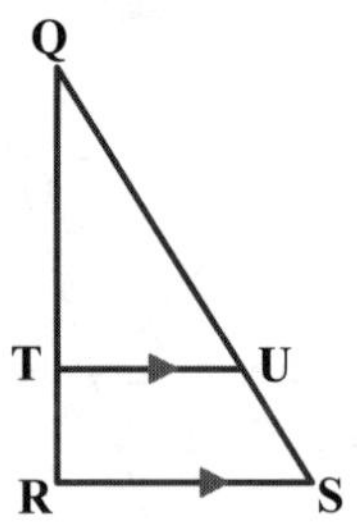

36. In the figure shown, $\overline{RS} \parallel \overline{TU}$. If the measure of $\angle QTU$ is 89° and the measure of $\angle Q$ is 37°, what is the measure of $\angle S$?

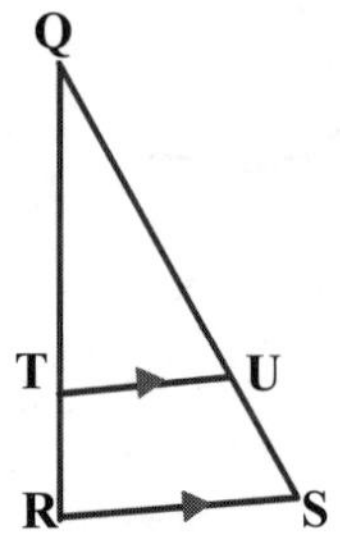

37. In the figure shown, $\overline{XZ} \parallel \overline{VW}$. If the measure of $\angle VWY$ is 93° and the measure of $\angle X$ is 43°, what is the measure of $\angle Y$?

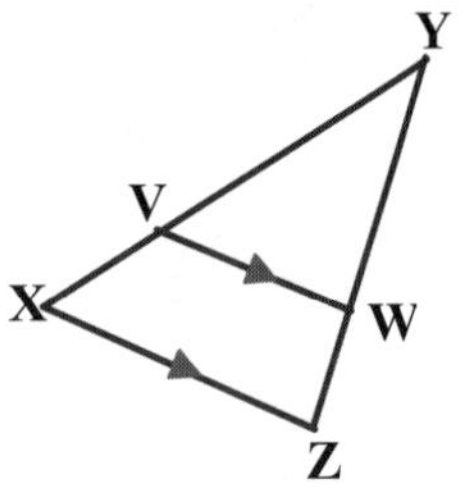

38. In the figure shown, $\overline{XZ} \parallel \overline{VW}$. If the measure of $\angle YVW$ is 47° and the measure of $\angle Y$ is 34°, what is the measure of $\angle Z$?

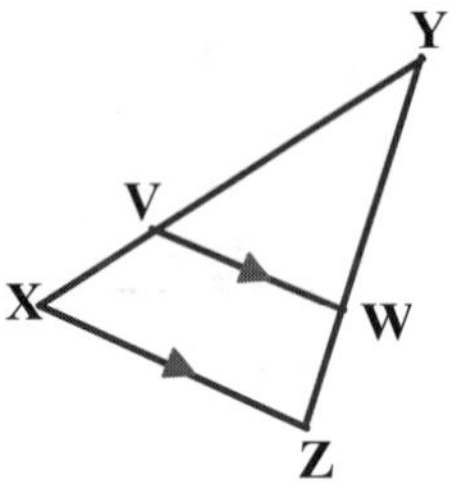

39. In the figure shown, $\overline{XZ} \parallel \overline{VW}$. If the measure of $\angle YVW$ is 101° and the measure of $\angle Y$ is 27°, what is the measure of $\angle Z$?

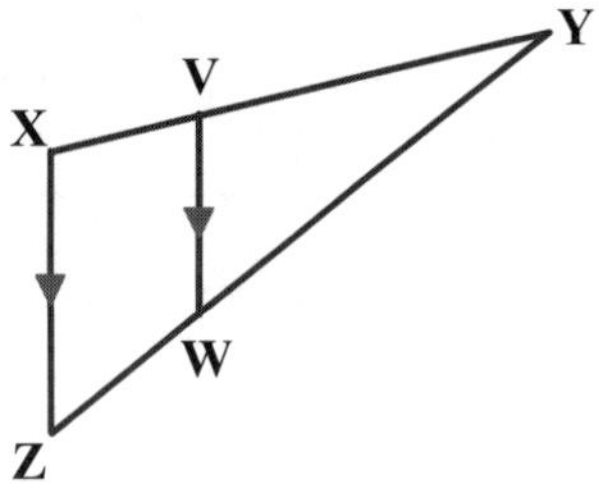

40. In the figure shown, $\overline{XZ} \parallel \overline{VW}$. If the measure of $\angle YWV$ is 55° and the measure of $\angle X$ is 110°, what is the measure of $\angle Y$?

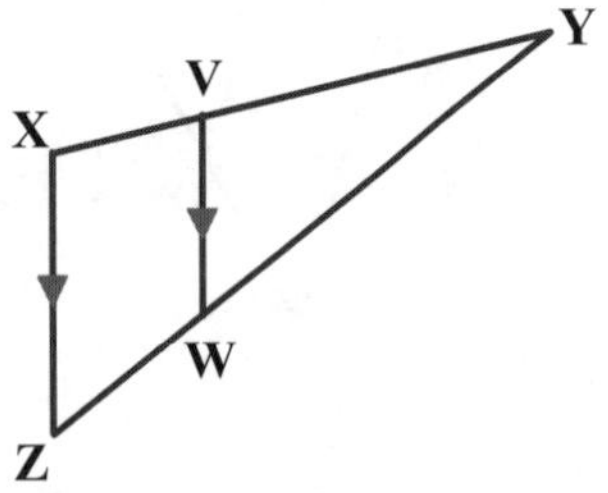

MPA-106: Identifying a Solid Figure from a Net

A **solid figure** is a three-dimensional shape composed of a region in space bounded by familiar geometric figures. One such solid figure is a triangular pyramid.

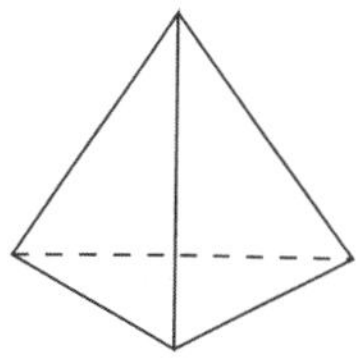

A triangular pyramid is a region of space bounded by three triangles that meet at a single point, all built on a base that is a triangle. Each of the triangles and the base are called the **faces** of the pyramid. The line segments that join these faces are called the **edges** of the pyramid. Finally, the points where the line segments meet are called the **vertices** of the pyramid. In this case, there are 4 faces, 6 edges, and 4 vertices.

Faces	Flat sides formed by polygons and their interiors.

Edges	Line segments that join the faces of polyhedrons.

Vertex	The point where the line segments of polyhedrons intersect.

Another standard example of a solid figure is a cube. It has 6 faces (squares), 12 edges, and 8 vertices.

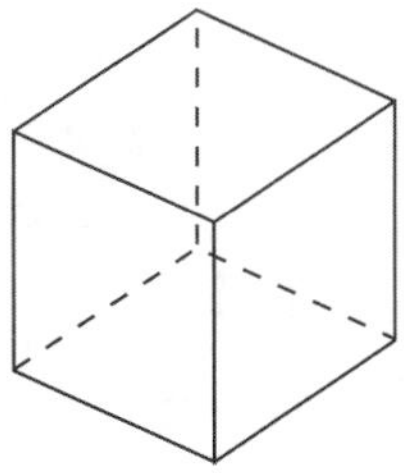

Imagine cutting a cardboard cube along its edges and laying it flat. The result is a two-dimensional figure known as a **net**. Nets are useful in visualizing the faces of a solid. They are also useful when calculating the surface area (the sum of the areas of all the faces) of the solid figure. For example, the net on the following page indicates that the surface area of the cube is the sum of the areas of the six faces.

Surface Area	The sum of the areas of all the faces of a three-dimensional solid.

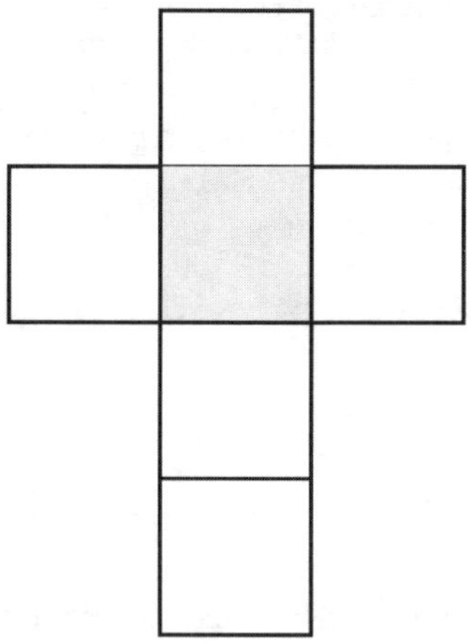

Understanding how to identify a solid object and find its properties from its net are important spatial reasoning skills that are useful in a variety of mathematical applications.

Let's consider an example of a solid and its net. Examine the cylinder below. In this case not all of the faces are polygons, as they were in the first two examples.

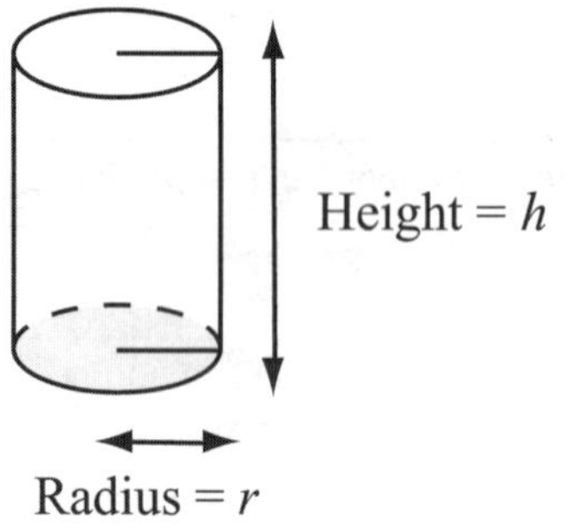

The net of the cylinder consists of two circles and a rectangle as shown below:

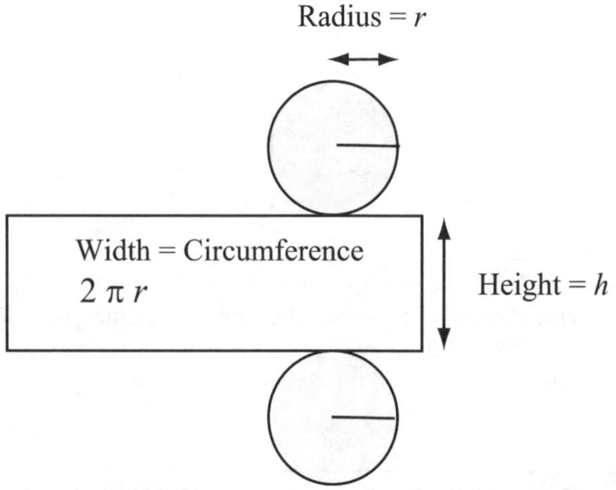

Note that the dimensions of the rectangle are calculated by multiplying the height of the original cylinder, h, by the circumference of the top of the cylinder, which is $2\pi r$.

Example 1 Consider the net of a pentagonal prism shown below. How many faces does the prism have?

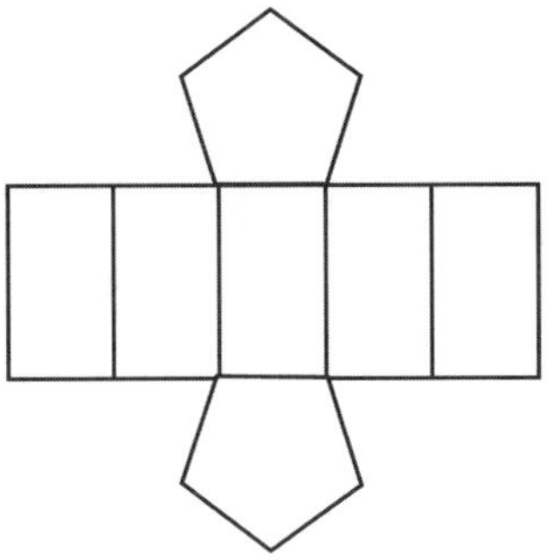

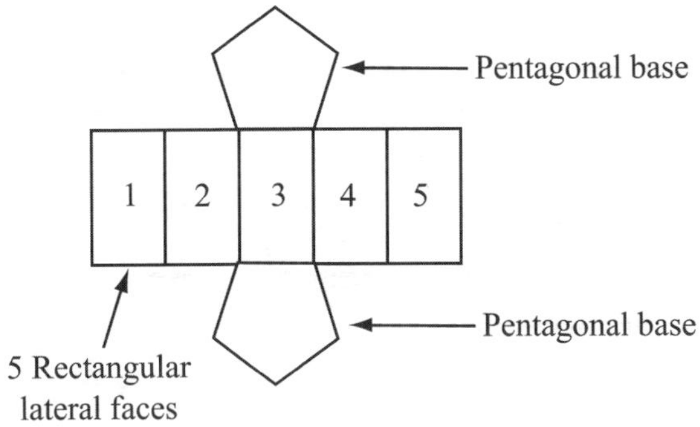

Step 1: The net shows exactly how many faces there are in the prism. There are 5 rectangular lateral faces and 2 pentagonal bases for a total of 7 faces.

Answer: A pentagonal prism has 7 faces (5 rectangles and 2 pentagons).

Example 2 How many faces, edges, and vertices does a triangular prism have?

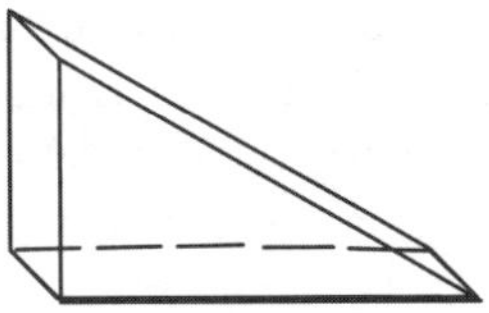

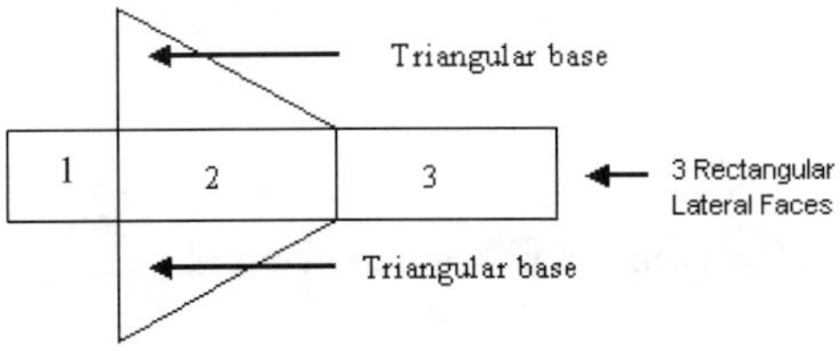

Step 1: Count the number of faces. The triangular prism has 2 bases that are triangles joined by 3 lateral faces that are rectangles. This means that there are 5 faces.

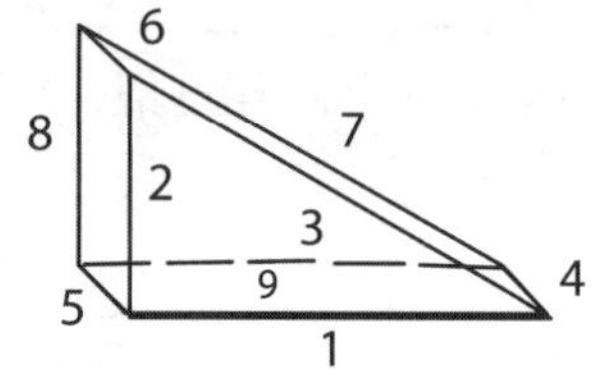

Step 2: Now count the edges. Each triangular face is bounded by 3 edges, making 6. Each vertex of the triangular faces is joined by an edge, adding 3 more for a total of 9 edges.

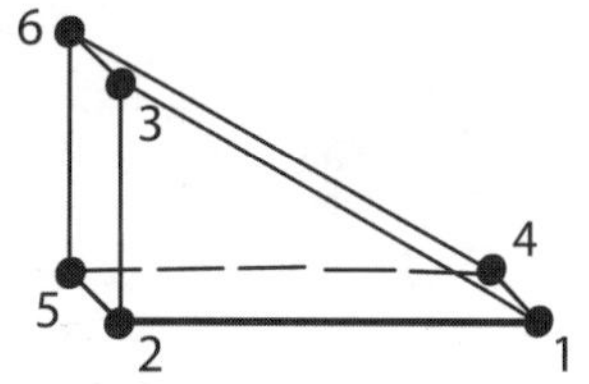

Step 3: Finally, count the number of vertices. Note that there are 3 vertices per triangular face for a total of 6 vertices.

Answer: A triangular prism has 5 faces, 9 edges, and 6 vertices.

Example 3 Name the solid figure formed from the net shown below.

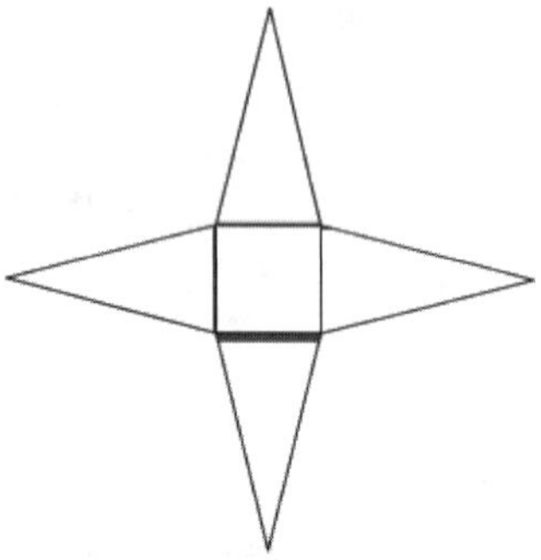

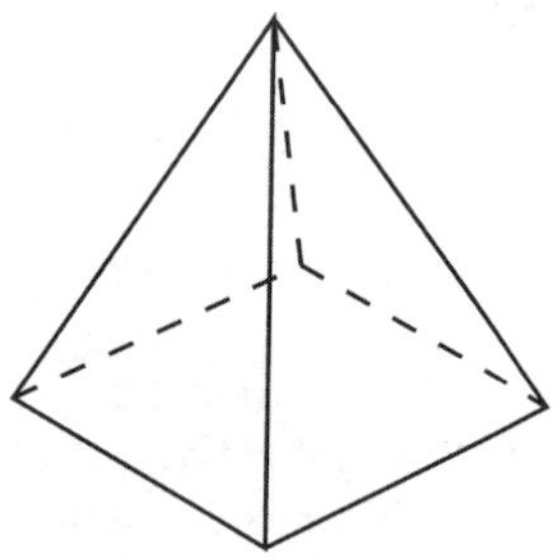

Step 1: This figure has one face that is a square and four faces that are triangles. Imagine folding the triangles along the line where they join the square. If this is done, the four triangles will meet at one point.

Answer: The figure formed from this net is a square pyramid.

Example 4 If the height of the cylinder is 5 centimeters and its radius is 2 centimeters, what is the surface area of the cylinder?

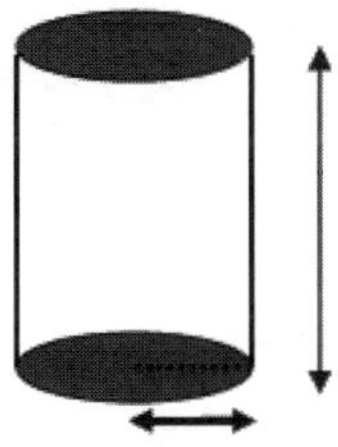

Step 1: Recall that the net of a cylinder is composed of three pieces: two circles and one rectangle.

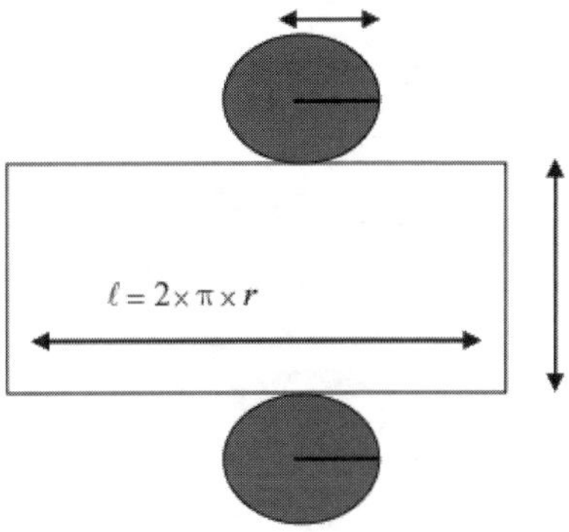

Area of each circle $= \pi \times r^2$

$= \pi \times 2^2$

$= 4\pi\ \text{cm}^2$

Step 2: The area of each circular piece is 4π square centimeters.

Length of the rectangle $= 2 \times \pi \times r$

$= 4\pi\ \text{cm}$

Step 3: The rectangular face has a width of 5 centimeters and a length that is the circumference of the circle. Since the radius is 2 centimeters, this makes the length of the circle 4π centimeters.

Area of rectangle $= \ell \times w$

$= 4\pi \times 5$

$= 20\pi\ \text{cm}^2$

Step 4: The area of the rectangular piece is the length times the width. The length is 4π centimeters and the width is 5 centimeters. Therefore, the area is 20π square centimeters.

Surface area $= 8\pi + 20\pi$

$= 28\pi\ \text{cm}^2$

Step 5: Adding the areas of the circles to the area of the rectangle gives us an area of 28π square centimeters.

Answer: The surface area of a cylinder with a height of 5 centimeters and a radius of 2 centimeters is 28π square centimeters.

Example 5 If the length of the sides of a cube is x centimeters, what is a formula for the surface area of the cube in square centimeters?

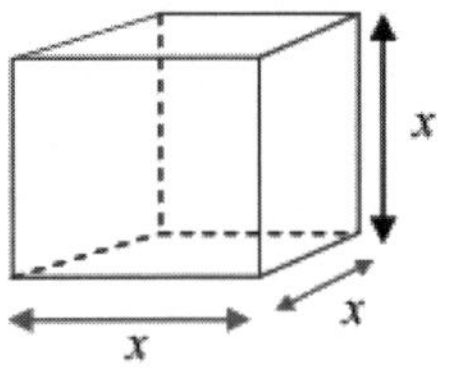

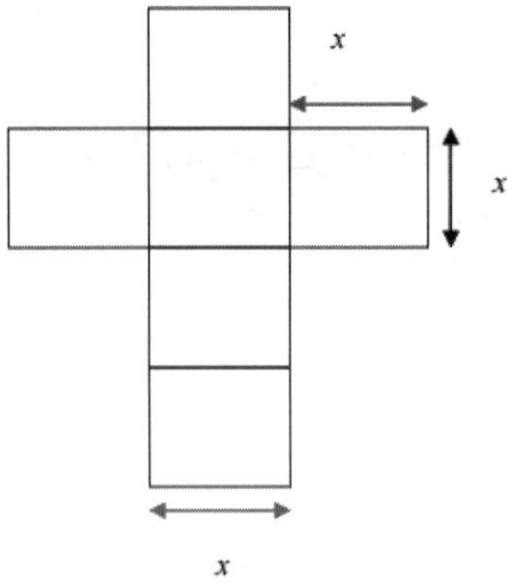

Step 1: Consider the drawing of the cube and its net.

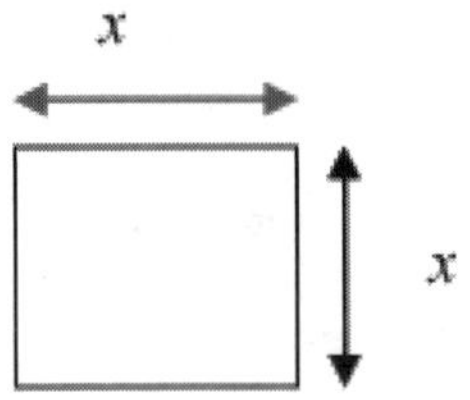

Step 2: Note that each of the faces of the cube is a square with sides of length x.

Area of square $= x \times x = x^2$ cm^2

Surface area $= 6x^2$ cm^2

Step 3: The area of each square is x^2 square centimeters. Since there are six identical faces, the surface area is $6x^2$ square centimeters.

Answer: The surface area of a cube with edges of length x centimeters is $6x^2$ square centimeters.

Problem Set

1. A tetrahedron has _____ faces, ______ vertices, and _______ edges.

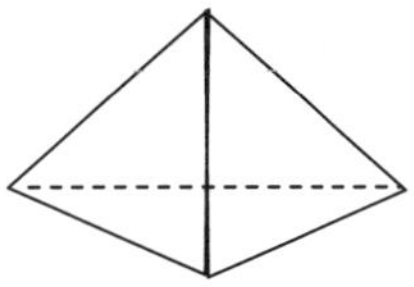

2. An octahedron has ____ faces, ____ vertices, and ____ edges.

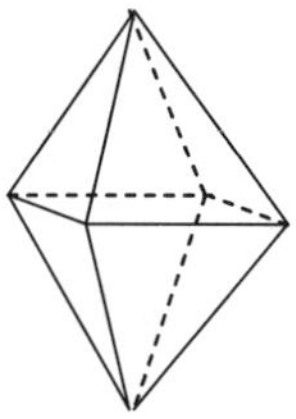

3. A hexagonal prism has ____ faces, ____ vertices, and ____ edges.

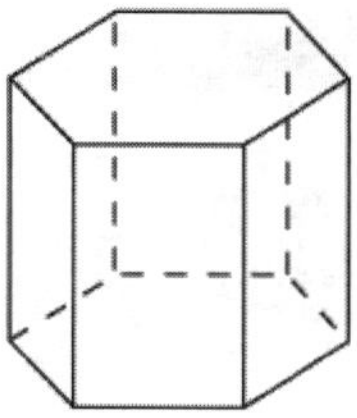

4. A cube has ____ faces, ____ vertices, and ____ edges.

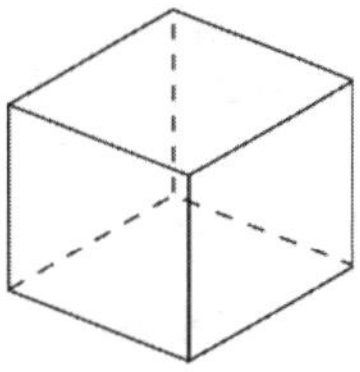

5. A square pyramid has ____ faces, ____ vertices, and ____ edges.

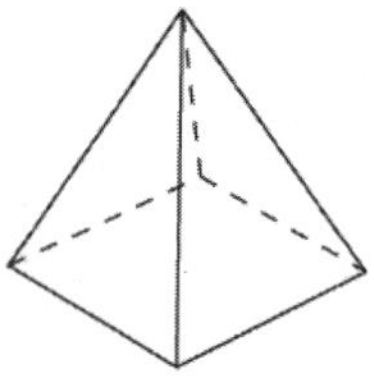

6. A rectangular prism has ____ faces, ____ vertices, and ____ edges.

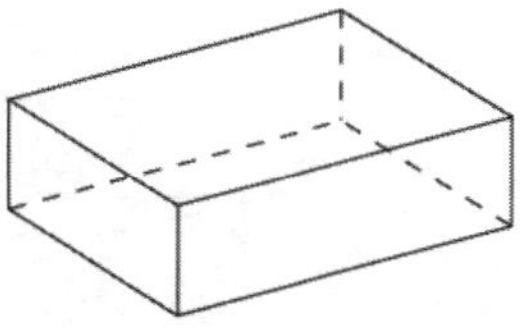

7. True or False:

   The altitude of a pyramid is the distance from the apex to the right hand corner of the base.

8. True or False:

   An octahedron has 6 faces.

9. True or False:

   There are 5 vertices in a triangular pyramid.

10. True or False:

    Vertex is the point where the line segments of polyhedrons intersect.

11. True or False:

    A pentagonal prism has 8 faces.

12. True or False:

    The sides of a pyramid are always triangles.

13. True or False:

    The size of the altitude affects the height of the pyramid.

14. True or False:

    Faces are line segments that join the faces of polyhedrons.

15. Identify the solid that corresponds to the following net.

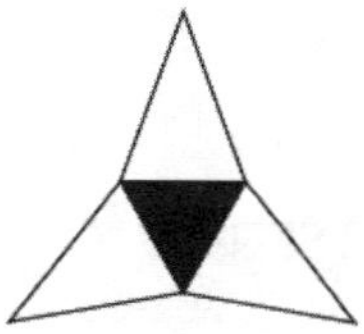

16. Identify the solid that corresponds to the following net.

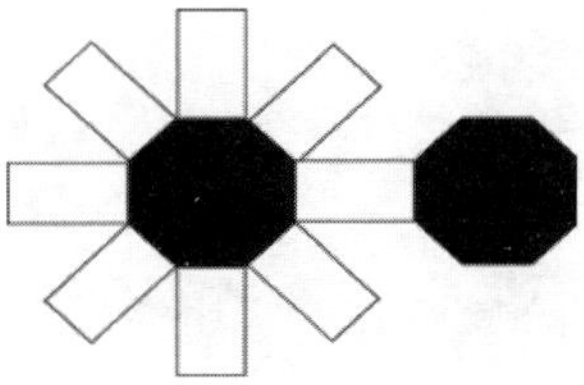

17. Identify the solid that corresponds to the following net.

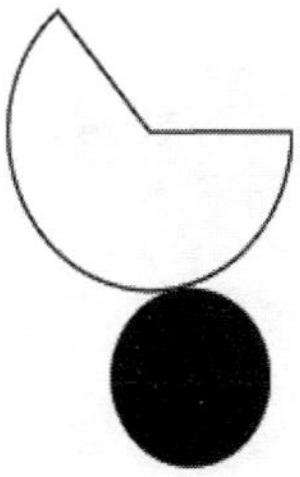

18. Identify the solid that corresponds to the following net.

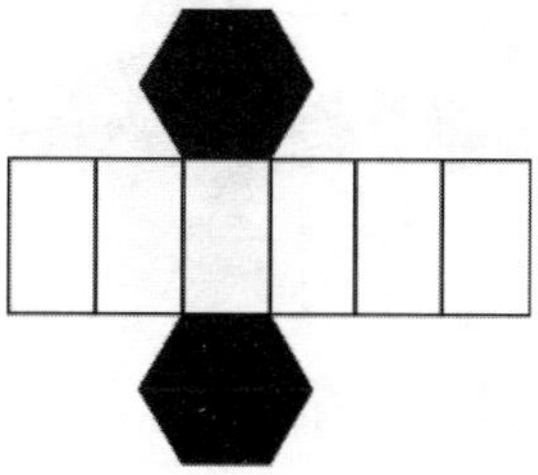

19. Find the surface area of the cube from the net shown below.

20. Find the surface area of the cylinder from the net shown below.

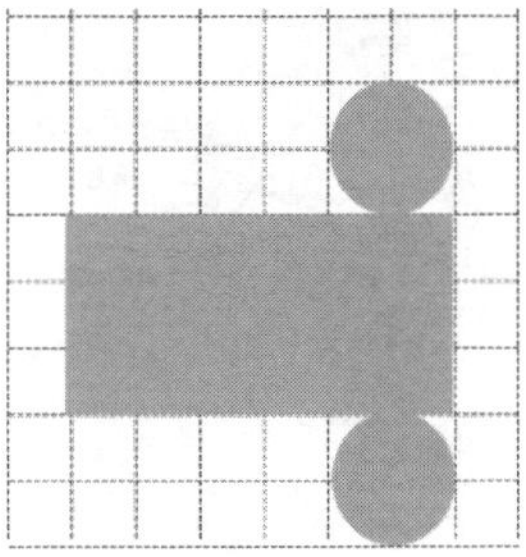

21. Find the surface area of the cylinder from the net shown below.

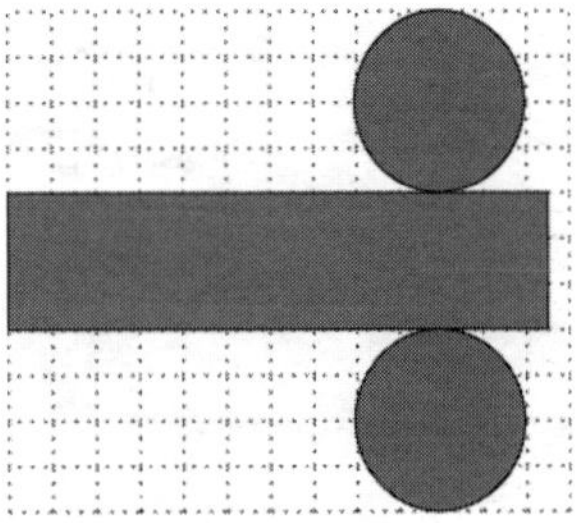

22. Find the surface area of the square pyramid from the net shown below.

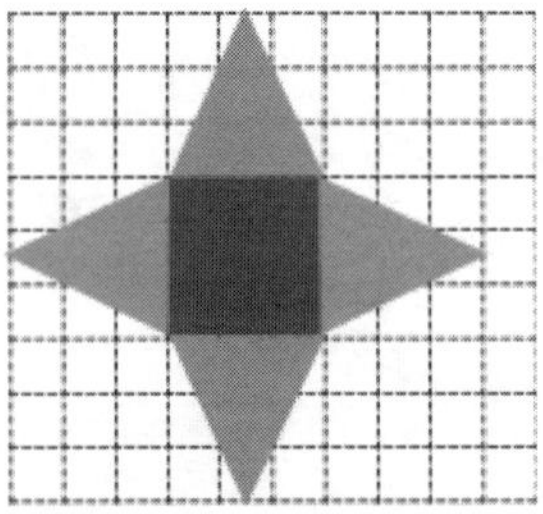

23. Find the surface area of the rectangular prism from the net shown below.

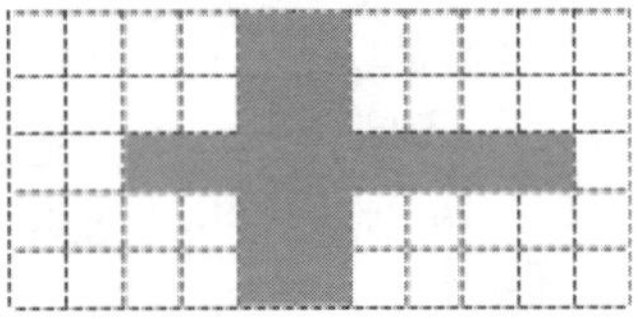

24. Find the surface area of the cube from the net shown below.

25. Find the surface area of the square pyramid from the net shown below.

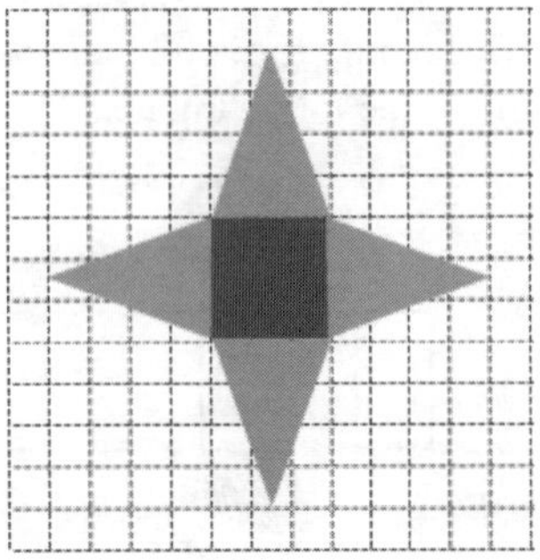

26. Find the surface area of the cylinder from the net shown below.

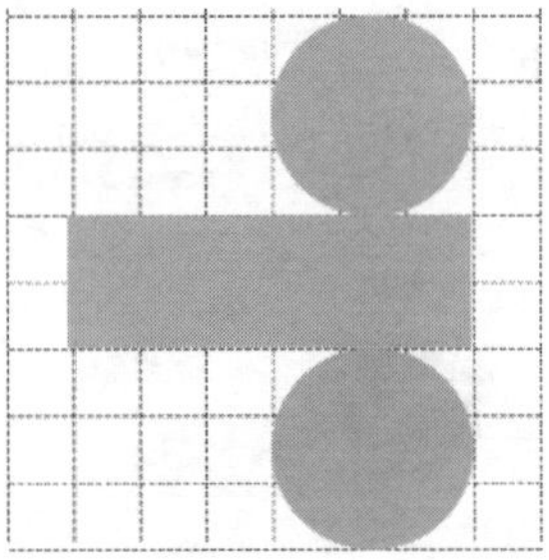

27. Find the surface area of the square pyramid from the net shown below.

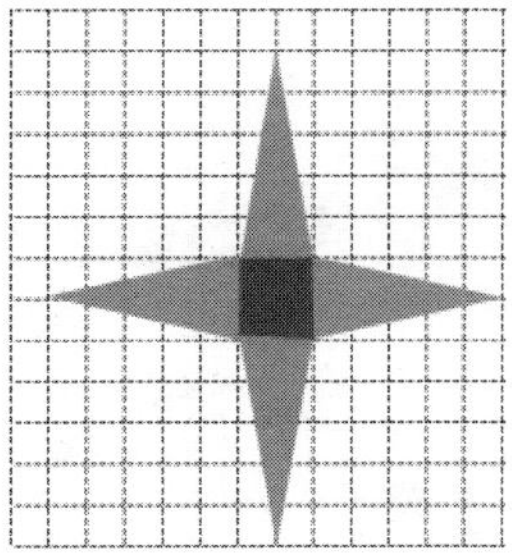

28. Find the surface area of the square pyramid from the net shown below.

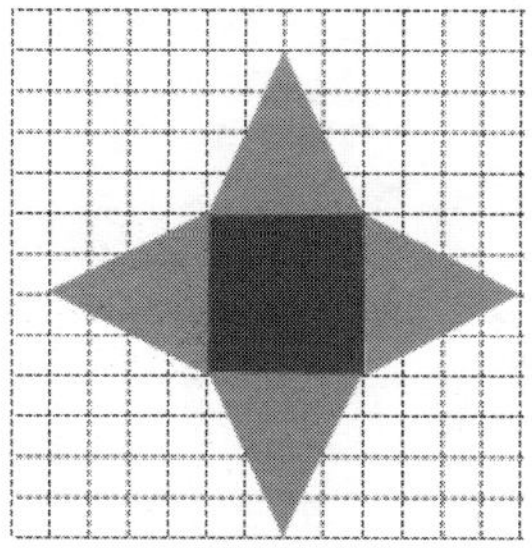

29. Find the surface area of the rectangular prism shown below. ______ in^2

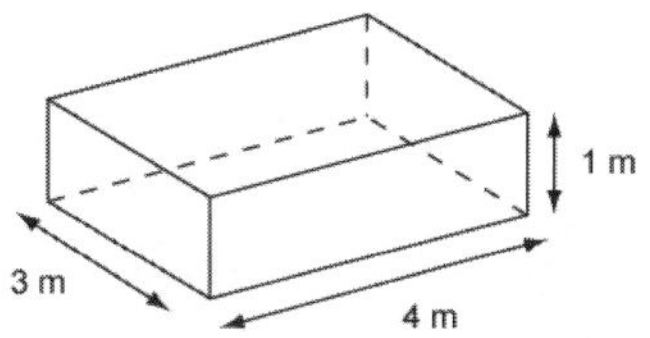

30. Find the surface area of the rectangular prism shown below. ______ in^2

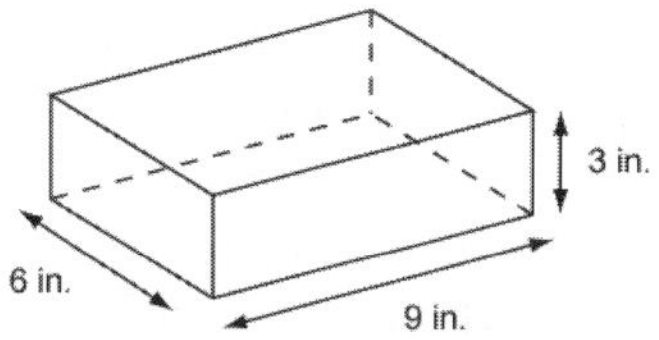

31. Find the surface area of the cube shown below.

______ m^2

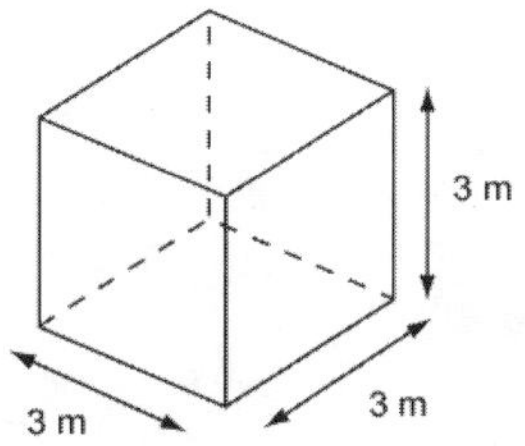

32. Find the surface area of the cube shown below.

______ m^2

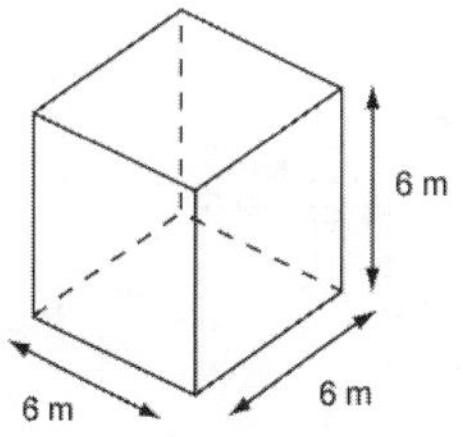

33. Find the surface area of the square pyramid shown below. ______ m^2

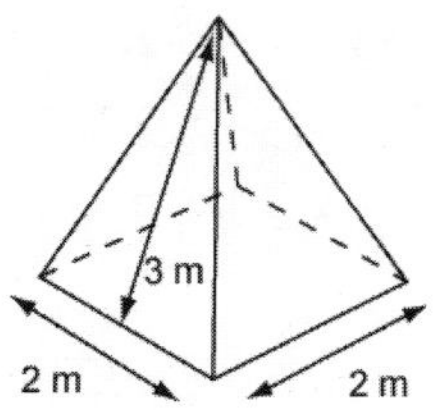

34. Find the surface area of the square pyramid shown below. ______ m^2

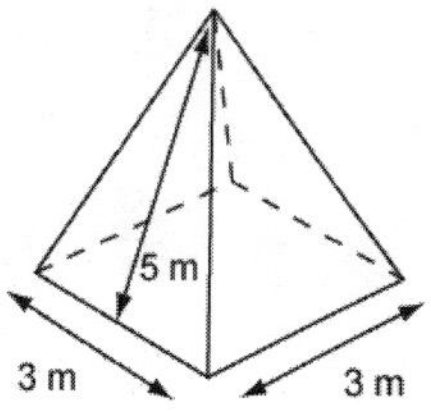

35. Find the surface area of the square pyramid shown below. ______ m^2

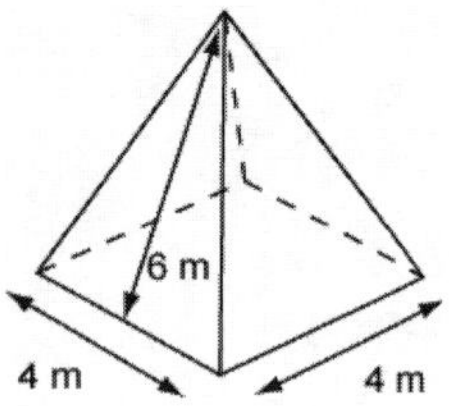

36. Find the surface area of the square pyramid shown below. ______ m^2

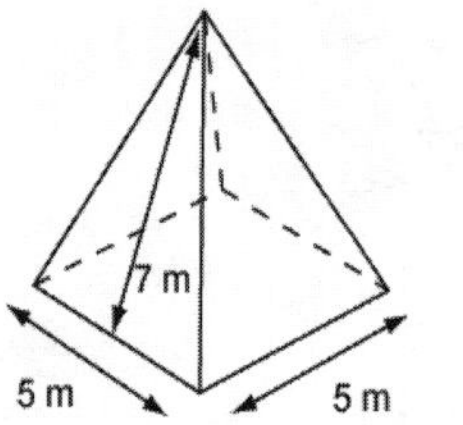

37. Find the surface area of the rectangular prism shown below. ______ m^2

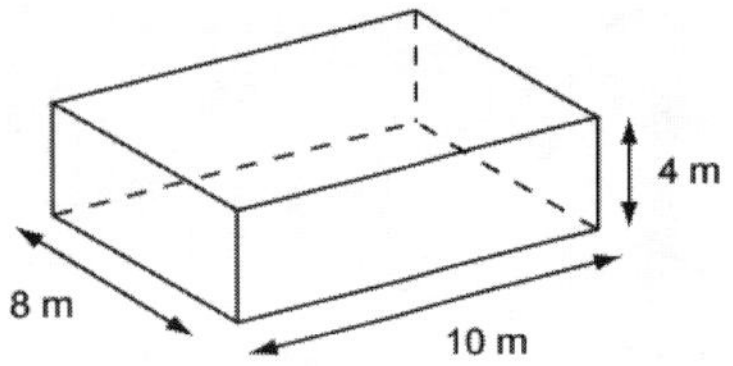

38. Find the surface area of the cube shown below.

______ m^2

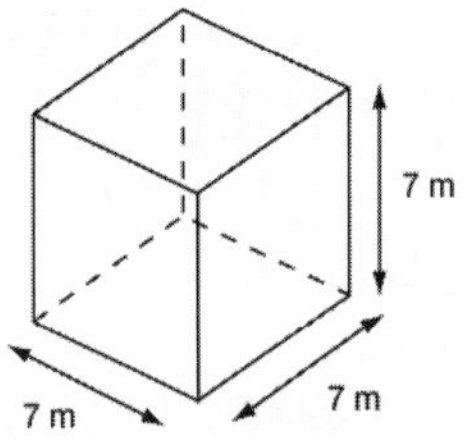

39. Find the surface area of the rectangular prism shown below. ______ m^2

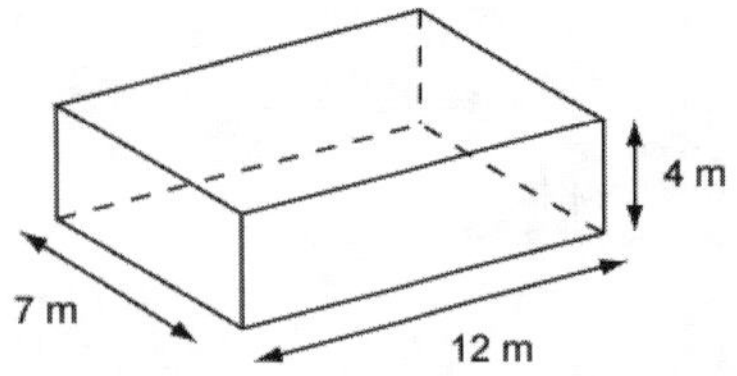

40. Find the surface area of the square pyramid shown below. ______ m^2

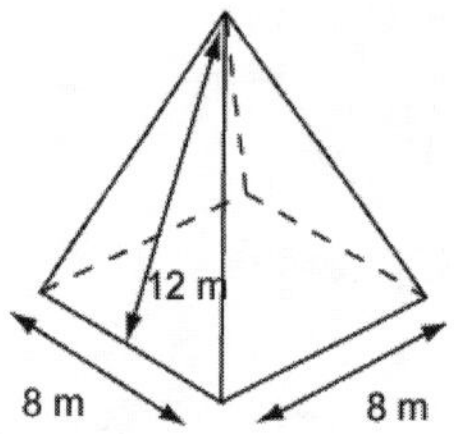

MPA-107: Constructing Three-Dimensional Figures and Examining Their Dimensions

A **solid figure** is a three-dimensional shape composed of a region in space bounded by familiar geometric figures. This lesson will explore prisms, pyramids, cones, cylinders, and spheres.

Solid Figure	A three-dimensional shape composed of a region in space bounded by familiar geometric figures.

A **prism** is a three-dimensional solid with bases that are identical polygons in parallel planes. Its edges are connected by parallelograms. The connecting sides are called the **lateral faces** of the solid.

Prism	A three-dimensional solid with bases that are identical polygons in parallel planes. Its edges are connected by parallelograms.

A prism is named for the polygon that makes its base. The figure below is called a hexagonal prism because its bases are hexagons.

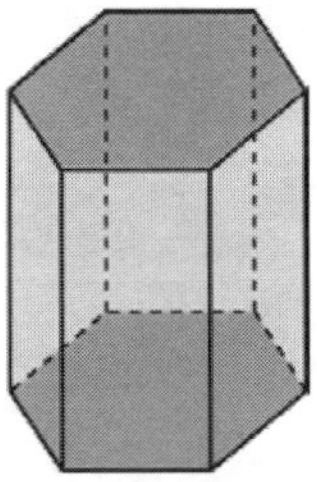

A **pyramid** is a three-dimensional solid with one polygonal base and triangular lateral faces that meet at one point called the **apex**.

Pyramid	A three-dimensional solid with one polygonal base and triangular lateral faces that meet at one point called the apex.

The figure below is called a square pyramid, since its base is a square.

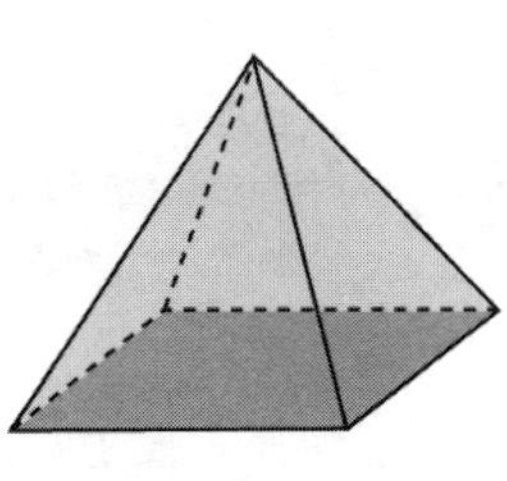

A **cone** is similar to a pyramid, except that it has a circular base. The lateral face is conical in shape.

Cone	A three-dimensional solid that has one circular base and a curved lateral face that meets at one point.

An example of a cone is shown below.

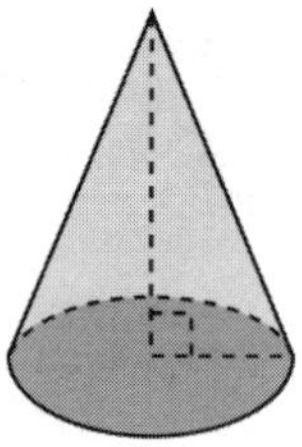

A **cylinder** is like a prism, except that its bases are circles.

Cylinder	A three-dimensional solid with two circular bases in parallel planes connected with a curved side.

A picture of a cylinder is shown below.

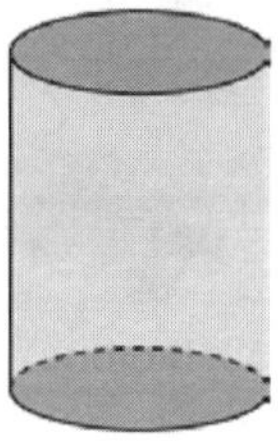

Spheres are like circles in three dimensions. Some common examples of spheres are basketballs and ball bearings.

Sphere	A three-dimensional solid that is in the shape of a circle.

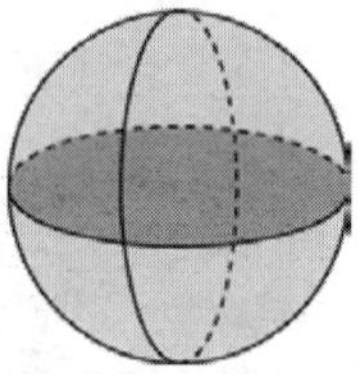

Example 1 Name the three-dimensional geometric figure shown below.

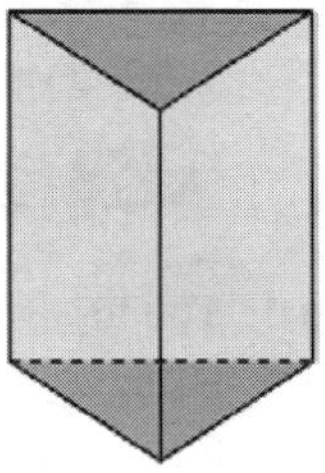

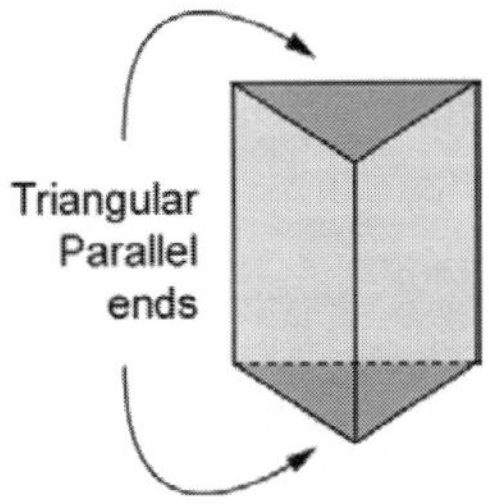

Step 1: Notice that this figure has two identical parallel bases (or ends) that are triangles. These are joined by rectangular faces. This figure is called a triangular prism.

Answer: Triangular prism

Example 2 A familiar solid figure is a soda can. Name the three-dimensional geometric figure that best describes the soda can shown below.

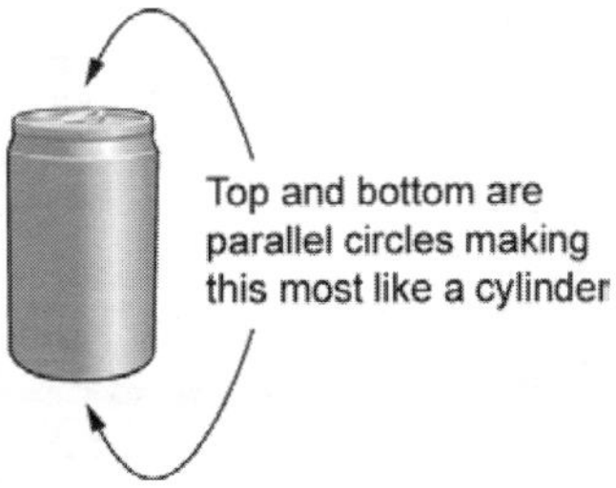

Step 1: Notice that this figure has a circular top and bottom which are both parallel. This type of figure is called a cylinder.

Answer: Cylinder

Example 3 Name the three-dimensional geometric figure that is shown below.

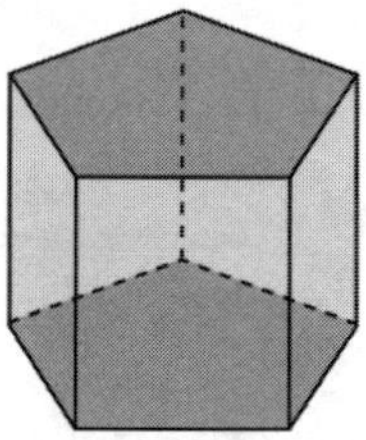

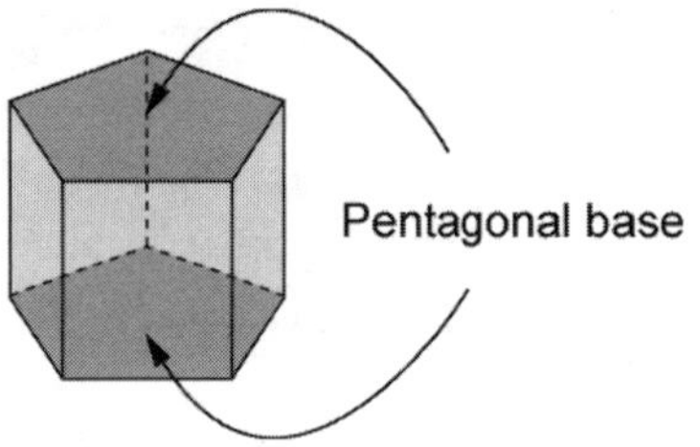

Step 1: The top and bottom of this figure are identical parallel pentagons joined by rectangular faces. This figure is called a pentagonal prism.

Answer: Pentagonal prism

Example 4 Name the three-dimensional geometric figure shown below.

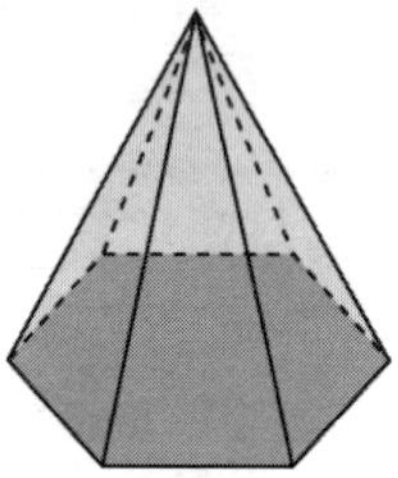

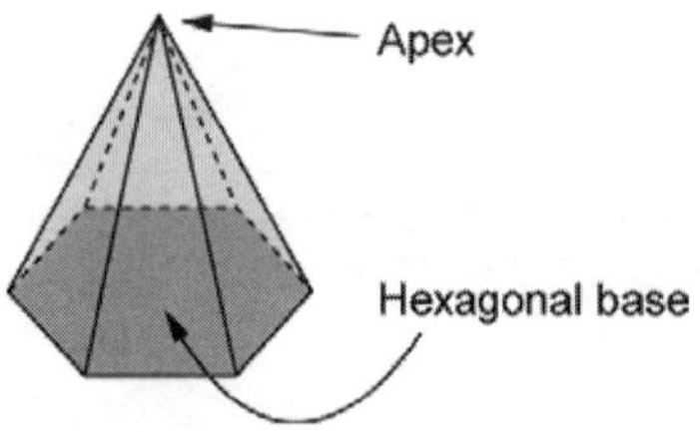

Step 1: This figure has a base that is a 6-sided polygon which is called a hexagon. The lateral faces are triangles that meet in a single point called the apex. This figure is called a hexagonal pyramid.

Answer: Hexagonal pyramid

Example 5 Name a prism with bases that are 8-sided polygons.

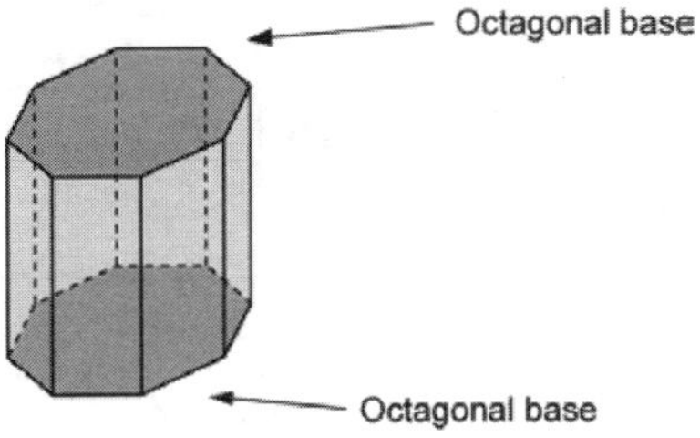

Step 1: Prisms are named for their bases. In this case, the bases are 8-sided polygons. An 8-sided polygon is called an octagon. This prism is called an octagonal prism.

Answer: Octagonal prism

Problem Set

1. Identify the three-dimensional geometric figure shown below.

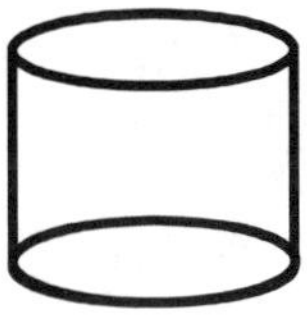

2. Identify the three-dimensional geometric figure shown below.

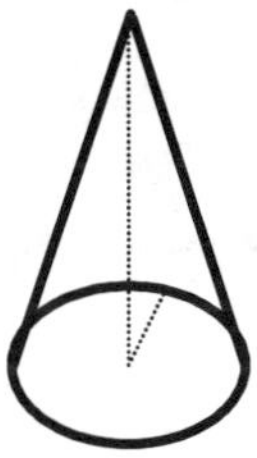

3. Identify the three-dimensional geometric figure shown below.

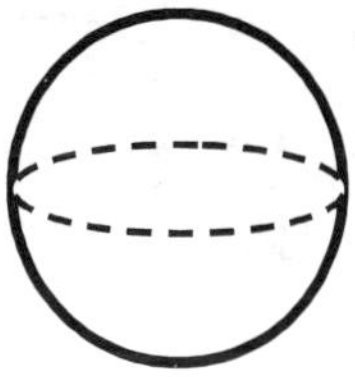

4. Identify the three-dimensional geometric figure shown below.

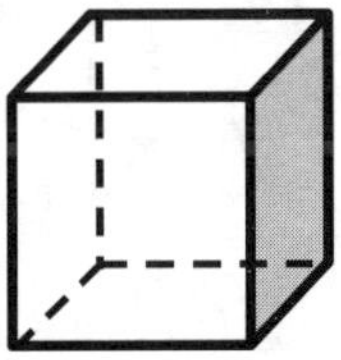

5. Identify the three-dimensional geometric figure shown below.

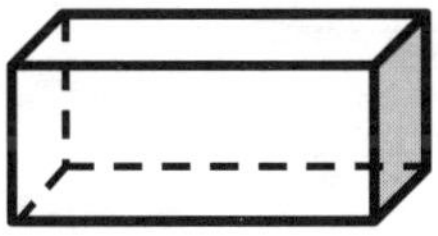

6. Identify the three-dimensional geometric figure shown below.

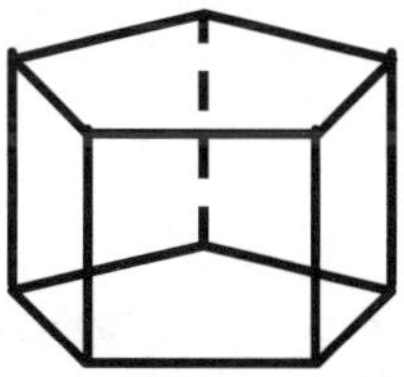

7. Identify the three-dimensional geometric figure shown below.

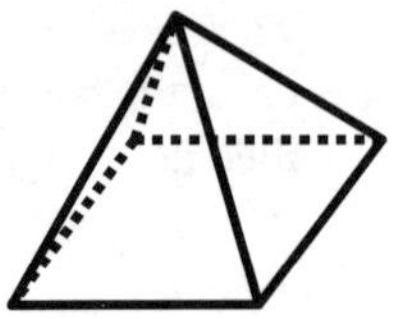

8. Which figure represents a triangular prism?

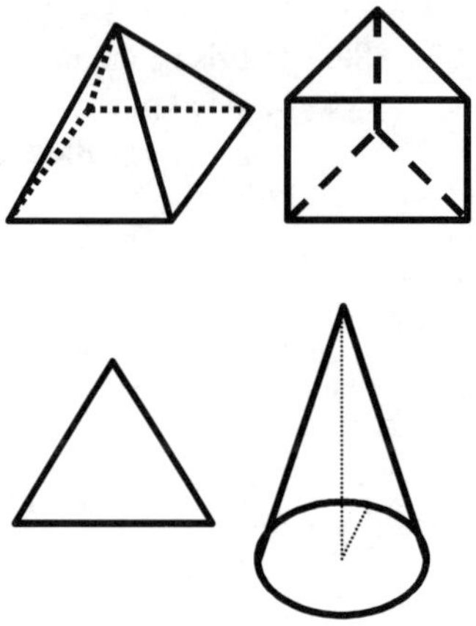

9. Which figure represents a hexagonal prism?

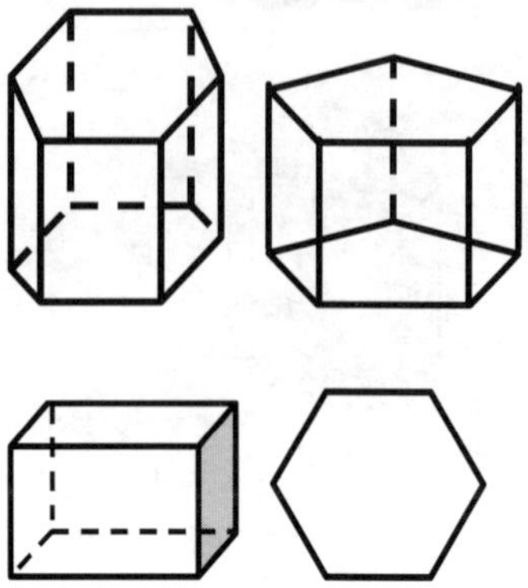

10. Which figure represents an octagonal prism?

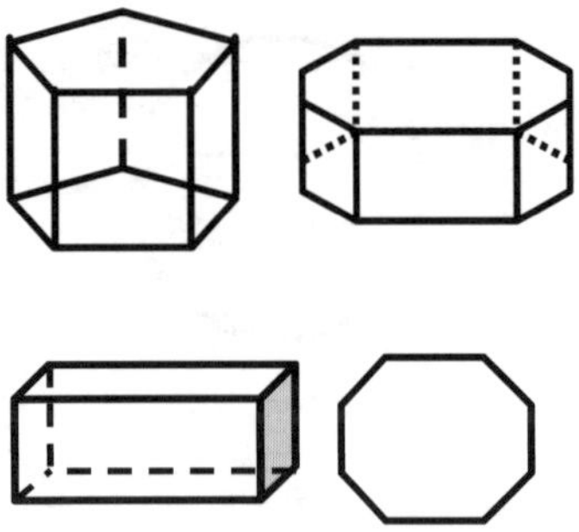

11. Which figure represents a triangular pyramid?

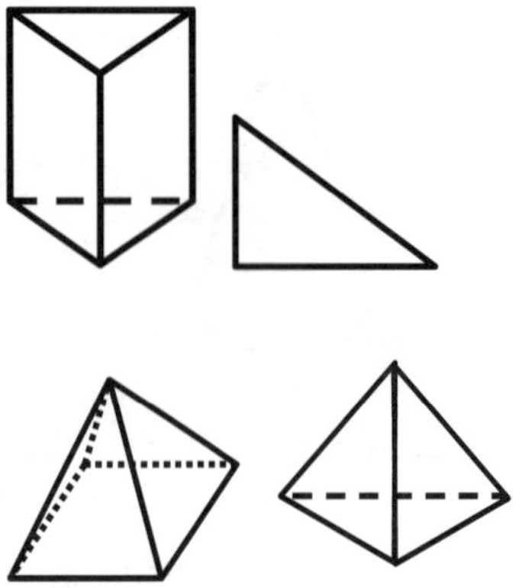

12. Which figure represents a pentagonal pyramid?

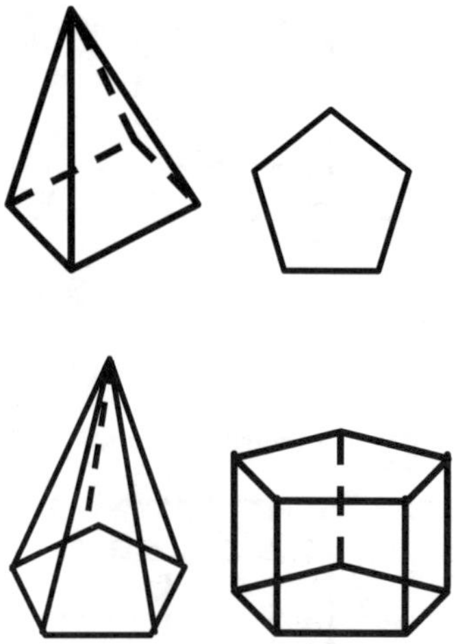

13. Which figure represents a cylinder?

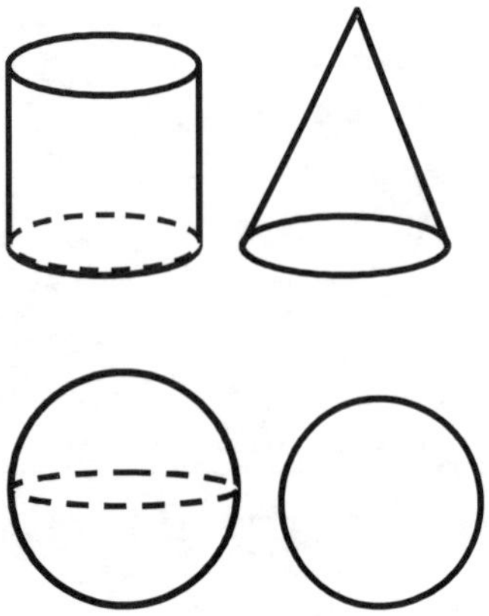

14. Which figure represents a square pyramid?

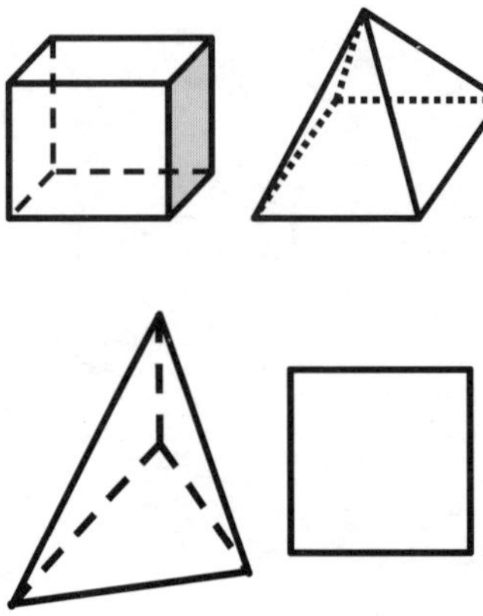

15. Which figure represents a hexagonal pyramid?

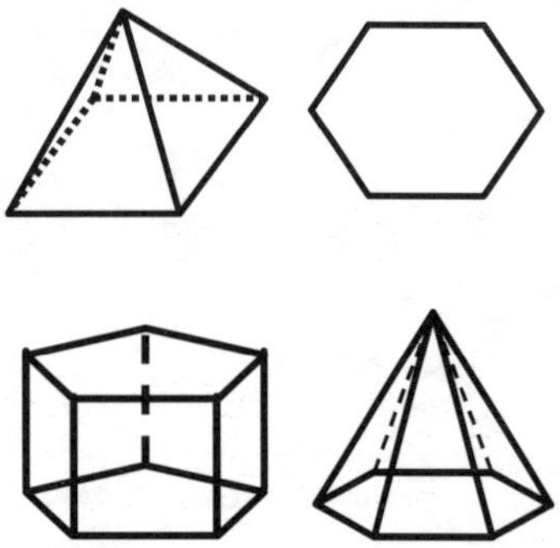

16. The given picture shows the intersection of a pentagonal prism and a plane parallel to its bases. If the plane cuts the prism into two pieces, what is the shape of the base of the top piece of the prism?

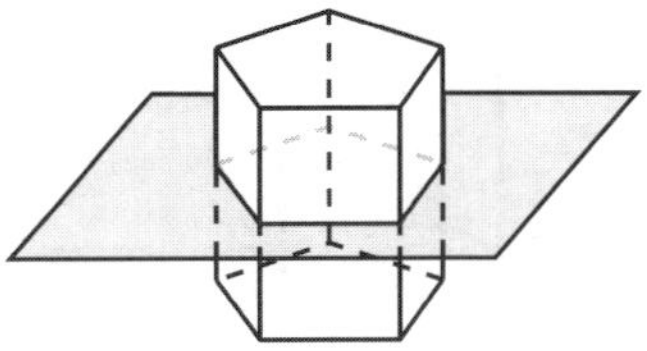

17. The given picture shows the intersection of a triangular prism and a plane parallel to its bases. If the plane cuts the prism into two pieces, what is the shape of the base of the top piece of the prism?

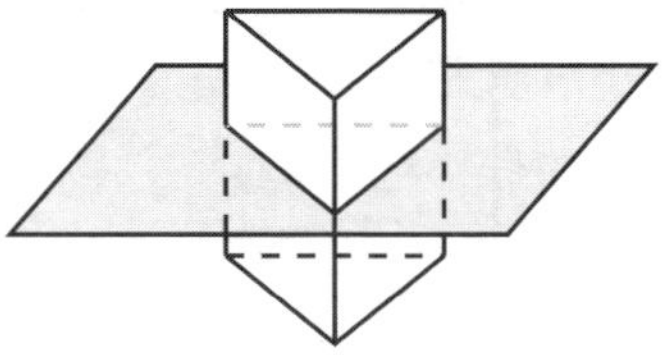

18. The given picture shows the intersection of a cylinder and a plane parallel to its bases. If the plane cuts the cylinder into two pieces, what is the shape of the base of the top piece of the cylinder?

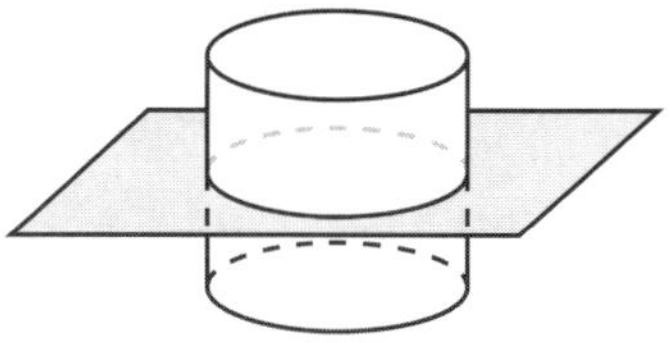

19. The given picture shows the intersection of a square pyramid and a plane parallel to its bases. If the plane cuts the pyramid into two pieces, what is the shape of the base of the top piece of the pyramid?

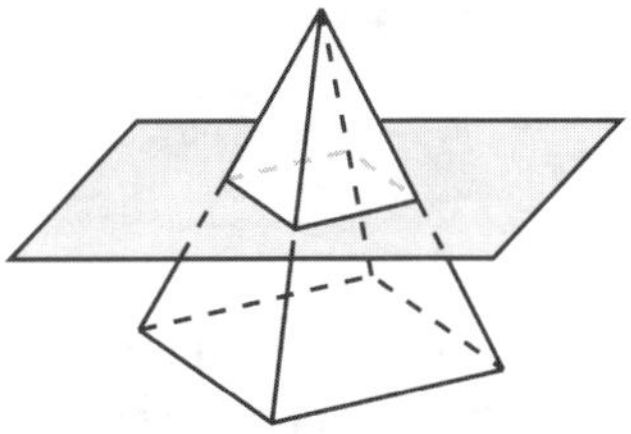

20. The given picture shows the intersection of a hexagonal prism and a plane parallel to its bases. If the plane cuts the prism into two pieces, what is the shape of the base of the top piece of the prism?

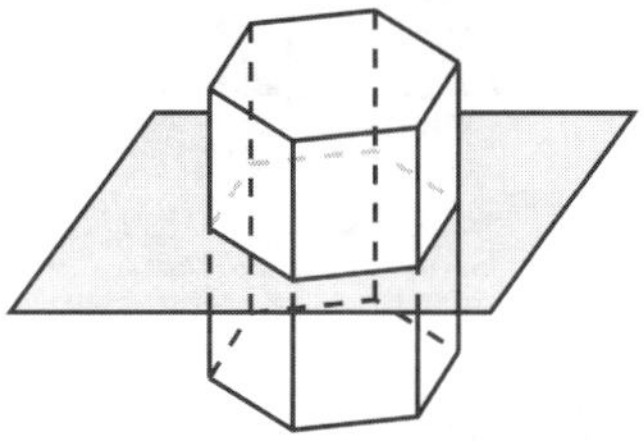

21. The given picture shows the intersection of an octagonal pyramid and a plane parallel to its bases. If the plane cuts the pyramid into two pieces, what is the shape of the base of the top piece of the pyramid?

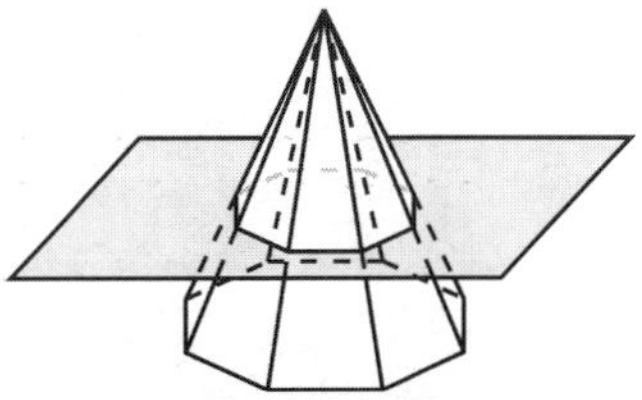

22. The given picture shows the intersection of a cube and a plane parallel to its bases. If the plane cuts the cube into two pieces, what is the shape of the base of the top piece of the cube?

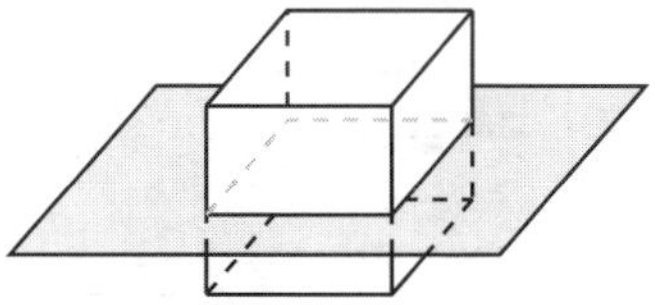

23. The given picture shows the intersection of a cone and a plane parallel to its bases. If the plane cuts the cone into two pieces, what is the shape of the base of the top piece of the cone?

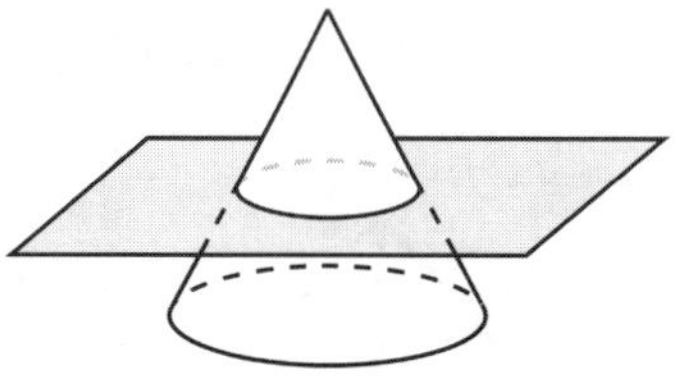

24. What would be the name of a prism whose bases are trapezoids?

25. What would be the name of a prism whose bases are rhombuses?

26. What would be the name of a prism whose bases are decagons?

27. What would be the name of a pyramid whose base is a decagon?

28. What would be the name of a pyramid whose base is a heptagon?

29. What would be the name of a prism whose bases are heptagons?

30. What would be the name of a pyramid whose base is a trapezoid?

31. Find the area of the base of the cylinder shown below.

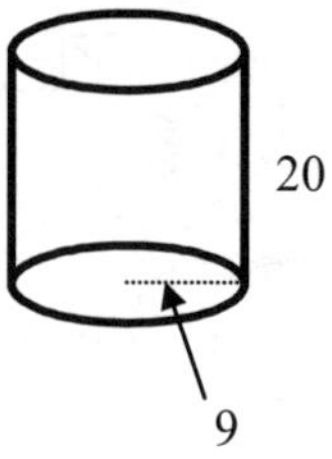

32. Find the area of the base of the cylinder shown below.

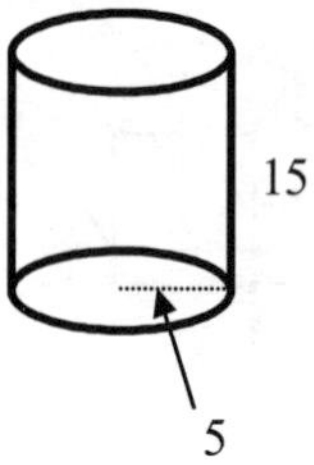

33. Find the area of the base of the cone shown below.

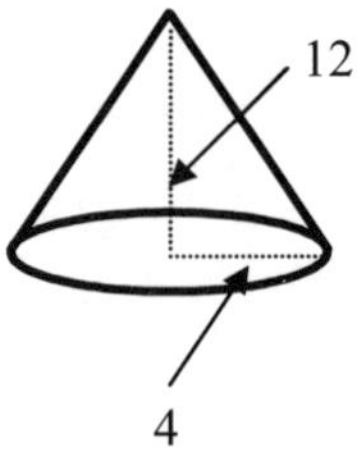

34. Find the area of the base of the square pyramid shown below.

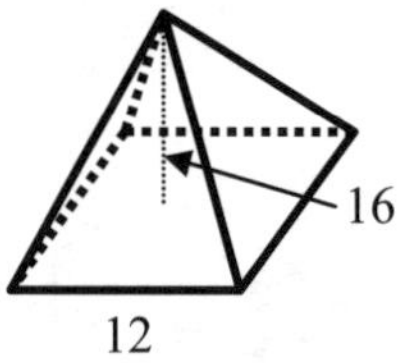

35. Find the area of the base of the square pyramid shown below.

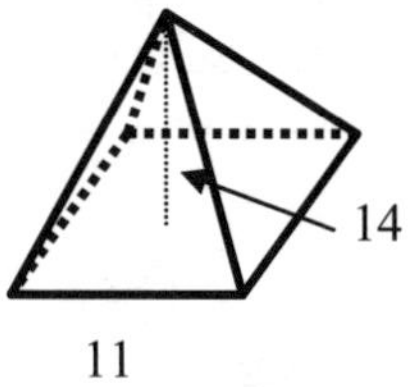

36. Find the area of the base of the rectangular prism shown below.

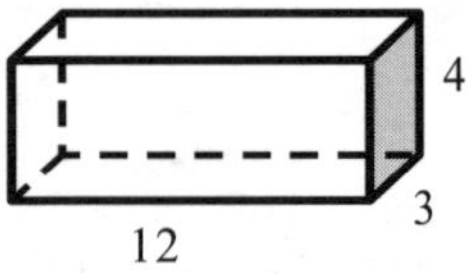

37. Find the area of the base of the rectangular prism shown below.

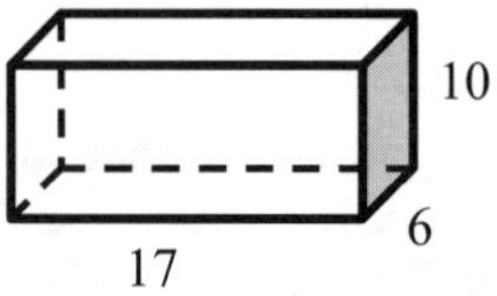

38. Find the area of the base of the triangular prism shown below.

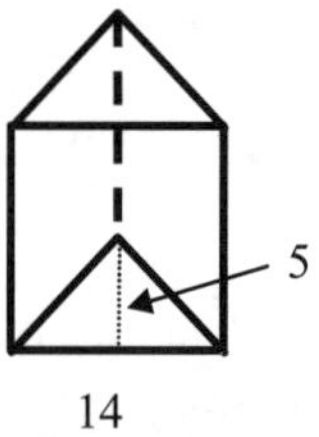

39. Find the area of the base of the triangular prism shown below.

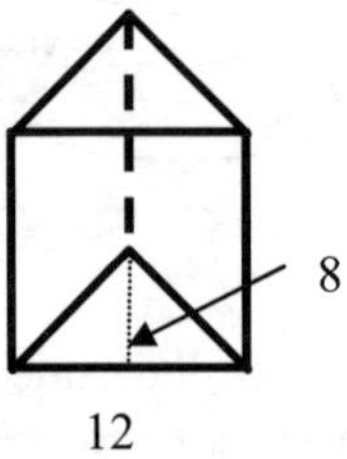

40. Find the area of the base of the triangular pyramid shown below.

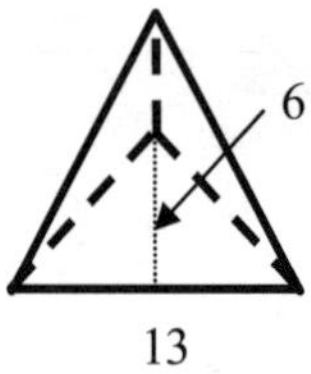

MPA-108: Graphing Translations and Reflections on the Coordinate Plane

The location, position, and size of geometric figures can be changed and distorted. This process is known as a **transformation**. This lesson explores two transformations called translations and reflections.

Transformation	An action that either distorts or changes the location, position or size of a geometric figure.

A **translation** of a geometric object is the movement of all points that make up that object along a straight path. For example, a triangle may be "translated" down and to the right as shown in the following diagram.

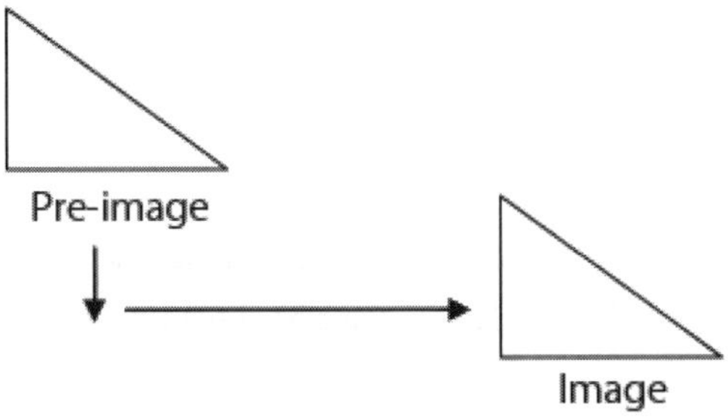

A translation is specified when we say "how far" and "in what direction" the object is to be moved. That is, we specify a distance and a direction to move all of the points in the object.

Translation	The movement of a geometric figure from one location to another along a straight path.

The original object is called the **pre-image** and the transformed object is called the **image**. Notice that the triangle did not change in size or orientation during the process of transformation.

Pre-Image and Image	When a geometric object is transformed, the original object is called the pre-image and the repositioned object is called the image.

A **reflection** is the movement of all points of a geometric figure over a line resulting in a mirror image of the figure. For example, a rectangle may be "reflected" over a line as indicated in the diagram on the next page. The line over which the reflection is made is called the **axis of reflection**.

Reflection	The movement of all of the points of a geometric figure over a line making a mirror image of the reflected figure.

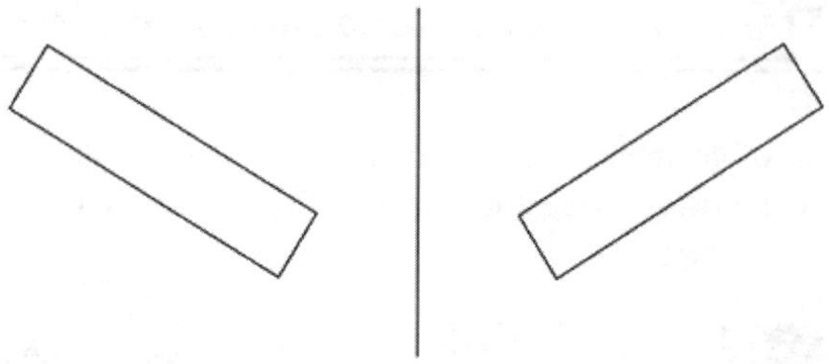

Axis of Reflection	The line over which the reflection is made.

In this case, the size of the object is the same but its orientation has changed. A reflection is specified by giving the axis of reflection.

Example 1 Given the figures below, state whether or not the image is a translation of the pre-image.

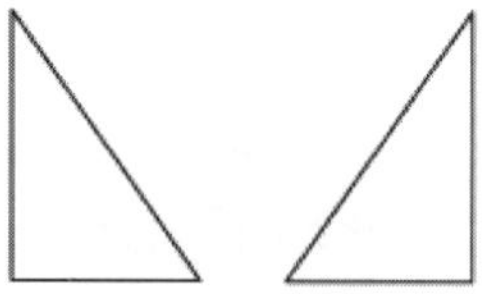

Step 1: A translation is defined as movement of a geometric figure from one location to another along a straight path.

Step 2: From the given figures, we can see that the transformation is not a translation, since the image is a reflection of the pre-image. The symmetry here implies that the transformation is a reflection.

Answer: The image is not a translation of the pre-image.

Example 2 Does the following figure show a translation, a reflection or neither?

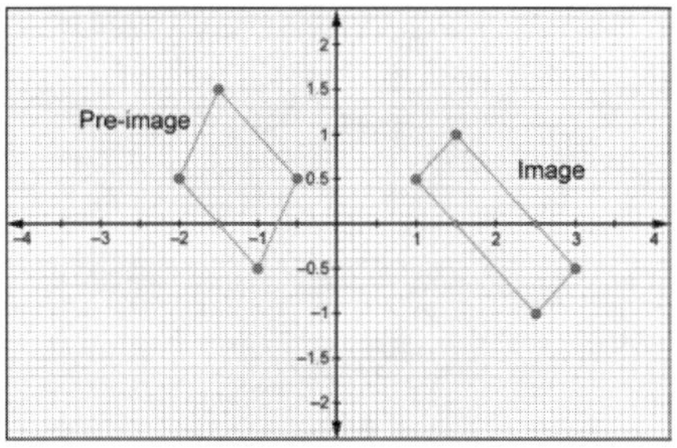

Step 1: The image has changed in size, so this transformation cannot be a translation or a reflection, since both of those transformations preserve size.

Answer: The transformation that is displayed is neither a translation nor a reflection.

Example 3 If a reflection is made across the *y*-axis in the following graph, what will be the coordinates of the corresponding vertex of (–2, 2)?

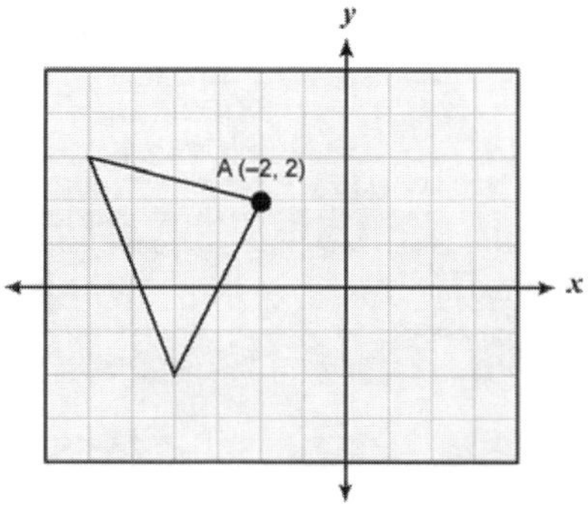

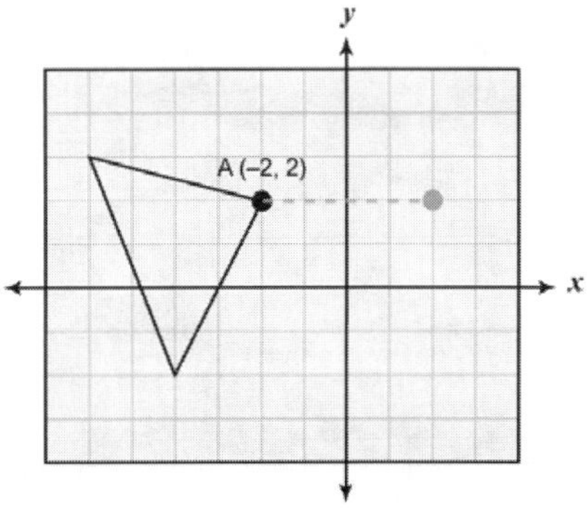

Step 1: The point (–2, 2) is exactly 2 units to the left of the *y*-axis. Its image will be exactly 2 units to the right of the *y*-axis.

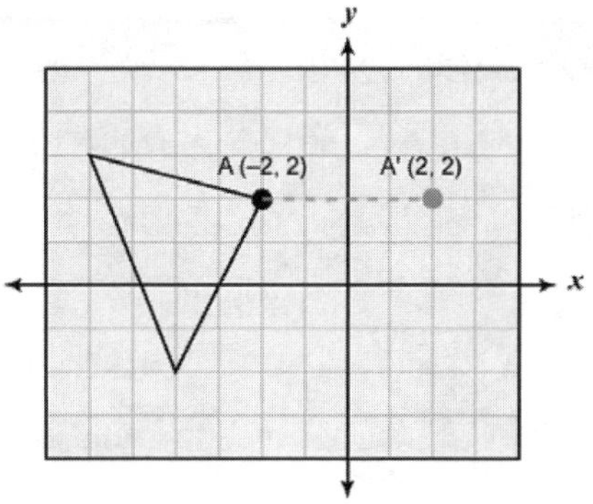

Step 2: Call this point A'. Since the x-coordinate of all the points on the y-axis is 0, the x-coordinate of the image point will be $0 + 2 = 2$. This makes the image point A' (2, 2).

Answer: The image of the point $A(-2, 2)$ reflected across the y-axis is the point A' (2, 2).

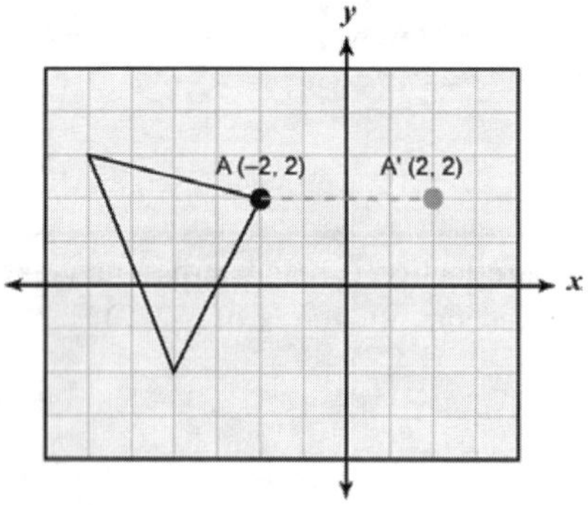

Example 4 Given the following triangle in the Cartesian Plane, list the coordinates of the image if it is translated two units down and five units to the right.

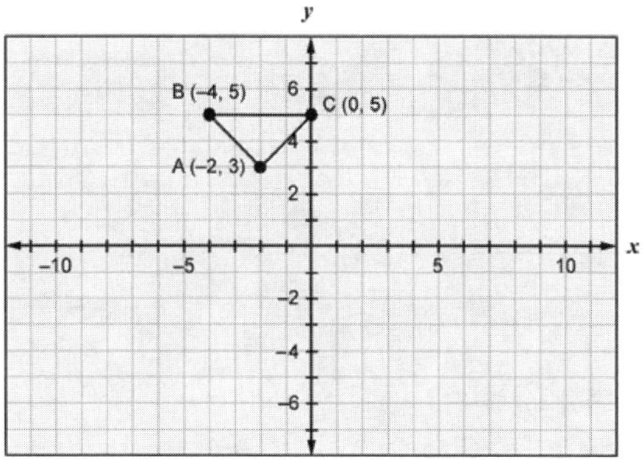

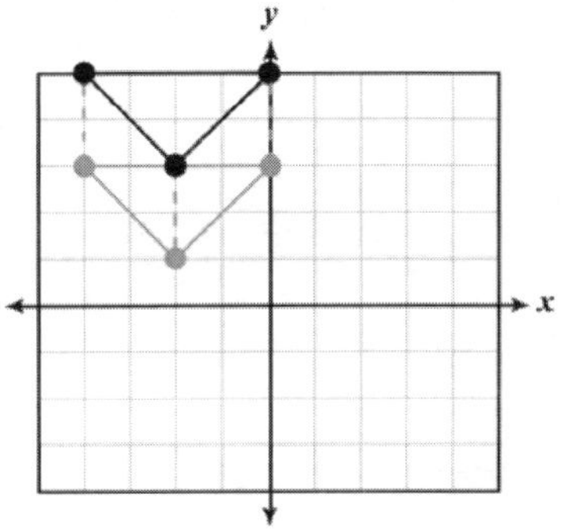

Step 1: The problem asks us to move the points of the figure two units down and five units to the right. To find the coordinates of the image, start by subtracting two from the *y*-coordinates of each point, since we are moving down two units.

$$A(-2, 3) \rightarrow (-2, 3 - 2)$$
$$B(-4, 5) \rightarrow (-4, 5 - 2)$$
$$C(0, 5) \rightarrow (0, 5 - 2)$$

Step 2: To find the x-coordinates of each point, add five to the x-coordinates. By subtracting two from the y-coordinates and adding five to the x-coordinates, we find the new translated image.

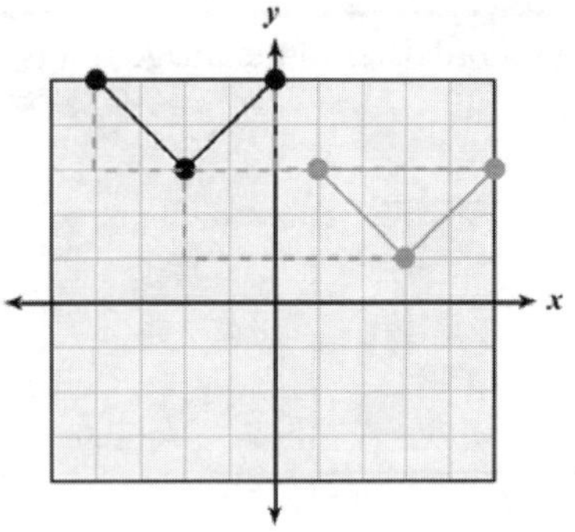

$$A(-2, 3) \rightarrow (-2+5, 3-2) = A'(3, 1)$$
$$B(-4, 5) \rightarrow (-4+5, 5-2) = B'(1, 3)$$
$$C(0, 5) \rightarrow (0+5, 5-2) = C'(5, 3)$$

Answer: The points of the new translated image are $A'(3, 1)$, $B'(1, 3)$, and $C'(5, 3)$.

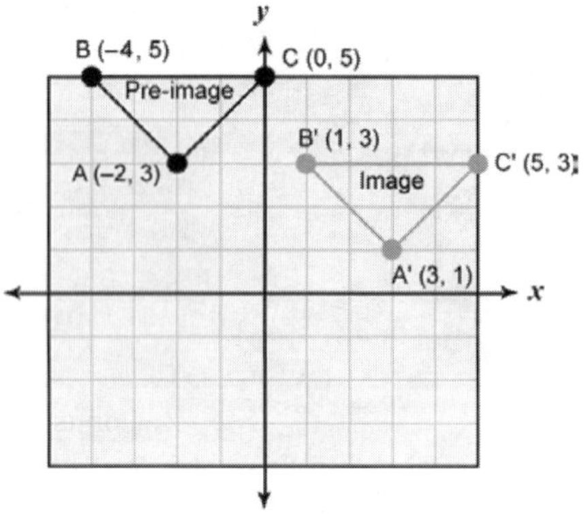

Example 5 What is the image of the triangle ABC shown in the figure below when reflected across the x-axis?

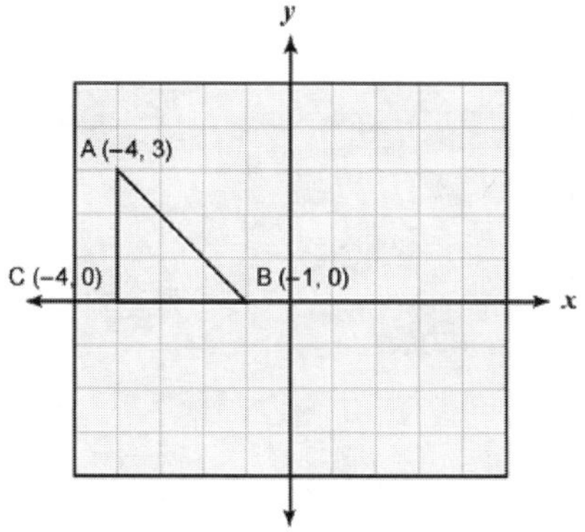

Step 1: Notice that point A is exactly 3 units above the x-axis. Its image will be exactly 3 units below the x-axis. This point is (–4, –3), which we will call A'.

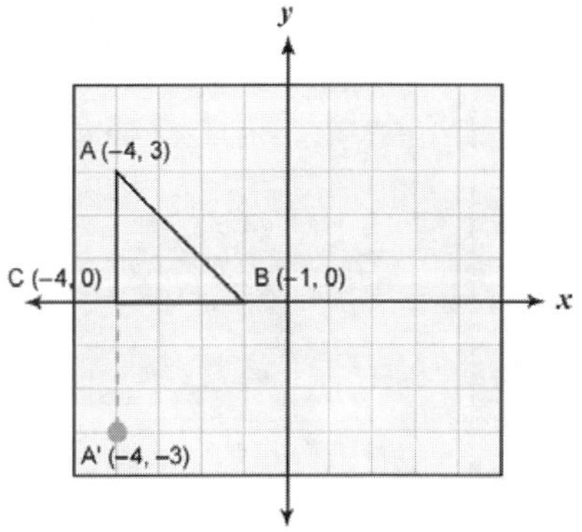

Step 2: Points B and C lie on the x-axis, so when the image is reflected, these points will remain on the x-axis. Call these points B' and C'.

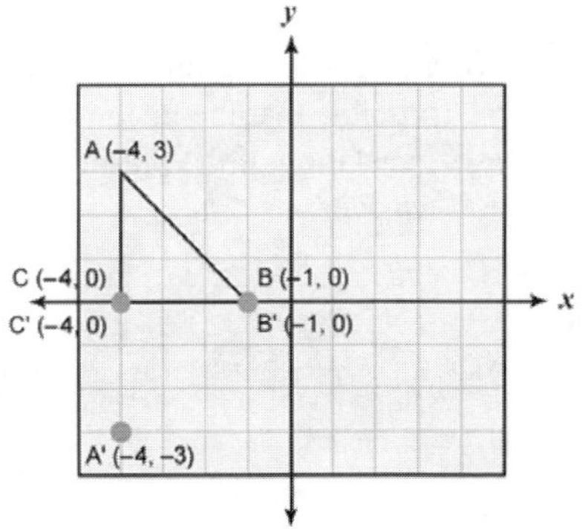

Step 3: Connect the vertices to see the image of the triangle.

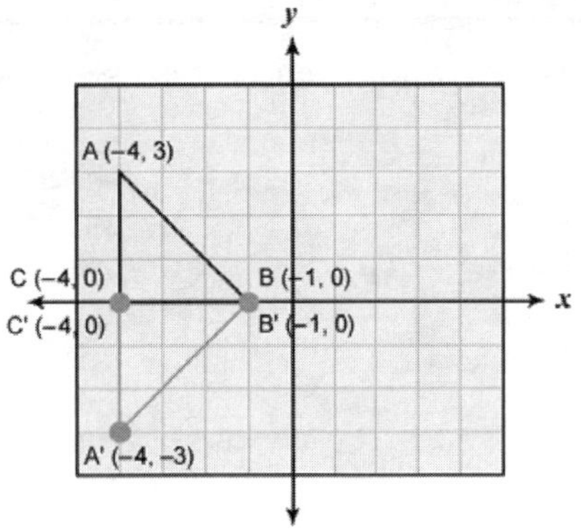

Answer: The image of triangle *ABC* reflected across the *x*-axis is the triangle A′B′C′.

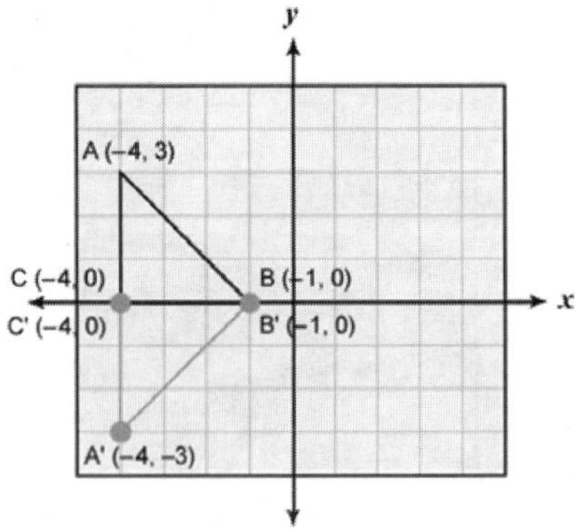

Problem Set

1. Given the diagram below, is it possible that the image is a translation of the pre-image? Answer yes or no.

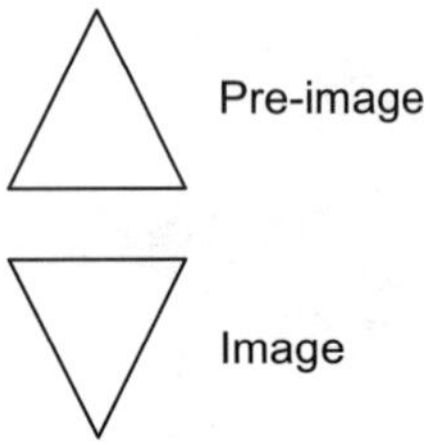

2. Given the diagram below, is it possible that the image is a translation of the pre-image? Answer yes or no.

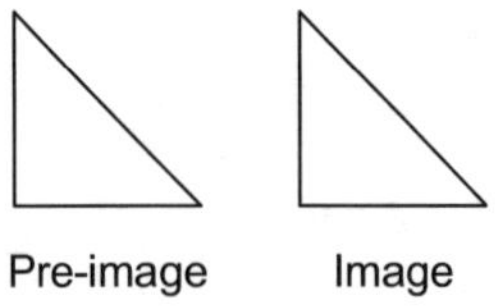

3. Given the diagram below, is it possible that the image is a translation of the pre-image? Answer yes or no.

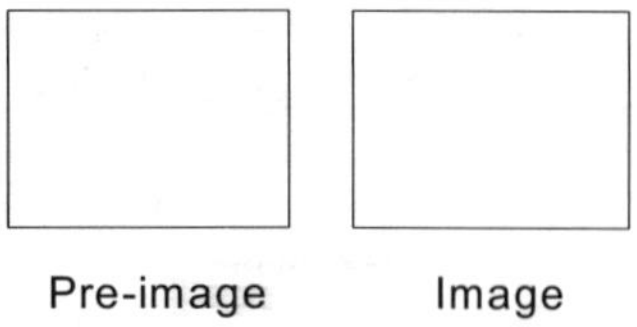

4. Given the diagram below, is it possible that the image is a translation of the pre-image? Answer yes or no.

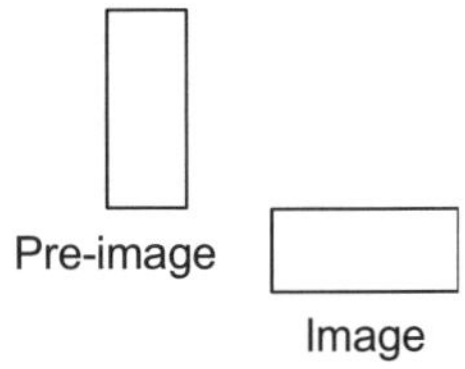

5. Given the diagram below, is it possible that the image is a translation of the pre-image? Answer yes or no.

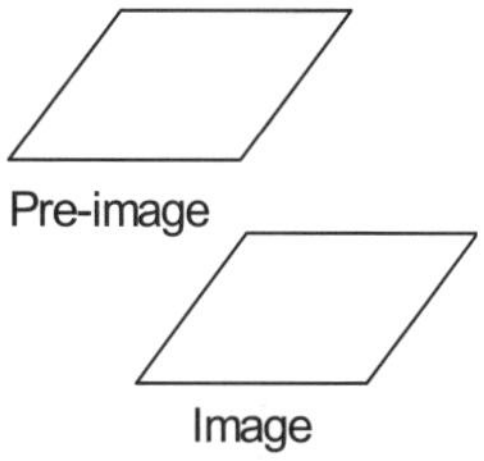

6. Given the diagram below, is it possible that the image is a translation of the pre-image? Answer yes or no.

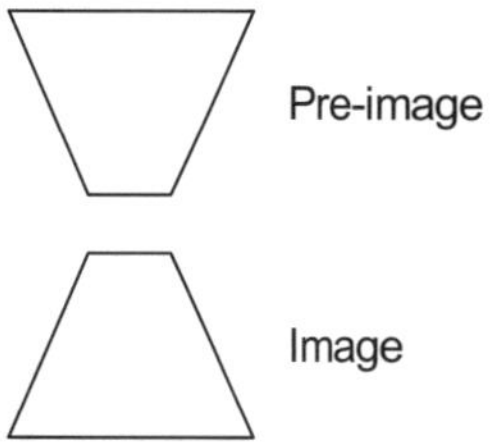

7. Given the diagram below, is it possible that the image is a translation of the pre-image? Answer yes or no.

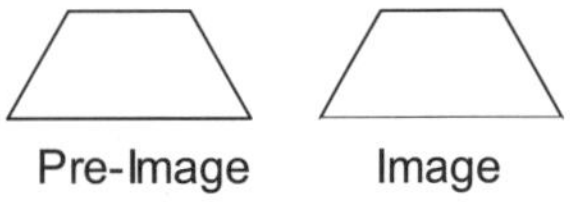

8. If the pre-image of a translation is a trapezoid, then the image will be a smaller trapezoid. True or False?

9. If the pre-image of a translation is a rectangle, then the image can be a congruent rectangle. True or False?

10. If the pre-image of a translation is an octagon, then the image can be a larger octagon. True or False?

11. If the pre-image of a translation is a hexagon, then the image can be a pentagon. True or False?

12. If the pre-image of a translation is a rhombus, then the image can be a congruent rhombus. True or False?

13. If the pre-image of a translation is an equilateral triangle, then the image can be an isosceles triangle. True or False?

14. If the pre-image of a translation is a parallelogram, then the image can be a congruent parallelogram. True or False?

15. If the pre-image of a translation is a rectangle, and the image is a rectangle that is one-third the area of the pre-image, then the image can be a translation. True or False?

16. If a reflection is made across the x-axis in the following graph, what would be the coordinates of the corresponding vertex of (–4, –2)?

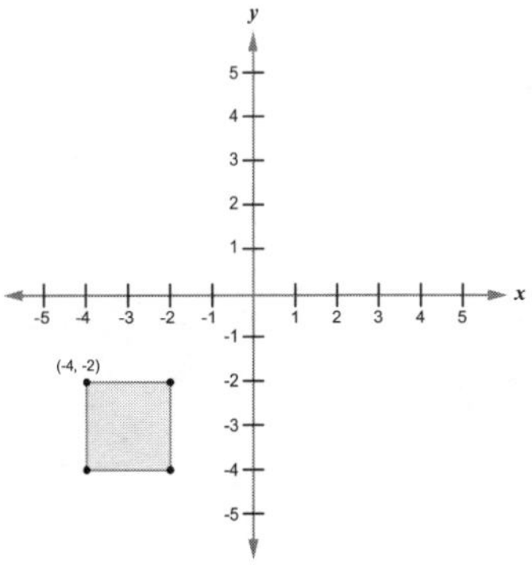

17. If a reflection is made across the y-axis in the following graph, what would be the coordinates of the corresponding vertex of (–1, –3)?

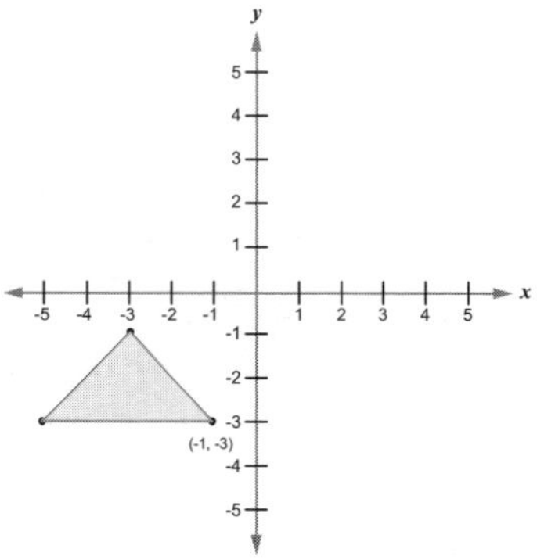

18. If a reflection is made across the x-axis in the following graph, what would be the coordinates of the corresponding vertex of (4, –3)?

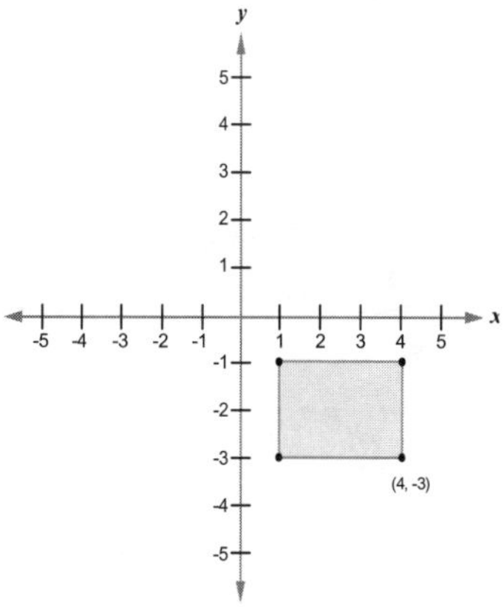

19. If a reflection is made across the y-axis in the following graph, what would be the coordinates of the corresponding vertex of $(-4, 4)$?

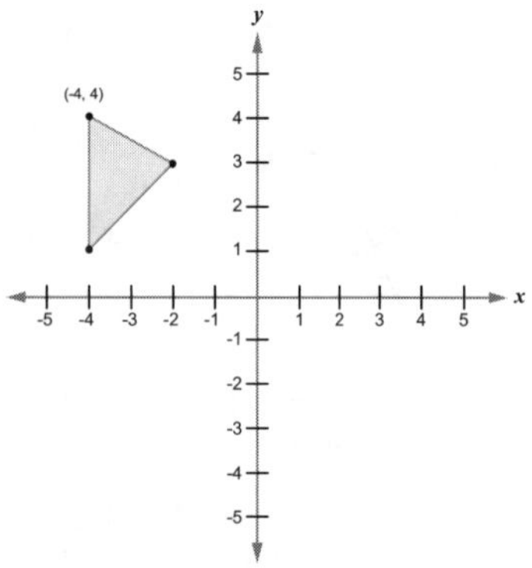

20. If a reflection is made across the x-axis in the following graph, what would be the coordinates of the corresponding vertex of $(5, -2)$?

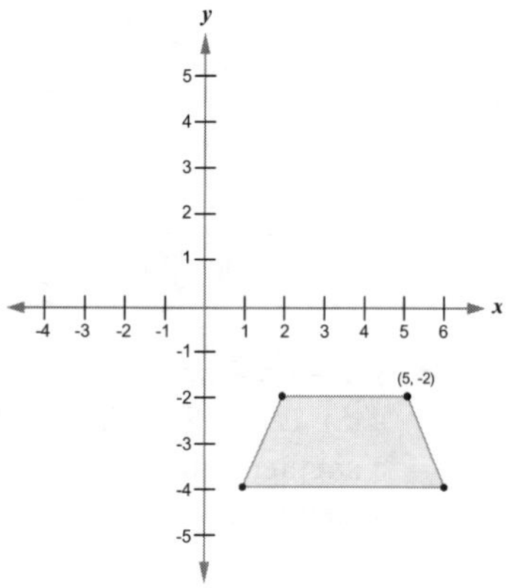

21. If a reflection is made across the y-axis in the following graph, what would be the coordinates of the corresponding vertex of $(5, 4)$?

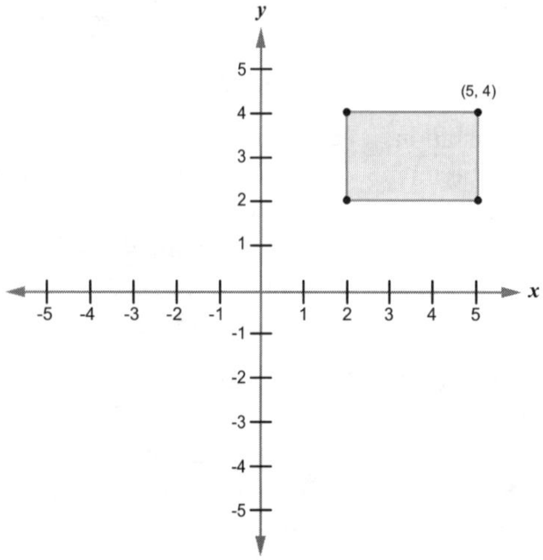

22. If a reflection is made across the y-axis in the following graph, what would be the coordinates of the corresponding vertex of $(4, -4)$?

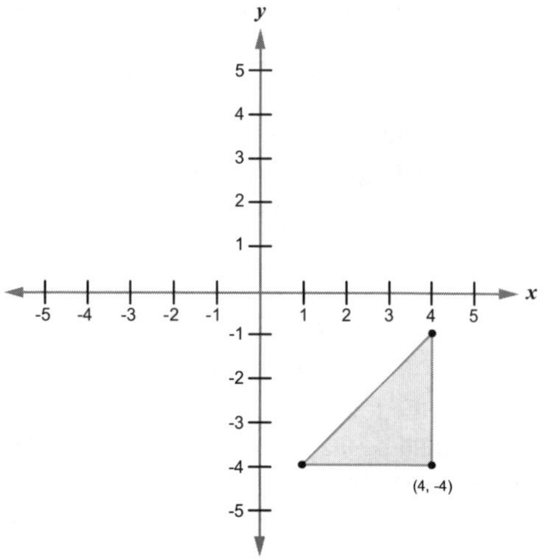

23. What would be the correct coordinates of the figure when it is translated 2 units right and 1 unit down?

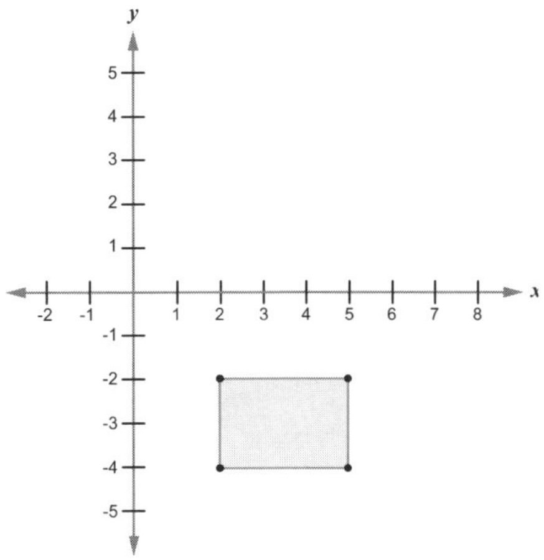

24. What would be the correct coordinates of the figure when it is translated 3 units to the right and 3 units up?

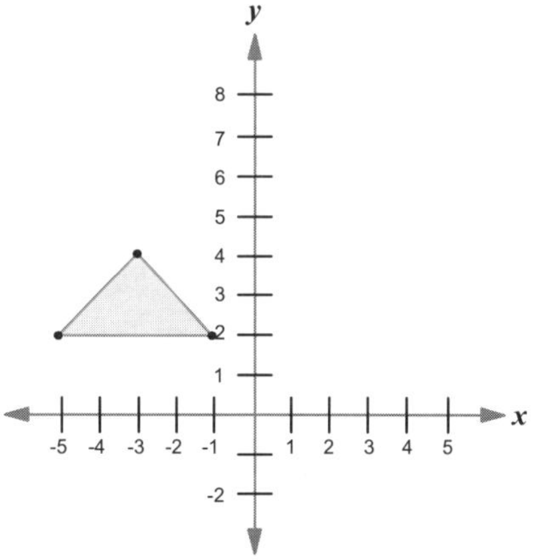

25. What would be the correct coordinates of the figure when it is translated 1 unit to the left and 5 units up?

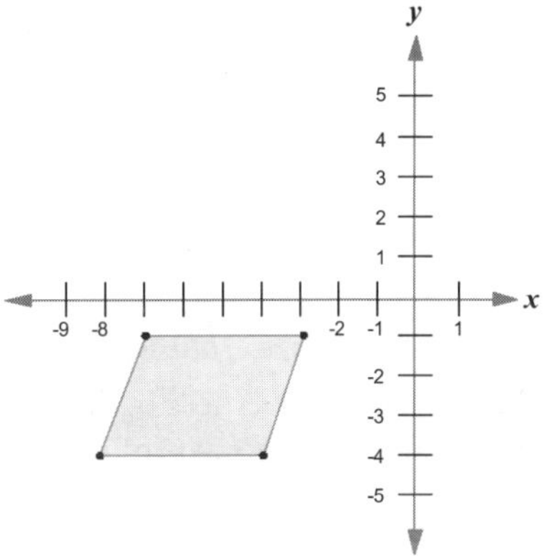

26. What would be the correct coordinates of the figure when it is translated 5 units down and 4 units to the left?

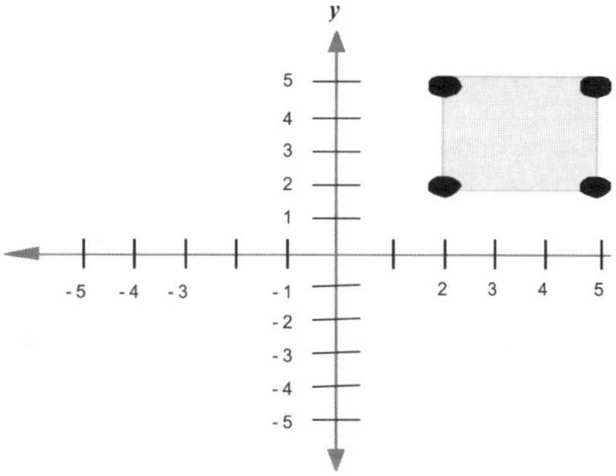

27. What would be the correct coordinates of the figure when it has been translated 6 units to the right and 2 units up?

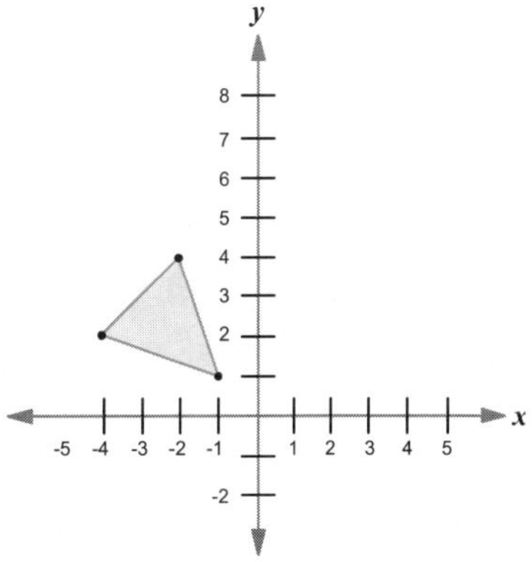

28. What would be the correct coordinates of the figure when it is translated 7 units to the left and 5 units up?

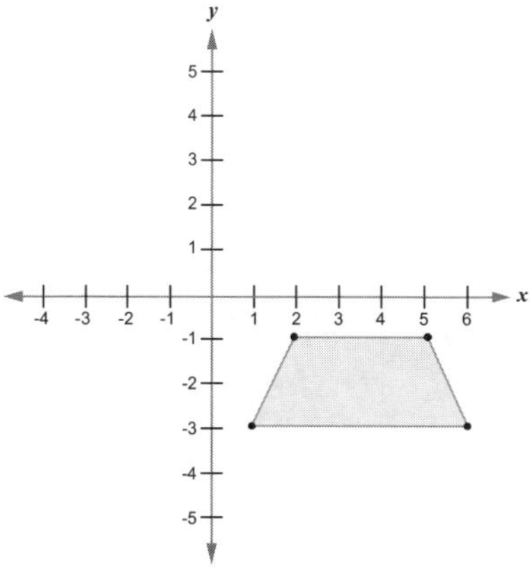

29. What would be the correct coordinates of the figure when it is translated 2 units down and 4 units to the left?

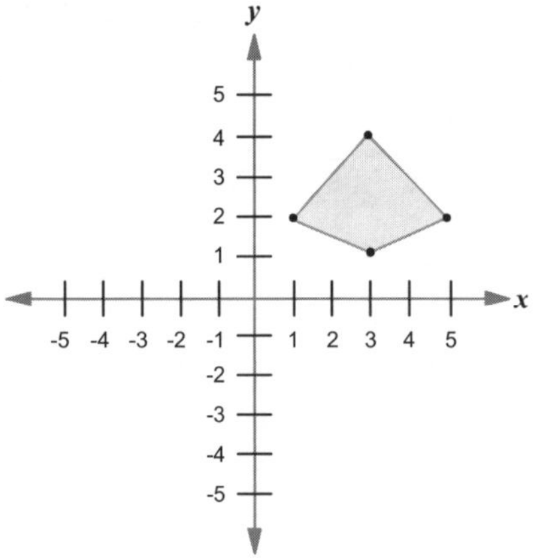

30. What would be the correct coordinates of the figure when it is translated 6 units to the right and 7 units down?

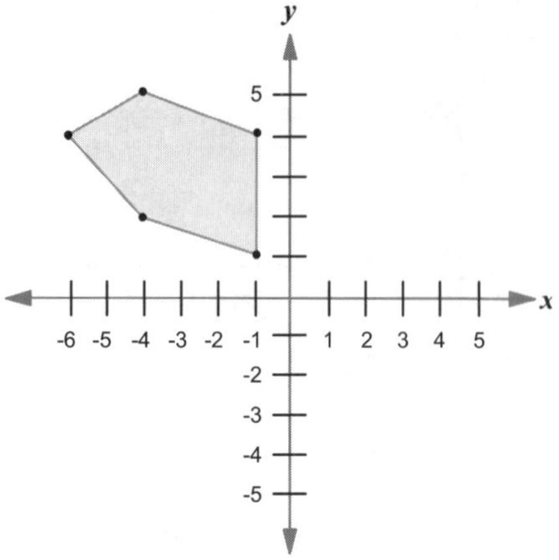

31. What would be the correct vertices of the square that is the reflection over the y-axis of a square with the vertices: (2, 1), (6, 1), (2, 5), and (6, 5)?

32. What would be the correct vertices of the rectangle that is a reflection over the x-axis of a rectangle with the vertices: (–1, –1), (–1, –3), (–6, –1), and (–6, –3)?

33. What would be the correct vertices of a triangle that is the reflection over the y-axis: (–2, –1), (–4, –3) and (–1, –4)?

34. What would be the correct vertices of the trapezoid that is the reflection over the x-axis of a trapezoid with the vertices: (–1, 2), (–2, 4), (–5, 2) and (–4, 4)?

35. What would be the correct vertices of the parallelogram that is the reflection over the x-axis of a parallelogram with the vertices: (2, –1), (5, –1), (1, –3) and (4, –3)?

36. What would be the correct vertices of the rectangle that is the reflection over the y-axis of a rectangle with the vertices: (3, –1), (5, –1), (3, –5), and (5, –5)?

37. What would be the correct vertices of the kite that is the reflection over the y-axis of a kite with the vertices: (2, 4), (3, 5), (4, 4) and (3, 2)?

38. What would be the correct vertices of the trapezoid that is the reflection over the x-axis of a trapezoid with the vertices: (3, 1), (3, 5), (5, 2) and (5, 4)?

39. What would be the correct vertices of the square that is the reflection over the x-axis of a square with the vertices: (–5, 0), (–2, 0), (–5, –3), (–2, –3)?

40. What would be the correct vertices of the pentagon that is the reflection over the y-axis of a pentagon with the vertices: (0, 0), (0, 4), (5, 2), (3, 0), (3, 4)?

MPA-109: Solving and Graphing Linear Inequalities on a Number Line

A **linear inequality** is an inequality that involves one variable and uses the symbols $\geq, >, <, \leq$ as seen in the problem displayed below:

$$3x - 5 \leq 13$$

Linear Inequality	An inequality in which each variable has an exponent of 1 and no variable in the denominator.

When solving an inequality, you will get a number of values for the variable that will satisfy the inequality. Notice that you can add 5 to both sides of this inequality to obtain the equivalent inequality:

$$3x - 5 + 5 \leq 13 + 5$$

$$3x \leq 18$$

$$\frac{3x}{3} \leq \frac{18}{3}$$

$$x \leq 6$$

We can now see that any value of x that is less than or equal to 6 will satisfy the inequality above.

Notice that unlike solving a linear equation, there are an infinite number of solutions to a linear inequality. To show the range of solutions, we draw a number line for the set of x-values.

In this case, we want to display all numbers that are less than or equal to 6. To do this, shade the part of the number line that is less than 6. Since the point $x = 6$ is part of the solution set, use a solid dot at the value of $x = 6$ to show that it is included.

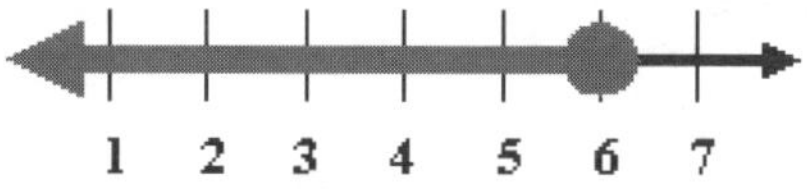

Unlike linear equations, when **multiplying or dividing an inequality by a negative number**, you must **reverse** the direction of the inequality. For example, look at the equation below.

$$-2x > 8$$

Divide both sides of this inequality by –2.

$$\frac{-2x}{-2} < \frac{8}{-2}$$

Notice that the inequality switched from "greater than" to "less than." Simplify the inequality to get the expression below.

$$x < -4$$

Now draw the solution on a number line.

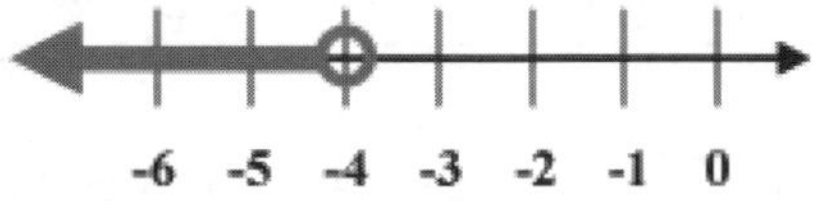

In this case, we want all points smaller than -4. To indicate that -4 is not included we draw an open circle at the point $x = -4$.

Example 1 Graph the linear inequality on a number line: $x > -2$

$x > -2$

Step 1: This inequality is read "x is greater than negative two."

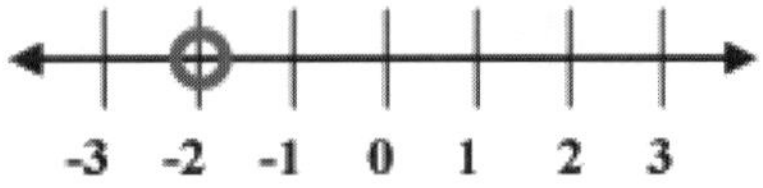

Step 2: To graph this inequality on a number line, start by drawing an open circle at -2 to show that -2 is not included as part of the solution.

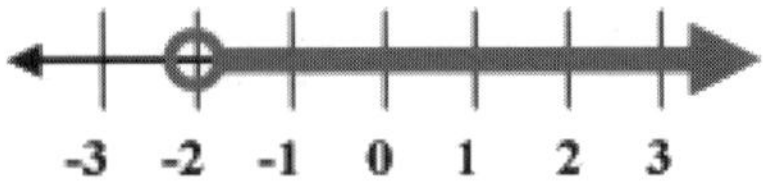

Step 3: Since the sign is "greater than," shade the number line for all values greater than -2.

Answer: The graph for the linear inequality of $x > -2$ is

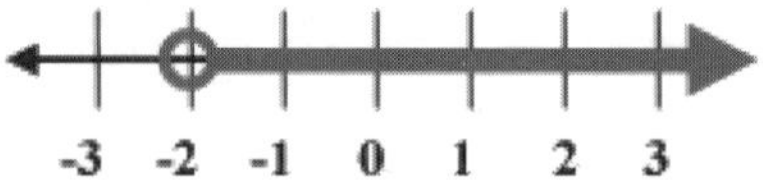

Example 2 Solve the linear inequality. Graph the solution on a number line.

$$x + 8 < 9$$

$$x + 8 < 9$$
$$x + 8 - 8 < 9 - 8$$
$$x < 1$$

Step 1: First solve the inequality by subtracting 8 from each side. The solution for x is all values that are less than one.

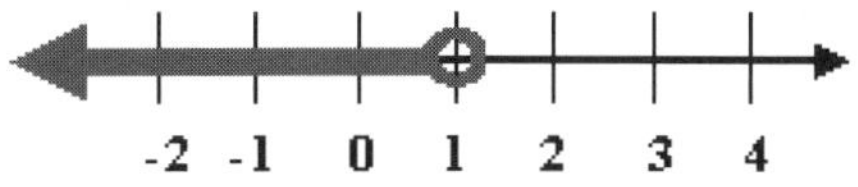

Step 2: The inequality is read "x is less than one," so place an open circle at 1 on the number line to show that 1 is not included in the solution.

Answer: The solution to the linear inequality $x + 8 < 9$ is the set of all values for x such that $x < 1$. The solution on a number line is:

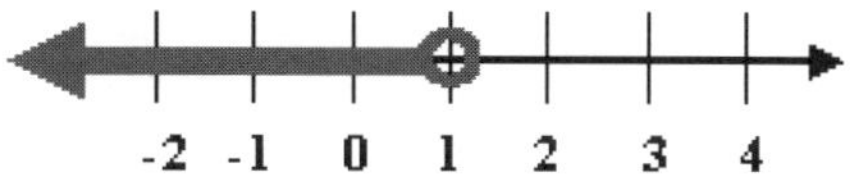

Example 3 Solve the linear inequality. Graph the solution on a number line.

$$5x + 2 \geq 12$$

$$5x + 2 \geq 12$$
$$5x + 2 - 2 \geq 12 - 2$$
$$5x \geq 10$$

Step 1: First solve the inequality by subtracting 2 from each side.

$$\frac{5x}{5} \geq \frac{10}{5}$$
$$x \geq 2$$

Step 2: Divide both sides of the inequality by 5.

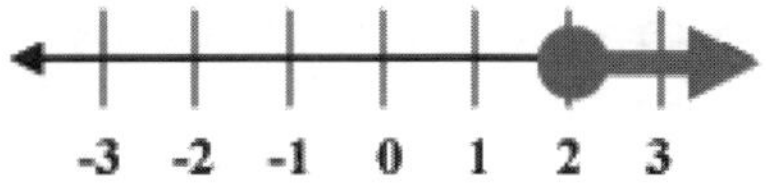

Step 3: Plot the solution on a number line showing all points greater than or equal to 2.

Answer: The solution to the linear inequality $5x + 2 \geq 12$ is the set of all values for x such that $x \geq 2$. The solution on a number line is:

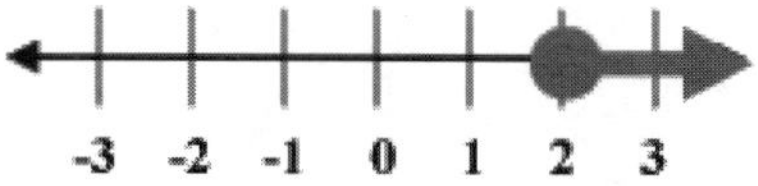

Example 4 Solve the linear inequality. Graph the solution on a number line.

$$4x + 2 < -6$$

$$4x + 2 < -6$$
$$4x + 2 - 2 < -6 - 2$$
$$4x < -8$$

Step 1: First solve the inequality by subtracting 2 from each side of the inequality.

$$\frac{4x}{4} < \frac{-8}{4}$$

$$x < -2$$

Step 2: Divide both sides of the inequality by 4.

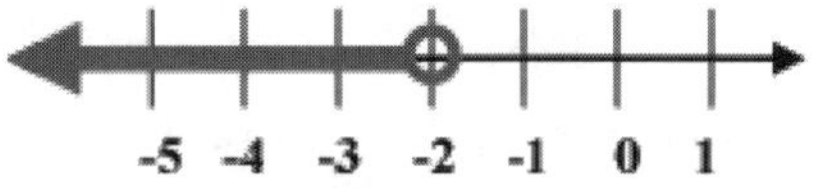

Step 3: Plot all of the points less than the number –2 on a number line. The number –2 has an open circle because it is not included in the solution set.

Answer: The solution to the inequality $4x + 2 < -6$ is the set of all values for x such that $x < -2$. The solution on a number line is:

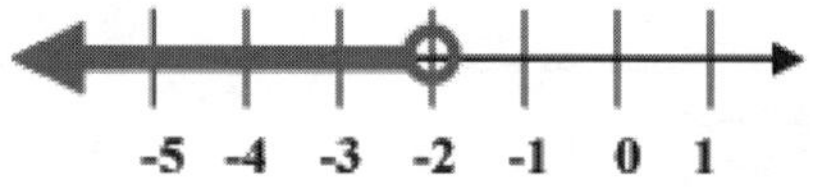

Example 5 Solve the linear inequality. Graph the solution on a number line.

$$12 - 3x \le 9$$

$$12 - 12 - 3x \le 9 - 12$$

$$-3x \le -3$$

Step 1: Start by subtracting 12 from each side of the inequality.

$$\frac{-3x}{-3} \ge \frac{-3}{-3}$$

$$x \ge 1$$

Step 2: Next divide both sides of the inequality by –3. Since we are dividing by a negative number, the direction of the inequality reverses.

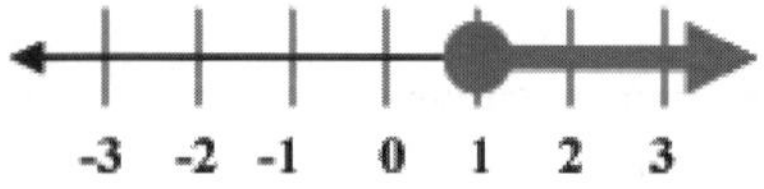

Step 3: Plot all of the points greater than and including the number 1 on a number line.

Answer: The solution to the inequality $12 - 3x \le 9$ is the set of all values for x such that $x \ge 1$. The solution on the number line is:

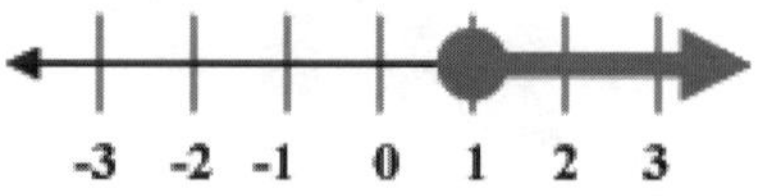

Problem Set

1. Write the inequality for the graph.

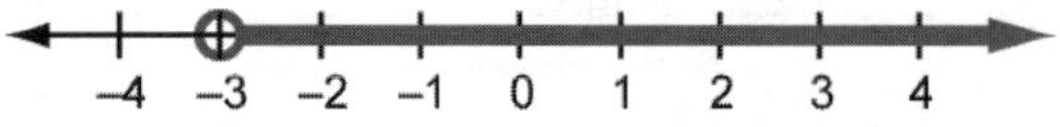

2. Write the inequality for the graph.

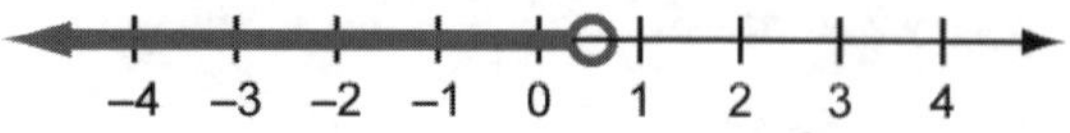

3. Write the inequality for the graph.

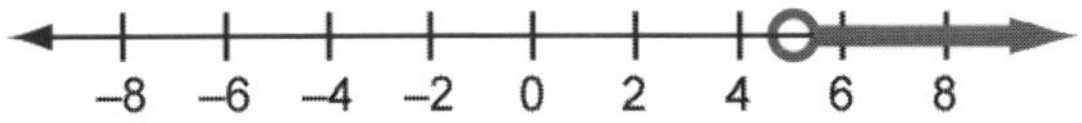

4. Write the inequality for the graph.

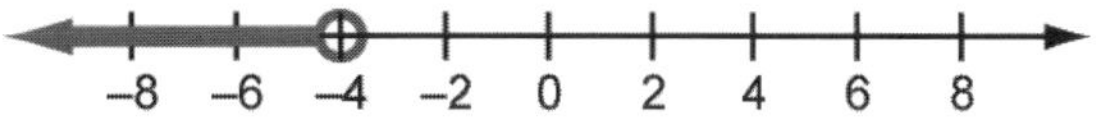

5. Write the inequality for the graph.

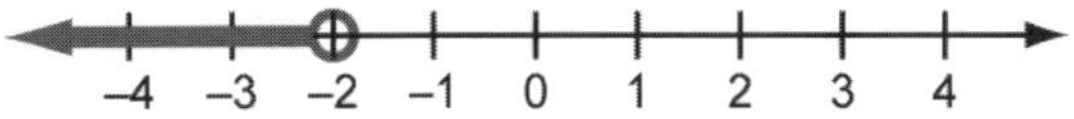

6. Draw the correct graph of the inequality:

$x > -8 + 9$

7. Draw the correct graph of the inequality:

$b < -\frac{9}{3}$

8. Write the inequality for the graph.

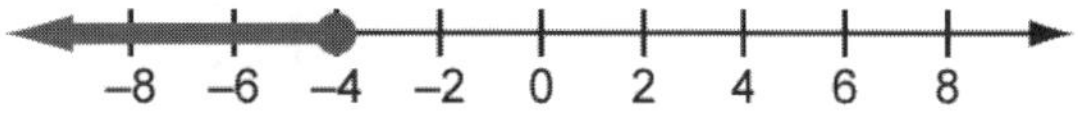

9. Write the inequality for the graph.

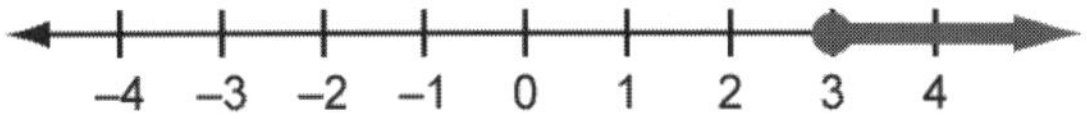

10. Write the inequality for the graph.

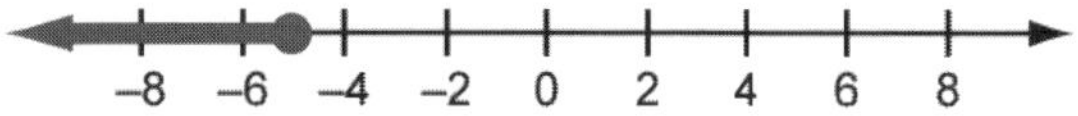

11. Draw the correct graph of the inequality:

$6\frac{1}{2} \le b$

12. Draw the correct graph of the inequality:

$m \ge 2 + (-4)$

13. Draw the correct graph of the inequality:
$r \le -6 + 7$

14. Draw the correct graph of the inequality:
$3(5) \le d$

15. Draw the correct graph of the inequality:
$-6 - 1 \ge y$

16. Draw the correct graph of the inequality:
$a + (-4) \le 3$

17. Draw the correct graph of the inequality:
$b - 7 \ge -4$

18. Draw the correct graph of the inequality:
$c + 4 < -15$

19. Draw the correct graph of the inequality:
$-8 + m > -8$

20. Draw the correct graph of the inequality:
$-10 + y \le -8$

21. Determine the graph of the inequality:

$7 + x \ge 12$

22. Determine the graph of the inequality:

$-15 \ge -9 + x$

23. Determine the graph of the inequality:

$9a \ge -36$

24. Determine the graph of the inequality:

$7c \le 21$

25. Determine the graph of the inequality:

$-45 \le 5m$

26. Determine the graph of the inequality:

$-28 < 4x$

27. Determine the graph of the inequality:

$-5c > -35$

28. Determine the graph of the inequality:

$-12y \le 36$

29. Determine the graph of the inequality:

$-15x \geq 75$

30. Determine the graph of the inequality:

$-90 \leq -10m$

31. Determine the graph of the inequality:

$4x + 4 > -24$

32. Determine the graph of the inequality:

$-6x - 5 > 37$

33. Determine the graph of the inequality:

$-8x + 12 \leq 76$

34. Determine the graph of the inequality:

$10 - 2x < 12$

35. Determine the graph of the inequality:

$5x + 7 \geq 92$

36. Determine the graph of the inequality:

$11x - 4 > -48$

37. Determine the graph of the inequality:

$-12x - 5 \leq 31$

38. Determine the graph of the inequality:

$-7x + 4 \geq -52$

39. Determine the graph of the inequality:

$-22 > -5x + 3$

40. Determine the graph of the inequality:

$-53 \leq -15x - 8$

MPA-110: Solving Problems Using Proportions, Scale Drawings, Models and Maps

Scale is a proportion between two sets of dimensions. For example, on a map, 20 kilometers of actual distance might be represented by 2 centimeters. Another example is a temperature scale where the height of a column of mercury is marked to correspond to the temperature as measured in degrees. In this second case, you can see that the two dimensions, height versus temperature, need not be related.

Scale	Scale is a proportion between two sets of dimensions.

Another familiar example of scale is a timeline, where time is represented as distance as in the drawing below.

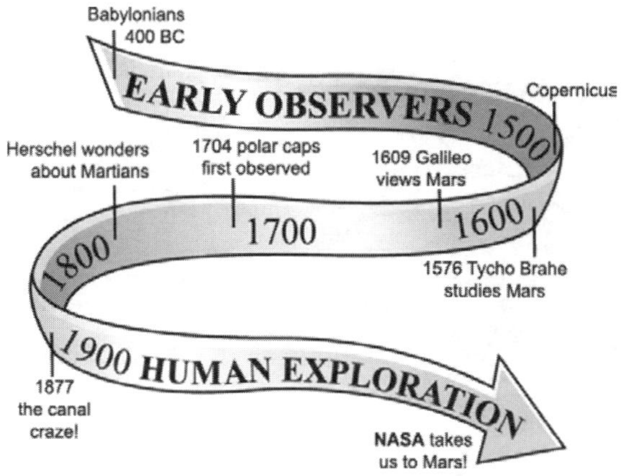

Problems that typically arise when using scales are of two types: (1) using scale to measure one dimension in terms of the other or (2) measuring two dimensions in a model to determine the scale used.

Example 1 Estimate the longest distance across the Ellipse using the map below. Note that the scale is given in the upper right-hand corner of the map.

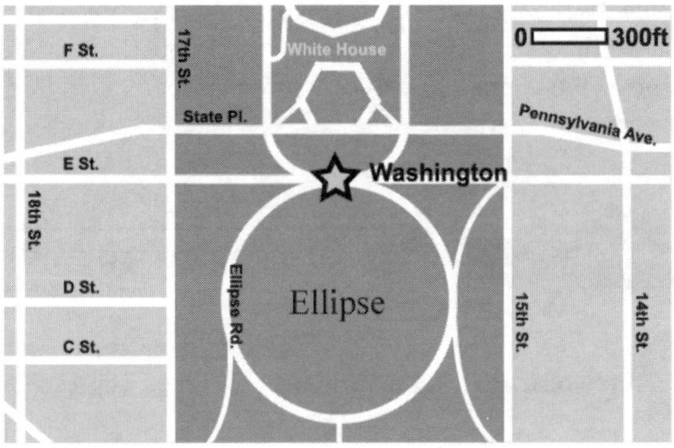

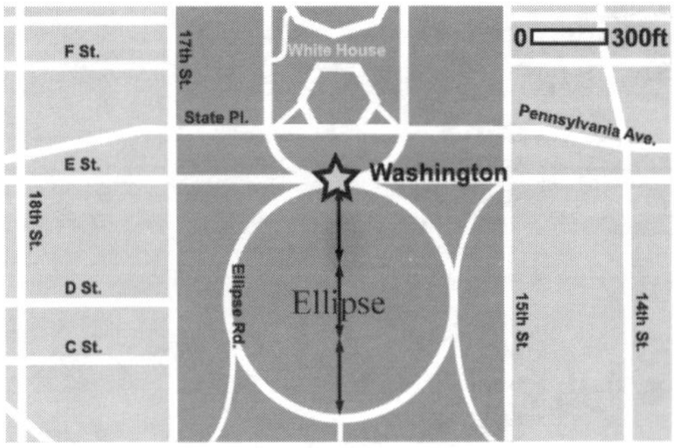

Step 1: Using a ruler or other marked measuring device, lay the map scale out along the longest distance across the Ellipse. The distance is approximately 3×300 feet, or 900 feet across.

Answer: The Ellipse is approximately 900 feet across at its widest point.

Example 2 Many people enjoy building or collecting model trains. One popular model called the HO employs a scale of $\frac{1}{87}$. That means 1 unit of measure in the model corresponds to 87 units in the real object. If a real train is 44 feet long, how many inches long will the model train be?

$$\frac{1}{87} = \frac{x}{44}$$

Step 1: This problem can be solved using proportions. Let x represent the length of the model train.

$$87x = 44$$

Step 2: Cross-multiply.

$$\frac{87x}{87} = \frac{44}{87}$$

Step 3: Divide both sides by 87.

$$x = \frac{44}{87}$$

$$\approx \frac{1}{2} \text{ foot}$$

$$\approx 6 \text{ inches}$$

Step 4: Solve for x.

Answer: The model train is about 6 inches long.

Example 3 Toy miniature soldiers are normally built on a $\frac{1}{35}$ scale. If an average man is 6 feet tall, how many inches tall would a typical toy soldier be?

$\frac{1}{35} = \frac{x}{72}$	**Step 1:** Let x represent the height of the toy soldier. Since an average man is 6 feet tall, and the answer needs to be in inches, convert 6 feet to 72 inches.
$35x = 72$	**Step 2:** Cross-multiply.
$\frac{35x}{35} = \frac{72}{35}$	**Step 3:** Divide by 35.
$x = \frac{72}{35}$ ≈ 2 inches	**Step 4:** Solve for x.

Answer: The toy soldier is approximately 2 inches tall.

Example 4 A floor plan is a scale model or drawing of a building or room. An example of a floor plan is shown below. The room dimensions are stated in the diagram. If the scale of the longer dimension of the carport is represented by 2.5 inches, what is the scale of the smaller dimension?

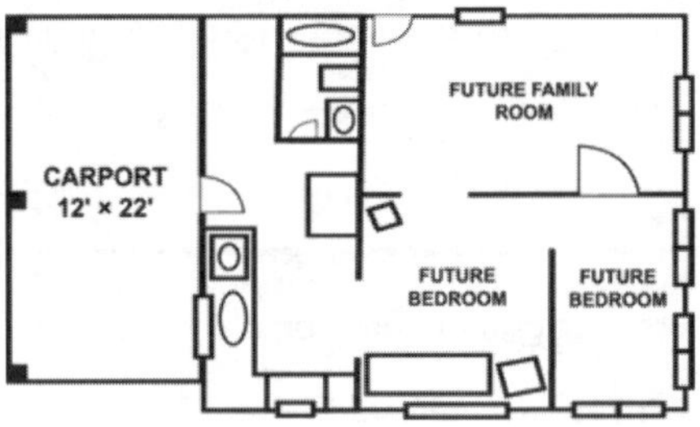

$\text{Scale} = \frac{2.5}{22}$	**Step 1:** Since the longer dimension of the carport is 22 feet, the scale of the diagram is 2.5 inches to 22 feet.
$\frac{2.5}{22} = \frac{x}{12}$	**Step 2:** Let x represent the other dimension of the carport.
$22x = 30$	**Step 3:** Cross-multiply.
$\frac{22x}{22} = \frac{30}{22}$	**Step 4:** : Divide both sides by 22.
$x = \frac{30}{22}$ ≈ 1.4 inches	**Step 5:** Solve for x.

Answer: The smaller dimension of the model will be approximately 1.4 inches long.

Example 5 A model designer wants to create a model of a World War II battleship that was 826 feet long. If he wants the scale to be 1 : 700, how long will the model be in inches?

$\frac{1}{700} = \frac{x}{826}$	**Step 1:** Since the scale is 1 : 700, let x represent the length of the model.
$700x = 826$	**Step 2:** Cross-multiply.
$\frac{700x}{700} = \frac{826}{700}$	**Step 3:** Divide both sides by 700.

$x = 1.18$ feet	**Step 4:** Solve for x.
$1.18 \text{ feet} \times \dfrac{12 \text{ inches}}{1 \text{ foot}} \approx 14 \text{ inches}$	**Step 5:** Convert to inches.
	Answer: The model will be approximately 14 inches long.

Problem Set

Solve:

1. The distance between Miami, Florida and Orlando, Florida is approximately 200 miles. On a map, the distance is 5 inches. What is the map scale?

2. The distance between Auburn, Alabama and Mobile, Alabama is approximately 200 miles. On a map, the distance is 4 inches. What is the map scale?

3. The distance between Madison, Wisconsin and New Orleans, Louisiana is approximately 910 miles. On a map, the distance is 5 inches. What is the map scale?

4. The distance between Seattle, Washington and Dallas, Texas is approximately 2,710 kilometers. On a map, the distance is 10 inches. What is the map scale?

5. The distance between Atlanta, Georgia and Las Vegas, Nevada is approximately 2,816 kilometers. On a map, the distance is 8 inches. What is the map scale?

6. The distance between Boston, Massachusetts and Jackson, Mississippi is approximately 1,260 miles. On a map, the distance is 18 centimeters. What is the map scale?

7. The distance between Philadelphia, Pennsylvania and Knoxville, Tennessee is approximately 549 miles. On a map, the distance is 3 inches. What is the map scale?

8. The distance between Louisville, Kentucky and Metairie, Louisiana is approximately 625 miles. On a map, the distance is 5 inches. What is the map scale?

9. On a city map, the distance between the sandwich shop and the mall is 8.2 centimeters. If the map scale is 2 centimeters to 400 meters, find the actual distance between the sandwich shop and the mall.

10. On a city map, the distance between the gas station and the mall is 6.4 centimeters. If the map scale is 4 centimeters to 600 meters, find the actual distance between the gas station and the mall.

11. On a city map, the distance between the hospital and the bank is 7.5 centimeters. If the map scale is 3 centimeters to 900 meters, find the actual distance between the hospital and the bank.

12. On a city map, the distance between the museum and the coffee shop is 9.6 centimeters. If the map scale is 3 centimeters to 900 meters, find the actual distance between the museum and the coffee shop.

13. On a city map, the distance between the bakery and the sandwich shop is 3.9 centimeters. If the map scale is 3 centimeters to 300 meters, find the actual distance between the bakery and sandwich shop.

14. On a city map, the distance between the coffee shop and the bakery is 6.8 centimeters. If the map scale is 2 centimeters to 500 meters, find the actual distance between the coffee shop and the bakery.

15. On a city map, the distance between the grocery store and the bank is 5.5 centimeters. If the map scale is 5 centimeters to 500 meters, find the actual distance between the grocery store and the bank.

16. In a picture, the height of the doors is 2 inches. If the drawing scale is 0.2 inches to 1 foot, approximately what is the actual height of the door?

17. Samantha drew a picture of a tree. The tree in her drawing has a height of 3 inches. If the drawing scale is 0.6 inches to 1 foot, approximately what is the actual height of the tree?

18. Angela drew a picture of the Eiffel Tower. Her drawing of the Eiffel Tower has a height of 3 inches. If the drawing scale is 0.003 inches to 1 foot, approximately what is the actual height of the Eiffel Tower?

19. Pete has a picture of his boat in his house. The length of the boat in the picture is 2 inches. If the drawing scale is 0.05 inches to 1 foot, approximately what is the actual length of the boat?

20. Christopher drew a picture of a ship. The ship in the drawing has a length of 4 inches. If the drawing scale is 0.02 inches to 1 foot, approximately what is the actual length of the ship?

21. Bailey drew a picture of a building. The building in her drawing has a height of 5 inches. If the drawing scale is 0.07 inches to 1 foot, approximately what is the actual height of the building? Round your answer to the nearest tenth.

22. Jamie drew a picture of the sand castle that he made at the beach. The drawing of his sand castle has a height of 2 inches. If the drawing scale is 0.04 inches to 1 foot, approximately what is the actual height of the sand castle?

23. Jim drew a picture of his ladder. The height of the ladder in the picture is 2 inches. If the drawing scale is 0.05 inches to 1 foot, approximately what is the actual height of the ladder?

24. Jay drew a bus. The length of the bus in his drawing is 3 inches. If the drawing scale is 0.2 inches to 1 foot, approximately what is the actual length of the bus?

25. Brandon drew a picture of a pyramid. The drawing of his pyramid has a height of 6 inches. If the drawing scale is 0.1 inches to 1 foot, approximately what is the actual height of the pyramid?

26. Patti painted a picture of a castle she saw on her trip. The drawing of the castle has a height of 4 inches. If the drawing scale is 0.1 inches to 1 foot, approximately what is the actual height of the castle?

27. Shannon sketched a picture of a ladder. The ladder in his drawing has a height of 5 inches. If the drawing scale is 0.2 inches to 1 foot, approximately what is the actual height of the ladder?

28. Kayla bought a picture of the Golden Gate bridge. The picture of the bridge has a length of 18 inches. If the drawing scale is 0.002 inches to 1 foot, approximately what is the actual length of the Golden Gate bridge?

29. An artist drew a picture of a courthouse. The drawing of the courthouse has a height of 2 inches. If the drawing scale is 0.07 inches to 1 foot, approximately what is the actual height of the courthouse? Round the answer to the nearest tenth.

30. The architects of a new church showed the community a drawing of what the new church will look like. The drawing of the new church has a length of 4 inches. If the drawing scale is 0.2 inches to 1 foot, approximately what will be the actual length of the new church?

31. The actual distance between two cities A and B is 500 miles. The distance between those cities on the map is 2 inches. If the distance, on the same map, between cities C and D is 4 inches, find the actual distance between cities C and D.

32. The actual distance between two cities A and B is 800 miles. The distance between those cities on the map is 4 inches. If the distance, on the same map, between cities C and D is 6 inches, find the actual distance between cities C and D.

33. The actual distance between two cities A and B is 700 miles. The distance between those cities on the map is 2 inches. If the distance, on the same map, between cities C and D is 9 inches, find the actual distance between cities C and D.

34. The actual distance between two cities A and B is 1,000 miles. The distance between those cities on the map is 5 inches. If the distance, on the same map, between cities C and D is 7 inches, find the actual distance between cities C and D.

35. The actual distance between two cities A and B is 300 miles. The distance between those cities on the map is 2 inches. If the distance, on the same map, between cities C and D is 6 inches, find the actual distance between cities C and D.

36. Paul wants to make a drawing of his rose garden using the scale 1 to 400. He measured that the actual length of the rose garden is 642 centimeters. How long should his drawing of the rose garden be? Round your answer to nearest hundredth.

37. Rhonda wants to make a drawing of her orange grove using the scale 1 to 3,000. She measured that the actual length of the grove is 12,000 centimeters. How long should her drawing of the orange grove be?

38. Janice wants to make a drawing of her garden using the scale 1 to 873. She measured that the actual length of the garden is 8,000 centimeters. How long should her drawing of the garden be? Round your answer to nearest hundredth.

39. Jillian wants to make a drawing of her rose garden using the scale 1 to 539. She measured that the actual length of the rose garden is 5,980 centimeters. How long should her drawing of the rose garden be? Round your answer to the nearest hundredth.

40. The actual distance between two cities A and B is 450 miles. The distance between those cities on the map is 2.5 inches. If the distance, on the same map, between cities C and D is 8 inches, find the actual distance between cities C and D.

MPA-111: Comparing Perimeters, Areas and Volumes of Similar Geometric Figures and Solids

Two geometric figures are similar if the lengths of their corresponding parts are in proportion. For example, consider the two triangles shown below.

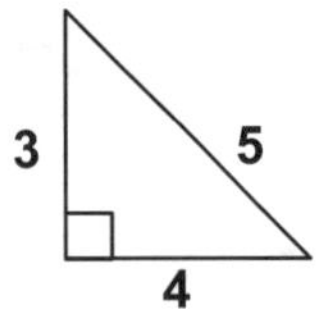

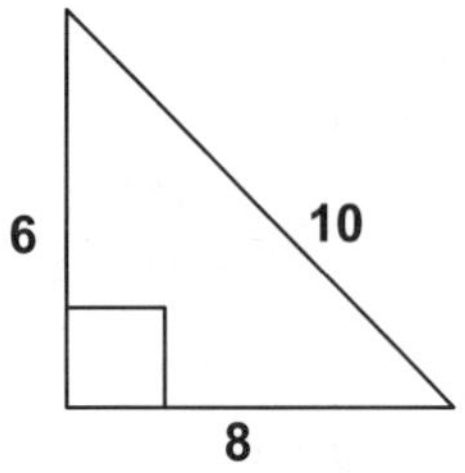

$\text{Perimeter} = 3 + 4 + 5 = 12$ units

$\text{Area} = \frac{1}{2} \times 3 \times 4 = 6$ square units

$\text{Perimeter} = 6 + 8 + 10 = 24$ units

$\text{Area} = \frac{1}{2} \times 6 \times 8 = 24$ square units

Similar Geometric Figures	Two geometric figures are similar if all of their corresponding parts are proportional.

The sides of the larger triangle are exactly two times the lengths of the smaller triangle. This means that the proportions, or ratios, of their side lengths are 1 : 2. The perimeters, which are 12 units and 24 units respectively, also have the same ratio as the sides.

The areas of the triangles, however, have a ratio of 6 : 24, or 1 : 4. This means that the ratio of the areas is the square of the ratio of the sides. To determine how the perimeter, area, or volume of a geometric figure or solid changes when its dimensions are proportionally increased or decreased, use the **scale factor**.

$$\text{Ratio of perimeters} = \text{Ratio of side lengths} = \text{Scale factor}$$

$$\text{Ratio of areas} = (\text{Ratio of side lengths})^2 = (\text{Scale factor})^2$$

$$\text{Ratio of volumes} = (\text{Ratio of side lengths})^3 = (\text{Scale factor})^3$$

Example 1 Two rectangles are similar. The scale factor is 8, and the perimeter of the larger rectangle is 96 centimeters. Find the perimeter of the smaller rectangle.

Let n represent the perimeter of the smaller rectangle.

Step 1: Set up the equation.

Ratio of perimeters = Scale factor

$$\frac{96}{n} = 8$$

Step 2: Cross-multiply.

$$\frac{96}{n} = \frac{8}{1}$$

$$96 \times 1 = 8 \times n$$

$$96 = 8n$$

$$\frac{96}{8} = \frac{8n}{8}$$

$$12 = n$$

Step 3: Divide both sides by 8.

Answer: The perimeter of the smaller rectangle is 12 centimeters.

Example 2 If each side of a triangle is increased 6 times, how much is the area of the triangle increased?

Scale factor = 6

Step 1: Find the ratio of the side lengths of the old and new triangles.

$$\text{Ratio of areas} = (\text{Scale factor})^2 = 6^2 = 36$$

Step 2: Find the ratio of the areas of the old and new triangles.

Answer: The area of the triangle is increased 36 times.

Example 3 Two rectangles are similar. The area of the small rectangle is 20 square inches. The area of the large rectangle is 180 square inches. Find the scale factor.

$$\text{Scale factor} = \text{Ratio of perimeters} = \frac{\text{Perimeter of large rectangle}}{\text{Perimeter of small rectangle}}$$

Step 1: Set up the equation.

$$\text{Scale factor} = \frac{180}{20} = 9$$

Step 2: Substitute the given values and solve.

Answer: The scale factor of the two rectangles is 9.

Example 4 Kenya has a small pond in her backyard. She dug a new pond that has the same shape but has dimensions that are 7 times larger. How much more water will be necessary to fill the new pond?

Scale = 7

Step 1: To solve this problem, find the ratio of volumes. In other words, find out how many times the volume of the new pond is bigger than the volume of the old pond. Since the dimensions of the new pond are 7 times larger than the old one, use 7 as the scale factor.

Ratio of volumes $= (\text{Scale factor})^3$ $= 7^3$ $= 343$	**Step 2:** Find the ratios of the volumes. Recall that this ratio is the cube of the scale factor.

Answer: It will take 343 times more water to fill the new pond.

Example 5 Jeff wants to fill two cans with his favorite juice. Both of the cans are the same shape but are different sizes. If the dimensions of the large can are 5 times the dimensions of the small can, how much more juice can be stored in the large can than in the small can?

$\frac{\text{Height of large can}}{\text{Height of small can}} = 5$ Scale factor $= 5$	**Step 1:** Find the scale factor.
Ratio of volumes $= (\text{Scale factor})^3$ $= 5^3$ $= 125$	**Step 2:** Find the ratio of the volumes. Recall that this ratio is the cube of the scale factor.

Answer: The large can will hold 125 times more juice than the small can.

Problem Set

Solve:

1. Two rhombi are similar. The perimeter of the large rhombus is 35 feet. The perimeter of the small rhombus is 7 feet. Find the scale factor.
2. Two squares are similar. The perimeter of the large square is 36 centimeters. The perimeter of the small square is 12 centimeters. Find the scale factor.
3. Two triangles are similar. The perimeter of the large triangle is 72 inches. The perimeter of the small triangle is 9 inches. Find the scale factor.
4. Two trapezoids are similar. The perimeter of the large trapezoid is 18 feet. The perimeter of the small trapezoid is 9 feet. Find the scale factor.
5. Two trapezoids are similar. The scale factor is 8 and the perimeter of the large trapezoid is 64 inches. Find the perimeter of the small trapezoid.
6. Two squares are similar. The scale factor is 10 and the perimeter of the large square is 60 centimeters. Find the perimeter of the small square.
7. Two squares are similar. The scale factor is 6 and the perimeter of the large square is 72 centimeters. Find the perimeter of the small square.
8. Two triangles are similar. The scale factor is 9 and the perimeter of the large triangle is 108 inches. Find the perimeter of the small triangle.
9. Two rhombi are similar. The scale factor is 7 and the perimeter of the larger rhombus is 91 centimeters. Find the perimeter of the smaller rhombus.
10. Two triangles are similar. The scale factor is 9 and the perimeter of the larger triangle is 135 inches. Find the perimeter of the smaller triangle.
11. If each side of a triangle is increased 8 times, how much is the area of the triangle increased?
12. If each side of a rectangle is increased 11 times, how much is the area of the rectangle increased?
13. If each side of a trapezoid is increased 12 times, how much is the area of the trapezoid increased?
14. If the radius of a circle is increased 3 times, how much is the area of the circle increased?

15. If the radius of a given circle is increased 5 times, how much is the area of the circle increased?

16. Two rhombi are similar. The area of the small rhombus is 6 square inches. The small rhombus has a perimeter of 12 inches and the large rhombus has a perimeter of 24 inches. Find the area of the large rhombus.

17. Two squares are similar. The area of the large square is 256 square centimeters. The small square has a perimeter of 32 centimeters and the large square has a perimeter of 64 centimeters. Find the area of the small square.

18. Two parallelograms are similar. The area of the small parallelogram is 2 square feet. The small parallelogram has a perimeter of 7 feet and the large parallelogram of 35 feet. Find the area of the large parallelogram.

19. Two right triangles are similar. The area of the small triangle is 16 square inches. The small triangle has a perimeter of 12 inches and the large triangle has a perimeter of 36 inches. Find the area of the large triangle.

20. Two trapezoids are similar. The area of the small trapezoid is 30 square inches. The small trapezoid has a perimeter of 30 inches and the large trapezoid has a perimeter of 150 inches. Find the area of the large trapezoid.

21. Two rectangles are similar. The area of the small rectangle is 2 square feet. The small rectangle has a perimeter of 6 feet and the large rectangle has a perimeter of 24 feet. Find the area of the large rectangle.

22. Two rectangles are similar. The area of the small rectangle is 12 square inches. The small rectangle has a perimeter of 35 inches and the large rectangle has a perimeter of 105 inches. Find the area of the large rectangle.

23. Two squares are similar. The area of the large square is 36 square feet. The small square has a perimeter of 8 feet and the large square has a perimeter of 24 feet. Find the area of the small square.

24. Susan and Cynthia each have cookies. Susan's cookie is 4 times the radius of Cynthia's cookie. How much more is the area of Susan's cookie than Cynthia's cookie?

25. Jason is buying tires. The radius of the large tire is 2 times the radius of the small tire. How much more is the area of the large tire than the small tire?

26. Thomas has two circular car speakers. The radius of the large speaker is 4 times the radius of the small speaker. How much more is the area of the large speaker than the small speaker?

27. Mary has drawn two pictures of the sky. One is an enlargement of the other. The pictures contain a sun. The larger sun is 5 times the radius of the smaller sun. How much more is the area of the large sun than the small sun?

28. A carpenter is cutting out circular holes in the ceiling for two different size lights. The radius of the large hole must be 6 times the radius of the small hole. How much greater will the area of the large hole be than the small hole?

29. A kitchen has two different sized windows. The radius of the large window is twice the radius of the small window. How much more is the area of the large window than the small window?

30. Martha has two buttons of different sizes to sew on a clown costume. The large button is 7 times the radius of the small button. How much more area will the large button take up than the small button on the costume?

31. Michael is moving into a new home. He has two boxes of different sizes. The dimensions of the large box are 4 times the dimensions of the small box. How much more room will Michael have to pack his belongings in the large box?

32. Darwin is selling chocolate covered raisins in two different size containers. The large container is 3 times the dimensions of the small container. How much more candy will fit in the large container?

33. Samantha and Lisa have doll houses with the same shape, however, the dimensions of Samantha's house are 2 times larger than the dimensions of Lisa's house. How much larger is the volume of Samantha's doll house than Lisa's doll house?

34. A company sells two different sizes of disc players. The dimensions of the larger disc player are 4 times the dimensions of the small disc player. How much more volume does the large disc player take up than the small disc player?

35. Tina and Lori are each going to make pumpkin pie using similar pumpkins. The dimensions of Tina's pumpkin are 7 times larger than the dimensions of Lori's pumpkin. How much more pumpkin can Tina use to make her pies?

36. Robert and his teacher have similar rectangular shaped color boxes. The dimensions of her box are 2 times larger than the dimensions of Robert's box. How much more storage capacity does his teacher's box store than Robert's?

37. The hardware store has paint cans of two sizes. The large can has the same shape but the dimensions are 8 times larger than the small can. How much more paint does the large can hold as compared to the small can?

38. A Pizzeria sells individual, small, medium and large pizzas. The dimensions of the large pizza box are 5 times larger than the dimensions of the individual pizza box. How much more volume does the large pizza box have than the individual pizza box?

39. John and Sam each purchased milk in similar containers. The dimensions of Sam's container are 3 times larger than the dimensions of John's milk container. How much more milk does Sam's container hold than John's container?

40. Dana has two rectangular storage containers of different sizes. Both containers are the same shape, but the dimensions of the larger container are 4 times that of the smaller container. How much more food can be filled in the larger container as compared to the smaller container?

MPA-112: Constructing Sample Spaces for Compound Events (Dependent and Independent)

The general formula for finding the probability of an event for a given experiment is:

$$\text{Probability (Event)} = \frac{\text{number of outcomes in event}}{\text{number of possible outcomes}}$$

If we are considering more than one event, we might ask questions like: What is the probability that Event A occurs given that Event B has occurred? If the occurrence of Event B has no effect on the probability of Event A occurring, we say the two events are **independent**. If the occurrence of Event B changes the probability of Event A, we say the two events are **dependent**.

Independent Events	Events for which the occurrence of one does not change the probability of the other.

Dependent Events	Events for which the occurrence of one changes the probability of the other.

For example, suppose we roll a number cube with six sides numbered 1, 2, 3, 4, 5, and 6. Let Event A be the result that the roll is an odd number $\{1, 3, 5\}$, and let Event B be the event that the roll is a prime number $\{2, 3, 5\}$. There are six total possible outcomes for this experiment.

Sometimes we want to know what the probability is that Event A occurs when we know that Event B has occurred. If Event B occurs, we know the result is one of these numbers $\{2, 3, 5\}$. Since Event A is the event that the result is an odd number, we see that:

$$\text{Probability (toss is odd when toss is prime)} = \frac{2}{3}$$

Note that since both A and B have three possible outcomes, we see that:

$$\text{Probability (toss is odd)} = \frac{1}{2}$$

Since these results are not equal, we see that Events A and B are dependent events.

Example 1 Determine whether the following event is dependent or independent: Taking a green marble from a bag with two green and two yellow marbles, replacing it, and then taking another.

Answer: The event is independent, since the outcome of the first draw does not affect the outcome of the second draw.

Example 2 A basket contains one apple (A), one orange (O), and one pear (P). Suppose a fruit is randomly picked from a basket and then, without replacing it, another fruit is picked. What is the sample space of this experiment?

Step 1: Use a tree diagram to find all the possible outcomes.

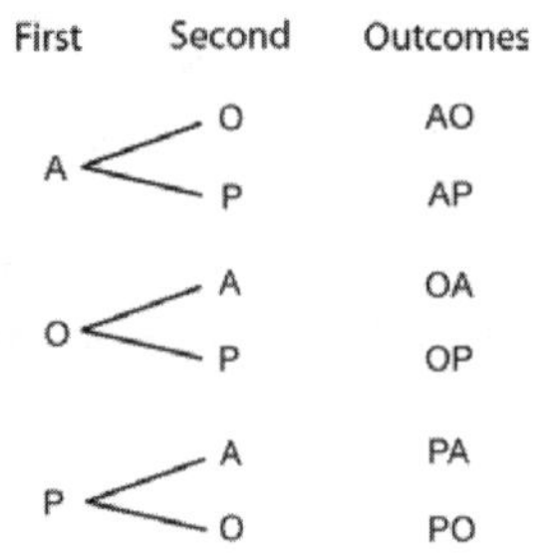

{AO, AP, OA, OP, PA, PO}

Step 2: List all the possible outcomes of the experiment.

Answer: The sample space of the experiment is {AO, AP, OA, OP, PA, PO}.

Example 3 A desk drawer contains one pencil, one red pen, and one blue pen. Ben randomly chooses a writing instrument from the drawer and, without replacing it, chooses another. What is the probability that the second writing instrument chosen was a pencil?

Step 1: Use a tree diagram to find all the possible outcomes. Let P represent a pencil, R represent a red pen, and B represent a blue pen.

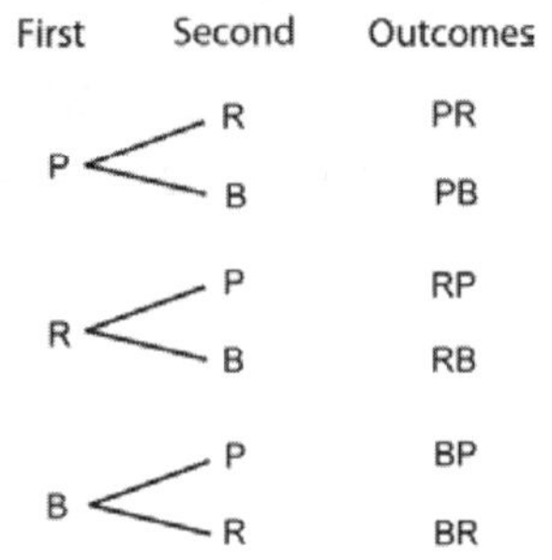

{PR, PB, RP, RB, BP, BR}

Step 2: List all the possible outcomes of the experiment.

{RP, BP}

Step 3: List all the outcomes from the sample space, found in Step 2, which list that the second writing instrument chosen is a pencil (P).

Step 4: Find the probability of choosing a pencil on the second draw.

$$P(2^{nd}\text{ pencil}) = \frac{2}{6}$$
$$= \frac{1}{3}$$

Answer: The probability of getting a pencil on the second choice is $\frac{1}{3}$.

Example 4 In Example 3, if the first writing instrument chosen is a red pen, what is the probability that the second writing instrument chosen is a pencil?

{PR, PB, RP, RB, BP, BR}	**Step 1:** List all the possible outcomes of the experiment.
{RP, RB}	**Step 2:** List all the possible outcomes found in Step 1 that lists a red pen as the first writing instrument chosen.
{RP}	**Step 3:** From Step 2, list all the possible outcomes that show the second instrument chosen is a pencil.
$P\begin{pmatrix} 2^{nd} \text{ pencil when} \\ 1^{st} \text{ red pen} \end{pmatrix} = \frac{1}{2}$	**Step 4:** Use Steps 2 and 3 to find the probability of choosing a red pen first, and then choosing a pencil second. Notice that the result is not the same as Example 3, so these two events are dependent.

Answer: The probability of getting a pencil on the second choice if a red pen is chosen first is $\frac{1}{2}$.

Example 5 Suppose a fair coin is tossed 2 times. If the first toss is heads (H), what is the probability that the second toss is tails (T)? State whether the experiment is dependent or independent.

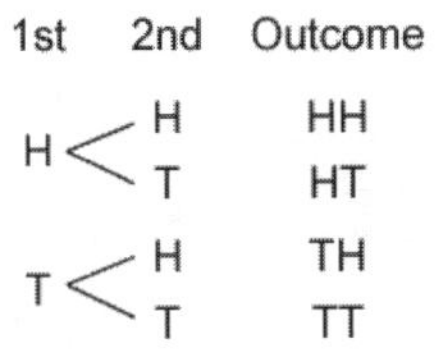

Step 1: Use a tree diagram to find all possible outcomes.

{HH, HT, TH, TT}	**Step 2:** List all the possible outcomes of the experiment.
{HH, HT}	**Step 3:** List all the possible outcomes, found in Step 2, that lists the first toss as heads (H).
{HT}	**Step 4:** List all the possible outcomes, found in Step 3, that lists the second toss as tails (T).
$P(2^{nd}\text{ T when } 1^{st}\text{ H}) = \frac{1}{2}$	**Step 5:** Use Steps 3 and 4 to find the probability of tossing heads first and tails second.

Answer: The two events are dependent because tossing heads first creates a smaller sample space for tossing tails second. Therefore, the probability that heads is tossed the second time is $\frac{1}{2}$.

Problem Set

1. Are the following events independent or dependent?

   Selecting a vowel from a deck of alphabet cards and then, without replacing it, selecting a consonant from the same deck of alphabet cards.

2. Are the following events independent or dependent?

   Rolling a 3 on a six-sided number cube and then rolling a 1 on the same number cube.

3. Are the following events independent or dependent?

   Selecting a boy from a group of ten boys and seven girls and then, without replacing him, selecting a girl from the same group.

4. Are the following events independent or dependent?

   Selecting a chocolate chip cookie from a tray of 3 cookies (2 chocolate chip and 1 peanut butter) and then, without replacing it, selecting a peanut butter cookie from the same tray of cookies.

5. Are the following events independent or dependent?

   Selecting the C from a deck of alphabet cards, replacing it, and then selecting the K from the same deck of cards.

6. Are the following events independent or dependent?

   Selecting a carrot from a vegetable tray which contains 20 carrots, 1 radish, and 19 celery sticks, eating it, and then selecting a radish from the same tray.

7. Are the following events independent or dependent?

   Flipping heads on a fair coin and then flipping tails on the same coin.

8. Suppose Anthony flips a fair coin two times. List all the possible outcomes of the two flips.

9. Suppose Sheila flips a fair coin three times. List all the possible outcomes of the three flips.

10. Suppose Jan picks a fruit from a basket containing an apple (A), an orange (O), and a banana (B), and then she picks a cookie from a tray containing a chocolate chip cookie (C), a peanut butter cookie (P), and a sugar cookie (S). List all the possible outcomes of the two choices.

11. Suppose Carlos picks a vegetable from a choice of corn (C), beans (B), and potatoes (P), and then he picks a meat from a choice of steak (S), lamb (L), and duck (D). List all the possible outcomes of the two choices.

12. Suppose Sally picks a number from 1 through 3, and then she picks a letter from A through C. List all the possible outcomes of the number and letter picks.

13. Suppose John picks a letter from A through C, and then he picks a letter from X through Z. List all the possible outcomes of the two letter picks.

14. Suppose Harold picks a shirt from a red (R), a green (G), and a blue (B), and then he picks a solid (S), paisley (P) or dotted (D) tie. List all the possible outcomes of the shirt and tie combinations.

15. Suppose Sierra picks a red (R), a green (G), or a blue (B) shirt, and then she picks a skirt (S) or pants (P). List all the possible outcomes of the top and bottom combinations.

16. Suppose we have a bowl of fruit containing 3 apples, 2 oranges, 4 bananas, and 3 pears. If Angie randomly selects and eats an apple, what is the probability that her second selection will be a banana?

17. Suppose we have a bowl of fruit containing 5 apples, 3 oranges, 2 bananas, and 1 pear. If Sheila randomly selects and eats an orange, what is the probability that her second selection will be a pear?

18. Suppose we have a dessert tray containing 6 brownies, 4 pieces of pie, 3 chocolate chip cookies, and 5 pieces of fudge. If Jared randomly selects and eats a brownie, what is the probability that his second selection will be a piece of fudge?

19. Suppose we have a dessert tray containing 5 brownies, 3 pieces of pie, 5 chocolate chip cookies, and 4 pieces of fudge. If Amy randomly selects and eats a chocolate chip cookie, what is the probability that her second selection will be a piece of pie?

20. Suppose we have a sack of marbles containing 6 red marbles, 4 blue, 5 green, and 3 yellow. If Robin randomly selects and replaces a yellow marble, what is the probability that her second selection will be a red marble?

21. Suppose we have a sack of marbles containing 2 red marbles, 3 blue, 4 green, and 6 yellow. If James randomly selects and replaces a green marble, what is the probability that his second selection will be a blue marble?

22. Suppose we have a basket of toys containing 3 dolls, 2 planes, 4 trucks, and 5 balls. If Nicolas randomly selects a ball and throws it back in, what is the probability that his second selection will be a truck?

23. Suppose we have a basket of toys containing 3 dolls, 2 planes, 4 trucks, and 5 balls. If Anna randomly selects a doll and throws it back in, what is the probability that her second selection will be a plane?

24. John is randomly picking what he will wear to work. He must wear a shirt and tie. His shirt choices are 1 red, 1 blue, and 1 black shirt. His tie choices are 1 solid, 1 striped, and 1 paisley tie. If he picks the red shirt, what is the probability that he will pick the paisley tie?

25. Bonnie is randomly picking what she will wear to work. She must wear a skirt and blouse. Her skirt choices are 1 black skirt and 1 brown skirt. Her blouse choices are 1 red, 1 white, 1 black, and 1 green blouse. If she picks the black skirt, what is the probability that she will pick the white blouse?

26. Chris is randomly picking what he will cook for dinner. He wants to have meat and a vegetable. In the refrigerator, he has hamburger, chicken, and pork chops. For vegetables, he has corn and potatoes. If he chooses to cook hamburgers, what is the probability that he will choose potatoes?

27. Jason is randomly picking what kind of sandwich he will eat for lunch. He must choose the meat and bread. The bread choices are Kaiser roll, wheat roll, French loaf, and toasted sourdough. The meat choices are turkey, ham, bacon, tuna salad, and chicken salad. If he chooses a wheat roll, what is the probability that he will choose tuna salad?

28. Stacey is randomly picking what kind of ice cream cone she will have for dessert. She must choose the cone and ice cream flavor. The cone choices are sugar, waffle, and cake. The ice cream choices are vanilla, chocolate, strawberry, fudge ripple, chocolate almond, and butter pecan crunch. If she chooses a waffle cone, what is the probability that she will choose chocolate almond ice cream?

29. Maggie is randomly picking what kind of ice cream sundae she will have for dessert. She must choose the ice cream flavor and topping. The ice cream choices are vanilla, chocolate, strawberry, fudge ripple, chocolate almond, and butter pecan crunch. The topping choices are hot fudge, caramel, strawberries, and butterscotch. If she chooses fudge ripple ice cream, what is the probability that she will choose caramel topping?

30. Brandon is taking Marcy on their first date. He does not know her dinner and outing preferences, so he randomly chooses what they will do. His restaurant choices are Italian, seafood, steak house, and Chinese. His outing choices are the movies, bowling, silly golf, dancing, and the amusement park. If he chooses to take her to the Italian restaurant, what is the probability that he will take her to the amusement park after dinner?

31. The Key Club is randomly drawing for their President, Vice President, Secretary, and Treasurer. There are 11 girls and 13 boys running for office. If a boy is drawn for President, find the probability of a boy being drawn for Vice President. State whether the events, of choosing a boy as president and then a boy as vice president, are dependent or independent.

32. The Math Club is randomly drawing for their President and Vice President. There are 12 girls and 9 boys running for office. If a girl is drawn for President, find the probability of a boy being drawn for Vice President. State whether the events, choosing a girl as president and then a boy as vice president, are dependent or independent.

33. McKinley High School is randomly drawing for its Homecoming King and Queen. There are 12 Seniors and 8 Juniors running for King, and 11 seniors and 5 Juniors running for Queen. If a Senior is drawn for King, find the probability of a Senior being drawn for Queen. State whether the events, choosing a Senior as King and then a Senior as Queen, are dependent or independent.

34. A normal deck of 52 cards contains four suits: spades, hearts, diamonds, and clubs. Each suit contains thirteen cards labeled 2 through 10, Jack, Queen, King, and Ace. Suppose we draw an Ace from the deck and keep it. Find the probability that our second draw will be another Ace. State whether the events, drawing an Ace and then drawing another Ace, are dependent or independent.

35. A normal deck of 52 cards contains four suits: spades, hearts, diamonds, and clubs. Each suit contains thirteen cards labeled 2 through 10, Jack, Queen, King, and Ace. Suppose we draw a heart from the deck and keep it. Find the probability that our second draw will be another heart. State whether the events, drawing a heart and then drawing another heart, are dependent or independent.

36. McGee High School is randomly drawing for its Homecoming King and Queen. There are 12 Seniors and 8 Juniors running for King, and 11 Seniors and 5 Juniors running for Queen. If a Junior is drawn for King, find the probability of a Junior being drawn for Queen. State whether the events, drawing a Junior as King and then a Junior as Queen, are dependent or independent.

37. Sandra is randomly selecting two pets for her children from the pound. There are 15 puppies and 14 kittens to choose from. If her first choice is a kitten, find the probability that her second choice will be a puppy. State whether the events, selecting a kitten and then a puppy, are dependent or independent.

38. Mark is randomly selecting two pets for his children from the pound. There are 13 birds and 12 iguanas to choose from. If his first choice is a bird, find the probability that his second choice will be an iguana. State whether the events, selecting a bird and then an iguana, are dependent or independent.

39. A normal deck of 52 cards contains 4 suits: spades, hearts, diamonds, and clubs. Each suit contains 13 cards labeled 2 through 10, Jack, Queen, King, and Ace. Suppose we draw the Ace of Spades from the deck and return it. Find the probability that our second draw will be the King of Spades. State whether the events, drawing the Ace of Spades and then the King of Spades, are dependent or independent.

40. A normal deck of 52 cards contains 4 suits: spades, hearts, diamonds, and clubs. Each suit contains 13 cards labeled 2 through 10, Jack, Queen, King, and Ace. Suppose we draw a Queen from the deck and return it. Find the probability that our second draw will be a Jack. State whether the events, drawing a Queen and then a Jack, are dependent or independent.

MPA-113: Finding the Probability of Compound Events Through Experimentation

Recall that to compute the probability of an outcome or set of outcomes from an experiment, we use the following formula:

$$\text{Probability (Event)} = \frac{\text{number of outcomes in event}}{\text{number of possible outcomes}}$$

This formula is sufficient when considering a single event, but what happens when we have two events and we want to combine them? We can combine events in a number of ways:

- What is the probability that Event A occurs when we know that Event B has occurred?
- What is the probability that Event A or Event B occurs?
- What is the probability that Event A and Event B occur?

Combining Events Using "When"

First, we will learn how to compute the probability of Event A when we know that Event B has occurred. For example, if we flip a fair coin three times, what is the probability that the second and third flips are tails when the first flip is tails?

In this case, there are two events to consider: Event A (the second and third flips are tails) and Event B (the first flip is tails). In the problem, we know that Event B has occurred.

In general, we consider the set of all possible outcomes of the three flips (H = heads; T = tails):

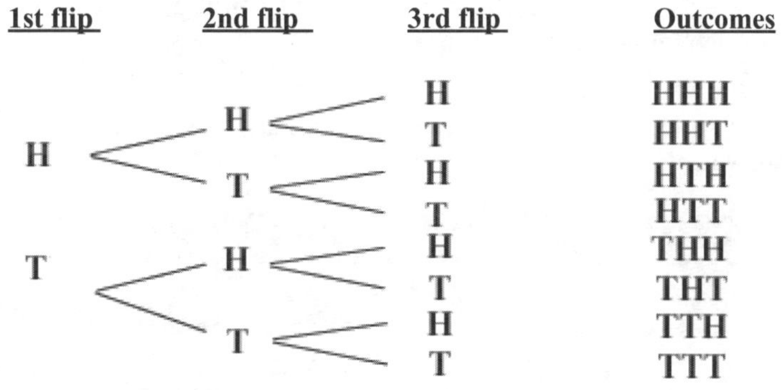

Total number of outcomes = 8

However, we know that Event B (first coin flip is tails) has occurred, so we restrict our attention to the following events.

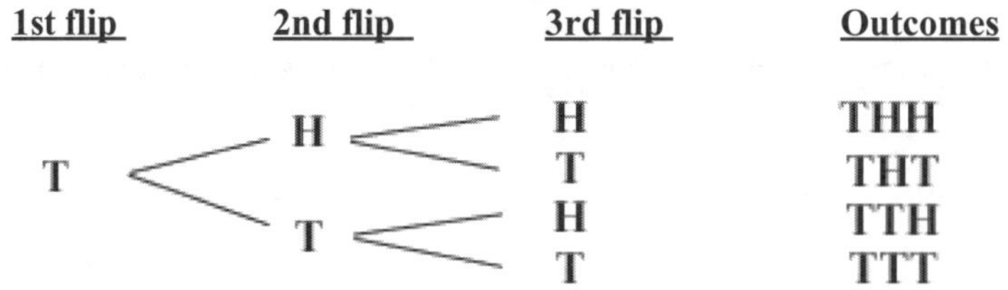

Total number of outcomes = 4

To compute the probability of Event A (the second and third flips are tails) we are only considering these four events. Since only one of them has all three flips as tails, the probability of getting the second and third flips as tails when the first flip is tails is:

$$\text{Probability (second and third T } \underline{\text{when}} \text{ first is T)} = \frac{1}{4}$$

In general, when computing the probability of a combination of events where one event is known to have occurred, consider only those events that make up the event that has occurred. This is called reducing the sample space or reducing the total number of outcomes.

Combining Events Using "Or"

Consider a combination of events in which you want to know whether Event A <u>or</u> Event B has occurred. For example, consider rolling two number cubes (one gray and the other white). What is the probability that the sum of the cubes is 7 <u>or</u> the gray cube is a 5?

Again, we have two events: Event A (the sum of the cubes is 7) and Event B (the gray cube is a 5). We want to know how many events have either a sum of 7 or a gray cube of 5. Consider the diagram of all possible outcomes of rolling the two number cubes shown below.

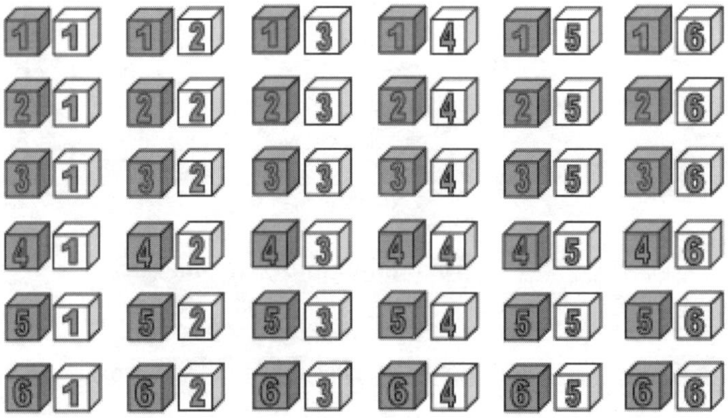

To simplify our writing we will represent these outcomes using ordered pair notation. So, (4, 3) stands for the outcome:

This means that we can write all of the outcomes in the ordered pair notation as shown below:

{(1,1), (1,2), (1,3), (1,4), (1,5), (1,6),
(2,1), (2,2), (2,3), (2,4), (2,5), (2,6),
(3,1), (3,2), (3,3), (3,4), (3,5), (3,6),
(4,1), (4,2), (4,3), (4,4), (4,5), (4,6),
(5,1), (5,2), (5,3), (5,4), (5,5), (5,6),
(6,1), (6,2), (6,3), (6,4), (6,5), (6,6)}

Now all of the outcomes in Event A (the sum of the cubes is 7) are shaded, and Event B (gray cube is 5) are surrounded by a box.

{(1,1), (1,2), (1,3), (1,4), (1,5), (1,6),
(2,1), (2,2), (2,3), (2,4), (2,5), (2,6),
(3,1), (3,2), (3,3), (3,4), (3,5), (3,6),
(4,1), (4,2), (4,3), (4,4), (4,5), (4,6),
(5,1), (5,2), (5,3), (5,4), (5,5), (5,6),
(6,1), (6,2), (6,3), (6,4), (6,5), (6,6)}

There are six shaded events and six events in the box. So, there are 11 distinct events that are either shaded (sum is 7) or in the box (gray is 5). This means we can compute the probability of either event occurring as:

$$\text{Probability (sum of 7 } \underline{\text{or}} \text{ gray 5)} = \frac{11}{36}$$

In general, when trying to compute the probability of two combined events by determining if either of the events occurred, take the union of all outcomes that could occur in either of the two events. Remember not to count any event twice if it occurs in both events, since it is really just one outcome.

Combining Events Using "And"

The last case we must consider is when we want to find that Event A and Event B have occurred. For example, let's consider our number cubes again. Let Event A be the event that the sum of the cubes is greater than 8, and let Event B be the event that the white cube is a 6.

In the diagram below, we have shaded all the outcomes which have a sum greater than 8 (Event A), and we have surrounded all the outcomes where the white cube is a 6 (Event B) by a box.

{(1,1), (1,2), (1,3), (1,4), (1,5), (1,6),
(2,1), (2,2), (2,3), (2,4), (2,5), (2,6),
(3,1), (3,2), (3,3), (3,4), (3,5), (3,6),
(4,1), (4,2), (4,3), (4,4), (4,5), (4,6),
(5,1), (5,2), (5,3), (5,4), (5,5), (5,6),
(6,1), (6,2), (6,3), (6,4), (6,5), (6,6)}

In this case, there are 10 outcomes that make up Event A (sum greater than 8) and 6 outcomes that make up Event B (white cube is a 6). Since we want to know the probability of both events occurring, consider the four events they have in common. This means we can compute the probability of both events occurring as:

$$\text{Probability (sum} > 8 \text{ } \underline{\text{and}} \text{ white 6)} = \frac{4}{36} = \frac{1}{9}$$

In general, when trying to compute the probability of two combined events by determining if both events occurred, take the intersection of all outcomes that occur in both events.

Example 1 Suppose you roll two number cubes, one gray and the other white. What is the probability that the numbers will be the same? In other words, what is the probability of rolling doubles?

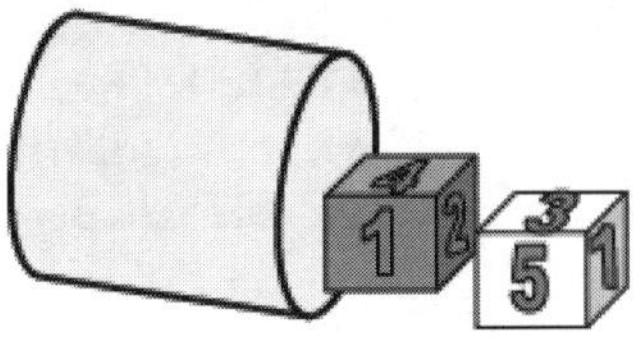

{(1,1), (1,2), (1,3), (1,4), (1,5), (1,6),
(2,1), (2,2), (2,3), (2,4), (2,5), (2,6),
(3,1), (3,2), (3,3), (3,4), (3,5), (3,6),
(4,1), (4,2), (4,3), (4,4), (4,5), (4,6),
(5,1), (5,2), (5,3), (5,4), (5,5), (5,6),
(6,1), (6,2), (6,3), (6,4), (6,5), (6,6)}

Step 1: First, note that there are 36 possible outcomes when rolling the two number cubes. The gray cube can land in 6 ways and for each of those ways the white cube can land 6 ways. This gives us a total of 36 possible outcomes for the experiment.

{(1,1), (1,2), (1,3), (1,4), (1,5), (1,6),
(2,1), (2,2), (2,3), (2,4), (2,5), (2,6),
(3,1), (3,2), (3,3), (3,4), (3,5), (3,6),
(4,1), (4,2), (4,3), (4,4), (4,5), (4,6),
(5,1), (5,2), (5,3), (5,4), (5,5), (5,6),
(6,1), (6,2), (6,3), (6,4), (6,5), (6,6)}

Step 2: There are 6 ways to roll doubles on the number cubes.

$$\text{Probability (doubles)} = \frac{6}{36} = \frac{1}{6}$$

Step 3: The probability of rolling doubles is the number of ways to get doubles divided by the total number of possible outcomes which is $\frac{6}{36}$ or $\frac{1}{6}$.

Answer: The probability of rolling doubles is $\frac{1}{6}$.

Example 2 Suppose you toss a fair coin followed by a number cube. What is the probability that a tail <u>or</u> an odd number is obtained?

{(H,1), (H,2), (H,3), (H,4), (H,5), (H,6),
(T,1), (T,2), (T,3), (T,4), (T,5), (T,6)}

Step 1: There are 2 sides on the coin and 6 numbers on the cube. Therefore, there are a total of 12 outcomes.

{(H,1), (H,2), (H,3), (H,4), (H,5), (H,6),
(T,1), (T,2), (T,3), (T,4), (T,5), (T,6)}

Step 2: Since there are a total of 12 outcomes in the sample space, and each outcome is equally likely to occur, we can highlight the outcomes with a "T" or an odd number.

$$\text{Probability (T or odd)} = \frac{9}{12} = \frac{3}{4}$$

Step 3: There are 9 ways in which flipping the coin and rolling the number cube can include either a tail or an odd number, so the probability that we will obtain a tail or an odd number is $\frac{9}{12}$ or $\frac{3}{4}$.

Answer: The probability of obtaining a tail or an odd number is $\frac{3}{4}$.

Example 3 Brad rolls two number cubes, one gray and one white. Each cube has sides numbered one through six. What is the probability that he rolls a white 4 <u>or</u> a sum less than 6?

{(1,1), (1,2), (1,3), (1,4), (1,5), (1,6),
(2,1), (2,2), (2,3), (2,4), (2,5), (2,6),
(3,1), (3,2), (3,3), (3,4), (3,5), (3,6),
(4,1), (4,2), (4,3), (4,4), (4,5), (4,6),
(5,1), (5,2), (5,3), (5,4), (5,5), (5,6),
(6,1), (6,2), (6,3), (6,4), (6,5), (6,6)}

Step 1: First, note that there are 36 possible outcomes when rolling the two number cubes. The gray cube can land 6 ways, and for each of those, the white cube can land 6 ways. This gives us a total of 36 possible outcomes for the experiment.

{(1,1), (1,2), (1,3), (1,4), (1,5), (1,6),
(2,1), (2,2), (2,3), (2,4), (2,5), (2,6),
(3,1), (3,2), (3,3), (3,4), (3,5), (3,6),
(4,1), (4,2), (4,3), (4,4), (4,5), (4,6),
(5,1), (5,2), (5,3), (5,4), (5,5), (5,6),
(6,1), (6,2), (6,3), (6,4), (6,5), (6,6)}

Step 2: Next, look at the column that contains a 4 as the second number in the number pair, which represents a 4 on the white number cube. There are 6 such outcomes. Now we will look at the cubes that have a sum less than 6. There are 10 such outcomes. We do not count any outcomes more than once.

$$\text{Probability (white 4 or sum} < 6) = \frac{15}{36} = \frac{5}{12}$$

Step 3: The probability of obtaining a white 4 or a sum less than 6 is $\frac{15}{36}$ or $\frac{5}{12}$.

Answer: The probability of obtaining a white 4 or a sum less than 6 is $\frac{5}{12}$.

Example 4 Given the two spinners below, suppose you first spin the spinner with the numbers and then spin the spinner with the letters. What is the probability of spinning a 3 and a C?

{(1,A), (1,B), (1,C), (1,D),
(2,A), (2,B), (2,C), (2,D),
(3,A), (3,B), (3,C), (3,D),
(4,A), (4,B), (4,C), (4,D)}

Step 1: The number spinner can land on 4 different numbers, and for each of those numbers, the letter spinner can land on 4 different letters. Therefore, there are 16 total possible outcomes.

{(1,A), (1,B), (1,C), (1,D),
(2,A), (2,B), (2,C), (2,D),
(3,A), (3,B), (3,C), (3,D),
(4,A), (4,B), (4,C), (4,D)}

Step 2: Next look at the row that contains a 3 which represents a 3 on the number spinner. There are 4 such outcomes. Now look at the column that contains a C. There are 4 such outcomes. The events are made up of all outcomes that spin both a 3 and a C. There is 1 such outcome.

Probability (3 and C) $= \frac{1}{16}$

Step 3: The probability that you will spin both a 3 and a C is $\frac{1}{16}$.

Answer: The probability that you will spin a 3 and a C is $\frac{1}{16}$.

Example 5 Allie rolls one letter cube which has six letters A, B, C, D, E, and F, and one number cube with numbers 1, 2, 3, 4, 5, and 6. What is the probability that she will roll a number greater than 4 when a vowel has been rolled?

{(1,A), (1,B), (1,C), (1,D), (1,E), (1,F),
(2,A), (2,B), (2,C), (2,D), (2,E), (2,F),
(3,A), (3,B), (3,C), (3,D), (3,E), (3,F),
(4,A), (4,B), (4,C), (4,D), (4,E), (4,F),
(5,A), (5,B), (5,C), (5,D), (5,E), (5,F),
(6,A), (6,B), (6,C), (6,D), (6,E), (6,F)}

Step 1: There are 36 possible outcomes when rolling the number cube and the letter cube. The number cube can land 6 ways, and for each of those ways the letter cube can land 6 ways.

New Sample Space: {(1,A), (1,E), (2,A), (2,E), (3,A), (3,E), (4,A), (4,E), (5,A), (5,E), (6,A), (6,E)}	**Step 2:** Since we know the letter cube is a vowel, we must look at a smaller sample space and find the outcomes with a number cube greater than 4. The word "when", used to combine events, always indicates that we will use a new (usually smaller) sample space. The first and forth columns of the list gives all the outcomes with a vowel as the letter cube, so this is our new sample space.
{(1,A), (1,E), (2,A), (2,E), (3,A), (3,E), (4,A), (4,E), (5,A), (5,E), (6,A), (6,E)}	**Step 3:** Of these 12 outcomes, 4 of them have a number greater than 4.
Probability (# > 4 when vowel) $= \frac{4}{12}$ $= \frac{1}{3}$	**Step 4:** The probability of rolling a number greater than 4 when a vowel has been rolled is $\frac{4}{12}$ or $\frac{1}{3}$.
	Answer: The probability of rolling a number greater than 4 when a vowel has been rolled is $\frac{1}{3}$.

Problem Set

Solve:

1. When you get to gym class, your teacher tells you to reach in two different bags and pick one item from each bag. The first bag has one white (W) and one blue (B) shirt, and the second bag has one red (R), one green (G), one yellow (Y), and one orange (O) pair of shorts. What is the probability that you will pick the blue shirt and the yellow shorts?

**{(B, R), (B, G), (B, Y), (B, O),
(W, R), (W, G), (W, Y), (W, O)}**

2. To win a game you must roll an odd number on a number cube with the numbers 1, 2, 3, 4, 5 and 6 and get heads on the flip of a coin. What is the probability that you will win?

**{(1, H), (2, H), (3, H), (4, H), (5, H), (6, H),
(1, T), (2, T), (3, T), (4, T), (5, T), (6, T)}**

3. Given the two spinners, what is the probability that you will spin a consonant and a multiple of 4?

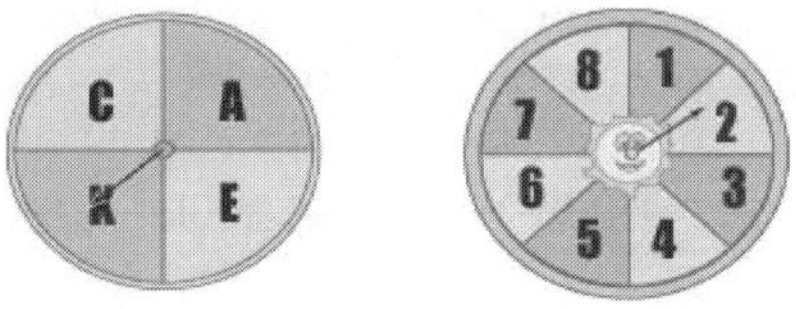

{(C, 1), (C, 2), (C, 3), (C, 4), (C, 5), (C, 6), (C, 7), (C, 8),
(A, 1), (A, 2), (A, 3), (A, 4), (A, 5), (A, 6), (A, 7), (A, 8),
(K, 1), (K,2), (K, 3), (K, 4), (K, 5), (K, 6), (K, 7), (K, 8),
(E, 1), (E, 2), (E, 3), (E, 4), (E, 5), (E, 6), (E, 7), (E, 8)}

4. Tudi spins the spinner and then randomly picks one of the following number cards. What is the probability that she will spin yellow (Y) and pick the number 3?

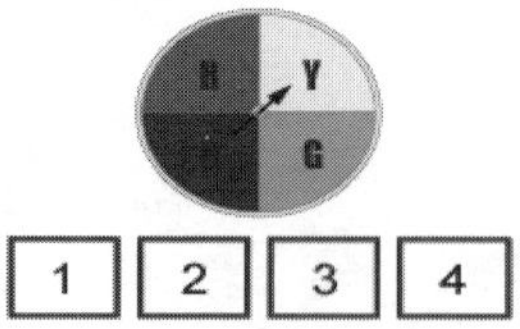

{(R, 1), (Y, 1), (G, 1), (B, 1)
(R, 2), (Y, 2), (G, 2), (B, 2)
(R, 3), (Y, 3), (G, 3), (B, 3)
(R, 4), (Y, 4), (G, 4), (B, 4)}

5. In Mo's wallet there is a quarter (Q), a dime (D), and a nickel (N). If he pulls out 2 coins a the same time without looking, what is the probability that he will pull out a quarter or a nickel?

{(Q, D), (Q, N),
(D, Q), (D, N),
(N, Q), (N, D)}

6. You roll two number cubes, each numbered 1, 2, 3, 4, 5, and 6. What is the probability that you will roll a 3 or a 5?

{(1, 1), (1, 2), (1, 3), (1, 4), (1, 5), (1, 6)
(2, 1), (2, 2), (2, 3), (2, 4), (2, 5), (2, 6)
(3, 1), (3, 2), (3, 3), (3, 4), (3, 5), (3, 6)
(4, 1), (4, 2), (4, 3), (4, 4), (4, 5), (4, 6)
(5, 1), (5, 2), (5, 3), (5, 4), (5, 5), (5, 6)
(6, 1), (6, 2), (6, 3), (6, 4), (6, 5), (6, 6)}

7. Jeremy has two boxes of donuts. Both boxes have one glazed (G), one chocolate (C), one jelly (J), and one sprinkle donut (S). If Jeremy randomly chooses one donut from each box, what is the probability that Jeremy will choose a glazed or a jelly donut?

{(G, C), (G, J), (G, S),(G, G)
(C, G), (C, J), (C, S), (C, C)
(J, G), (J, C), (J, S), (J, J)
(S, G), (S, L), (S, C), (S, S)}

8. Jane has a spinner with the letters X, Y, and Z, and a number cube with the numbers 1, 2, 3, 4, 5, and 6. What is the probability of the spinner landing on the Z when a 4 has been rolled?

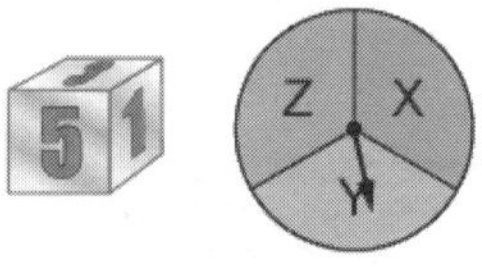

{(1, X), (2, X), (3, X), (4, X), (5, X), (6, X)
(1, Y), (2, Y), (3, Y), (4, Y), (5, Y), (6, Y)
(1, Z), (2, Z), (3, Z), (4, Z), (5, Z), (6, Z)}

9. Becky rolls two number cubes with numbers 1, 2, 3, 4, 5, and 6. What is the probability that the sum of the cubes is greater than 7 or one of the cubes is 5?

{(1, 1), (1, 2), (1, 3), (1, 4), (1, 5), (1, 6)
(2, 1), (2, 2), (2, 3), (2, 4), (2, 5), (2, 6)
(3, 1), (3, 2), (3, 3), (3, 4), (3, 5), (3, 6)
(4, 1), (4, 2), (4, 3), (4, 4), (4, 5), (4, 6)
(5, 1), (5, 2), (5, 3), (5, 4), (5, 5), (5, 6)
(6, 1), (6, 2), (6, 3), (6, 4), (6, 5), (6, 6)}

10. James has to pick a chip from a cup that contains one green (G), one black (B), one white (W), and one red (R). He then spins a spinner with the numbers 1, 2, 3, 4, and 5. Find the probability of James randomly picking the red chip or spinning a 5.

{(G, 1), (G, 2), (G, 3), (G, 4), (G, 5)
(B, 1), (B, 2), (B, 3), (B, 4), (B, 5)
(W, 1), (W, 2), (W, 3), (W, 4), (W, 5)
(R, 1), (R, 2), (R, 3), (R, 4), (R, 5)}

11. Given the two spinners below, what is the probability of getting a heart or the color purple (P)?

{(♥, B), (♥, P), (♥, G), (♥, R),
(♪, B), (♪, P), (♪, G), (♪, R),
(☼, B), (☼, P), (☼, G), (☼, R),
(♦, B), (♦, P), (♦, G), (♦, R)}

12. At a birthday party, there was a box of cupcakes containing one strawberry (S), one chocolate (C), one vanilla (V), and one white chocolate (W). In another box, there were two party hats, one red (R) and one blue (B). If Brandy closed her eyes and picked one of the four cupcakes and one of the two hats, what is the probability that Brandy picked the chocolate cupcake or the blue hat?

{(S, R), (S, B),
(C, R), (C, B),
(V, R), (V, B),
(W, R), (W, B)}

13. Tyree arrives at a party to find four gifts, one red (R), one blue (B), one green (G), and one yellow (Y), and three balloons, one purple (P), one orange (O), and one teal (T). He was told to close his eyes and pick one of the four gifts and one of the three balloons. Find the probability of Tyree randomly selecting the green gift or the purple balloon.

{(R, P), (R, O), (R, T),
(B, P), (B, O), (B, T),
(G, P), (G, O), (G, T),
(Y, P), (Y, O), (Y, T)}

14. At a bowling alley, Paul has to pick a pair of shoes from one box and a bowling ball from another. The box of shoes contains one pair of blue shoes (B), one pair of tan shoes (T), and one pair of red shoes (R). The box of bowling balls contains one orange (O), one yellow (Y), and one pink (P). If Paul randomly picks a pairs of shoes and a ball, what is the probability that he picks the tan shoes or the pink ball?

{(B, O), (B, Y), (B, P),
(T, O), (T, Y), (T, P),
(R, O), (R, Y), (R, P)}

15. Pam rolls a number cube with the numbers 1, 2, 3, 4, 5, and 6 and then tosses a coin. Find the probability of Pam rolling a 3 or getting tails.

{(1, H), (2, H), (3, H), (4, H), (5, H), (6, H),
(1, T), (2, T), (3, T), (4, T), (5, T), (6, T)}

16. Given the two spinners, suppose you first spin the spinner with the letters and then the spinner with the shapes. What is the probability that you will spin a 2 and a circle?

17. Patricia has two sets of cards. One set has 7 cards each with a different day of the week on them. The other set has two cards, one with the letter A and the other with the letter B. If Patricia randomly chooses one card from each set, what is the probability that she will pick a day that begins with the letter T and the letter A?

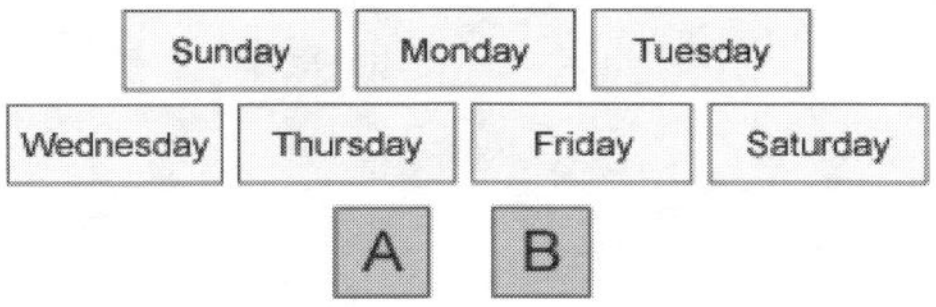

18. Mewelde rolls two different number cubes with the numbers 1, 2, 3, 4, 5, and 6. What is the probability that both numbers rolled will be odd?

19. A department store is giving away a free shirt (blue, white, or gray) and a free pair of shorts (red, orange, or khaki) with any purchase. If the shirt and shorts are given to you at random, what is the probability that you are given a white shirt and red shorts?

20. Nikki has 2 shades of lipstick (maroon and plum) and 5 shades of nail polish (clear, burgundy, red, toast, and lavender). If she reaches in her drawer without looking and pulls out one of each, what is the probability it will be the plum lipstick and the clear nail polish?

21. In order to win the game, you need to land on the color red on the first spinner and "Move Forward 2" on the second spinner. What is the probability that you will win the game?

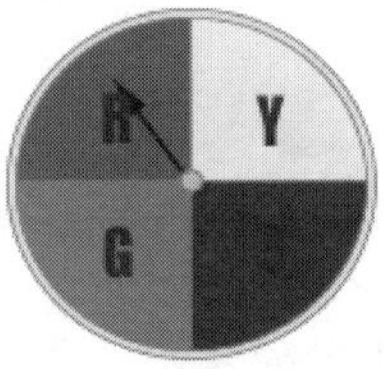

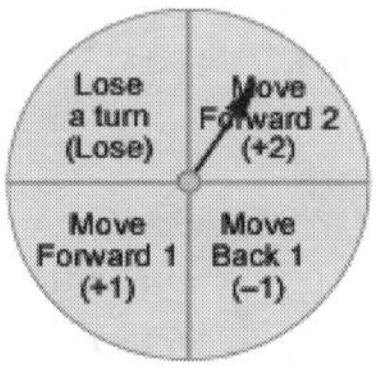

22. Tom has three ties, one green, one red, and one blue, and three pairs of pants, one tan, one black, and one brown. Tom reaches into his closet without looking and picks out a tie and a pair of pants to wear to work. What is the probability that he picks the red tie or the tan pants?

23. Ben tells his friend to get him something from the ice cream parlor. The flavors are vanilla, swirl, and chocolate, and the toppings are nuts and sprinkles. If, Ben's friend randomly chooses one flavor and one topping, what is the probability Ben will get swirl and sprinkles?

24. Suppose you spin the spinner and randomly pick a card. What is the probability of spinning a multiple of 2 or picking a 10?

25. Dave's suitcase contains 4 sweatshirts, one blue, one green, one white, and one yellow and 3 pairs of sweatpants, one black, one red, and one orange. If Dave randomly chooses one sweatshirt and one pair of sweatpants, what is the probability that he will choose the blue sweatshirt or the red sweatpants?

26. Beth has 3 cans of paint, one red, one yellow, and one blue and 5 crayons, one green, one orange, one black, one brown, and one purple. Beth randomly picks one can of paint and one crayon. What is the probability she will pick a blue can of paint or a black crayon?

27. Lisa has 6 cards with the numbers 1, 2, 3, 4, 5 and 6. If she randomly chooses two cards, what is the probability that she will choose a card numbered 4 or a card numbered 5?

28. Jennifer rolls a pair of number cubes with the numbers 1, 2, 3, 4, 5, and 6. What is the probability that the sum of the cubes will be 11 or 12?

29. Bryce's mother bought 2 variety packs of snacks that each contains one bag of chips, one candy bar, and one lollipop. If Bryce randomly chooses one snack from each pack of snacks, what is the probability that he will pick a bag of chips or a lollipop?

30. Sydney rolls two number cubes with number 1, 2, 3, 4, 5 and 6. What is the probability that the sum of the cubes is 8 or one of the cubes is 4?

31. A jar contains a red, a yellow, and a blue marble. Another jar contains a green and a white marble. Find the probability of randomly picking the white marble when the yellow marble has been picked.

32. If Paullina rolls the number cube with numbers 1, 2, 3, 4, 5 and 6 and then spins the spinner, what is the probability that the spinner will land on 12 when a 6 has been rolled?

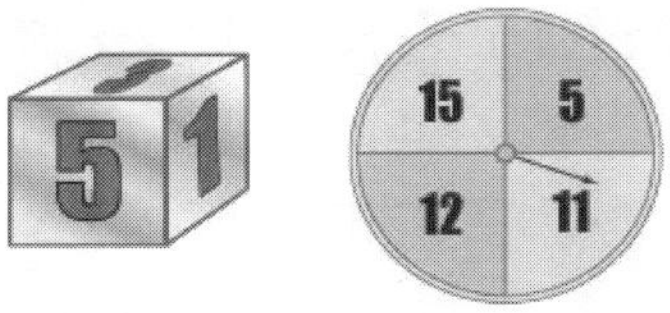

33. If you flip a coin and then spin a spinner with the numbers 1 through 8, what is the probability of the spinner landing on the number 8 when the coin lands on heads?

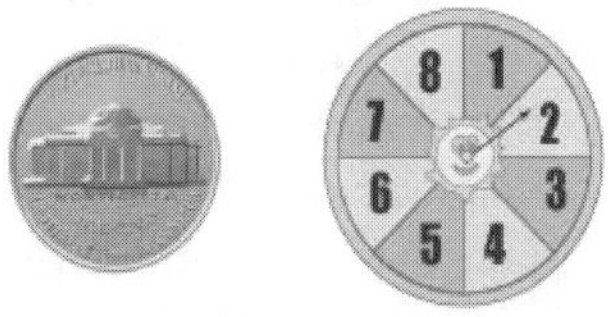

34. Malcolm rolls a letter cube with the letters A, B, C, D, E, and F and a number cube with the numbers 1, 2, 3, 4, 5, and 6. Find the probability that he will roll an even number when a consonant has been rolled.

35. Jose has 3 pairs of pants (tan, white, and black) and 4 ties (green, yellow, purple, and red) in his closet. He reaches in his closet with the lights off and grabs one pair of pants and one tie at random. What is the probability that he will grab the yellow tie when he first grabs the black pants?

36. Billy's teacher has given him poster boards and a box of markers to start his art project. There are 5 different colored poster boards (yellow, orange, green, purple, and red) and six markers (violet, blue, black, pink, brown, and maroon). If Billy chooses one poster board and one marker at random, what is the probability that Billy will pick a pink marker when he has already chosen the yellow poster board?

37. Jamie and her friends are getting ready to play a round of golf. There are two golf clubs (one driver and one wedge) and three golf balls (one red, one yellow, and one blue) in her bag. If Jamie randomly chooses one golf club and one ball from the bag, what is the probability that she will select the red ball when she has already chosen the wedge?

38. The dentist has a one drawer with 2 toothbrushes (one pink and one blue) and another drawer with 5 lollipops (one green, one yellow, one red, one violet, and one orange). The nurse tells you to randomly pick one toothbrush and one lollipop. What is the probability that you will choose the violet lollipop when you first choose the pink toothbrush?

39. A box contains one orange, one yellow, and one blue cube. Another box contains one red, one white, and one green cube. Find the probability of randomly picking the red cube when the blue cube is picked first.

40. Hailey rolls two number cubes with the numbers 1, 2, 3, 4, 5 and 6. What is the probability that Hailey will roll a sum that is even when a 2 has been rolled?

MPA-114: Finding the Odds of Events and Experimental Probability from a Math Model

In previous lessons, you learned how to compute the probability of an event from an experiment using the following formula:

$$\text{Probability (Event)} = \frac{\text{number of outcomes in event}}{\text{number of possible outcomes}}$$

There are times when we do not directly need the probability of an event but are interested in the likelihood that an event will occur compared to the likelihood that it will not occur. The ratios of these two likelihoods are called the **odds**.

Odds in Favor	The ratio of the number of favorable outcomes to the number of unfavorable outcomes.

Odds in favor of event = Number of favorable outcomes : Number of unfavorable outcomes

Consider rolling a number cube. What are the odds in favor of rolling a 6? The odds in favor of rolling a six are 1 : 5, since there is one way to get an outcome of 6, and five ways not to get an outcome of 6.

Odds Against	The ratio of the number of unfavorable outcomes to the number of favorable outcomes.

Odds against an event = Number of unfavorable outcomes : Number of favorable outcomes

For example, there are 6 teams in the Eastern division, and 5 teams in the Western division of a professional sports league. If a team is chosen at random, what are the odds against the choice being from the Eastern division?

The number of favorable outcomes is 6, while the number of unfavorable outcomes is 5, so the odds against choosing an Eastern division team are 5 : 6.

In many experiments, there is no theoretical way to know the likelihood of an event, so we must estimate its value. One way to do this is to find the **experimental probability** of an event.

Experimental Probability	Probabilities that are found by repeating an experiment and calculating the results.

Experimental probability is found by repeating an experiment and calculating the results. If we measure the number of times the event occurs in our experiment, and we know how many times we have run the experiment, then we call the ratio of favorable outcomes to the total number of trials the experimental probability.

$$\text{Experimental probability (event)} = \frac{\text{number of favorable outcomes}}{\text{number of experimental trials}}$$

For example, if we want to estimate what percentage of people in a school with 600 students were born in Texas, we could estimate that number by randomly asking how many students in one of the classes were born in Texas, and then use that answer to make an estimate.

Suppose we ask a class of 30 students to raise their hands if they were born in Texas; this is like running an experiment 30 times. Suppose two were born in Texas. We then have:

$$\text{Experimental probability (born in Texas)} = \frac{2}{30} = \frac{1}{15}$$

This allows us to estimate that $\frac{1}{15}$ or 7% of the school population is from Texas. Because 7% of 600 is $0.07 \times 600 = 42$, we estimate that 42 of the students in the school were born in Texas.

Example 1 A drawer contains 4 pairs of white socks, 3 pairs of black socks, and 5 pairs of brown socks. What are the odds in favor of reaching into the drawer and selecting a brown pair of socks?

$\text{Odds in favor of (brown)} = \underbrace{\text{brown}}_{5} : \text{not brown}$

Step 1: Since there are 5 pairs of brown socks in the drawer and you are equally likely to choose any pair, the total number of favorable outcomes is 5.

$\text{Odds in favor of (brown)} = \underbrace{\text{brown}}_{5} : \underbrace{\text{not brown}}_{7}$

Step 2: Since there are 7 other pairs of socks in the drawer, the total number of unfavorable outcomes is 7. The odds in favor of selecting a pair of brown socks is 5 : 7.

Answer: The odds in favor of selecting a pair of brown socks is 5 : 7.

Example 2 Ben has a bag of marbles containing 2 yellow, 2 red, and 1 green marble. If he chooses one marble at random, what are the odds against choosing a red marble?

Odds against (red) = number not red : number red

Step 1: To find the odds against choosing a red marble, find the ratio of the number of marbles that are not red to the number of marbles that are red.

$\text{Odds against (red)} = \underbrace{\text{number not red}}_{3} : \text{number red}$

Step 2: There are 3 marbles that are not red.

$\text{Odds against (red)} = \underbrace{\text{number not red}}_{3} : \underbrace{\text{number red}}_{2}$

Step 3: There are 2 red marbles. Therefore, the odds against choosing a red marble are 3 : 2.

Answer: The odds against choosing the red marble are 3 : 2.

Example 3 If 6 potential guest speakers for graduation are female and 9 are male, what are the odds in favor of getting a female guest speaker at graduation? Assume that the choice of speaker is random and each is equally likely.

Odds in favor of (female) = $\underbrace{\text{female}}_{6}$: not female

Step 1: Since there are 6 females, the number of favorable outcomes is 6.

Odds in favor of (female) = $\underbrace{\text{female}}_{6} : \underbrace{\text{not female}}_{9}$

Step 2: Since there are 9 males, the number of unfavorable outcomes is 9.

Odds in favor of (female) = $\underbrace{\text{female}}_{6} : \underbrace{\text{not female}}_{9}$

$= 2 : 3$

Step 3: The odds of 6 : 9 can be reduced, since $\frac{6}{9} = \frac{2}{3}$.

Therefore, we have 6 : 9 = 2 : 3

Answer: The odds in favor of getting a female speaker for graduation are 2 : 3.

Example 4 A class of 24 students was asked to name their favorite type of food. The results appear in the table below. What is the experimental probability that a randomly selected student will say his or her favorite food is a hamburger?

Favorite Food	Number of Students
Pizza	12
Hamburger	8
Chicken Nuggets	2
Hot Dog	2

Total surveyed = 12 + 8 + 2 + 2 = 24

Step 1: The total number of people surveyed is 24.

Number who favor hamburgers = 8

Step 2: The number of students who favor hamburgers is 8.

Probability (hamburger) = $\frac{8}{24} = \frac{1}{3}$

Step 3: The experimental probability of randomly selecting a student who prefers hamburgers is $\frac{8}{24}$ or $\frac{1}{3}$.

Answer: The experimental probability of randomly selecting a student who prefers hamburgers is $\frac{1}{3}$.

Example 5 A high school offers two foreign language courses to its students (Spanish and French). Each student must study at least one language during high school. The principal randomly selected 188 students and asked them to choose which language they would like to study. What is the experimental probability that a randomly selected student chose French as his or her only foreign language course?

Language	Number of Students
French	80
Spanish	98
Other	10

Total surveyed = 80 + 98 + 10 = 188

Step 1: The total number of students surveyed is 188.

Number who chose French = 80

Step 2: The total number of students who chose French is 80.

$\text{French} = \frac{80}{188} = \frac{20}{47}$

Step 3: The experimental probability of selecting a student who chose French is $\frac{80}{188}$ or $\frac{20}{47}$.

Answer: The experimental probability that a student selected French is $\frac{20}{47}$.

Problem Set

1. There are 3 red, 4 blue and 1 green gumdrops in a bag. If one gumdrop is selected at random, what are the odds in favor of selecting a green?

2. There are 33 students who want to participate in a bike rodeo. Each student has been assigned a number from 1 to 33. One number will be randomly selected to decide what person will be the first in line. What are the odds in favor of selecting the number 22?

3. Mark has collected a maple, an oak, a beech and a dogwood leaf in a bag for his school project. If he randomly selects one leaf from the bag, what are the odds in favor of selecting a maple leaf?

4. The brass section of the band is having an election to select a leader. There is 1 tuba, 1 trombone and 1 trumpet player on the ballot. If a voter randomly selects a leader from the ballot, what are the odds in favor of selecting a trumpet player as the leader?

5. There are 15 almonds, 8 cashews and 1 pecan in a nut dish. If one nut is randomly selected, what are the odds in favor of selecting a pecan?

6. The names of the nine planets in the solar system (Mercury, Venus, Earth, Mars, Jupiter, Saturn, Uranus, Neptune and Pluto) are posted on Mrs. Smith's bulletin board. If one planet is selected at random for further study, what are the odds in favor that it will begin with the letter S?

7. Beverly's tape collection has 1 jazz, 6 classical, 7 country and 5 rap tapes. If one tape is selected at random, what are the odds in favor of selecting a jazz tape?

8. Janet placed 1 clam, 4 scallop and 7 oyster shells in her seashell box. If she randomly selects one shell, what are the odds against her selecting a clam shell?

9. The math lab has models of three-dimensional figures stored in a box. There is 1 cylinder, 8 prisms, 9 pyramids, and 4 cubes in the box. If one model is randomly selected, what are the odds against selecting a cylinder?

10. A box of donuts has 1 glazed, 6 chocolate and 5 jelly-filled donuts. If one donut is randomly selected, what are the odds against selecting a glazed donut?

11. A spinner is divided evenly into 8 sections. Three of the sections have stars, 4 of the sections have stripes and 1 section is solid blue. What are the odds against the spinner landing on a section that is solid blue?

12. A gumball machine has 8 blue, 9 green, 12 red and 1 yellow gumballs. If the gumballs drop out of the machine randomly, what are the odds against the next gumball being a yellow one?

13. Michael places 8 cards with hearts, 1 with a club, and 9 with diamonds randomly face-down on a table. If he chooses one card at random, what are the odds against choosing a club?

14. A six-sided cube is tossed. On 3 of the faces there are odd numbers, 2 of the faces have even numbers, and 1 face is blank. What are the odds against tossing the blank face?

15. Paul's video collection consists of 1 movie, 1 cartoon and 8 documentaries. If he randomly selects a video, what are the odds against selecting a movie?

16. There are 2 A's, 7 B's and 3 C's printed on tiles that are randomly arranged. If one tile is selected, what are the odds in favor of selecting the letter C?

17. On a shelf, there are 10 cans of tomato soup, 5 cans of clam chowder, and 10 cans of chicken noodle soup randomly arranged. If Michelle selects one can without looking, what are the odds in favor of selecting a can of clam chowder?

18. The numbers 1 through 10 are painted on small balls that are placed in a spinner. If one ball is randomly selected after the spinner has spun the balls, what are the odds in favor it will be a multiple of 5?

19. In a school cafeteria cooler, there are 5 chocolate, 6 skim, and 11 white whole milk pints. If they are randomly arranged and Tommy selects one without looking, what are the odds in favor of Tommy selecting a pint of skim milk?

20. Dr. Jones gives each patient a tube of toothpaste. There are 17 peppermint, 10 wintergreen, and 8 spearmint flavored tubes in a bag. If the next tube is selected at random, what are the odds in favor of selecting wintergreen?

21. A bottle of chewable vitamins contains 10 lions, 10 tigers, 18 elephants, 12 giraffes, and 8 rhino-shaped vitamins. If a vitamin is randomly selected, what are the odds in favor of selecting a tiger-shaped vitamin?

22. Mr. King has randomly arranged the puzzles at the math station. There are 5 decimal, 13 probability, and 8 geometry puzzles. If one puzzle is selected, what are the odds in favor of it being a geometry puzzle?

23. Sixty students were surveyed about their favorite subject. Eighteen students selected math, 15 selected language arts, 12 selected history and 15 chose science. Find the experimental probability that a student's favorite subject is math.

24. A survey asked fifty students to name how many hours of sleep they averaged each night. The results were recorded in the table below. Find the experimental probability of a student averaging 8 hours of sleep each night.

Hours of Sleep	Number of students
6	4
7	12
8	18
9	8
10	8

25. The school librarian asked forty students to name the most recent type of reading material that they had read. The results were recorded in the table below. Find the experimental probability of a student having last read a newspaper.

Type of reading material	Number of students
Book	20
Newspaper	6
Magazine	12
Comic Book	2

26. The 26 students in Mrs. Smith's homeroom have earned a reward day. They have a choice of going to the movies, a sporting event, or a boat trip. Seven students voted for the movie, 10 voted for the boat trip and 9 voted for the sporting event. Find the experimental probability that the class will go on a boat trip.

27. A pizza restaurant surveyed 100 customers asking them to name their favorite pizza. Thirty stated pepperoni, 22 stated sausage, 8 stated plain cheese, 12 said veggie, and 28 voted for supreme. Find the experimental probability of supreme pizza being the favorite pizza.

28. The owners of a movie theatre asked 70 teenagers to name their favorite type of movie. The results were recorded in the table below. Find the experimental probability that a teenager's favorite type of movie is comedy.

Type of Movie	Number of votes
Comedy	15
Horror	15
Drama	12
Action	28

29. In art class, Mr. Stephens surveyed 30 students and asked them to name their favorite type of art activity. The results were recorded in the table below. Find the experimental probability of a student selecting pottery making.

Art Activity	Number of votes
Watercolor painting	7
Pottery making	15
Charcoal drawing	5
Oil pastels	3

30. The math department asked 42 students to name their favorite unit of study in math class. The results were recorded in the table below. Find the experimental probability that a student's favorite unit of study is geometry.

Unit of study in Math	Number of votes
Algebra	8
Calculus	8
Geometry	12
Statistics	14

31. During a unit of study on nutrition, the students in the seventh grade were asked to name the last piece of fruit they had eaten. The results were recorded in the table below. What is the experimental probability of a student not choosing an apple or banana?

Fruit	Number of students
Apple	20
Pear	11
Banana	14
Grapes	15

32. The computer club president asked each member of the club to select their favorite color to use for the lettering on the new school web page. Three members chose green, 2 chose red, 7 chose blue, and 4 chose black. What is the experimental probability of a member not choosing green or blue?

33. The recreation department asked 50 students to name their favorite type of game. The results were recorded in the table below. What is the experimental probability for students not selecting computer or card games?

Favorite type of game	Number of votes
Board games	13
Computer games	9
Sports	21
Card games	7

34. A high school librarian asked the last 40 students who visited the library to name the primary reason for their visit. Ten students came to study, 8 came to read, another 8 came to use the computers, and 14 came to do research for a project. What is the experimental probability of students not selecting reading and research?

35. The school cafeteria manager surveyed 120 students about their favorite item on the menu. Forty-two said pizza, 14 said hamburger, 30 stated tacos, and 34 chose chicken. What is the experimental probability of a student not selecting a hamburger or tacos?

36. Mrs. Michaels asked all of the students in the eighth grade to select the activity that they would like to participate in for the end of the year picnic. The results were recorded in the table below. What is the experimental probability of not selecting canoeing or basketball?

Activity	Number of votes
Canoeing	32
Dancing	15
Basketball	18
Swimming	45

37. The after-school program leader, Mr. Bailey, surveyed parent volunteers to see which subject they would prefer to tutor. Fifteen parents said English, 20 parents said Math, 10 said History, and 11 wanted to tutor Science. What is the experimental probability of a parent not selecting History or Science?

38. The school improvement team surveyed students to find out what students wanted to do first in regards to making changes at school. The results were recorded in the table below. What is the experimental probability of a student not selecting new lockers or cafeteria tables first?

Improvement	Number of students
Painting the gym	46
New lockers	68
New cafeteria tables	32
Add more computers	22

39. Parents of the members of the school marching band were asked to select one item they would be willing to do. The results of this survey showed 42 parents would work the concession stand at football games, 32 would be willing to chaperone on trips, 24 parents would help with fundraising, and 2 would help repair uniforms. What is the experimental probability of a parent not selecting to be a chaperone on trips or repairing uniforms?

40. Mrs. Harp, the school guidance counselor, sent a questionnaire to each homeroom to survey students about their after-school activities that occurred before dinnertime. The results were recorded in the table below. What is the experimental probability of a student not selecting chores at home, watching T.V., or working?

Activity	Number of students
Chores at home	22
Watching T.V.	33
Study/homework	50
Sports	40
Working	17

MPA-115: Finding the Volumes of Prisms, Cylinders, Pyramids, and Cones Using Models

The **volume** of a solid figure is the amount of space that it occupies measured in cubic units. We can compute the volume using standard formulas.

The areas of prisms and cylinders are similar. The volume of a prism or a cylinder is computed by multiplying the area of the base of the object by its height.

Volume	The amount of space that a solid figure occupies measured in cubic units.

Volume of a Prism or Cylinder

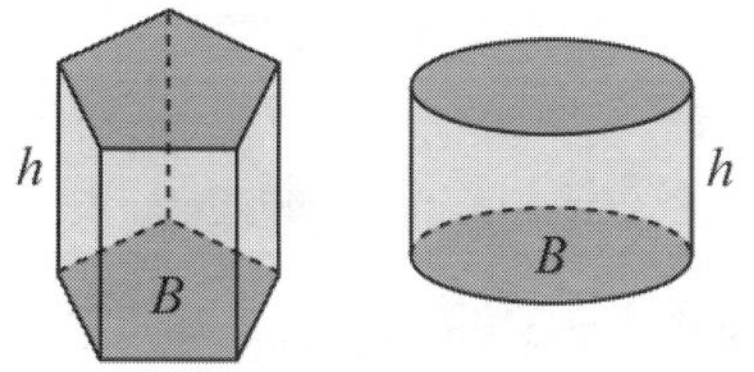

$V = Bh$, where B is the area of the base and h is the height.

In the case where we know the radius of the circle that forms the base of the cylinder, we know that $B = \pi \times r^2$ where r is the radius of the base of the cylinder.

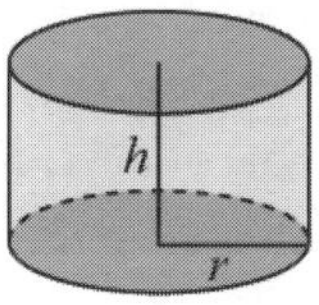

$V = \pi \times r^2 \times h$, **where r is the radius of the base and h is the height of the cylinder.**

Measuring the volume of a pyramid or a cone is similar to measuring the volume of a prism or a cylinder. The volume of a pyramid or a cone is computed by multiplying one-third times the area of the base of the object by its height.

Volume of a Pyramid

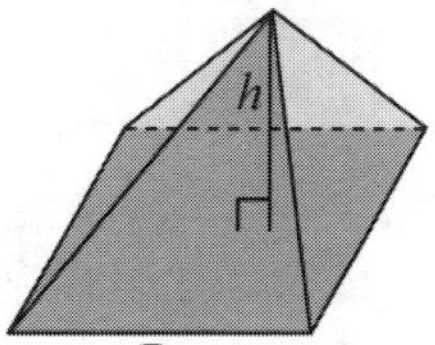

$V = \frac{1}{3} \times Bh$, **where B is the area of the base and h is the height.**

In cases where we know the radius of the circle that forms the base of a cone, we know that $B = \pi \times r^2$ where r is the radius of the base of the cone.

Volume of a Cone

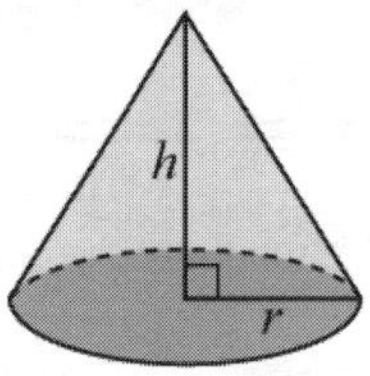

$V = \frac{1}{3}\pi \times r^2 \times h$, **where** r **is the radius of the base and** h **is the height of the cone**

Example 1 Find the volume of the rectangular prism shown below.

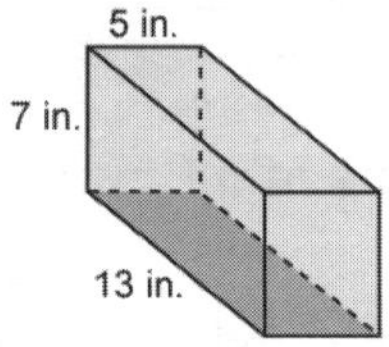

$V = Bh$	**Step 1:** The volume of the rectangular prism is the area of the base times the height.
$B = \ell \times w$ $= 13 \times 5$ $= 65 \text{ in.}^2$	**Step 2:** In this case, the base is a rectangle that is 13 inches by 5 inches. This makes the area of the base 65 square inches.
$V = 65 \times 7$ $= 455 \text{ in.}^3$	**Step 3:** The height of the prism is 7 inches. Therefore, the volume is 455 cubic inches.

Answer: The volume of the rectangular prism is 455 cubic inches.

Example 2 Find the volume of the triangular prism shown below.

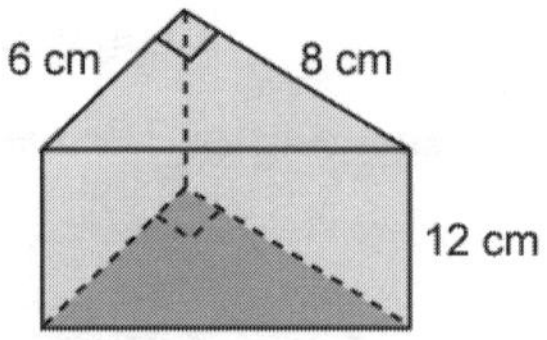

$V = Bh$	**Step 1:** The volume of the triangular prism is the area of the base times the height.
$B = \frac{1}{2} \times b \times h$ $= \frac{1}{2} \times 6 \times 8$ $= 24 \text{ cm}^2$	**Step 2:** In this case, the base is a right triangle that has a base of 6 centimeters and a height of 8 centimeters. The area of the base is 24 square centimeters.

$V = 24 \times 12$ $= 288 \text{ cm}^3$	**Step 3:** The height of the prism is 12 centimeters. Therefore, the volume is 288 cubic centimeters.

Answer: The volume of the triangular prism is 288 cubic centimeters.

Example 3 A soft drink can is a cylinder with a diameter of 10 centimeters and a height of 25 centimeters. What is the volume of the can?

$V = Bh$	**Step 1:** The volume of the cylinder is the area of the base times the height.
$r = \frac{1}{2} \times 10$ $= 5 \text{ cm}$	**Step 2:** The radius of the base is half of its diameter of 10 centimeters, or 5 centimeters.
$B = \pi \times 5^2$ $= 25\pi \text{ cm}^2$	**Step 3:** The area of the base is $\pi \times r^2$, where the radius is 5 centimeters. The area of the base is 25π square centimeters.
$V = 25\pi \times 25$ $= 625\pi$ $\approx 1,963 \text{ cm}^3$	**Step 4:** The volume of the can is the area of the base 25π times the height 25, or 625π cubic centimeters. The value 3.14 can be used to approximate π and the volume.

Answer: The volume of the cylinder is approximately 1,963 cubic centimeters.

Example 4 Find the volume of the square pyramid shown below.

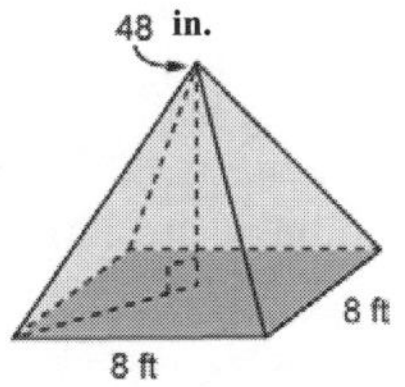

$V = \frac{1}{3}Bh$	**Step 1:** The volume of a pyramid is $\frac{1}{3}$ times the area of the base times the height of the pyramid.
$B = \ell \times w$ $= 8 \times 8$ $= 64 \text{ ft}^2$	**Step 2:** The base of the pyramid is a square with a side length of 8 feet. This makes the area of the base 64 square feet.
$h = 48 \div 12$ $= 4 \text{ ft}$	**Step 3:** The height of the pyramid is 4 feet since there are 4 feet in 48 inches.

$V = \frac{1}{3} \times 64 \times 4$

$= \frac{256}{3}$

$\approx 85 \text{ ft}^3$

Step 4: The volume of the square pyramid is approximately 85 cubic feet.

Answer: The volume of the square pyramid is approximately 85 cubic feet.

Example 5 Find the volume of the cone shown in the figure below.

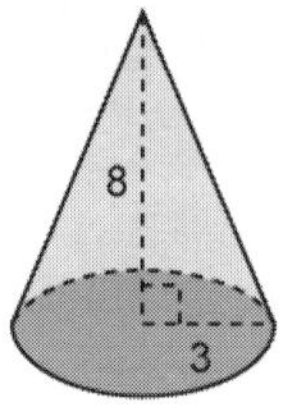

$V = \frac{1}{3}Bh$

Step 1: The volume of a cone is $\frac{1}{3}$ times the area of the base times the height of the cone.

$B = \pi \times r^2$

$= \pi \times 3^2$

$= 9\pi \text{ units}^2$

Step 2: The area of the base is πr^2, or 9π square units.

$V = \frac{1}{3} \times 9\pi \times 8$

$= 24\pi$

$\approx 75 \text{ units}^3$

Step 3: The volume of the cone is approximately 75 cubic units. The value 3.14 can be used to approximate π and the volume.

Answer: The volume of the cone is approximately 75 cubic units.

Problem Set

Solve:

1. Find the volume of a rectangular prism with a length of 7 centimeters, a width of 5 centimeters, and a height of 8 centimeters.
2. Find the volume of a rectangular prism with a height of 6 inches, a width of 4 inches, and a length of 5 inches.
3. Find the volume of a rectangular prism that has a height of 5 meters, a width of 3 meters, and a length of 6 meters.
4. Find the volume of a rectangular prism with a length of 12 centimeters, a width of 10 centimeters, and a height of 15 centimeters.
5. Find the volume of a rectangular prism with a height of 6 feet, a width of 7 feet, and a length of 8 feet.
6. Find the volume of a rectangular prism with a length of 9 yards, a width of 9 yards, and a height of 7 yards.

7. Find the volume of a rectangular prism with a length of 20 meters, a width of 30 meters, and a height of 40 meters.

8. Find the volume of the triangular prism.

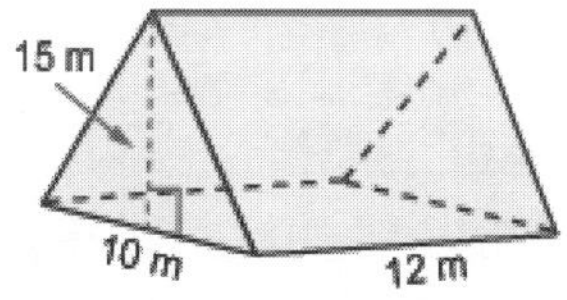

9. Find the volume of the triangular prism.

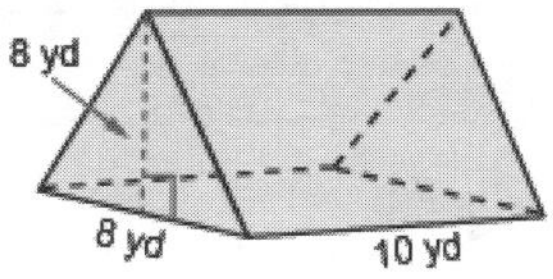

10. Find the volume of the triangular prism.

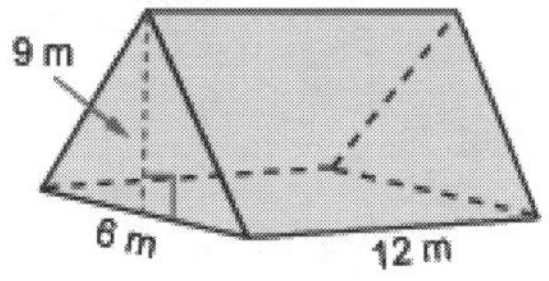

11. Find the volume of the triangular prism.

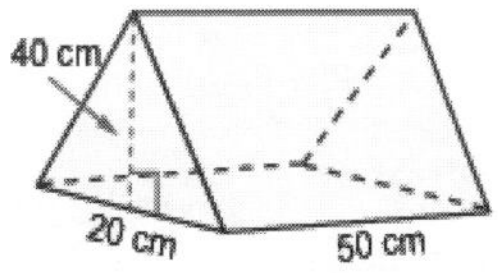

12. Find the volume of the triangular prism.

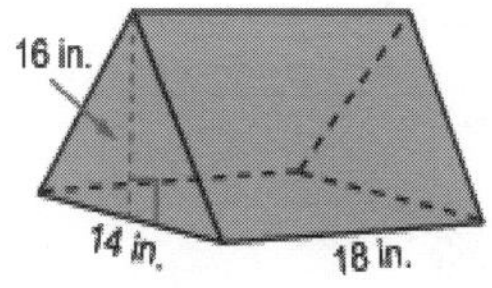

13. Find the volume of the triangular prism.

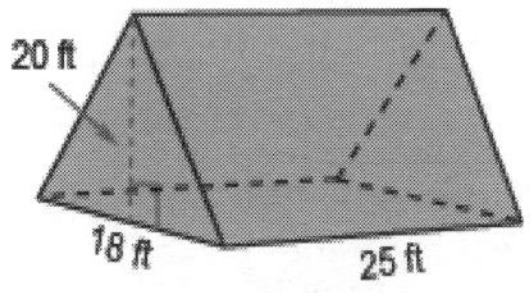

14. Find the volume of the triangular prism.

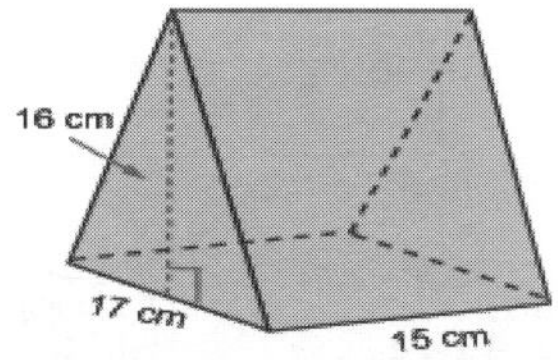

15. Find the volume of the triangular prism.

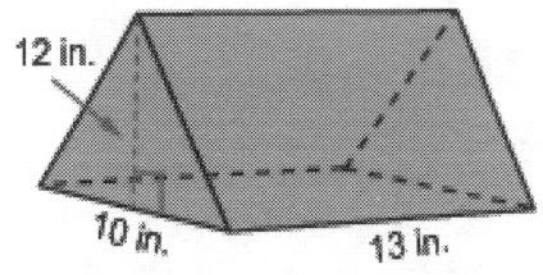

16. A soup can has a radius of 3 inches and a height of 7.5 inches. What is the volume of the soup can? Round the volume to the nearest tenth. (Use $\pi = 3.14$)

17. A coffee can has a radius of 5.5 inches and a height of 10 inches. What is the volume of this can? Round the volume to the nearest tenth. (Use $\pi = 3.14$)

18. A cardboard cylinder that holds documents such as blueprints and photographs has a radius of 4.5 centimeters and a height of 12.5 centimeters. What is the volume of this cylinder? Round the volume to the nearest tenth. (Use $\pi = 3.14$)

19. A trash can has a radius of 0.5 feet and a height of 3 feet. What is the volume of the trash can? Round the volume to the nearest tenth. (Use $\pi = 3.14$)

20. A thermos has a radius of 4 inches and a height of 9.5 inches. What is the volume of the thermos? Round the volume to the nearest tenth. (Use $\pi = 3.14$)

21. A bank in the shape of a cylinder that has a radius of 2.5 inches and a height of 7 inches. What is the volume of the bank? Round the volume to the nearest tenth. (Use $\pi = 3.14$)

22. A cylinder has a radius of 4 inches and a height of 10.5 inches. What is the volume of the cylinder? Round the volume to the nearest tenth. (Use $\pi = 3.14$)

23. Find the volume of the square pyramid.

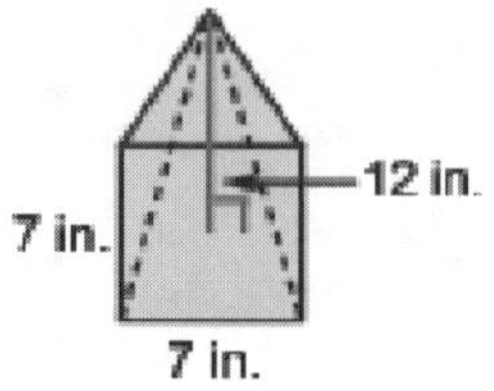

24. Find the volume of a rectangular pyramid that has a length of 10 feet, a width of 5 feet, and a height of 15 feet.

25. Find the volume of a triangular pyramid with the following dimensions.

Base of triangle: 7 inches

Height of triangle: 6 inches

Height of pyramid: 8 inches

26. Find the volume of a square pyramid that has sides of $5\frac{1}{2}$ yards and a height of $10\frac{1}{2}$ yards. Round the volume to the nearest tenth.

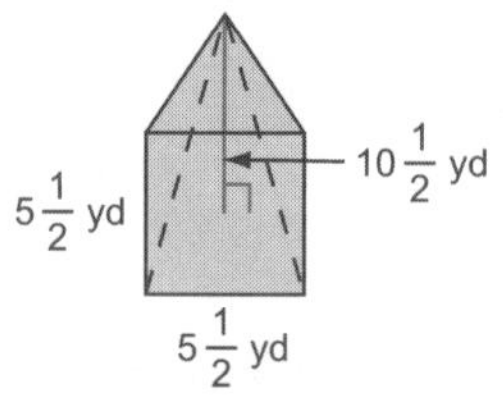

27. Find the volume of a rectangular pyramid that has the following dimensions. Round the volume to the nearest tenth.

Length of rectangle: $6\frac{1}{4}$ meters

Width of rectangle: 3 meters

Height of pyramid: 9 meters

28. Find the volume of a rectangular pyramid with the following dimensions. Round the volume to the nearest tenth.

Length of rectangle: 11 kilometers

Width of rectangle: $12\frac{3}{4}$ kilometers

Height of rectangle: 15 kilometers

29. Find the volume of a triangular pyramid with the following dimensions.

Base of triangle: 4 inches

Height of triangle: 6 inches

Height of pyramid: 9 inches

30. Find the volume of the square pyramid.

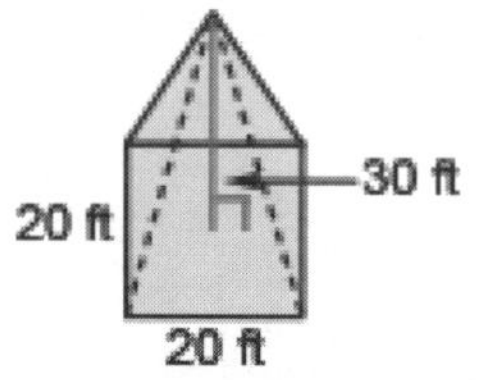

31. Find the volume of a cone that has a radius of 7 inches and a height of 15 inches. Round the volume to the nearest tenth. (Use $\pi = 3.14$)

32. Find the volume of a cone with a radius of 9 centimeters and a height of 10 centimeters. Round the volume to the nearest tenth. (Use $\pi = 3.14$)

33. Find the volume of a cone with a radius of 8 feet and a height of 9 feet. Round the volume to the nearest tenth. (Use $\pi = 3.14$)

34. Find the volume of a cone with a radius of 6 yards and a height of 8 yards. Round the volume to the nearest tenth. (Use $\pi = 3.14$)

35. Find the volume of a cone with a radius of 3 meters and a height of 6 meters. Round the volume to the nearest tenth. (Use $\pi = 3.14$)

36. Find the volume of a cone with a radius of 5 centimeters and a height of 12 centimeters. (Use $\pi = 3.14$)

37. Find the volume of a cone with a radius of 10 feet and a height of 21 feet. (Use $\pi = 3.14$)

38. Find the volume of a cone with a radius of 15 inches and a height of 25 inches. Round the volume to the nearest tenth. (Use $\pi = 3.14$)

39. Find the volume of a cone with a radius of 14 yards and a height of 27 yards. Round the volume to the nearest tenth. (Use $\pi = 3.14$)

40. Find the volume of a cone with a radius of 30 inches and a height of 40 inches. (Use $\pi = 3.14$)

MPA-116: Solving Real-Life Problems by Using Guess-and-Check and Working Backwards

Mathematical problems can be solved in a number of ways. One general strategy for solving problems was developed by the Hungarian mathematician George Polya. His strategy has four parts and is named Polya's Four-Step Problem Solving Process.

Polya's Four-Step Problem Solving Process

1. Understand the problem
2. Devise a plan
3. Carry out the plan
4. Look back

Each of these steps can be applied to all mathematical problems you solve. Below is a brief summary of each step.

Understand the problem

In order to understand a problem thoroughly, it may be necessary to read it several times and look up the meaning of unfamiliar words or procedures. It might be helpful to restate the problem or to reorganize the information. Sometimes a problem contains extra information that should be disregarded.

Devise a plan

Some of the most common plans or techniques to solve problems include looking for a pattern, drawing and labeling a diagram, making an orderly list, relating the problem to a similar problem, working backwards, using guess-and-check, writing and solving an equation, and applying a formula. In many cases, it is possible to use more than one plan to solve a problem.

Carry out the plan

Follow your plan to solve the problem. Several attempts may be necessary. Be careful to avoid computational errors.

Look back

Check your answer thoroughly. Make sure the answer gives a reasonable solution to the problem. The problem is solved only when all the questions are answered. Make sure the proper units are used for numerical answers.

This lesson focuses on the problem-solving techniques guess-and-check and working backwards. Working backwards is a good method to use when a problem describes a process that can be undone. Guess-and-check is best used when the problem gives only a small number of possible answers. With guess-and-check, we guess an answer that has the properties required by the problem, and then we check to see if it is correct.

Example 1 Before Jim got his first paycheck, his employer told him that \$20 was taken out for social security and \$18 was taken out for income taxes. If his check was \$182, how much money did he earn?

Reread the problem	**Step 1:** In this case, Jim does not know how much money he earned. He only knows what he received after paying social security and taxes. He does know that \$20 was taken out of his check for social security and \$18 was taken in taxes.
Undo the subtraction: Wages $= 182 + 18 + 20 = 220$	**Step 2:** Since money was subtracted from his wages twice, he can find out how much he made by adding these amounts back to his check.
Look back and check: Wages $= 220 - 20 - 18 = 182$	**Step 3:** If his wages were \$220, subtracting \$20 for social security leaves \$200. Now subtract \$18 for taxes, which leaves \$182.

Answer: Jim earned \$220.

Example 2 Brenda invested money in the stock market. In the first month her money earned \$12, the next month it lost \$8, and in the third month it made \$40. If Brenda now has \$494, how much did she invest in the beginning?

Reread the problem	**Step 1:** In this case, you know that whatever amount Brenda started with, \$12 was added, then \$8 was subtracted, and finally \$40 was added. These adjustments left Brenda with \$494.
Undo the adjustments: $494 - 40 = 454$ $454 + 8 = 462$ $462 - 12 = 450$	**Step 2:** Undo these adjustments to her investment by starting with \$494, subtracting \$40, adding \$8, and then subtracting \$12.
Look back and check: $450 + 12 = 462$ $462 - 8 = 454$ $454 + 40 = 494$	**Step 3:** If Brenda started with \$450 and made \$12, lost \$8, and made \$40, she would have \$494.

Answer: Brenda invested \$450.

Another plan for solving problems is guess-and-check. Guess-and-check is best used when the problem has only a small number of possible answers. We guess an answer that has the properties required by the problem and then check to see if it is correct. Even when the guess is not correct, it may yield some information that will make the next guess better.

Example 3 Taylor has \$0.97 to spend on candy. A piece of licorice costs \$0.07 and a piece of chocolate costs \$0.11. How many pieces of each candy can Taylor buy?

Reread the problem.

Step 1: We want to know how many pieces of licorice and how many pieces of chocolate Taylor can buy with \$0.97.

Guess #	Amount of licorice	Amount of chocolate	Total spent
1	$10 \times 0.07 = \$0.70$	$2 \times 0.11 = \$0.22$	\$0.92

Step 2: Let's begin by guessing that he bought 10 pieces of licorice and 2 pieces of chocolate. The total is too low.

Guess #	Amount of licorice	Amount of chocolate	Total spent
1	$10 \times 0.07 = \$0.70$	$2 \times 0.11 = \$0.22$	\$0.92
2	$11 \times 0.07 = \$0.77$	$2 \times 0.11 = \$0.22$	\$0.99

Step 3: Next, let's guess 11 pieces of licorice and 2 pieces of chocolate. The total is too high.

Guess #	Amount of licorice	Amount of chocolate	Total spent
1	$10 \times 0.07 = \$0.70$	$2 \times 0.11 = \$0.22$	\$0.92
2	$11 \times 0.07 = \$0.77$	$2 \times 0.11 = \$0.22$	\$0.99
3	$6 \times 0.07 = \$0.42$	$5 \times 0.11 = \$0.55$	\$0.97

Step 4: Next, let's guess 6 pieces of licorice and 5 pieces of chocolate. The total is \$0.97.

Look back and check:

$\$0.42 + \$0.55 = \$0.97$

Step 5: Now look back and check your answer.

Answer: Taylor can buy 6 pieces of licorice and 5 pieces of chocolate.

Example 4 Three consecutive positive integers have a sum of 27. Find these integers.

Reread the problem.

Step 1: We are looking for 3 consecutive numbers that add up to 27.

Guess #	1st #	2nd #	3rd #	Sum of Numbers
1	9	10	11	30

Step 2: Let's guess that the smallest integer is 9. This makes the three consecutive integers 9, 10, and 11. This sum is too large.

Guess #	1st #	2nd #	3rd #	Sum of Numbers
1	9	10	11	30
2	7	8	9	24

Step 3: Since our sum was too large, start with a smaller number. Let's guess 7. That makes the three consecutive integers 7, 8, and 9. This sum is too small.

Guess #	1st #	2nd #	3rd #	Sum of Numbers
1	9	10	11	30
2	7	8	9	24
3	8	9	10	27

Step 4: Since the last guess was too low, try the smallest integer, 8. This gives us three consecutive integers as 8, 9, 10. This is the correct sum.

$8 + 9 + 10 = 27$

Step 5: Now look back and check your answer.

Answer: Three consecutive integers whose sum is 27 are 8, 9, and 10.

Example 5 The high school cheerleading squad is having a fundraiser by selling boxes of doughnuts. Normally the squad charges $5.00 per dozen and sells 1,000 dozen doughnuts. For every $0.25 the price is raised, the squad will lose 30-dozen doughnut sales. What should they charge to make the most money?

Reread the problem.

Step 1: If the price is $5.00 per dozen, the squad will sell 1,000 dozen and make $5,000. If they raise the price to $5.25, they will only sell 970 dozen doughnuts, but they will make $5,092.50. This is more money, so they should raise the price.

Guess #	Price	Sold	Income
1	5.25	970	5,092.50
2	5.50	940	5,170.00

Step 2: What if the price goes to $5.50? Now they will only sell 940 dozen doughnuts. Notice each time the price goes up $0.25 the number of dozens sold goes down 30. In this case, they will make $5,170. Again, there is an increase.

Guess #	Price	Sold	Income
3	5.75	910	5,232.50
4	6.00	880	5,280.00

Step 3: Let's keep going. If the price is $5.75 they will sell 910 dozen doughnuts. This gives an income of $5,232.50. Now try $6.00 per dozen. They will only sell 880 dozen for an income of $5,280 — still an increase.

Guess #	Price	Sold	Income
5	6.25	850	5,312.50
6	6.50	820	5,330.00

Step 4: If the price goes to $6.25 they will sell 850 dozen for an income of $5,312.50. Another increase of $0.25 takes the price to $6.50 and the sales to 820 dozen for an income of $5,330.

Guess #	Price	Sold	Income
7	6.75	790	5,332.50
8	7.00	760	5,320.00

Step 5: Another increase brings the price to $6.75 with sales of 790 dozen. At this level they will make $5,332.50. The next increase to $7.00 cuts sales to 760 dozen giving them an income of $5,320 — a decrease. We see the greatest income when the price is $6.75 per dozen.

Answer: The most money can be made if the price is raised to $6.75 per dozen. At that price the squad will sell 790 dozen and make $5,332.50.

Problem Set

Solve:

1. Danielle sells lemonade. She sold 6 cups between 2:00 P.M. and 3:00 P.M. She sold another 10 cups between 3:00 P.M. and 4:00 P.M. She now has 5 cups left. How many cups did she start with?

2. Don collects miniature cars. For his birthday his uncle gave him 2 cars and his father gave him 6 cars. How many cars did he have before he received his birthday gifts if he now has 31 cars?

3. James drives to work. When he leaves his home, he stops for coffee before going to work. It takes him 8 miles to get to the coffee shop. It then takes him another 12 miles to get to work. Today when he got to work, his speedometer read 12,354 miles. What did his speedometer read before he left his house?

4. Amy is selling Mother's Day gift baskets. She makes all her baskets before she starts selling them. The first week she sold 7 baskets. The second week she sold 12 baskets. The third week she sold 15 baskets. The week before Mother's Day Amy had 6 baskets left. How many baskets did she make in the beginning?

5. Tammy is buying Christmas gifts for her family today. She first went to a home store and spent $140. Tammy then went to a clothing store and purchased $77 in gifts. She now has $35 left in her purse. How much money did she have before she began her day shopping?

6. Soledad is thinking of a number. If she subtracts 2, adds 9, and then multiplies by 3, she gets 36. What number is Soledad thinking of?

7. The public transportation bus was holding a number of people when it left Sunset Street. At the first stop, 3 more people got on. The second stop 5 people got off. The third stop 1 person got off and 2 people got on. If there were 17 people left on the bus, how many people were on the bus when it left Sunset Street?

8. On Tuesday, Jonah spent $7 on food. Taylor gave him $15 on Wednesday which left him with $18. How much money did Jonah have to start?

9. Tamara had bags of candy to divide up equally to her class. She gave away 5 bags to her friends that were not in the class. Tamara then gave 2 bags to each of her 12 classmates. When Tamara got home, she had 7 bags left. How many bags of candy did Tamara begin with?

10. Tony is saving money to go on vacation. At the end of March, he had doubled the amount of money in his account. At the end of April, Tony saved an additional $55 and added it to his account. Tony now has $175 in his savings account. How much money did Tony start with in his savings account?

11. Dominique is monitoring her dog Lucky's weight. The first week Lucky lost 5 pounds. The second week she gained 2 pounds. The third week she lost 4 pounds. Lucky now weighs 50 pounds. How much did Lucky initially weigh?

12. Joseph is thinking of a number. If he multiplies by 5, adds 7, then subtracts 3 he gets 19. What number is Joseph thinking of?

13. The Gators are playing in a football game. During the second play in the second quarter, they gained 7 yards, the third play they lost 3 yards, and the fourth play they gained 11 yards. If they were on the 50 yard line after the fourth play, what yard line were they on after the second play?

14. Thomas went to the store and purchased apples, oranges, lettuce, and tomatoes. He spent $4 on apples, $5 on oranges, and $3 on lettuce. His bill was $15. How much did Thomas spend on tomatoes?

15. Sandy gave Jason $15. Jason spent $25 at the store. Jason then gave Sandy $3. If Jason has $12 left in his pocket, how much did Jason have before Sandy gave him money?

16. JoAnn shops at a clothing store that gave her a $10 discount if she spends over $50. She purchased 3 items from the list below. How many of each item did she purchase if she spent $70?

Item	Cost
Dress	$30
Sweater	$25

17. Jessica is 4 years older than Brandon. Brandon is twice as old as Justin. The sum of their ages is 39. How old is Jessica?

18. Jose is 5 times as old as Juan. Juan is half as old as Hector. The sum of their ages is 24. How old is Jose?

19. In a basketball game, a ball in the basket is worth 2 or 3 points, depending on the distance from the basket. Foul shots are worth 1 point. The Crocodiles have twice as many points earned from 2-point shots as 3-point shots. They have 4 more points in 3-point shots than foul shots. The Crocodiles have a total of 32 points. If they have scored at least 1 of each type of basket, how many points were scored from 2-point shots?

20. Janet's age is a multiple of 5 and next year her age will be a multiple of 6. Her current age is less than 50. How old is Janet?

21. Cory has 1 dollar, 5 dollar and 10 dollar bills. He has a total of 42 dollars. He has at least 1 of each bill. How many 10 dollar bills does Cory have?

22. Sam has a total of 90 dollars. He has 5 dollar, 10 dollar, and 20 dollar bills. His total is made up of at least 1 of each. He has twice as many 10 dollar bills as he does 20 dollar bills. He has the same number of 5 dollar bills and 20 dollar bills. How many 10 dollar bills does Sam have?

23. Jaime has the same number of quarters, nickels, and dollar coins. Their combined value is $11.70. How many of each coin does Jaime have?

24. Sherry went to an annual shoe sale. The sale gave her $5 off for every $20 purchase she made. Their shoes are priced by the type of shoe. She purchased 3 different pairs of shoes from the list below. If her total was $100, which 3 pairs of shoes did she purchase?

Shoe Type	Price
Walking shoes	$60
Running shoes	$70
Sandals	$20
Dress shoes	$50
Loafers	$40

25. Two television companies, one selling cable and the other selling satellite, charge a set up fee and a monthly fee for every month of service performed. After how many months of service will the costs of each service be equal?

Television company	Set-up fee	Cost per month
Cable	$45	$40
Satellite	$60	$35

26. On a farm, there are 90 chickens and cows. There are 5 times as many chickens as cows. How many cows are on the farm?

27. The Danburry's are having a family reunion. They are expecting to have 3 times as many adults than children. There are a total of 48 people. How many adults are expected at the family reunion?

28. Derrick has 10 coins. He has quarters and nickels. If their combined value is $1.30, how many of each coin are there?

29. On a particular day, the same number of people got on or off the bus. The number of people who got on or off the bus is divisible by 3 and 6 and is less than 40. What is the maximum possible number of people getting on or off the bus?

30. In Mrs. William's class, there are 28 students. There are 3 times as many girls as boys. How many girls are in Mrs. William's class?

31. Kathryn is thinking of a whole number that is both a multiple of 6 and 8. When it is divided by 5, the remainder is 2. What is the smallest whole number possible that Kathryn could have in mind?

32. Angelle's mother is 22 years older than Angelle. In 12 years, her mother will be 2 times as old as Angelle. How old is Angelle's mother now?

33. Maria went to a café for lunch. She selected 3 different items from the following menu. If Maria paid with a $20 bill and received $7 in change, which 3 items did she order?

Item	Cost
Hot dog	$4.00
Hamburger	$6.00
Salad	$8.00
Water	$1.00
Soda	$2.00

34. Bruce is 20 years younger than his mother. In 8 years, his mother will be three times as old as Bruce. How old is Bruce now?

35. Jefferson, Washington and Roosevelt High Schools were competing in a quiz bowl. Jefferson High answered 2 times as many correct questions as Washington High. Washington answered 3 more questions correctly than Roosevelt. If there were 29 questions answered correctly, how many did Jefferson High School answer?

36. Tina, Lisa, and Linda went to a sweet shop to buy candy. Tina spent $7 more than Lisa. Lisa spent 3 times as much as Linda. Linda spent $3. Among the 3 girls, they had $7 left over. How much money did they have together before they went into the sweet shop?

37. For homework, the teacher assigned different problems to each of 3 students. Roberta completed 3 more of her problems than Doris. Nola completed 7 fewer problems than Doris. Nola completed 5 problems. There were 13 problems that were not completed. How many total problems were assigned to the 3 students?

38. Jared has a marble collection. If he adds 20 to the number of marbles, divides by 4, and then multiplies by 2, he will get 26. How many marbles does Jared have?

39. Leroy, Jonathan, and Marvin were offered cookies by Ms. Landon. She told them to take as many as they would like. They took half the cookies and left the rest. Leroy took 2 more than Jonathan. Jonathan took half as many as Marvin. Marvin took 8 cookies. How many cookies did Ms. Landon have to offer the boys?

40. Katie keeps coins in her piggy bank. She has 3 times as many dimes as quarters, she has 5 fewer nickels than dimes, and she has 13 nickels. How many quarters does Katie have?

MPA-117: Modeling Integer Arithmetic Using Cups and Counters

Adding and subtracting algebraic expressions can be modeled using cups and counters. Use a cup to represent a variable. Think of a cup as containing an unknown number of counters. A yellow counter will represent positive one, and a red counter will represent negative one.

= a variable

= positive one (yellow counter)

= negative one (red counter)

To model the number 7, use seven yellow counters.

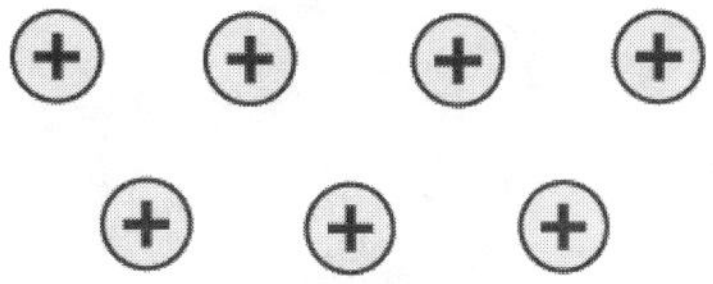

To model the number –5, use five red counters.

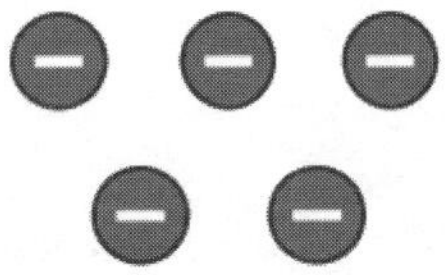

To model the expression $2x$, use two cups.

Example 1 Write an algebraic expression for the model below.

$3x$

Step 1: The model contains three cups, so the expression begins with $3x$ using x as the variable.

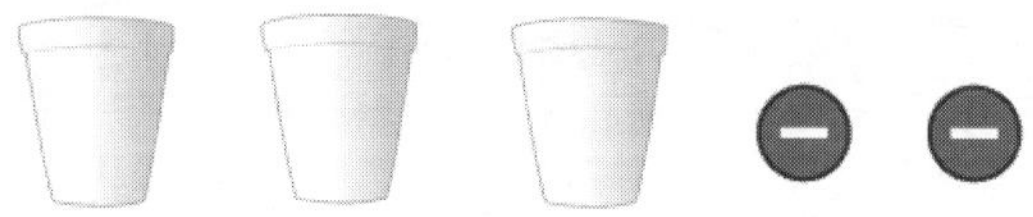

$3x - 2$

Step 2: There are two red counters, so the number –2 is in the equation.

Answer: The model represents the algebraic expression $3x - 2$.

Example 2 Create a model using cups and counters to add the addition problem $3 + 8 = 11$.

Step 1: Since all three numbers are positive, we will use only yellow counters to model.

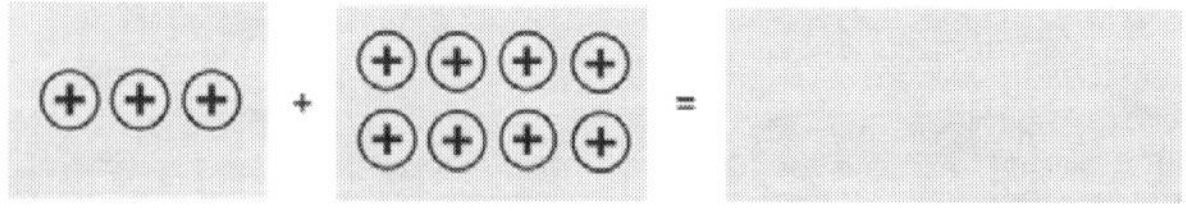

Step 2: Represent the first addend, 3, with three yellow counters and the second addend, 8, with eight yellow counters.

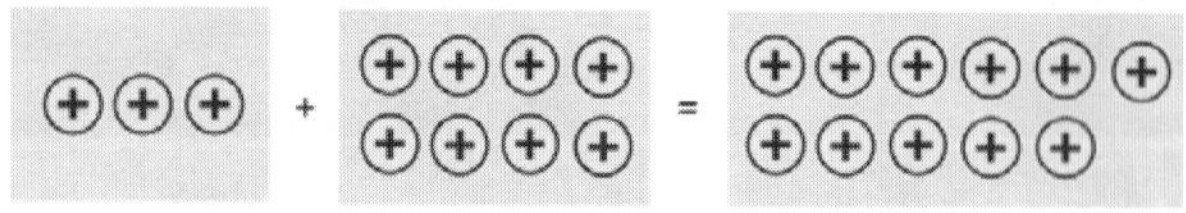

Step 3: Represent the sum, 11, with eleven yellow counters.

Answer: The model that represents $3 + 8 = 11$ is:

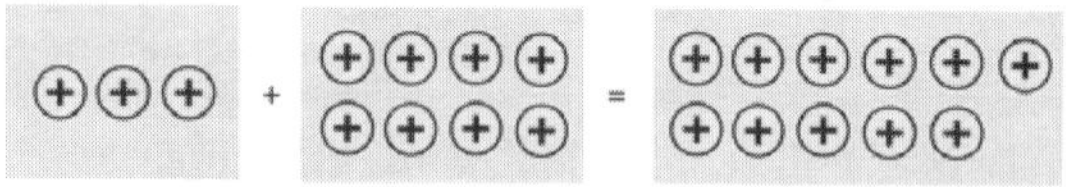

Example 3 Write an addition problem to match the model below.

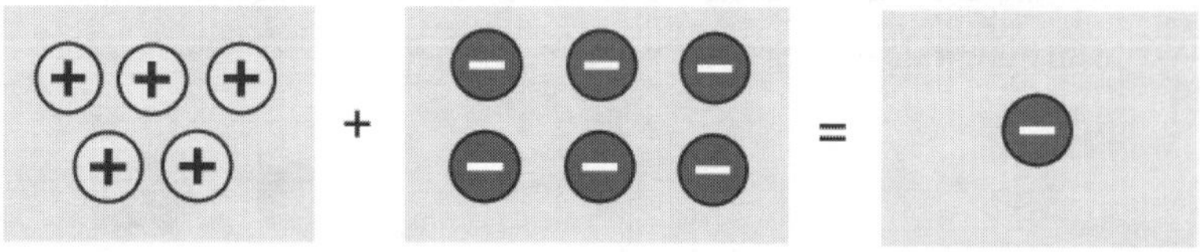

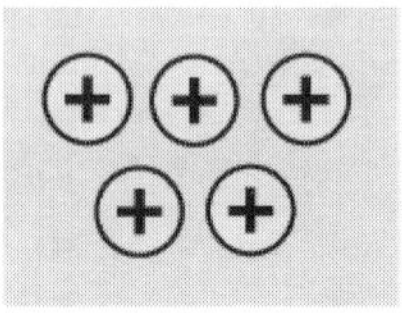

5

Step 1: Since the first addend in this sentence is represented by five yellow counters, the number is 5.

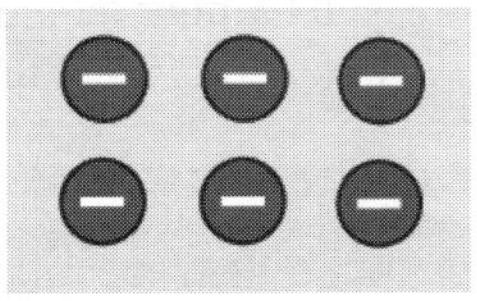

−6

Step 2: The second addend is represented by six red counters, so the number is −6.

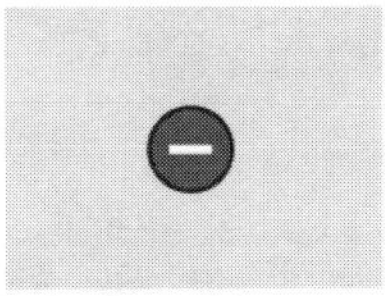

−1

Step 3: Since the sum of these numbers is represented by one red counter, the sum is −1.

Answer: The addition problem that matches this model is $5 + (-6) = -1$.

Example 4 Use cups and counters to subtract the following expression:

$(-3) - 4 = ?$

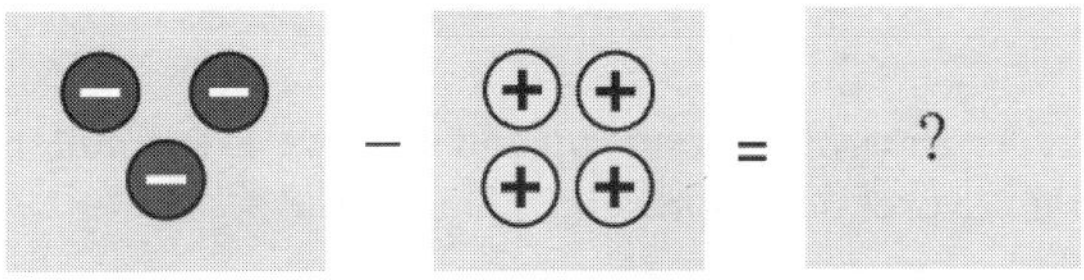

(-3) − 4 = ?

Step 1: Represent −3 with three red counters and 4 with four yellow counters.

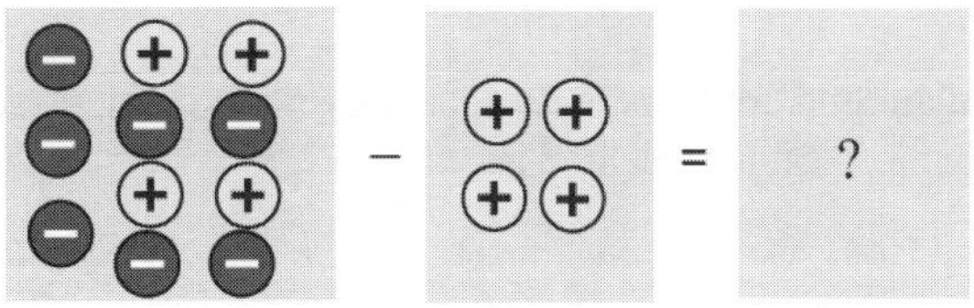

(-3) − 4 = ?

Step 2: While addition is done by combining, subtraction is done by taking away. Since there are no yellow counters in the minuend, add four yellow counters and four red counters to it. Each red and yellow pair represents zero, so the represented number remains −3.

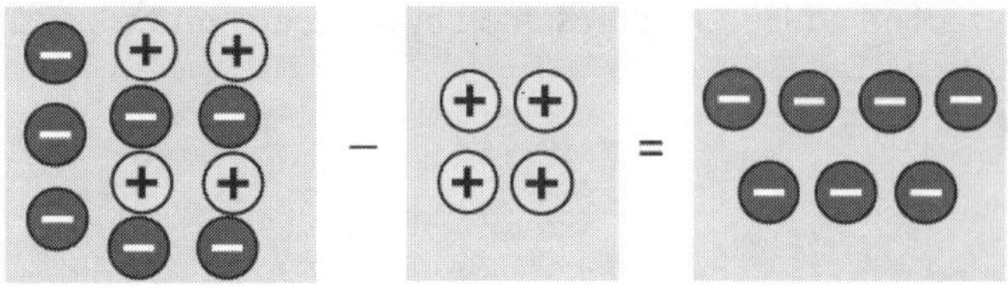

(-3) – 4 = -7

Step 3: After taking away four yellow counters, the remaining seven red counters represent the difference.

Answer: The model below shows the expression $(-3) - 4 = -7$.

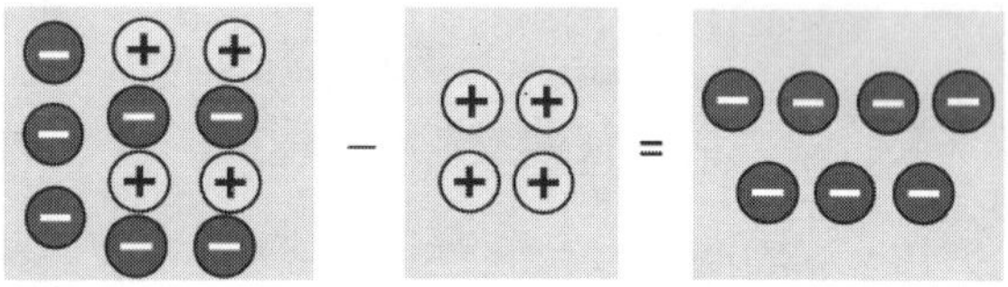

Example 5 Write an addition problem that matches the model below.

$x + 3$

Step 1: The first addened contains one cup and three yellow counters, so the algebraic expression represented here is $x + 3$.

$x - 2$

Step 2: The second addend contains one cup and two red counters, so the algebraic expression represented here is $x - 2$.

$2x + 1$

Step 3: The sum contains two cups and one yellow counter, so the algebraic expression represented here is $2x + 1$.

Answer: The model represents the equation $(x+3)+(x-2) = (2x+1)$.

Problem Set

1. Write an algebraic expression to match the model below.

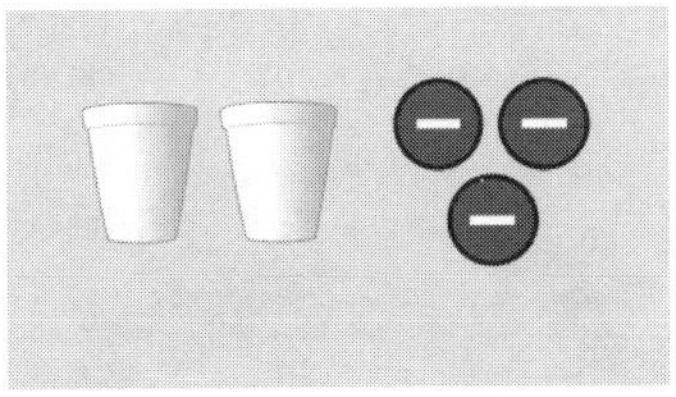

2. Write an algebraic expression to match the model below.

3. Write an algebraic expression to match the model below.

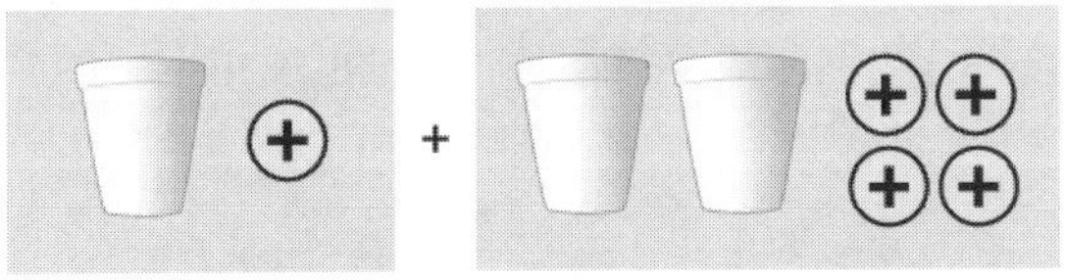

4. Write an algebraic expression to match the model below.

5. Model the addition problem below using cups and counters.

   $5 + 5 = 10$

6. Model the addition problem below using cups and counters.

   $-9 + (-1) = -10$

7. Model the problem below using cups and counters.

   $4 - 3 = 1$

8. Model the problem below using cups and counters.

   $5 - 3 = 2$

9. Write an algebraic expression to match the model below.

10. Write an algebraic expression to match the model below.

11. Write an addition problem to match the model below.

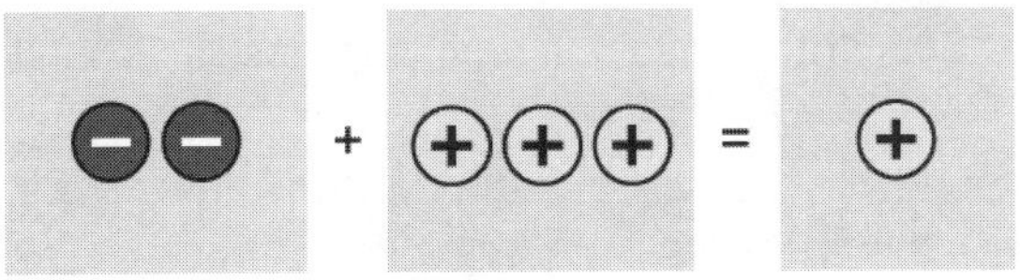

12. Write a subtraction problem to match the model below.

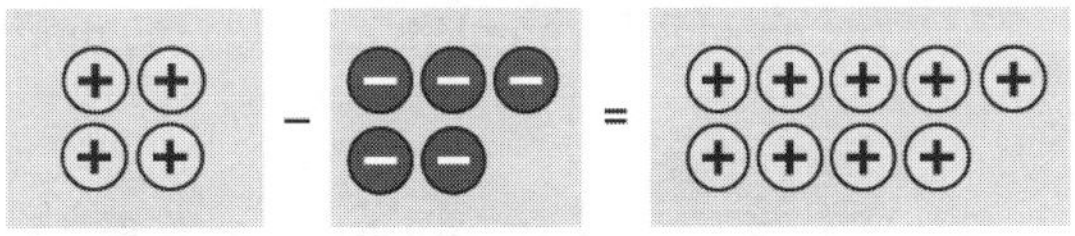

13. Subtract the integers using cups and counters.

    $4 - (-2) = ?$

14. Add the integers using cups and counters.

    $-4 + 5 = ?$

15. Add the integers using cups and counters.

    $-5 + 2 = ?$

16. Write an addition problem to match the model below.

17. Write a subtraction problem to match the model below.

18. Solve the equation modeled below.

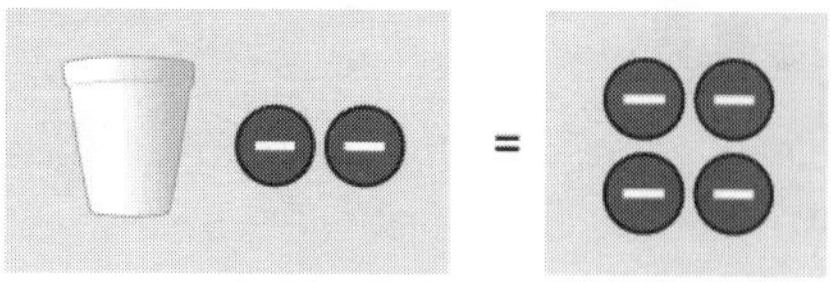

19. Solve the equation modeled below.

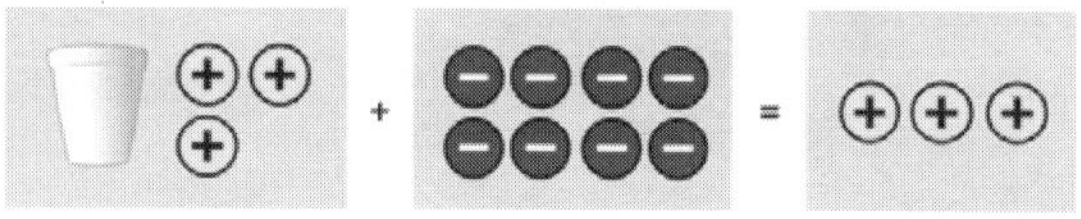

20. Solve the equation modeled below.

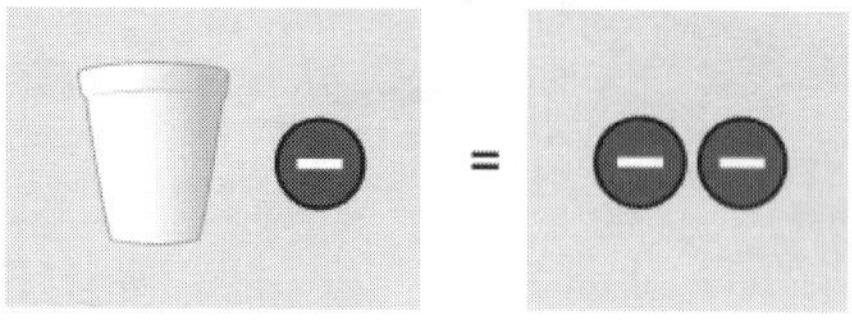

21. Write an algebraic expression to match the model below.

22. Write an algebraic expression to match the model below.

23. Write an algebraic expression to match the model below.

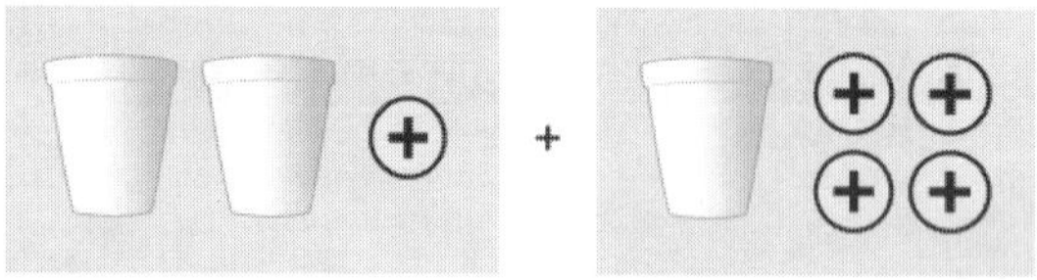

24. Write an algebraic expression to match the model below.

25. Model the addition below using cups and counters.

$-7 + (-3) = -10$

26. Write an addition problem to match the model below.

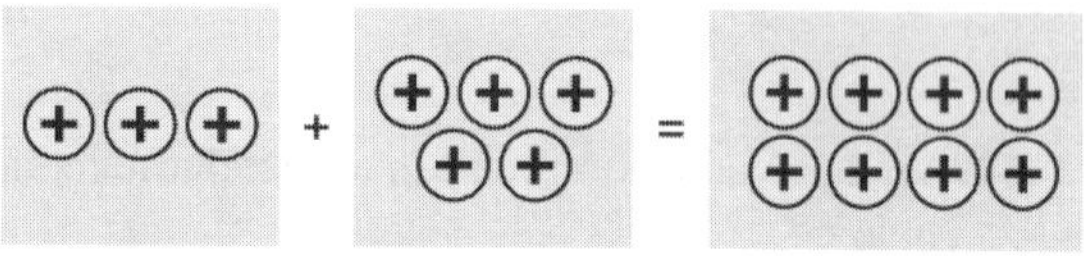

27. Write an addition problem to match the model below.

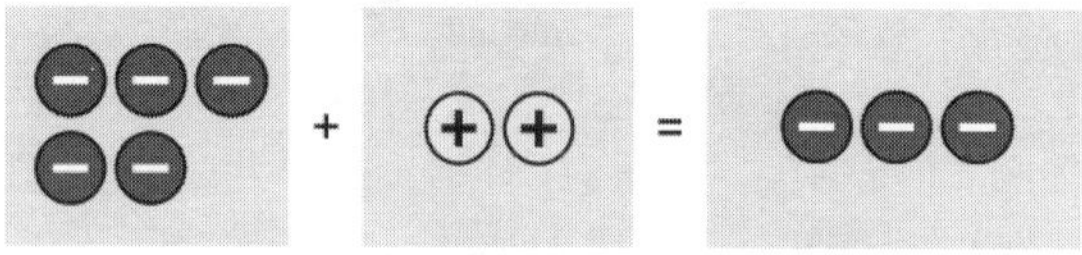

28. Write an addition problem to match the model below.

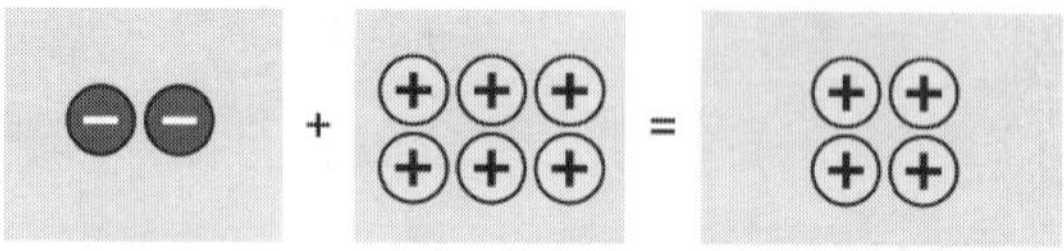

29. Write an addition problem to match the model below.

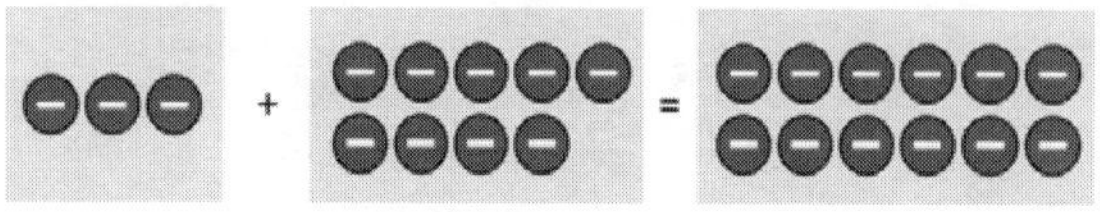

30. Subtract the integers using cups and counters.

$2 - (-4) = ?$

31. Add the integers using cups and counters.

$1 + (-5) = ?$

32. Add the integers using cups and counters.

$-4 + 6 = ?$

33. Subtract the integers using cups and counters.

$-7 - 4 = ?$

34. Subtract the integers using cups and counters.

$-3 - 2 = ?$

35. Write a subtraction problem to match the model below.

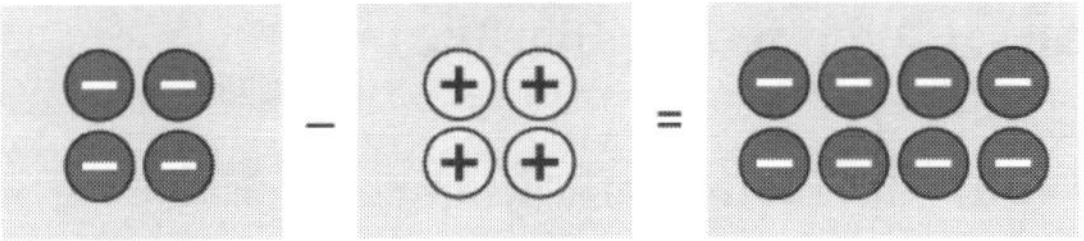

36. Solve the equation modeled below.

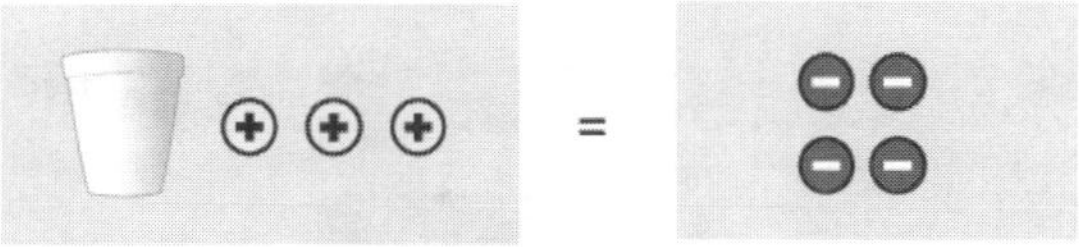

37. Solve the equation modeled below.

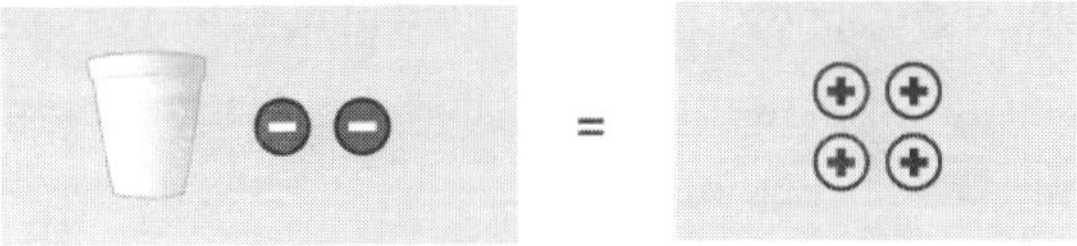

38. Solve the equation modeled below.

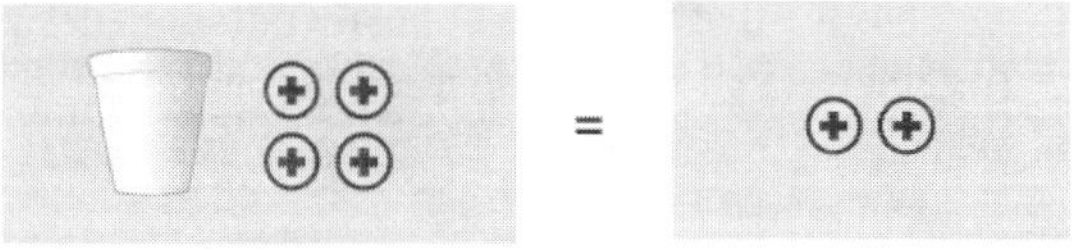

39. Solve the equation modeled below.

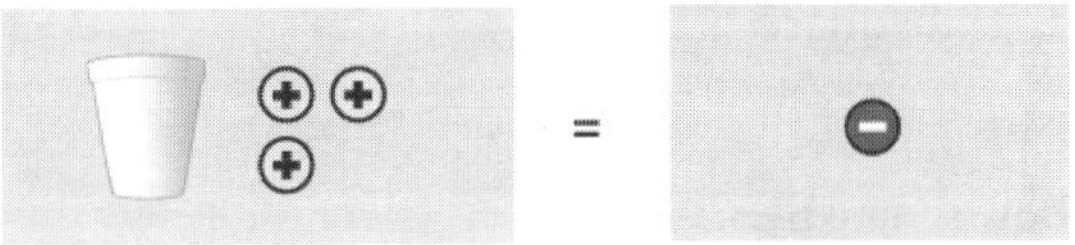

40. Solve the equation modeled below.

MPA-118: Modeling Integer Arithmetic Using Algebra Tiles

Algebra tiles can be used to model operations performed with algebraic expressions. When using algebra tiles, quantities are represented by a specified set of tiles. Like using a cup with cups and counters, a green rectangular tile represents an unknown quantity represented by the variable x. A red rectangular tile represents a negative unknown quantity.

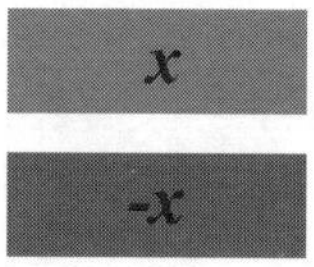

= **positive variable** (green rectangular tile)

= **negative variable** (red rectangular tile)

A yellow square tile represents one unit, or positive one. A red square tile represents one negative unit, or negative one. Notice that there are opposite number tiles just like there are opposite counters used in cups and counters. Together, a yellow tile and a red tile represent zero.

= **positive one** (yellow square tile)

= **negative one** (red square tile)

To use algebra tiles to model the number 7, use seven yellow square tiles.

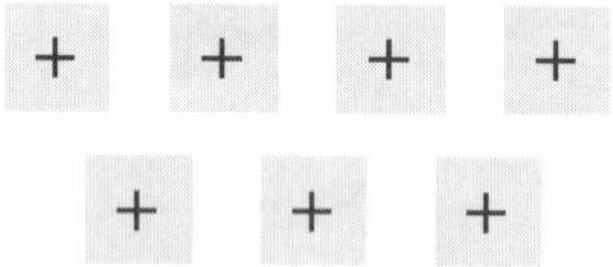

To use algebra tiles to model the number –5, use five red square tiles.

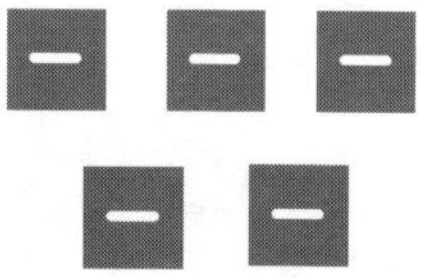

To model the expression $2x$, use two green rectangular tiles.

To model the expression $-2x$, use two red rectangular tiles.

Example 1 Write an addition problem to match the model below.

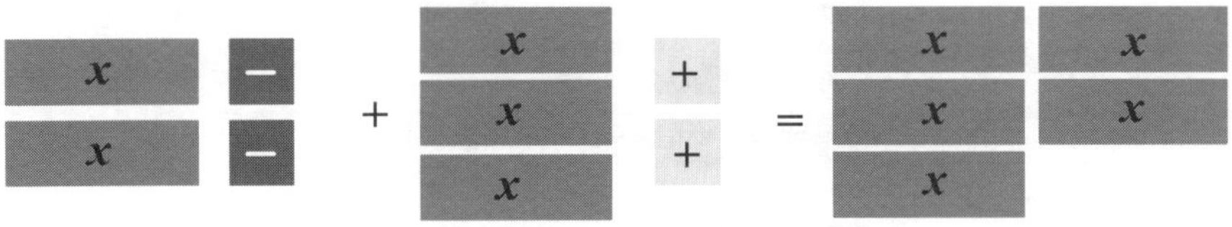

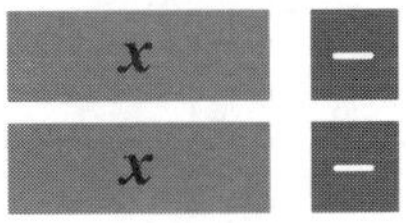

$2x$ - 2

Step 1: Start with the first group of tiles, which is called the addend. When two green rectangular tiles are added to two small red square tiles, the result is $2x - 2$.

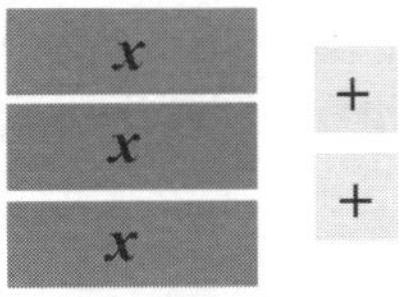

$3x + 2$

Step 2: The second addend contains three green rectangular tiles and two yellow square tiles, which represents the expression $3x + 2$.

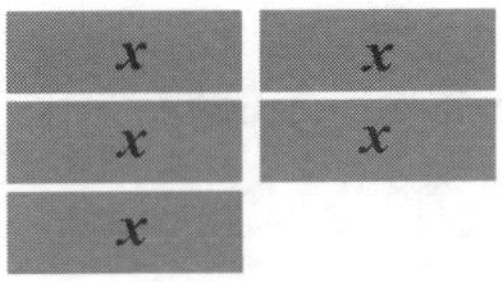

$5x$

Step 3: The sum of the addition problem contains five green rectangular tiles, which represents $5x$.

Answer: The model represents the addition problem $(2x - 2) + (3x + 2) = 5x$.

Algebra tiles can also be used to model the multiplication of algebraic expressions. For this operation, new tiles will be used. The term x^2 will be represented by a large green square tile that has the same length as the rectangular tile representing x. A red rectangular tile will represent the opposite of the variable x, and a large red square tile will represent the opposite of x^2.

= large green square tile

= green rectangular tile

= large red square tile

= red rectangular tile

Example 2 Write an algebraic expression to match the model below.

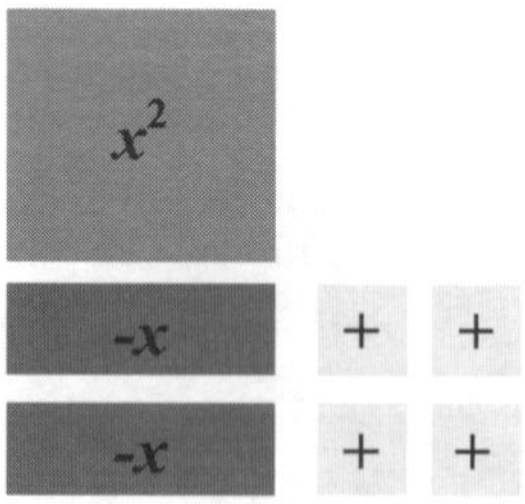

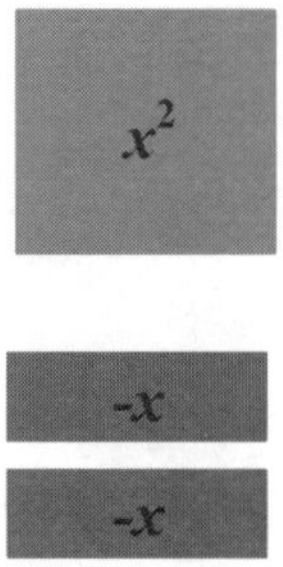

$= x^2$

$= -2x$

Step 1: Each green square tile represents x^2. Each red rectangular tile represents $-x$.

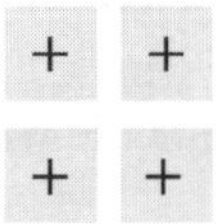

$= 4$

Step 2: Each yellow square tile represents positive one, so four yellow square tiles represent positive 4.

Answer: The model represents the algebraic expression $x^2 - 2x + 4$.

To multiply, begin with the two expressions represented with groups of tiles arranged in rows. These factors are placed at right angles to each other, leaving a rectangular space between them for the product. To find the product, place new tiles into the rectangle that match the lengths and widths of the tiles in the factors.

Example 3 Write the algebraic expression represented by the product of the multiplication model below.

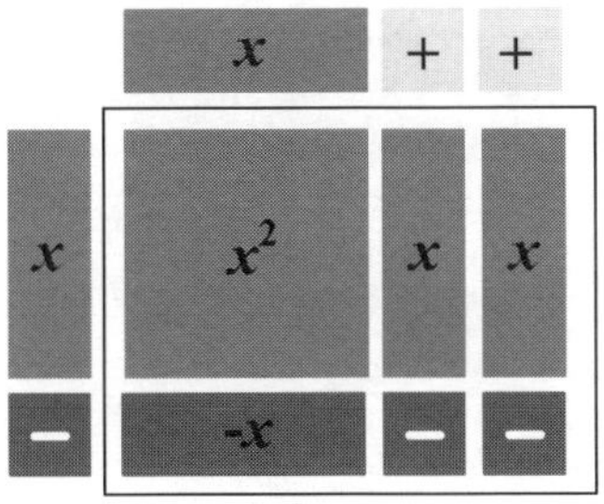

Step 1: Recall when multiplying with algebra tiles, the tiles along the top row and the left-most column, that are outside the rectangle, represent the factors.

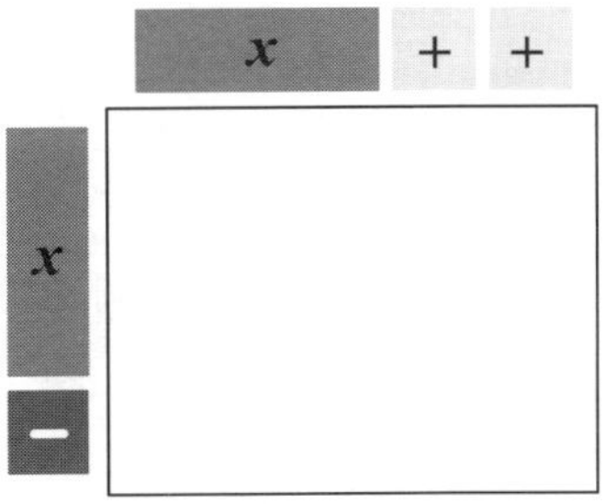

Step 2: The expression for the product is represented by the tiles that are inside the rectangle.

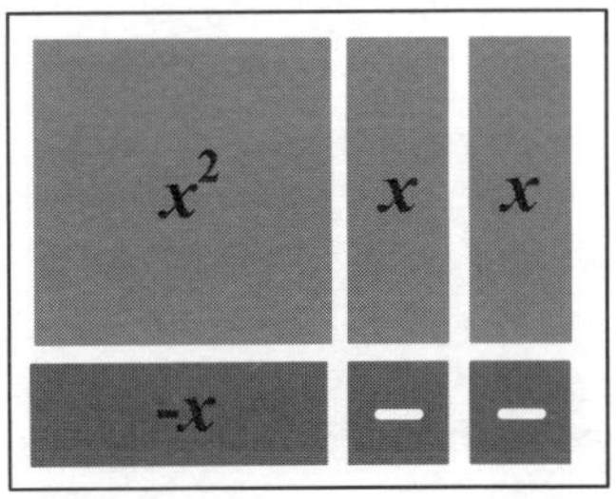

Step 3: Notice that the rectangle contains one green square tile, two green rectangular tiles, one red rectangular tile and two red squares. Since red and green tiles are opposites, they cancel each other out.

Therefore, the product of the multiplication model is $x^2 + x - 2$.

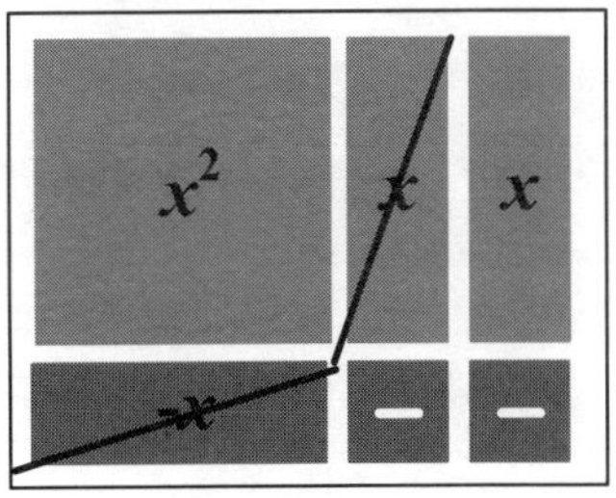

$x^2 + x - 2$

Answer: The product of the multiplication model is $x^2 + x - 2$.

Example 4 Use algebra tiles to model the multiplication of $(2x - 1)(x + 3)$.

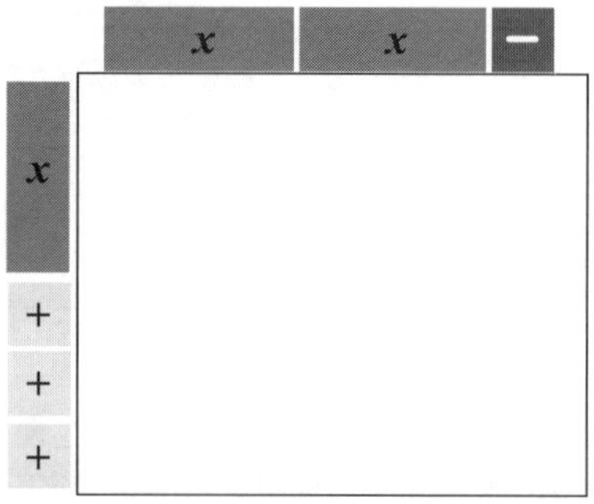

Step 1: Arrange the two factors, $2x - 1$ and $x + 3$, at right angles as shown.

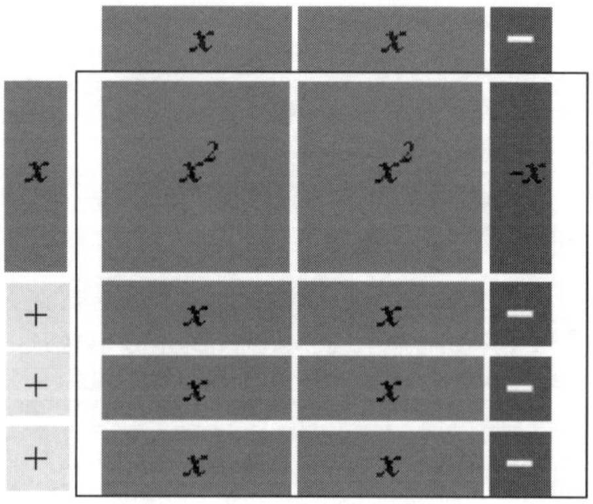

Step 2: The box can be filled with tiles as follows. Remember that when multiplying a positive and negative, the result is a negative.

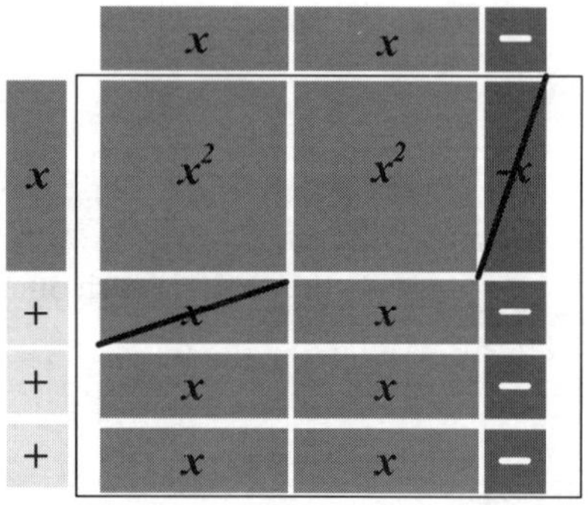

Step 3: One of the green rectangular tiles (positive variable) cancels one of the red rectangular tiles (negative variable), leaving the solution $(2x - 1)(x + 3) = 2x^2 + 5x - 3$.

Answer: The model for the solution of the multiplication problem (2x – 1) (x + 3) is: $2x^2 + 5x - 3$.

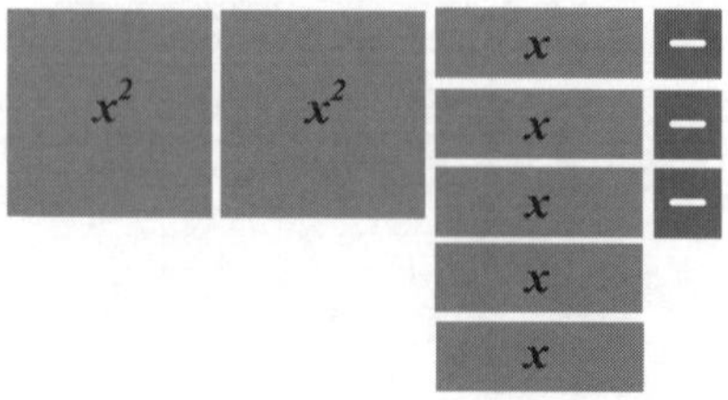

Algebra tiles can be used for other mathematical representations like solving equations. Let's look at an example. Starting with two quantities, we want to find out what number is represented by one green rectangle.

Example 5 Solve the linear equation $3x - 2 = 7$ using algebra tiles.

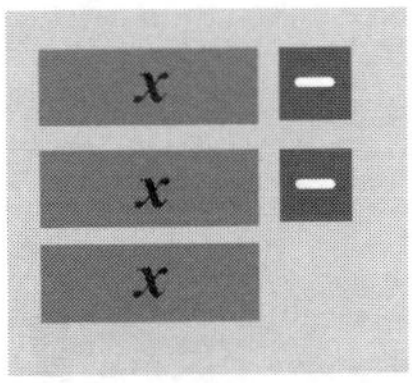

Step 1: Model $3x - 2$.

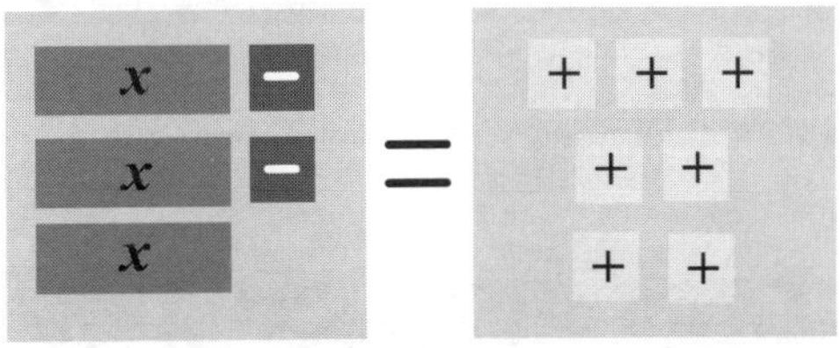

Step 2: Model 7 and place an equal sign between the two models, creating the model of the equation.

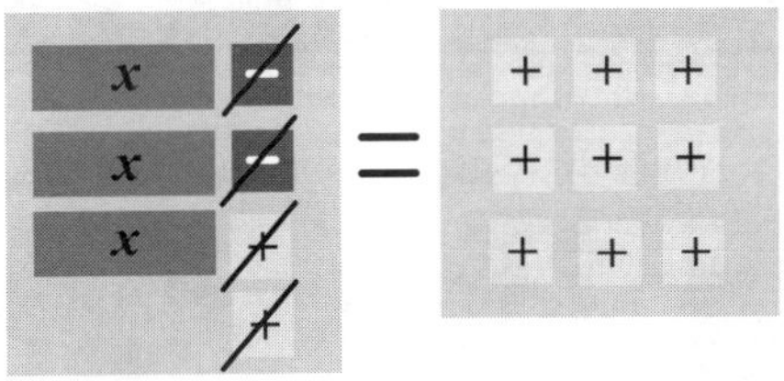

Step 3: To get one green rectangular tile (positive variable) on the left side of the equation, start by removing the two red square tiles. To do this, add two yellow square tiles to both sides of the equal sign. Remember when performing an action to one side of the equation, the same action must be done to the other side. Remove the red and yellow pairs on the left side of the equal sign.

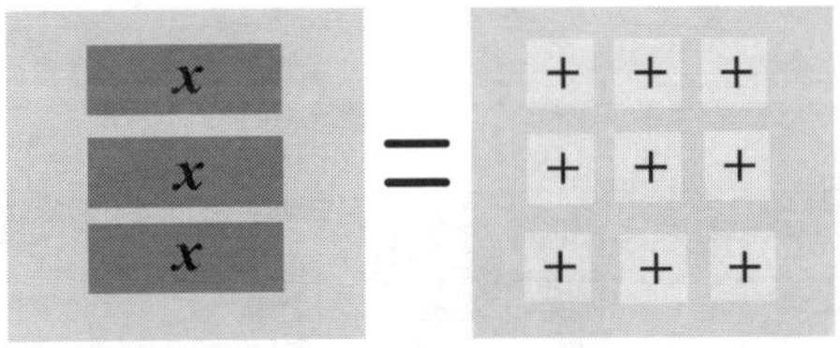

Step 4: Now we have $3x = 9$.

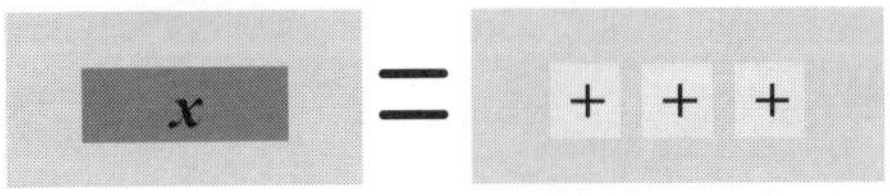

Step 5: Notice that the left-hand side is arranged in three equal groups. The right-hand side is also arranged into three equal groups. Identifying the groups, we see that $x = 3$.

Answer: The solution to the equation $3x - 2 = 7$ is $x = 3$.

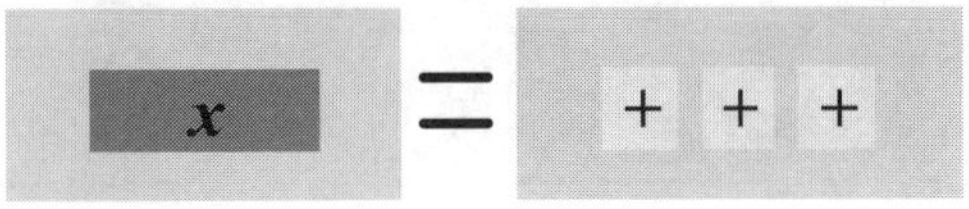

Problem Set

1. Write an algebraic expression to match the model below.

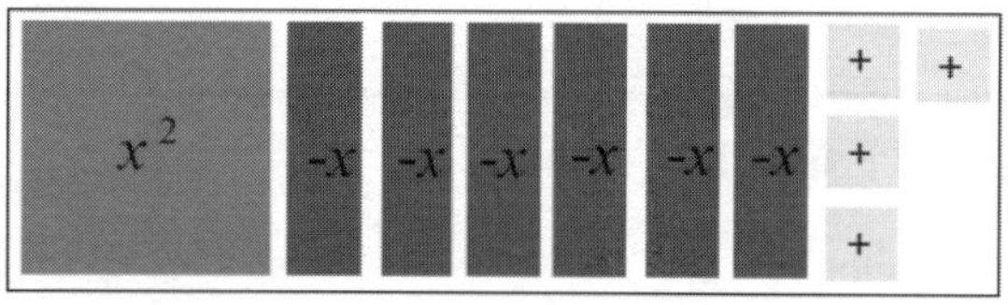

2. Write an algebraic expression to match the model below.

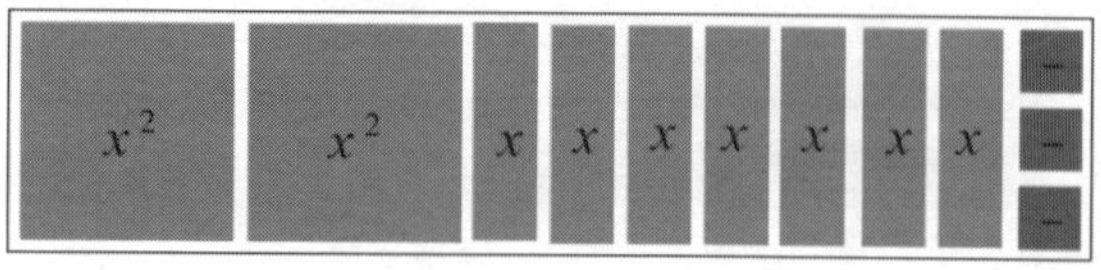

3. Write an algebraic expression to match the model below.

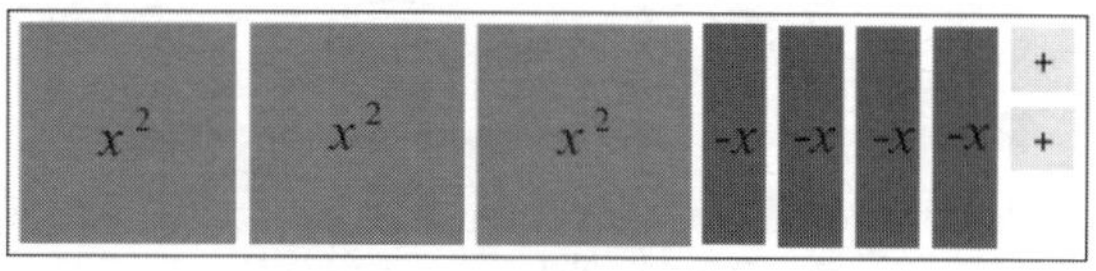

4. Write an algebraic expression to match the model below.

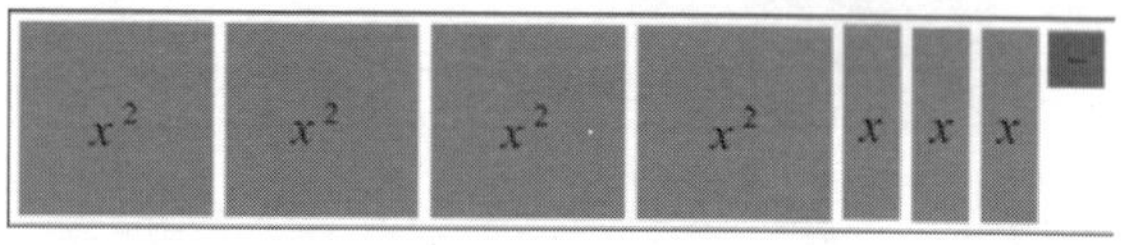

5. Write an algebraic expression to match the model below.

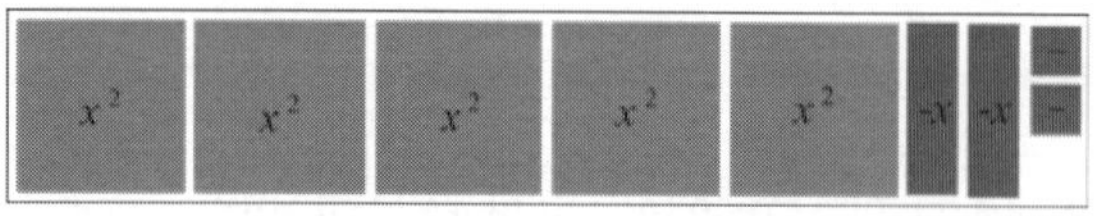

6. Write an algebraic expression to match the model below.

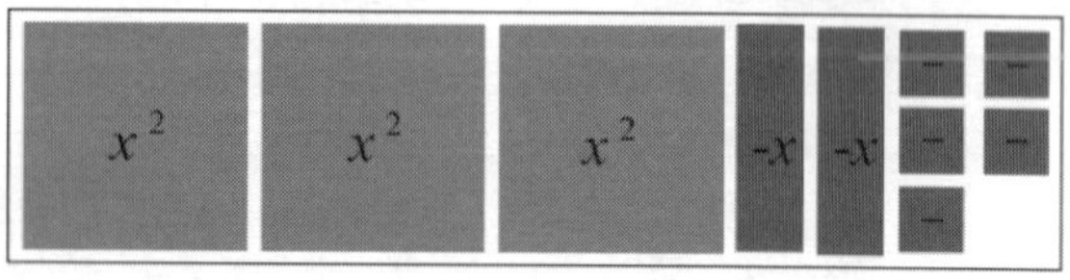

7. Write an algebraic expression to match the model below.

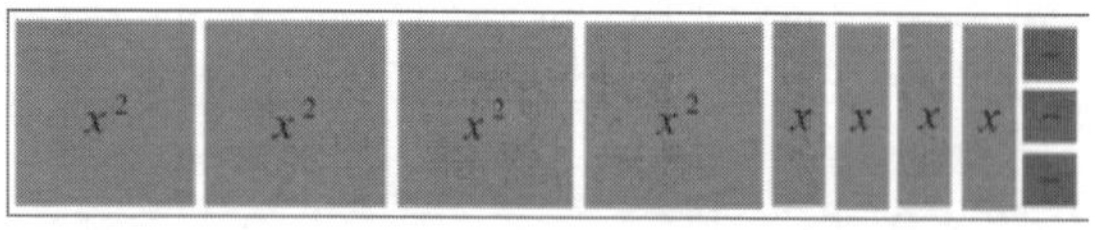

8. Write a problem to match the model below.

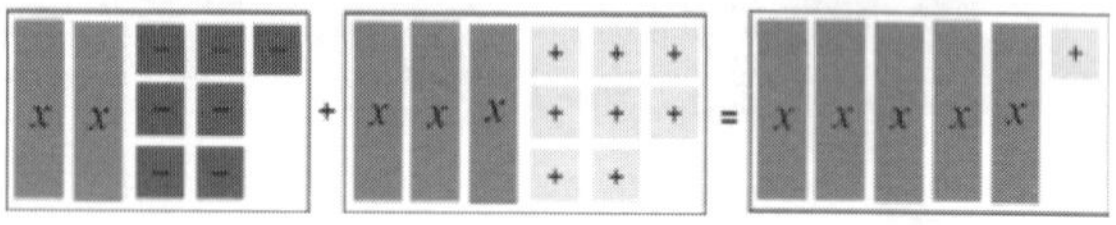

9. Write a problem to match the model below.

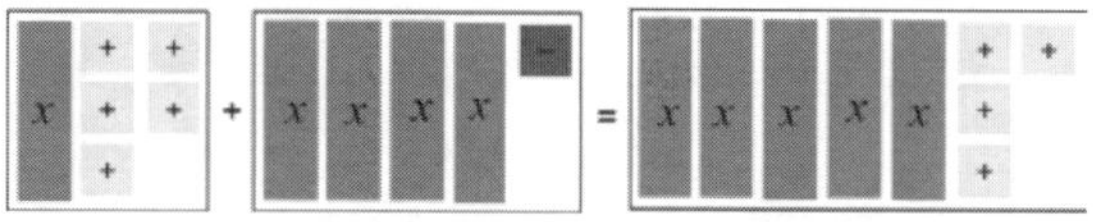

10. Write a problem to match the model below.

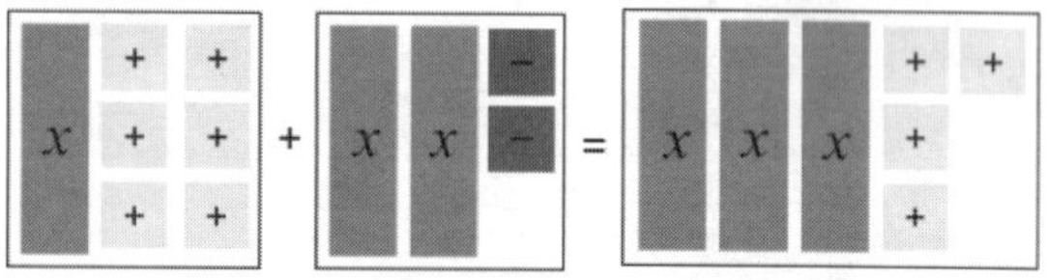

11. Write a problem to match the model below.

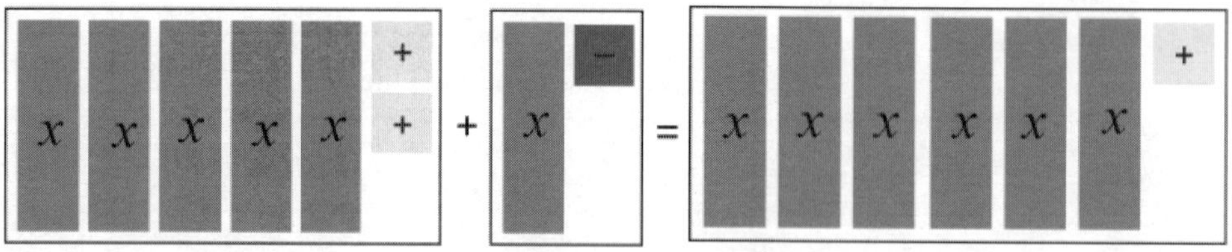

12. Write a problem to match the model below.

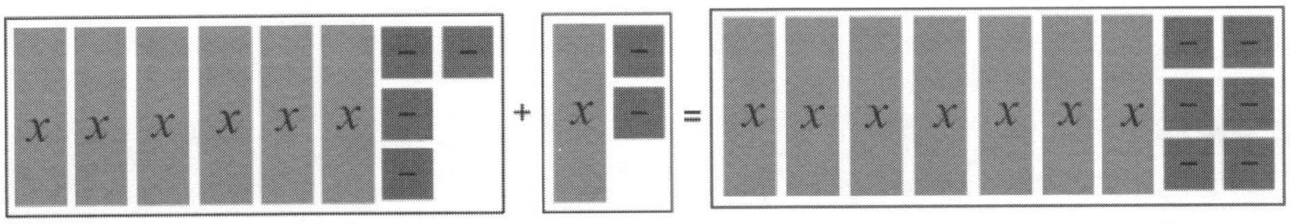

13. Write a problem to match the model below.

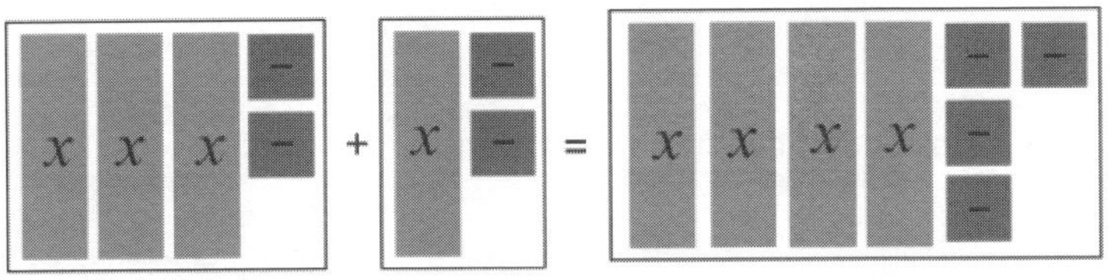

14. Write a problem to match the model below.

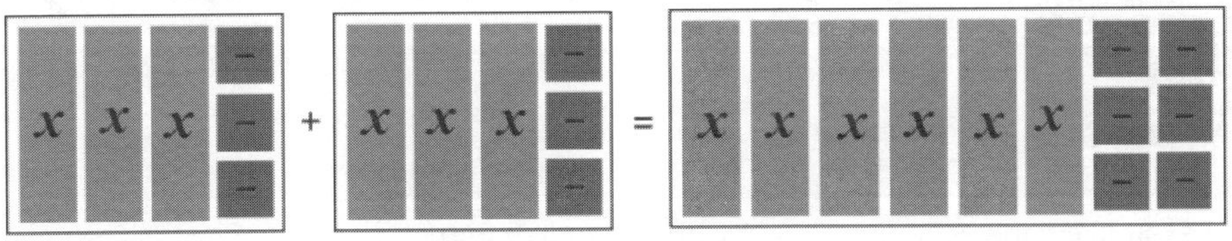

15. Write a problem to match the model below.

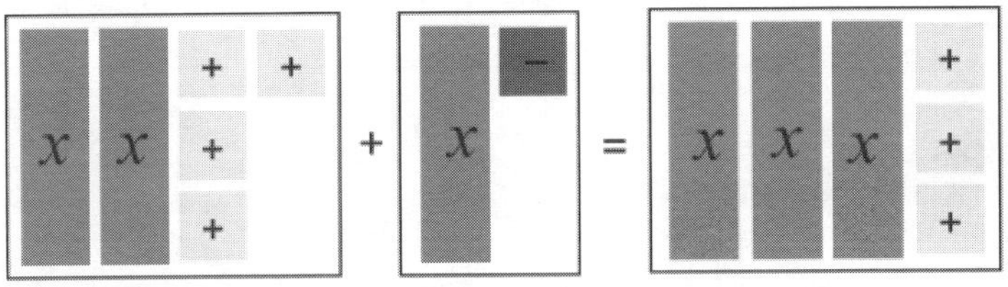

16. Write the algebraic expression that is represented by the product of the multiplication model below.

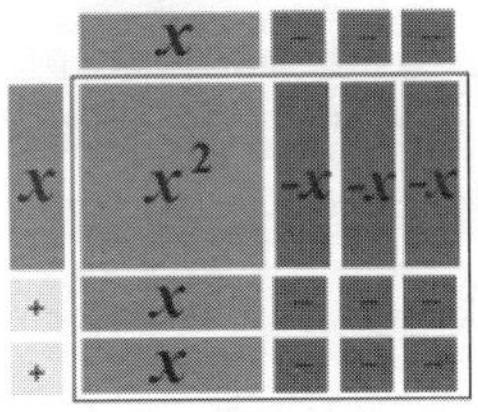

17. Write the algebraic expression that is represented by the product of the multiplication model below.

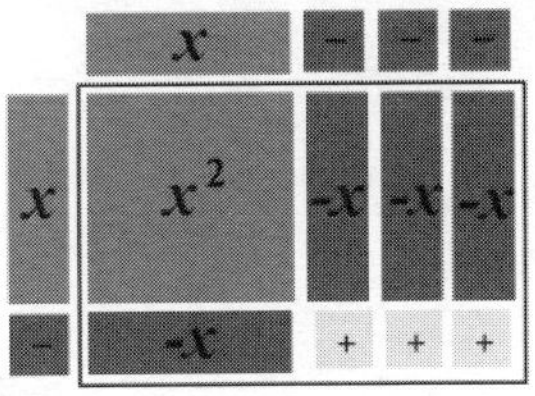

18. Write the algebraic expression that is represented by the product of the multiplication model below.

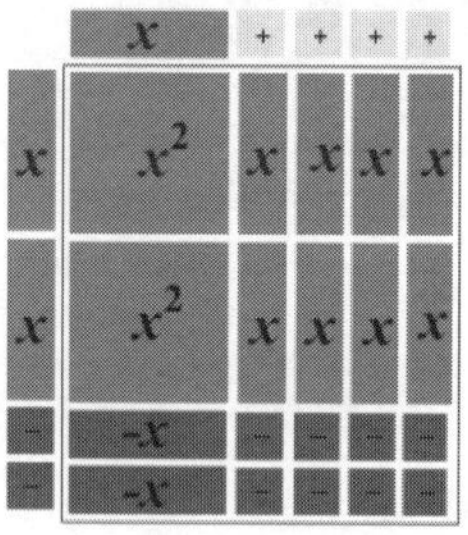

19. Write the algebraic expression that is represented by the product of the multiplication model below.

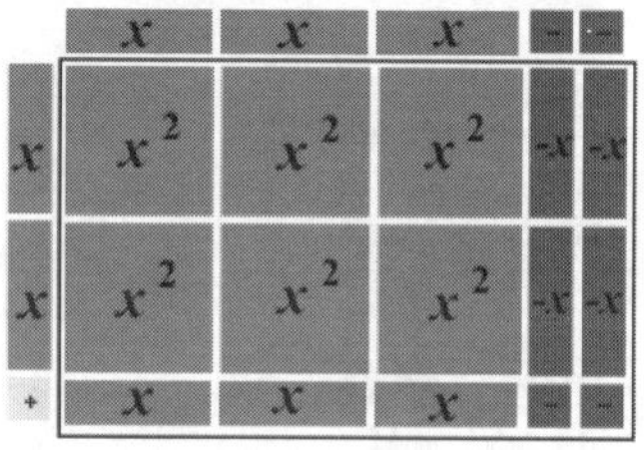

20. Write the algebraic expression that is represented by the product of the multiplication model below.

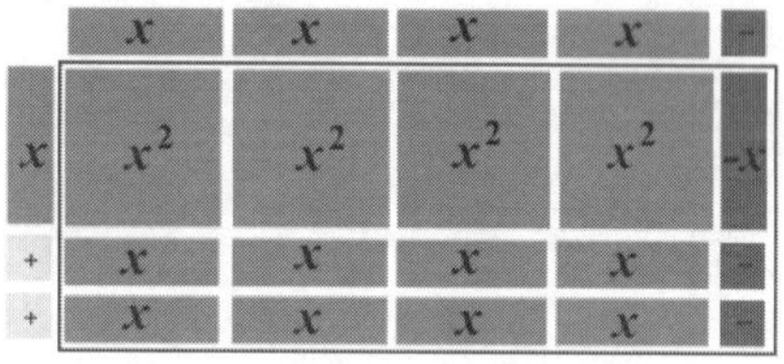

21. Write the algebraic expression that is represented by the product of the multiplication model below.

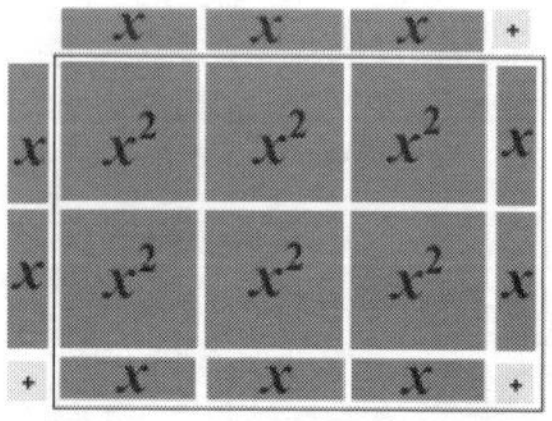

22. Write the algebraic expression that is represented by the product of the multiplication model below.

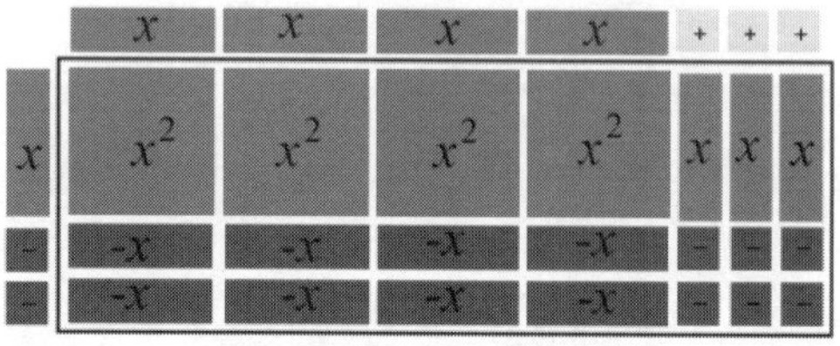

23. Write the algebraic expression that is represented by the product of the multiplication model below.

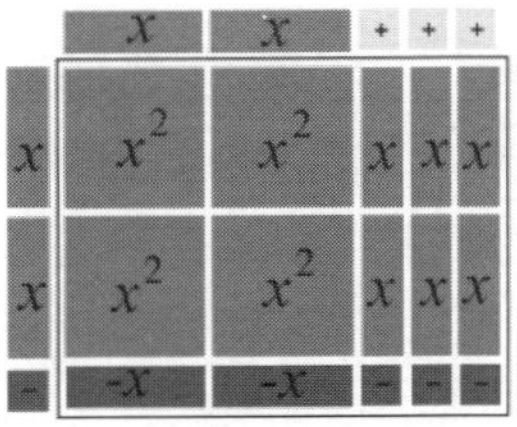

24. Model the product below.

$(x - 3)(2x - 2)$

25. Model the product below.

$(2x + 2)(2x - 1)$

26. Model the product below.

$(3x + 2)(2x - 1)$

27. Model the product below.

$(3x - 4)(x + 1)$

28. Model the product below.

$(4x + 2)(2x - 2)$

29. Model the product below.

$(4x + 3)(x - 1)$

30. Model the product below.

$(x + 3)(2x + 2)$

31. Solve the equation modeled below with algebra tiles.

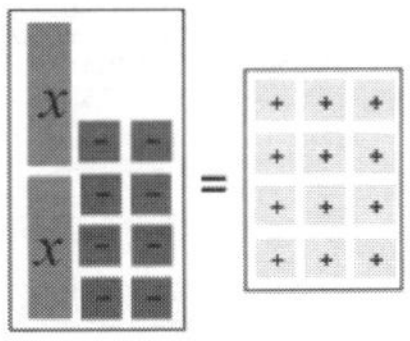

32. Solve the equation modeled below with algebra tiles.

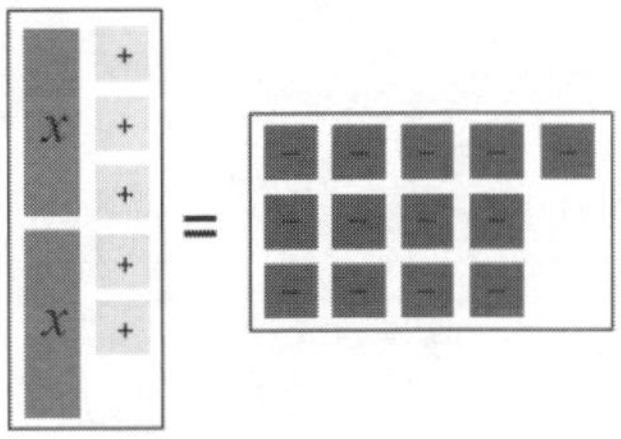

33. Solve the equation modeled below with algebra tiles.

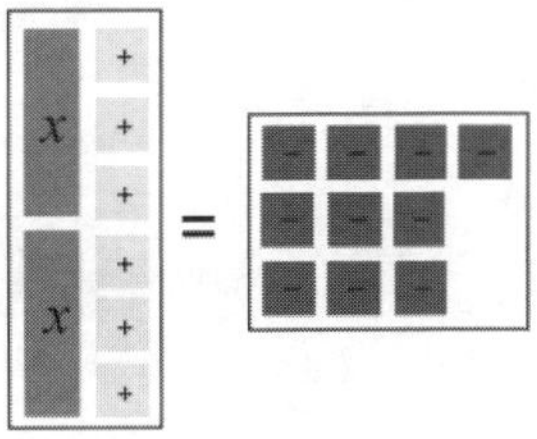

34. Solve the equation modeled below with algebra tiles.

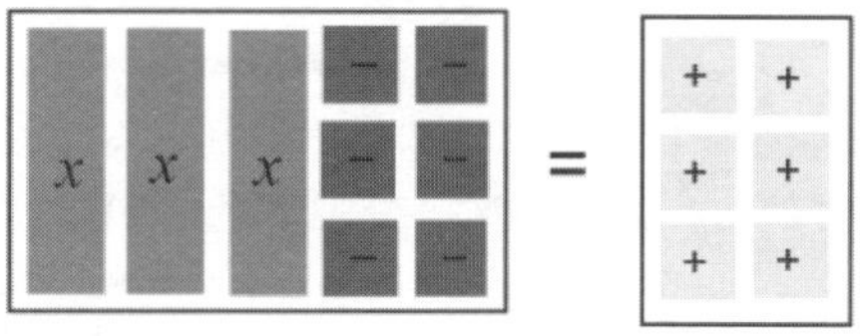

35. Solve the equation modeled below with algebra tiles.

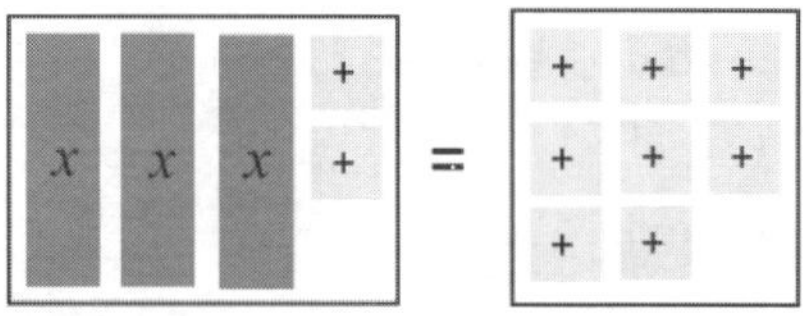

36. Solve the equation modeled below with algebra tiles.

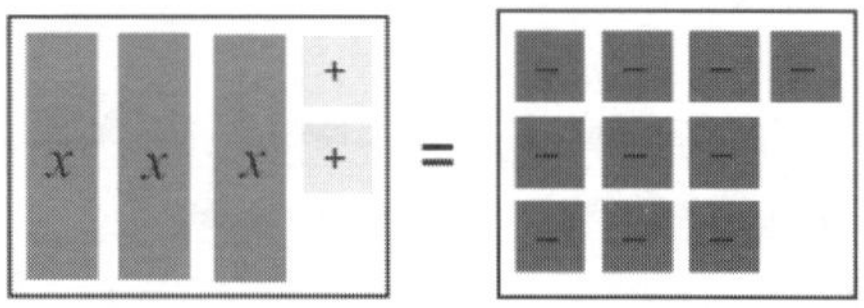

37. Solve the equation modeled below with algebra tiles.

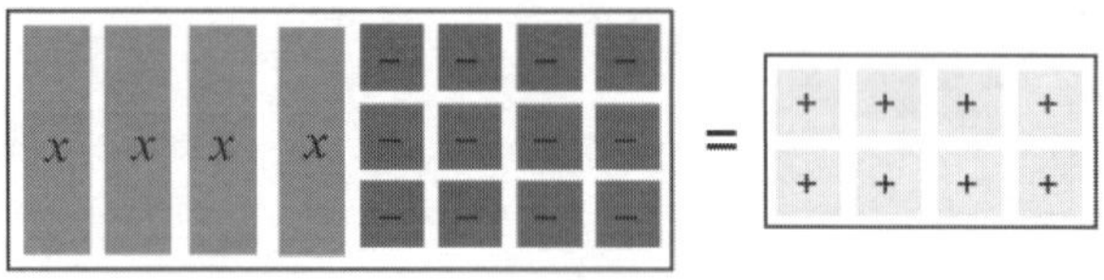

38. Solve the equation modeled below with algebra tiles.

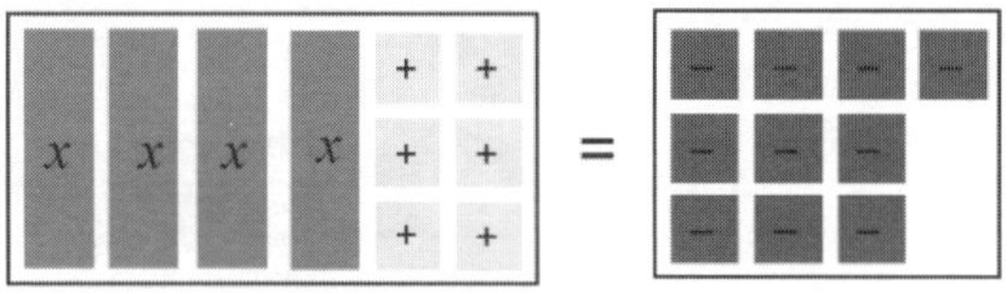

39. Solve the equation modeled below with algebra tiles.

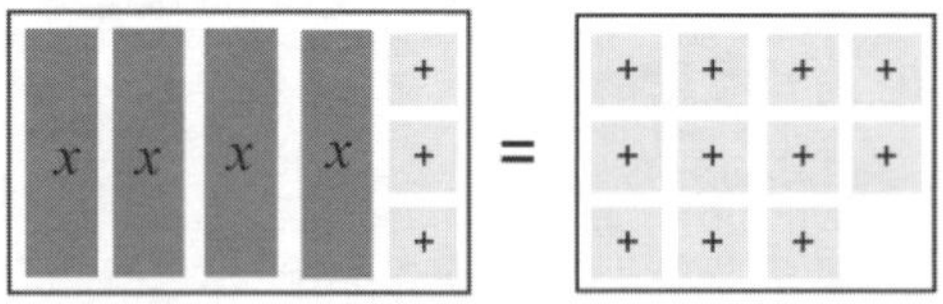

40. Solve the equation modeled below with algebra tiles.

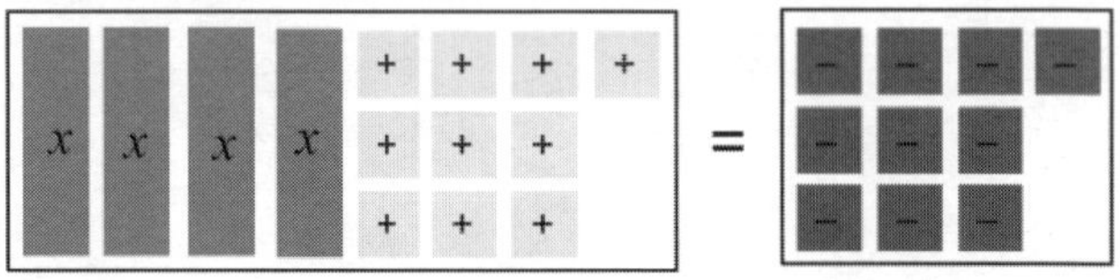

MPA-119: Dividing Decimals (an Introduction)

In this lesson, we will learn to divide decimal numbers. Let's start by reviewing the step-by-step process of division with whole numbers.

1. Divide.
2. Multiply.
3. Subtract.
4. Bring down the next digit.

When dividing a decimal by a whole number, follow these steps:

1. Place a decimal point in the quotient directly above the decimal point in the dividend.
2. Divide as usual.

When dividing a decimal by a decimal, follow these steps:

1. If the divisor is a decimal, move the decimal point in the divisor all the way to the right to make it a whole number.
2. Move the decimal point in the dividend the same number of places.
3. Place the decimal point in the quotient directly above the decimal point in the dividend.
4. Divide as usual.

Example 1 Divide: 25.2 ÷ 6

$$6\overline{)25.2}$$

Step 1: Write in standard form.

$$\begin{array}{r} . \\ 6\overline{)25.2} \end{array}$$

Step 2: When dividing a decimal by a whole number, place the decimal point in the quotient directly above the decimal point in the dividend.

$$\begin{array}{r} 4. \\ 6\overline{)25.2} \\ -\underline{24}\downarrow \\ 12 \end{array}$$

Step 3: Divide 6 into 25.

$6\overline{)25.2}$ quotient 4.2; -24 ↓; 12; -12; 0	**Step 4:** Bring down the 2, then divide 6 into 12.

Answer: 25.2 ÷ 6 = 4.2

Example 2 Divide: 8.06 ÷ 2.6

$2.6\overline{)8.06}$	**Step 1:** Write in standard form.
$26\overline{)80.6}$	**Step 2:** Move the decimal point in the divisor and the dividend one place to the right.
quotient: . $26\overline{)80.6}$	**Step 3:** Place the decimal point in the quotient directly above the decimal point in the dividend.
quotient: 3. $26\overline{)80.6}$ -78 2	**Step 4:** Divide 26 into 80.
quotient: 3.1 $26\overline{)80.6}$ -78 ↓ 26 -26 0	**Step 5:** Bring down the 6, then divide 26 into 26.

Answer: 8.06 ÷ 2.6 = 3.1

Example 3 Divide: 36.05 ÷ 7

$7\overline{)36.05}$

Step 1: Write in standard form.

$\begin{array}{r} . \\ 7\overline{)36.05} \end{array}$

Step 2: Place the decimal point in the quotient directly above the decimal in the dividend.

$\begin{array}{r} 5. \\ 7\overline{)36.05} \\ -35 \\ \hline 1 \end{array}$

Step 3: Divide 7 into 36.

$\begin{array}{r} 5.1 \\ 7\overline{)36.05} \\ -35 \\ \hline 10 \\ -07 \\ \hline 3 \end{array}$

Step 4: Bring down the 0, then divide 7 into 10.

$\begin{array}{r} 5.15 \\ 7\overline{)36.05} \\ -35 \\ \hline 10 \\ -07 \\ \hline 35 \\ -35 \\ \hline 0 \end{array}$

Step 5: Bring down the 5, then divide 7 into 35.

Answer: 36.05 ÷ 7 = 5.15

Example 4 Divide: 45.5 ÷ 0.25

$0.25\overline{)45.5}$

Step 1: Write in standard form.

$25\overline{)4550}$

Step 2: Move the decimal points in the divisor and the dividend two places to the right.

$$\begin{array}{r} 1 \\ 25\overline{)4550} \\ -25 \\ \hline 20 \end{array}$$

Step 3: Divide 25 into 45.

$$\begin{array}{r} 1 \\ 25\overline{)4550} \\ -25\downarrow \\ \hline 205 \end{array}$$

Step 4: Bring down the 5, then divide 25 into 205.

$$\begin{array}{r} 182 \\ 25\overline{)4550} \\ -25\downarrow \\ \hline 205 \\ -200\downarrow \\ \hline 50 \\ -50 \\ \hline 0 \end{array}$$

Step 5: Bring down the 0, then divide 25 into 50.

Answer: $45.5 \div 0.25 = 182$

Example 5 How many quarters are there in $41.25?

$0.25\overline{)41.25}$

Step 1: A quarter is $0.25 and we want to know how many are in $41.25. Write the division problem.

$25\overline{)4125}$

Step 2: Move the decimal points in the dividend and the divisor two places to the right.

$$\begin{array}{r} 1 \\ 25\overline{)4125} \\ -25 \\ \hline 16 \end{array}$$

Step 3: Divide 25 into 41.

```
    16
25)4125
  -25↓
   162
  -150
    12
```

Step 4: Bring down the 2. Then divide 25 into 162.

```
    165
25)4125
  -25↓
   162
  -150↓
    125
   -125
      0
```

Step 5: Bring down the 5, then divide 25 into 125.

Answer: There are 165 quarters in $41.25.

Problem Set

Solve:

1. Divide: 18.8 ÷ 4

2. Divide: 24.5 ÷ 7

3. Divide: 87.3 ÷ 9

4. Divide: 96.8 ÷ 8

5. Divide: 100.5 ÷ 15

6. Divide: 134.4 ÷ 24

7. Divide: 174.6 ÷ 18

8. Divide: 140.8 ÷ 8.8

9. Divide: 182.4 ÷ 7.6

10. Divide: 158.4 ÷ 13.2

11. Divide: 188.1 ÷ 5.7

12. Divide: 291.2 ÷ 2.6

13. Divide: 916.5 ÷ 3.9

14. Divide: 119.7 ÷ 6.3

15. Divide: 1,657.5 ÷ 25.5

16. Divide: 58.95 ÷ 9

17. Divide: 321.5 ÷ 5

18. Divide: 76.48 ÷ 8

19. Divide: 789.04 ÷ 14

20. Divide: 382.5 ÷ 15

21. Divide: 394.8 ÷ 6

22. Divide: 827.2 ÷ 11

23. Divide: 14.4 ÷ 0.96

24. Divide: 81.6 ÷ 0.85

25. Divide: 22.1 ÷ 0.26

26. Divide: 14.76 ÷ 0.328

27. Divide: 10.35 ÷ 0.138

28. Divide: 62.1 ÷ 0.46

29. Divide: 52.29 ÷ 0.415

30. Divide: 29.21 ÷ 0.254

31. For a science experiment, Janice needs to divide 75.6 milliliters of water equally among 9 test tubes. How much water should she pour into each test tube?

32. Chris has 16.64 ounces of pepperoni to divide equally among 4 pizzas that he is making. How many ounces of pepperoni should he put on each pizza?

33. Mrs. Marks has 59.8 ounces of candy to divide equally among 13 gift bags. How many ounces of candy should she place in each bag?

34. A farmer has 42.72 pounds of chicken feed to divide equally among 12 containers. How many pounds of feed should he place in each container?

35. A factory produces 1,567.2 meters of twine that needs to be divided equally among 24 spools. How much twine should be wrapped around each of the spools?

36. Mr. Martin, owner of Handy Hardware, has 192.4 pounds of nails to be divided equally among 37 packages. How many pounds of nails should he place in each package?

37. The Band Booster club is going to sell popcorn to raise money. They have 195.2 pounds of popcorn to divide equally among 122 boxes. How many pounds of popcorn should the club members place in each box?

38. The art club wants to paint a mural on a wall that is 71.72 meters long. There are 11 members in the club and each member will be assigned a section of the mural. If the wall is divided equally among the 11 members, how long is each section of the mural?

39. Judy set a goal of walking 50.6 miles in 23 days. If she wants to walk the same amount every day to reach her goal, how many miles should she walk each day?

40. Mr. Wallace's three sons want to buy him a telescope that costs $180.93. If they split the cost equally, how much should each son contribute for the telescope?

MPA-120: Applying Dilations in the Coordinate Plane

A **dilation** is a transformation that changes a geometric figure by changing its size but not its shape. The description of a dilation includes the center of the dilation and the scale factor.

Dilation	A transformation that changes a geometric figure by changing its size but not its shape.

The **scale factor** tells us how much to enlarge or shrink a figure. If the scale factor is greater than one, the image is larger than the pre-image. If the scale factor is between zero and one, the image is smaller than the pre-image.

Note: In a dilation, the pre-image and image are similar instead of congruent.

Since a dilation centered at the origin with scale factor s is moving points in the plane, we can describe its effect in terms of the coordinates. If the point (a, b) is on the pre-image, then the point (sa, sb) is its image after dilation. That is, we simply multiply the coordinates by the scale factor.

In this lesson, the origin will be used as the center of the dilation.

Example 1 Given the dilation shown on the graph, compare the scale factor with 1. Assume the origin is the center of the dilation.

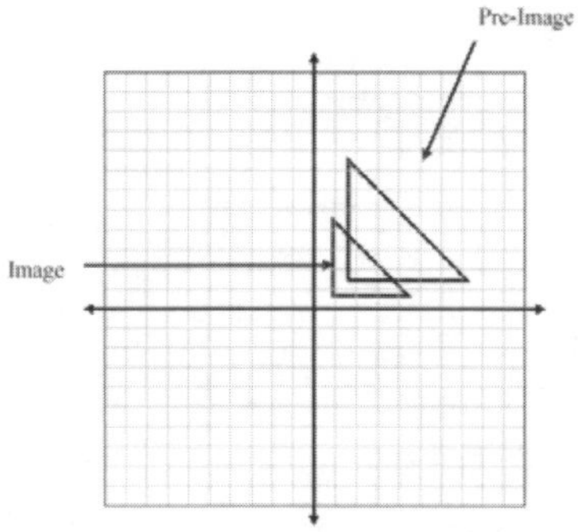

Step 1: Remember, a scale factor is a number which describes the ratio of an image to a pre-image.

The scale factor is between 0 and 1.

Step 2: Since the image is smaller than the pre-image, the scale factor must be between 0 and 1.

Answer: The scale factor is between 0 and 1.

Example 2 What is the image of the point $(2, -3)$ for the dilation centered at the origin with a scale factor of 4?

$(2 \times 4, -3 \times 4)$

Step 1: Multiply the coordinates of the point by the scale factor. In this case, the scale factor is 4.

$(8, -12)$

Step 2: Simplify.

Answer: The image of the point $(2, -3)$ is $(8, -12)$.

Example 3 Consider the triangle with vertices $A(-2, 0)$, $B(0, 4)$, and $C(1, -3)$. If we dilate this triangle with a dilation whose center is the origin, and with a scale factor of 3, what are the vertices of the dilated triangle?

$$A(-2,0) \to (-2\times 3, 0\times 3) = A'(-6,0)$$
$$B(0,4) \to (0\times 3, 4\times 3) = B'(0,12)$$
$$C(1,-3) \to (1\times 3, -3\times 3) = C'(3,-9)$$

Step 1: To obtain the image, multiply each point by the scale factor of 3.

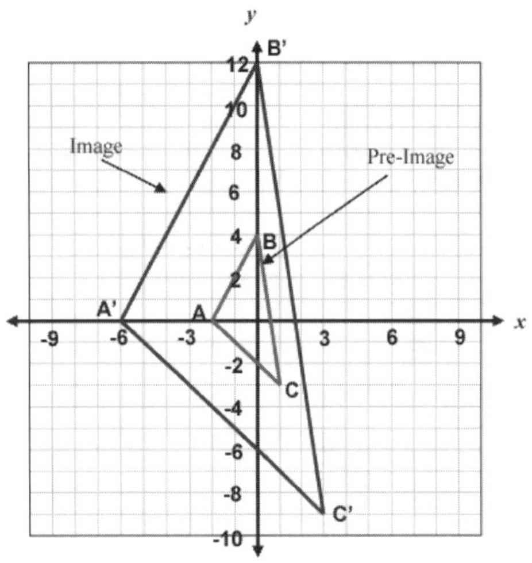

Step 2: Draw both the image and the pre-image on a coordinate plane.

Answer: The vertices of the dilated triangle are $(-6, 0)$, $(0, 12)$, and $(3, -9)$.

Example 4 A rectangle is defined by the vertices $A(-1,-1)$, $B(3,-1)$, $C(3,5)$ and $D(-1,5)$. Find the image of the rectangle that is dilated with the center at the origin and a scale factor of $\frac{1}{4}$.

$$A(-1,-1)\rightarrow\left(-1\times\frac{1}{4},-1\times\frac{1}{4}\right)=A'\left(-\frac{1}{4},-\frac{1}{4}\right)$$
$$B(3,-1)\rightarrow\left(3\times\frac{1}{4},-1\times\frac{1}{4}\right)=B'\left(\frac{3}{4},-\frac{1}{4}\right)$$
$$C(3,5)\rightarrow\left(3\times\frac{1}{4},5\times\frac{1}{4}\right)=C'\left(\frac{3}{4},\frac{5}{4}\right)$$
$$D(-1,5)\rightarrow\left(-1\times\frac{1}{4},5\times\frac{1}{4}\right)=D'\left(-\frac{1}{4},\frac{5}{4}\right)$$

Step 1: To obtain the image, multiply each point by the scale factor of $\frac{1}{4}$.

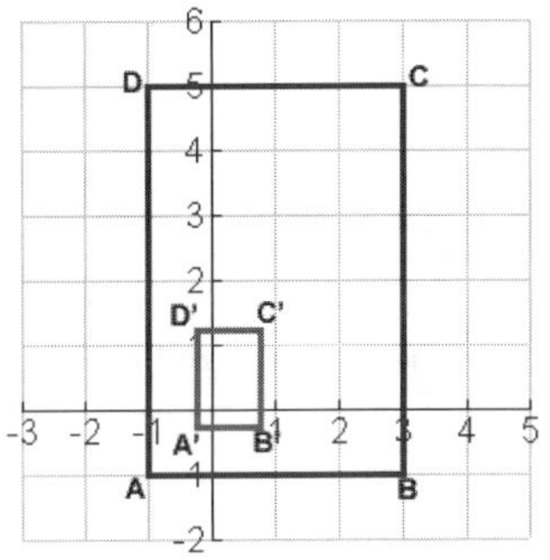

Step 2: Draw both the image and the pre-image on a coordinate plane.

Answer: The vertices of the dilated rectangle are $\left(-\frac{1}{4},-\frac{1}{4}\right)$, $\left(\frac{3}{4},-\frac{1}{4}\right)$, $\left(\frac{3}{4},\frac{5}{4}\right)$, and $\left(-\frac{1}{4},\frac{5}{4}\right)$.

Example 5 Suppose the point $(3,6)$ is transformed by a dilation centered at the origin to the point $(1,2)$. What is the scale factor of the dilation?

$$(3,6)\rightarrow(3\times s,6\times s)=(1,2)$$

Step 1: Let s represent the scale factor. We need to multiply the pre-image by a common number that represents the scale factor to obtain the image.

$$(3,6)\rightarrow\left(3\times\frac{1}{3},6\times\frac{1}{3}\right)=(1,2)$$

Step 2: The image is produced when we multiply the pre-image by a scale factor of $\frac{1}{3}$.

Answer: The scale factor of the dilation is $\frac{1}{3}$.

Problem Set

1. Given the dilation on the graph, compare the scale factor with one. Assume the origin as the center of dilation.

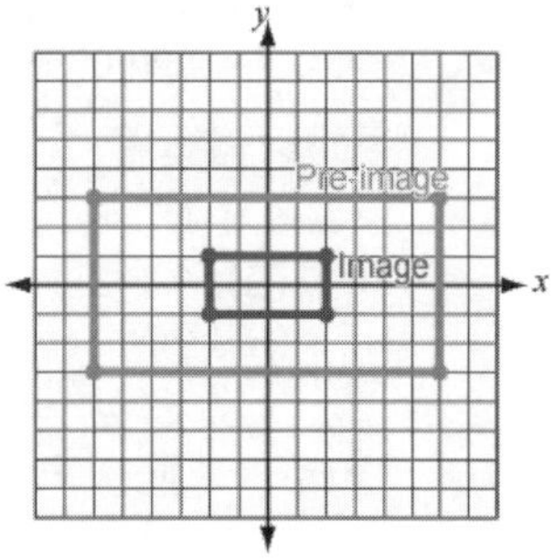

2. Given the dilation on the graph, compare the scale factor with one. Assume the origin as the center of dilation.

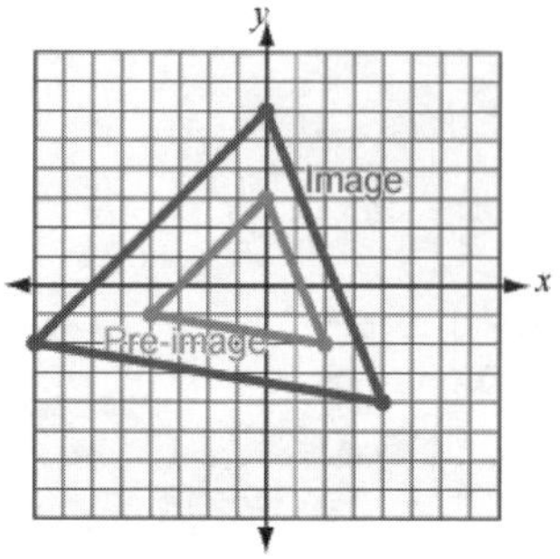

3. Given the dilation on the graph, compare the scale factor with one. Assume the origin as the center of dilation.

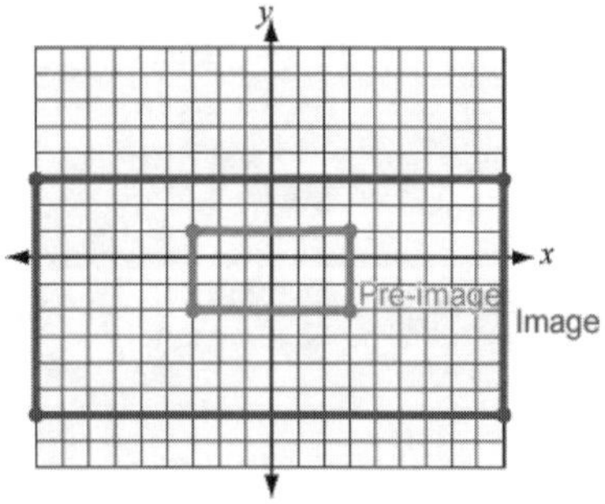

4. Given the dilation on the graph, compare the scale factor with one. Assume the origin as the center of dilation.

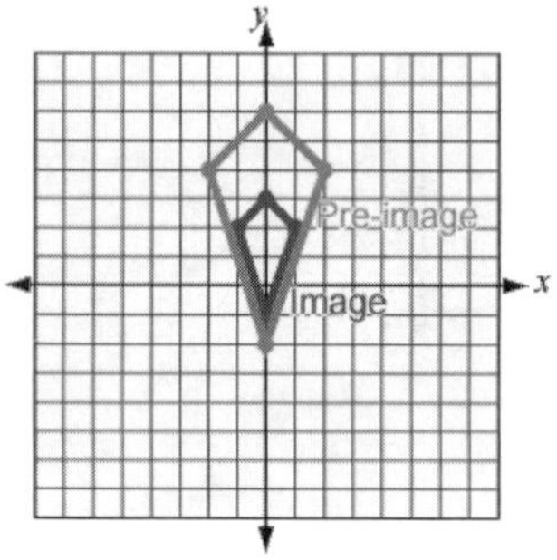

5. Given the dilation on the graph, compare the scale factor with one. Assume the origin as the center of dilation.

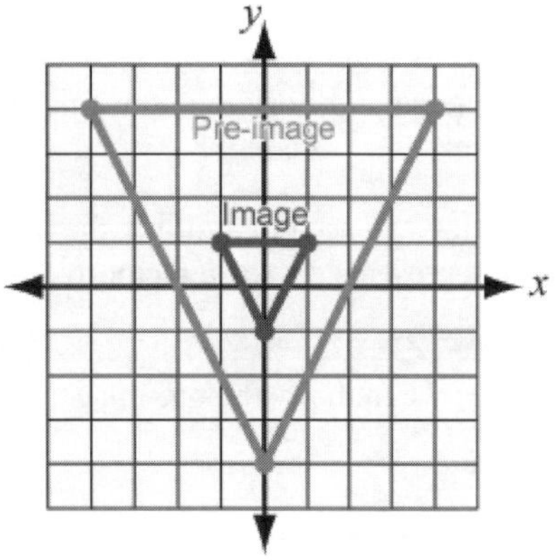

6. Which of the following pair of images represents a dilation with the center of dilation at the origin?

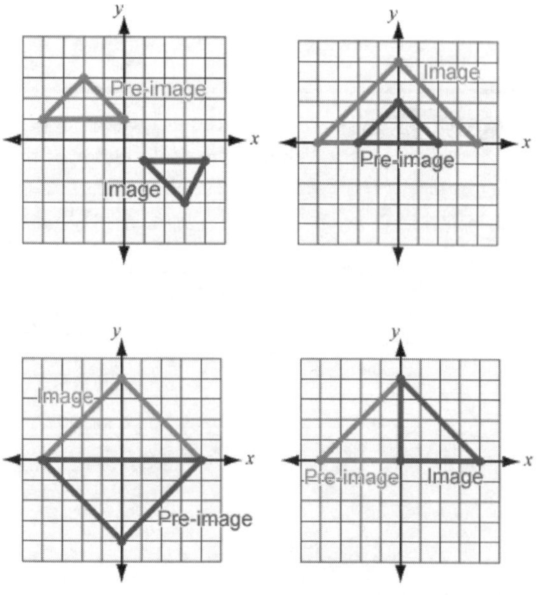

7. Which of the following pair of images represents a dilation with the center of dilation at the origin?

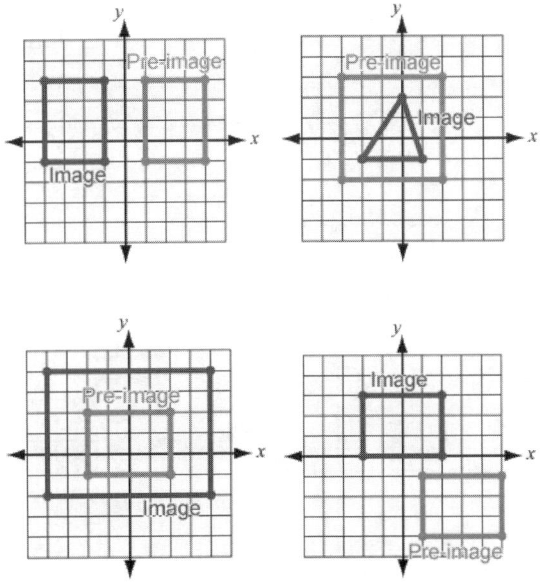

8. Which of the following pair of images represents a dilation with the center of dilation at the origin?

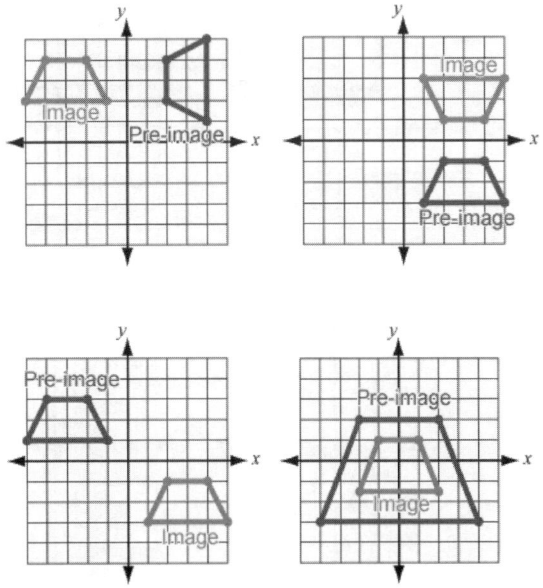

9. Which of the following pair of images represents a dilation with the center of dilation at the origin?

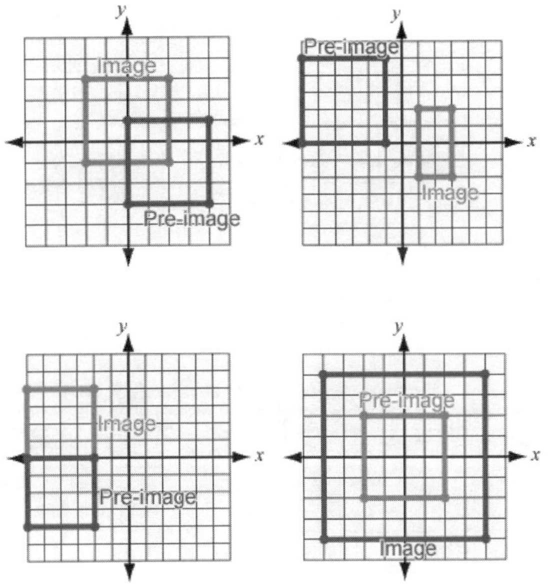

10. Which of the following pair of images represents a dilation with the center of dilation at the origin?

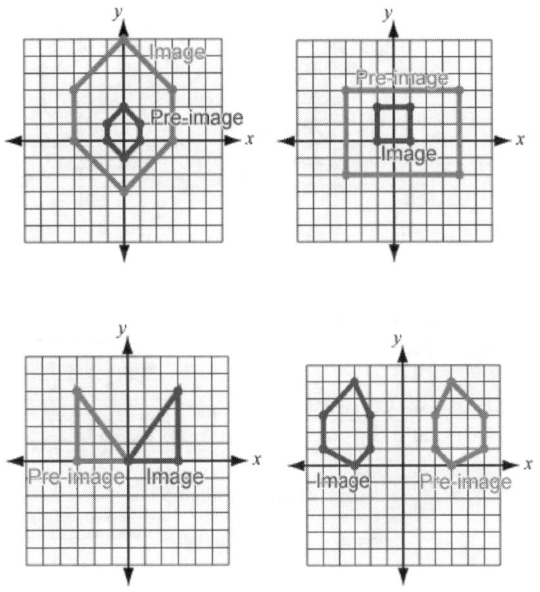

11. Which of the following pair of images represents a dilation with the center of dilation at the origin?

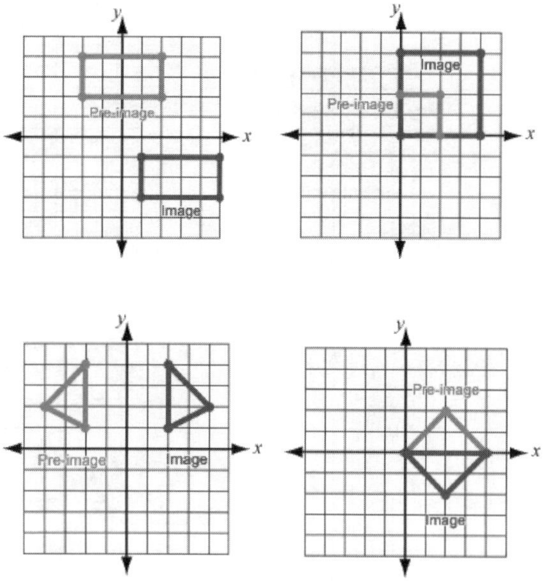

12. Which of the following pair of images represents a dilation with the center of dilation at the origin?

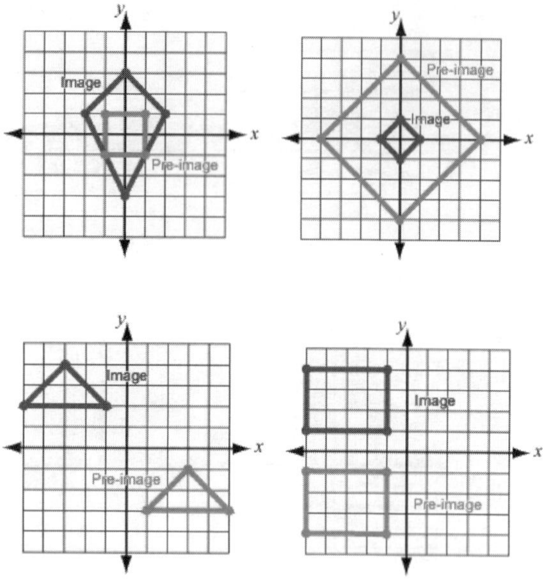

13. Which of the following pair of images represents a dilation with the center of dilation at the origin?

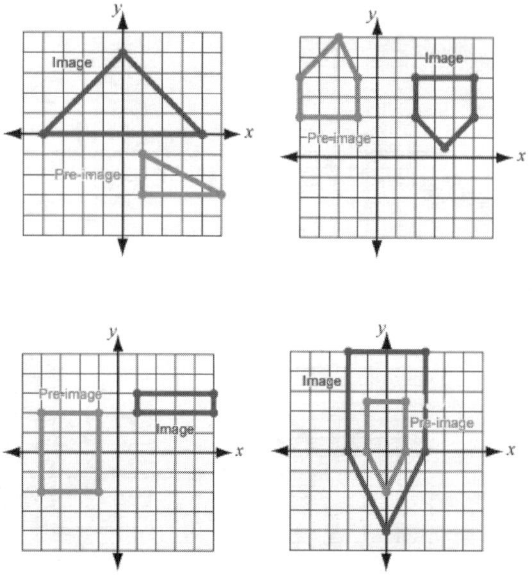

14. Which of the following pair of images represents a dilation with the center of dilation at the origin?

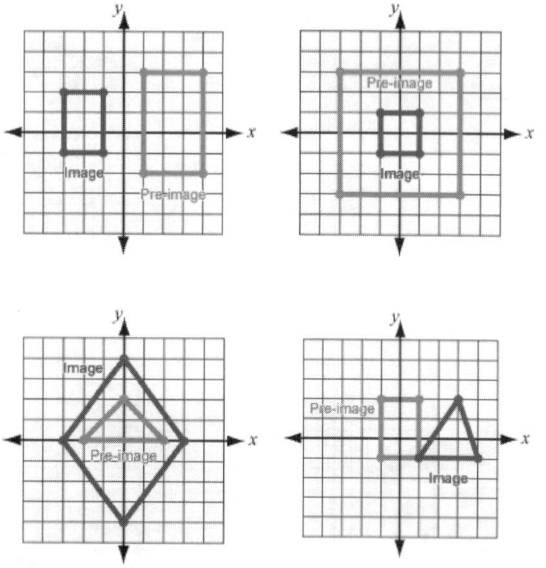

15. Which of the following pair of images represents a dilation with the center of dilation at the origin?

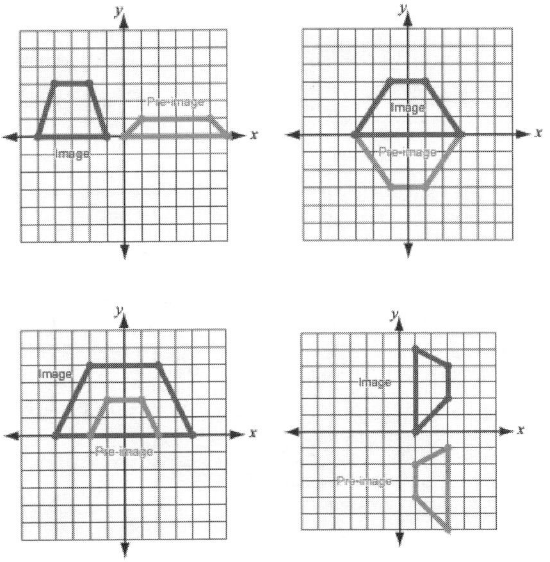

16. Which of the following pair of images represents a dilation with the center of dilation at the origin?

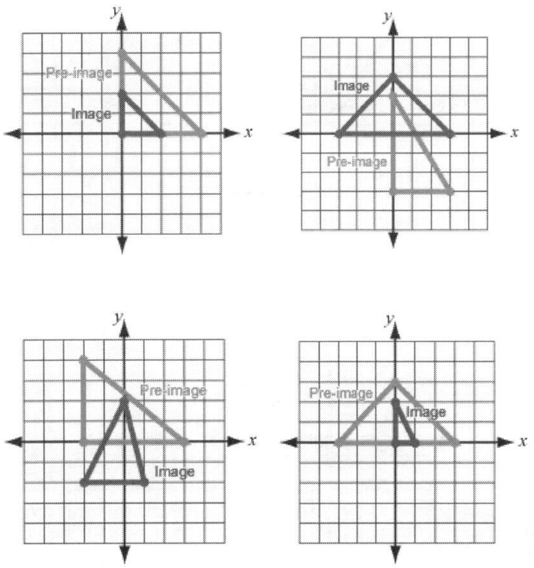

17. Which of the following pair of images represents a dilation with the center of dilation at the origin?

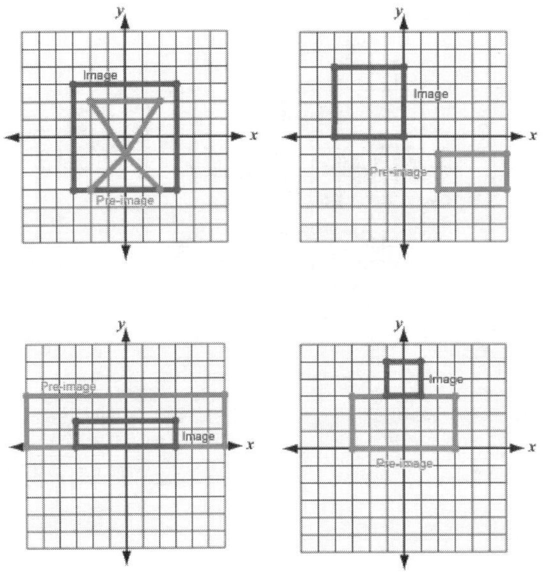

18. Which of the following pair of images represents a dilation with the center of dilation at the origin?

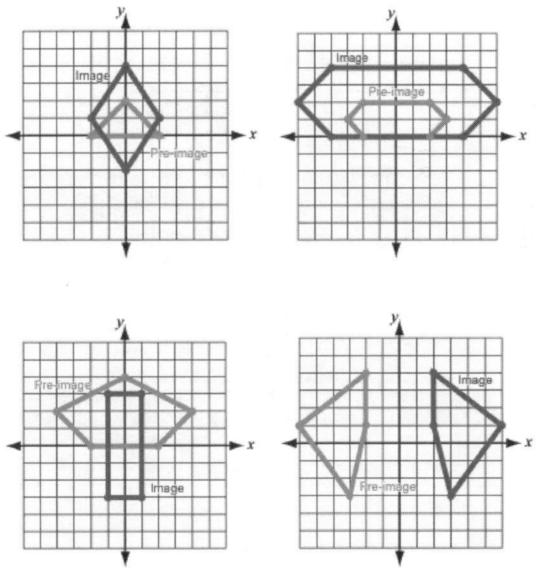

19. Which of the following pair of images represents a dilation with the center of dilation at the origin?

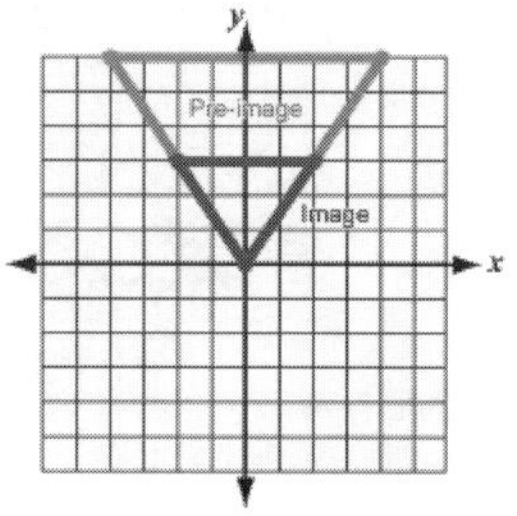

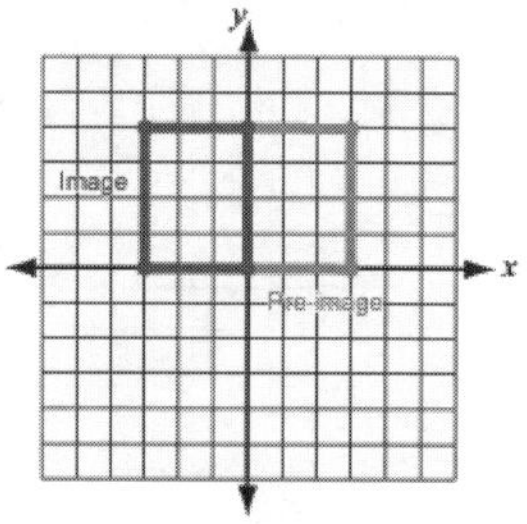

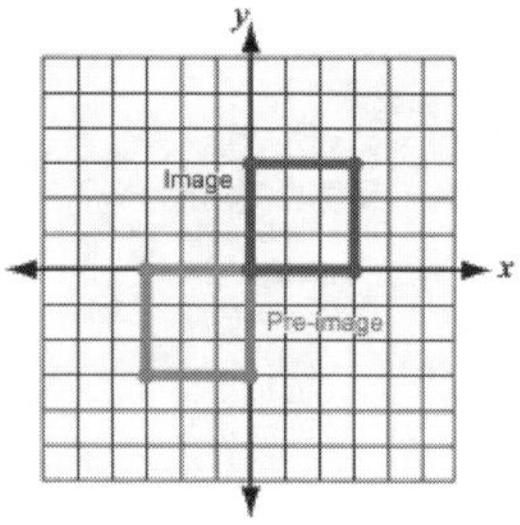

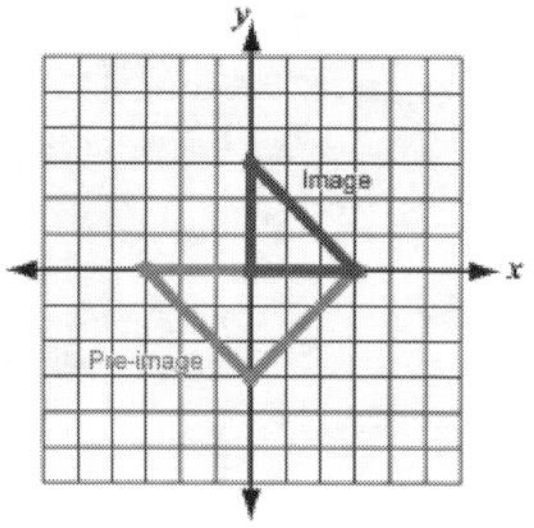

20. Which of the following pair of images represents a dilation with the center of dilation at the origin?

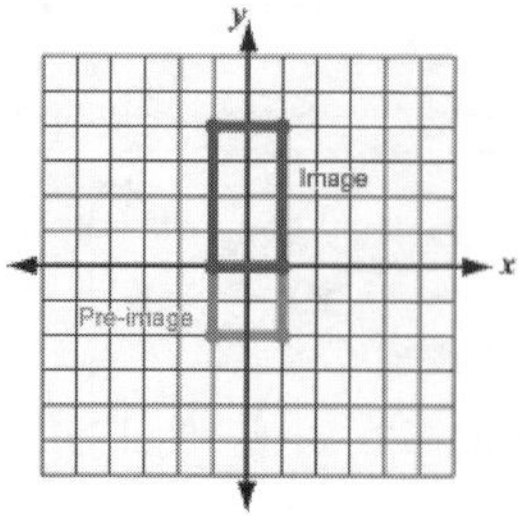

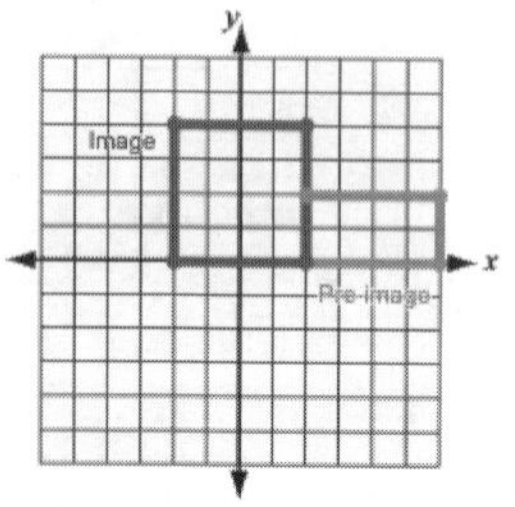

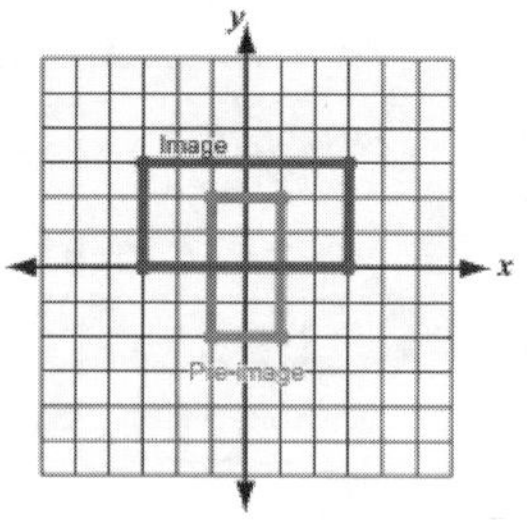

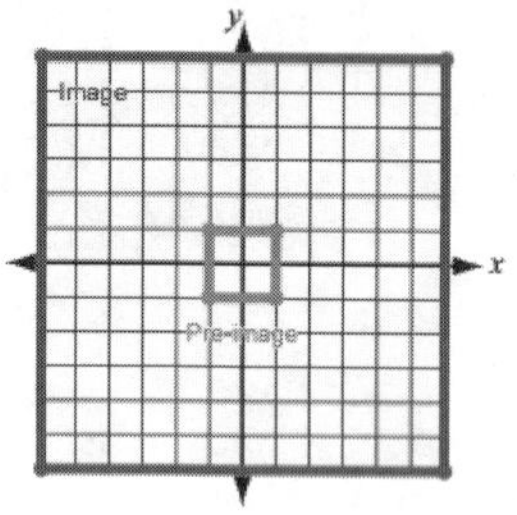

21. Draw the dilation image of trapezoid *CDEF* with the center at the origin and a scale factor of 2.

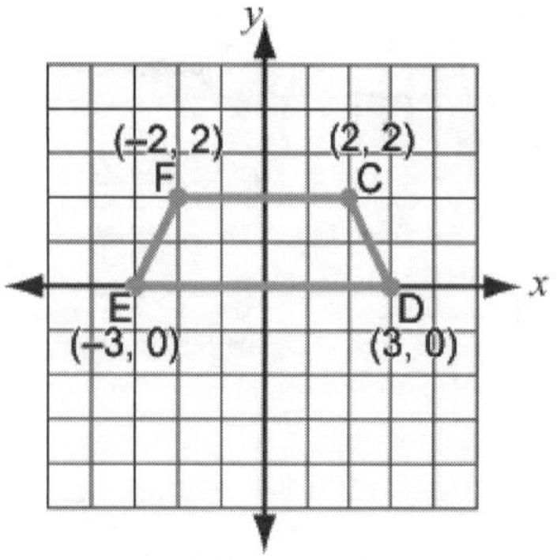

22. Draw the dilation image of rectangle *FGHJ* with the center of dilation at the origin and a scale factor of 3.

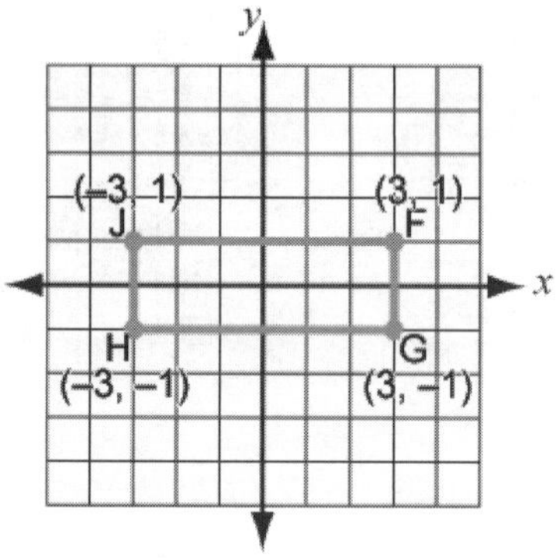

23. Draw the dilation image of hexagon *ABCDEF* with the center of dilation at the origin and a scale factor of $\frac{2}{3}$.

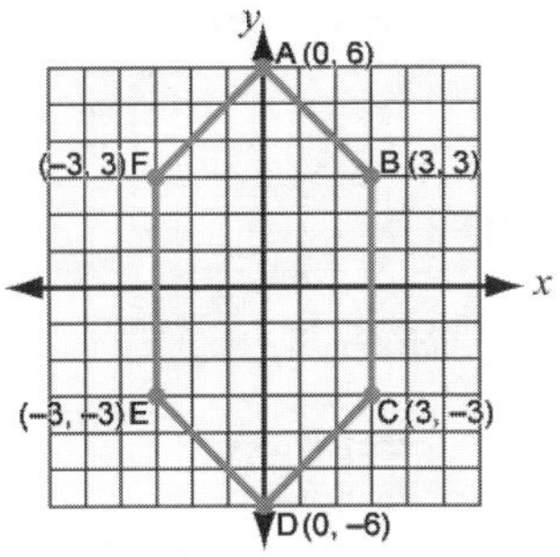

24. Draw the dilation image of kite *HJKL* with the center of dilation at the origin and a scale factor of $\frac{3}{4}$.

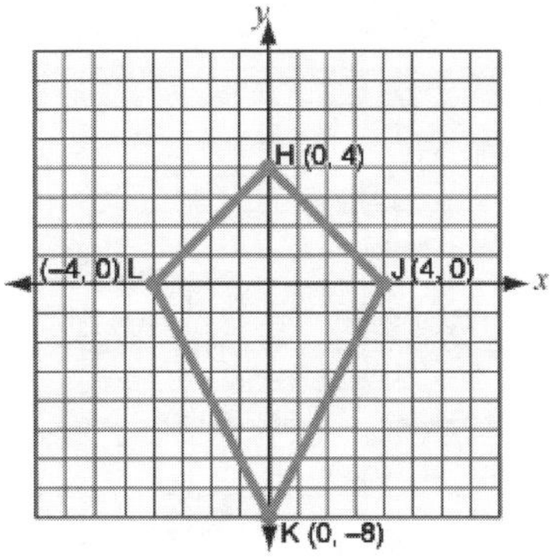

25. Draw the dilation image of square *WXYZ* with the center of dilation at the origin and a scale factor of $\frac{3}{8}$.

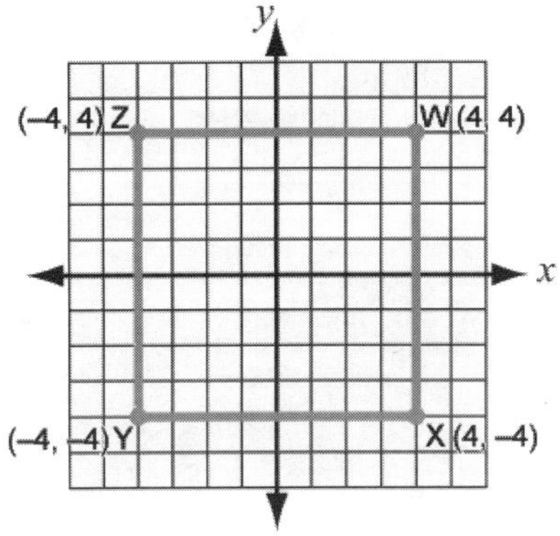

26. What is the scale factor of the dilation? Assume that the center of dilation is at the origin.

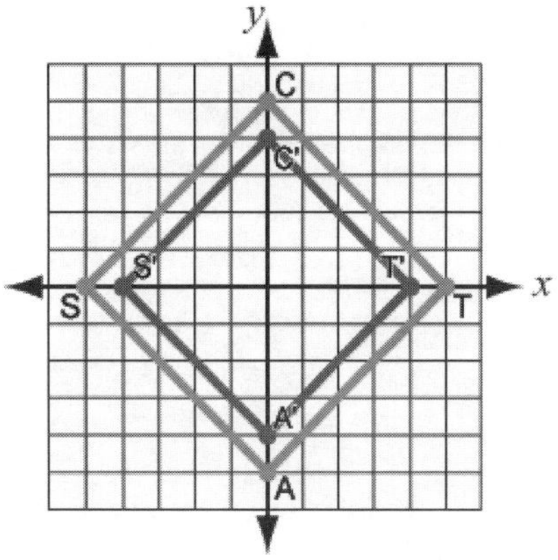

27. What is the scale factor of the dilation? Assume that the center of dilation is at the origin.

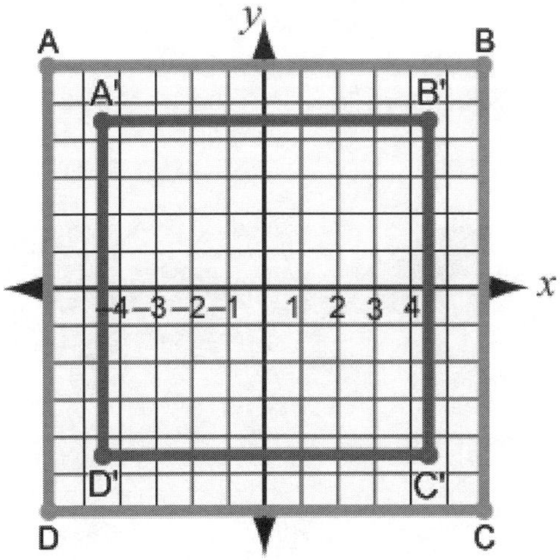

28. What is the scale factor of the dilation? Assume that the center of dilation is at the origin.

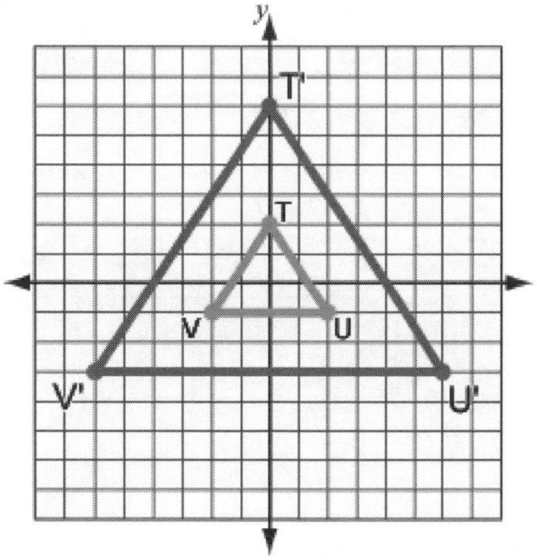

29. What is the scale factor of the dilation? Assume that the center of dilation is at the origin.

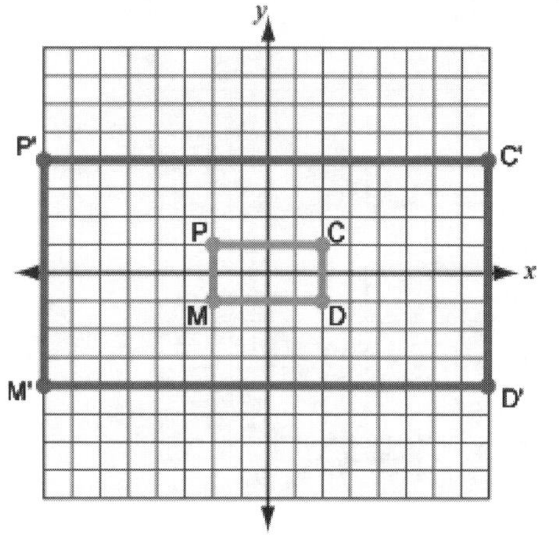

30. What is the scale factor of the dilation? Assume that the center of dilation is at the origin.

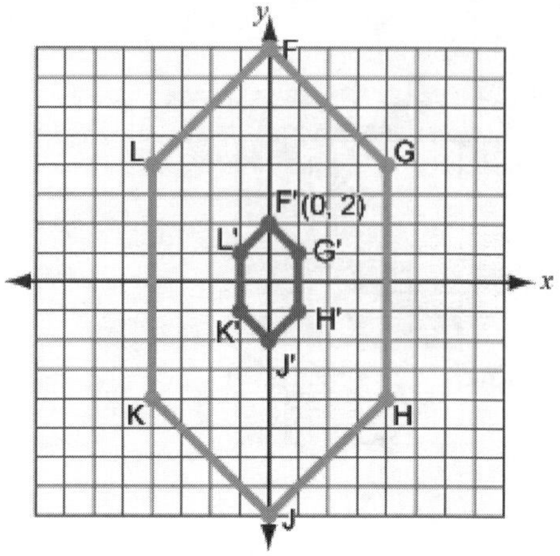

MPA-121: Identifying Similar and Congruent Polygons Using Proportions

Two polygons are **similar** when they have the same shape but not necessarily the same size. Specifically, in order to be similar, two polygons must have congruent, corresponding angles and proportional, corresponding sides. An example of two similar quadrilaterals is shown below.

Similar Polygons	Two polygons are similar if they have congruent, corresponding angles, and proportional, corresponding sides.

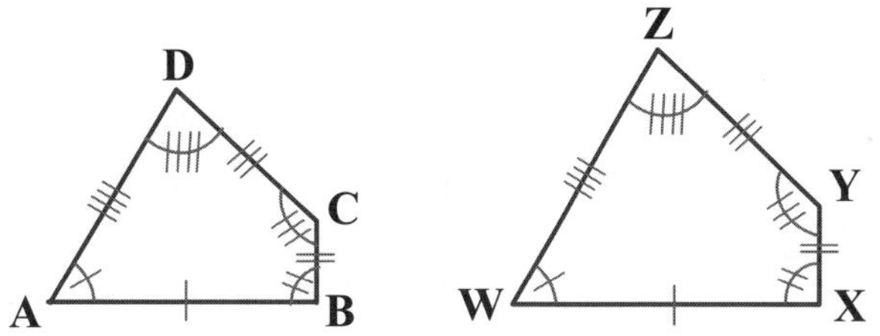

Corresponding Angles	Corresponding Sides
$\angle A \leftrightarrow \angle W$	$\overline{AB} \leftrightarrow \overline{WX}$
$\angle B \leftrightarrow \angle X$	$\overline{BC} \leftrightarrow \overline{XY}$
$\angle C \leftrightarrow \angle Y$	$\overline{CD} \leftrightarrow \overline{YZ}$
$\angle D \leftrightarrow \angle Z$	$\overline{AD} \leftrightarrow \overline{WZ}$

Congruent Angles: $\angle A \cong \angle W$, $\angle B \cong \angle X$, $\angle C \cong \angle Y$, $\angle D \cong \angle Z$

Proportional Corresponding Sides: $\frac{AB}{WX} = \frac{BC}{XY} = \frac{CD}{YZ} = \frac{AD}{WZ}$

Two polygons are **congruent** when they have the same shape and the same size. Specifically, two polygons are congruent if they have congruent, corresponding angles and congruent, corresponding sides. When a rectangle is cut into two pieces along its diagonal, two congruent triangles are formed as shown below.

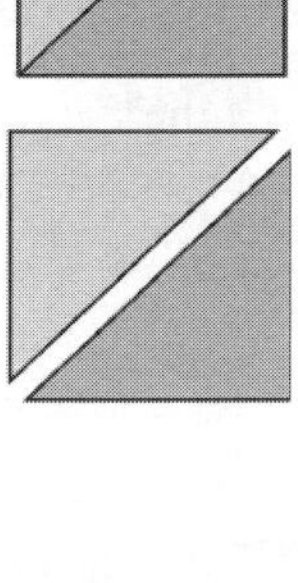

Congruent Polygons	Two polygons are congruent if they have congruent, corresponding angles and congruent, corresponding sides.

Example 1 Show that the rectangles *ABCD* and *EFGH* are similar.

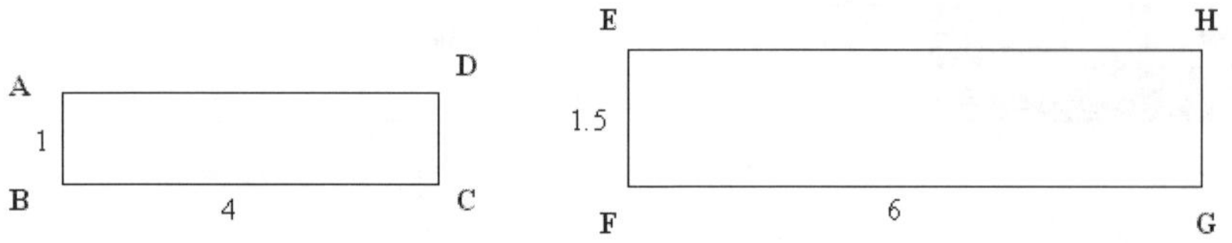

$\angle A \cong \angle E$
$\angle B \cong \angle F$
$\angle C \cong \angle G$
$\angle D \cong \angle H$

Step 1: Notice that all of the angles are right angles since the two figures are rectangles. So, the corresponding angles are congruent.

$$\frac{AB}{EF} = \frac{BC}{FG} = \frac{CD}{GH} = \frac{AD}{EH}$$

$$\frac{1}{1.5} = \frac{4}{6} = \frac{1}{1.5} = \frac{4}{6}$$

Step 2: Since we know that the corresponding angles of the two rectangles are congruent, we must determine whether the corresponding sides are proportional.

$$\frac{1}{1.5} = \frac{4}{6}$$

$$6 = 6$$

Step 3: Use cross-multiplication to check if the ratios are equal. Since the cross products are equal, the corresponding sides are proportional.

Answer: Rectangle *ABCD* and rectangle *EFGH* are similar.

Example 2 State whether the triangles are similar. If so, what angle corresponds to angle *A*?

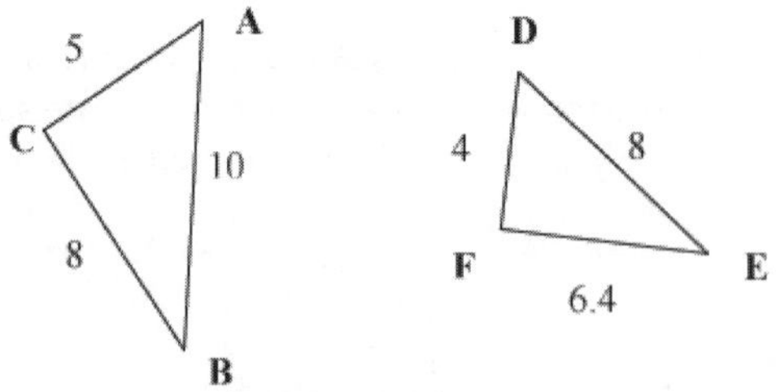

$$\frac{AB}{DE} = \frac{BC}{EF} = \frac{AC}{DF}$$

Step 1: If these triangles are similar, the ratio of the corresponding sides are equal.

$$\frac{10}{8} = \frac{8}{6.4} = \frac{5}{4}$$

$$\frac{10}{8} = \frac{8}{6.4} \qquad \frac{8}{6.4} = \frac{5}{4}$$

$$10 \times 6.4 = 8 \times 8 \qquad 8 \times 4 = 6.4 \times 5$$

$$64 = 64 \qquad 32 = 32$$

Step 2: Write the ratios using the lengths of the sides and cross-multiply. Since the cross products are equal, the ratios are equal. Therefore, the triangles are similar.

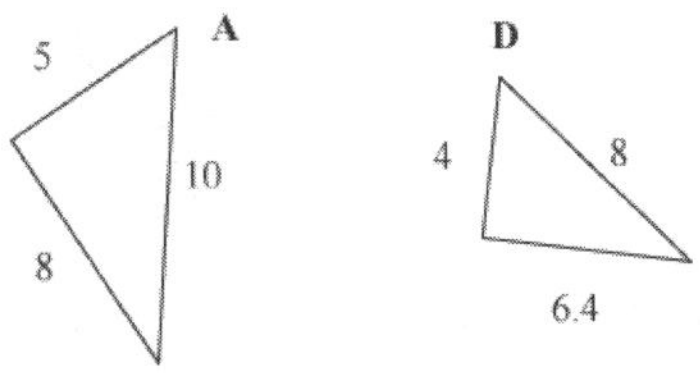

Step 3: To find which angle corresponds to $\angle A$, notice that $\angle A$ lies between the longest and shortest sides of the triangle. In the similar triangle, the corresponding angle is $\angle D$.

Answer: Triangle *ABC* and triangle *DEF* are similar. The angle that corresponds to $\angle A$ is $\angle D$.

Example 3 Explain why these two quadrilaterals are not similar.

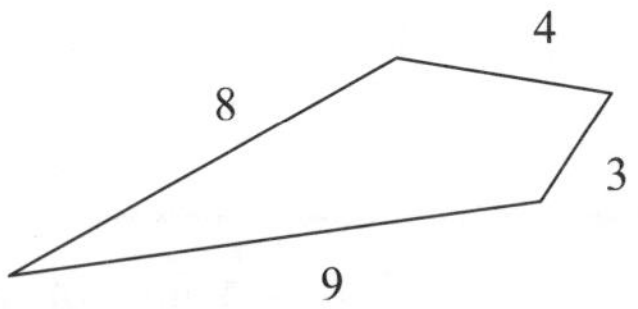

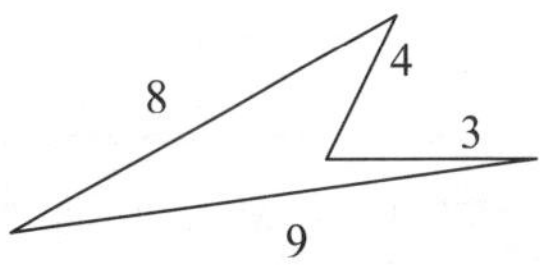

Step 1: The quadrilaterals have four sides of the same length, but they don't have congruent angles.

Step 2: Notice that the angle between the sides of length 8 and 4, in one case, is obtuse (left), and in the other, it is acute (right).

Step 3: Since the corresponding angles are not congruent, the figures are not similar.

Answer: The quadrilaterals do not have congruent corresponding angles, therefore they are not similar.

Example 4 The two right triangles shown are similar with side AB corresponding to side DE. Find the length of the missing side.

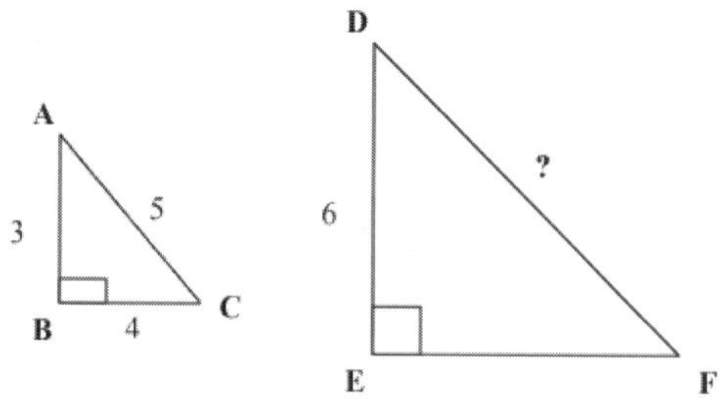

$\frac{DE}{AB} = \frac{DF}{AC}$

Step 1: Since these triangles are similar, the ratios of corresponding sides are equal.

$\frac{6}{3} = \frac{n}{5}$

Step 2: Write the ratios using the lengths of the sides. Let n represent the missing side length.

$6 \times 5 = 3n$

$\frac{30}{3} = \frac{3n}{3}$

$10 = n$

Step 3: Cross-multiply and solve for n.

Answer: The length of the missing side $\overline{DF}$ is 10 units.

Example 5 If trapezoid $ABCD$ is similar to trapezoid $EFGH$, find the length of GH.

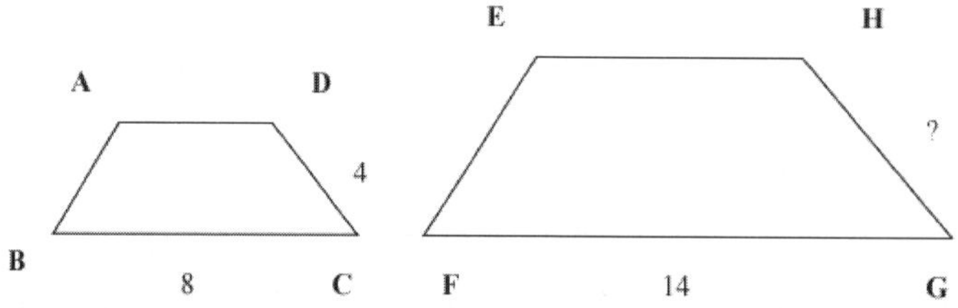

$\frac{FG}{BC} = \frac{GH}{CD}$

Step 1: Since the trapezoids are similar, the ratios of the corresponding sides are equal.

$\frac{14}{8} = \frac{n}{4}$

Step 2: Write the ratios using the lengths of the sides. Let n represent the missing side length.

$14 \times 4 = 8n$

$\frac{56}{8} = \frac{8n}{8}$

$7 = n$

Step 3: Cross-multiply and solve for n.

Answer: The length of the missing side $\overline{GH}$ is 7 units.

Problem Set

1. Label the angles of the similar polygons to show that the corresponding angles are congruent.

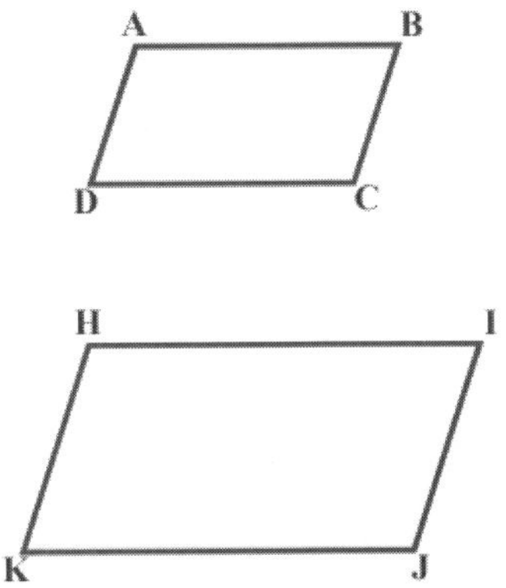

2. Label the angles of the similar polygons to show that the corresponding angles are congruent.

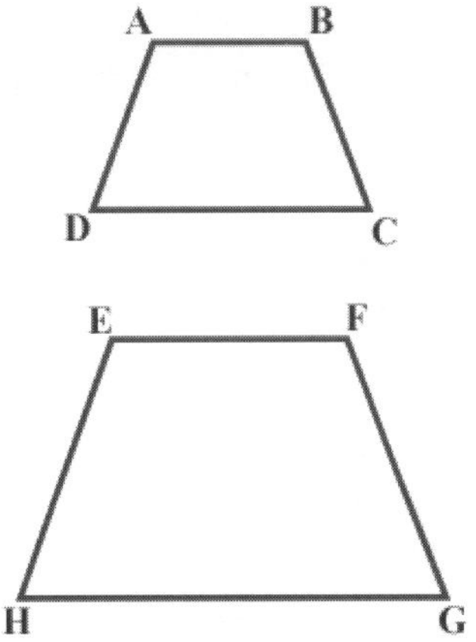

3. Label the angles of the similar polygons to show that the corresponding angles are congruent.

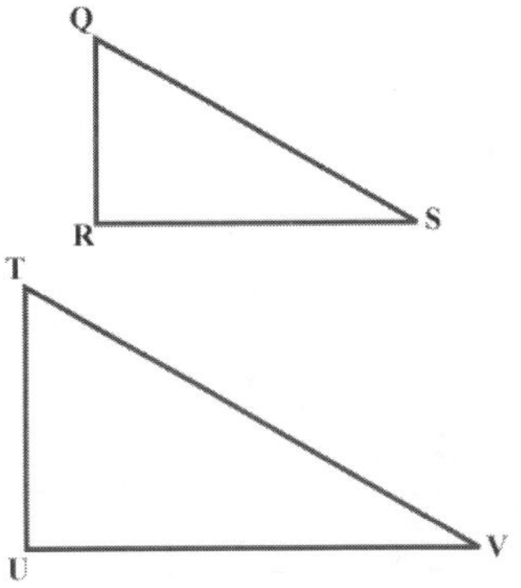

4. Label the angles of the similar polygons to show that the corresponding angles are congruent.

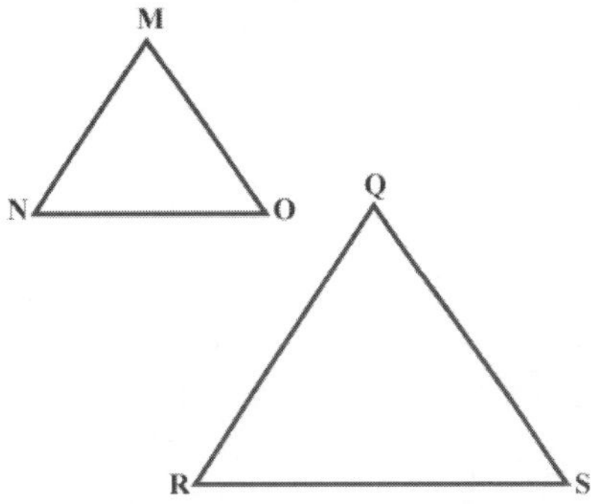

5. Label the angles of the similar polygons to show that the corresponding angles are congruent.

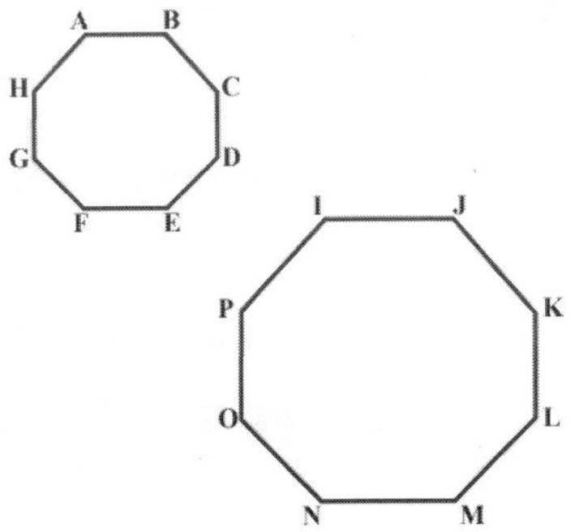

6. Interpret the mathematical statement below.

$\Delta ABC \sim \Delta MNO$

7. Interpret the mathematical statement below.

$\angle A \Leftrightarrow \angle X$

8. Interpret the mathematical statement below.

$\angle B \cong \angle Y$

9. Interpret the mathematical statement below.

$\overline{XY} \Leftrightarrow \overline{MN}$

10. Interpret the mathematical statement below.

$\overline{AB} \Leftrightarrow \overline{JK}$

11. Given that ▱*ABCD* ~ ▱*HIJK*, write a set of proportions given by the similarity.

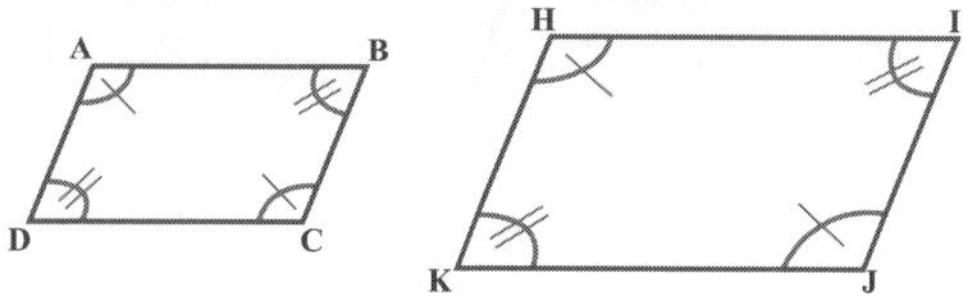

12. Given that ⏢*ABCD* ~ ⏢*EFGH*, write a set of proportions given by the similarity.

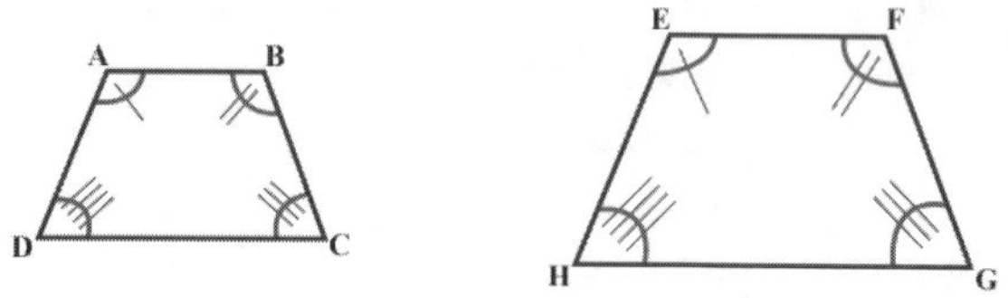

13. Given that $\Delta QRS \sim \Delta TUV$, write a set of proportions given by the similarity.

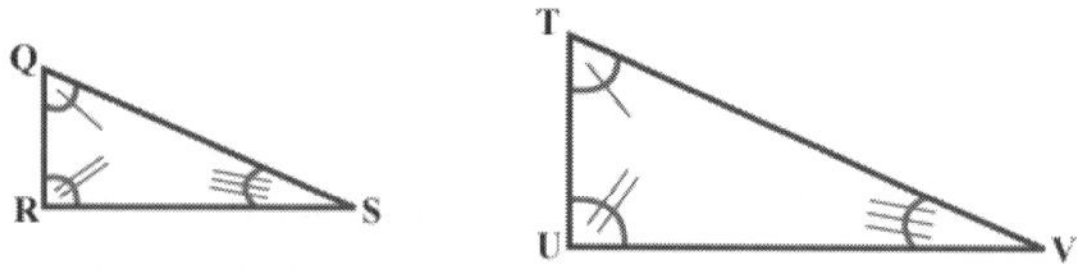

14. Which of the following figures is similar to rectangle *ABCD*?

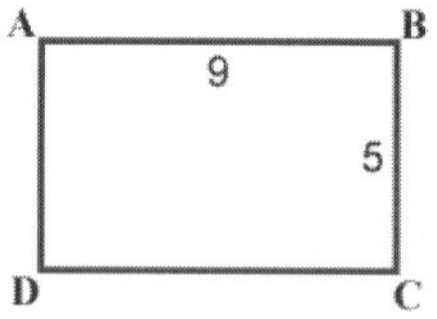

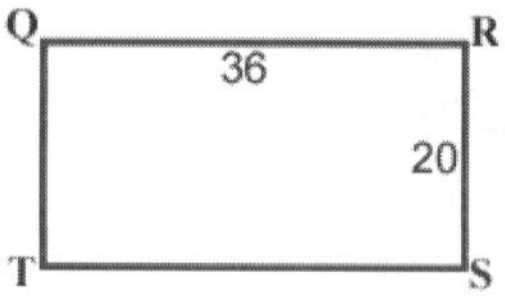

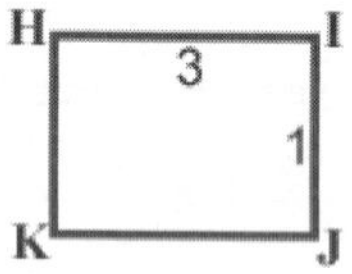

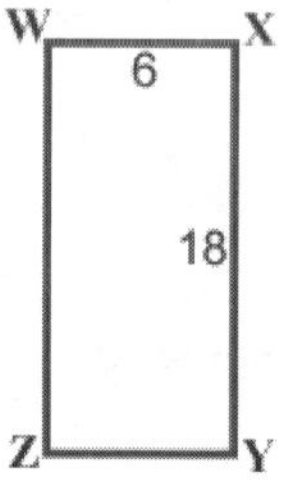

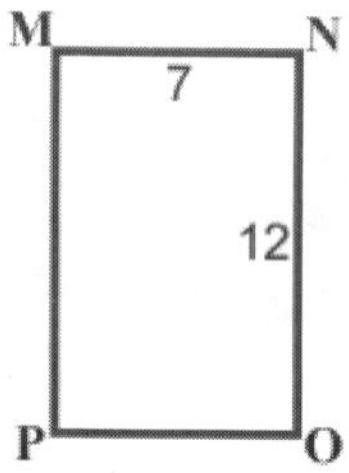

15. Which of the following figures is similar to rectangle *ABCD*?

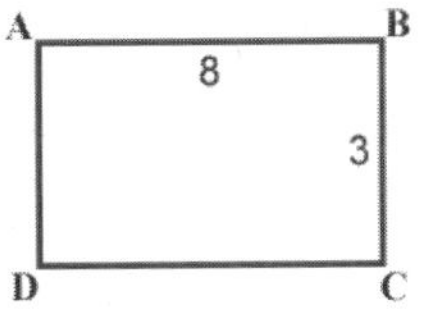

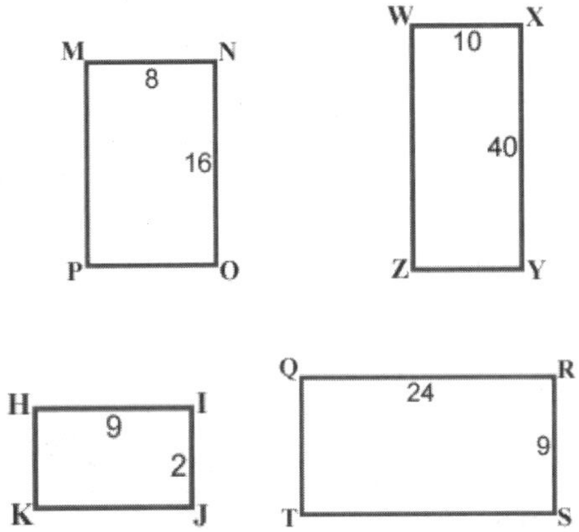

16. Find the missing length in the two similar figures below.

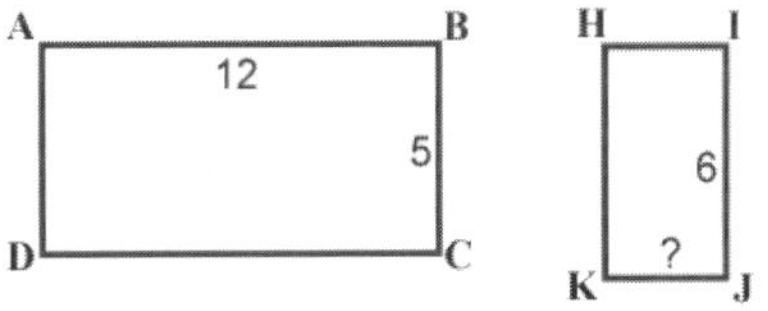

17. Find the missing length in the two similar figures below.

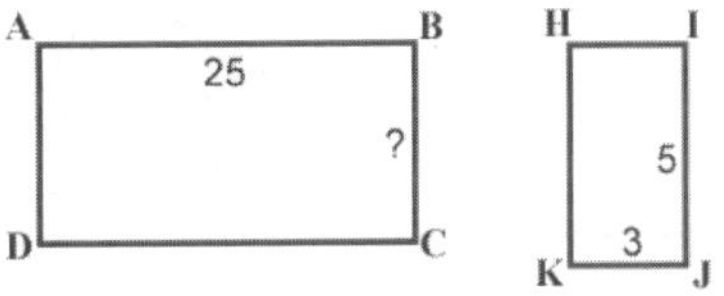

18. Find the missing length in the two similar figures below.

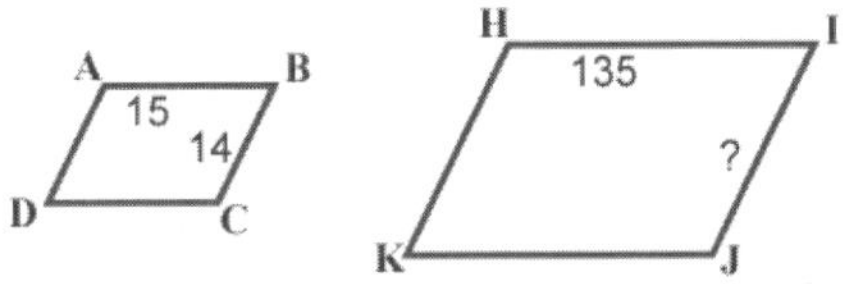

19. Find the missing length in the two similar figures below.

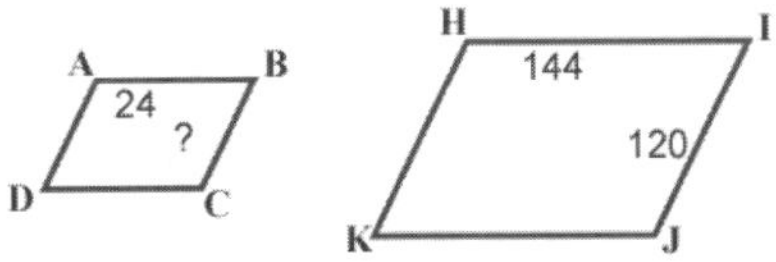

20. Find the missing length in the two similar figures below.

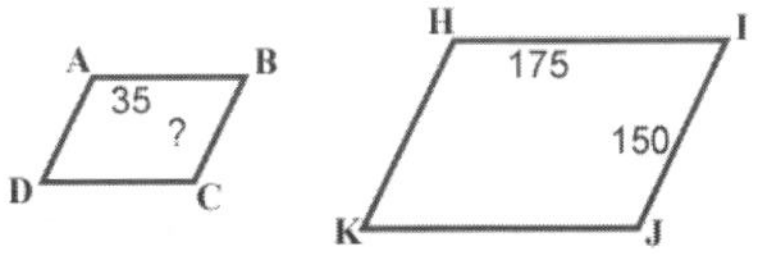

MPA-122: Modeling Multiplication and Division of Decimals

This lesson explores how to model the multiplication and division of decimals. To multiply decimals, imagine that two numbers represent the lengths of the sides of a rectangle. Think of the product as the area of the rectangle, as shown in the example below. Each division is 0.1 units of length, making each small square 0.1 units long by 0.1 units wide. Therefore, the area of each small square is 0.01 square units.

Multiply Using a Model: 0.8×1.3

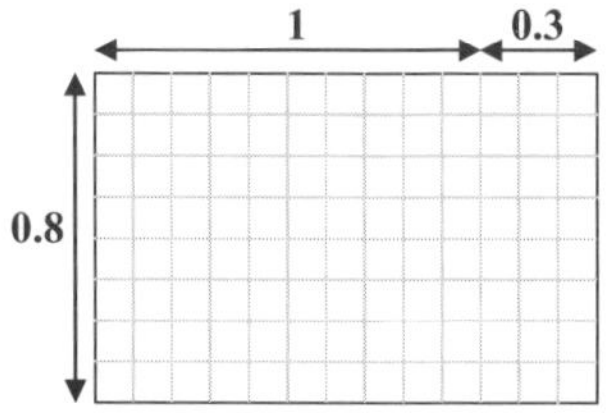

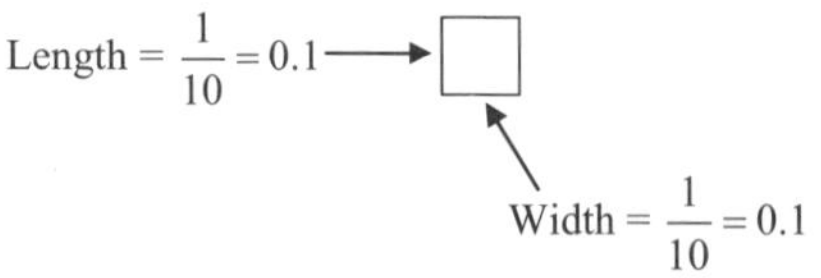

$$\begin{aligned} \text{Area} &= \text{Height} \times \text{Width} \\ &= 0.1 \times 0.1 \\ &= 0.01 \text{ units}^2 \end{aligned}$$

Example 1 Draw an area model that represents 1.3×2.1 .

Step 1: Since each square of an area model represents one hundredth of a unit, the width 1.3 is 13 squares wide and the length 2.1 is 21 squares long.

Answer:

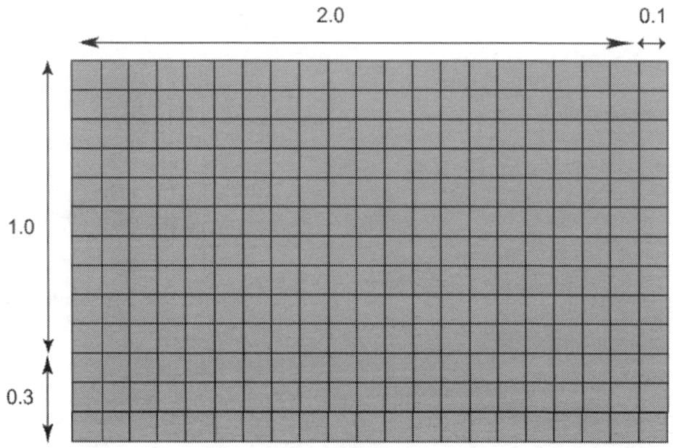

Example 2 An area model represents the product of 0.8×0.3. If the length and width are made up of 8 and 3 small squares respectively, how many small squares will there be in the area of the rectangle? What is the product of 0.8×0.3?

$8 \times 3 = 24$ **Step 1:** Multiply.

$24 \times 0.01 = 0.24$ **Step 2:** Each small square has an area of 0.01 square units.

Answer: The model will consist of a rectangle that is composed of 24 small squares and the product of 0.8×0.3 is 0.24.

Example 3 Use an area model to find the product of 2.1×3.6.

Step 1: Create a rectangle with a width of 2.1 and a length of 3.6 marked in tenths of a unit.

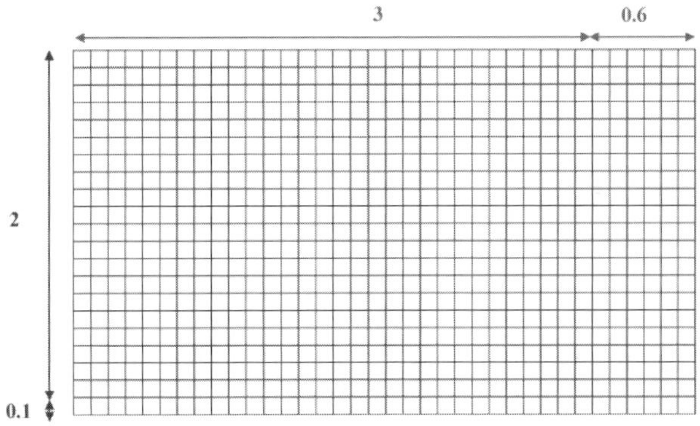

Step 2: Now we can see that the product 2.1 times 3.6 can be thought of as the area of a rectangle with a length of 3.6 units and a width of 2.1 units. The area is the sum of the areas of the four pieces shown in the diagram. Compute each of the areas separately.

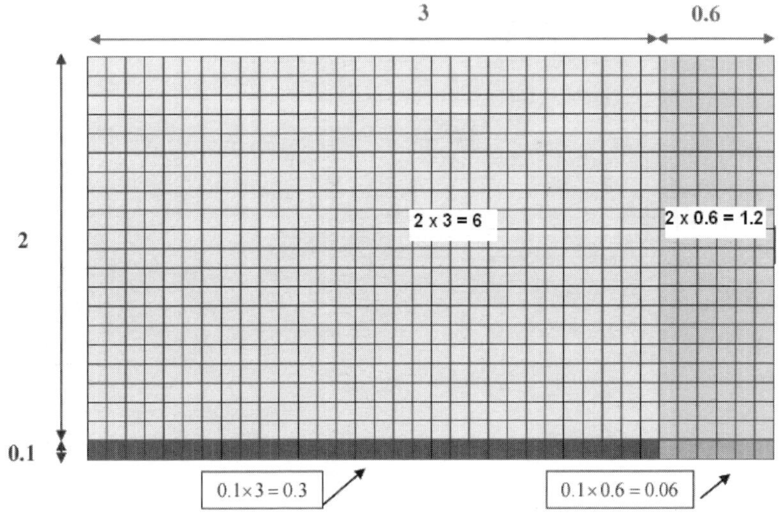

$20 \times 30 = 600$ squares Area = 6 units^2 $20 \times 6 = 120$ squares Area = 1.2 units^2	**Step 3:** The 2×3 area is 20 squares by 30 squares, for a total of 600 squares. Since each square is 0.01 units of area, this represents 6 square units of area. The 2×0.6 area is 20 squares by 6 squares, for a total of 120 squares. This represents 1.2 square units of area.
$1 \times 30 = 30$ squares Area = 0.3 units^2 $1 \times 6 = 6$ squares Area = 0.06 units^2	**Step 4:** The 0.1×3 area is 1 square by 30 squares, for a total of 30 squares. This represents 0.3 square units of area. The 0.1×0.6 area is 1 square by 6 squares, for a total of 6 squares. This represents 0.06 square units of area.
$6 + 1.2 + 0.3 + 0.06 = 7.56$	**Step 5:** The total area is the sum of the areas of the sections. This total area is the product of 2.1×3.6.

Answer: The product of 2.1×3.6 is 7.56.

This method can also be adapted to model the division of decimals. In this case, we know the area (dividend) and one of the sides (divisor), and we are looking for the other side (quotient). For example, the division problem $2.55 \div 1.7$ can be modeled as a rectangle with an area of 2.55 square units, and one side of length 1.7 units. With this information, the length of the other side can be determined.

Example 4 Use the area model to find the quotient of $2.55 \div 1.7$.

Step 1: Start with a rectangle of width 1.7. How wide must this rectangle be to have an area of 2.55? Since the area is 2.55, and each small square has an area 0.01, the rectangle is composed of 255 small squares.

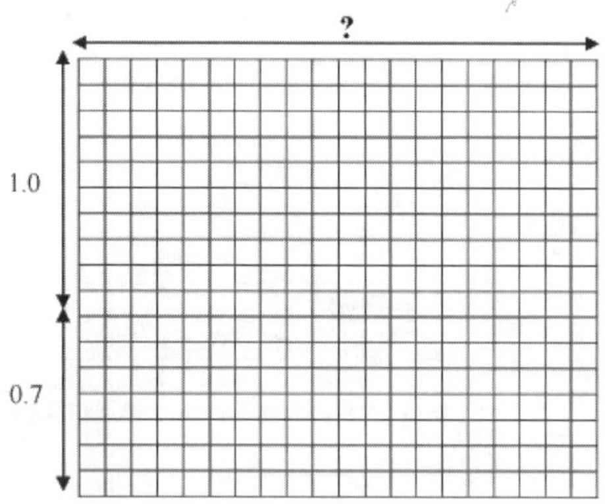

Step 2: Since the rectangle is 17 squares tall, each column added to make the area will add 0.17 square units of area. Since 255 squares are needed, the answer is greater than 10 squares long, since $10 \times 17 = 170$ is less than the required 255 small squares.

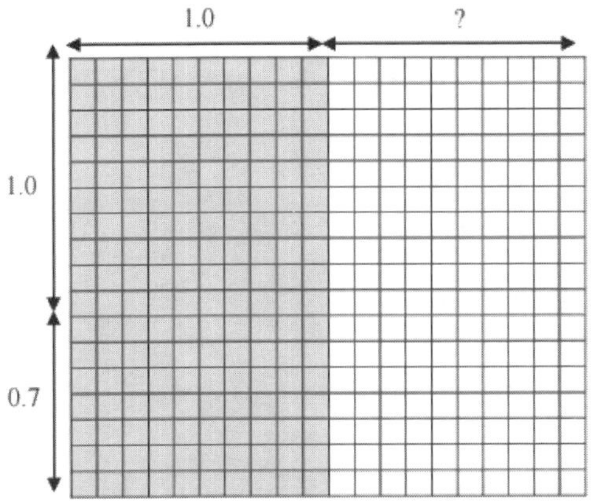

Step 3: Each column adds 17 more small squares. Constructing a chart, we see that the number of columns needed is 15. This makes the length of the rectangle 1.5 units long.

Number of Columns	Total number of small squares
10	170
11	187
12	204
13	221
14	238
15	255

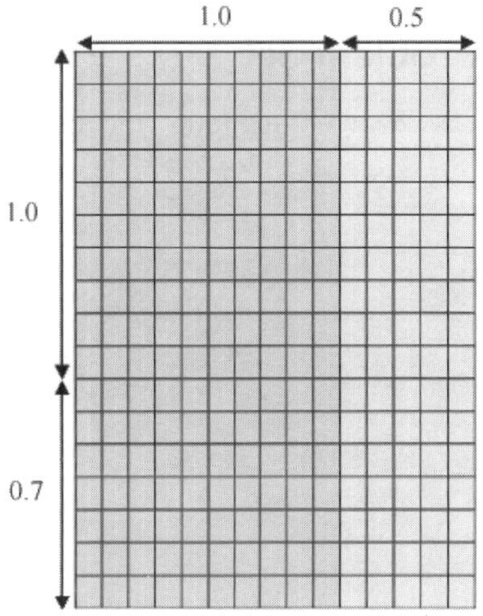

Answer: The quotient of $2.55 \div 1.7$ is 1.5.

Example 5 What decimal multiplication problem could be represented by the figure shown below?

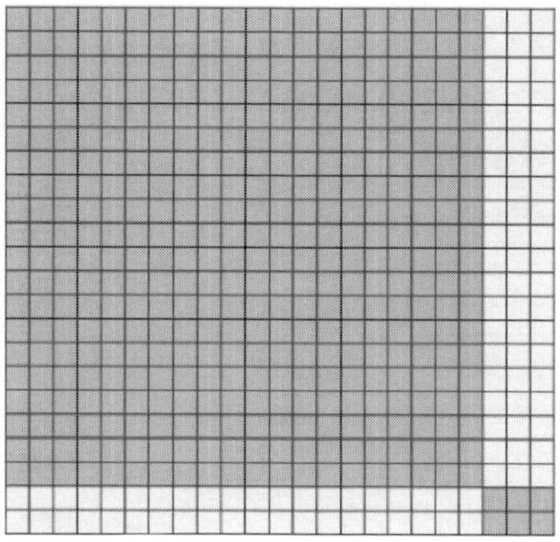

Step 1: By counting squares, we see that this rectangle is 23 squares long, and 22 squares wide. Since each square is 0.01 unit of area, this represents 2.3×2.2.

Answer: The model represents 2.3×2.2.

Problem Set

1. Which of the following models represents 1.3×1.6?

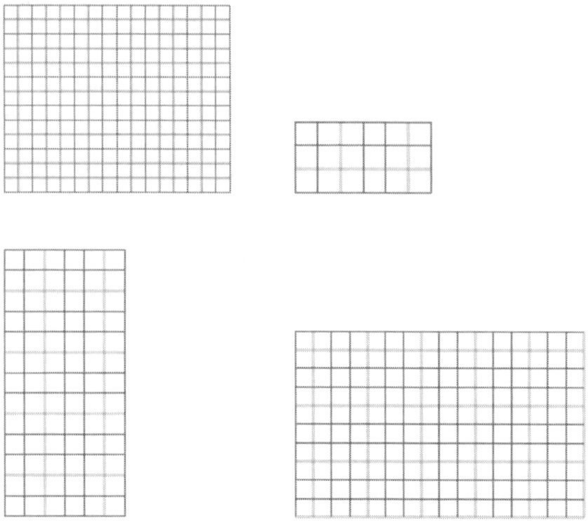

2. Which of the following models represents 2.2×1.5?

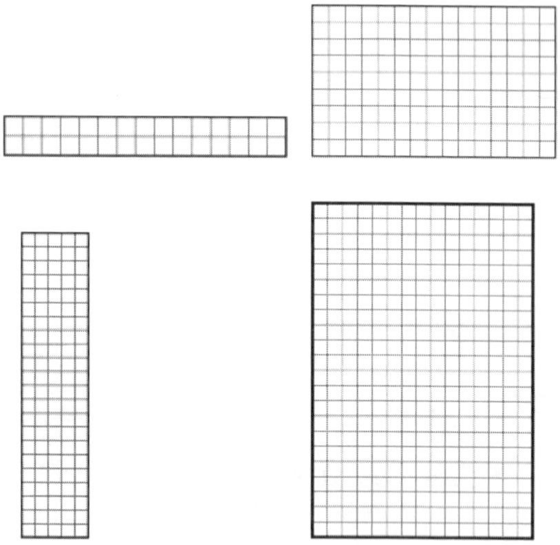

3. Which of the following models represents 3.2 × 1.8?

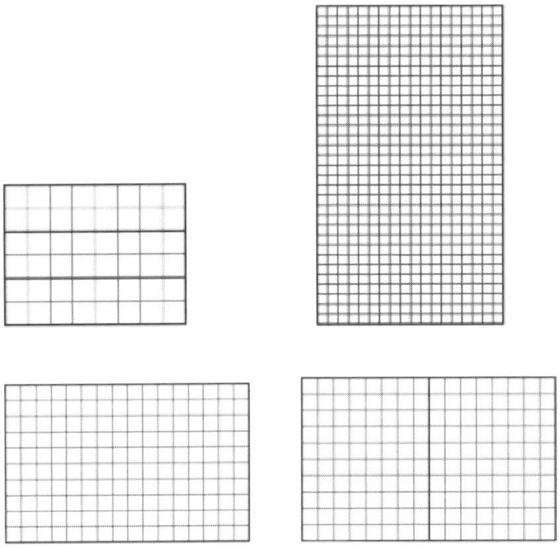

4. Which of the following models represents 1.4 × 2.5?

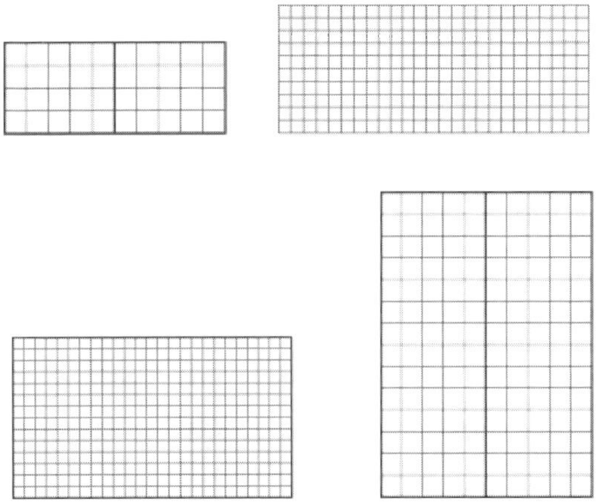

5. Which of the following models represents 3.1 × 1.3?

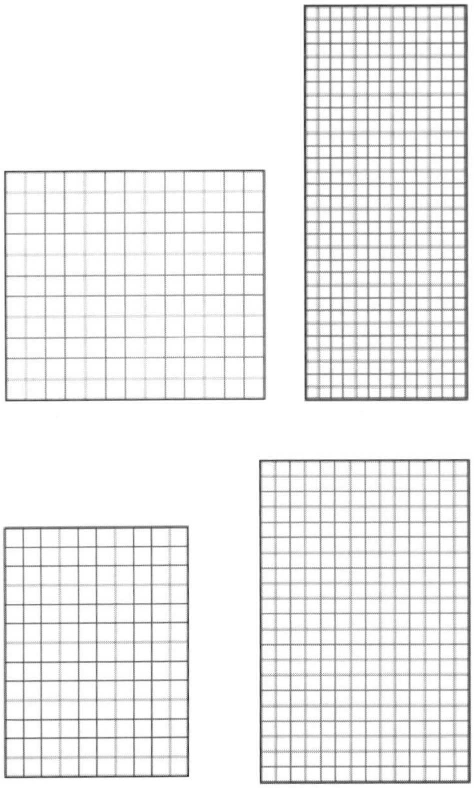

6. Which of the following models represents 2.9 × 1.7?

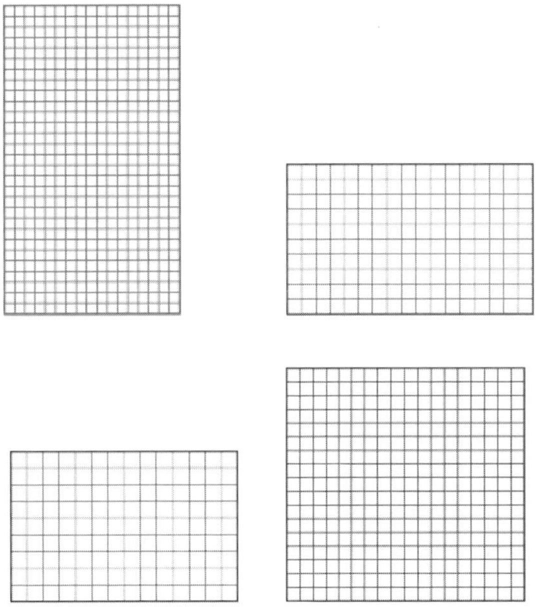

7. Which of the following models represents 2.4×1.1?

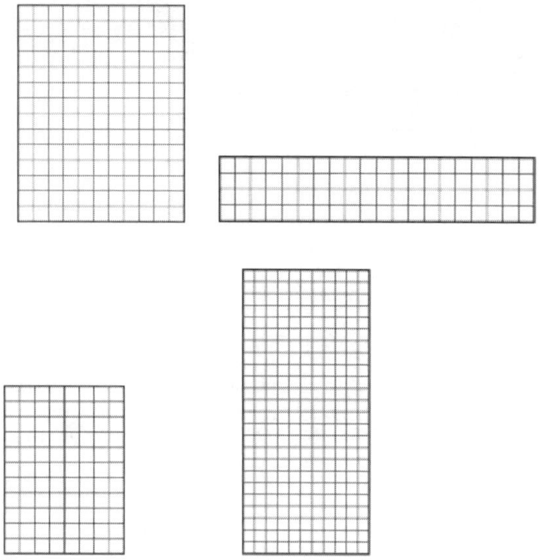

8. If a rectangular model of the division problem below has a width of 17 squares, how long should the model be?

$4.25 \div 1.7 = 2.5$

9. If a rectangular model of the division problem below has a width of 14 squares, how long should the model be?

$4.48 \div 1.4 = 3.2$

10. If a rectangular model of the division problem below has a length of 34 squares, how wide should the model be?

$5.1 \div 3.4 = 1.5$

11. If a rectangular model of the division problem below has a width of 28 squares, how long should the model be?

$8.12 \div 2.8 = 2.9$

12. If a rectangular model of the division problem below has a length of 23 squares, how wide should the model be?

$4.14 \div 2.3 = 1.8$

13. If a rectangular model of the division problem has a length of 31 squares, how wide should the model be?

$6.82 \div 3.1 = 2.2$

14. If a rectangular model of the division problem below has a length of 42 squares, how wide should the model be?

$6.72 \div 4.2 = 1.6$

15. If a rectangular model of the division problem below has a length of 52 squares, how wide should the model be?

$11.96 \div 5.2 = 2.3$

16. Which of the following models represents the product of 3.1 × 1.4?

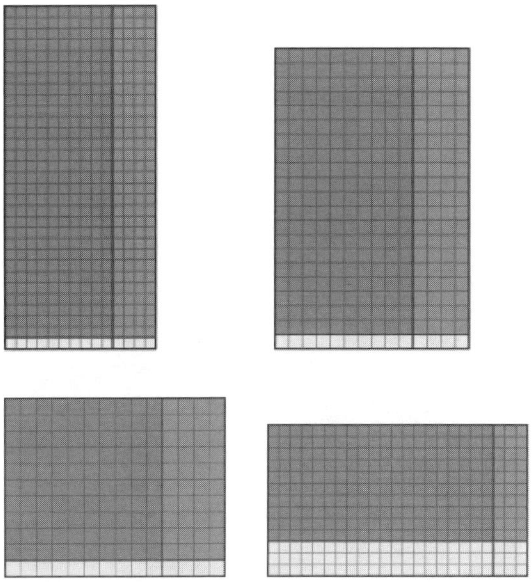

17. Which of the following models represents the product of 2.6 × 1.3?

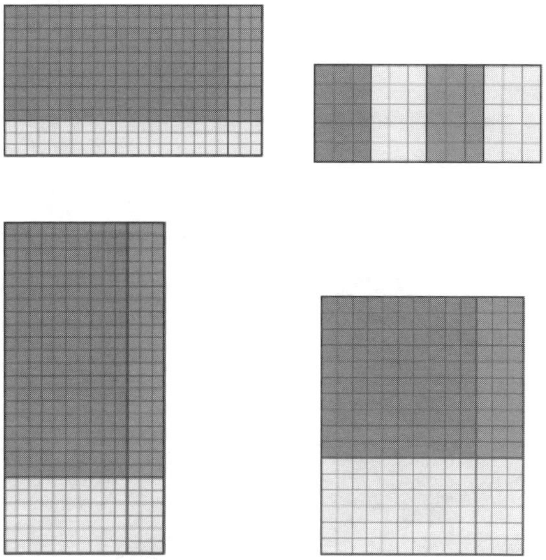

18. Which of the following models represents the product of 1.9 × 1.1?

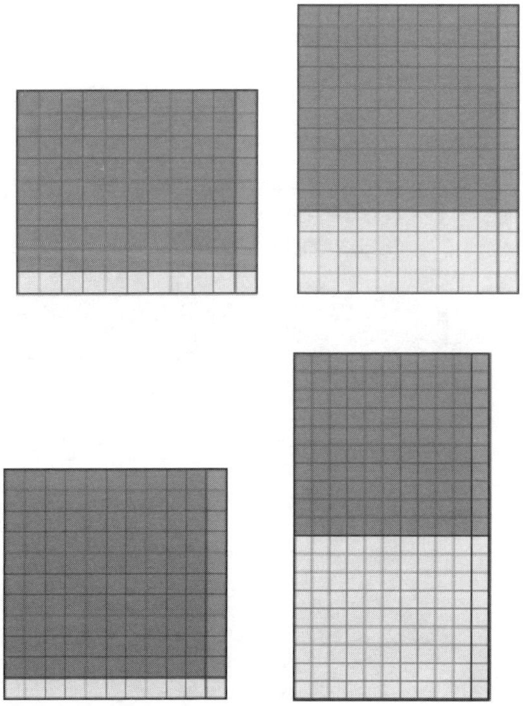

19. Which of the following models represents the product of 2.5 × 1.2?

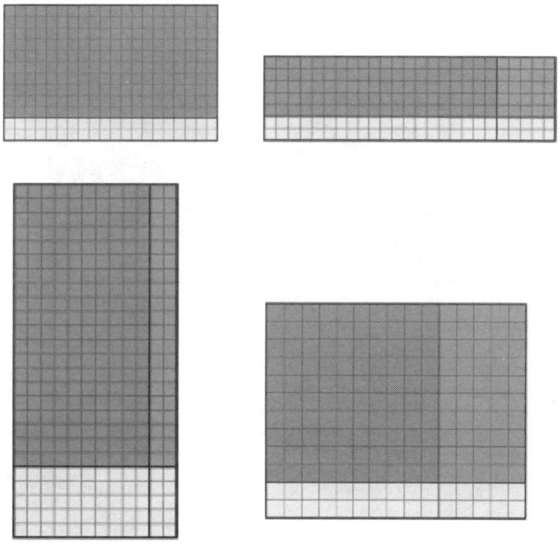

20. Which of the following models represents the product of 3.2 × 1.3?

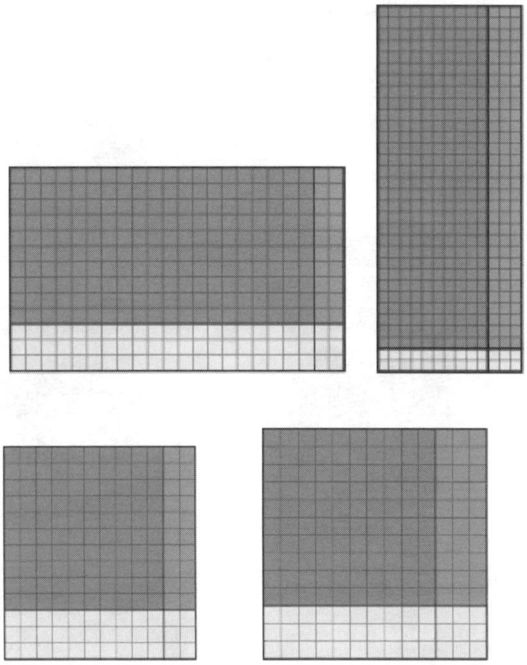

21. Which of the following models represents the product of 2.8 × 1.8?

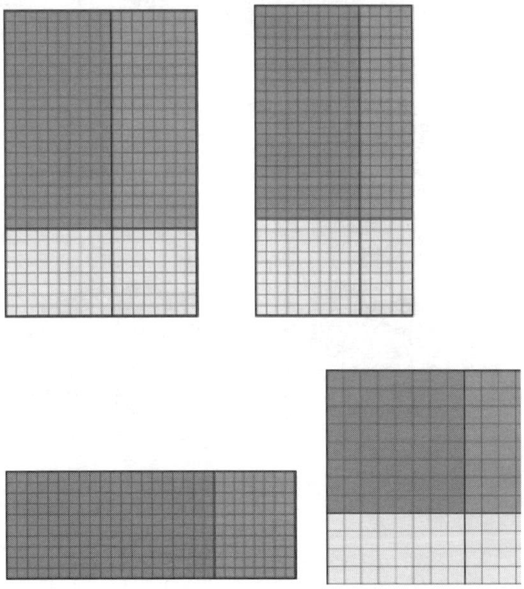

22. Which of the following models represents the product of 2.2 × 1.7?

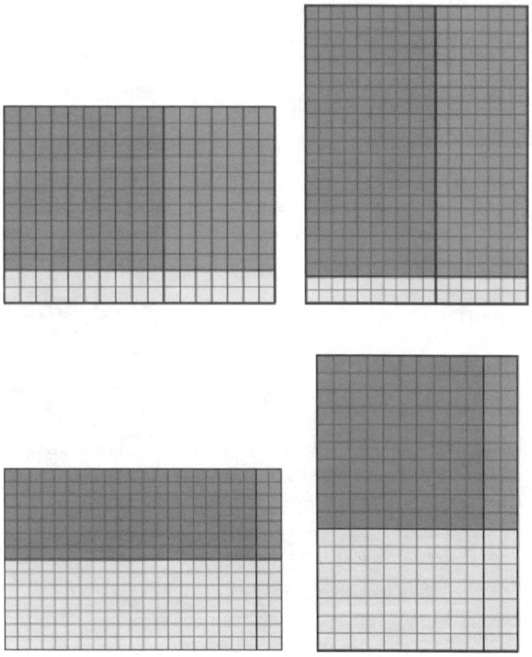

23. Which of the models represents the division problem below if each square has an area equal to one hundredth?

1.95 ÷ 1.3

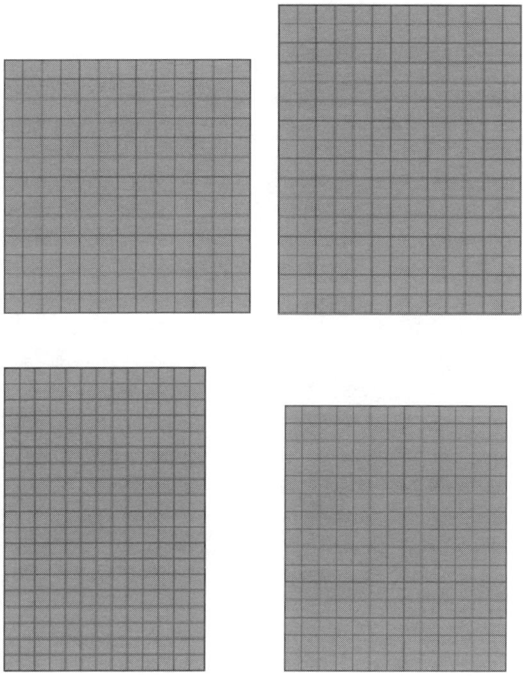

24. Which of the models represents the division problem below if each square has an area equal to one hundredth?

3.57 ÷ 1.7

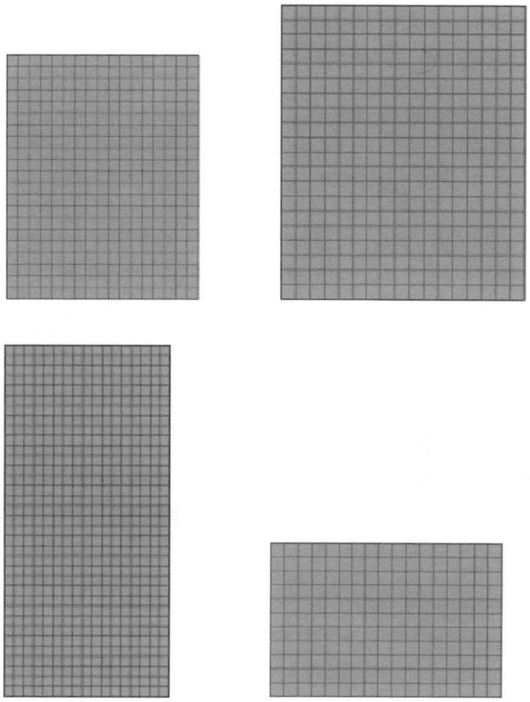

25. Which of the models represents the division problem below if each square has an area equal to one hundredth?

4.18 ÷ 2.2

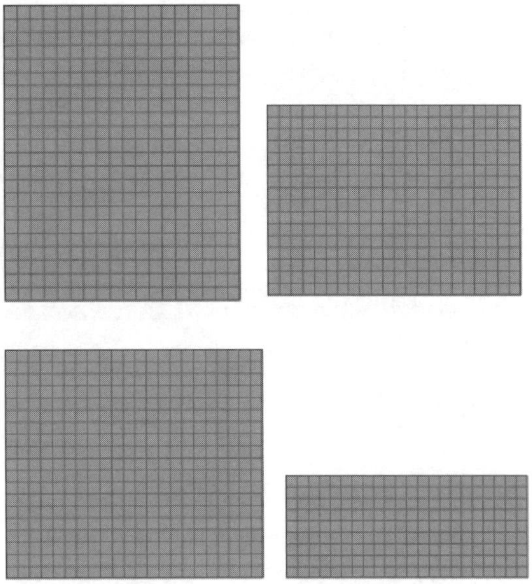

26. Which of the models represents the division problem below if each square has an area equal to one hundredth?

4.14 ÷ 1.8

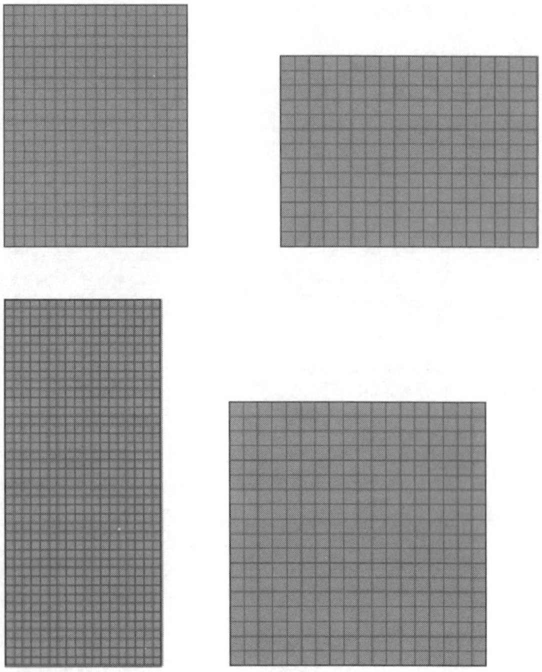

27. Which of the models represents the division problem below if each square has an area equal to one hundredth?

3.52 ÷ 3.2

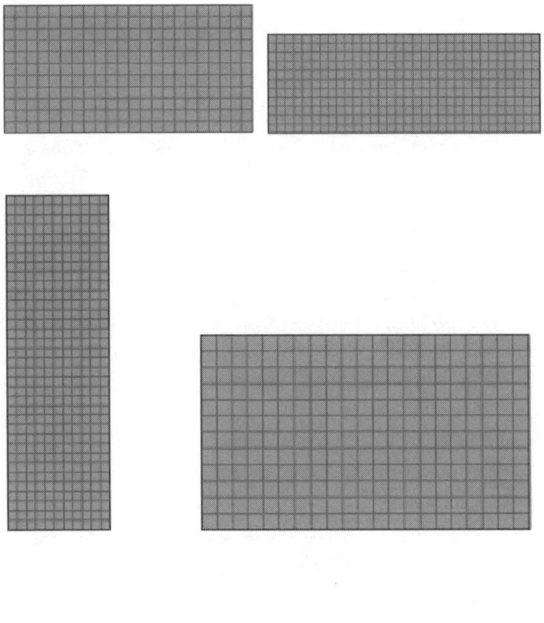

28. Which of the models represents the division problem below if each square has an area equal to one hundredth?

3.48 ÷ 1.2

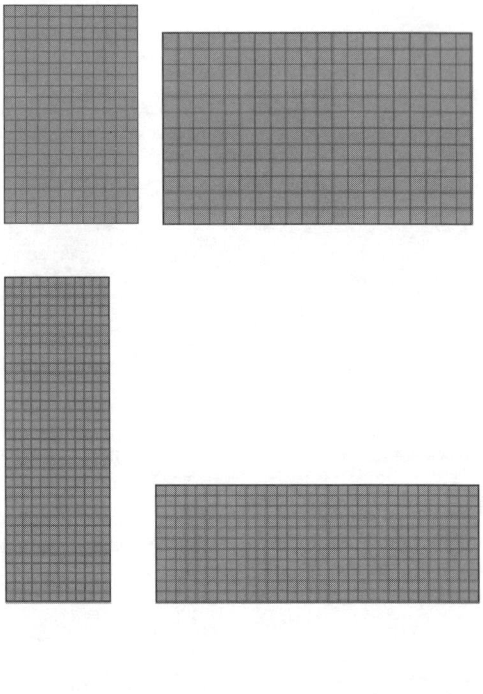

29. Which of the models represents the division problem below if each square has an area equal to one one hundredth?

$4.29 \div 3.3$

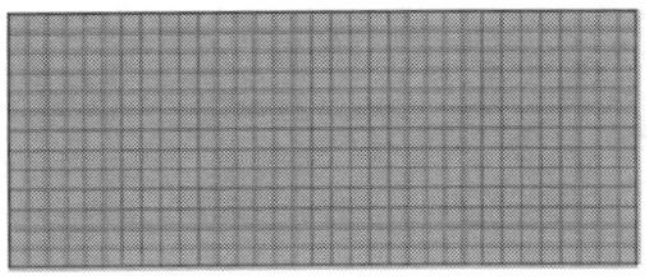

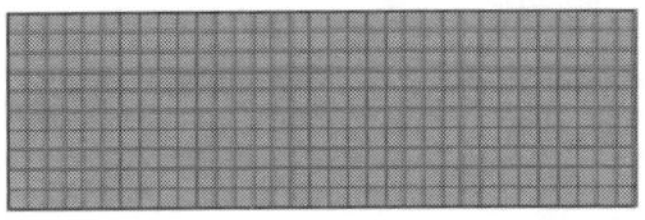
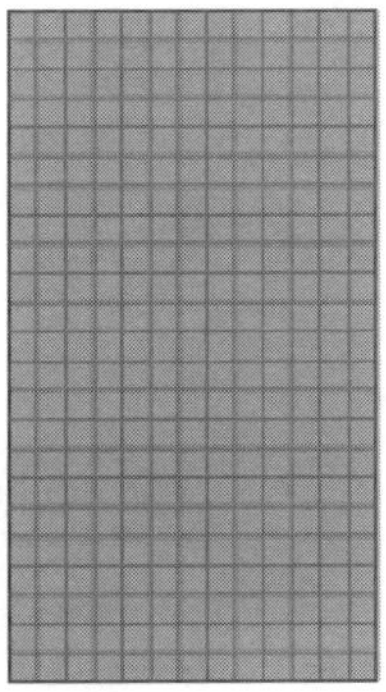

30. Which of the models represents the division problem below if each square has an area equal to one one hundredth?

$4.48 \div 2.8$

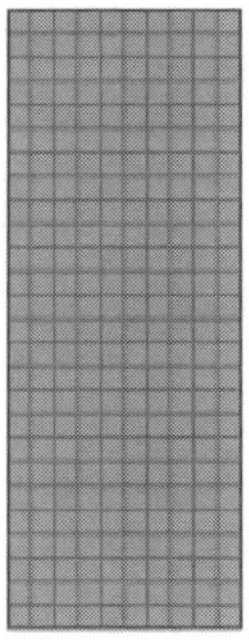
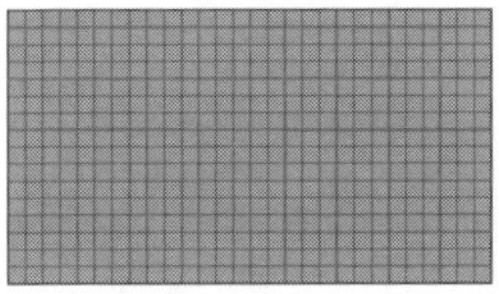
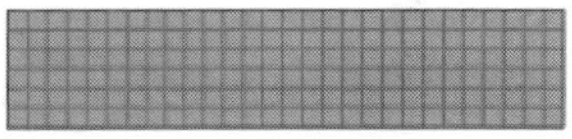

31. What decimal multiplication problem could be represented by the model?

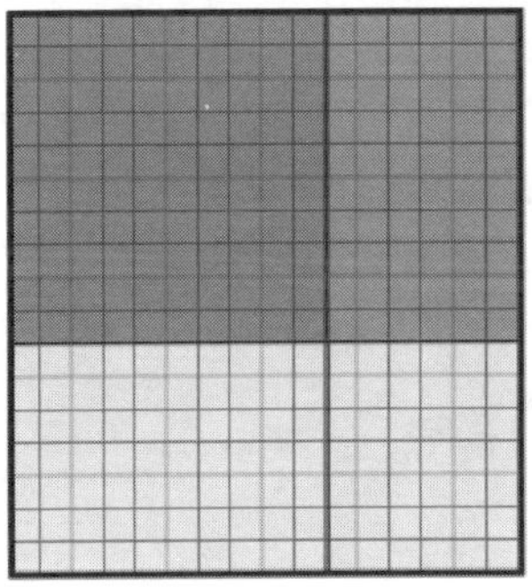

32. What decimal multiplication problem could be represented by the model?

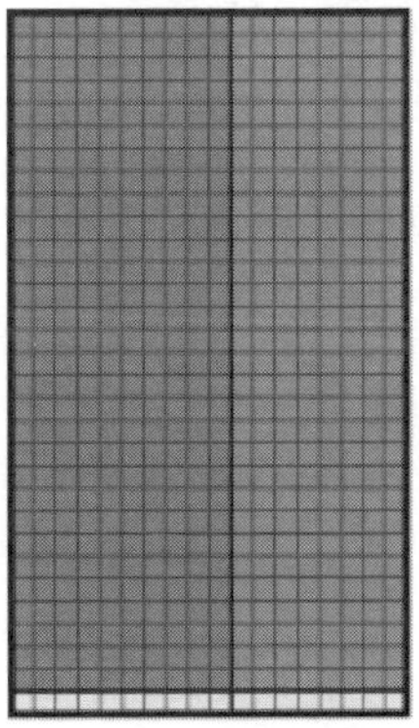

33. What decimal multiplication problem could be represented by the model?

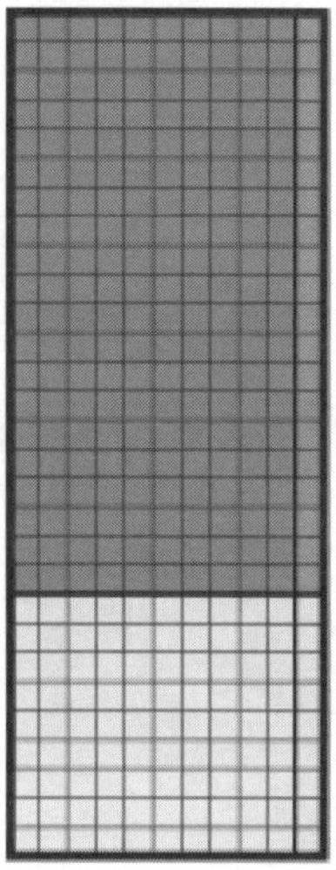

34. What decimal multiplication problem could be represented by the model?

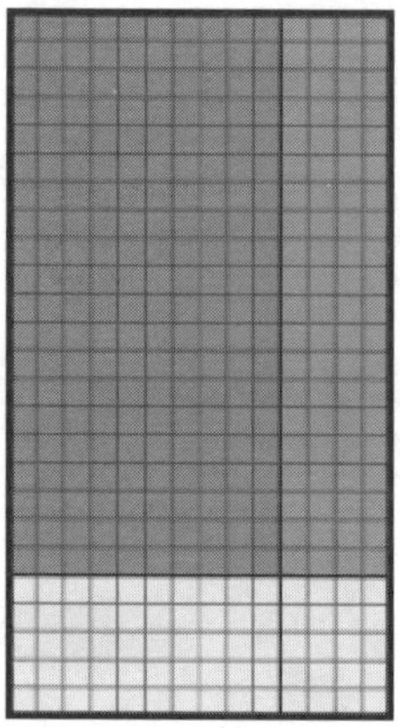

35. What decimal multiplication problem could be represented by the model?

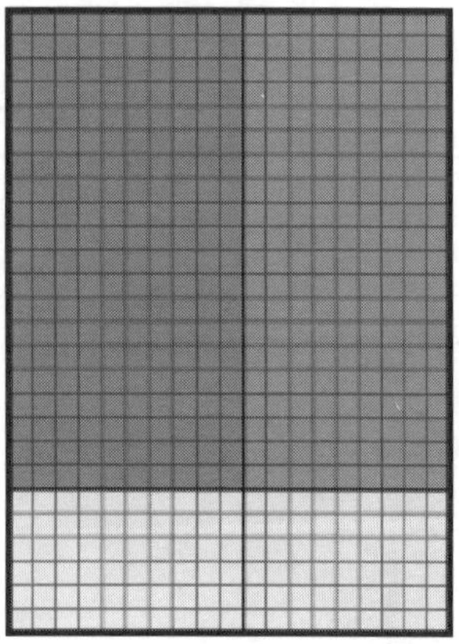

36. What decimal multiplication problem could be represented by the model?

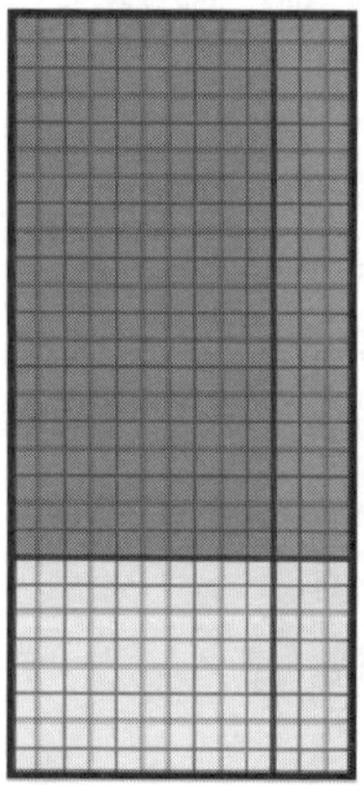

37. What decimal multiplication problem could be represented by the model?

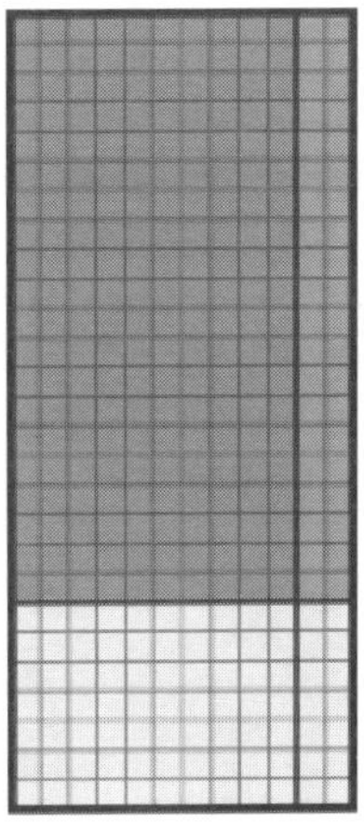

38. What decimal multiplication problem could be represented by the model?

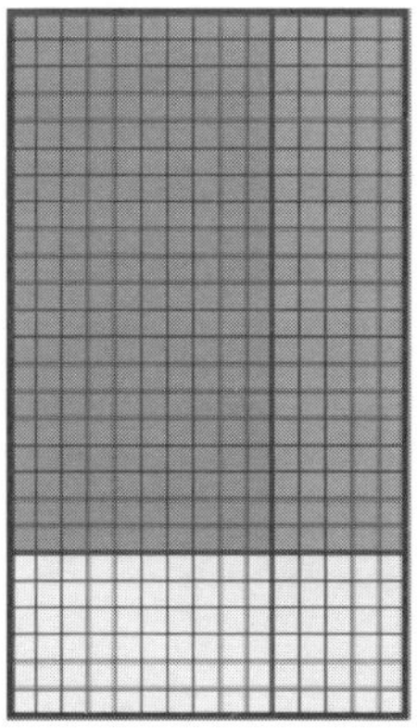

39. What decimal multiplication problem could be represented by the model?

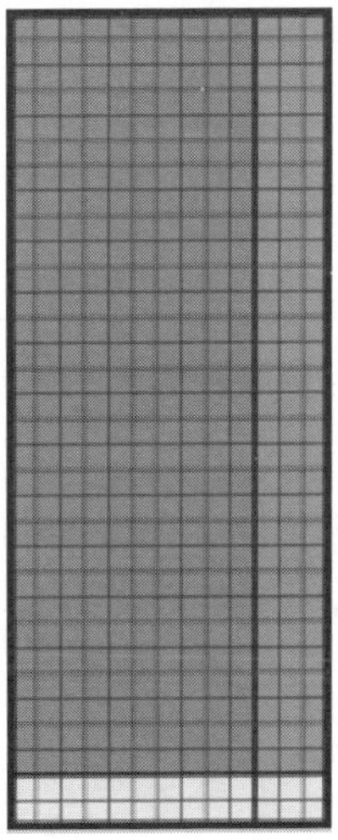

40. What decimal multiplication problem could be represented by the model?

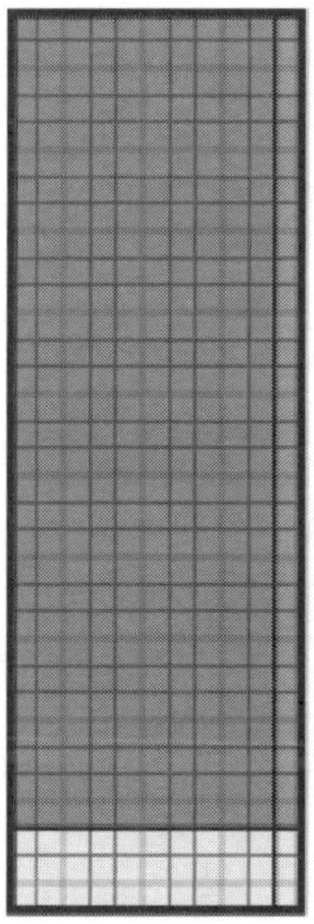

MPA-123: Modeling Multiplication and Division of Fractions

The ability to multiply and divide fractions is needed in many real-world applications of mathematics. It is important to learn how to visualize models used for these operations so that you can understand your answers. In this lesson, you will learn how to model the multiplication and division of fractions.

Example 1 Find the product of $\frac{3}{4} \times 2$.

$\frac{3}{4} \times 2 = \frac{3}{4} \times \frac{2}{1}$

Step 1: Change the whole number to a fraction.

$\frac{3}{4} \times \frac{2}{1} = \frac{6}{4}$

Step 2: Multiply.

$\frac{6}{4} = \frac{3}{2} = 1\frac{1}{2}$

Step 3: Simplify.

Answer: The product of $\frac{3}{4} \times 2$ is $1\frac{1}{2}$.

Example 2 Create a visual model of the fraction $\frac{5}{4}$.

Step 1: To create a visual model use a rectangle that is broken into fourths. Each part represents $\frac{1}{4}$ of the total rectangle. By shading all four parts we have $\frac{4}{4}$.

Step 2: To represent $\frac{5}{4}$ we will need 5 parts.

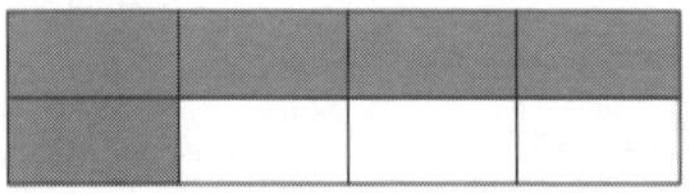

Step 3: Add another rectangle, divide it into four parts, and shade one of those parts.

Answer: The fraction $\frac{5}{4}$ is represented by the shaded areas in the model below.

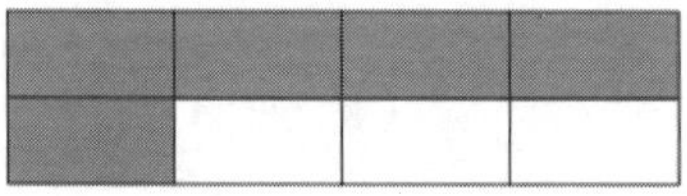

The next example presents a visual model of the process for multiplying a fraction by an integer.

Example 3 Model the product of $\frac{2}{3} \times 4$.

$\frac{2}{3} \times 4 = \frac{2}{3} + \frac{2}{3} + \frac{2}{3} + \frac{2}{3}$

Step 1: We can interpret $\frac{2}{3} \times 4$ as adding together 4 copies of the fraction $\frac{2}{3}$.

Step 2: Model $\frac{2}{3}$ by splitting a rectangle into 3 equal parts and crossing out 2 of the parts with an X.

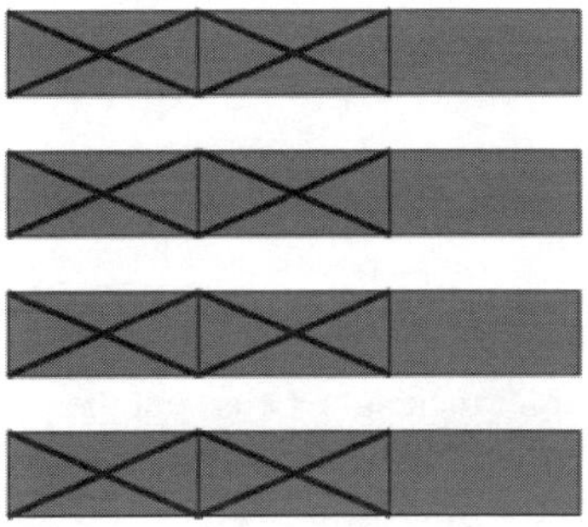

Step 3: Make 4 copies of the model. There are now 8 boxes marked with an X.

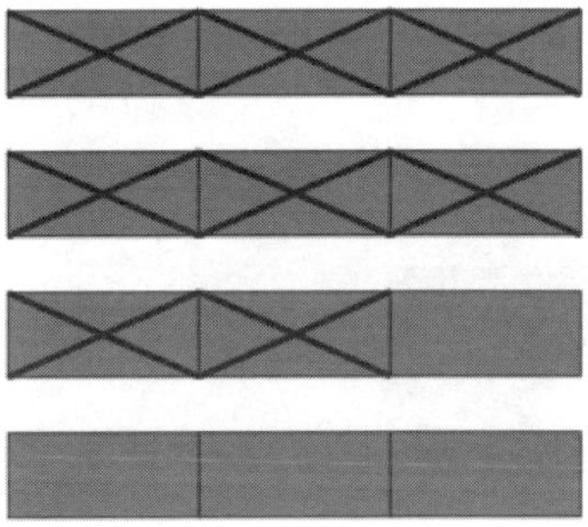

Step 4: Rearrange the boxes to show that $\frac{2}{3} \times 4$ is the same as $2\frac{2}{3}$. We can also write the answer as $\frac{8}{3}$.

Answer: The model for $\frac{2}{3} \times 4 = \frac{8}{3} = 2\frac{2}{3}$ is:

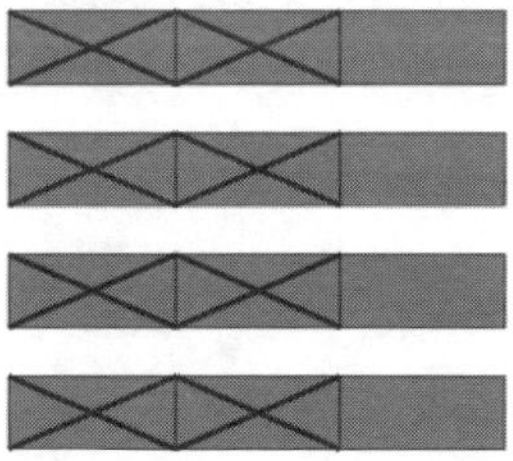

or

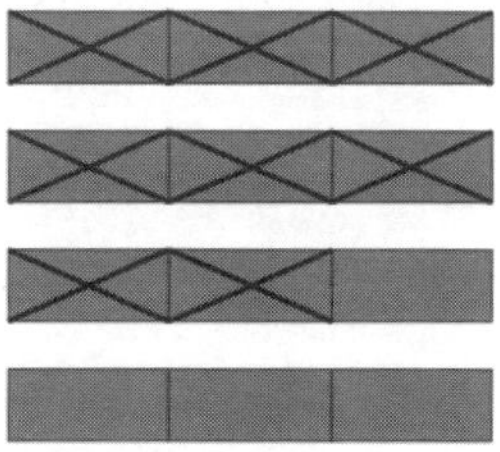

The next example models multiplying a fraction by a fraction. Instead of replicating the fraction as in the previous example, here we think of the multiplier as taking "part of" the multiplicand. So we will think of $\frac{3}{4} \times \frac{2}{3}$ as taking three-fourths of $\frac{2}{3}$.

Example 4 Model the product of $\frac{3}{4} \times \frac{2}{3}$.

Step 1: To model $\frac{2}{3}$, draw a rectangle that is divided into 3 columns. Shade 2 of the columns.

Step 2: To model $\frac{3}{4}$, draw a rectangle that is divided into 4 rows. Shade 3 of the rows.

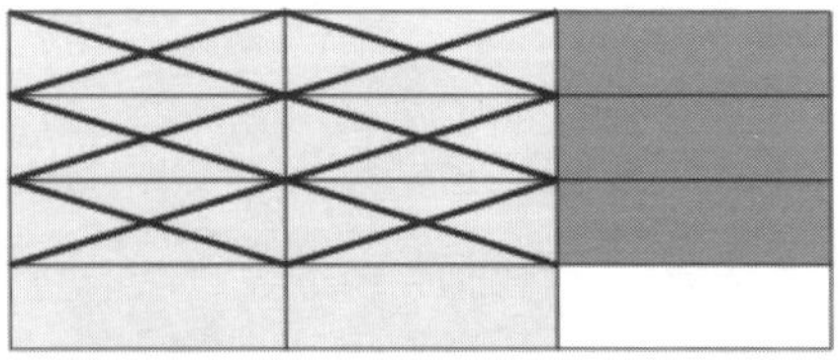

Step 3: Combine the 2 rectangles. Now there are a total of 12 smaller rectangles within the larger rectangle. Cross out the parts of the rectangle that have the combined shadings. These 6 smaller rectangles represent $\frac{6}{12}$ or $\frac{1}{2}$ of the entire rectangle.

Answer: The model for $\frac{3}{4} \times \frac{2}{3} = \frac{6}{12} = \frac{1}{2}$ is:

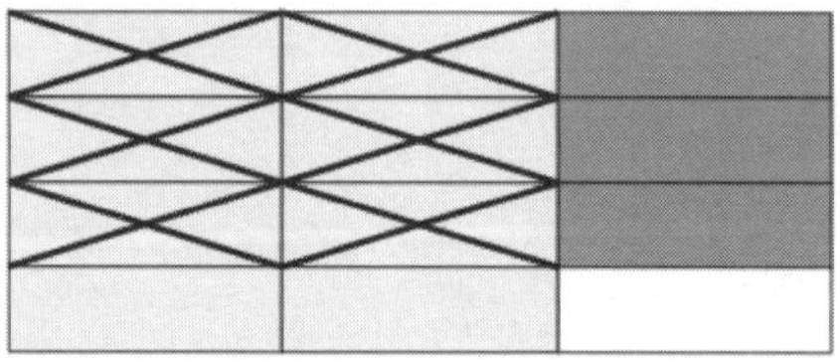

The next example models a fraction divided by an integer. Begin with a model of the original fraction, or the dividend, and subdivide it into as many parts as defined by the divisor. The quotient is the size of one of these parts.

Example 5 Model the quotient of $\frac{3}{4} \div 5$.

Step 1: To visualize this problem, create a model for $\frac{3}{4}$ and divide each of its pieces into 5 equal parts.

Step 2: To model $\frac{3}{4}$, draw a rectangle with four equal rows and shade three of them.

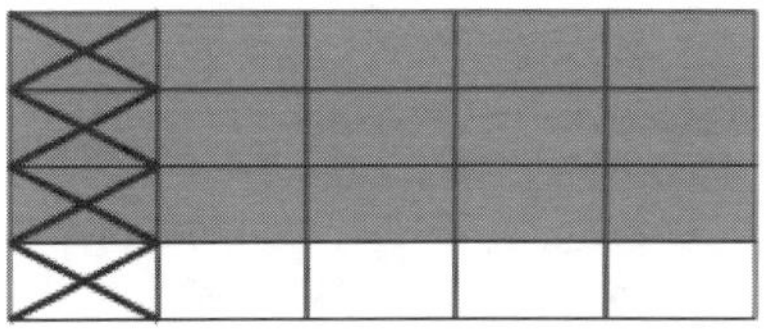

Step 3: To model the quotient of $\frac{3}{4} \div 5$, divide the model into five equal columns and place X's in one column.

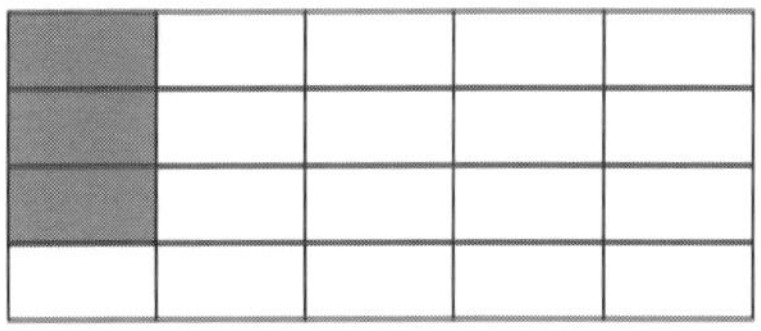

Step 4: The parts that are both shaded and marked with an X, which are three of the twenty parts, represent the quotient of $\frac{3}{4} \div 5$.

Answer: The model shows that the quotient of $\frac{3}{4} \div 5 = \frac{3}{20}$.

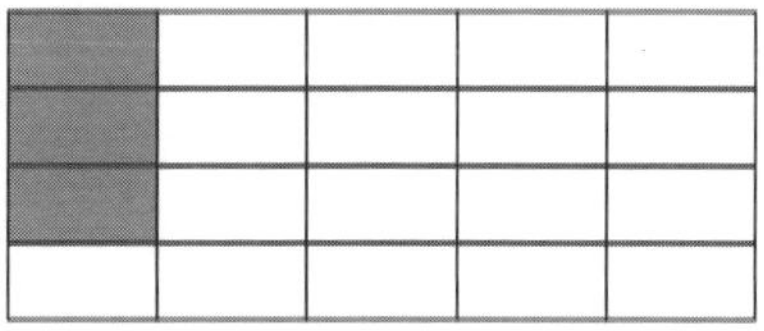

Problem Set

Solve:

1. Find the product of $\frac{1}{8} \times 7$.

2. Find the product of $\frac{2}{9} \times 7$.

3. Find the product of $\frac{3}{8} \times 5$.

4. Find the product of $\frac{4}{7} \times 3$.

5. Find the product of $\frac{5}{6} \times 6$.

6. Find the product of $\frac{3}{4} \times 8$.

7. Find the product of $\frac{2}{3} \times 12$.

8. Find the product of $\frac{5}{7} \times 7$.

9. Model the fraction $\frac{6}{5}$.

10. Model the fraction $\frac{9}{7}$.

11. Model the fraction $\frac{7}{4}$.

12. Model the fraction $\frac{11}{10}$.

13. Model the fraction $\frac{4}{5}$.

14. Model the fraction $\frac{3}{4}$.

15. Model the fraction $\frac{2}{3}$.

16. Model the product of $\frac{1}{4} \times 3$.

17. Model the product of $\frac{1}{8} \times 7$.

18. Model the product of $\frac{1}{11} \times 5$.

19. Model the product of $\frac{1}{5} \times 3$.

20. Model the product of $\frac{2}{7} \times 4$.

21. Model the product of $\frac{3}{8} \times 5$.

22. Model the product of $\frac{3}{5} \times 4$.

23. Model the product of $\frac{2}{9} \times 7$.

24. Model the product of $\frac{1}{9} \times \frac{2}{3}$.

25. Model the product of $\frac{2}{5} \times \frac{3}{4}$.

26. Model the product of $\frac{3}{7} \times \frac{1}{2}$.

27. Model the product of $\frac{1}{3} \times \frac{4}{5}$.

28. Model the product of $\frac{5}{6} \times \frac{2}{3}$.

29. Model the product of $\frac{6}{7} \times \frac{3}{5}$.

30. Model the product of $\frac{7}{8} \times \frac{3}{5}$.

31. Model the quotient of $\frac{2}{5} \div 5$.

32. Model the quotient of $\frac{5}{8} \div 3$.

33. Model the quotient of $\frac{7}{8} \div 4$.

34. Model the quotient of $\frac{8}{9} \div 3$.

35. Model the quotient of $\frac{9}{10} \div 4$.

36. Model the quotient of $\frac{5}{6} \div 4$.

37. Model the quotient of $\frac{6}{7} \div 5$.

38. Model the quotient of $\frac{3}{4} \div 4$.

39. Model the quotient of $\frac{2}{9} \div 3$.

40. Model the quotient of $\frac{5}{9} \div 2$.

MPA-124: Classifying Numbers in the Real Number System

In this lesson, you will explore the real number system and learn to classify numbers into their appropriate set. The diagram below shows the relationship between different sets of numbers.

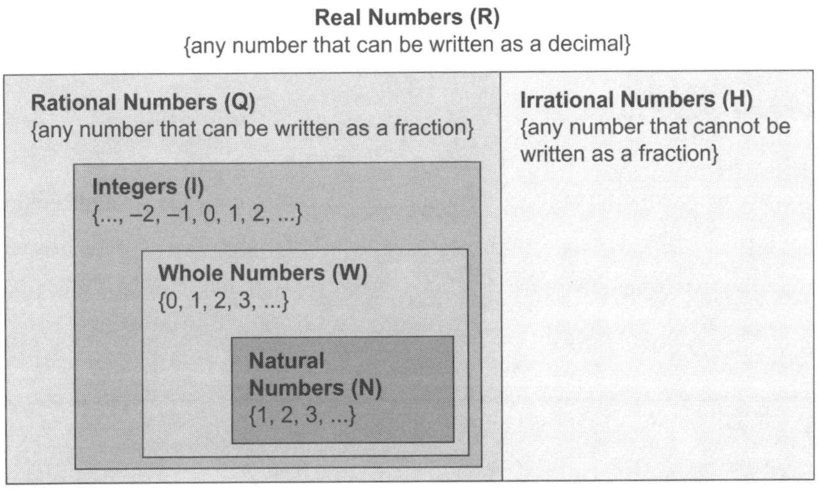

Natural Numbers	The set of numbers used for counting objects. {1, 2, 3, 4, 5, ...}
Whole Numbers	The set of natural numbers and zero. {0, 1, 2, 3, 4, 5, ...}
Integers	The set of whole numbers and negative natural numbers. {..., –3, –2, –1, 0, 1, 2, 3, ...}
Rational Numbers	Any number that can be expressed as a quotient $\frac{a}{b}$, where *a* and *b* are integers and $b \neq 0$.

In other words, any number that can be written as a fraction is a rational number. We can then conclude that all integers, whole numbers, and natural numbers are rational numbers, because we can simply convert any of those numbers into a fraction by writing the number over a denominator of 1. For example, the integer –5 can be written as $-\frac{5}{1}$

A rational number can also be expressed as a decimal.

1. **Terminating decimal**

   A decimal that ends and can be written as a fraction.

   Ex. 0.4, 0.75, and 0.125

2. Repeating decimal

A decimal in which a sequence of digits repeats endlessly and which can also be written as a fraction.

Ex. 0.151515... or $0.\overline{15}$

Irrational Numbers	Any number that *cannot* be expressed as the quotient $\frac{a}{b}$, where *a* and *b* are integers and $b \neq 0$.

Since a rational number in decimal form either terminates or repeats a sequence of digits endlessly, an **irrational number** in decimal form does *not* terminate or repeat a sequence of digits. An example of such a number is $\sqrt{6}$. As a decimal, $\sqrt{6}$ is approximately equal to 2.449489742... This decimal does not terminate and does not repeat a sequence of digits.

Example 1 Which set below is the set of all natural numbers?

(A) {0, 1, 2, 3, ...} (B) {..., −3, −2, −1, 0, 1, 2, 3, ...}
(C) {3, 4, 5, ...} (D) {1, 2, 3, ...}

N = {1, 2, 3, 4, 5, ...}	**Step 1:** Define the set of natural numbers.
(A) {0, 1, 2, 3, ...}	**Step 2:** This set does consist of the natural numbers. However, it also contains the number 0. Zero is not a natural number.
(B) {..., −3, −2, −1, 0, 1, 2, 3, ...}	**Step 3:** This set consists of 0 and negative numbers. The natural numbers do not contain 0 or negative numbers.
(C) {3, 4, 5, ...}	**Step 4:** This set contains only natural numbers. However, it does not contain *all* the natural numbers.
(D) {1, 2, 3 ...}	**Step 5:** This set contains all the natural numbers.

Answer: {1, 2, 3 ...} is the set of all natural numbers.

Example 2 Which of the following numbers is a natural number, a whole number, *and* an integer?

(A) $\frac{1}{3}$ (B) 0

(C) –11 (D) 27

	0	$\frac{1}{3}$	–11	27
Natural Numbers (N) {1, 2, 3, 4, ...}				
Whole Numbers (W) {0, 1, 2, 3, ...}				
Integers (I) {..., –2, –1, 0, 1, 2, ...}				

Step 1: Make a table.

	0	$\frac{1}{3}$	–11	27
Natural Numbers (N) {1, 2, 3, 4, ...}				√
Whole Numbers (W) {0, 1, 2, 3, ...}	√			√
Integers (I) {..., –2, –1, 0, 1, 2, ...}	√		√	√

Step 2: Zero is not a natural number because the natural numbers are counting numbers starting with one. Zero is a whole number and an integer.

$\frac{1}{3}$ is not a natural number, a whole number, or an integer because it is a fractional part.

–11 is not a natural number or a whole number because it is negative. –11 is an integer.

27 is a natural number, a whole number, and an integer.

Answer: 27 is a natural number, a whole number, and an integer.

Example 3 List all the sets to which –9 does *not* belong.

	–9
Natural Numbers	
Whole Numbers	
Integers	
Rational Numbers	
Irrational Numbers	
Real Numbers	

Step 1: Make a table.

	–9
Natural Numbers	×
Whole Numbers	×
Integers	√
Rational Numbers	√
Irrational Numbers	×
Real Numbers	√

Step 2: –9 is neither a natural number nor a whole number because it is negative.

Integers consist of the whole numbers and the negatives of the natural numbers. Therefore, –9 is an integer.

–9 is a rational number because it can be written as the fraction $-\frac{9}{1}$.

Since –9 is a rational number, it cannot be an irrational number.

–9 is a real number because it can be written as the decimal –9.0

Answer: –9 does not belong to the set of natural numbers, whole numbers, or irrational numbers.

Example 4 Is $\sqrt{5}$ a rational or an irrational number?

A rational number is any number that can be expressed in the form $\frac{a}{b}$, where a and b are integers and $b \neq 0$.

A rational number terminates or repeats a sequence of digits in decimal form.

Step 1: Define rational number.

$\sqrt{5} \approx 2.2360679$

Step 2: Convert the given number into a decimal.

2.2360679... does not terminate nor does it repeat a sequence of digits.

Step 3: Check to see if the decimal terminates or repeats a sequence of digits.

Answer: $\sqrt{5}$ is an irrational number.

Example 5 List all the sets to which 0.5 belongs.

	0.5
Natural Numbers	
Whole Numbers	
Integers	
Rational Numbers	
Irrational Numbers	
Real Numbers	

Step 1: Make a table.

	0.5
Natural Numbers	×
Whole Numbers	×
Integers	×
Rational Numbers	√
Irrational Numbers	×
Real Numbers	√

Step 2: 0.5 is not a natural number, a whole number, or an integer because it is a decimal.

Rational numbers are any numbers that can be written as a fraction. 0.5 can be written as $\frac{1}{2}$. Therefore, it is a rational number.

Since 0.5 is a rational number, it cannot be an irrational number.

Real numbers are any numbers that can be written as a decimal. 0.5 is a decimal. Therefore, it is also a real number.

Answer: 0.5 belongs to the set of rational numbers and real numbers.

Problem Set

1. Which of the sets listed below is the set of *all* natural numbers?

$\{1, 2, 3, 4, ...\}$ $\{..., -2, -1, 0, 1, 2, ...\}$
$\{0, 1, 2, 3, 4, ...\}$ $\{..., -8, -4, 0, 4, 8, ...\}$

2. Which of the sets listed below is the set of *all* integers?

$\{1, 2, 3, 4, ...\}$ $\{..., -3, -2, -1, 0, 1, 2, 3, ...\}$
$\{0, 1, 2, 3, 4, ...\}$ $\{..., -1.5, -0.5, 0, 0.5, 1.5, ...\}$

3. Name the set of numbers that cannot be expressed as the quotient $\frac{a}{b}$, where a and b are integers and $b \neq 0$.

4. Name the set of numbers that contains only the number 0 and the set of natural numbers.

5. Which set of numbers below belongs to the set of irrational numbers?

$\{..., -6, -3, 0, 3, 6, ...\}$
$\{..., -4.5, -2.5, 0, 2.5, 4.5, ...\}$
$\{\frac{1}{3}, \frac{2}{3}, \frac{3}{8}, \frac{7}{8}\}$
$\{3.4545512..., 67.321234...\}$

6. Which of the following does not contain the number 0?

Natural Numbers Integers
Rational Numbers Irrational Numbers

7. Which of the following contains both rational and irrational numbers?

Integers Whole Numbers
Real Numbers Natural Numbers

8. Which of the following numbers is both a natural number and an integer?

$\frac{1}{4}$ 0
4 -8

9. Which of the following numbers is a whole number and an integer?

-6 0

$-\frac{3}{4}$ 6.2

10. Which of the following numbers is an integer, a whole number, and a natural number?

9 $-\frac{7}{9}$

-9 0.9

11. Which of the following numbers is both an integer and a natural number?

7 -7

0 0.7

12. Which of the following numbers is an integer, a whole number, and a natural number?

-80 $\frac{1}{8}$

0.8 80

13. Which of the following numbers is both a natural number and a whole number?

-6 0.6

6 $\frac{1}{6}$

14. Which of the following numbers is a whole number, a natural number, and an integer?

-0.7 $\frac{1}{7}$

17 -7

15. Which of the following numbers is a whole number, a natural number, and an integer?

5.4 -54

$\frac{1}{54}$ 54

16. Choose the set in which –12 does *not* belong.

Whole Numbers   Rational Numbers

Integers   Real Numbers

17. Choose the set in which 0 does *not* belong.

Rational Numbers   Integers

Real Numbers   Natural Numbers

18. Which of the following sets includes the number –18.5?

Natural Numbers   Whole Numbers

Integers   Rational Numbers

19. Choose the set in which –44 does *not* belong.

Natural Numbers   Integers

Real Numbers   Rational Numbers

20. The number 5.6 belongs in which of the following sets?

Rational Numbers   Integers

Natural Numbers   Whole Numbers

21. Choose the set in which –25 does *not* belong.

Integers   Whole Numbers

Rational Numbers   Real Numbers

22. Choose the set in which 7.8 belongs.

Natural Numbers   Integers

Real Numbers   Whole Numbers

23. Which of the following is an irrational number?

-7 $\quad$ 6.5

$\sqrt{25}$ $\quad$ $\sqrt{6}$

24. Which of the following is an irrational number?

$15.\overline{14}$ $\quad$ -8

-10.1 $\quad$ $\sqrt{17}$

25. Which of the following is an irrational number?

99.09 $\quad$ $\sqrt{36}$

$2.\overline{145}$ $\quad$ $\sqrt{56}$

26. Which of the following is an irrational number?

12.5 $\quad$ -14.7

$2.871543...$ $\quad$ 0

27. Which of the following is a rational number?

147 $\quad$ $\sqrt{7}$

$\sqrt{14}$ $\quad$ $14.177711...$

28. Which of the following is an irrational number?

$\frac{20}{2}$ $\quad$ -2.2

$12.221112...$ $\quad$ $0.\overline{2}$

29. Which of the following is a rational number?

$0.13511511...$ $\quad$ $\sqrt{5}$

$\sqrt{3}$ $\quad$ 0.35

30. Which of the following is an irrational number?

$0.\overline{45}$ $\quad$ -4.5

0.45 $\quad$ $\sqrt{45}$

31. Which statement *best* describes the number 7?

It is a real number and an irrational number.

It is an integer, a real number, a rational number, a whole number, and a natural number.

It is an integer, a real number, and a rational number.

It is an integer, a real number, a rational number, and a natural number.

32. Which statement *best* describes $\sqrt{12}$?

It is a real number and an irrational number.

It is a real number and a rational number.

It is a real number.

It is a real number, a rational number, and an integer.

33. Which statement *best* describes the number 0?

It is a real number, a rational number, a whole number, and an integer.

It is a real number, a rational number, and an integer.

It is a real number, a rational number, an integer, a natural number, and a whole number.

It is a real number, a rational number, an integer, and a whole number.

34. Which statement *best* describes the number -14?

It is a real number, a rational number, and an integer.

It is a real number and a rational number.

It is a real number and an integer.

It is a real number, a rational number, an integer, and a natural number.

35. Which statement *best* describes the number 0.6?

It is a real number and an irrational number.
It is a real number, a rational number, and a natural number.
It is a real number and a rational number.
It is a rational number.

36. Which statement *best* describes the number $0.\overline{18}$?

It is a real number and an irrational number.
It is a real number and a rational number.
It is a real number and an integer.
It is a real number.

37. Which statement *best* describes the number $-3.\overline{8}$?

It is a real number and an irrational number.
It is a real number, a rational number, and an integer.
It is a real number, an irrational number, and an integer.
It is a real number and a rational number.

38. Which of the following *best* describes the number 150?

It is a real number, a rational number, an integer, and a whole number.
It is a real number, a rational number, a whole number, and a natural number.
It is a real number, a rational number, and an integer.
It is a real number, a rational number, an integer, a natural number, and a whole number.

39. Which statement *best* describes the number 3.12754...?

It is a real number and a rational number.
It is a real number and an irrational number.
It is a real number.
It is a real number and an integer.

40. Which of the following *best* describes the number –8?

It is a real number and an irrational number.
It is a real number and an integer.
It is a real number, a rational number, and an integer.
It is a real number, an irrational number, and an integer.

MPA-125: Formulating a Possible Problem Situation Given an Equation

In this lesson, we will formulate a possible problem situation when given an equation. Remember that in previous lessons, we were given a problem situation, and in order to solve the problem we had to formulate an equation. In this lesson we will learn how to do the reverse.

The formulas used in this lesson include those for the perimeter and area of parallelograms and the volume of a cube.

Perimeter of a Parallelogram	$P = 2\ell + 2w$

Area of a Parallelogram	$A = \ell \times w$

Volume of a Cube	$V = s^3$

Example 1 Formulate a possible problem situation that can be solved using the equation $y = x + 2$.

$y = x + 2$	**Step 1:** The equation shows that the quantity y is equal to the sum of x and 2.
y = Tom's age x = Samantha's age	**Step 2:** Let y represent Tom's age. Let x represent Samantha's age.
	Answer: Samantha is twenty years old. Tom is two years older than Samantha. How old is Tom?

Example 2 Devise a possible problem situation that requires the formula for the volume of a cube.

$V = s^3$	**Step 1:** The formula for the volume of a cube is volume equals the length of one side cubed.
Let s = 6 inches	**Step 2:** A problem should ask for the volume of a cube-shaped container, and give the length of one side or dimension.
	Answer: A cube has sides that are 6 inches long. What is the volume of the cube?

Example 3 Formulate a possible problem situation that can be solved using the following equation: $y = 5x + 4$

$y = 5x + 4$	**Step 1:** The equation shows that the quantity y is equal to the sum of 4 and the product of 5 and x.
y = amount of weekly paycheck x = number of hours worked per week	**Step 2:** Let y represent the amount of Tammy's weekly paycheck. Let x represent the number of hours Tammy works per week.
	Answer: Tammy makes \$5 per hour, plus a weekly bonus of \$4. How much is her weekly pay?

Example 4 Formulate a possible problem situation that uses the formula for the area of a rectangle:

$A = \ell \times w$

$A = \ell \times w$	**Step 1:** The problem situation should ask for one variable, and give values for the other two.
Let A = 80 square feet Let ℓ = 2 feet more than w	**Step 2:** Create a problem that depends on this equation.
	Answer: The area of a rectangular box is 80 square feet. If the length of the box is two feet longer than its width, what are the dimensions of the rectangular box?

Example 5 Write and solve a problem that matches the following equation: $y = 3x + 25$

$y = 3x + 25$	**Step 1:** The equation shows that the quantity y is equal to the sum of 25 and the product of 3 and x.
y = amount Ross spent x = amount Brenda spent	**Step 2:** Let y represent the amount of money that Ross spent at the mall. Let x represent the amount of money that Brenda spent at the mall.
$y = 3x + 25$ $y = 3(15) + 25$ $y = 45 + 25$ $y = 70$	**Step 3:** If the problem tells us that Brenda spent \$15, solve for y.
	Answer: Problem: Ross spent \$25 more than 3 times the amount that Brenda spent at the mall. How much did Ross spend if Brenda spent \$15? Solution: Ross spent \$70.

Problem Set

1. Choose the possible problem situation that best fits the equation:

   $y = 3.25x$

   The cost of riding a taxi is $3.25 per mile.

   The cost of a taxi is $3.25 more than the city bus.

   The cost of a taxi is $3.25 less than the limousine.

   It costs $3.25 to take a trip in a taxi.

2. Choose the possible problem situation that best fits the equation:

   $y = 0.25x$

   The game costs $0.25 more than the movie.

   The game costs $0.25 less than the soda.

   The game costs $10 plus $0.25 per game played.

   Each time the game is played the cost is $0.25.

3. Choose the possible problem situation that best fits the equation:

   $y = 60x$

   Sue can type 60 words per minute more than Veronica.

   Sue can type 60 words per minute.

   The number of words typed is 60 times the number of minutes typed.

   Veronica typed 60 minutes longer than Sue.

4. Choose the possible problem situation that best fits the equation:

   $y = 0.75x$

   Candy bars cost $0.75 more than potato chips.

   Candy bars cost $0.75 each.

   Candy bars cost $0.75 less than soda.

   Candy bars cost $75.

5. Choose the possible problem situation that best fits the equation:

   $y = 0.1x$

   The time it takes to empty an auditorium is 0.1 minutes per person.

   The width of the auditorium is 0.1 meters less than its length.

   It takes 0.1 minutes to get everyone out of the auditorium.

   The auditorium is 0.1 meters longer than the cafeteria.

6. Choose the possible problem situation that best fits the equation:

   $y = 200x$

   The population of one town is 200 more than the population of an adjacent town.

   The town has a population of 200 people.

   The population of a town is increasing at the rate of two hundred people per year.

   One town is 200 miles from another town.

7. Choose the possible problem situation that best fits the equation:

   $y = 40x$

   One row of a field has 40 fewer trees than another row.

   The total number of trees in a field is 40 times the number of rows of trees.

   One row of a field has 40 more trees than another row.

   There are 40 trees in a field.

8. Choose the possible problem situation that best fits the equation:

   $y = 0.95x$

   Brand X gum costs $0.95 more than Brand Z gum.

   You can save $0.95 if you purchase 10 packs of gum.

   The gum has decreased in price by $0.95 per pack.

   The packages of gum are $0.95 each.

9. Choose the possible problem situation that requires the formula for the volume of a cube.

Franco needs to line the bottom of his aquarium with netting. The sides of the aquarium all measure 2 feet. How much netting does he need?

Franco has an aquarium that is 2 feet on each side. If he wants to fill it with water, how much water will it hold?

Franco is putting a border around the top edge of an aquarium whose sides all measure 2 feet. How much border does he need?

Franco has an aquarium that is 2 feet long, 1 foot wide and 1 foot deep. If he wants to fill it with water, how much water does he need?

10. Choose the possible problem situation that requires the formula for the volume of a cube.

Gertrude has a clothes basket that is 4 feet on each side. How many cubic feet of clothes does it hold?

Gertrude is putting a border around the top edge of her clothes basket whose sides all measure 4 feet. How much border does she need?

Gertrude has a clothes basket that is 4 feet long, 3 feet wide, and 2 feet deep. How many cubic feet of clothes does it hold?

Gertrude needs to line the bottom of her clothes basket with cloth. The sides of the basket all measure 4 feet. How much cloth does she need?

11. Choose the possible problem situation that requires the formula for the volume of a cube.

Hank needs to line the bottom of his tool box with foil. The sides of the box all measure 18 inches. How much foil does he need?

Hank has a tool box that measures 18 inches on each side. How many cubic inches of rock will it hold?

Hank is putting a piece of rope around the top edge of a tool box whose sides all measure 18 inches. How much rope does he need?

Hank has a tool box that is 18 inches long, 9 inches wide, and 6 inches deep. How many cubic inches of rock will it hold?

12. Choose the possible problem situation that requires the formula for the volume of a cube.

Joey needs to line the bottom of a tub with plastic. The sides of the tub all measure 4 feet. How much plastic does he need?

Joey has a tub that measures 4 feet on each side. How much concrete will the tub hold?

Joey is putting a border around the top edge of a tub whose sides all measure 4 feet. How much border does he need?

Joey has a tub that is 4 feet long, 4 feet wide, and 3 feet deep. If he wants to fill it with concrete, how much concrete does he need?

13. Choose the possible problem situation that requires the formula for the area of a square.

Nancy's room measures 14 feet on each side. How much tile does she need to cover her floor?

Nancy needs to fill a box with water that is 14 inches on each side. How much water will the box hold?

Nancy is putting a strip of decorated paper around the top edge of her room. How much paper does she need?

Nancy's room is 14 feet by 12 feet. How much tile would she need to cover the floor?

14. Choose the possible problem situation that requires the formula for the area of a square.

Otis needs to tar the surface of a parking lot that is 50 yards on each side. How much tar does he need to cover the surface of the parking lot?

Otis wants to put a fence around a parking lot that is 50 yards on each side. How much fence does Otis need?

Otis needs to put tar on the surface of a parking lot that is 50 yards long and 40 yards wide. How much tar does he need?

Otis wants to put a fence around a parking lot that is 50 yards long and 40 yards wide. How much fence does he need?

15. Choose the possible problem situation that requires the formula for the area of a square.

Pablo has a container of paint that measures 9 inches on each side. How much paint does the container hold?

Pablo is putting edging around the floor of his room which measures 9 feet on each side. How much edging does Pablo need?

Pablo wants to paint a wall in his room that is 9 feet by 12 feet. Find the area of the surface that he wants to paint.

Pablo wants to paint a wall in his room that is 9 feet on each side. Find the area of the surface that he wants to paint.

16. Choose the possible problem situation that best fits the equation:

$y = 0.12x + 50$

A rental car company charges \$12 plus \$0.50 per mile the car is driven.

A rental car company charges \$50 per person plus \$0.12 per mile the car is driven.

A rental car company charges \$50 plus \$0.12 per mile the car is driven.

A rental car company charges \$0.12 per mile the car is driven but gives a \$50 discount.

17. Choose the possible problem situation that best fits the equation:

$y = 800x + 400$

An apartment complex charges a \$800 deposit plus \$400 per month.

An apartment complex charges \$400 per person plus \$800 per month.

An apartment complex charges a \$400 deposit plus \$800 per month.

An apartment complex charges \$800 per month but gives a \$400 discount.

18. Choose the possible problem situation that best fits the equation:

$y = 1,200x + 500$

The weekly salary of a real estate agent is \$1,200 plus a \$500 commission on each house that she sells.

The weekly salary of a real estate agent is \$500 plus a \$1,200 commission on each house that she sells.

The weekly salary of a real estate agent is \$500 plus a \$1,200 commission.

The weekly salary of a real estate agent is \$500 per day plus a \$1,200 commission on each house that she sells.

19. Choose the possible problem situation that best fits the equation:

$y = 45x + 125$

A cable company charges a \$125 installation fee plus \$45.

A cable company charges a \$45 installation fee plus \$125 for each month of service.

A cable company \$45 for each month of service and then gives a \$125 refund when the service is discontinued.

A cable company charges a \$125 installation fee plus \$45 for each month of service.

20. Choose the possible problem situation that best fits the equation:

$y = 0.2x + 30$

The monthly charge of a certain cell phone company is \$30 plus \$0.20 for each minute used.

The monthly charge of a certain cell phone company is \$30 per day plus \$0.20 for each minute used.

The monthly charge of a certain cell phone company is \$20 plus \$0.30 for each minute used.

The monthly charge of a certain cell phone company is \$20 per day plus \$0.30 for each minute used.

21. Choose the possible problem situation that best fits the equation:

$y = 0.05x + 1.5$

At the beginning of the year gasoline was priced at $5 per gallon. Since then it has increased in price at about $0.15 per month.

At the beginning of the year gasoline was priced at $1.50 per gallon. Since then it has increased in price at about $0.05 per month.

At the beginning of the year gasoline was priced at $0.50 per gallon. Since then it has increased in price at about $1.50 per month.

At the beginning of the year gasoline was priced at $1.50 per gallon. Since then it has decreased in price at about $0.05 per month.

22. Choose the possible problem situation that best fits the equation:

$y = 1.2x + 55$

The temperature at sunrise was 1.2 degrees. Since that time it has risen at 55 degrees per hour.

The temperature at sunrise was 55 degrees. Since that time it has risen 1.2 degrees.

The temperature at sunrise was 55 degrees. Since that time it has dropped at 1.2 degrees per hour.

The temperature at sunrise was 55 degrees. Since that time it has risen at 1.2 degrees per hour.

23. Choose the possible problem situation that requires the formula for the area of a rectangle.

Salene would like to cover a table with a piece of cloth. The table is 6 feet long and 3 feet wide. How much cloth does Salene need?

Salene wants to put a decorative border around her table that is 6 feet long and 3 feet wide. How much border does she need?

Salene has a jewelry box that measures 12 inches long, 8 inches wide, and 6 inches deep. How much jewelry will it hold?

Salene's table top is twice the size of her friend's. Her friend's table top is 9 square feet. How many square feet is Salene's table top?

24. Choose the possible problem situation that requires the formula for the area of a rectangle.

Travis bought a piece of property that is 250 feet long and 100 feet wide. How much fence would he need to completely enclose his property?

Travis has an aquarium that measures 18 inches long, 8 inches wide, and 12 inches deep. How much water will it hold?

Travis bought a piece of property measuring 25,000 square feet. If his neighbor's property is half the size, how large is his neighbor's property?

Travis bought a piece of property measuring 25,000 square feet. If the length of the piece of property is 250 feet, what is the width of the piece of property?

25. Choose the possible problem situation that requires the formula for the area of a rectangle.

Uma is building a flower bed that is 40 feet long and 3 feet wide. How much material does she need to make a border for her flower bed?

Uma is building a sidewalk that is 50 feet long, 3 feet wide and 4 inches deep. How much concrete does she need to build the sidewalk?

Uma is building a sidewalk that is 50 feet long and 3 feet wide. What is the surface area of the sidewalk?

The sidewalk in front of Uma's house is 50 feet long and 3 feet wide. Her neighbor's sidewalk is twice as long. How long is her neighbor's sidewalk?

26. Choose the possible problem situation that requires the formula for the area of a rectangle.

A bulletin board is 15 feet long and 3 feet wide. If the teacher wants to put a border around the bulletin board, how much border material is needed?

A classroom is 21 feet long, 18 feet wide, and 8 feet high. What is the volume of the classroom?

A bulletin board contains 45 square feet of space but it is half the size of the bulletin board next door. How large is the bulletin board next door?

A bulletin board contains 45 square feet of space. The width of the bulletin board is 3 feet. What is the length of the bulletin board?

27. Choose the possible problem situation that requires the formula for the perimeter of a rectangle.

Victor's room is 14 feet by 16 feet. He wants to put a new baseboard around the edges of his room. How much baseboard does he need?

Victor is carpeting his room that is 14 feet by 16 feet. How much carpet does he need?

Victor wants to fill a carton with lemonade. The carton is 4 inches wide, 4 inches long and 8 inches high. How much lemonade will it hold?

Victor's room has an area of 192 square feet. The length of the room is 16 feet. What is the width of his room?

28. Choose the possible problem situation that requires the formula for the perimeter of a rectangle.

Hector wants to frame his diploma. If the diploma measures 8 inches by 10 inches, how much framing material is needed?

Hector is carpeting his room that is 8 feet by 10 feet. How much carpet does he need?

Hector wants to fill a carton with water. The carton is 4 inches wide, 4 inches long and 8 inches high. How much water will it hold?

Hector's room has an area of 80 square feet. The length of the room is 10 feet. What is the width of his room?

29. Choose the possible problem situation that requires the formula for the perimeter of a rectangle.

Melissa is carpeting her sun room that is 10 feet by 8 feet. How much carpet does she need?

Melissa's sun room is 10 feet by 8 feet. She wants to place a string of lights along the edge of the room. How long will the string of lights be?

Melissa wants to humidify her sun room which is 8 feet wide, 10 feet long and 8 feet high. How much air is in the sun room for a humidifier to work on?

Melissa's sun room has an area of 180 square feet. The length of the room is 18 feet. What is the width of the sun room?

30. Choose the possible problem situation that requires the formula for the perimeter of a rectangle.

Andre is planning to cover his swimming pool that is 30 meters by 10 meters. How many square meters does the cover need to be?

Andre wants to fill his swimming pool. The pool is 10 meters wide, 30 meters long, and 2 meters deep. How much water is needed to fill the pool?

Andre is planning to tile the edge of his swimming pool. If the pool is 30 meters by 10 meters, what is the tile measurement around the pool?

Andre's pool deck has an area of 300 square meters. The length of the deck is 10 meters. What is its width?

31. Choose a story problem and answer that best matches the equation given below.

$y = 15x + 20$

Problem: A kennel charges \$15 per night for boarding a dog plus a \$20 fee for bathing the dog. If the dog spends 7 nights at the kennel, what is the total coat of boarding?

Answer: The cost of boarding the dog is \$105.

Problem: A kennel charges \$20 per night for boarding a dog plus a \$15 fee for bathing the dog. If the dog spends 7 nights at the kennel, what is the total coat of boarding?

Answer: The cost of boarding the dog is \$155.

Problem: A kennel charges \$15 times the sum of 20 and the number of nights the dog is boarded. If the dog spends 7 nights at the kennel, what is the total coat of boarding?

Answer: The cost of boarding the dog is \$405.

Problem: A kennel charges \$15 per night for boarding a dog plus a \$20 fee for bathing the dog. If the dog spends 7 nights at the kennel, what is the total coat of boarding?

Answer: The cost of boarding the dog is \$125.

32. Choose a story problem and answer that best matches the equation given below.

$y = 0.05x + 25$

Problem: A cell phone company charges a monthly fee of \$25.00 plus \$0.05 per minute used. If 400 minutes are used one month, what is the cost of the cell phone for the month?

Answer: The cost of the cell phone for the month is \$20.

Problem: A cell phone company charges a monthly fee of \$25.00 plus \$0.05 per minute used. If 400 minutes are used one month, what is the cost of the cell phone for the month?

Answer: The cost of the cell phone for the month is \$45.

Problem: A cell phone company charges a monthly fee of \$5.00 plus \$0.25 per minute used. If 400 minutes are used one month, what is the cost of the cell phone for the month?

Answer: The cost of the cell phone for the month is \$105.

Problem: A cell phone company charges a fee of \$0.25 times the sum of 5 and the number of minutes used. If 400 minutes are used one month, what is the cost of the cell phone for the month?

Answer: The cost of the cell phone for the month is \$101.25.

33. Choose a story problem and answer that best matches the equation given below.

$y = 250x + 200$

Problem: The weekly income of a car salesman is $200 plus a $250 commission on each car sold. If he sells 5 cars in a week, what is his income for the week?

Answer: The car salesman earned $1,250 for the week.

Problem: The weekly income of a car salesman is $200 plus a $250 commission on each car sold. If he sells 5 cars in a week, what is his income for the week?

Answer: The car salesman earned $1,450 for the week.

Problem: The weekly income of a car salesman is $250 plus a $200 commission on each car sold. If he sells 5 cars in a week, what is his income for the week?

Answer: The car salesman earned $1,250 for the week.

Problem: The weekly income of a car salesman is $200 plus a $250 bonus if he works every day of the week. If the car salesman works every day of the week, what is his income for the week?

Answer: The car salesman earned $450 for the week.

34. Choose a story problem and answer that best matches the equation given below.

$y = 0.1x + 16$

Problem: The monthly charge of a certain Internet provider is $16 plus $0.10 per hour of Internet use. If a person uses the Internet 60 hours during a month, what is the cost of service for that month?

Answer: The cost of Internet service for the month was $6.

Problem: The monthly charge of a certain Internet provider is $10 plus $0.16 per hour of Internet use. If a person uses the Internet 60 hours during a month, what is the cost of service for that month?

Answer: The cost of Internet service for the month was $19.60.

Problem: The monthly charge of a certain Internet provider is $16 plus $1 per hour of Internet use. If a person uses the Internet 60 hours during a month, what is the cost of service for that month?

Answer: The cost of Internet service for the month was $76.

Problem: The monthly charge of a certain Internet provider is $16 plus $0.10 per hour of Internet use. If a person uses the Internet 60 hours during a month, what is the cost of service for that month?

Answer: The cost of Internet service for the month was $22.

35. Choose a story problem and answer that best matches the equation given below.

$y = 30x + 75$

Problem: A workout facility charges a $75 membership fee plus $30 per month of membership. If Becky joined the facility 10 months ago, how much has she spent on her membership?

Answer: Becky has spent $375 on membership.

Problem: A workout facility charges a $75 membership fee plus $30 per month of membership. If Becky joined the facility 10 months ago, how much has she spent on her membership?

Answer: Becky has spent $300 on membership.

Problem: A workout facility charges a $30 membership fee plus $75 per month of membership. If Becky joined the facility 10 months ago, how much has she spent on her membership?

Answer: Becky has spent $780 on membership.

Problem: A workout facility charges $75 times 30 plus the number of months of membership. If Becky joined the facility 10 months ago, how much has she spent on her membership?

Answer: Becky has spent $3,000 on membership.

36. Choose a story problem and answer that best matches the equation given below.

$y = 25x + 800$

Problem: A golf club charges an initial membership fee of $800 plus $25 for each round of golf that you play. If Sheila joined the club and has played 20 rounds of golf, how much money has she spent?

Answer: Sheila has spent $500 on golf.

Problem: A golf club charges an initial membership fee of $25 plus $800 for each round of golf that you play. If Sheila joined the club and has played 20 rounds of golf, how much money has she spent?

Answer: Sheila has spent $16,025 on golf.

Problem: A golf club charges an initial membership fee of $800 plus $25 for each round of golf that you play. If Sheila joined the club and has played 20 rounds of golf, how much money has she spent?

Answer: Sheila has spent $1,300 on golf.

Problem: A golf club charges an initial membership fee of $800 plus $25 a month. If Sheila joined the club and has played 20 rounds of golf, how much money has she spent?

Answer: Sheila has spent $825 on golf.

37. Choose a story problem and answer that best matches the equation given below.

$y = 2.5x + 50$

Problem: An Internet CD club charges \$50 to join plus \$2.50 for each CD purchased. If Hamilton joined the club and then purchased 15 CD's, how much money did he spend?

Answer: Hamilton spent \$37.50 on the CD's.

Problem: An Internet CD club charges \$25 to join plus \$5 for each CD purchased. If Hamilton joined the club and then purchased 15 CD's, how much money did he spend?

Answer: Hamilton spent \$100 on the CD's.

Problem: An Internet CD club charges \$50 to join plus \$2.50 for each CD purchased. If Hamilton joined the club and then purchased 15 CD's, how much money did he spend?

Answer: Hamilton spent \$87.50 on the CD's.

Problem: An Internet CD club charges \$25 to join plus \$0.50 for each CD purchased. If Hamilton joined the club and then purchased 15 CD's, how much money did he spend?

Answer: Hamilton spent \$32.50 on the CD's.

38. Choose a story problem and answer that best matches the equation given below.

$y = 65x + 110$

Problem: A satellite company charges a \$110 installation fee plus \$65 per month for service. If Tasha had the service for 6 months, how much money did the satellite service cost her?

Answer: Tasha spent \$500 on the satellite service.

Problem: A satellite company charges a \$110 installation fee plus \$65 per month for service. If Tasha had the service for 6 months, how much money did the satellite service cost her?

Answer: Tasha spent \$390 on the satellite service.

Problem: A satellite company charges a \$65 installation fee plus \$110 per month for service. If Tasha had the service for 6 months, how much money did the satellite service cost her?

Answer: Tasha spent \$725 on the satellite service.

Problem: A satellite company charges a \$175 installation fee plus \$65 per month for service. If Tasha had the service for 6 months, how much money did the satellite service cost her?

Answer: Tasha spent \$565 on the satellite service.

39. Choose a story problem and answer that best matches the equation given below.

$y = 0.75x + 20$

Problem: A toll road offers a pass for $20 so that a driver pays a discount amount of $0.75 at each toll booth. If a driver purchased the pass and has paid tolls at 40 toll booths since he bought the pass, how much money has he spent on tolls?

Answer: The driver has spent $30 on tolls.

Problem: A toll road offers a pass for $75 so that a driver pays a discount amount of $0.20 at each toll booth. If a driver purchased the pass and has paid tolls at 40 toll booths since he bought the pass, how much money has he spent on tolls?

Answer: The driver has spent $83 on tolls.

Problem: A toll road offers a pass for $75 so that a driver pays a discount amount of $2 at each toll booth. If a driver purchased the pass and has paid tolls at 40 toll booths since he bought the pass, how much money has he spent on tolls?

Answer: The driver has spent $155 on tolls.

Problem: A toll road offers a pass for $20 so that a driver pays a discount amount of $0.75 at each toll booth. If a driver purchased the pass and has paid tolls at 40 toll booths since he bought the pass, how much money has he spent on tolls?

Answer: The driver has spent $50 on tolls.

40. Choose a story problem and answer that best matches the equation given below.

$y = 2x + 10$

Problem: A cafeteria offers a card for $10 which allows a person to purchase a sandwich for only $2 each. If Karla buys the card and purchased 25 sandwiches during a month, how much money did she spend on the sandwiches that month?

Answer: Karla spent $50 on sandwiches that month.

Problem: A cafeteria offers a card for $2 which allows a person to purchase a sandwich for only $10 each. If Karla buys the card and purchased 25 sandwiches during a month, how much money did she spend on the sandwiches that month?

Answer: Karla spent $252 on sandwiches that month.

Problem: A cafeteria offers a card for $10 which allows a person to purchase a sandwich for only $2 each. If Karla buys the card and purchased 25 sandwiches during a month, how much money did she spend on the sandwiches that month?

Answer: Karla spent $70 on sandwiches that month.

Problem: A cafeteria offers a card for $10 which allows a person to purchase a sandwich for only $2 each. If Karla buys the card and purchased 25 sandwiches during a month, how much money did she spend on the sandwiches that month?

Answer: Karla spent $60 on sandwiches that month.

MPA-126: Solving Real-World Problems Involving Sales Tax

The word percent means "per hundred", and simply expresses a ratio of some number to one hundred. For example, fifteen hundredths is the same as fifteen percent:

$$0.15 = \frac{15}{100} = 15\%$$

When we buy things, we usually have to pay a sales tax on each item purchased. The amount of sales tax that you are charged depends on where you are. Different cities charge different percentages. When solving problems involving sales tax, it is important to know how to solve the following types of problems:

- A percent of a total is what number?
- What percent of a total is a number?
- A percent of what total is a number?

Example 1 What is the dollar amount of sales tax on an item priced at $32 if the sales tax is 6% of the original price?

$n = 6\% \times 32$

$n = 0.06 \times 32$

$n = 1.92$

Step 1: The question is asking us "what is 6% of $32?" Let n represent the dollar amount of sales tax.

Answer: The dollar amount of sales tax is $1.92.

Example 2 Write an equation that represents the question below using n to represent the unknown.
"What percent of $422 is $37.98?"

"What percent of $422 is $37.98?"

$n \quad \times \quad =$

Step 1: First, look for key words. The verb "is" means "is equal to." The preposition "of", used with percent, indicates multiplication. The interrogative "what" indicates an unknown quantity.

Answer: The question is represented by the following equation:

$$n \times 422 = 37.98$$

Example 3 Becca wants to buy a new bicycle that costs $98 with a sales tax rate of 8%. How much money does Becca need to buy the bicycle?

$n = 8\% \times 98$

$n = 0.08 \times 98$

$n = 7.84$

Step 1: First, compute the sales tax. Let n represent the dollar amount of sales tax.

$98 + 7.84 = 105.84$

Step 2: To find the total cost of the bicycle, add the dollar amount of sales tax to the original price.

Answer: Becca will need $105.84 to buy the bicycle.

Example 4 Shannon cannot remember what she paid for her car, but she knows that the sales tax was $1,000 and the sales tax rate was 8%. How much was the original price of the car before tax?

Step 1: The problem asks us, "8% of what total is $1,000?"

$$8\% \times n = 1,000$$
$$0.08 \times n = 1,000$$

Step 2: Let n represent the original price of the car.

$$\frac{0.08 \times n}{0.08} = \frac{1,000}{0.08}$$
$$n = 12,500$$

Step 3: Divide both sides of the equation by 0.08.

Answer: The original price of Shannon's car was $12,500.

Example 5 Chris wants to buy a pair of pants that costs a total of $32.70. If the price before tax was $30, what was the sales tax rate?

$32.70 - 30 = 2.70$

Step 1: First find the dollar amount of sales tax by subtracting the original price from the total price.

$2.70 = n \times 30$

Step 2: Now answer the question, "2.70 is what percent of 30?" Let n represent the sales tax rate.

$$\frac{2.70}{30} = \frac{n \times 30}{30}$$
$$0.09 = n$$

Step 3: Divide both sides of the equation by 30.

$$0.09 = \frac{9}{100} \text{ or } 9\%$$

Step 4: Convert 0.09 into a percent.

Answer: If the total price is $32.70 for a $30 item, the sales tax rate is 9%.

Problem Set

Solve:

1. What is the dollar amount of sales tax on a $33 item if the sales tax rate is 6% of the original price?

2. What is the dollar amount of sales tax on a $65 coat if the sales tax rate is 4% of the original price?

3. If the sales tax rate is 12% of the original price, what is the dollar amount of sales tax on a $25 sweater?

4. A bag of chips costs $2 and the sales tax rate is 5.5% of this price. What is the dollar amount of sales tax on the bag of chips?

5. If the sales tax rate is 3.25% of the original price of an item, what is the dollar amount of sales tax on a $12 item?

6. Write an equation that represents the question below using n to represent the unknown.

"What percent of $315 is $28.35?"

7. Write an equation that represents the question below using n to represent the unknown.

"What is 13% of $25?"

8. Write an equation that represents the question below using n to represent the unknown.

"What percent of $28 is $1.96?"

9. Write an equation that represents the question below using n to represent the unknown.

"What percent of $250 is $37.50?"

10. Write an equation that represents the question below using n to represent the unknown.

"What percent of $115 is $6.90?"

11. Jacob is looking to purchase a sweater for his mother's birthday present. He sees a sweater he likes priced at $22.50. The sales tax rate is 9%. How much money does Jacob need to purchase the sweater for his mother?

12. Joshua wants to purchase a guitar case for his guitar. The guitar case costs $32.22. The sales tax rate is 8.75% of the original price. How much money does Joshua need to purchase the guitar case?

13. Dan went to the store to purchase a pint of strawberries. The sales tax rate is 4.5% of the original price. If Dan pays $0.18 in sales tax, what is the original price of the pint of strawberries?

14. Steven is buying a combo meal at a local fast food restaurant. The sales tax rate is 3.75% of the original price. How much is the original price of Steven's meal if he pays $0.12 in tax?

15. Mr. Berger wants to purchase a car. The sales tax rate is 6% of the original price. If Mr. Berger pays a tax of $900, what is the original price of the car?

16. Tony wants to buy a bicycle. The original price of the bicycle is $85. If Tony pays a total price of $102, what was the sales tax rate?

17. Mr. Rodney wants to buy a sofa. The price of the sofa is $115. If Mr. Rodney pays a total $118.45, what was the sales tax rate?

18. Samantha went to the grocery store. She selected $35 in groceries. If Samantha pays a total of $36.50, what was the sales tax rate on her groceries?

19. Mr. Prod went to the drug store to pick up his medicine. His medicine costs $50. Mr. Prod paid a total of $51 for his medicine. What was the sales tax rate?

20. Tina wants to purchase a pair of shoes. The original price of the pair of shoes is $45. If Tina pays a total of $46.35, what is the sales tax rate?

21. What is the dollar amount of sales tax on an item originally priced at $28 if the sales tax rate is 3.25% of the original price?

22. What is the dollar amount of sales tax on an item originally priced at $73 if the sales tax rate is 5% of the original price?

23. Write an equation that represents the question below using n to represent the unknown.

"What percent of $35 is $1.93?"

24. Write an equation that represents the question below using n to represent the unknown.

"What is 7% of $25?"

25. Write an equation that represents the question below using n to represent the unknown.

"What is 5.25% of $52?"

26. Dominic wants to purchase a new hat that is $7.50. The sales tax rate is 6.25% of the original price. How much money will Dominic need to purchase the hat?

27. Donna wants to purchase a baby doll for her daughter that is $19.99. The sales tax rate is 5.25% of the original price. How much money will Donna need to purchase the baby doll?

28. Paula wants to purchase make-up that costs $8.50. The sales tax rate is 4.75% of the original price. How much money will Paula need to purchase the make-up?

29. Paul wants to purchase a camcorder that costs \$80.50. The sales tax rate is 5.25% of the original price. How much money will Paul need to purchase the camcorder?

30. Pablo wants to purchase tools that cost \$78.50. The sales tax rate is 8% of the original price. How much money will Pablo need to purchase the tools?

31. Stan wants to buy a computer printer that is on sale. The sales tax rate is 6.25% of the sale price. If Stan pays a sales tax of \$3, what is the sale price of the computer printer?

32. Tammy wants to buy a computer that is on sale. The sales tax rate is 5% of the sale price. If Tammy pays a sales tax of \$41.25, what is the sale price of the computer?

33. Tom wants to buy a gift for his mother that is on sale. The sales tax rate is 5.25% of the sale price. If Tom pays a sales tax of \$3.50, what is the sale price of the gift?

34. Sandy wants to buy an outfit which is on sale for her baby. The sales tax rate is 3.25% of the sale price. If Sandy pays a sales tax of \$0.84, what is the sale price of the outfit?

35. Tony wants to buy a pair of jeans. The sales tax rate is 4.5% of the original price. If Tony pays a sales tax of \$1.35, what is the original price of the jeans?

36. Hector wants to buy a VCR. The original price of the VCR is \$35. If Hector pays a total price of \$36.75, what is the sales tax rate?

37. Mr. Gonzales wants to buy a new briefcase. The original price of the briefcase is \$65. If Mr. Gonzales pays a total price of \$71.01, what is the sales tax rate?

38. Linda wants to buy an umbrella. The original price of the umbrella is \$6.25. If Linda pays a total price of \$6.75, what is the sales tax rate?

39. Peter wants to buy a baseball bat. The original price of the bat is \$9.77. If Peter pays a total price of \$10.60, what is the sales tax rate?

40. Ms. Van needs to buy a new set of tires. The original price of the set of tires is \$525. If Ms. Van pays a total price of \$557.81, what is the sales tax rate?

MPA-127: Solving Real-World Problems Involving Discounts

Percentages are often used to determine discounts on merchandise. In this lesson, we will learn to evaluate the total price of an item using both discounts and sales tax. **Discount** is the amount by which the original price of an item is reduced.

Discount	The amount by which the original price of an item is reduced.

$$\text{Discount} = \text{rate of discount} \times \text{original price}$$

Example 1 Suppose your favorite shoe store is giving a 20% discount on all pairs of shoes. What is the dollar amount of the discount on a pair of shoes that costs $49.95?

Discount = rate of discount × original price = 20% × 49.95	**Step 1:** Substitute the given values into the equation.
= 0.20 × 49.95	**Step 2:** Convert 20% to a decimal.
= 9.99	**Step 3:** Multiply.
	Answer: The discount is $9.99.

Example 2 Estimate how much a video game will cost if its original price of $59 is discounted 19%.

Discount = rate of discount × original price = 20% × 59	**Step 1:** 19% is almost 20%. Substitute the given values into the equation.
= 0.20 × 59	**Step 2:** Convert 20% to a decimal.
= 11.80	**Step 3:** Multiply.
total price = original price – discount = 59.00 – 11.80 = 47.20	**Step 4:** Find the total price of the video game.
	Answer: The total price of the video game with the discount is approximately $47.20.

Example 3 If a clock originally priced at \$194 is discounted 25%, what is the price of the clock after the discount?

$$\begin{aligned}\text{Discount} &= \text{rate of discount} \times \text{original price}\\ &= 25\% \times \$194\\ &= 0.25 \times 194\\ &= 48.50\end{aligned}$$

Step 1: First, find the dollar amount of the discount.

$$\begin{aligned}\text{Sale price} &= \text{original price} - \text{discount}\\ &= \$194 - \$48.50\\ &= \$145.50\end{aligned}$$

Step 2: Next, find the sale price by subtracting the discounted amount from the original price.

Answer: The price of the clock after the discount is \$145.50.

Example 4 If the owner of a new coffeehouse reduces the cost of a large coffee from \$2.95 to \$2.49, what is the rate of the discount rounded to the nearest percent?

$$\begin{aligned}\text{Discount} &= 2.95 - 2.49\\ &= 0.46\end{aligned}$$

Step 1: Find the dollar amount of the discount.

$$\begin{aligned}\text{Discount} &= \text{rate of discount} \times \text{original price}\\ 0.46 &= n \times 2.95\\ \frac{0.46}{2.95} &= \frac{n \times 2.95}{2.95}\\ 0.1559 &= n\\ 0.16 &\approx n\end{aligned}$$

Step 2: Find the rate of the discount. Let n represent the rate of the discount.

$$16\% = n$$

Step 3: Convert 0.16 to a percent.

Answer: The rate of the discount is 16%.

Example 5 Danny bought his groceries on the day the store gives a 5% discount off the total price. If the sales tax at the store is 8%, what will Danny's final bill be if he bought \$112.84 worth of groceries?

$$\begin{aligned}\text{Discount} &= \text{rate of discount} \times \text{original price}\\ &= 5\% \times 112.84\\ &= 0.05 \times 112.84\\ &= 5.64\end{aligned}$$

Step 1: Find the amount of the discount.

Sale price = original price – discount $= 112.84 - 5.64$ $= 107.20$	**Step 2:** Find the sale price.
Sales tax = tax rate × sale price $= 8\% \times 107.20$ $= 0.08 \times 107.20$ $= 8.58$	**Step 3:** Find the amount of sales tax.
Total price = sale price + sales tax $= 107.24 + 8.58$ $= 115.82$	**Step 4:** Find the total price of the groceries.

Answer: The bill, after a 5% discount and tax, is $115.82.

Problem Set

Solve.

1. If an item priced at $94 is discounted 30%, what is the dollar amount of the discount?

2. If an item priced at $168 is discounted 15%, what is the dollar amount of the discount?

3. If an item priced at $112.50 is discounted 42%, what is the dollar amount of the discount?

4. Jackson's Appliance Store has an $8\frac{1}{2}\%$ discount on a washing machine that is priced at $398. What is the dollar amount of this discount?

5. What is the dollar amount of a 33% discount applied to an item that was originally priced at $220?

6. During a sale, a sweater originally priced at $45.70 is discounted 11%. What is the dollar amount of this discount? Round to the nearest cent.

7. Clara's Coffeehouse is offering a 21.5% discount this week. If a coffee maker was originally priced at $48.60, what is the dollar amount of the discount? Round to the nearest cent.

8. Travis wants to buy a jacket that was originally priced at $60. Estimate how much money Travis will save if a 38% discount is offered on this jacket.

9. Wayne is being offered an 18% discount on a beach umbrella that is normally priced at $125. Estimate the amount of this discount.

10. Lamar selects a video game system to buy that is originally priced at $160. It has a 27% discount. Estimate how much money Lamar will save on this purchase.

11. During the spring sale, Marine World is offering a 31% discount on a boat radio that is originally priced at $96. Estimate the amount of this discount.

12. Jordan's plane tickets were originally priced at $880. His travel agent is offering him an 8% discount. Estimate how much money Jordan will save on his tickets.

13. Haley wants to buy a designer label purse that has been discounted 52%. If the purse was originally priced at $122, estimate the amount of money Haley will save on the purse.

14. At a Going Out of Business Sale, Alex finds a car stereo originally priced at $200 with a 73% discount. Estimate how much Alex will save on this purchase.

15. Conner wants to buy an $80 tent for camping this summer. The tent he selects has a 43% discount. Estimate the amount of this discount.

16. If an item originally priced at $32 is discounted 22%, what is the price of the item after discount?

17. If Mr. Sam's bicycle is originally priced at $124 and is discounted 17%, what is the price of the bicycle after discount?

18. If an item originally priced at $165 is discounted 32%, what is the price of the item after discount?

19. A shirt originally priced at $18.50 is now on sale with a $7\frac{1}{2}$% discount. What is the price of the shirt after the discount?

20. If an item originally price at $96.25 is discounted 48%, what is the price of the item after discount?

21. If an item originally priced at $212.50 is discounted at 63%, what is the price of the item after discount?

22. A $451 item is discounted 25%. What is the price of this item once the discount has been applied?

23. What is the percent of discount off the original price on a pair of jeans reduced from $42 to $35.70?

24. What is the percent of discount off the original price on a rug reduced from $56.25 to $38.25?

25. What is the percent of discount off the original price on an item reduced from $92.20 to $59.93?

26. What is the percent of discount off the original price on an item reduced from $88.80 to $53.28?

27. Mr. Johnson bought an item originally priced at $110 for $89.65. What percent of discount off the original price was applied to Mr. Johnson's purchase?

28. What is the percent of discount off the original price on an item reduced from $54.50 to $16.35?

29. Megan finds a leather coat that has been reduced from $255 to $214.20. What is the percent of discount off the original price of the coat?

30. What is the percent of discount off the original price on an item reduced from $565 to $412.45?

31. The original price of a wallet is $25.50. A 15% discount is being offered on the wallet. The sales tax is 4%. What is the total cost of the wallet?

32. Travis wants to buy a backpack that has a 21% discount. It was originally priced at $37.68. The sales tax is 5.5%. What is the total cost of the backpack?

33. At the pet store, Ryan selects an aquarium set that is $57.50. The owner offers him an 18% discount. The sales tax where the store is located is 6%. What is the total cost Ryan will pay for the aquarium?

34. Mr. Huffman selects a new pair of eyeglasses to purchase that are originally priced at $112. He will receive a 12.5% discount on this purchase. He will also need to pay a 4.25% sales tax. What is the total cost of the eyeglasses?

35. Michelle finds a pair of skates originally priced at $65.58. The skates are now being discounted 38%. Michelle will still need to pay a 3.5% sales tax on this purchase. What is the total price of the skates?

36. Melinda wants to buy a charm bracelet that was originally priced at $155. The jewelry store is having a sale and the bracelet is now being discounted 42%. Melinda will still need to pay a 6.25% sales tax on her purchase. What is the total cost of the bracelet?

37. A set of golf clubs that was originally priced at $312.50 is now discounted 16%. If the sales tax is $7\frac{1}{2}$%, what will be the total cost of the clubs?

38. For his new office, Mr. Carey wants to buy a desk. He finds one that is originally priced at $187.50 but has now been discounted 25%. If the sales tax is 8%, how much will Mr. Carey pay for the desk?

39. Mrs. Hicks wants to replace the microwave in the teacher's lounge. She finds one originally priced at $98.25 but now has a 37% discount. The sales tax is 6.5%. What is the total cost of the microwave?

40. Cathleen wants to purchase a $575 microscope that has been discounted 65%. The sales tax would be 9%. What would be the total cost of the microscope?

MPA-128: Solving Real-World Problems Involving Simple Interest

When you borrow money from a bank, you are charged **simple interest**. You can earn interest when you let others use your money, such as when you put your money into a savings account. The amount of money borrowed or deposited is called the **principal**. The **interest rate** is the percentage of the principal owed or earned.

Simple Interest	Money paid for the use of money for a certain period of time.
Principal	The amount of money borrowed or deposited.
Interest Rate	The percentage of the principal owed or earned.

An interest rate is usually expressed as a percentage per year. **Time** is the duration for which the money is borrowed or deposited, measured in years. In a simple interest formula, the amount of interest equals the principal times the rate times the length of time. Use the letter I for the amount of interest, p for the principal, r for the interest rate and t for the time.

$$I = p \times r \times t$$

Time	The duration for which money is borrowed or deposited, measured in years.

Example 1 Diane borrowed $2,500 for 2 years from a bank charging 8.5% simple interest. How much interest will Diane have to pay back at the end of 2 years?

$I = p \times r \times t$

$I = 2,500 \times 8.5\% \times 2$

Step 1: Set up the equation and let p = $2,500, t = 2 years, and r = 8.5%.

$I = 2,500 \times 0.085 \times 2$

Step 2: Change the percent to a decimal.

$I = \$425$

Step 3: Multiply.

Answer: The amount of simple interest owed after 2 years is $425.

Example 2 If Joe invests $1,000 with a bank that pays 1.75% simple interest for 48 months, how much interest will he earn at the end of 48 months?

$I = p \times r \times t$

$I = 1,000 \times 1.75\% \times 4$

Step 1: Set up the equation and let p = $1,000 and r = 1.75%. Since time must be in years, convert 48 months to years. Twelve months equals one year, so 48 months is the same as 4 years.

$I = 1,000 \times 0.0175 \times 4$

Step 2: Change the percent to a decimal.

$I = 70$	**Step 3:** Multiply.
	Answer: Joe's investment will earn $70 in simple interest.

Example 3 What simple interest rate will allow a $2,000 investment to earn $1,000 interest in 5 years?

$I = p \times r \times t$ $1,000 = 2,000 \times r \times 5$	**Step 1:** Set up the equation and let I = $1,000, p = $2,000, and t = 5 years.
$1,000 = 10,000 \times r$	**Step 2:** Simplify.
$\frac{1,000}{10,000} = \frac{10,000 \times r}{10,000}$ $0.10 = r$	**Step 3:** Divide both sides by 10,000.
$10\% = r$	**Step 4:** Change the decimal to a percent.
	Answer: The simple interest rate is 10%.

Example 4 How much money would have to be invested at 8% simple interest to make $500 in interest in a 3-year period?

$I = p \times r \times t$ $500 = p \times 8\% \times 3$	**Step 1:** Set up the equation and let I = $500, r = 8%, and t = 3 years.
$500 = p \times 0.08 \times 3$	**Step 2:** Change the percent to a decimal.
$500 = p \times 0.24$	**Step 3:** Simplify.
$\frac{500}{0.24} = \frac{p \times 0.24}{0.24}$ $\$2,083.33 \approx p$	**Step 4:** Divide both sides by 0.24.
	Answer: The investment would be approximately $2,083.33.

Example 5 Stephanie invests $3,000 in a mutual fund that pays 5.6% in simple interest. How long will it take Stephanie to make $1,344 in interest?

$I = p \times r \times t$ $1,344 = 3,000 \times 5.6\% \times t$	**Step 1:** Set up the equation and let I = $1,344, p = $3,000, and r = 5.6%.
$1,344 = 3,000 \times 0.056 \times t$	**Step 2:** Change the percent to a decimal.

$1,344 = 168 \times t$	**Step 3:** Simplify.
$\frac{1,344}{168} = \frac{168 \times t}{168}$ $8 = t$	**Step 4:** Divide both sides by 168.

Answer: The money will have to be invested for 8 years.

Problem Set

Solve:

1. If $200 is invested for 5 years earning a 4% rate of interest, how much interest will be earned?
2. If $300 is invested for 3 years earning a 5% rate of interest, how much interest will be earned?
3. If $500 is invested for 6 years earning a 7% rate of interest, how much interest will be earned?
4. If $750 is invested for 4 years earning an 8% rate of interest, how much interest will be earned?
5. If $1,250 is invested for 2 years earning a 6% rate of interest, how much interest will be earned?
6. If $1,500 is invested for 4 years earning a 5.5% rate of interest, how much interest will be earned?
7. If $900 is invested for 7 years earning a 3% rate of interest, how much interest will be earned?
8. If $1,000 is invested for 10 years earning a 10% rate of interest, how much interest will be earned?
9. Doug borrowed $2,000 for 2 years from a bank charging 7% simple interest. What is the interest rate of Doug's loan?
10. Alise borrowed $5,500 for 3 years from a bank charging 7.25% simple interest. What is the interest rate of Alise's loan?
11. Christine borrowed $1,500 for 1 year from a bank charging 6.75% simple interest. What is the time of Christine's loan?
12. Benny borrowed $3,500 for 4 years from a bank charging 6.25% simple interest. What is the principal of Benny's loan?
13. Jeanne invested $4,500 for 5 years to a bank paying 5.5% simple interest. What is the interest rate earned on Jeanne's investment?
14. Adam invested $5,000 for 6 years to a bank paying 4.5% simple interest. What is the time of Adam's investment?
15. Herb invested $6,250 for 7 years to a bank paying 4.25% simple interest. What is the principal of Herb's investment?
16. If $1,250 is invested for 72 months earning a 4.25% rate of interest, how much interest will be earned?
17. If $1,575 is invested for 96 months earning a 5% rate of interest, how much interest will be earned?
18. If $2,000 is invested for 102 months earning a 5.75% rate of interest, how much interest will be earned?
19. If $12,500 is invested for 84 months earning a 5.5% rate of interest, how much interest will be earned?
20. If $10,000 is invested for 90 months earning a 6% rate of interest, how much interest will be earned?
21. If $8,000 is invested for 90 months earning a 6.25% rate of interest, how much interest will be earned?
22. If $6,500 is invested for 78 months earning a 7% rate of interest, how much interest will be earned?
23. If $5,500 is invested for 72 months earning a 6.5% rate of interest, how much interest will be earned?
24. Chris borrows $5,500 from a bank. The bank charges 5.25% simple interest. How much interest would Chris have to pay if he borrows the money for 5 years?
25. Julie borrows $14,000 from a bank. The bank charges 6.25% simple interest. How much interest would Julie have to pay if she borrows the money for 6 years?
26. Alethia borrows $10,000 from a bank. The bank charges 6.75% simple interest. How much interest would Alethia have to pay if she borrows the money for 3 years?

27. Lisa deposits $500 in a savings account. The account pays 5% simple interest. How much interest would accrue if Lisa leaves the money in the account for 5 years?

28. Dawn deposits $750 in a savings account. The account pays 4% simple interest. How much interest would accrue if Dawn leaves the money in the account for 6 years?

29. Dave deposits $1,000 in a savings account. The account pays 6% simple interest. How much interest would accrue if Dave leaves the money in the account for 4 years?

30. Jon deposits $1,750 in a savings account. The account pays 5.5% simple interest. How much interest would accrue if Jon leaves the money in the account for 8 years?

31. Juliet deposits $2,500 in an account that pays 5% simple interest per year. How long will it take Juliet to earn $500 in interest?

32. Romeo deposits $8,000 in an account that pays 6.5% simple interest per year. How long will it take Romeo to earn $3,120 in interest?

33. Jet deposits $4,200 in an account that pays 5.5% simple interest per year. How long will it take Jet to earn $1,617 in interest?

34. Joan deposits $5,600 in a savings account and earns $3,920 interest in 10 years. What simple interest rate is Joan earning on her account?

35. Paul deposits $7,800 in a savings account and earns $4,992 interest in 8 years. What simple interest rate is Paul earning on his account?

36. Hank earns $702 in interest from a savings account in 6 years. The simple interest rate is 4.5%. What is the principal of Hank's deposit?

37. William earns $1,125 in interest from a savings account in 5 years. The simple interest rate is 6%. What is the principal of William's deposit?

38. Jan deposits $3,500 in a savings account. The account pays 7.25% simple interest. What total amount of money would be in the account if Jan leaves the money in the account for 3 years?

39. Clark deposits $5,000 in a savings account. The account pays 6.5% simple interest. What total amount of money would be in the account if Clark leaves the money in the account for 5 years?

40. Lewis deposits $6,000 in a savings account. The account pays 8% simple interest. What total amount of money would be in the account if Lewis leaves the money in the account for 7 years?

MPA-129: Choosing Appropriate Scales and Intervals for Data (an Introduction)

When making a graph or a chart, units of scale and labels explaining the data are needed for both the horizontal and vertical axes. This lesson explains how to label the axes of both line graphs and bar graphs.

In the line graph shown below, the horizontal axis is labeled "Number of Days" and is scaled in units of 20. The vertical axis is labeled "Total Number of Accidents" and is scaled in units of 5.

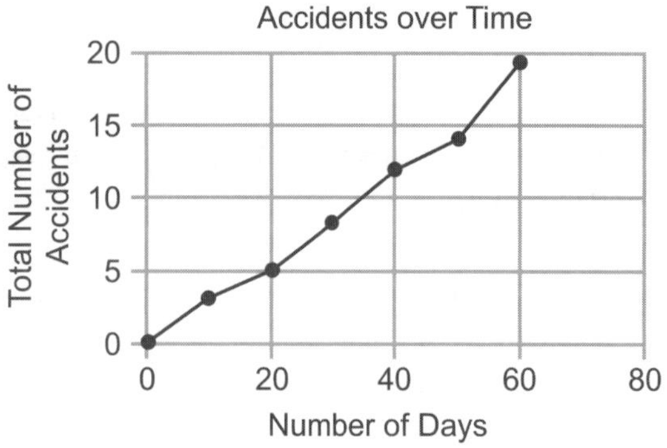

Total Number of Accidents	Number of Days
0	0
3	10
5	20
8	30
12	40
14	50
19	60

The values on the vertical axis must range from the lowest number in the data through the highest number in the data. When deciding how to label the vertical axis, calculate the range of the data. Recall that the range is the difference between the highest value and the lowest value.

$$\text{Range} = \text{highest value} - \text{lowest value}$$

When labeling the vertical axis, you must decide what values to include and whether to use units or multiples of units.

Example 1 Consider the chart of data below. What is the range of the test scores?

Student	Test Score
Tim	88
Joan	86
Sandy	93
Ralph	78
Sigmund	90
Hanna	74
Arthur	72
Charles	99

$$\begin{aligned}\text{Range} &= \text{highest value} - \text{lowest value}\\ &= 99 - 72\\ &= 27\end{aligned}$$

Step 1: The range is the difference between the highest and lowest values. The highest value is 99 and the lowest value is 72.

Answer: The range of the test scores is 27.

Example 2 For the data in the chart below, if we want to use 10 divisions on the "Test Score" axis starting at 72 and ending at 99, approximately how large would each division be?

Student	Test Score
Tim	88
Joan	86
Sandy	93
Ralph	78
Sigmund	90
Hanna	74
Arthur	72
Charles	99

Range = 27

Step 1: From the last example, we found that the range of the data is 27.

Division = 10

$\frac{27}{10} = 2.7 \approx 3$

Step 2: If we want to use 10 divisions, each mark on the vertical axis should be 2.7 points wide. Since we are measuring test scores, we can use a division of 3 points, since 2.7 is approximately 3.

Answer: Each division should be 3 points wide.

Example 3 Create a bar graph for the data in the chart below that uses the range from 0 to 35 for the mileage.

Automobile	Gas Mileage
Van	30
Car	25
Truck	32

Range = highest value – lowest value

= 35 – 0

= 35

Step 1: The range in the gas mileage is 35.

Division = 10

$\frac{35}{7} = 5$

Step 2: Since the range has to include 35 units, we will use 7 divisions. This makes each division 5 units wide.

Answer:

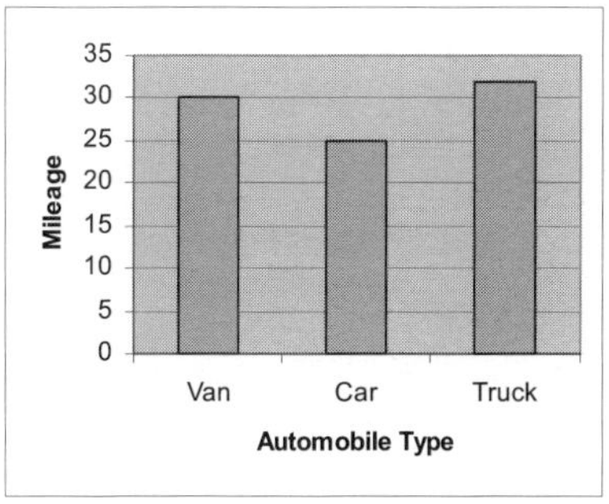

Example 4 Jo Ann, the class president, has gathered information from a survey and recorded the data in the table below.

Sport	Average Attendance
Basketball	4,850 spectators
Baseball	2,000 spectators
Tennis	1,500 spectators
Soccer	2,125 spectators

Jo Ann wants to compare the number of spectators of each sport with a bar graph using 10 divisions. How wide should each division be?

$$\begin{aligned}\text{Range} &= \text{highest value} - \text{lowest value}\\ &= 4,850 - 1,500\\ &= 3,350\end{aligned}$$

Step 1: The range in the data is 3,350.

Division = 10

$$\frac{3,350}{10} = 335 \approx 350$$

Step 2: If there are to be 10 divisions, each should be 350 units wide.

Answer: The divisions should be 350 units wide.

Example 5 Steve wants to cut back on energy expenses in his monthly budget. He logged his monthly expenditures on energy in the table below.

Month	Energy Bills
Jan.	$327
Feb.	$330
Mar.	$297
Apr.	$288
May	$252
June	$266

Make a line graph to represent Steve's monthly energy expenses. Label the axes appropriately.

$$\begin{aligned}\text{Range} &= \text{highest value} - \text{lowest value}\\ &= 330 - 252\\ &= 78\end{aligned}$$

Step 1: The range of the data is approximately 78 units.

Division = 10

$$\frac{78}{10} = 7.8 \approx 10$$

Step 2: If we make 10 divisions, each will be 10 units wide.

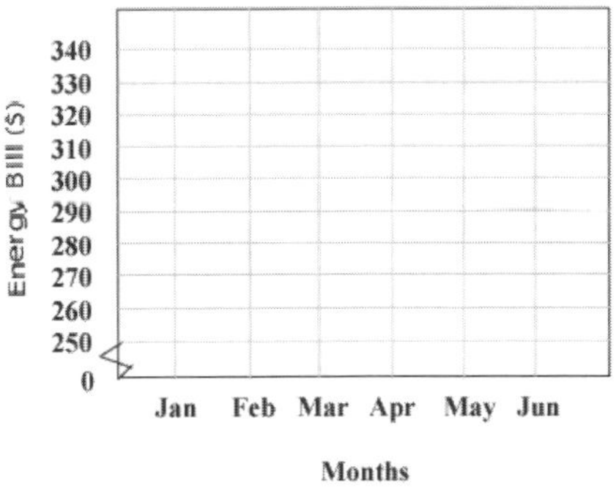

Step 3: Since our lowest value is 252, we will start our vertical axis with 250. To do this we must include a break symbol to show that there are missing values on the axis.

Answer:

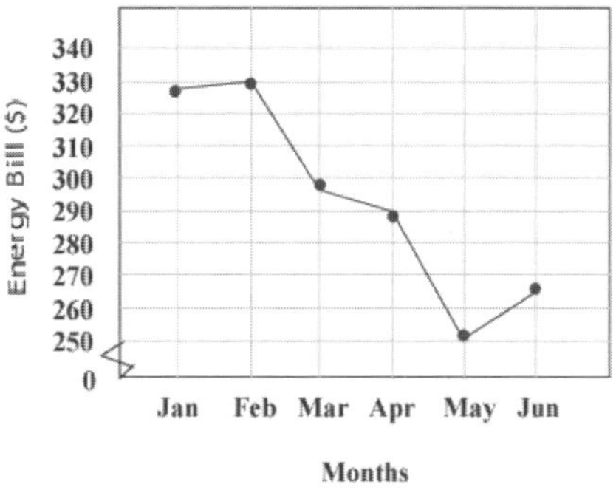

Problem Set

1. What is the range of data shown in this table?

Vending Machine Snack	Amount of Sales
Pretzels	$15
Potato Chips	$34
Juice	$56
Water	$42
Yogurt	$28

2. A dry cleaning store recorded the number of items cleaned for six months. This data is shown in the table below. What is the range of data?

Month	Number of Items Cleaned
February	231
March	198
April	334
May	269
June	421
July	362

3. The school cafeteria decided to offer salads for lunch. Mrs. Saunders, the cafeteria manager, recorded the income for salads for the first week below. What is the range of data?

Day of the Week	Amount Sold ($)
Monday	$46
Tuesday	$59
Wednesday	$78
Thursday	$89
Friday	$139

4. For a class project, Alice surveyed her classmates about their T.V. show favorites. This information is recorded in the table below. What is the range of data?

Favorite Type of T.V. Show	Number of Students
News	17
Detective	36
Comedy	61
Reality	52
Music Videos	111

5. Joe opened a new business and recorded the amount of income that he had for the first seven weeks. This data is recorded in the table below. What is the range of data for the income?

Week	Amount of Income
1	$524
2	$598
3	$608
4	$578
5	$679
6	$712
7	$653

6. A gym recorded the number of customers who used each area of the gym for the past week. What is the range of this data?

Area of Gym	Number Who Used This Area
Weights	171
Pool	103
Aerobics Class	113
Racquetball	46
Tennis Courts	52
Steam Room	62

7. The science department at the college recorded the number of students enrolled in each course. This information is recorded in the table shown below. What is the range of data?

Science Class	Number Enrolled
Chemistry	157
Biology	245
Geology	68
Astronomy	125
Oceanography	59

8. A travel agency conducted a survey concerning family vacations. Each person was asked to identify where they spent their last vacation. This information is shown in the table below. What is the range of this data?

Place	Number Who Visited
Beach	234
Mountains	89
Amusement Park	66
Historical Site	78
Other Family Members	198

9. Eddie is creating a bar graph to display the results of his survey. He recorded data concerning the favorite school lunches served in the cafeteria. Eddie will label the horizontal axis with the lunches and the vertical axis with the student amounts. If he decides to start the vertical axis at 0 and end it at 100 using a total of 10 marks on the vertical axis, how much separation should there be between each mark?

Favorite Lunch	Number of Students
Pizza	92
Cheeseburger	72
Taco	59
Spaghetti	48
Corndog	32

10. Curtis, a grocery store manager, recorded the number of cans sold of each type of soup this past year in the table shown below. He will create a bar graph from this information. He will label the horizontal axis with the type of soup and label the vertical axis with the amount sold. If Curtis decides to start his vertical axis at 0 and end it at 700 using a total of 10 marks on the axis, how much separation should there be between each mark?

Soup Type	Cans Sold
Chicken Noodle	421
Clam Chowder	320
Vegetable	683
Potato	220
Minestrone	129

11. Deanna is creating a line graph to display the data she collected concerning the number of service calls that a plumbing company answered for the first six months of this year. She will label the horizontal axis with the months and the vertical axis with the amount of service calls. If she decides to start her vertical axis at 0 and end it at 100 using a total of 10 marks on the axis, how much separation should there be between each mark?

Month	Number of Service Calls
January	78
February	96
March	52
April	33
May	49
June	67

12. A mall counted the number of shoppers for a period of six months. John is creating a line graph to represent this data. He will label the horizontal axis with the months and the vertical axis with the amount of shoppers. If he decides to start his vertical axis at 3,000 and end it at 8,000 using a total of 11 marks on the axis, how much separation should there be between each mark?

Month	Number of Shoppers
July	3,990
August	6,003
September	5,666
October	4,500
November	6,505
December	7,899

13. Margaret is creating a line graph to show the number of visitors at the aquarium for each day this week. She will label the horizontal axis with the day of the week and the vertical axis with the number of visitors. If Margaret decides to start her vertical axis at 100 and end it at 650 using a total of 12 marks on the vertical axis, how much separation should there be between each mark?

Day of the Week	Number of Visitors
Sunday	589
Monday	123
Tuesday	238
Wednesday	422
Thursday	378
Friday	488
Saturday	649

14. Heidi is creating a bar graph that shows the costs for 5 different summer camps that she is interested in attending. Each camp's weekly price is recorded in the table below. She will label the horizontal axis with the name of the camp and the vertical axis with the cost. If she decides to start her vertical axis at 500 and end it at 1,400 using a total of 10 marks on the vertical axis, how much separation should there be between each mark?

Camp	Weekly Cost
Day Camp	$556
Lake Camp	$889
Canoe Camp	$1,367
Basketball Camp	$780
Softball Camp	$690

15. Amber, the school yearbook editor, is creating a bar graph to show the number of yearbooks ordered by each class at the high school. The information she will use is recorded in the table below. Amber will label the horizontal axis with the name of each class and the vertical axis with the number of yearbooks sold. If she decides to start her vertical axis at 30 and end it at 160 using a total of 8 marks on the axis, how much separation should there be between each mark?

Class	Number of Yearbooks Sold
Freshman	42
Sophomore	98
Junior	136
Senior	158

16. Mrs. Garrett, a realtor, recorded data concerning the number of customers who visited her realty company this month in the following table. Use this information to label the axes on the following graph.

Reason For Visit	Number of Customers
Rent a Home	10
Buy a Home	54
Sell a Home	45
Buy a Condo	32

Realty Company Visits

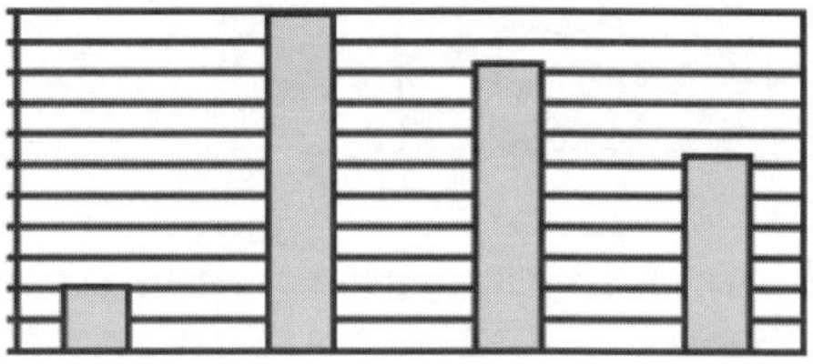

17. The Barnes family recorded the number of miles they drove each day on their family vacation in the following table. Use this information to label the axes on the following graph.

Day of the Week	Miles Driven
Tuesday	312
Wednesday	241
Thursday	152
Friday	75
Saturday	280

The Barnes' Family Vacation

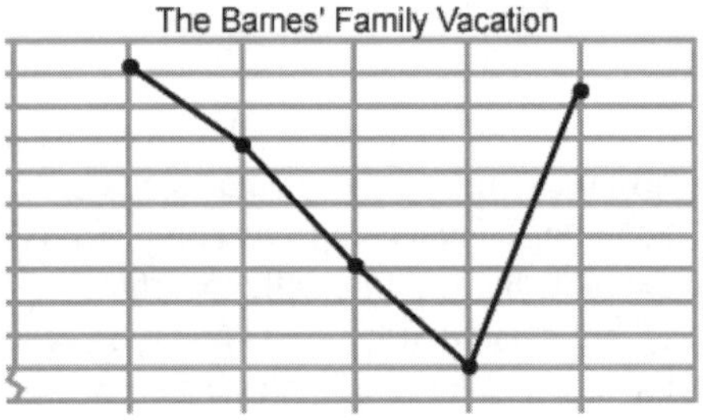

18. A convenience store has recorded data concerning the items sold during one evening in the following table. Use this information to label the axes on the following graph.

Types of Item Sold	Dollar Amount
Gas	$446
Newspapers	$54
Beverages	$145
Food Items	$295
Other	$120

Evening Sales

19. Amber is mowing lawns this summer to earn money. She recorded the amount of money she earned for five months in the following table. Use this information to label the axes on the following graph.

Month	Dollar Amount
May	$65
June	$188
July	$250
August	$345
September	$220

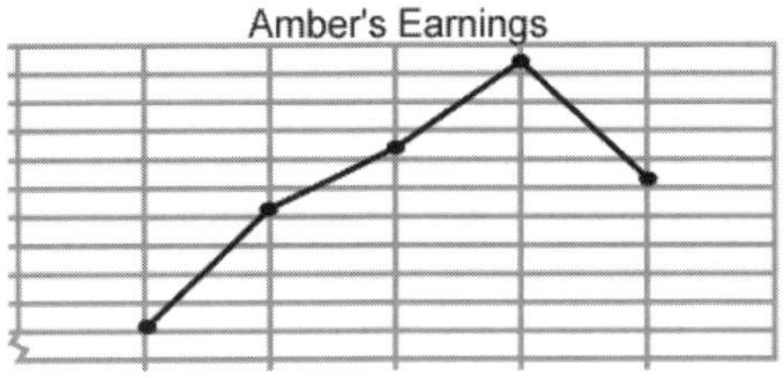

20. The athletic club for a middle school is selling items at a concession stand during basketball season. They recorded the number of items sold for each type of item in the following table. Use this information to label the axes on the following graph.

Items	Units Sold
Popcorn	234
Nachos	85
Hotdogs	155
Soda	225
Candy	410

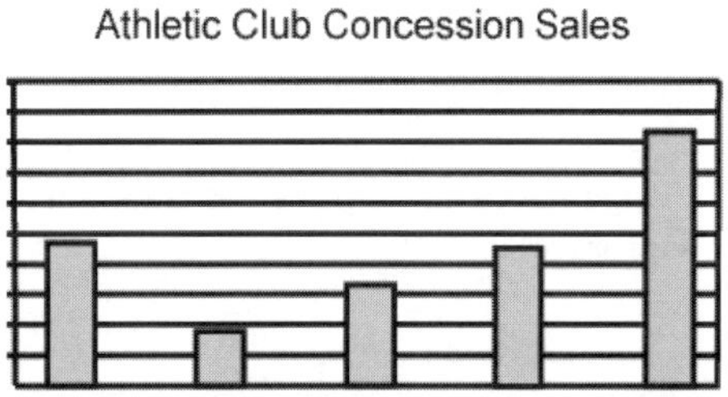

21. The drama club recorded the number of tickets sold for each night of their play performance in the following table. Use this information to label the axes on the following graph.

Performance	Number of tickets sold
Thursday	78
Friday	93
Saturday Matinee	56
Saturday Evening	130
Sunday Matinee	65

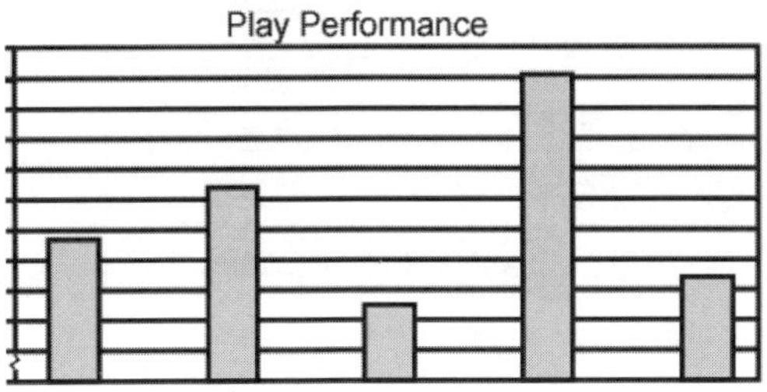

22. Chip measured six rooms in his house. He recorded the square feet of each room in the following table. Use this information to label the axes on the following graph.

Room	Square Feet
Living Room	225
Kitchen	120
Closest	15
Bathroom	56
Bedroom #1	144
Bedroom #2	160

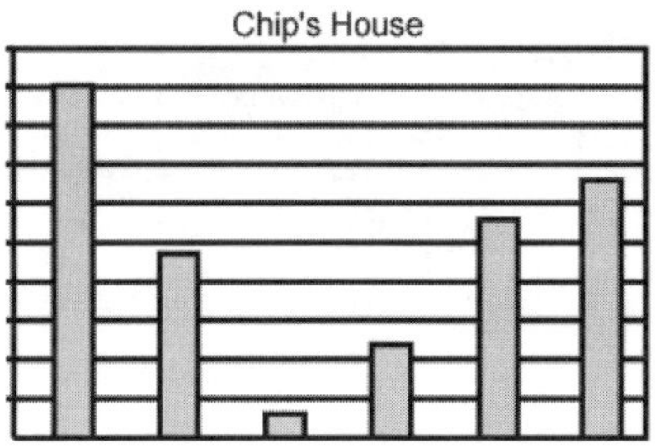

23. Mr. Williams, the owner of several car dealerships, has recorded data about the number of cars sold at each of his car lots during the past year in the following table. Mr. Williams wants to compare the number of cars sold with a bar graph. Draw the graph that best helps Mr. Williams interpret this information.

Car Lot	Number of Cars Sold
North Lot	324
South Lot	457
East Lot	223
West Lot	670

24. Dale has been recording this week's restaurant sales in the following table. Dale wants to compare the number of items sold with a bar graph. Draw the graph that best interprets this information.

Type of Item	Number Sold
Breakfast	456
Lunch	678
Dinner	550
Beverages	2,000
Desserts	350

25. Jane kept a record of her travel expenses for the past six months and recorded the data in the following table. Jane wants to see how her expenses have changed over time. Draw the graph that best interprets Jane's data.

Month	Amount of Expenses
March	$1,200
April	$1,000
May	$1,550
June	$1,708
July	$2,000
August	$1,650

26. A basketball team kept a record of the total number of points scored each season for the past 5 years in the following table. The coach wants to evaluate the team's progress over time. Draw the graph that best helps the coach to do this.

Year	Total Number of Points in Season
1999	560
2000	593
2001	615
2002	630
2003	646

27. Mrs. Hinson, the school board clerk, recorded the average daily attendance for the five middle schools in the school district in the following table. Mrs. Hinson wants to compare the average daily attendance of each middle school with a bar graph. Draw the graph that best helps Mrs. Hinson interpret the data.

School	Average Daily Attendance
A	266
B	359
C	460
D	420
E	410

28. Latonya just opened her own catering business. She recorded her personnel costs for the first six months in the following table. Latonya wants to see how the personnel costs of her business have changed over the past six months by using a line graph. Draw the graph that best interprets this information.

Month	Personnel Costs
April	$125
May	$275
June	$375
July	$435
August	$545
September	$580

29. Mr. Harriston, the owner of a new apartment rental complex, recorded the number of new apartment rentals for the first 5 months in the following table. Mr. Harriston wants to see how the number of new rentals has changed over the first five months. Draw a line graph that will best help Mr. Harriston interpret his information.

Month	Number of New Rentals
June	36
July	57
August	35
September	22
October	10

30. The chamber of commerce for a beach resort town is collecting data concerning the amount of revenue each hotel generated for this past summer. This information is recorded in the table below. The chamber wants to compare the amount of revenue generated at each hotel. Draw a bar graph that will best help them interpret this information.

Hotel	Amount of Revenue
Sand Dollar	$25,000
Beach Ball Inn	$16,500
Sand Castle	$17,000
Sunny Day	$30,000

31. Cameron's family recorded how far each family member traveled in order to attend his graduation. This data is recorded in the following table. Draw the graph that best helps Cameron's family compare the relative distance each family member had to travel.

Family Members	Miles Traveled
Sister	490
Brother	400
Uncle	460
Aunt	315
Cousin	340

32. The cheerleaders are raising money to attend a summer camp. For five Saturdays, they washed cars. The captain of the squad recorded their earnings in the following table. Draw the best graph that will help the cheerleaders compare the amount raised on each Saturday.

Saturdays	Amount Raised
1st Saturday	$58
2nd Saturday	$167
3rd Saturday	$212
4th Saturday	$328
5th Saturday	$280

33. Molly has several audio tapes of novels to listen to as she travels long distances in her car. She recorded the length in minutes for five of these tapes. Draw the best graph that will help Molly compare the length of each tape.

Tape	Length in Minutes
A	320
B	120
C	160
D	180
E	240

34. An after-school tutoring program has expanded into 6 different schools. The enrollment data is recorded in the following table. Draw the best graph that will help the owner of the program compare the enrollments at each school.

School	Enrollment
Bayside	230
Wood Acre	115
Jones Middle	90
Davis Middle	150
South High	185
Forest High	133

35. Mike is a traveling salesman. He recorded the number of miles that he traveled for the first seven weeks of the year in the following table. Draw the best graph that will help Mike compare the number of miles traveled each week.

Week #	Miles Traveled
1	452
2	421
3	380
4	350
5	330
6	330
7	315

36. Mr. Thomas, a business owner, is analyzing the cost of office supplies. He recorded the monthly expenses for these for the past seven months in the following table. Draw the best graph that will help Mr. Thomas compare the amount spent of office supplies each month.

Month	Amount of Office Expenses
March	$ 875
April	$ 880
May	$ 950
June	$ 990
July	$1,025
August	$1,080
September	$1,110

37. Alicia is investigating buying a new computer. She has recorded the prices of five computers in the following table. Draw the best graph that will help Alicia compare the prices of each computer.

Computer	Price
A	$499
B	$512
C	$678
D	$990
E	$1,305

38. A roofer has to repair five different roofs. Each roof will need a different number of shingles to complete the repair. The number of shingles for each roof is recorded in the following table. Draw the best graph that will help the roofer compare the number of shingles needed for each roof.

Roof	Number of Shingles
Roof A	825
Roof B	1,235
Roof C	300
Roof D	750
Roof E	480

39. Marcy has worked at the same company for 6 years. She recorded her annual salary for each year that she has been employed in the following table. Draw the best graph that will help Marcy compare her annual salary for each year employed.

Year	Salary
1	$16,000
2	$18,500
3	$20,800
4	$27,000
5	$35,000
6	$39,000

40. Tony, a chef, is reading new recipes. He has recorded the baking time in minutes of five of these recipes in the following table. Draw the best graph that will help Tony compare the baking time of each recipe.

Recipe	Baking time (minutes)
A	30
B	45
C	55
D	90
E	180

MPA-130: Developing a Sense of Relative Sizes of Measures

One of the earliest inventions by humans was a system of weights and measures. Far back in time, people began to measure and weigh things by using parts of their body as units of measure. For instance, a foot was actually the length of a person's foot. This early system of measurement is known as the **customary system**. Obviously, this type of system soon caused difficulties as the measurements were clearly different for different people.

Customary System	Older measurement system that is still in use today, originating from using body parts as comparison for measures.

As a result of these complications, another measuring system was created. This system is called the **metric system**. Today, the U.S. uses both the customary and metric system.

Metric System	Measurement system developed in 1790 by the French Academy of Sciences which uses consistent units for the measurement of length, capacity, and mass.

Using both the customary system and metric system, we will explore three different types of measurements: the length, the capacity, and the weight or mass of an object.

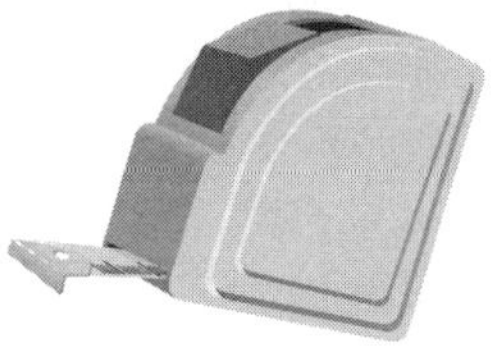

	Measurement of Length (How long an item is)	**Measurement of Capacity** (How much an item holds)	**Measurement of Weight / Mass** (How heavy an item is)
Customary Units	Inches Feet Yards Miles	Teaspoons Tablespoons Fluid Ounces Cups Pints Quarts Gallons	Ounces Pounds Tons

	Measurement of Length (How long an item is)	Measurement of Capacity (How much an item holds)	Measurement of Weight / Mass (How heavy an item is)
Metric Units	Millimeters (mm) Centimeters (cm) Decimeters (dm) **Meters (m)** Dekameters (dam) Hectometers (hm) Kilometers (km)	Milliliters (mL) Centiliters (cL) Deciliters (dL) **Liters (L)** Dekaliters (daL) Hectoliters (hL) Kiloliters (kL)	Milligrams (mg) Centigrams (cg) Decigrams (dg) **Grams (g)** Dekagrams (dag) Hectograms (hg) Kilograms (kg)

In this lesson, we will identify appropriate customary and metric units in relation to their real-life applications. We will begin with the customary units.

Customary Units

Measurement of Length

Inch (in.): Approximately the height of the middle joint of the index finger. Used to measure short lengths (ex. pen, stapler, and notebook).

Foot (ft): **= 12 inches**
Approximately the length of a man's foot. Used to measure such things as the length of a room, a car, and the height of filing cabinet.

Yard (yd): **= 3 feet = 36 inches**
Approximately the length from a chin to the end of an out-stretched arm. Used to measure longer lengths (ex. football field, purchasing fabric).

Mile (mi): **= 1,760 yards**
Used to measure long lengths (ex. distance between two cities).

Measurement of Capacity

Teaspoon (tsp): Holds approximately the capacity of a bottle cap. Used to measure very small quantities of liquid (ex. cooking ingredients and cough syrup).

Measurement of Capacity

Tablespoon (tbsp):	= 3 Teaspoons Used to measure small quantities of liquid (ex. cooking).
Fluid Ounce (fl oz):	= 2 Tablespoons = 6 Teaspoons Common units for small drinks and cosmetic items.
Cup (c):	= 8 Fluid Ounces Approximately the size of a small can of soup.
Pint (pt):	= 2 Cups = 16 Fluid Ounces Approximately the size of a small ice cream container or water bottle.
Quart (qt):	= 2 Pints = 4 Cups Approximately one-half of a milk carton.
Gallon (gal):	= 4 Quarts = 8 Pints = 16 Cups Used to measure large capacities (ex. gasoline and paint).

Measurement of Weight

Ounce (oz):	Approximately the weight of a greeting card. Used to weigh light masses (ex. pencils, cups, and small fruits).
Pound (lb):	= 16 ounces Used to weigh medium masses (ex. bags of fruit and a person's weight).
Ton (t):	= 2,000 lbs = 32,000 oz Used to weigh very heavy masses (ex. elephants and cars).

Now that we have identified common units in the customary system, let's look at units in the metric system. As stated earlier, the metric system uses consistent units throughout its measurements. Notice that all the prefixes among length, capacity, and mass are identical. Only the roots are different, depending on the type of measurement. The metric system is also more consistent than the customary system because its *prefixes are multiples of ten*. In other words, 10 milli-units is equal to 1 centi-unit, and 10 centi-units is equal to 1 deci-unit.

Metric Units

Measurement of Length

Millimeter (mm):	Approximately the width of a dime. Measures very small lengths (ex. length of an ant, paper clip, or rubber band).
Centimeter (cm):	= 10 mm Approximately the width of the tip of a little finger. Measures small lengths (ex. length of banana and diameter of CD).
Decimeter (dm):	= 10 cm = 100 mm
Meter (m):	= 10 dm A little longer than a yard. Approximately the length of a broom handle. Used to measure longer lengths (ex. height of a refrigerator, building, or house).
Dekameter (dam):	= 10 m
Hectometer (hm):	= 10 dam
Kilometer (km):	= 10 hm Equal to a little less than a mile. Approximately equal to 10 football fields or 7 city blocks. Used to measure very long distances (ex. distances between two cities).

Measurement of Capacity

Milliliter (mL):	Smallest common unit of capacity. Approximately the amount of ink inside a fountain pen / 1 sugar cube. Measures very small capacities (ex. sugar pack and cough medicine).
Centiliter (cL):	= 10 mL = 2 Teaspoons Approximately the amount of 10 sugar cubes / ink inside 10 fountain pens.
Deciliter (dL):	= 10 cL = 100 mL Approximately half a small glass of juice.

Measurement of Capacity

Liter (L):	= 10 dL = 100 cL = 1,000 mL Usually half the size of a large soda bottle.
Dekaliter (daL):	= 10 L Approximately half a bathtub.
Hectoliter (hL):	= 10 daL A baby pool is approximately 10 hL.
Kiloliter (kL):	= 10 hL Measures very large amounts of liquid.

Measurement of Mass

Milligram (mg):	Smallest common unit of mass. Approximately the mass of one grain of salt. Used to weigh very small masses (ex. the wings of a housefly).
Centigram (cg):	= 10 mg 10 seeds is approximately 1 cg.
Decigram (dg):	= 10 cg One vitamin tablet is approximately 3 dg.
Gram (g):	= 10 dg Approximately one paper clip.
Dekagram (dag):	= 10 g A mouse is approximately 7.5 dag.
Hectogram (hg):	= 10 dag
Kilogram (kg):	= 10 hg A little over two pounds. Approximately the mass of a textbook, hammer, or a baseball bat.

Example 1 Determine the most appropriate unit for measuring the weight of a watermelon.

(A) Inches (B) Centimeters

(C) Liters (D) Pounds

Determine the most appropriate unit for measuring the **weight** of a watermelon.	**Step 1:** Notice the key word is "weight." We are finding a measurement of weight.
(A) Inches	**Step 2:** Inch is a measurement of length in the customary system.
(B) Centimeters	**Step 3:** Centimeter is a measurement of length in the metric system.
(C) Liters	**Step 4:** Liter is a measurement of capacity in the metric system.
(D) Pounds	**Step 5:** Pound is a measurement of weight in the customary system.
	Answer: The most appropriate unit for measuring the weight of a watermelon is **pounds**.

Example 2 What is the most appropriate unit, in the customary system, for measuring the length of a pencil?

Units of length in the customary system: Inches (*middle joint of index finger*) Feet (*approximate length of a man's foot*) Yards (*chin to out-stretched arm of a man*) Miles (*measures distances between two cities*)	**Step 1:** List the units of **length** in the customary system, and recall common items for each unit.
What is the **most appropriate** unit, in the customary system, for measuring the length of a pencil?	**Step 2:** You can use any of the above units to measure the length of a pencil. However, notice that we are asked to find the most appropriate unit. Feet would be longer than a pencil. Therefore, yards and miles are both longer than a pencil.
	Answer: The most appropriate unit for measuring the length of a pencil in customary system is inches.

Example 3 Which unit of measure will give you the smallest measurement?

(A) Deciliter (B) Dekaliter

(C) Hectoliter (D) Centiliter

Milliliter
Centiliter
Deciliter
Liter
Dekaliter
Hectoliter
Kiloliter

Step 1: List the units of capacity for the metric system in order from smallest to largest.

Milliliter
Centiliter
Deciliter
Liter
Dekaliter
Hectoliter
Kiloliter

Step 2: Highlight the choices given.

Answer: **Centiliter** is the smallest measurement.

Example 4 Which of the following is the best estimation for the height of a closet door?

(A) 7 inches (B) 7 yards

(C) 7 miles (D) 7 feet

(A) 7 inches

Step 1: 7 inches is approximately the length of a pen. A closet door is much taller than this.

(B) 7 yards

Step 2: 7 yards is the height of approximately 4 men. A closet door is much smaller than this.

(C) 7 miles

Step 3: 7 miles is approximately the distance between two neighborhoods. A closet door is much shorter than this distance.

Answer: The best estimation for the height of a closet door is **7 feet**.

Example 5 What is the best estimation for the weight of a cheeseburger?

(A) 6 milligrams (B) 6 decigrams

(C) 6 pounds (D) 6 ounces

(A) 6 milligrams

Step 1: 6 milligrams is approximately the mass of 6 grains of salt. A cheeseburger weighs much more than 6 grains of salt.

(B) 6 decigrams

Step 2: 6 decigrams is approximately the mass of 2 vitamin tablets. A cheeseburger weighs more than 2 vitamin tablets.

(C) 6 pounds

Step 3: 6 pounds is approximately the mass of a healthy newborn baby. A cheeseburger weighs much less than a newborn baby.

Answer: The best estimation for the weight of a cheeseburger is **6 ounces**.

Problem Set

1. Determine the most appropriate unit for measuring the length of a bookshelf.

meters grams

quarts pounds

2. Determine the most appropriate unit for measuring the length of a gym floor.

yards tons

hectograms gallons

3. Determine the most appropriate unit for measuring the weight of a hamster.

cups feet

millimeters ounces

4. Determine the most appropriate unit for measuring the weight of a toolbox.

kilograms quarts

pounds yards

5. Determine the most appropriate unit for measuring the capacity of a milk carton.

pounds fluid ounces

grams centimeters

6. Determine the most appropriate unit for measuring the capacity of a coffee pot.

cups feet

hectograms tons

7. Determine the most appropriate unit for measuring the mass of an orange.

pounds inches

hectograms tablespoons

8. What is the most appropriate unit for measuring the mass of a pencil?

kilograms centigrams

grams milligrams

9. What is the most appropriate unit for measuring the length of a laptop computer?

millimeters centimeters

kilometers meters

10. What is the most appropriate unit for measuring the length of a calculator?

yards miles

inches feet

11. What is the most appropriate unit for measuring the capacity of a bottle of liquid paper?

liters centiliters

kiloliters hectoliters

12. What is the most appropriate unit for measuring the thickness of 10 loose leaf sheets of paper?

millimeters hectometers

kilometers meters

13. What is the most appropriate unit for measuring the mass of a safety pin?

milligrams grams

centigrams kilograms

14. What is the most appropriate unit for measuring the mass of a slice of bread?

tons ounces

pounds

15. What is the most appropriate unit to measure the length of a sofa in your living room?

centimeters kilometers

meters hectometers

16. Which unit of measure will allow you to accurately measure the smallest weight?

dekagram kilogram

hectogram decigram

17. Which unit of measure will allow you to accurately measure the longest length?

yard mile

inch foot

18. Which unit of measure will allow you to accurately measure the smallest capacity?

fluid ounce cup

tablespoon teaspoon

19. Which unit of measure will allow you to accurately measure the largest weight?

decigram dekagram

centigram hectogram

20. Which unit of measure will allow you to accurately measure the smallest capacity?

deciliter dekaliter

hectoliter centiliter

21. Which unit of measure will allow you to accurately measure the largest capacity?

cup quart

gallon pint

22. Which unit of measure will allow you to accurately measure the smallest weight?

centigram decigram

milligram hectogram

23. Which of the following is the best estimate for the length of a loose leaf sheet of paper?

$10\frac{1}{2}$ inches $\frac{1}{2}$ yard

$\frac{1}{3}$ mile 2 feet

24. Which of the following is the best estimate for the distance between Houston, TX and Dallas, TX?

100 yards 3 miles

350 kilometers 100 meters

25. Which of the following is the best estimate for the mass of a backpack with a 70-sheet notebook inside?

10 milligrams 1 kilogram

40 decigrams 4 dekagrams

26. Which of the following is the best estimate for the length of a pickup truck?

22 inches 11 feet

$\frac{1}{3}$ mile 9 yards

27. Which of the following is the best estimate for the capacity of a pitcher of iced tea?

2 quarts   32 cups

4 gallons   16 fluid ounces

28. Which of the following is the best estimate for the mass of 2 horseflies?

5 grams   2 dekagrams

1 kilogram   8 milligrams

29. Which of the following is the best estimate for the capacity of a bottle of dish soap?

4 liters   5 hectoliters

3 kiloliters   400 milliliters

30. Which of the following is the best estimate for the weight of a brick?

2 pounds   2 tons

2 ounces

31. What is the best estimate for the capacity of a medium-sized bottle of liquid laundry detergent?

1 kiloliter   3 gallons

3 liters   5 hectoliters

32. What is the best estimate for the weight of a student's school desk?

5 kilograms   10 dekagrams

50 ounces $\frac{1}{2}$ ton

33. What is the best estimate for the length of a canoe?

$4\frac{1}{2}$ yards   15 meters

50 centimeters $\frac{1}{8}$ mile

34. What is the best estimate for the capacity of a small bowl used to feed a house cat?

2 centiliters   1 pint

5 teaspoons   5 hectoliters

35. What is the best estimate for the weight of a bowling ball?

$\frac{1}{2}$ kilogram   80 centigrams

12 pounds   50 grams

36. What is the best estimate for the length of a slide found on a playground?

5 meters   20 inches

10 yards   50 dekameters

37. What is the best estimate for the capacity of a can of hair spray?

10 milliliters   2 dekaliters

10 fluid ounces   5 gallons

38. What is the best estimate for the mass of a refrigerator?

6 kilograms   250 pounds

500 milligrams   1 ton

39. What is the best estimate for the height of a giraffe?

6 meters   1 yard

60 inches   2 kilometers

40. What is the best estimate for the capacity of a bottle of pancake syrup?

$\frac{1}{2}$ hectoliter   24 fluid ounces

3 kiloliters   10 pints

MPA-131: Interpreting and Creating Histograms

There are many different types of bar graphs. One type is called a histogram. In this lesson, we will learn how to interpret and create a histogram.

Histogram	A type of bar graph with adjacent bars that are used to represent numerical data in equal intervals.

CHARACTERISTICS OF A HISTOGRAM:

- specific type of bar graph
- has no spaces between the bars
- categories are numbers
- data is displayed in equal intervals

INTERPRETING A HISTOGRAM:

Histograms are a good way to visually represent data, making it simpler to interpret the data.

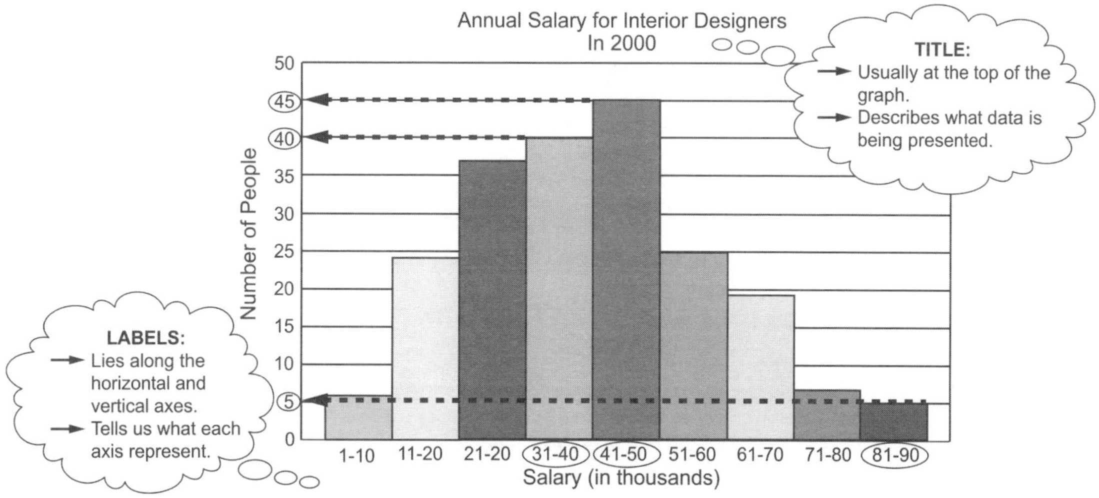

1. **What is the most common annual salary for interior designers in 2000, and how many interior designers made that salary?**
   In 2000, 45 interior designers made between $41,000 and $50,000.

2. **What is the next most common annual salary for interior designers in 2000, and how many interior designers made that salary?**
   In 2000, 40 interior designers made between $31,000 and $40,000.

3. **What is the least common annual salary for interior designers in 2000, and how many interior designers made that salary?**
   In 2000, only 5 interior designers made between $81,000 and $90,000.

<u>Steps in Creating a Histogram:</u>

1. Collect the numerical data.

Heights of Sunflowers (in inches) of a Particular School

42	43	45	45	47
76	72	67	49	64
53	66	55	60	58
57	59	54	50	58
53	78			

2. Organize the data in a frequency table.

a. Find the range (Range = largest number – smallest number).

$$\text{Range} = 78 - 42 = 36$$

b. Set the intervals.

c. Tally the data.

d. Record the frequency.

Frequency Table

Heights (inches)	Tally	Frequency
40 – 44	||	2
45 – 49	||||	4
50 – 54	||||	4
55 – 59	~~||||~~	5
60 – 64	||	2
65 – 69	||	2
70 – 74	|	1
75 – 79	||	2

3. Construct the histogram.

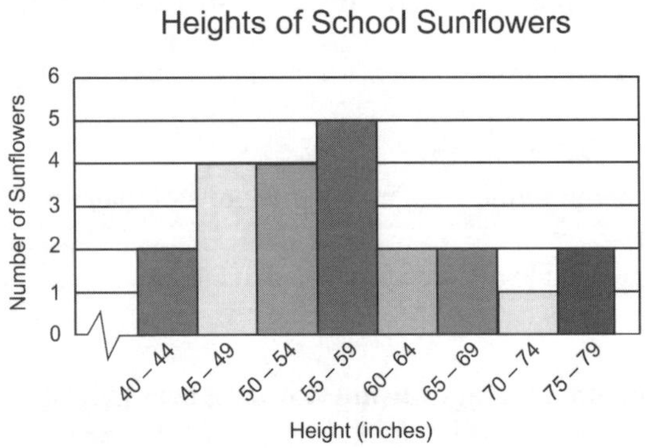

Example 1 The math scores for 20 students from Ms. Thai's Geometry class are listed below. Find the range of this set of data.

85	99	59	80	77
65	92	90	88	91
77	81	83	69	89
76	78	73	80	85

The largest number is 99.
The smallest number is 59.

Step 1: Identify the largest number and the smallest number.

Range = 99 – 59 = 40

Step 2: Subtract the smallest number from the largest number.

Answer: The range is 40.

Example 2 Using the same data given above in Example 1, construct a frequency table from the data.

Scores	Tally	Frequency
50 – 59		
60 – 69		
70 – 79		
80 – 89		
90 – 99		

Step 1: Draw a table with the intervals you have chosen.

Recall that the smallest number is 59 and largest number is 99. Therefore, we will choose intervals of tens, starting with interval 50 – 59.

Scores	Tally	Frequency
50 – 59	|	
60 – 69	||	
70 – 79	卌	
80 – 89	卌 |||	
90 – 99	||||	

Step 2: Tally the number of scores in each interval.

Scores	Tally	Frequency
50 – 59	|	1
60 – 69	||	2
70 – 79	卌	5
80 – 89	卌 |||	8
90 – 99	||||	4

Step 3: Count the number of tallies for each interval, and record the count in the column labeled "Frequency."

Answer:

Scores	Tally	Frequency
50 – 59	|	1
60 – 69	||	2
70 – 79	卌	5
80 – 89	卌 |||	8
90 – 99	||||	4

Example 3 The ages of the children at Angie's Day Care were tallied and graphed into the histogram below. What is the least common age range, and how many children are in that age range?

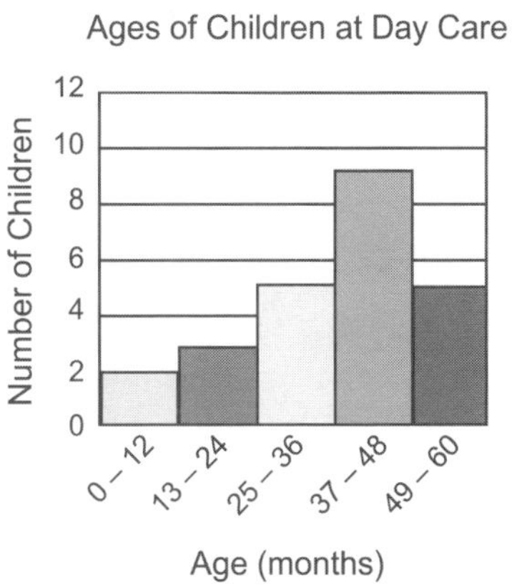

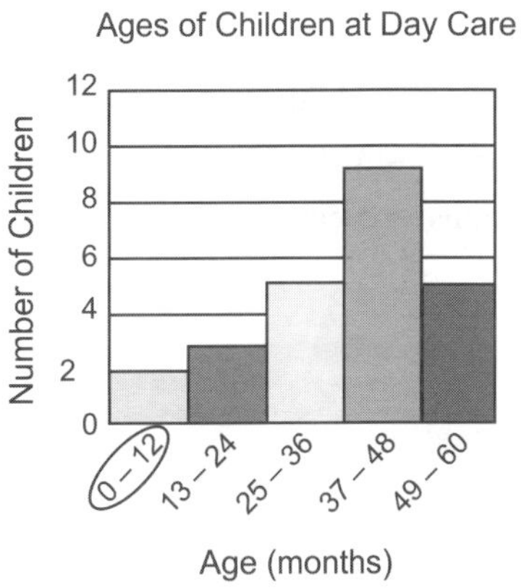

Step 1: Look for the shortest bar. The shortest bar has the fewest number of students. Therefore, it is the least common.

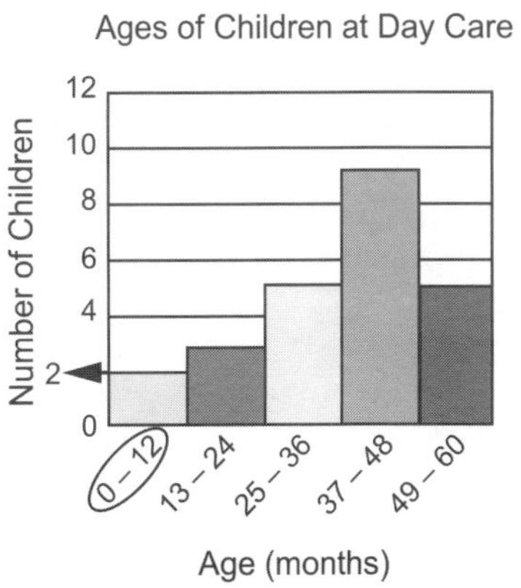

Step 2: Find its corresponding categories on the vertical and horizontal axes.

Answer: There are 2 children who are in the age range from 0 – 12 months.

Example 4 Using the same table from Example 3, how many children are 36 months or younger?

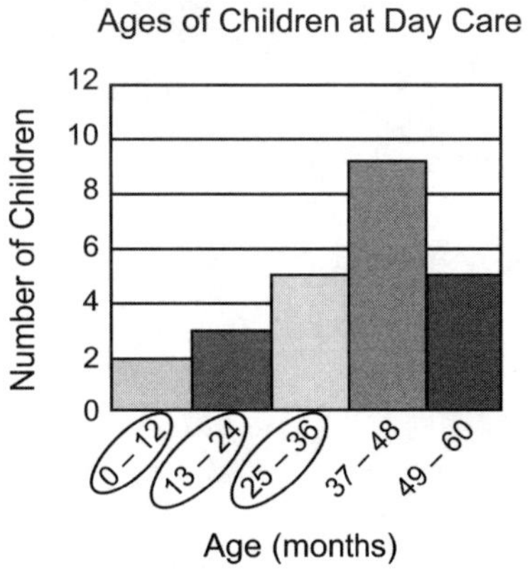

Step 1: Look for all the categories of children 36 months or younger.

There are 3 categories in which children are 36 months or younger.

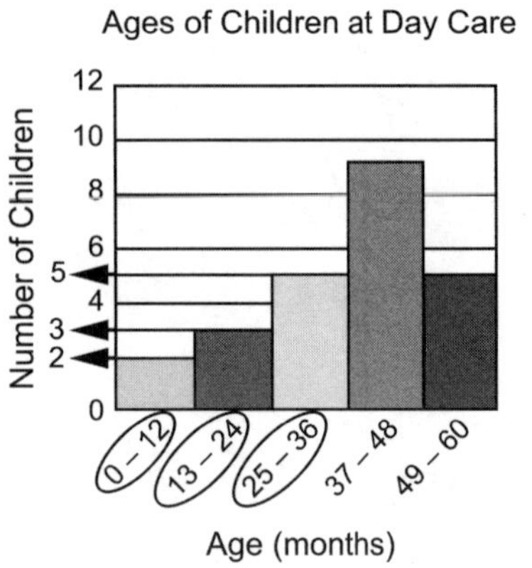

Step 2: Find the number of children in each category.

0 – 12 months → 2 children
13 – 24 months → 3 children
25 – 36 months → 5 children

$2 + 3 + 5 = 10$

Step 3: Add the total number of children that are 36 months or younger.

Answer: There are 10 children 36 months or younger.

Example 5 Construct a histogram using the table given below.

Math Scores in Ms. Thai's Geometry Class

Scores	Frequency
50 – 59	1
60 – 69	2
70 – 79	5
80 – 89	8
90 – 99	4

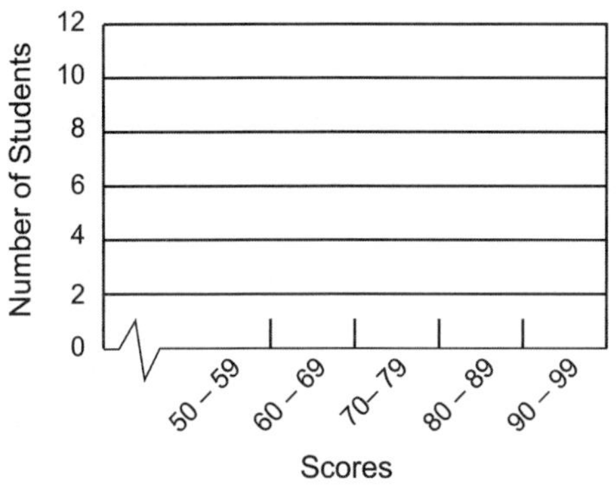

Step 1: Draw the horizontal and vertical axes and label them appropriately.

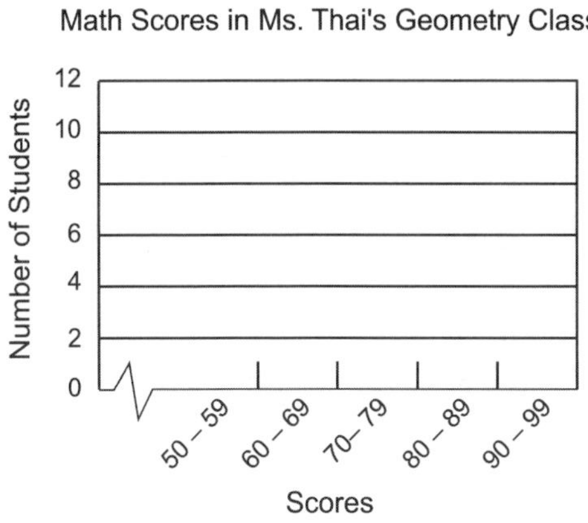

Step 2: Title the histogram.

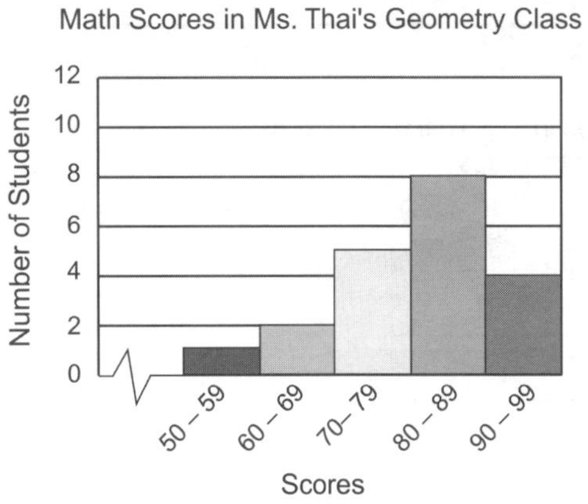

Step 3: Using the frequency, draw the *adjacent* bars to its appropriate heights.

Answer:

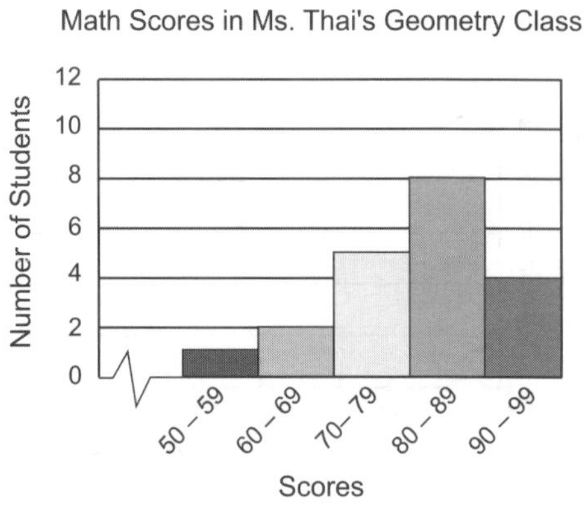

Problem Set

1. The heights of 15 students, in inches, participating in the after-school sports club are listed below. Find the range of this set of data.

63 60 67 50 55
48 62 49 56 57
57 51 65 63 60

2. The science quiz scores for Mr. Johnson's first period class are listed below. Find the range of this set of data.

88 80 78 99 90
77 91 95 96 99
88 86 89 90 82
92 84 85 95 83

3. A restaurant manager recorded the number of customers during the lunch shift for two weeks. Find the range of this set of data.

55 63 69 85 64 63 82
45 95 63 75 81 52 63

4. A school bus driver recorded the number of miles he drove for 20 days. Find the range of this set of data.

52 78 44 50 65
34 45 68 55 85
61 34 42 52 43
36 47 68 74 52

5. A video store recorded the amount of videos rented for three weeks. Find the range of this set of data.

68 88 56 45 55 93 78
90 77 66 75 76 61 88
54 53 69 82 89 78 66

6. Ryan recorded the amount of money (in dollars) that he deposited in his savings account each month for twelve months. Find the range of this set of data.

135 188 201 177
111 221 189 322
101 124 266 245

7. A basketball team recorded the number of points it scored during the fourteen games of this past season. Find the range of this set of data.

32 45 77 66 55
44 66 56 54 38
47 70 59 67 66

8. The number of books checked out at the local library was recorded by the librarian for 18 days.
Construct a frequency table from this data set.

34 56 66 103 110 120
78 49 88 155 90 99
152 78 89 123 134 156

9. For a class project, Susan recorded the number of minutes each of her classmates spent watching television over the weekend. She recorded this data below.

Construct a frequency table from this data set.

80 235 160 200 45
120 90 210 230 80
60 70 30 237 100
80 100 45 180 90

10. A fast-food restaurant with a drive-thru recorded the number of customers that went through the drive-thru each day for four weeks.
Construct a frequency table from the data set shown below.

45 90 67 88 99 102 105
98 49 50 77 88 110 120
65 78 80 130 96 115 119
81 97 78 112 70 75 130

11. A magazine company recorded the number of new subscriptions that were sold in 20 weeks. This data is recorded below.
Construct a frequency table from this data set.

250 450 500 325 230
310 267 321 440 190
280 480 280 430 200
235 475 525 345 222

12. A tax service company recorded the amount (in dollars) of tax refunds given for the first 20 customers this week.
Construct a frequency table from this data set.

330 445 106 125 230
80 260 100 239 270
76 389 378 108 200
50 175 488 550 167

13. The surface area, in square feet, of 20 prisms is recorded below.
Construct a frequency table from the data set.

65	100	134	156	188
90	120	220	280	225
178	156	145	135	190
345	320	300	290	168

14. A physical education teacher at Jackson Middle School recorded the number of sit-ups each student in her fourth period class could complete during a physical fitness test.
Construct a frequency table from this data set.

6	34	39	23	40	22
7	15	18	55	20	19
11	48	44	10	18	30

15. A pizza delivery service recorded the number of pizzas that were delivered weekly for 15 weeks.
Construct a frequency table from the data set.

125	188	104	121	193
75	90	90	150	200
83	126	108	119	179

16. The number of miles each teacher at Jackson Middle School drives to work was tallied and graphed in the given histogram.

Which mile range is the least common, and how many teachers are in this range?

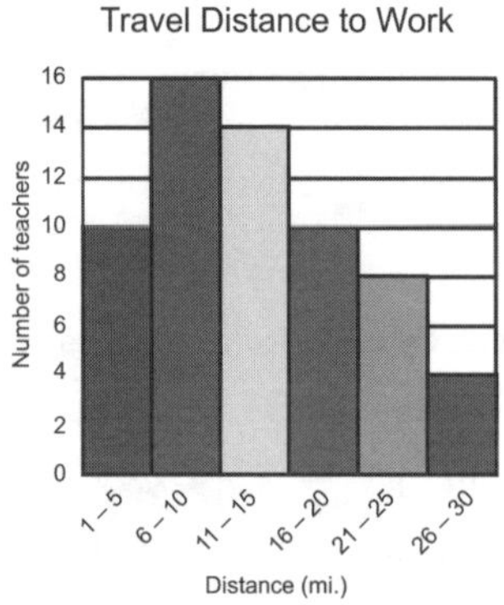

17. The salaries of employees at the City Newspaper were tallied and graphed in the given histogram.

Which salary range is the most common, and how many employees are in this range?

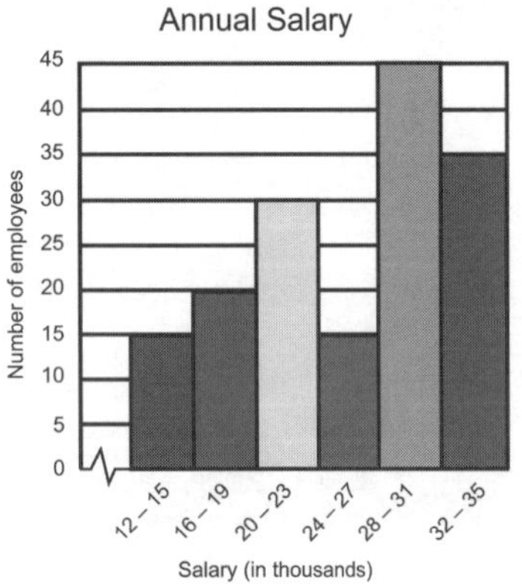

18. The number of pages in each book published by the Royal Book Company was tallied and graphed in the given histogram.

Which range of page numbers is the most common, and how many books are in this range?

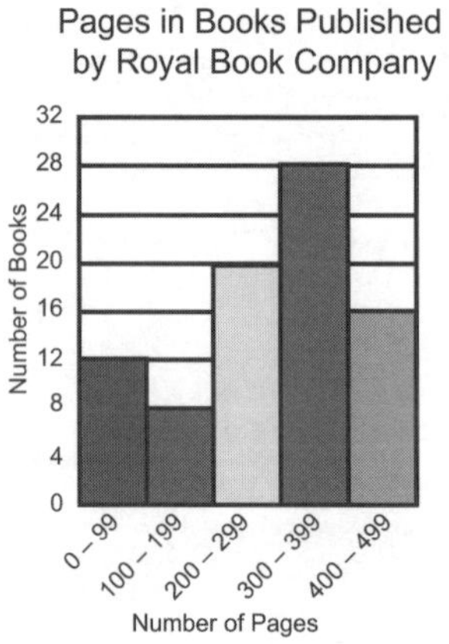

19. A realty company gathered information about monthly rental rates for apartments in the city that were available. This information was tallied and graphed in the given histogram.

Which rental fee range is the least common, and how many apartments are in this range?

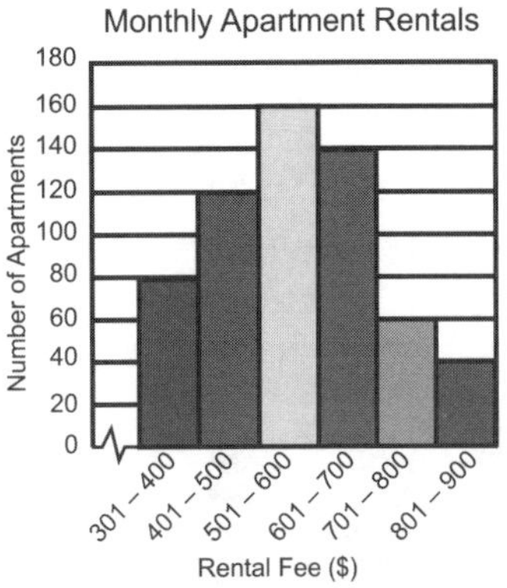

20. A survey of selected college students concerning their average monthly phone bill was tallied and graphed in the given histogram.

Which monthly phone bill range is the most common, and how many students are in this range?

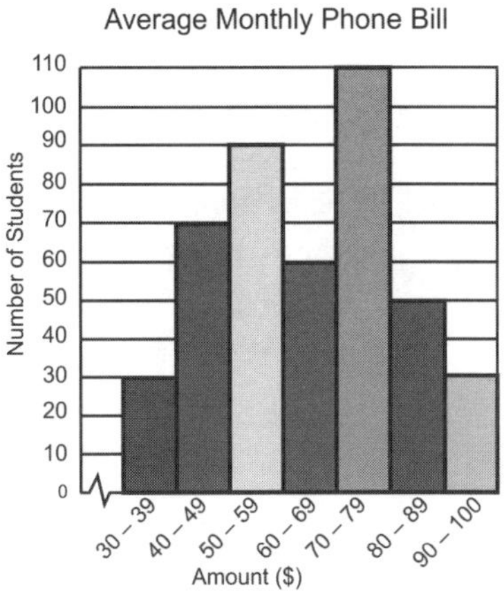

21. A travel agency conducted a survey in which families were asked to report the number of miles that they traveled on their last vacation. The results were tallied and graphed in the given histogram.

Which mile ranges show the same number of families, and how many families are in each range?

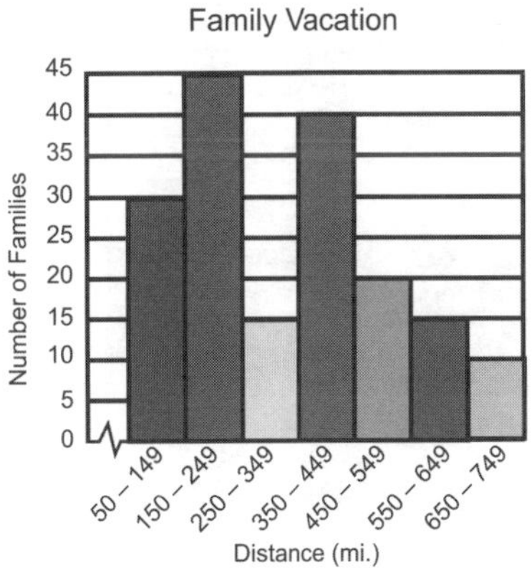

22. A restaurant manager recorded the amount each customer spent on dinner over a two-week period. These records were tallied and graphed in the given histogram.

Which dollar range is the most common, and how many customers are in this range?

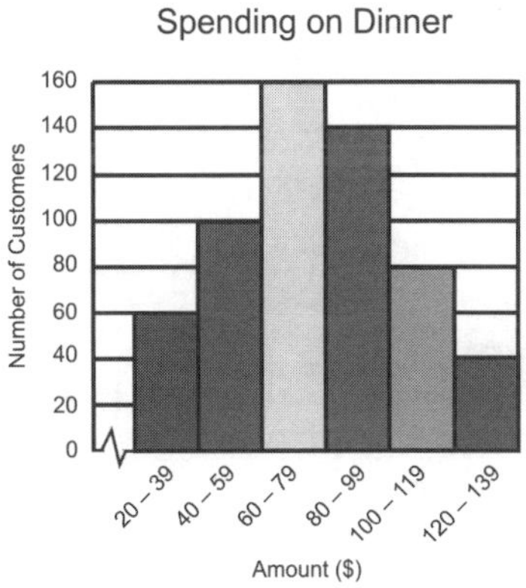

23. Refer to the given histogram to answer the following question:

How many employees are younger than age 42?

24. A boat manufacturer surveyed several owners of recreational boats to find out how many times they used their boat in the past three months. The results of this survey were tallied and graphed in the given histogram.

How many boaters used their boats more than 29 times?

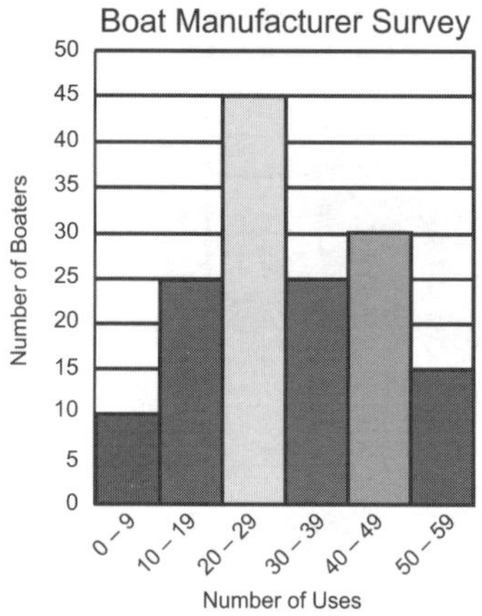

25. The number of miles driven by visitors to the Science Museum was tallied and graphed in the given histogram.

How many visitors drove 44 miles or less to visit the museum?

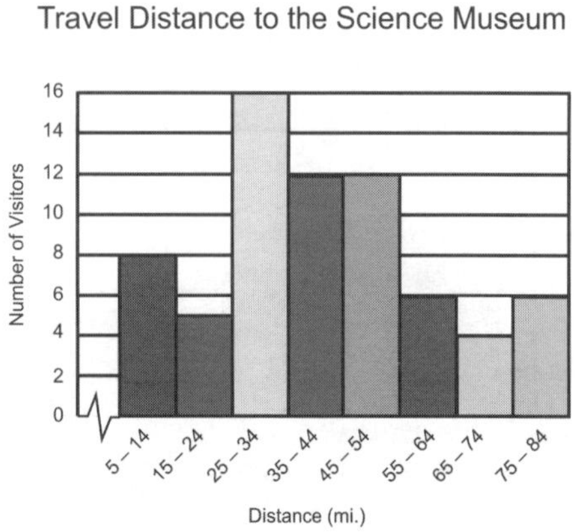

26. The ticket prices to a baseball game were tallied and graphed in the given histogram.

How many tickets sold cost less than $25?

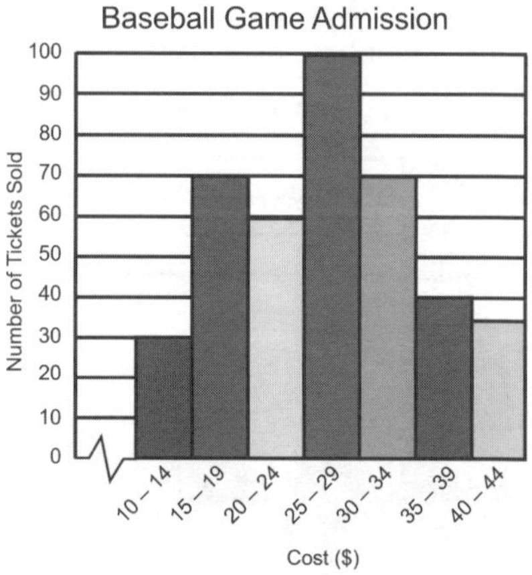

27. A photo developing counter at a local store conducted a survey asking each customer how many rolls of film they had developed in the past year. These responses were tallied and graphed in the given histogram.

How many customers developed more than 15 rolls of film in the past year?

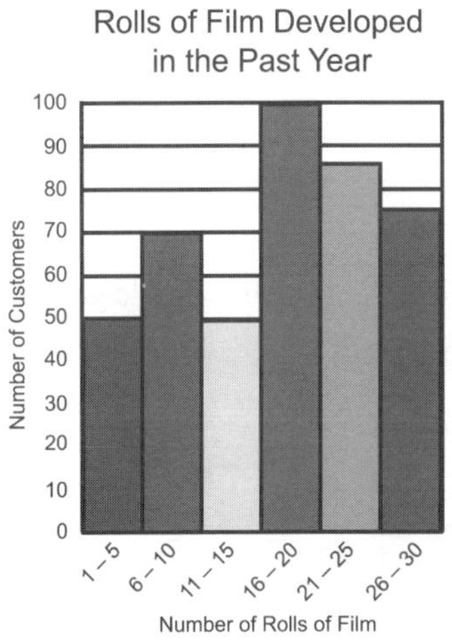

28. The school secretary at Millwood School tallied the number of absences of the students. This data was graphed in the given histogram.

How many students had more than 9 absences?

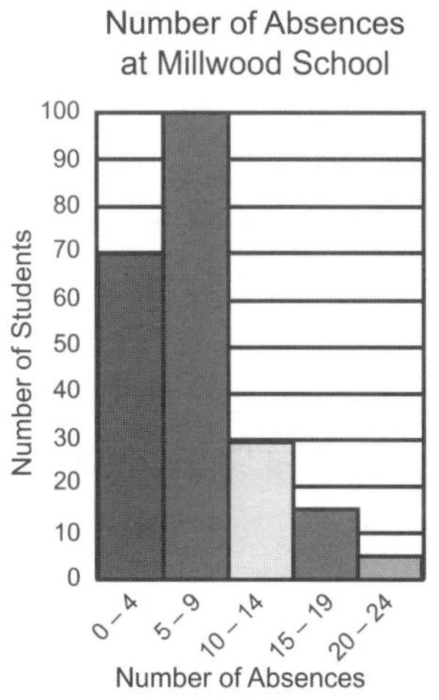

29. The car prices of cars sold at the Auto Barn this past month were tallied and graphed in the given histogram.

How many cars were priced $25,000 or less?

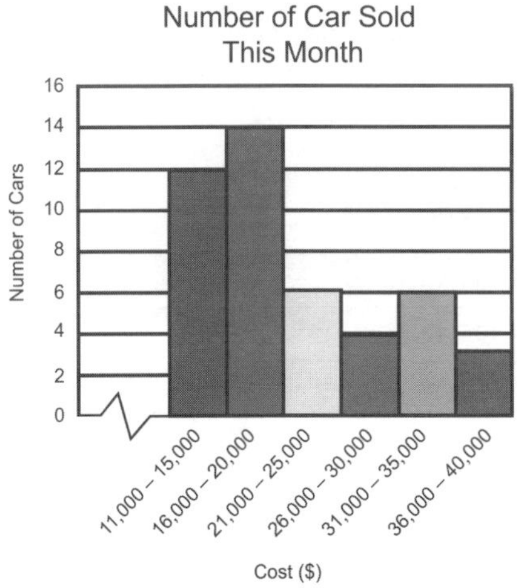

30. Mrs. Smith tallied the scores from her last spelling test and graphed them in the given histogram.

How many students scored greater than an 85?

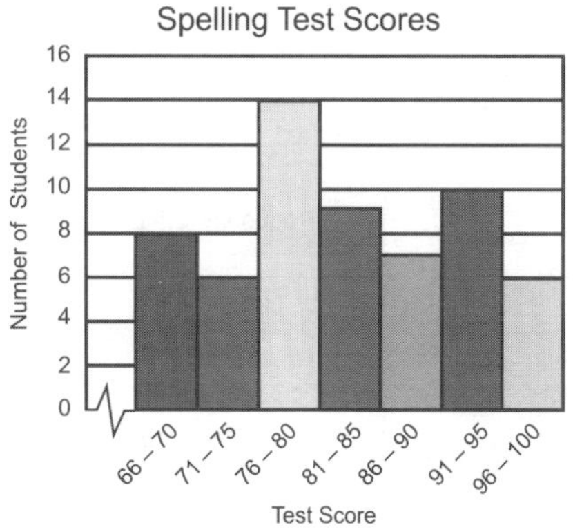

31. Construct a histogram using the table given below.

Miles Walked this Month
by the Walking Club

Number of Miles Walked	Frequency
0 – 9	12
10 – 19	8
20 – 29	16
30 – 39	2
40 – 49	10

32. Construct a histogram using the table given below.

Pizza Deliveries During
Dinner Shift

Number of Pizza Deliveries	Frequency
0 – 19	10
20 – 39	15
40 – 59	35
60 – 79	40
80 – 99	25

33. Construct a histogram using the table given below.

Students' Daily
Study Time

Number of Minutes	Frequency
0 – 14	10
15 – 29	60
30 – 44	50
45 – 59	40
60 – 74	35
75 – 89	25

34. Construct a histogram using the table given below.

Cost of Lunch
at Danny's Diner

Cost ($)	Frequency
5.00 – 6.99	6
7.00 – 8.99	4
9.00 – 10.99	10
11.00 – 12.99	16
13.00 – 14.99	12

35. Construct a histogram using the table given below.

Weekly CD Sales

Number of CDs	Frequency
20 – 39	10
40 – 59	40
60 – 79	70
80 – 99	65
100 – 119	55
120 – 139	60

36. Construct a histogram using the table given below.

Books Read this Year
by Library Patrons

Number of Books Read	Frequency
0 – 9	12
10 – 19	24
20 – 29	20
30 – 39	36
40 – 49	32

37. Construct a histogram using the table given below.

Students Playing Video Games

Number of Minutes	Frequency
0 – 29	30
30 – 59	60
60 – 89	64
90 – 119	80
120 – 149	15
150 – 179	10

38. Construct a histogram using the table given below.

Students' Baseball Card Collection

Number of Cards	Frequency
0 – 39	45
40 – 79	60
80 – 119	90
120 – 159	120
160 – 199	15

39. Construct a histogram using the table given below.

Attending After-School Programs

Number of Students	Frequency
0 – 4	2
5 – 9	5
10 – 14	3
15 – 19	4
20 – 24	8

40. Construct a histogram using the table given below.

Weekly Travel Distance by Biking Club Members

Number of Miles	Frequency
0 – 9	3
10 – 19	9
20 – 29	18
30 – 39	12
40 – 49	6

MPA-132: Interpreting and Creating Scatterplots

There are many different types of graphs that are used in math. Some graphs are used to show a relationship between two set of data. One such graph is called a **scatterplot**.

Scatterplot	A graph that shows the relationship between two sets of data.

A scatterplot is made by plotting two sets of data as ordered pairs on the coordinate plane.

EXAMPLE OF A SCATTERPLOT:

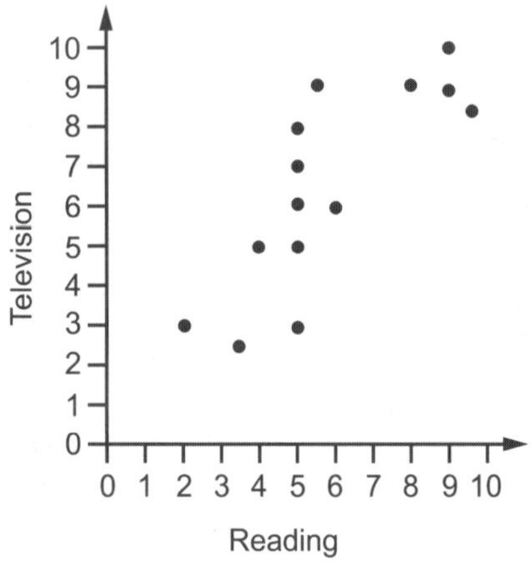

The points on the scatterplot may reveal a pattern. If a pattern is revealed, it could help determine a correlation (relationship) between two sets of data.

Positive Correlation: Reading from left to right, the points show an *upward trend.*

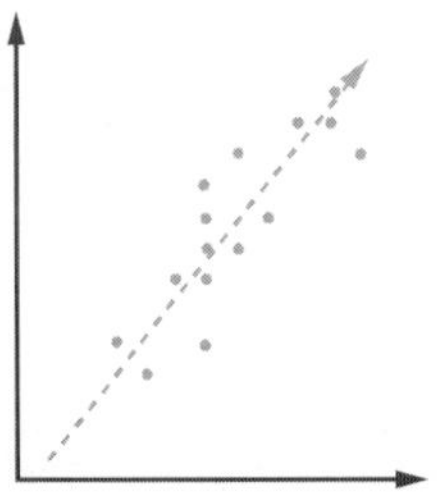

As x increases ⟶ y increases.

Negative Correlation: Reading from left to right, the points show a *downward trend.*

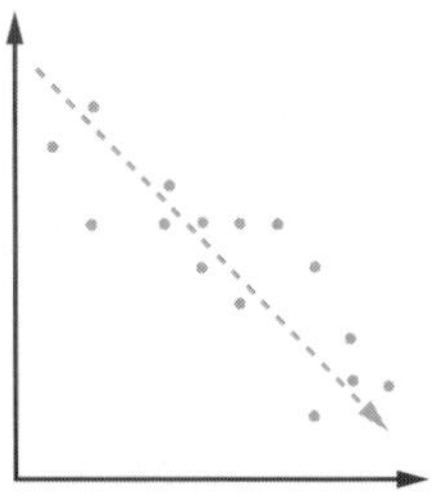

As x increases ⟶ y decreases.

No Correlation: Reading from left to right, the points show *no trend.*

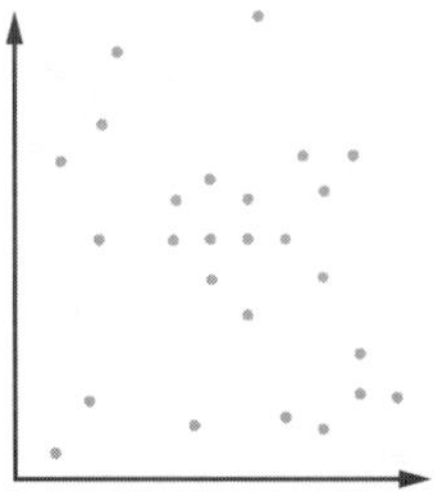

Creating a Scatterplot:

1. Collect the numerical data
2. Set up and label the axes and title the graph
3. Plot the data points

Homework Hours and Grades

	Hours on Homework	Test Grade
Unit 1	10	96
Unit 2	9	97
Unit 3	4	70
Unit 4	7	86
Unit 5	3	63
Unit 6	4	74
Unit 7	6.5	80
Unit 8	1	45
Unit 9	8	97
Unit 10	10	99
Unit 11	5	78
Unit 12	8.5	92

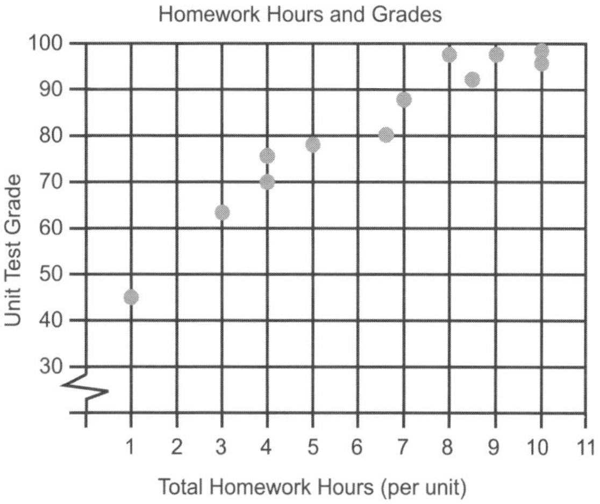

1. **Is there a correlation between homework hours and test grades? If so, what type of correlation exists?**
   There is a positive correlation between homework hours and test grades.
2. **Make a conclusion about the relationship between the set of data using the graph.**
   The more hours spent on homework, the higher the test grade.

When a scatterplot has a correlation, we can draw a **line of best fit** to help predict other values that might not be included on the graph.

Line of Best Fit	Line drawn on a scatterplot that passes close to most of the data points.

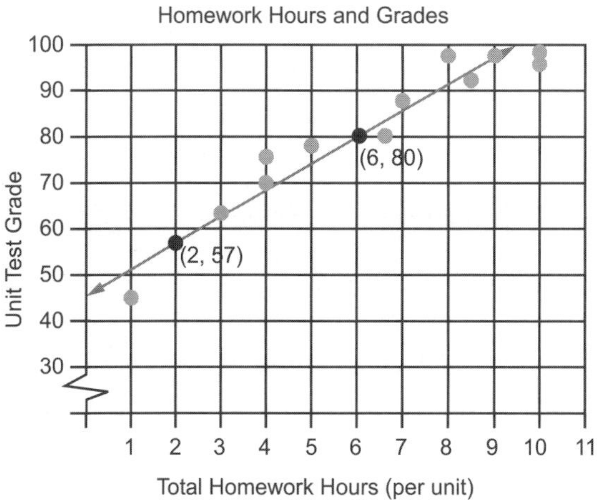

1. **Using the line of best fit, what can you predict a student will make on the test if the student spends 2 hours on homework?**
   The student will make approximately a 57 on the test.
2. **If a student makes an 80 on the next test, about how many hours do you think the student spent on homework for that unit?**
   The student spent approximately 6 hours on homework.

Example 1 In the scatterplot given, is there a correlation and, if so, what type?

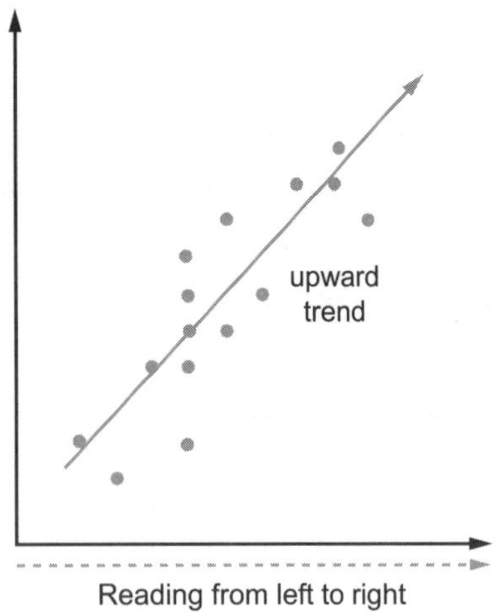

Step 1: Reading from left to right, the points have an upward trend.

Answer: The points on the scatterplot show a positive correlation.

Example 2 Using the scatterplot given, describe the relationship between the number of shopping visits in a week and the amount of money left.

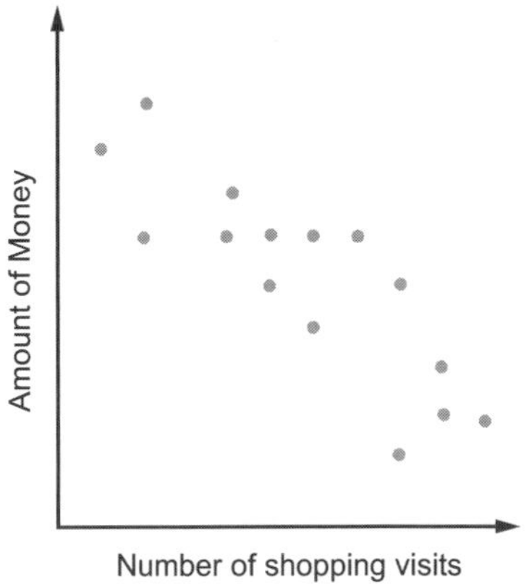

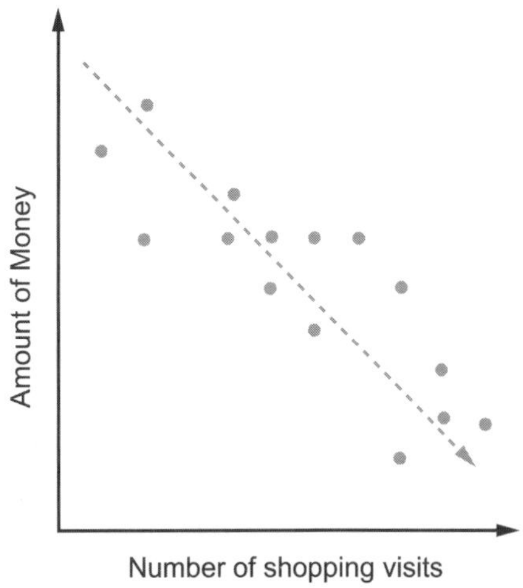

Step 1: Reading from left to right, the points show a downward trend. This means it has a negative correlation.

When a correlation is negative, as x increases, y decreases.

In this example, x is the number of shopping visits and y is the amount of money.

Answer: As the number of shopping visits increases, the amount of money decreases.

Example 3 Given the data below, create a scatterplot representing the data.

Value of Antique over Time

Age (years)	Value (dollars)
0	5
5	5
10	10
15	20
20	50
25	70
30	100
35	105
40	110
45	150

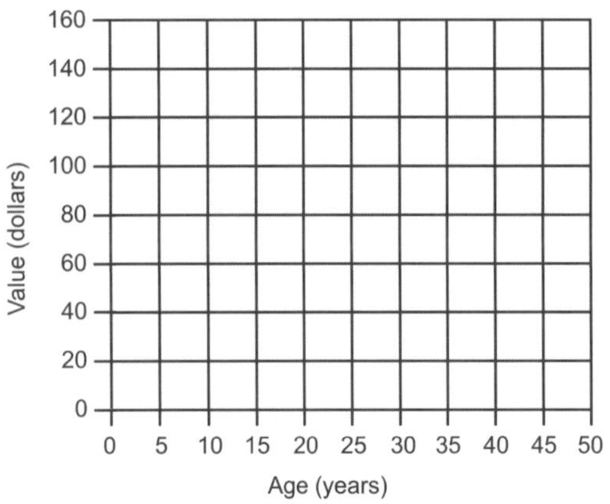

Step 1: Draw and label the horizontal and vertical axes.

The horizontal label is "Age (years)". Since the age ranges from 0 to 45, draw the axis with increments of 5.

The vertical label is "Value (dollars)". Since the value ranges from \$5 to \$150, draw this axis with increments of 20.

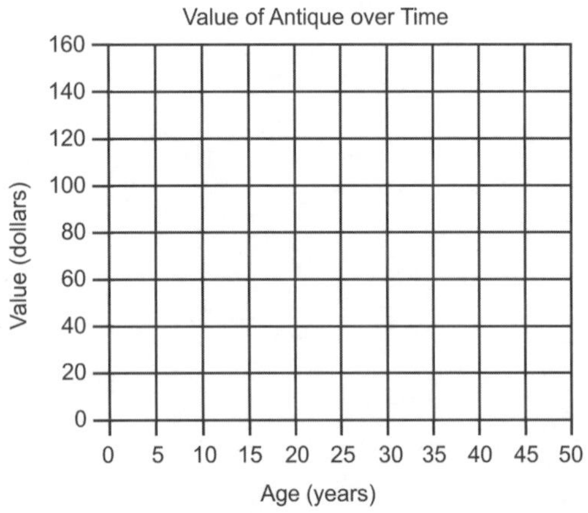

Step 2: Title the scatterplot.

The scatterplot is about the value of an antique over time.

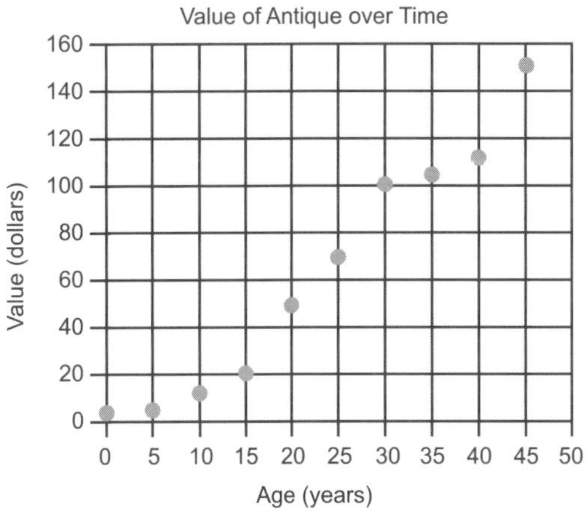

Step 3: Plot each data point.

Answer:

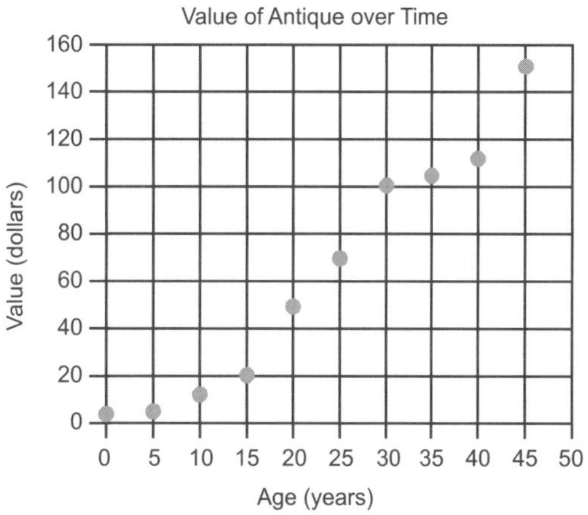

Example 4 The scatterplot below shows the number of pages in a hardcover textbook and its weight. Draw a line of best fit for the data, then use it to predict the approximate weight of a textbook that has 600 pages.

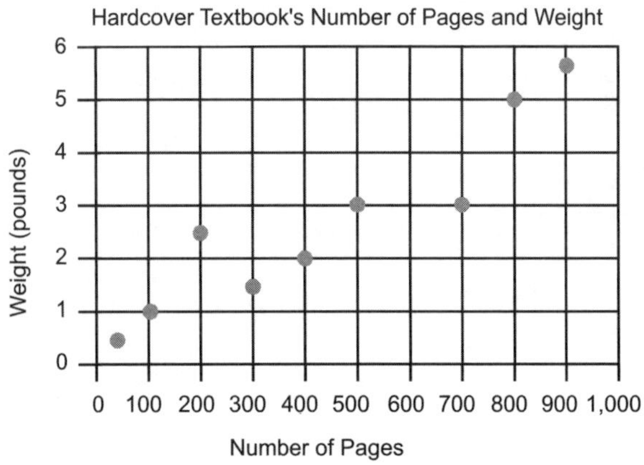

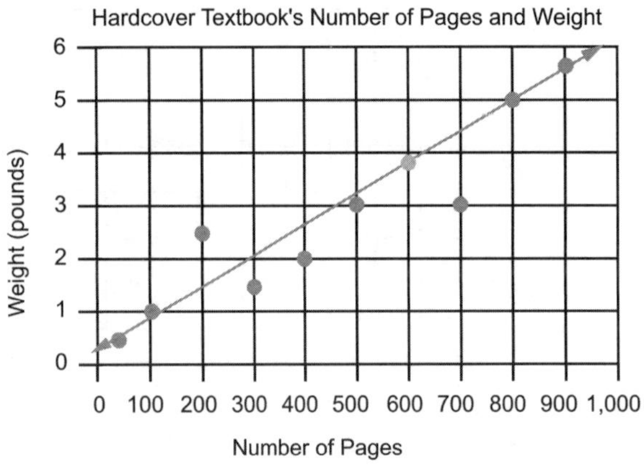

Step 1: Draw a line of best fit.

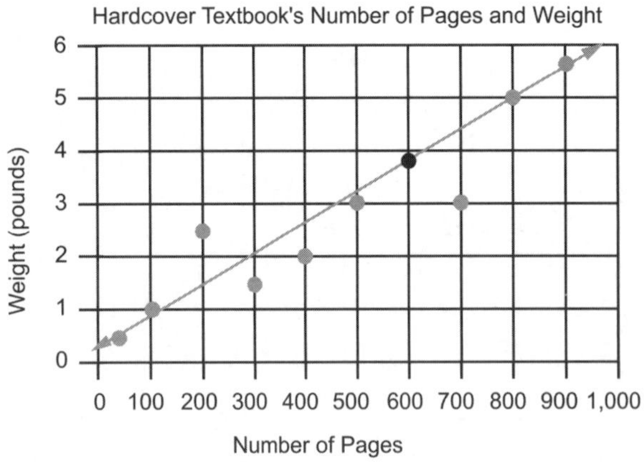

Step 2: Since 600 is the number of pages, locate 600 on the horizontal axis.

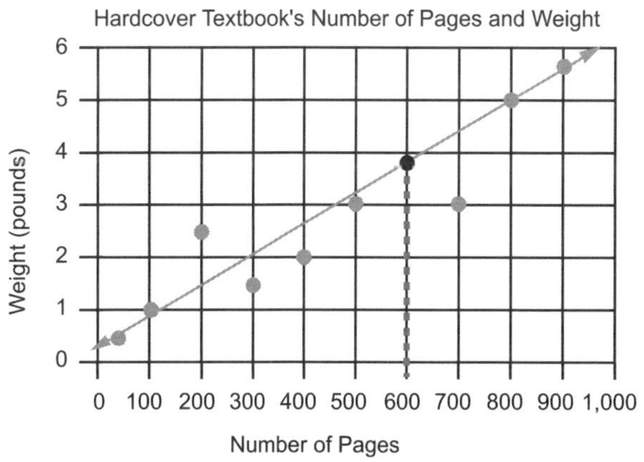

Step 3: Find the point on the line of best fit that is directly above 600.

The point is located at $\left(600, 3\frac{3}{4}\right)$.

Answer: A 600-page book would weigh about $3\frac{3}{4}$ pounds.

Example 5 Using the same scatterplot from Example 4, predict how much a 1,200-page textbook would weigh.

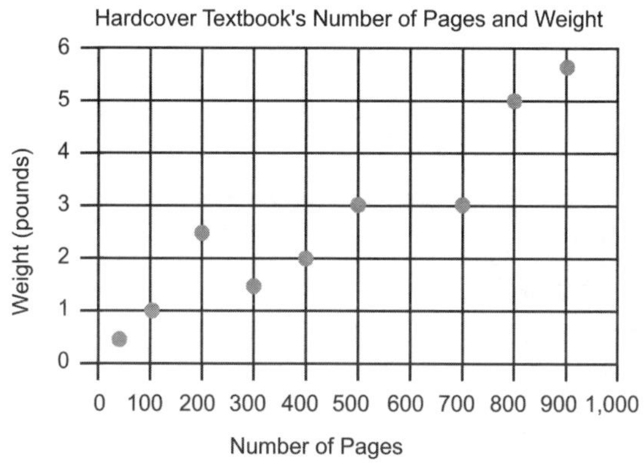

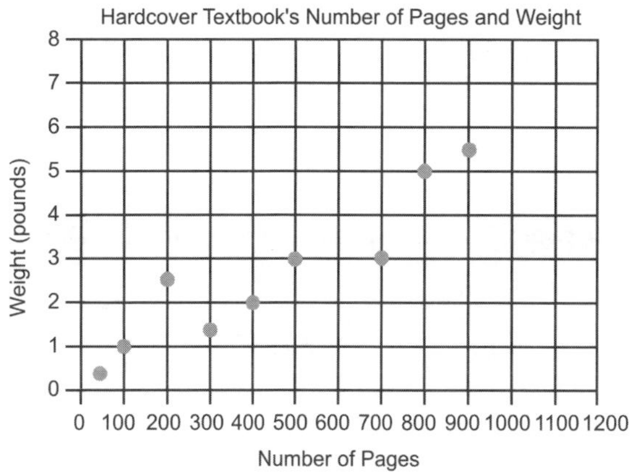

Step 1: Expand the graph to include 1,200 pages.

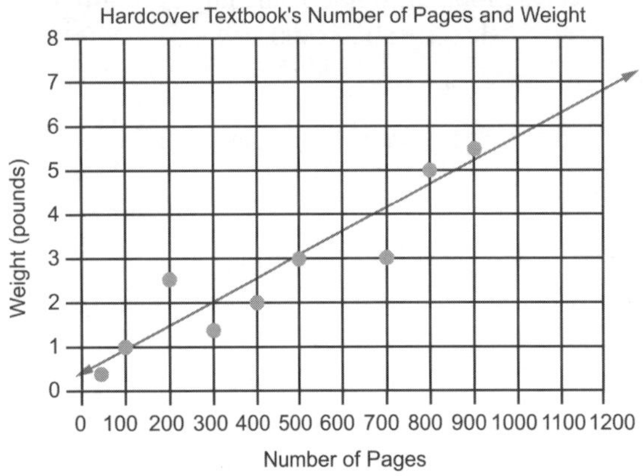

Step 2: Draw a line of best fit.

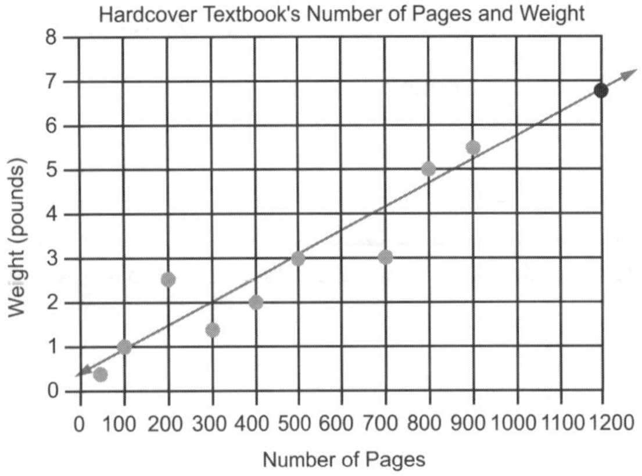

Step 3: Since 1,200 is the number of pages, locate 1,200 on the horizontal axis.

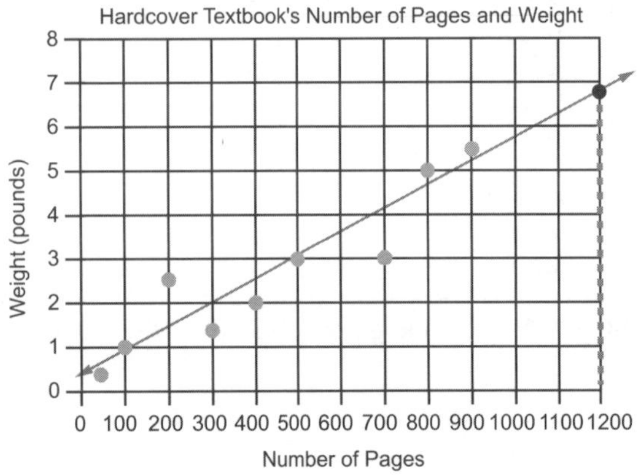

Step 4: Find the point on the line of best fit that is directly above 1,200.

The point is located at $\left(1200, 6\frac{3}{4}\right)$.

Answer: A 1,200-page book would weigh about $6\frac{3}{4}$ pounds.

Problem Set

1. Determine the type of correlation shown in the given scatterplot.

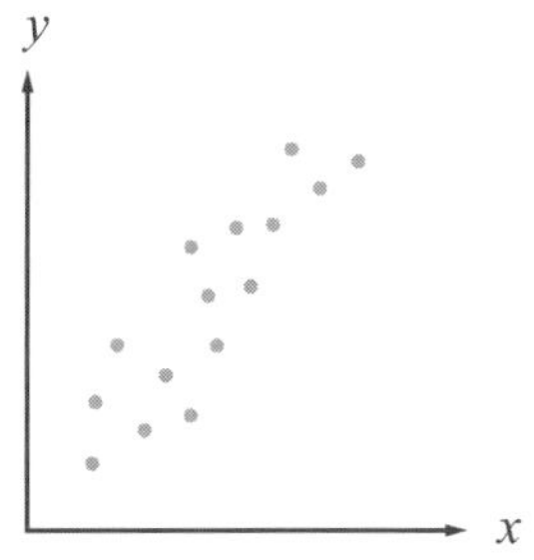

2. Determine the type of correlation shown in the given scatterplot.

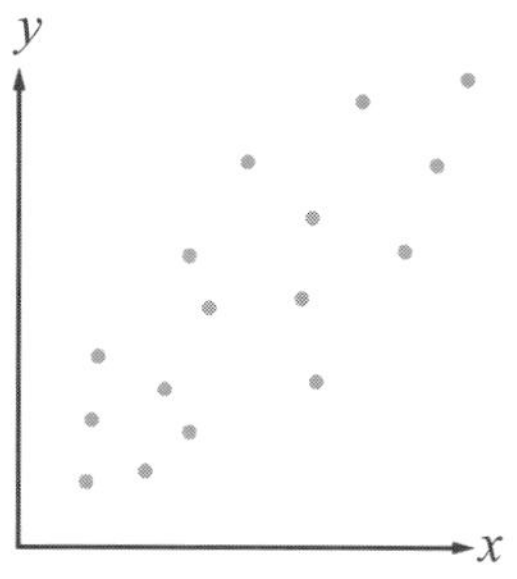

3. Determine the type of correlation shown in the given scatterplot.

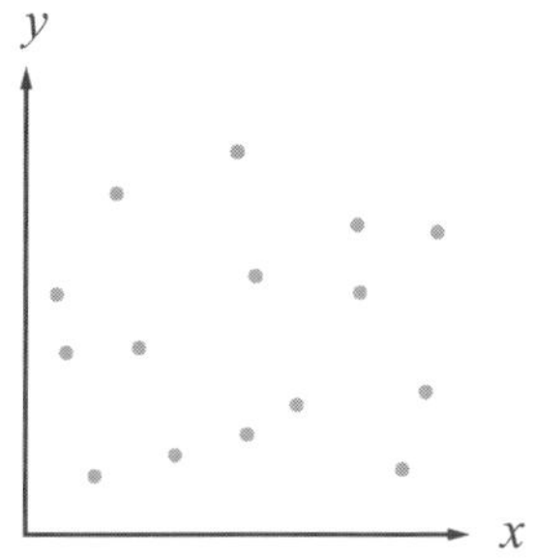

4. Determine the type of correlation shown in the given scatterplot.

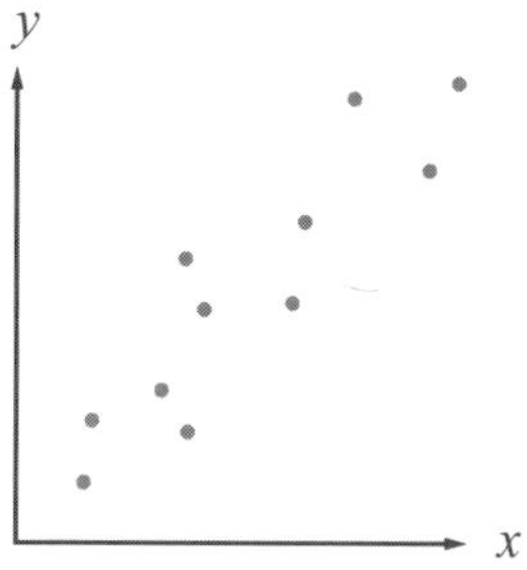

5. Determine the type of correlation shown in the given scatterplot.

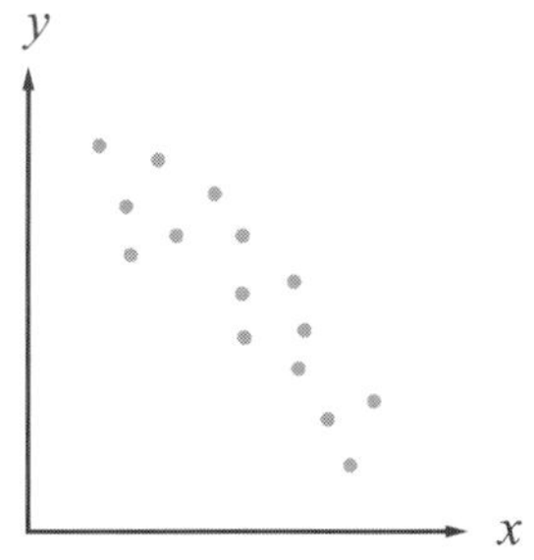

6. Determine the type of correlation shown in the given scatterplot.

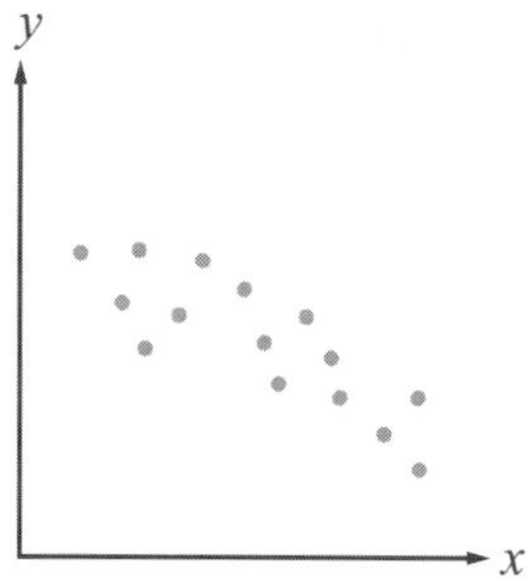

7. Determine the type of correlation shown in the given scatterplot.

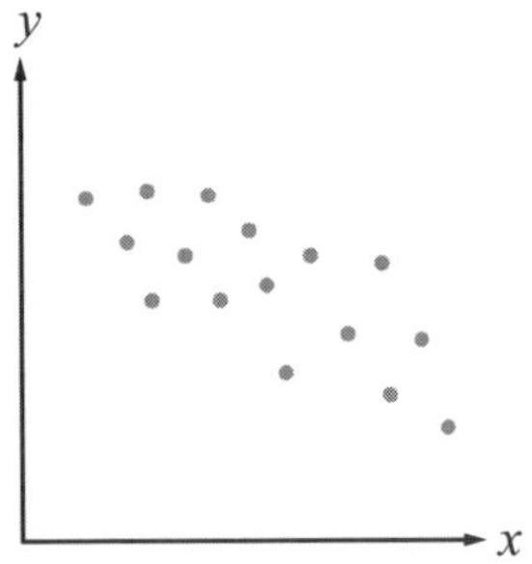

8. Determine the type of correlation shown in the given scatterplot.

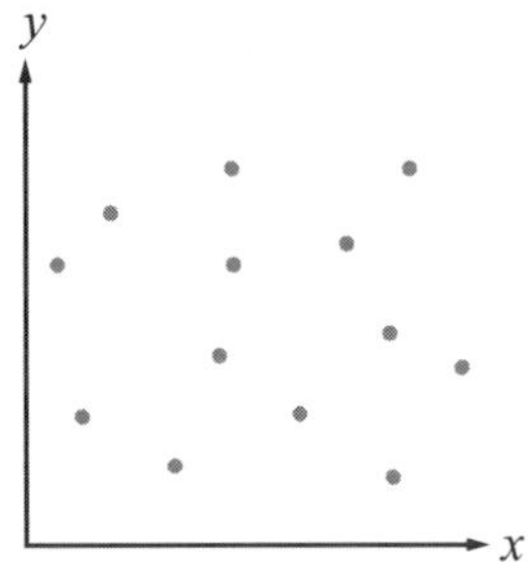

9. Using the given scatterplot, describe the relationship between the interest rates and the selling prices of homes.

10. Using the given scatterplot, describe the relationship between time spent watching television and test scores.

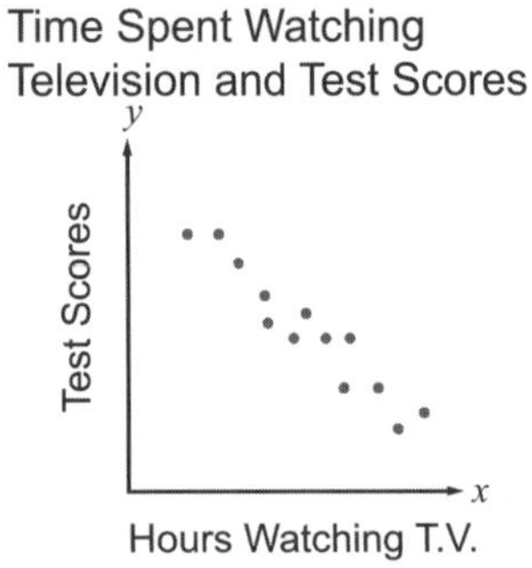

11. Using the given scatterplot, describe the relationship between monthly rainfall and the number of mosquito bites on a person.

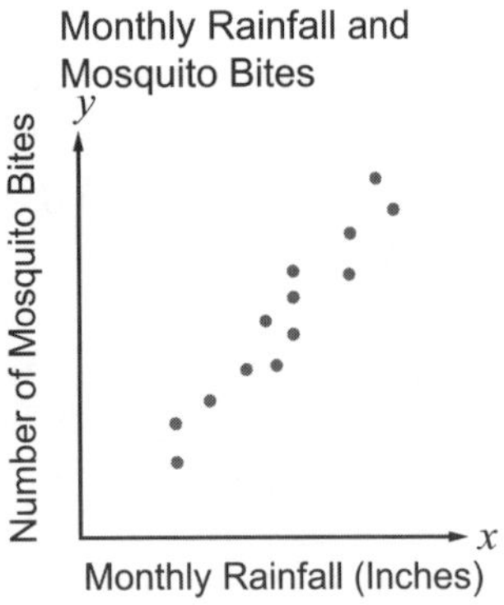

12. Using the given scatterplot, describe the relationship between time spent at work and salary.

13. Using the given scatterplot, describe the relationship between the amount of clouds in the sky and the number of cars in a lot.

Clouds in the Sky and Cars in Lot

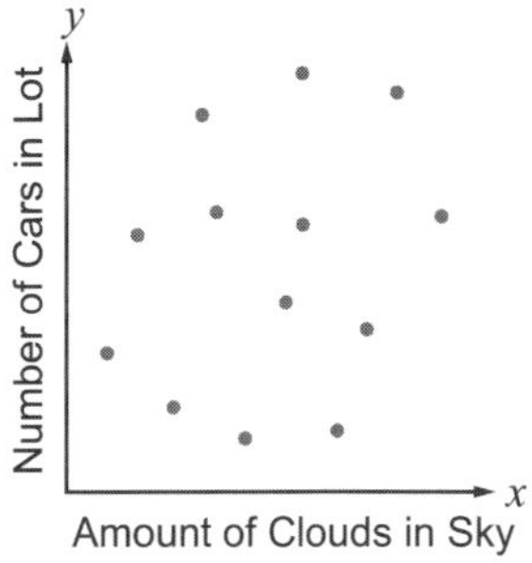

14. Using the given scatterplot, describe the relationship between the age of a car and the car insurance rate.

Age of a Car and Car Insurance Rate

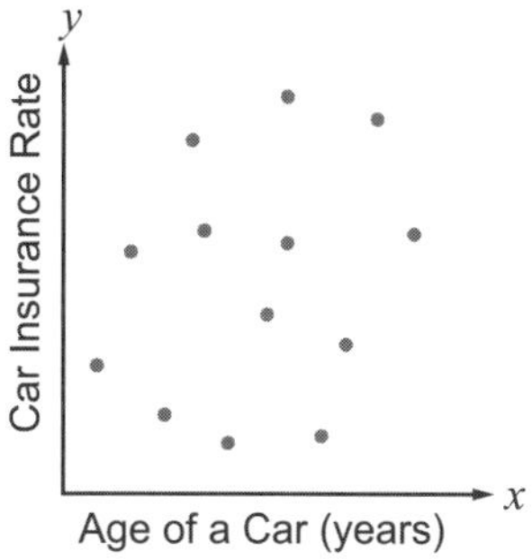

15. Using the given scatterplot, describe the relationship between the years of education and the salary.

Years of Education and Salary

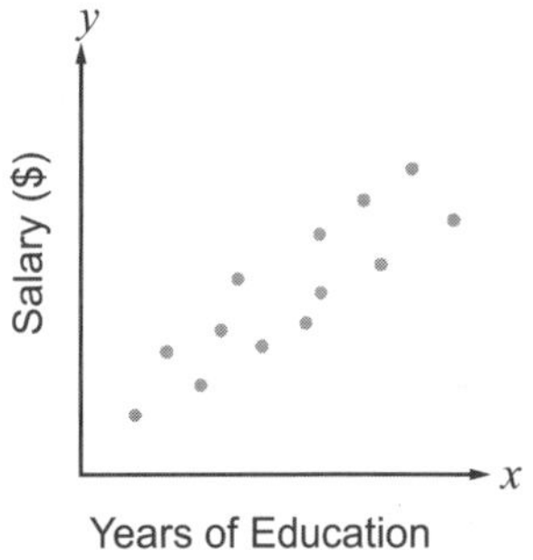

16. Use the given table to create a scatterplot.

Track Meets

Number of Hours Practicing	Number of Medals Won
4	0
5	1
6	2
6	1
7	2
7.5	3
7.5	4
8	5
9	4
10	5

17. Use the given table to create a scatterplot.

Math Grades

Number of Hours on Homework	Test Grade
0.5	58
1	53
1	59
2	67
3	73
3.5	76
4	89
4	82
5	97
5	98

18. Use the given table to create a scatterplot.

Lawn Cutting Service

Number of Lawns Cut	Money Made (In Dollars)
2	40
3	55
4	60
5	100
6	90
6	100
8	140
9	160
10	180
10	200

19. Use the given table to create a scatterplot.

Housing

Square Feet (in hundreds)	Price of House (in thousands)
7	50
9	70
15	95
17	120
18	115
21	165
24	180
24	190
26	200
28	230

20. Use the given table to create a scatterplot.

Base Salary

Years of Education	Salary (in thousands)
8	9
10	11
11	10
12	15
13	16
16	25
17	27
18	28
20	35
21	36

21. Use the given table to create a scatterplot.

Time Watching Television

Number of Homework Items	Hours Spent Watching Television
1	4
2	4
2	3.5
3	3.5
4	3
4	2.5
5	2
6	1.5
7	1
7	0.5

22. Use the given table to create a scatterplot.

Days Swimming

Number of Days Raining	Number of Days Swimming
1	20
2	18
3	16
5	15
6	14
9	10
9	8
10	8
11	6
12	3

23. What is the type of correlation between the number of pairs of shoes sold and the amount of money earned on shoe sales?

24. What is the type of correlation between the number of tickets sold and the number of tickets remaining?

25. What is the type of correlation between the number of boxes of candy sold and the amount of money raised from the boxes of candy?

26. What is the type of correlation between a person's height and the amount of money he or she makes in a year?

27. What is the type of correlation between a person's weight and the length of his or her hair?

28. What is the type of correlation between the number of times a pen is used and the amount of ink remaining?

29. What is the type of correlation between a person's age and the number of siblings he or she has?

30. What is the type of correlation between the amount of time spent watching television and the amount of the phone bill?

31. What is the type of correlation between the number of meals a person eats and his height?

32. What is the type of correlation between the number of days it rains and the number of times the plants need to be watered?

33. The given scatterplot was created to show the relationship between the number of tickets sold and the amount of money the theater earned that day.

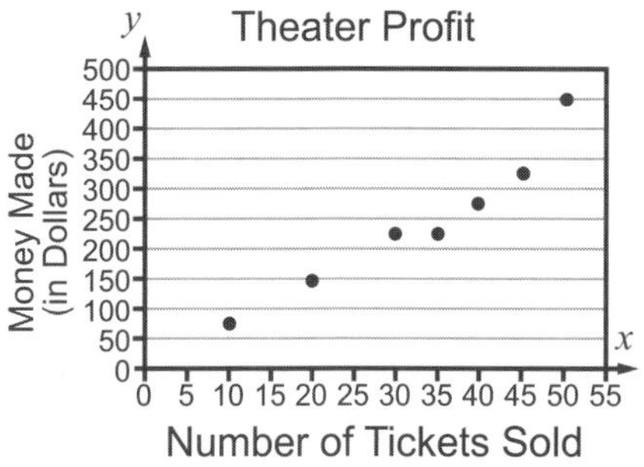

If the theater sells 25 tickets, approximately how much money will it make?

34. The given scatterplot was created to show the relationship between the number of hours Mark exercises and the amount of weight he lost.

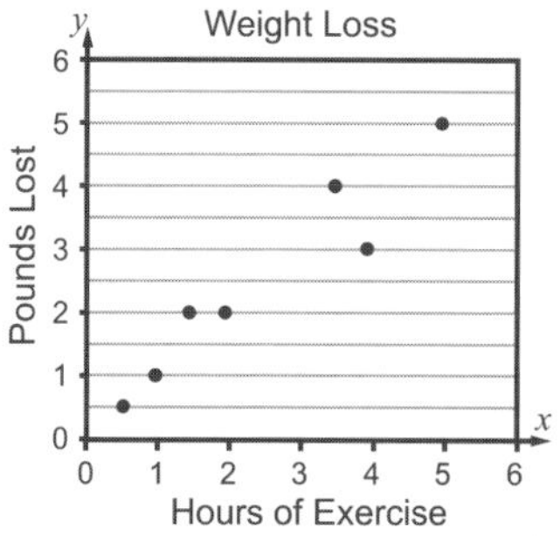

If Mark exercises for 6 hours each day, approximately how much weight will he lose?

35. Lee wanted to improve her math grade, so she decided to record the amount of hours she spent studying for her tests. The given scatterplot shows the relationship between the hours she spent studying and her test grades.

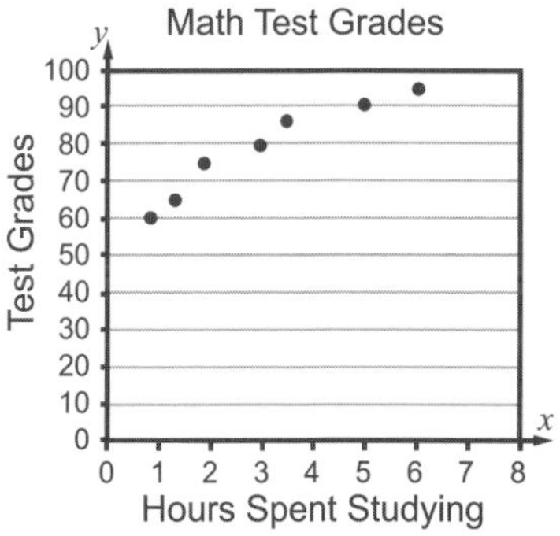

If Lee studies for 4 hours before her test, what test grade should she expect to make?

36. Tina sold lemonade to save money to buy toys. The given scatterplot shows the relationship between how many cups of lemonade she sold and how much money she made.

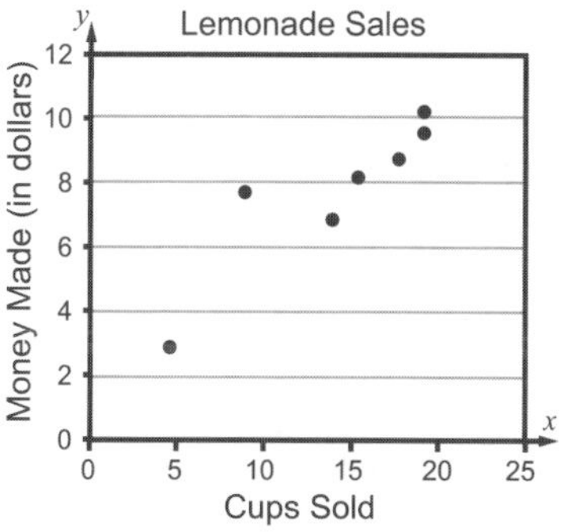

If she sells 15 cups of lemonade, approximately how much money will she make?

37. A barber created a scatterplot to illustrate the relationship between how many haircuts he gives and how much money he makes each day.

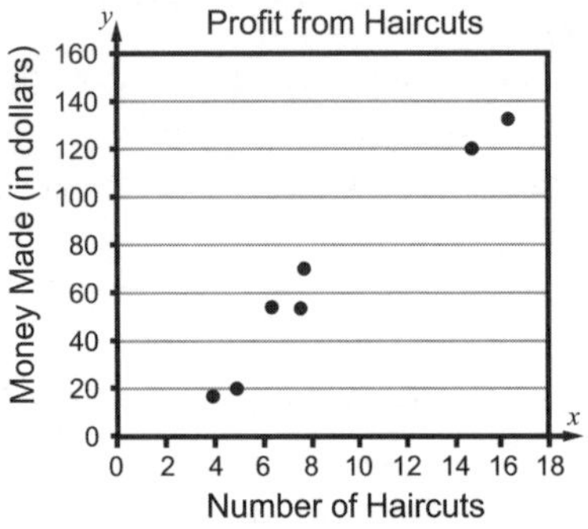

How much money would he expect to make if he gives 10 haircuts tomorrow?

38. The given scatterplot shows the relationship between the number of CDs sold and the amount of money a particular singer made.

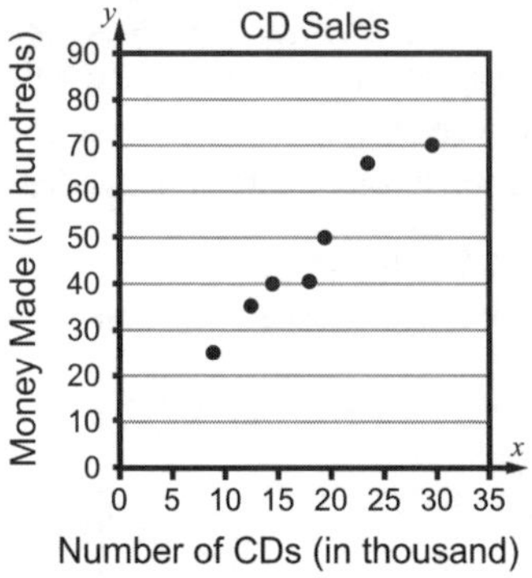

If the singer sells 3,500 CDs, how much money should she expect to make?

39. Joe is trying to save money for a pair of shoes, but he is not satisfied with the amount of money he is saving. Joe created a scatterplot to show the relationship between the snacks he bought during the week and the amount of money he had left for his shoes.

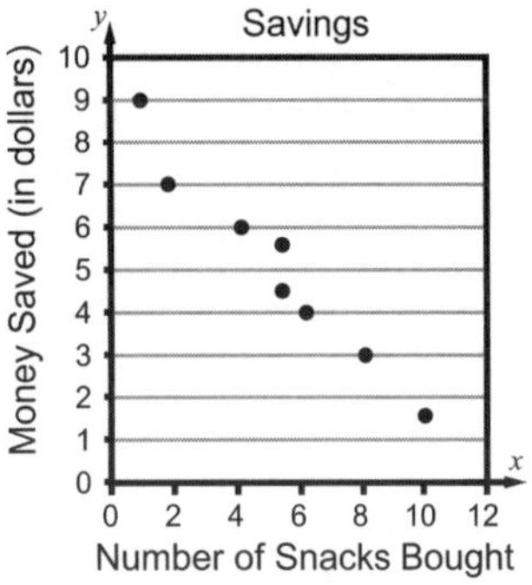

If Joe buys 3 snacks during the week, approximately how much money will he have saved?

40. Derrick saves change he finds around the house. The given scatterplot shows the relationship between the number of coins he finds and the amount of money he has collected.

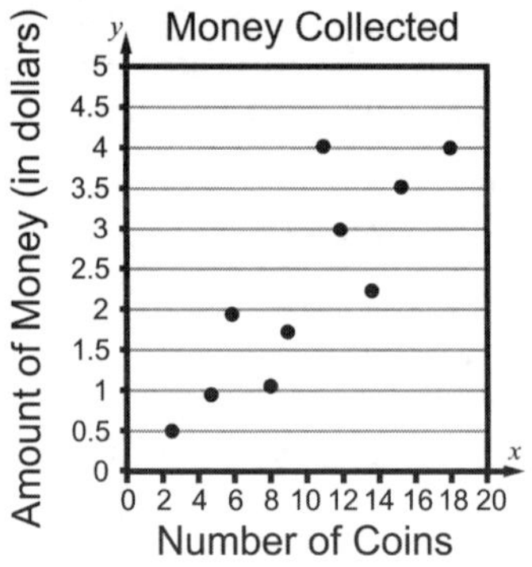

If Derrick finds 10 coins, approximately how much money will he have collected?

MPA-133: Distinguishing Between Exact and Approximate Answers

There are two types of computational answers: exact answers and approximate answers. With the vast number of math computations we use daily, both types of answers play significant roles in our lives. Descriptions of the two types of answers are given below.

Exact Answer	The actual answer. No error exists.
Approximate Answer	Answer that is close to exact. Small error may exist.

Depending on the situation, sometimes an exact answer is necessary and other times an approximate answer is necessary. Below is a list of some characteristics and situations that will help you to determine whether an exact or an approximate answer is necessary.

Exact Answer	Approximate Answer
Uses an "equal to" sign ($=$)	Uses an "approximately equal to" sign ($\approx$)
When we want an exact number	When using mental calculations
When accuracy is needed	To save time
	To obtain a rough idea of cost
	When using measurement instruments
	When rounding or truncating decimals

Measurement instruments yield only approximate answers because all measurement instruments have limitations. The limitations may be due to human error or due to the precision of the tool being used. Consider the example below that presents two different approximate answers for the same measurement. Since the ruler units on the right are smaller, it would yield a different and more precise measurement.

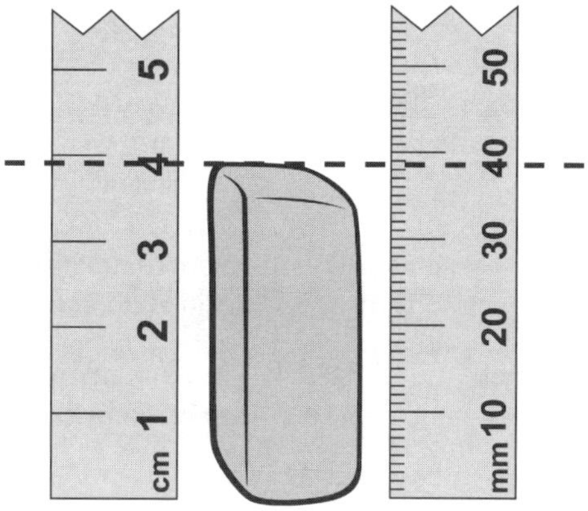

- According to the left ruler, the eraser is approximately 4 cm tall.
- According to the right ruler, the eraser is approximately 3.9 cm tall.

Example 1 For the problem $\frac{1}{3} \cdot \frac{2}{5}$, is the answer $\frac{2}{15}$ an approximate or exact answer?

$\frac{1}{3} \cdot \frac{2}{5} = \frac{2}{}$

Step 1: To multiply fractions, first multiply the numerators. One times two is two.

$\frac{1}{3} \cdot \frac{2}{5} = \frac{2}{15}$

Step 2: Next, multiply the denominators. Three times five is fifteen.

The answer $\frac{2}{15}$ is an exact answer.

Step 3: The answer $\frac{2}{15}$ is an actual unmodified answer that has not been rounded or truncated.

Answer: $\frac{2}{15}$ is an exact answer for $\frac{1}{3} \cdot \frac{2}{5}$.

Example 2 Which of the situations given below *best* describes when an exact answer is possible?

A. Measuring two cups of water to bake a cake.

B. Reading the thermometer outside.

C. Deciding how much concrete is needed to build a basketball court in the backyard.

D. Paying the apartment's monthly rent.

A. Measuring two cups of water to bake a cake.

Step 1: Anytime a measuring instrument is used, only an approximate answer can be given. Measuring two cups of water is an approximation because the actual amount yielded will vary depending on the units of the measuring instrument and human error.

B. Reading the thermometer outside.

Step 2: A thermometer is also a measuring instrument. It measures the temperature. The number read on a thermometer may be different from one individual to another due to a number of reasons, such as the perspective of the reader's eyes or the accuracy of the thermometer.

C. Deciding how much concrete is needed to build a basketball court in the backyard.

Step 3: To try to determine the amount of concrete needed for a backyard, we can multiply the backyard's length, width, and the depth of how thick the concrete should be to get an answer. However, the measurement for each dimension is an approximation. Therefore, the amount of concrete decided upon is also an approximation.

D. Paying the apartment's monthly rent.

Step 4: The amount owe for rent each month is an exact amount agreed on by the tenets and the landlord. This is an exact answer.

Answer: "Paying the apartment's monthly rent" is the situation that *best* describes when an exact answer is possible.

Example 3 Which of the problems given below will yield an approximate answer?

A. Counting the number of marbles in a bag.

B. The amount of change a register states should be given back to you after an excess payment.

C. The number of pages in a particular textbook.

D. Calculating the time it will take to get from one city to another.

A. Counting the number of marbles in a bag.	**Step 1:** Counting items will give an exact number.
B. The amount of change a register states should be given back to you after an excess payment.	**Step 2:** The register will subtract the amount you owe from the amount you give the cashier. This will result in an exact amount to be given back to you.
C. The number of pages in a particular textbook.	**Step 3:** Textbooks have an exact number of pages.
D. Calculating the time it will take to get from one city to another.	**Step 4:** Due to unforeseen situations, such as traffic and weather, calculating the time it will take to get from one city to another can only be an approximation.

Answer: "Calculating the time it will take to get from one city to another" will yield an approximate answer.

Example 4 Using a centimeter ruler, measure the length of the rectangle below to the nearest centimeter. Make sure to use the correct symbol ($=$ or $\approx$).

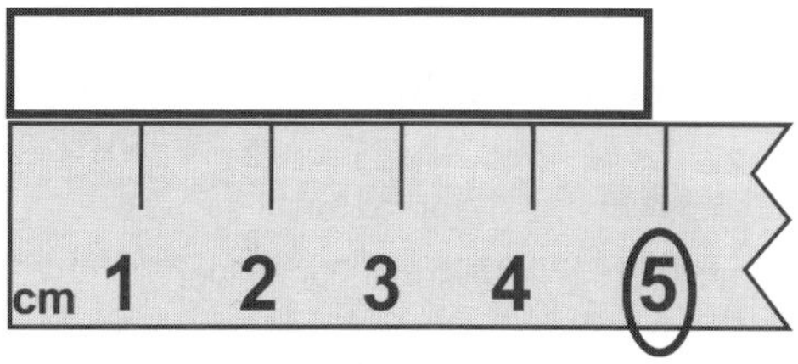

Step 1: The end of the rectangle sits at about the 5 centimeter marking.

The rectangle's length is approximately equal to 5 cm.	**Step 2:** Since all measurements are approximations, we use the "approximately equal to" sign.

Answer: The length of the rectangle is ≈ 5 centimeters.

Example 5 Find the area of the shaded region. Each box is equal to one square unit.

A. 30 square units C. 36 square units

B. 18 square units D. 15 square units

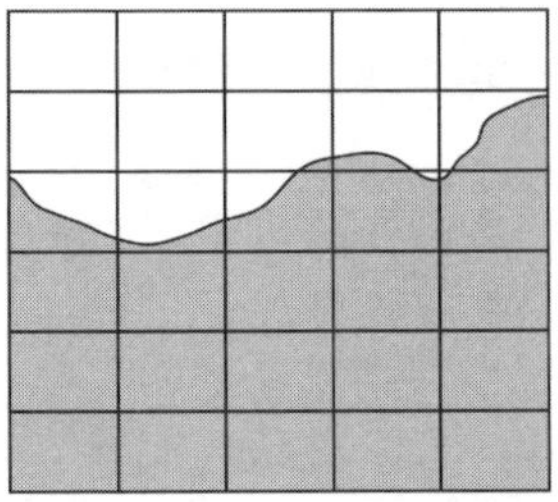

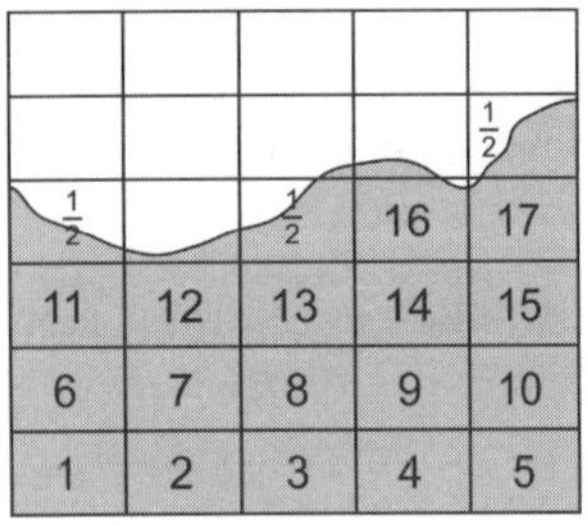

Step 1: Counting the squares, the shaded region covers approximately $18\frac{1}{2}$ squares.

Step 2: Since the half-squares are approximations, choose the answer that is closest to the approximation.

Answer: The area of the shaded region is about 18 square units.

Problem Set

1. One divided by 4 is 0.25.
What type of answer is 0.25?

Wrong answer Approximate answer
Exact answer Estimated answer

2. The answer given for 5.7 multiplied by 6.3 is 36.
What type of answer is 36?

Wrong answer Approximate answer
Exact answer Estimated answer

3. Jack said the answer to 12.65 – 5.75 is 6.9.
What type of answer is 6.9?

Wrong answer Approximate answer
Exact answer Estimated answer

4. The answer given for twelve divided by 7 is 1.7.
What type of answer is 1.7?

Wrong answer Approximate answer
Exact answer Estimated answer

5. When Jill added 3.5 + 12.9 + 7.6, she said the answer was 24. What type of answer is 24?

Wrong answer Approximate answer
Exact answer Estimated answer

6. The answer given for two and five-tenths divided by 8 is 0.31. What type of answer is 0.31?

Wrong answer Approximate answer
Exact answer Estimated answer

7. When multiplying 45.5 by 7.25, Jack gave the answer as 330. What type of answer is 330?

Wrong answer	Approximate answer
Exact answer	Estimated answer

8. The answer given for one hundred sixty divided by 3 is 53.3. What type of answer is 53.3?

Wrong answer	Approximate answer
Exact answer	Estimated answer

9. Which of the following situations best describes when an exact answer is possible?

- Sandra needs to fill a basket with homemade cookies to give as a gift.
- George has to determine the amount of change that Mr. Martin should receive back from his purchases totaling $13.39 when Mr. Martin hands him a $20.00 bill.
- Mrs. Jones is purchasing pizza for the baseball team. She needs to determine how many pizzas to buy.
- Melissa needs to squirt an amount of toothpaste on her toothbrush.

10. Which of the following situations best describes when an approximate answer is possible?

- Marica is following a recipe when making 24 cupcakes. She needs to divide the cupcakes into 3 groups.
- The cafeteria manager at Jackson Middle School needs to determine how many cartons of milk to order for this week's lunches.
- A babysitter charges $3.25 for 5 hours of babysitting.
- A carpenter counts the number of boards he will need to make bookshelves.

11. Which of the following situations best describes when an exact answer is possible?

- Jason is trying to determine how many basketballs will fit into his locker.
- Morgan has to determine how many days it will take her to read a book about basketball.
- Cameron must pay for her ticket to the basketball game.
- Michelle is trying to determine how many times she can make 3-point shots during this season's tournament games.

12. Which of the following situations best describes when an approximate answer is possible?

- Cameron must write a check for the amount of his purchases in a store.
- A business owner must calculate his employees' paychecks.
- Michael needs to determine how many hours he will need to work in order to buy a new video game system.
- The coach must select how many baseball players to place on the field during the game.

13. Which of the following situations best describes when an exact answer is possible?

- A gardener is determining how many bags of mulch will be needed to cover the area of a flower bed.
- Steven is determining how many apples will fit into a bushel basket.
- A restaurant manager is determining how many cups of coffee will be served during the dinner shift tonight.
- A waitress must calculate the amount of a bill for dinner in a restaurant.

14. Which of the following situations best describes when an approximate answer is given?

- A banker is determining how much money is in a person's savings account.
- A biologist determines the number of eggs an ostrich has in its nest.
- A person must pay for the gas that they pump into their car.
- Gordon is determining how much birdseed he will need to buy to put in his bird feeders this month.

15. Which of the following situations best describes when an exact answer is possible?

- Jason needs to determine how long it will take him to swim the length of the pool.
- A bank teller must make change for a $20 bill using quarters.
- Margie needs to determine the cost of punch she will serve at her party if she expects 20 people will attend the party.
- The pool manager must determine how much chlorine he will need to treat his pool over the summer.

16. Which of the given problems will yield an exact answer?

- Find the diameter of a circle using an inch ruler.
- Find the length of your bedroom wall using a meter stick.
- Find the percent of red candy in a bag of 20 pieces in which 5 are red.
- Find the volume of a baby pool using a measuring cup.

17. Which of the given problems will yield an approximate answer?

- Find the square root of 49.
- Find the factors of the number 25.
- Find the number of cubes in this rectangular prism:

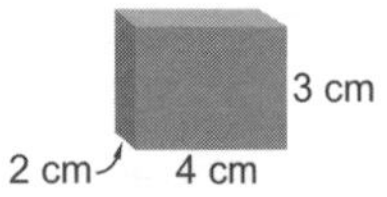

- Find the area of your hand using a tape measure:

18. Which of the given problems will yield an exact answer?

- Determine the amount of gas it will take to drive from your house to school.
- Find the perimeter of your backyard by counting your footsteps as you walk around the edges of your yard.
- Find the number of halves in 6 wholes.
- Find the surface area of a basketball using a meter stick.

19. Which of the given problems will yield an approximate answer?

- Find the sales tax for a bill that is $20 and tax that is 5%.
- Find the percent of red candies in a bag that has 25 pieces and 5 of these are red.
- Find the surface area of a large butterfly using a ruler.
- Find the number of red candies in a bag that has 30 pieces and 20% of these are red.

20. Which of the given problems will yield an exact answer?

- Determine the number of cars that are not red if an automobile factory produced 11,000 cars last year and 4,122 cars were red.
- Determine the percentage of children at ABC Elementary who ride in cars to school if it is observed that the first 23 out of 96 students are brought to school in a car.
- Find the square footage of rain forests in South America using a map.
- Find the amount of money Nina spends on clothes each month if she budgets 15% of her paycheck for clothing.

21. Which of the given problems will yield an approximate answer?

- Determine the cost of 10 pounds of peanuts if they are \$1.25 per pound.
- Find the percent correct on a 20-word spelling test if you get 15 correct.
- Find the height of the triangle by comparing it to the base of the triangle:

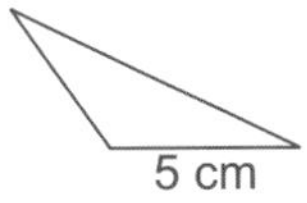

- Determine the product of $-\frac{3}{5}$ and its reciprocal:

$$-\frac{3}{5} \times -\frac{5}{3}$$

22. Which of the given problems will yield an exact answer?

- Determine the length of a bridge by comparing it to the length of a 16-foot sailboat.
- Find the volume of a box that will hold 18 books.
- Find the cost of 35 baseballs if they are priced near \$3.00.
- Find the value of the expression $4x + 5$, if $x = 12$.

23. Using the centimeter ruler provided, measure the height of the rectangle. Choose the best answer from the choices on the right.

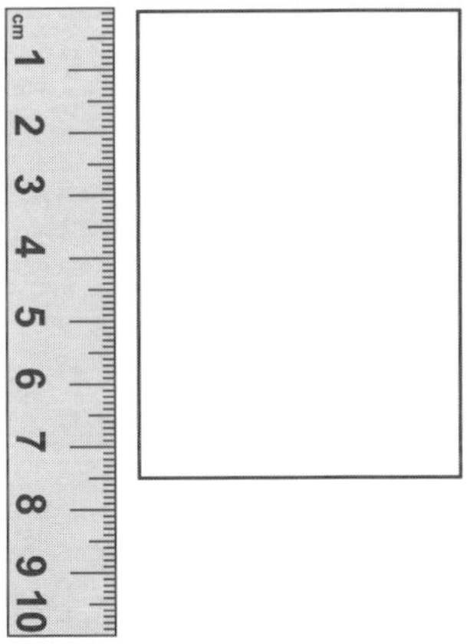

= 8.5 centimeters

≈ 7.5 centimeters

≈ 10 centimeters

= 7 centimeters

24. Using the inch ruler provided, measure the length of the book. Choose the best answer from the choices on the right.

$= 9\frac{1}{2}$ inches

≈ 9 inches

$= 8\frac{3}{4}$ inches

≈ 10 inches

25. Using the inch ruler provided, measure the length of the calculator. Choose the best answer from the choices on the right.

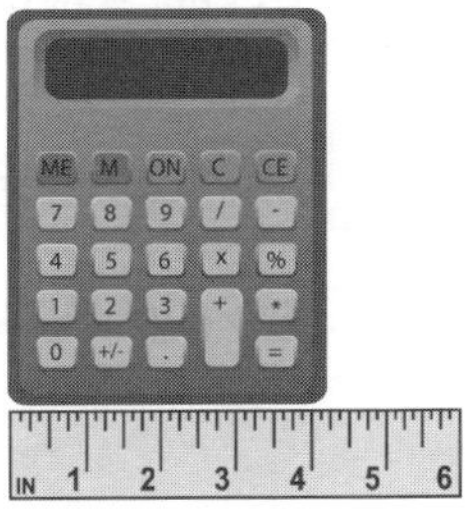

$\approx 5\frac{1}{4}$ inches

= 5 inches

= 4 inches

$\approx 4\frac{1}{4}$ inches

26. Using the centimeter ruler provided, measure the length of the diagonal of this square. Choose the best answer from the choices on the right.

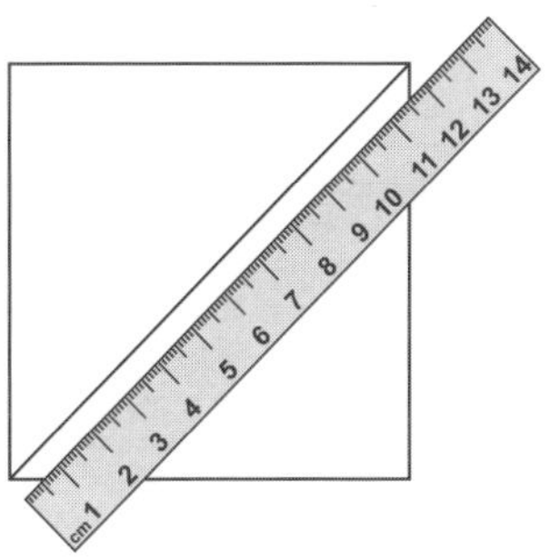

≈ 10.5 centimeters

≈ 12 centimeters

= 13 centimeters

= 11 centimeters

27. Using the inch ruler provided, measure the perimeter of this regular hexagon. Choose the best answer from the choices on the right.

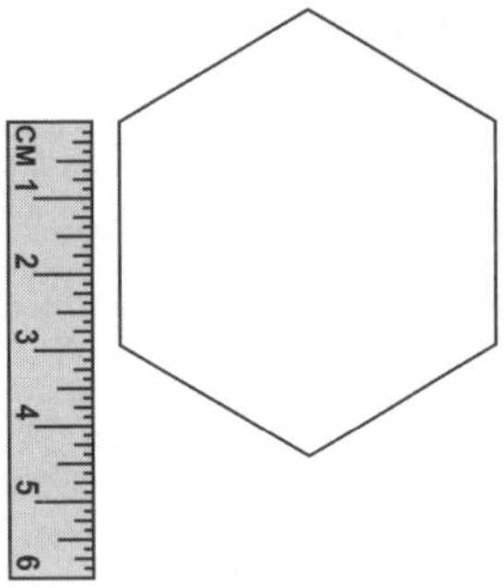

= 17 centimeters

= 18.6 centimeters

≈ 18 centimeters

≈ 16 centimeters

28. Using the inch ruler provided, measure the length of this water bottle. Choose the best answer from the choices on the right.

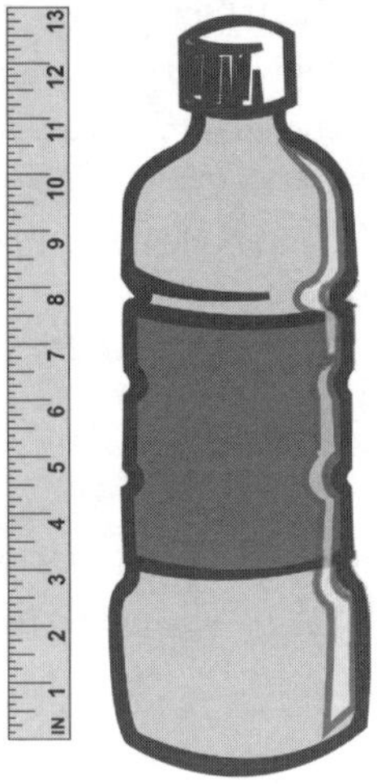

= 13 centimeters

≈ 11.5 centimeters

≈ $12\frac{1}{4}$ inches

≈ 11 centimeters

29. Using the centimeter ruler provided, measure the diameter of this circle. Choose the best answer from the choices on the right.

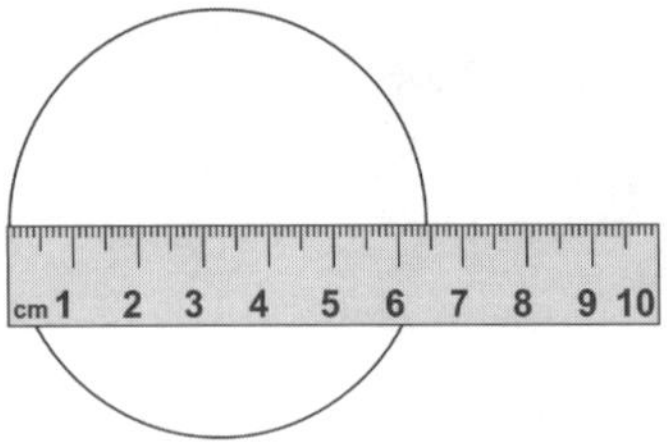

= 6 centimeters

≈ 6.5 centimeters

= 7 centimeters

≈ 5.5 centimeters

30. Using the inch ruler provided, measure the length of this beetle. Choose the best answer from the choices on the right.

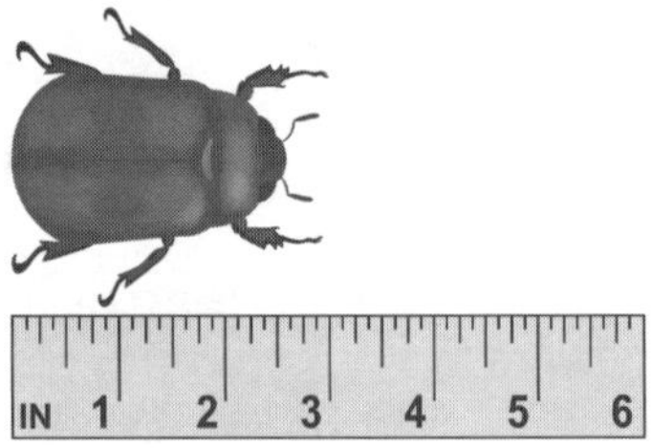

= $2\frac{3}{4}$ inches

= $3\frac{1}{2}$ inches

≈ 3 inches

≈ 4 inches

31. Find the area of the shaded region. Each box is equal to one square unit. Choose the best answer from the choices on the right.

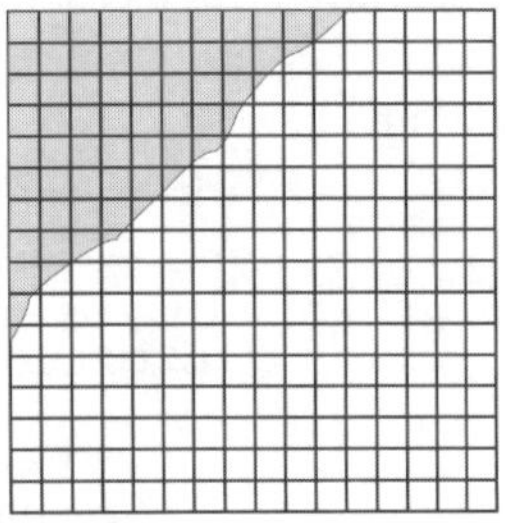

= 50 square units

≈ 90 square units

≈ 256 square units

≈ 64 square units

32. Find the area of the shaded region. Each box is equal to one square unit. Choose the best answer from the choices on the right.

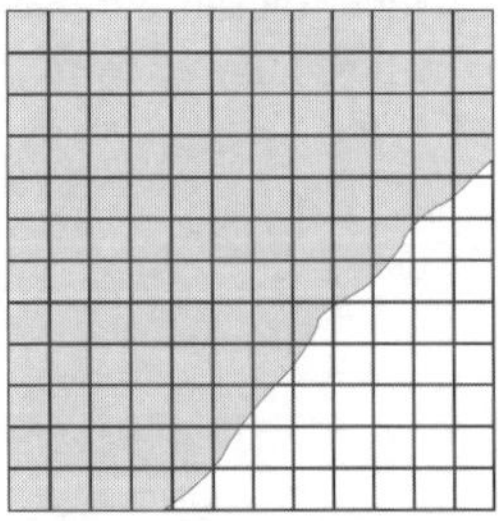

= 72 square units

≈ 108 square units

≈ 120 square units

= 100 square units

33. Find the area of the shaded region. Each box is equal to one square unit. Choose the best answer from the choices on the right.

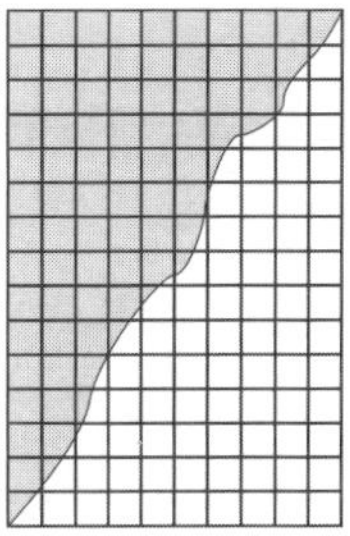

= 60 units

≈ 75 square units

= 90 square units

≈ 100 square units

34. Find the area of the shaded region. Each box is equal to one square unit. Choose the best answer from the choices on the right.

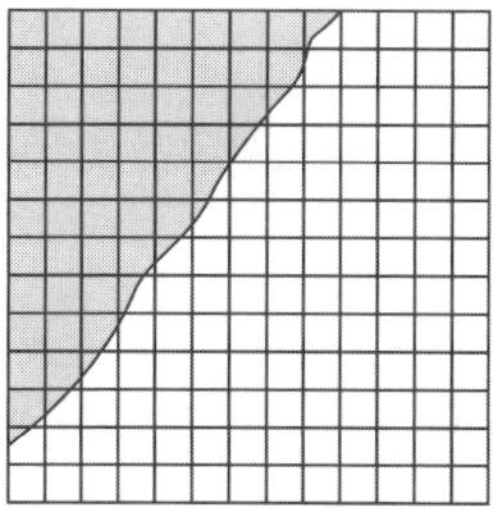

= 65 square units

= 80 square units

≈ 50 square units

≈ 70 square units

35. Find the area of the shaded region. Each box is equal to one square unit. Choose the best answer from the choices on the right.

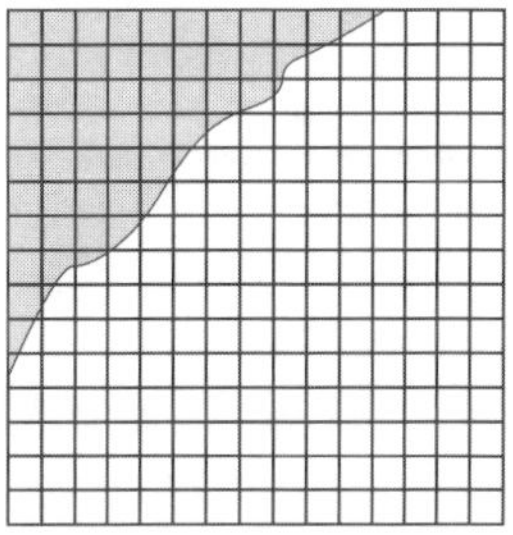

≈ 35 square units

≈ 55 square units

= 75 square units

= 50 square units

36. Find the area of the shaded region. Each box is equal to one square unit. Choose the best answer from the choices on the right.

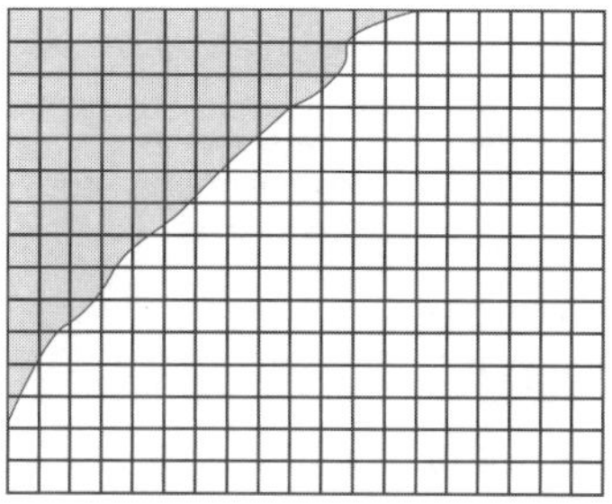

= 100 units

= 90 units

≈ 60 square units

≈ 75 square units

37. Find the area of the shaded region. Each box is equal to one square unit. Choose the best answer from the choices on the right.

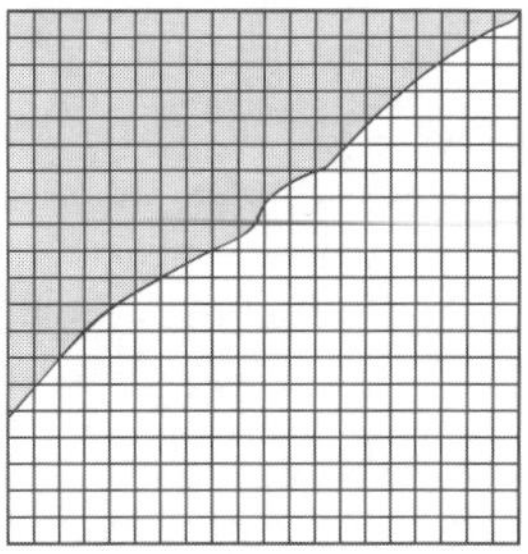

= 150 square units

= 120 square units

≈ 180 square units

≈ 135 square units

38. Find the area of the shaded region. Each box is equal to one square unit. Choose the best answer from the choices on the right.

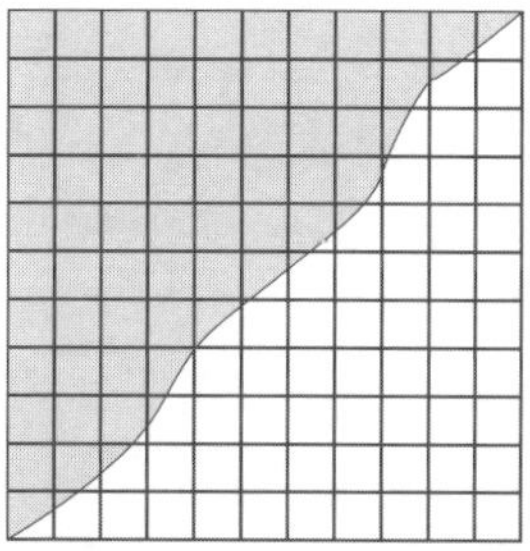

= 55 square units

= 65 square units

≈ 61 square units

≈ 75 square units

39. Find the area of the shaded region. Each box is equal to one square unit. Choose the best answer from the choices on the right.

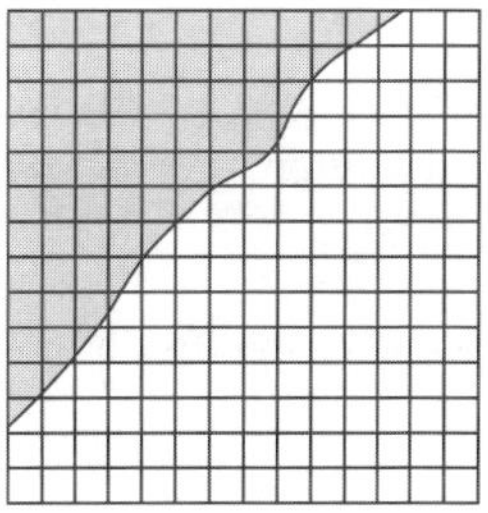

≈ 66 square units

≈ 100 square units

= 55 square units

= 90 square units

40. Find the area of the shaded region. Each box is equal to one square unit. Choose the best answer from the choices on the right.

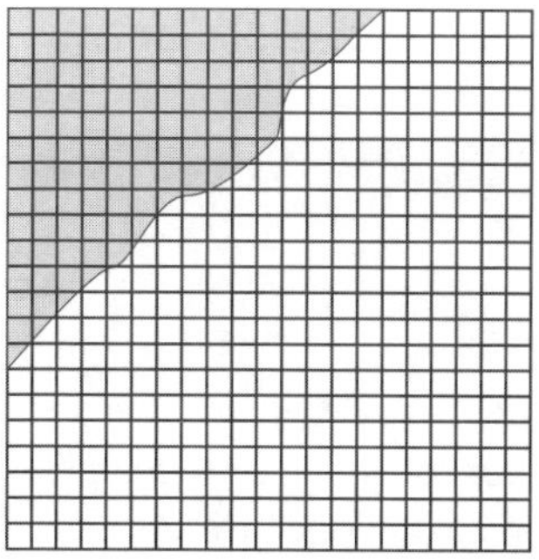

≈ 110 square units

≈ 85 square units

= 108 units

= 150 square units

MPA-134: Calculating with Precision, Accuracy, and Significant Digits

Accuracy and Precision

In a previous lesson, we learned that all physical measurements are approximations. How good an approximation is can be described by its accuracy and precision.

Accuracy	How close the measurement is to the actual value.
Precision (of several measurements)	How close the individual measurements are to each other.

Example 1 In a science experiment, students were asked to weigh a 3.011 pound piece of iron. Their results are as follows:

3.612 lb

3.613 lb

3.614 lb

Are the measurements accurate, precise, or both?

$3.612 - 3.011 = 0.601$ lb
$3.613 - 3.011 = 0.602$ lb
$3.614 - 3.011 = 0.603$ lb

Step 1: Since accuracy is how close the measure is to the actual value, subtract the actual value from each measurement.

3.612 – 3.011 = **0.6**01 lb
3.613 – 3.011 = **0.6**02 lb
3.614 – 3.011 = **0.6**03 lb

The measurements are not accurate.

Step 2: The difference is how far the measurement is from the actual value. In this problem, all the measurements are more than 0.6 lb away from the actual value, which is measured to the thousandths place. Therefore, the measurements are not accurate.

3.612 lb ← smallest measurement
3.613 lb
3.614 lb ← largest measurement
$3.614 - 3.612 = 0.002$ lb

Step 3: Since precision is how close the measurements are to each other, find the difference between the largest and smallest measurements.

Step 4: In this problem, the largest difference between any two measurements is only 0.002 lb. Therefore, the measurements are precise.

Answer: The measurements are not accurate, but they are precise.

The description of precision given earlier applies to the precision of several measurements. However, an individual measurement can also be described by its precision.

Precision (of an individual measurement)	The more partitions or decimal places, the more precise the measurement.

More Partitions → More Precise

On the two rulers below, the ruler to the right is more precise since it has more partitions.

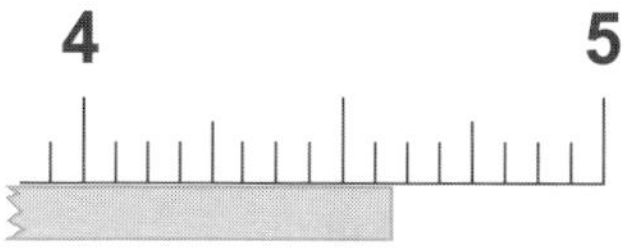

On this ruler, we may say the strip is about $4\frac{9}{16}$ inches.

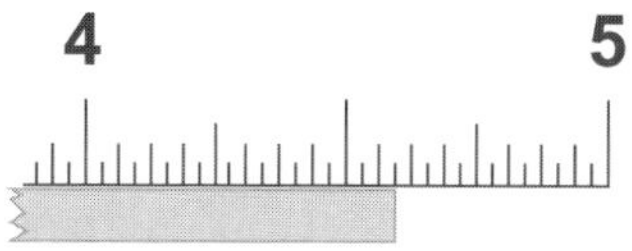

On this ruler, we may say the strip is $4\frac{19}{32}$ inches.

More Decimal Places → More Precise

The actual value of pi is non-repeating and non-terminating. Using more decimal places to approximate pi will result in greater precision.

Actual value of pi: 3.1415926535897932384626433832795…

Values of Pi used	Level of Precision
3.14	least precise
3.141592	
3.1415926535	
3.141592653589	
3.141592653589793	most precise

Significant Digits

To express answers with as much accuracy as the precision of the numbers allows, we follow significant digit rules.

Significant Digits	Significant digits give us a systematic method to express answers accurately.

Significant Digits are:
• All non-zero digits • All embedded zeros • Trailing zeros if the decimal is specified • Leading zeros are **not** significant

Example 2 How many significant digits does the number 0.038700810 have?

0.0**38**700**81**0	**Step 1:** All non-zero digits are significant. Therefore, there are 5 significant non-zero digits.
0.0387**00**810	**Step 2:** All imbedded zeros are significant. Therefore, there are two significant imbedded zeros.
0.03870081**0**	**Step 3:** Trailing zeros are significant if the decimal is specified. In this number, the decimal is specified. Therefore, the trailing decimal is significant.
0.0**38700810**	**Step 4:** Leading zeros are not significant. Therefore, the total number of significant digits is 8.

Answer: 0.038700810 has 8 significant digits.

Rules For Significant Digits in Computational Problems

Multiplication and Division:

The product or quotient has as many *significant digits* as the number with the fewest number of *significant digits.*

Addition and Subtraction:

The sum or difference has as many *decimal places* as the number with the fewest number of *decimal places.*

Example 3 Give the answer to the correct number of significant digits: 3.211 + 19.2 + 20.5

```
   3.211
  19.2
+ 20.5
  42.911
```

Step 1: Line up the decimals and add.

```
   3.211  ←  3 decimal places
  19.2    ←  1 decimal place
+ 20.5    ←  1 decimal place
  42.911
```

Step 2: Determine the number of decimal places for each number in the problem.

```
   3.211  ←  3 decimal places
  19.2    ←  1 decimal place
+ 20.5    ←  1 decimal place
  42.911
```

Step 3: Round the answer to the fewest number of decimal places. The fewest number of decimal places is 1. Therefore, round 42.911 to one decimal place.

Answer: 3.211 + 19.2 + 20.5 = 42.9

Example 4 Give the answer to the correct number of significant digits: 2.113 × 4.01

$$\begin{array}{r} 2.113 \\ \times\ 4.01 \\ \hline 8.47313 \end{array}$$

Step 1: Multiply.

$$\begin{array}{rl} 2.113 & \leftarrow \text{4 significant digits} \\ \times\ 4.01 & \leftarrow \text{3 significant digits} \\ \hline 8.47313 & \end{array}$$

Step 2: Determine the number of significant digits for each number in the problem.

$$\begin{array}{rl} 2.113 & \leftarrow \text{4 significant digits} \\ \times\ 4.01 & \leftarrow \textbf{3 significant digits} \\ \hline 8.47313 & \end{array}$$

Step 3: Round the answer to the fewest number of significant digits. The fewest number of significant digits is 3. Therefore, round 8.47313 to 3 significant digits.

Answer: 2.113 × 4.01 = 8.47

Example 5 A discus was thrown 8.805 feet. Three individuals measured the distance to be 8.005, 7.030, and 8.100 feet. The measurements:

A. are precise

B. are accurate

C. each have 2 significant digits

D. each have 4 significant digits

8.005 ft

7.030 ft ← smallest measure

8.100 ft ← largest measure

Step 1: Since precision is how close the measures are to each other, find the difference of the largest and smallest measure.

8.100 – 7.030 = 1.130 ft

Step 2: In this problem, the largest difference between two measurements is 1.130 ft. Therefore, the measurements are not precise. We can eliminate choice A.

8.805 – 8.005 = 0.800 ft

8.805 – 7.030 = 1.775 ft

8.805 – 8.100 = 0.705 ft

Step 3: Since accuracy is the closeness of a measure to the actual value, subtract the measurements from the actual value.

8.805 – 8.005 = 0.800 ft

8.805 – 7.030 = 1.775 ft

8.805 – 8.100 = 0.705 ft

Step 4: The difference is how far the measurement is from the actual value. In this problem, all the measurements are more than 0.705 feet away from the actual value. Therefore, the measurements are not accurate. We can eliminate choice B.

8.005 ft ← 4 significant digits 7.030 ft ← 4 significant digits 8.100 ft ← 4 significant digits	**Step 5:** Since the decimal is specified, all non-zero digits, imbedded zeros, and trailing zeros are significant. Each of the numbers contains 4 significant digits.

Answer: The measurements each have 4 significant digits.

Problem Set

1. Three cooking students were asked to measure the temperature of a roast that was 96.25 degrees. Their results are as follows:

96.24°

96.25°

96.26°

How can their measurements best be described?

Accurate and precise
Accurate but not precise
Not accurate but precise
Not accurate and not precise

2. Three meteorology students were asked to measure the temperature of a room that was 37.89 degrees. Their results are as follows:

37.65°

37.63°

37.64°

How can their measurements best be described?

Accurate and precise
Accurate but not precise
Not accurate but precise
Not accurate and not precise

3. The winner of a race finished in 25.962 seconds. Four clockers timed the race. Their results are as follows:

25.975 sec.

25.973 sec.

25.980 sec.

25.951 sec.

How can their measurements best be described?

Accurate and precise
Accurate but not precise
Not accurate but precise
Not accurate and not precise

4. Four bank tellers counted a drawer that contained $592.49. Their results are as follows:

$592.49

$592.48

$592.47

$592.48

How can their measurements best be described?

Accurate and precise
Accurate but not precise
Not accurate but precise
Not accurate and not precise

5. Jacob measured the growth of his tree four times. The actual tree growth is 34.0555 centimeters. His results are as follows:

34.0690 cm

34.0691 cm

34.0691 cm

34.0692 cm

How can their measurements best be described?

Accurate and precise
Accurate but not precise
Not accurate but precise
Not accurate and not precise

6. Four surveyor trainees were asked to measure the area of a yard that is 42.754 square yards. Their results are as follows:

42.652 square yards

42.596 square yards

42.714 square yards

42.850 square yards

How can their measurements best be described?

Accurate and precise
Accurate but not precise
Not accurate but precise
Not accurate and not precise

7. Four math students measure the diameter of a quarter. The actual diameter is 24.26 millimeters. Their results are as follows:

24.26 mm

24.24 mm

24.25 mm

24.26 mm

How can the measurements best be described?

Accurate and precise
Accurate but not precise
Not accurate but precise
Not accurate and not precise

8. How many significant digits does the number 0.0630 have?

9. How many significant digits does the number 0.0009041 have?

10. How many significant digits does the number 0.0073 have?

11. How many significant digits does the number 3.04010 have?

12. How many significant digits does the number 9,100 have?

13. How many significant digits does the number 10,001 have?

14. How many significant digits does the number 5,204,000 have?

15. How many significant digits does the number 6 have?

16. Give the answer to the correct number of significant digits: $34.001 + 6.9 + 75.0145$

17. Give the answer to the correct number of significant digits: $8.605 + 19.12 + 14.0087$

18. Give the answer to the correct number of significant digits: $106.45 + 254.892 + 100.0009$

19. Give the answer to the correct number of significant digits: $564.5 - 194.113$

20. Give the answer to the correct number of significant digits: $987.65 - 65.3256 - 98.741$

21. Give the answer to the correct number of significant digits: $54.006 - 22.0007 - 19.21456$

22. Give the answer to the correct number of significant digits: $64.0968 - 12.064 - 14.19564$

23. Give the answer to the correct number of significant digits: 8.1×9.304

24. Give the answer to the correct number of significant digits: 10.6×48.207

25. Give the answer to the correct number of significant digits: 25.04×3.1245

26. Give the answer to the correct number of significant digits: 50.92×4.0065

27. Give the answer to the correct number of significant digits: $62.005 \div 25$

28. Give the answer to the correct number of significant digits: $78.25 \div 6.25$

29. Give the answer to the correct number of significant digits: $83.3286 \div 17.4$

30. Give the answer to the correct number of significant digits: $477.78126 \div 32.01$

31. One meter is equal to 1.0936 yards. Three students are asked to measure the number of yards in one meter. Their results are 1.0939 yards, 1.0940 yards, and 1.0938 yards. How can the students' measurements best be described?

The students' measurements are accurate but not precise.
The students' measurements are accurate and precise.
Each measurement has 3 significant digits.
Each measurement has 4 significant digits.

32. The circumference of a newborn baby's head is 45.08 centimeters. Three nurses measured this circumference and got 44.97 centimeters, 44.71 centimeters, and 44.85 centimeters. How can the nurses' measurements best be described?

Each measurement has 4 significant digits.
The nurses' measurements are accurate and precise.
The nurses' measurements are accurate.
Each measurement has 2 significant digits.

33. The volume of a penny is 0.360 cubic centimeters. Three students measured this volume and got 0.345 cubic centimeters, 0.347 cubic centimeters, and 0.344 cubic centimeters. How can the students' measurements best be described?

Each measurement has 4 significant digits.
The students' measurements are precise.
Each measurement has 5 significant digits.
The students' measurements are accurate.

34. The winner of the long jump contest jumped 10.56 feet. Three other contestants measured the length and got 10.30 feet, 10.40 feet, and 10.20 feet. How can the contestants' measurements best be described?

Each measurement has 2 significant digits.
The contestants' measurements are accurate and precise.
Each measurement has 4 significant digits.
The contestants' measurements are accurate but not precise.

35. The thickness of a penny is 1.27 millimeters. Three students measured this thickness and got 1.25 millimeters, 1.26 millimeters, and 1.26 millimeters. How can the students' measurements best be described?

Each measurement has 2 significant digits.
Each measurement has 4 significant digits.
The students' measurements are not accurate and not precise.
The students' measurements are accurate and precise.

36. A flagpole stands 29.85 meters high. Three math students calculated this height using geometry and got 29.70 meters, 29.72 meters, and 29.73 meters. How can the students' measurements best be described?

The students' measurements are accurate but not precise.
The students' measurements are precise.
Each measurement has 3 significant digits.
Each measurement has 2 significant digits.

37. The area of a painting is 37.95 square inches. Three art students measured this area and got 36.55 square inches, 36.95 square inches, and 35.35 square inches. How can the students' measurements best be described?

The students' measurements are accurate and precise.
Each measurement has 4 significant digits.
Each measurement has 2 significant digits.
The students' measurements are precise.

38. The weight of a golden dollar coin is 8.10 grams. Three scientists measured this weight and got 8.09 grams, 8.09 grams, and 8.08 grams. How can the scientists' measurements best be described?

Each measurement has 2 significant digits.
Each measurement has 4 significant digits.
The scientists' measurements are accurate and precise.
The scientists' measurements are accurate but not precise.

39. An arrow is shot and travels 19.56 yards. Three hunters measured this distance and got 19.35 yards, 19.37 yards, and 19.17 yards. How can the hunters' measurements best be described?

The hunters' measurements are precise.
Each measurement has 2 significant digits.
The hunters' measurements are accurate and precise.
Each measurement has 4 significant digits.

40. The temperature of a patient is 99.785 degrees. Three nurses measure this temperature and got 99.780 degrees, 99.781 degrees, and 99.782 degrees. How can the nurses measurements best be described?

The nurses' measurements are accurate and precise.
Each measurement has 3 significant digits.
The nurses' measurements are accurate but not precise.
Each measurement has 4 significant digits.

MPA-140: Examining Linear Equations in Slope-Intercept Form

In this lesson, we will learn about linear equations that are written in slope-intercept form. Recall that the slope of a line is the ratio of the change in the y-coordinates to the corresponding change in the x-coordinates as you move from one point on the line to another. The y-intercept is the y-coordinate of the point where the line passes through the y-axis. The x-intercept is the x-coordinate of the point where the line passes through the x-axis.

y-intercept	The y-coordinate of the point where the line passes through the y-axis
x-intercept	The x-coordinate of the point where the line passes through the x-axis

The slope-intercept form uses the slope and the y-intercept. With this information, linear equations can be easily graphed.

Slope-intercept Form	Equations written in the form $y = mx + b$, where m is the slope and b is the y-intercept

Example 1 Write an equation that describes the given graph.

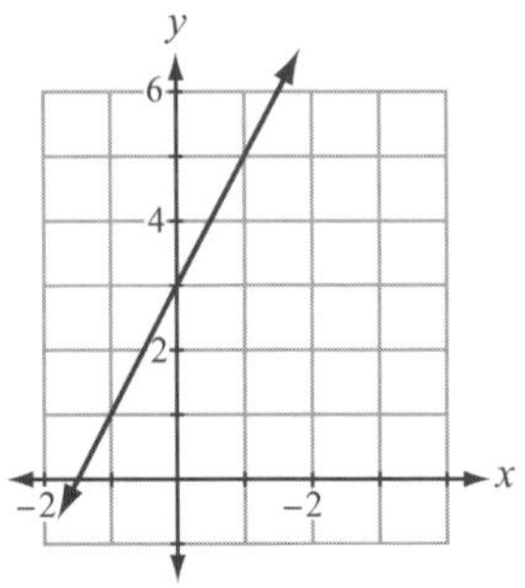

Step 1: Find the y-intercept, b.

$b = 3$

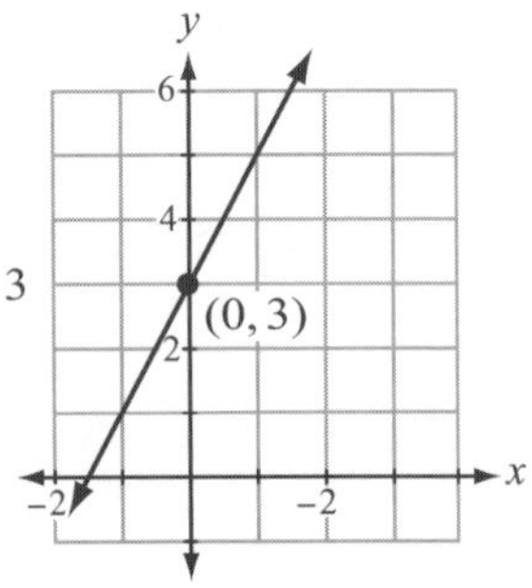

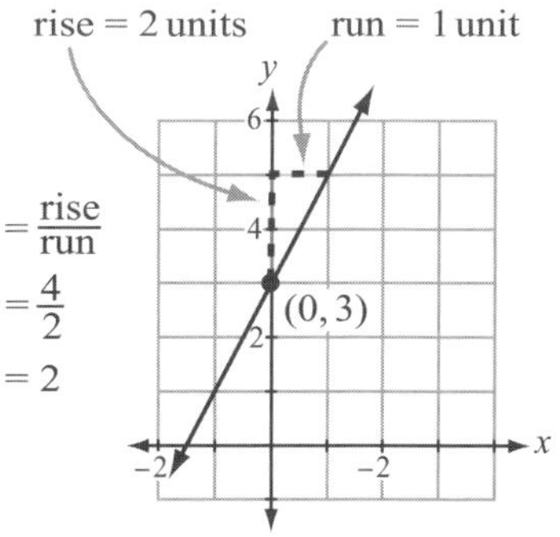

$m = \frac{\text{rise}}{\text{run}}$

$= \frac{4}{2}$

$= 2$

Step 2: Find the slope, m.

$y = mx + b$

$y = 2x + 3$

Step 3: Use the slope-intercept form, $y = mx + b$, to write the equation of the line.

Answer: The equation of the line is $y = 2x + 3$.

Example 2 Find the slope and y-intercept of a line from the linear equation:

$$y = 5x - 12$$

$y = \boldsymbol{mx} + \boldsymbol{b}$, where m is the slope of the line and b is the y-intercept.

Step 1: Define slope-intercept form.

$y = \mathbf{5}x - \mathbf{12}$

Step 2: Write the equation in slope-intercept form.

$y = \mathbf{5}x + (\mathbf{-12})$

$\boldsymbol{m} = \mathbf{5}$ and $\boldsymbol{b} = \mathbf{-12}$

Step 3: Identify the slope and y-intercept.

Answer: The slope of the line is 5, and the y-intercept is –12.

Example 3 Write the equation $4x + y = 8$ in slope-intercept form.

$y = \boldsymbol{mx} + \boldsymbol{b}$, where m is the slope of the line and b is the y-intercept.

Step 1: Define slope-intercept form.

$4x + y = 8$

$4x - 4x + y = -4x + 8$

$y = -4x + 8$

Step 2: Write the equation in slope-intercept form.

Answer: The slope-intercept form of the equation is $y = -4x + 8$.

Example 4 Find the slope and y-intercept of a line from the linear equation:

$$y - 7x = -2$$

$y = \boldsymbol{mx} + \boldsymbol{b}$, where m is the slope of the line and b is the y-intercept.

Step 1: Define slope-intercept form.

$y - 7x = -2$ $y - 7x + 7x = 7x - 2$ $y = 7x - 2$	**Step 2:** Write the equation in slope-intercept form.
$y = \mathbf{7}x + (\mathbf{-2})$ $\boldsymbol{m} = 7$ and $\boldsymbol{b} = -2$	**Step 3:** Identify the slope and y-intercept.

Answer: The slope of the line is 7, and the y-intercept is –2.

Example 5 Claudia is shopping for T-shirts for her students. She looked at three shops and found the prices listed in the given table. If Claudia plans to buy 30 T-shirts, which shop would charge her the least?

Shop	Artist Fee	Cost per T-shirt
A	\$12	\$3
B	\$24	\$2
C	none	\$5

Shop A: $y = 3x + 12$ Shop B: $y = 2x + 24$ Shop C: $y = 5x$	**Step 1:** Write an equation in slope-intercept form that describes each of the shops' charges. The y-intercept is the artist fee and the slope is the cost per T-shirt.
Shop A: $y = 3x + 12$ $y = 3(30) + 12$ $y = 90 + 12$ $\boldsymbol{y} = \mathbf{102}$ Shop B: $y = 2x + 24$ $y = 2(30) + 24$ $y = 60 + 24$ $\boldsymbol{y} = \mathbf{84}$ Shop C: $y = 5x$ $y = 5(30)$ $\boldsymbol{y} = \mathbf{150}$	**Step 2:** To find the amount Claudia will pay at each shop, substitute the number of T-shirts she plans to buy, 30, for x.

Answer: Shop B would charge Claudia the least.

Problem Set

1. Which equation describes the given graph?

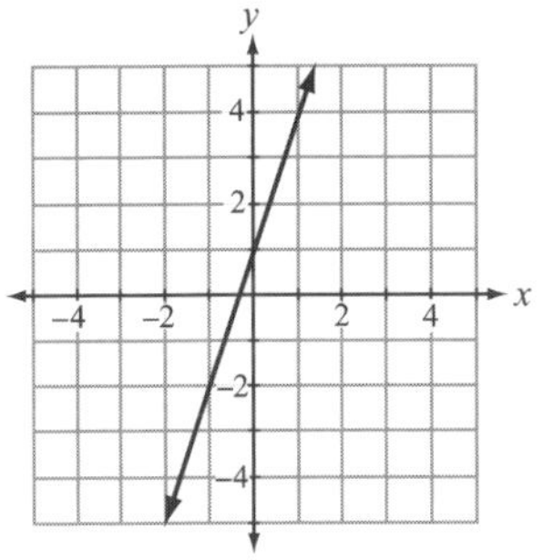

$y = 3x + 1$ $\qquad y = \frac{1}{3}x + 1$

$y = -3x + 1$ $\qquad y = -\frac{1}{3}x + 1$

2. Which equation describes the given graph?

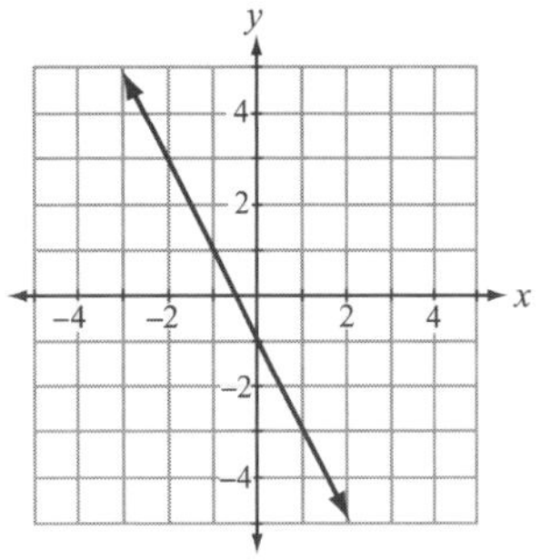

$y = -\frac{1}{2}x - 1$ $\qquad y = 2x - 1$

$y = -2x - 1$ $\qquad y = -2x + 1$

3. Which equation describes the given graph?

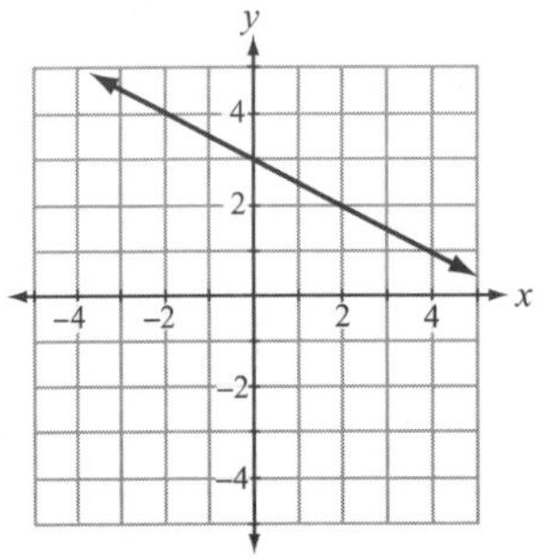

$y = -\frac{1}{2}x - 3$ $\qquad y = -\frac{1}{2}x + 3$

$y = -2x + 3$ $\qquad y = \frac{1}{2}x + 3$

4. Which equation describes the given graph?

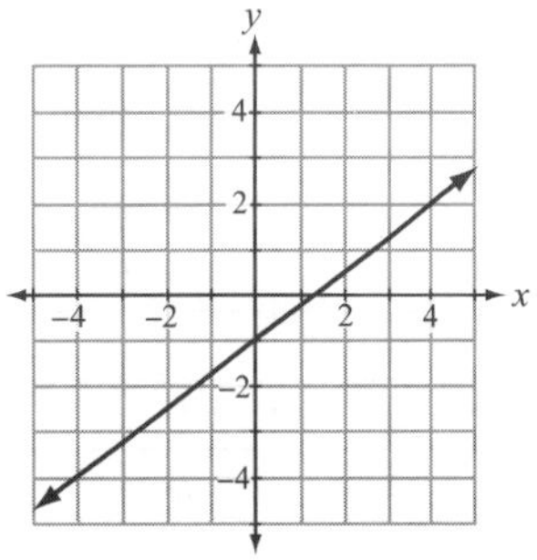

$y = \frac{3}{4}x - 1$ $\qquad y = -\frac{3}{4}x - 1$

$y = \frac{4}{3}x - 1$ $\qquad y = -\frac{4}{3}x - 1$

5. Find the slope and y-intercept of a line from the linear equation:

$$y = 2x - 5$$

6. Find the slope and y-intercept of a line from the linear equation:

$$y = -4x + 7$$

7. Find the slope and y-intercept of a line from the linear equation:

$$y = \frac{3}{5}x + 6$$

8. Find the slope and y-intercept of a line from the linear equation:

$$y = -\frac{2}{3}x - 8$$

9. Write the equation $3x + y = -7$ in slope-intercept form.

10. Write the equation $6x - y = 8$ in slope-intercept form.

11. Write the equation $5x - 2y = -4$ in slope-intercept form.

12. Write the equation $3x + 4y = 16$ in slope-intercept form.

13. Find the slope and y-intercept of a line from the linear equation:

$$2x - y = -3$$

14. Find the slope and y-intercept of a line from the linear equation:

$$2x + y = 7$$

15. Find the slope and y-intercept of a line from the linear equation:

$$4x + 3y = 12$$

16. Find the slope and y-intercept of a line from the linear equation:

$$5x - 2y = -8$$

17. Alex is trying to choose a cell phone company to use. The following table shows four different companies and their rates. If Alex uses 700 minutes per month, which company would be the cheapest for him to use?

Company	Amount per month	Amount per minute
A	\$75	\$0.06
B	\$50	\$0.08
C	\$125	none
D	none	\$0.20

18. Alta is trying to choose a car rental company to use for a day. The following table shows four different companies and their rates. If Alta plans to drive 500 miles, which company would be the cheapest for her to use?

Company	Amount per day	Amount per mile
A	\$45	\$0.10
B	\$60	\$0.08
C	\$100	none
D	none	\$0.20

19. Mr. Meyer is trying to choose a printing company to use. The following table shows four different companies and their rates. If Mr. Meyer plans to have 2,000 pamphlets printed, which company would be the cheapest for him to use?

Company	Setup fee	Amount per pamphlet
A	\$100	\$1.50
B	\$150	\$1.00
C	\$175	\$0.75
D	none	\$1.75

20. Jason is trying to choose a mail order DVD rental company to use. The following table shows four different companies and their rates. If Jason plans to rent 30 DVD's in the next year, which company would be the cheapest for him to use?

Company	Yearly membership fee	Amount per DVD
A	\$40	\$2.25
B	\$15	\$2.75
C	\$150	none
D	none	\$2.90

MPA-142: Solving Problems with Linear Functions and Direct Variation

In this lesson, we will learn how to solve real-life situations with linear functions and direct variation. In some problems, slope-intercept form can be used to represent the linear function. Recall that the slope-intercept form of a function is $y = mx + b$, where m is the slope or the rate of change and b is the y-intercept.

Example 1 It costs a flat fee of \$15 to ice skate, plus an additional \$5 per hour for ice skate rentals. If x represents the number of hours the ice skates are rented, write a linear function that represents the data.

$y = mx + b$

Step 1: Recall the slope-intercept form.

$y = 5x + b$

Step 2: The slope, m, is the rate of change. In this problem, the rate of change is \$5 per hour.

$y = 5x + 15$

Step 3: The y-intercept, b, is the initial amount. In this problem, the initial charge to ice skate is \$15.

Answer: Therefore, the linear function that represents this data is $y = 5x + 15$.

Sometimes real-life situations are given in a table. To write a function from a table, you must identify the slope as the rate of change in the table and the y-intercept as the initial amount.

Example 2 The following table shows the cost of a college math tutor. Write a function that represents the relationship between the number of hours, x, and the total cost, y.

# of hours	x	1	2	3	4
Total cost (\$)	y	60	110	160	210

+1 +1 +1

# of hours	x	1	2	3	4
Total cost (\$)	y	60	110	160	210

+50 +50 +50

Step 1: Find the slope by calculating the change in y over the change in x.

$50(1) = 50 \quad 50(2) = 100$

$50(3) = 150 \quad 50(4) = 200$

Step 2: Check to see if the function, $y = 50(x)$, represents the table.

$y = 50x + 10$

Step 3: Notice that each of our calculations do not result in y. In each calculation there is a difference of 10. Therefore, add 10 to our function.

$50(1) + 10 = 60$ $50(2) + 10 = 110$ $50(3) + 10 = 160$ $50(4) + 10 = 210$	**Step 4:** Check to see if the function, $y = 50x$, represents the table.

Answer: The function that represents the relationship given in the table is $y = 50x + 10$.

Another type of linear function is called direct variation. In direct variation, an increase or decrease in one variable results in an increase or decrease in the other variable. Another way of stating this is "*y* varies directly with *x*." In the equation for direct variation, $y = kx$, k is the constant of variation and k cannot equal zero.

Direct Variation	A function in the form $y = kx$, where $k \neq 0$, and y is proportional to x. This can also be stated as "y varies directly with x."
Constant of Variation	The constant k in the direct variation equation $y = kx$

Example 3 Suppose y varies directly with x, and $y = 30$ when $x = 5$. Find the constant of variation.

$y = kx$

Step 1: The constant of variation is represented by k in the direct variation equation.

$30 = k(5)$

$\frac{30}{5} = \frac{k(5)}{5}$

$6 = k$

Step 2: Substitute the given values for x and y, and solve the equation for k.

Answer: The constant of variation is 6.

Example 4 Determine if the given table lists values of y that vary directly with the values of x.

x	10	18	28	40
y	5	6	7	8

$k = \frac{5}{10}$ $k = \frac{6}{18}$ $k = \frac{7}{28}$ $k = \frac{8}{40}$

$k = \frac{1}{2}$ $k = \frac{1}{3}$ $k = \frac{1}{4}$ $k = \frac{1}{5}$

Step 1: The direct variation formula, $y = kx$, can be written as $k = \frac{y}{x}$. Substitute the values in the chart for y and x, and solve for k.

Answer: Since the values for k are not constant, y does not vary directly with x.

Example 5 A bus driver's earnings vary directly with the amount of time driving. If the bus driver earns \$135 for 15 hours of driving, what will the bus driver earn for driving 40 hours?

$y = kx$ or $k = \frac{y}{x}$

Step 1: Since the question states that a bus driver's earnings vary directly with the amount of time driving, this is a direct variation.

$x_1 = 15$ hours

$x_2 = 40$ hours

$y_1 = \$135$

$y_2 =$ unknown

Step 2: Use proportions to solve this situation. Four values are needed to set up a proportion.

$$\frac{y_1}{x_1} = \frac{y_2}{x_2}$$

$$\frac{135}{15} = \frac{y_2}{40}$$

$$15y_2 = 135(40)$$

$$\frac{y_2}{15} = \frac{5{,}400}{15}$$

$$y_2 = 360$$

Step 3: Set up a proportion using the given information and solve.

Answer: The bus driver will earn \$360 after 40 hours of driving.

Problem Set

1. It costs a flat fee of \$75 plus an additional \$15 per hour to rent a Jet Ski at the beach. If x represents the number of hours the Jet Ski is rented, which linear equation represents the data?

$y = 60x$ $\quad$ $y = 90x$

$y = 15x + 75$ $\quad$ $y = 75x + 15$

2. The cost to rent a cabin in the mountains is an application fee of \$45 plus an additional \$85 per day. If x represents the number of days the cabin is rented, which linear equation represents the data?

$y = 130x$ $\quad$ $y = 40x$

$y = 85x + 45$ $\quad$ $y = 45x + 85$

3. The cost to print engraved invitations is \$36 for the setup fee plus an additional \$1.50 per engraved invitation. If x represents the number of engraved invitations that need to be printed, which linear equation represents the data?

$y = 36x + 1.5$ $\quad$ $y = 1.5x - 36$

$y = 1.5x + 36$ $\quad$ $y = 37.5x$

4. The cost to rent a bus for the class field trip is an initial charge of \$70 plus an additional charge of \$5 per passenger. If x represents the number of passengers, which linear equation represents the data?

$y = 75x$ $\quad$ $y = 70x + 5$

$y = 5x - 70$ $\quad$ $y = 5x + 70$

5. The following table shows the cost of phone calls made from a cellular phone. Which equation represents the relationship between the number of minutes, x, and the amount paid for the call, y?

# of minutes	100	200	300	400
Amount Paid ($)	19	38	57	76

$y = 0.19x$ $y = 0.38x - 19$

$y = 19x$ $y = 0.38x$

6. The following table shows the cost of mileage driven in a rental car. Which equation represents the relationship between the number of miles, x, and the amount paid for the mileage, y?

# of miles	100	200	300	400
Amount Paid ($)	24	48	72	96

$y = 0.48x - 24$ $y = 24x$

$y = 0.24x$ $y = 0.48x$

7. The given table shows the price for bowling at the neighborhood bowling alley. Let x represent the time in hours, and let y represent the total cost of the bowling round. If the total cost includes a rental fee for one pair of shoes, find the total cost of the shoe rental and write an equation that models the situation.

# of hours	1	2	3	4
Total Cost ($)	17	22	27	32

$y = 17x$ $y = 12x + 5$

$y = 17x - 5$ $y = 5x + 12$

8. The given table shows the price for printing pamphlets. Let x represent the number of pamphlets, and let y represent the total cost for printing the pamphlets. If the total cost includes a setup fee, find the cost of the setup fee and write an equation that models the situation.

# of pamphlets	100	200	300	400
Total Cost ($)	205	380	555	730

$y = 30x + 1.75$ $y = 1.75x + 30$

$y = 1.9x + 15$ $y = 3x - 95$

9. Suppose y varies directly with x, and $y = 65$ when $x = -5$. Find the constant of variation.

10. Suppose y varies directly with x, and $y = -3$ when $x = 42$. Find the constant of variation.

11. Suppose y varies directly with x, and $y = 75$ when $x = 5$. Find the constant of variation.

12. Suppose y varies directly with x, and $y = -16$ when $x = -4$. Find the constant of variation.

13. Choose the table that gives values of y that vary directly with the values of x.

x	1	2	3	4
y	–8	–16	–24	–32

x	1	2	3	4
y	–3	–7	–11	–15

x	1	2	3	4
y	–2	1	4	7

x	1	2	3	4
y	–5	–11	–17	–23

14. Choose the table that gives values of y that vary directly with the values of x.

x	8	12	16	20
y	–2	–3	–4	–5

x	8	12	16	20
y	5	7	9	11

x	8	12	16	20
y	–5	–7	–9	–11

x	8	12	16	20
y	4	5	6	7

15. Choose the table that gives values of y that vary directly with the values of x.

x	–12	–9	–6	–3
y	5	4	3	2

x	–12	–9	–6	–3
y	7	6	5	4

x	–12	–9	–6	–3
y	–8	–6	–4	–2

x	–12	–9	–6	–3
y	0	–3	–6	–12

16. Choose the table that gives values of y that vary directly with the values of x.

x	–12	–9	–6	–3
y	–21	–15	–9	–3

x	–12	–9	–6	–3
y	3	2	1	0

x	–12	–9	–6	–3
y	–72	–54	–36	–18

x	–12	–9	–6	–3
y	0	–3	–6	–9

17. Peggy rode her bike at an average speed for 65 miles. Her trip lasted 5 hours. The number of miles, m, varies directly with the number of hours she rides, h. How many miles could Peggy ride in 8 hours?

18. Adam drove his car at an average speed for 165 miles. His trip lasted 3 hours. The number of miles, m, varies directly with the number of hours he drove, h. How many hours would it take Adam to drive 385 miles?

19. A printer can print 700 copies in 2 hours. The number of copies, c, varies directly with the number of hours, h. How many hours would it take the printer to print 2,100 copies?

20. Margaret can drive 308 miles on 11 gallons of gasoline. The number of miles, m, varies directly with the number of gallons of gasoline, g. How many miles could she drive on 17 gallons of gasoline?

MPA-150: Identifying and Graphing Linear and Nonlinear Functions

In this lesson, you will learn how to determine whether a function is linear or nonlinear. You will also learn how to graph nonlinear functions. First, recall the definition of a linear function. A linear function is an equation written in the form $y = mx + b$ and whose graph is a straight line. A linear function has a constant rate of change.

Linear Function	A linear function is an equation written in the form $y = mx + b$ and whose graph is a straight line. A linear function has a constant rate of change.
Constant Rate of Change	As the x-values change by the same number each time, the y-values change by the same number each time.

The equation $y = x - 1$ is a linear function because it is written in the form $y = mx + b$. When we graph the function, we can see that the graph is a straight line. In the table below, as the x-values increase by 1 each time, the y-values increase by 1 each time. This shows a constant rate of change.

x	y
0	–1
1	0
2	1

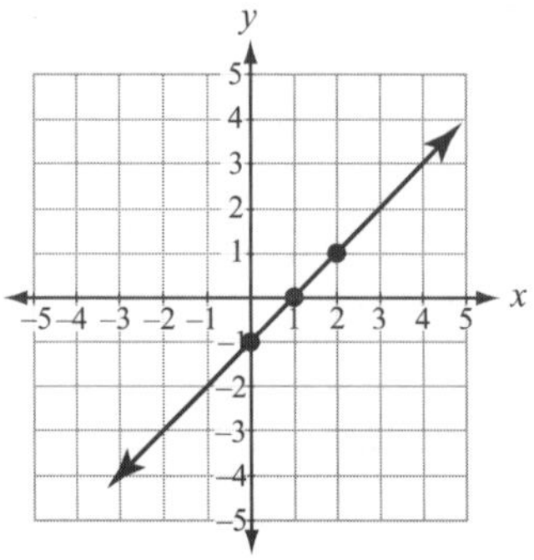

A nonlinear function cannot be written in the form $y = mx + b$ and the graph is not a straight line. A nonlinear function does not have a constant rate of change.

Nonlinear Function	A nonlinear function is an equation that cannot be written in the form $y = mx + b$ and whose graph is a not a straight line. A nonlinear function has a non-constant rate of change.
Non-constant Rate of Change	As the x-values change by the same number each time, the y-values change by a different number each time

There are two types of nonlinear functions that we will cover in this lesson: quadratic functions and cubic functions. A quadratic function is written in the form $y = ax^2$, where $x \neq 0$. The graph of a quadratic function is called a parabola. A cubic function is written in the form $y = ax^3$, where $a \neq 0$. The graph of a cubic function is an S-shaped curve.

Quadratic Function	An equation written in the form $y = ax^2$, where $a \neq 0$, and whose graph is called a parabola	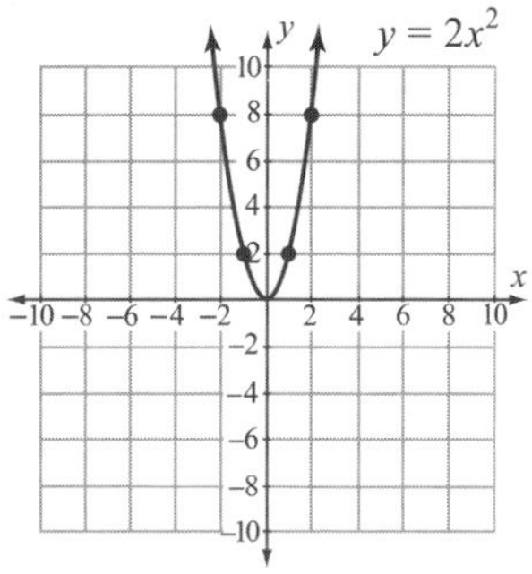
Cubic Function	An equation written in the form $y = ax^3$, where $a \neq 0$, and whose graph is an S-shaped curve .	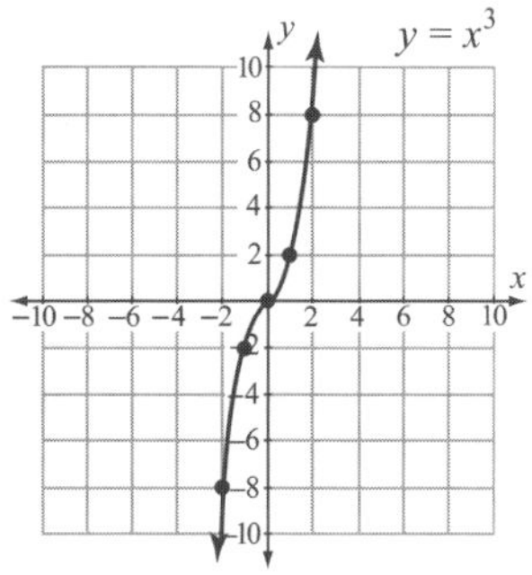

Example 1 Determine whether the following equation represents a linear or nonlinear function:

$$y = x^3 + 2$$

$y = x^3 + 2$

A cubic function is nonlinear.

Step 1: The equation is a cubic function because x is raised to the power of 3. It cannot be written in the form $y = mx + b$.

Answer: The equation $y = x^3 + 2$ represents a nonlinear function.

Example 2 Determine whether the following table represents a linear or nonlinear function:

x	y
3	6
7	5
11	3
15	0

Change in x	x	y	Change in y
	3	6	
+4	7	5	−1
+4	11	3	−2
+4	15	0	−3

Step 1: Calculate the change in the x-values and the change in the y-values.

As the x-values increase by four each time, the y-values decrease by a different value each time.	**Step 2:** In a non-constant rate of change, as the x-values change by the same number each time, the y-values change by a different number each time.
A nonlinear function has a non-constant rate of change.	**Step 3:** Recall the definition of a nonlinear function.

Answer: The table represents a nonlinear function.

When graphing quadratic and cubic functions, we can plot the points using a table of values. However, there are some very important characteristics of these nonlinear functions that will help in graphing. The value of a in the function $y = ax^2$ determines the width of the parabola and whether the graph opens upward or downward.

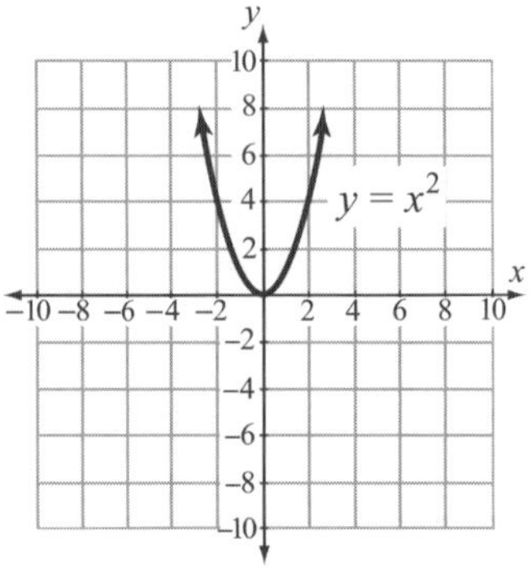

When $a > 0$, the parabola opens upward.

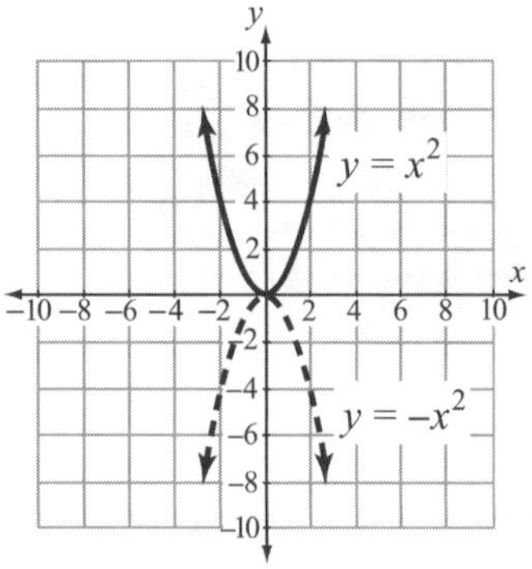

When $a < 0$, the parabola opens downward.

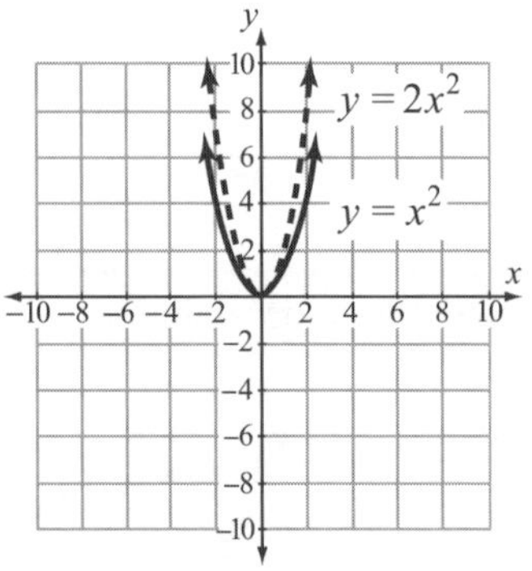

When $|a| > 1$, the parabola is narrower than the graph of the parent function, $y = x^2$.

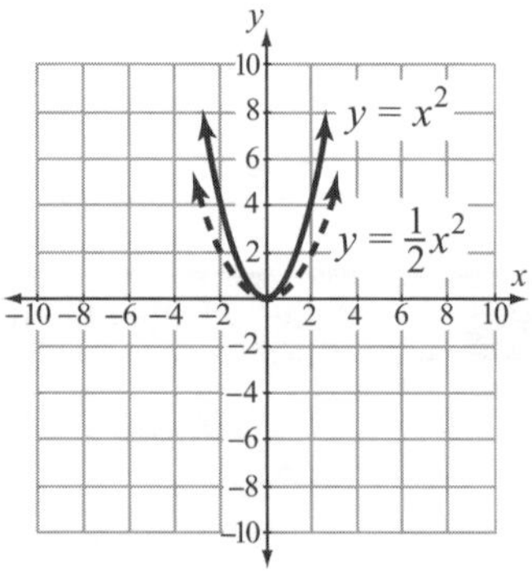

When $|a| < 1$, the parabola is wider than the graph of the parent function, $y = x^2$.

The value of a in the function $y = ax^3$ determines the width of the S-shaped curve and the quadrants in which the points on the graph lie.

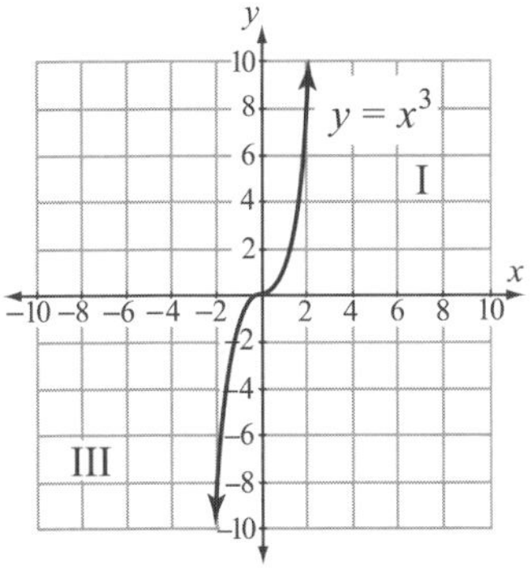

When a > 0, the graph is an S-shaped curve that lies in Quadrants I and III.

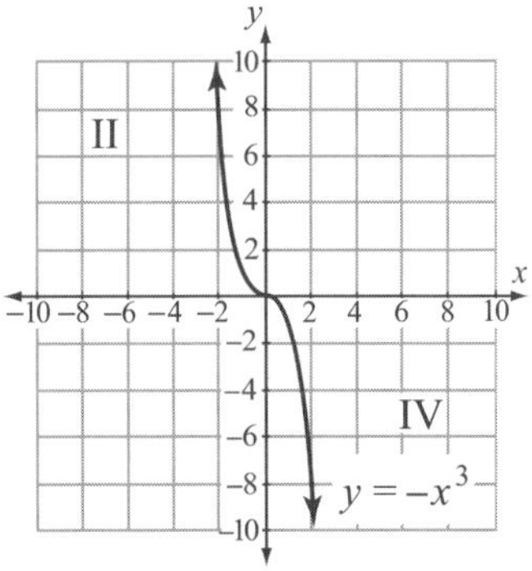

When a < 0, the graph is an S-shaped curve that lies in Quadrants II and IV.

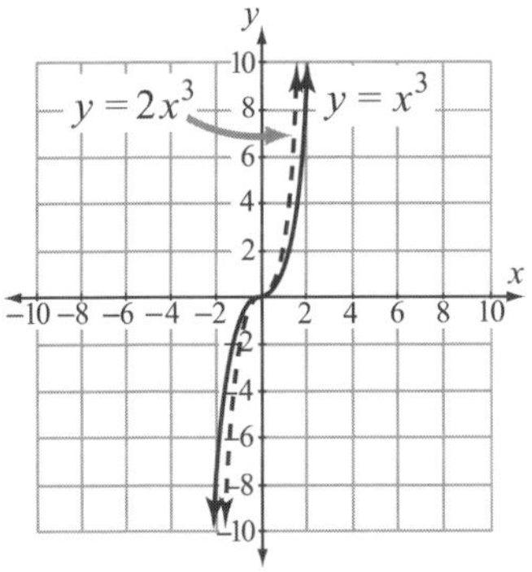

When $|a| > 1$, the graph is narrower than the graph of the parent function, $y = x^3$.

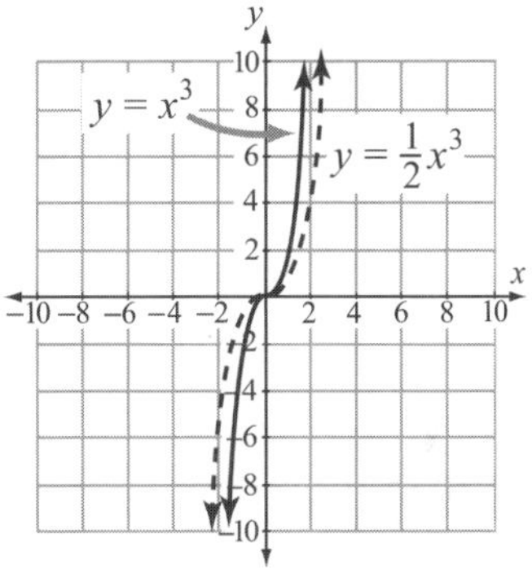

When $|a| < 1$, the graph is wider than the graph of the parent function, $y = x^3$.

Example 3 Describe the graph of the equation $y = -3x^3$.

The graph of $y = -3x^3$ is an S-shaped curve.	**Step 1:** The equation is in the form $y = ax^3$, which is a cubic function. The graph of a cubic function is an S-shaped curve.
Since $a = -3$, the points of the graph lie in quadrants II and IV.	**Step 2:** When $a < 0$, the graph is contained in quadrants II and IV.

Since the absolute value of a is greater than one, the graph of $y = -3x^3$ is narrower than the graph of $y = x^3$.

Step 3: When $|a| > 1$, the graph is narrower than the graph of the parent function, $y = x^3$.

Answer: The graph of $y = -3x^3$ is an S-shaped curve that lies in quadrants II and IV, and is narrower than the graph of $y = x^3$.

Example 4 Graph the equation $y = 4x^2$.

Step 1: Determine the characteristics of the graph.

Since the equation is in the form $y = ax^2$ and a is positive, the graph is a parabola that opens upward.

Step 2: Make a table of values.

x	$y = 4x^2$	(x, y)
–2	$4(-2)^2 = 16$	(–2, 16)
–1	$4(-1)^2 = 4$	(–1, 4)
0	$4(0)^2 = 0$	(0, 0)
1	$4(1)^2 = 4$	(1, 4)
2	$4(2)^2 = 16$	(2, 16)

Step 3: Plot the points on the graph and connect the points.

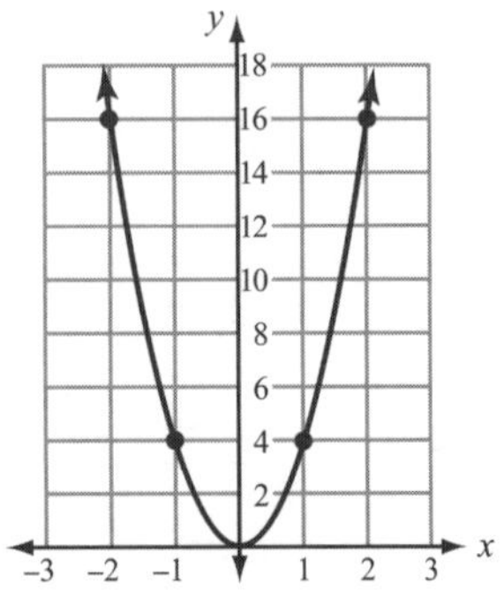

Answer:

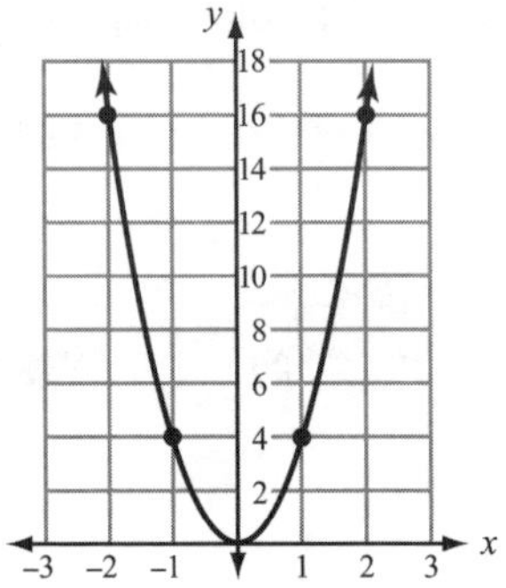

Example 5 The height of a tree is given by $y = \frac{1}{2}x^2$ where y is the height in feet and x is the number of years since the tree was planted. Graph the function and determine the height of the tree after four years.

Step 1: Determine the characteristics of the graph.

Since the equation is in the form $y = ax^2$, the graph is a parabola that opens upward.

Step 2: Make a table of values. Since the number of years cannot be negative, we can use only positive values for x.

x	$y = \frac{1}{2}x^2$	(x, y)
0	$\frac{1}{2}(0)^2 = 0$	(0, 0)
1	$\frac{1}{2}(1)^2 = 0.5$	(1, 0.5)
2	$\frac{1}{2}(2)^2 = 2$	(2, 2)
3	$\frac{1}{2}(3)^2 = 4.5$	(3, 4.5)
4	$\frac{1}{2}(4)^2 = 8$	(4, 8)
5	$\frac{1}{2}(5)^2 = 12.5$	(5, 12.5)

Step 3: Plot the points on the graph and connect the points.

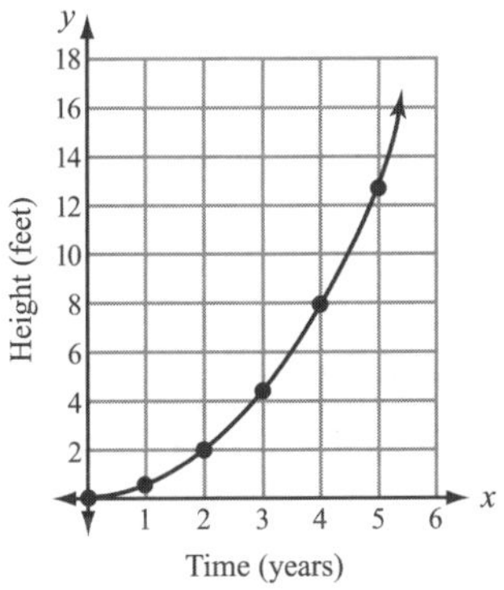

Step 4: Look at the graph to determine the height of the tree after 4 years.

According to the graph, when $x = 4$ years, $y = 8$ feet. Therefore, the tree will be 8 feet tall after 4 years.

Answer:

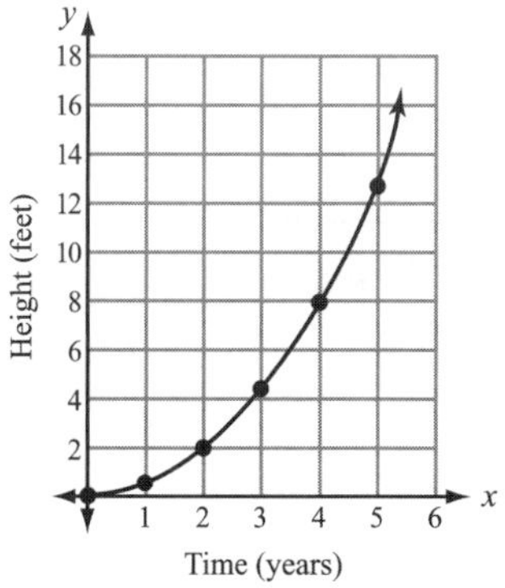

The tree is 8 feet tall after 4 years.

Problem Set

1. Choose the equation that represents a nonlinear function:

$y = 7x$ $\quad y = 7x^2$

$y = 2x - 3$ $\quad y = x + 4$

2. Choose the equation that represents a nonlinear function:

$y = -4x^3$ $\quad y = -2x - 3$

$y = -4x$ $\quad y = 3x + 1$

3. Choose the equation that represents a nonlinear function:

$y = \frac{1}{8}x^2$ $\quad y = \frac{1}{8}x$

$y = x + 5$ $\quad y = -3x + 2$

4. Choose the equation that represents a nonlinear function:

$y = \frac{1}{7}x^3$ $\quad y = 4x - 5$

$y = 2x + 1$ $\quad y = \frac{1}{7}x$

5. Choose the table that represents a nonlinear function:

x	y
1	3
2	6
3	9
4	12

x	y
1	–2
2	–5
3	–8
4	–11

x	y
1	3
2	12
3	27
4	48

x	y
1	–2
2	0
3	4
4	8

6. Choose the table that represents a nonlinear function:

x	y
–1	6
0	0
1	–6
2	–12

x	y
–1	–6
0	–4
1	–2
2	0

x	y
–1	2
0	–1
1	–4
2	–7

x	y
–1	6
0	0
1	–6
2	–48

7. Choose the table that represents a linear function:

x	y
1	–1
2	–8
3	–27
4	–48

x	y
1	3
2	12
3	27
4	48

x	y
1	–2
2	–5
3	–8
4	–11

x	y
1	–2
2	–8
3	–18
4	–32

8. Choose the table that represents a linear function:

x	y
2	–1
4	–8
6	–27
8	–64

x	y
2	2
4	3
6	18
8	32

x	y
2	–1
4	–4
6	–9
8	–16

x	y
2	0
4	1
6	2
8	3

9. Determine which description matches the equation $y = -8x^2$.

A parabola that opens downward and is narrower than $y = x^2$
A parabola that opens downward and is wider than $y = x^2$
A parabola that opens upward and is narrower than $y = x^2$
A parabola that opens upward and is wider than $y = x^2$

10. Determine which description matches the equation $y = 9x^2$.

A parabola that opens upward and is narrower than $y = x^2$
A parabola that opens downward and is wider than $y = x^2$
A parabola that opens upward and is wider than $y = x^2$
A parabola that opens downward and is narrower than $y = x^2$

11. Determine which description matches the equation $y = -6x^3$.

An S-shaped curve that is narrower than $y = x^3$ with points in quadrants II and IV
An S-shaped curve that is wider than $y = x^3$ with points in quadrants I and III
An S-shaped curve that is wider than $y = x^3$ with points in quadrants II and IV
An S-shaped curve that is narrower than $y = x^3$ with points in quadrants I and III

12. Determine which description matches the equation $y = 0.25x^3$.

An S-shaped curve that is wider than $y = x^3$ with points in quadrants I and III
An S-shaped curve that is narrower than $y = x^3$ with points in quadrants II and IV
An S-shaped curve that is narrower than $y = x^3$ with points in quadrants I and III
An S-shaped curve that is wider than $y = x^3$ with points in quadrants II and IV

13. Graph the equation $y = -3x^2$.

14. Graph the equation $y = -0.25x^2$.

15. Graph the equation $y = -\frac{1}{8}x^3$.

16. Graph the equation $y = 4x^3$.

17. Rhoda dropped a ball from a cliff 1,600 feet above the ground. The height (h) of the ball is given by the equation $h = -16t^2 + 1{,}600$, where t represents time in seconds. Draw the graph of the function and determine when the ball struck the ground.

18. Steve is a skydiver. The distance he falls when skydiving is expressed by the equation $d = 16t^2$, where d is the distance in feet and t is the time in seconds. Steve opens his parachute at 30 seconds. Draw the graph of the function, and determine how far he will have fallen when the parachute opens.

19. Jessica dropped a ball from a tower 78.4 meters above the ground. The height (h) of the ball is given by the equation $h = -4.9t^2 + 78.4$, where t represents time in seconds. Draw the graph of the function and determine when the ball struck the ground.

20. Alice is a skydiver. The distance she falls when skydiving is expressed by the equation $d = 4.9t^2$, where d is the distance in meters and t is the time in seconds. Alice opens her parachute at 15 seconds. Draw the graph of the function, and determine how far she will have fallen when the parachute opens.

MPA-155: Comparing and Converting Rates

In this lesson, you will learn how to find unit rates, and compare and convert these rates. You must first recall that a ratio is a comparison of two quantities by division. For example, in the statement "3 out of 12 people live in a rural area," the ratio is "3 out of 12" or $\frac{3}{12}$ written as a fraction. This ratio can be reduced to $\frac{1}{4}$.

Ratio	a comparison of two quantities by division

A rate is a ratio with different units of measure. For example, in the statement "A car travels 586 miles in 8 hours," the rate is $\frac{586 \text{ miles}}{8 \text{ hours}}$. A unit rate is a rate with a denominator of one. In this example, the unit rate would be found by dividing the numerator and the denominator by 8.

$$\frac{586 \text{ miles}}{8 \text{ hours}} \xrightarrow{\div 8} \frac{73.25 \text{ miles}}{1 \text{ hour}}$$

$$\frac{586 \text{ miles}}{8 \text{ hours}} = \frac{73.25 \text{ miles}}{1 \text{ hour}}$$

The unit rate is 73.25 miles per hour. Therefore, we can state that the car's average speed is 73.25 miles per hour.

Rate	a comparison of two quantities with different units of measure
Unit Rate	a rate with a denominator of one

Example 1 Maya's typing test determined that she can type 240 words in 3 minutes. How many words can she type per minute?

$\frac{240 \text{ words}}{3 \text{ minutes}}$

Step 1: Find the rate.

$\frac{240 \text{ words}}{3 \text{ minutes}} = \frac{80 \text{ words}}{1 \text{ minute}}$

Step 2: To express the rate as a unit rate, divide the numerator and denominator by 3.

Answer: Maya can type 80 words per minute.

Example 2

Read the advertisement and determine which bag is the best buy.

$\frac{\$1.49}{6.4 \text{ oz}} = \$0.23/\text{oz}$

$\frac{\$2.39}{12 \text{ oz}} = \$0.20/\text{oz}$

$\frac{\$2.99}{13.75 \text{ oz}} = \$0.22/\text{oz}$

Step 1: Find the unit rate for each bag of chips.

The 12 oz bag of chips has the lowest unit rate at \$0.20/oz.	**Step 2:** Compare the unit rates and find the bag with the lowest unit rate.
	Answer: The 12 oz bag of chips is the best buy.

To convert rates, we must use the concept of dimensional analysis, also called unit analysis. Dimensional analysis allows us to use conversion factors to convert a rate with one unit of measure to another rate with a different unit of measure without changing the value. For example, we may want to convert "15,840 feet per minute" to "miles per minute". To do this, we will need to know how many feet are in a mile. Since 5,280 feet equals one mile, the conversion factor would be $\frac{15,840 \text{ ft}}{1 \text{ min.}}$ or $\frac{1 \text{ mi}}{5,280 \text{ ft}}$. The conversion factor is multiplied by the rate you are converting.

$$\frac{15,840 \cancel{\text{ft}}}{1 \text{ min.}} \times \frac{1 \text{ mi}}{5,280 \cancel{\text{ft}}} = \frac{15,840 \text{ ft}}{5,280 \text{ min.}} = \frac{3 \text{ mi}}{1 \text{ min.}}$$

Therefore, 15,840 feet per minute = 3 miles per minute.

Dimensional Analysis (Unit Analysis)	the method used to convert rates

Example 3 Convert 6 miles per minute to miles per hour.

$\frac{6 \text{ mi}}{1 \text{ min.}} = \frac{? \text{ mi}}{1 \text{ hr.}}$	**Step 1:** We are converting 6 miles per minute to miles per hour.
Since 60 minutes = 1 hour, the conversion factor is $\frac{60 \text{ min}}{1 \text{ hr.}}$.	**Step 2:** Find the conversion factor.
$\frac{6 \text{ mi}}{1 \cancel{\text{min.}}} \times \frac{60 \cancel{\text{min}}}{1 \text{ hr.}} = \frac{360 \text{ mi}}{1 \text{ hr.}}$	**Step 3:** Multiply $\frac{6 \text{ mi}}{1 \text{ min.}}$ by the conversion factor and simplify.
	Answer: 6 mi/min. = 360 mi/hr.

Example 4 Convert the following rate: 32 yd/min = ________ in./sec.

$\frac{32 \text{ yd}}{1 \text{ min.}} = \frac{? \text{ in.}}{1 \text{ sec.}}$	**Step 1:** We are converting 32 yards per minute to inches per second.
Since 36 inches = 1 yard and 60 seconds = 1 minute, the conversion factors are $\frac{36 \text{ in.}}{1 \text{ yd}}$ and $\frac{1 \text{min.}}{60 \text{ sec.}}$.	**Step 2:** Find the conversion factors.
$\frac{32 \cancel{\text{yd}}}{1 \cancel{\text{min.}}} \times \frac{36 \text{ in.}}{1 \cancel{\text{yd}}} \times \frac{1 \cancel{\text{min.}}}{60 \text{ sec.}} = \frac{1,152 \text{ in.}}{60 \text{ sec.}}$	**Step 3:** Multiply $\frac{32 \text{ yd}}{1 \text{ min.}}$ by both conversion factors and simplify.

$$\frac{1{,}152 \text{ in.}}{60 \text{ sec.}} = \frac{19.2 \text{ in.}}{1 \text{ sec.}}$$

Step 4: Divide the numerator and denominator by 60 to get a unit rate in inches per second.

Answer: 32 yd/min. = 19.2 in./sec.

Example 5 Mrs. Payton's art class is making a tile mural. The tile costs $1.97 per square foot. How much is the tile per square yard?

$$\frac{\$1.97}{1 \text{ sq ft}} = \frac{\$\,?}{1 \text{ sq yd}}$$

Step 1: We are converting $1.97 per square foot to dollars per square yard.

Since 3 feet = 1 yard,
9 square feet = 1 square yard.

The conversion factor is $\frac{9 \text{ sq ft}}{1 \text{ sq yd}}$.

Step 2: Find the conversion factor.

$$\frac{\$1.97}{1 \cancel{\text{sq ft}}} \times \frac{9 \cancel{\text{sq ft}}}{1 \text{ sq yd}} = \frac{\$17.73}{1 \text{ sq yd}}$$

Step 3: Multiply $\frac{\$1.97}{1 \text{ sq ft}}$ by the conversion factor and simplify.

Answer: The tile costs $17.73 per square yard.

Example 6 A local pizza chain in Texas advertises that a large pizza costs 60 pesos. If the currency exchange rate is 12 pesos per US dollar, how much is the pizza in US dollars?

60 pesos = ? US dollars

Step 1: We are converting 60 pesos to US dollars.

The currency exchange rate is 12 pesos per US dollar, so the conversion factor is

$\frac{12 \text{ pesos}}{1 \text{ US dollar}}$ or $\frac{1 \text{ US dollar}}{12 \text{ pesos}}$.

Step 2: Find the conversion factor.

$$60 \cancel{\text{pesos}} \times \frac{1 \text{ US dollar}}{12 \cancel{\text{pesos}}} = \frac{\$60}{12} = \$5$$

Step 3: Multiply 60 pesos by the conversion factor and simplify.

Answer: The pizza costs $5.

Problem Set

1. If Keith drove 585 miles in 9 hours, what was his average rate of speed?
2. If a model rocket can travel 4,500 feet in 9 seconds, what is the average rate of speed for the model rocket?
3. If Rhoda drove 550 miles using 25 gallons of gasoline, what was her average miles per gallon?
4. If Boyce's heart beats 432 times in 6 minutes, what is his average heartbeat per minute?

5. In a local grocery store, your favorite cereal comes in four different sizes. Which box is the best buy? Round to the nearest cent.

Box	Size	Price
A	24 oz	$2.75
B	36 oz	$3.99
C	48 oz	$5.75
D	60 oz	$6.25

6. In a local grocery store, your favorite mixed nuts come in four different sizes. Which package is the best buy?
Round to the nearest cent.

Package	Size	Price
A	8 oz	$1.85
B	12 oz	$2.99
C	16 oz	$3.85
D	20 oz	$4.75

7. In a local grocery store, your favorite trail mix comes in four different sizes. Which package is the best buy?
Round to the nearest cent.

Package	Size	Price
A	8 oz	$1.55
B	12 oz	$2.25
C	16 oz	$2.75
D	20 oz	$3.85

8. In a local grocery store, your favorite fruit drink comes in four different sizes. Which bottle is the best buy?
Round to the nearest cent.

Bottle	Size	Price
A	6 fl oz	$1.39
B	10 fl oz	$2.20
C	16 fl oz	$3.00
D	36 fl oz	$7.25

9. Convert the following rate:

140 ft/sec. to ____ ft/min.

10. Convert the following rate:

0.75 mi/min. to ____ mi/hr.

11. Convert the following rate:

60 gal/min. to ____ gal/hr.

12. Convert the following rate:

$0.65/qt to $____ /gal

13. Jeremiah is carpeting a floor. The carpet costs $1.75 per square foot. How much is this per square yard?

14. Tiffany is installing a hardwood floor. The hardwood floor costs $5 per square foot. How much is this per square yard?

15. Lucinda is installing a tile floor. The tile costs $24.75 per square yard. How much is this per square foot?

16. Rickey is installing a granite countertop. The tile costs $359.55 per square yard. How much is this per square foot?

17. Convert the following rate:

48 gal/hr. to ____ qt/min.

18. Convert the following rate:

480 ft/sec. to ____ yd/min.

19. Bryan is traveling to Japan for the summer and wants to exchange $1,800 to yen. If the currency exchange rate is 120.36 yen per US dollar on the day he exchanges the money, how many yen will he receive?
Round to the nearest whole dollar.

20. Dorren is traveling to Mexico for the spring and wants to exchange $475 to pesos. If the currency exchange rate is 10.979 pesos per US dollar on the day she exchanges the money, how many pesos will she receive?
Round to the nearest whole number.

MPA-160: Plotting Polygons and Finding the Area

Recall the definition of area:

Area	The number of square units that are needed to cover the figure

Recall the following area formulas from previous lessons:

Rectangle	$A = \ell \times w$	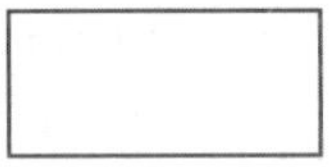
Parallelogram	$A = b \times h$	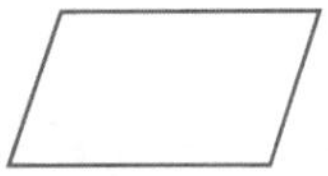
Triangle	$A = \frac{1}{2} \times b \times h$	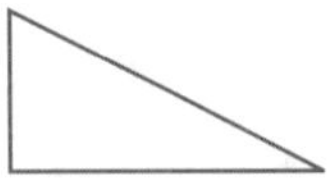
Trapezoid	$A = \frac{1}{2}(b_1 + b_2)h$	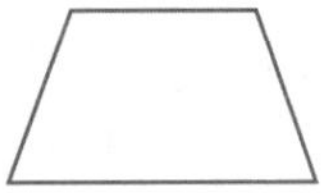

Example 1 Find the area of rectangle *EFGH* if the vertices are *E* (0, 0), *F* (5, 0), *G* (5, 3), and *H* (0, 3).

Key: 1 square = 1 unit

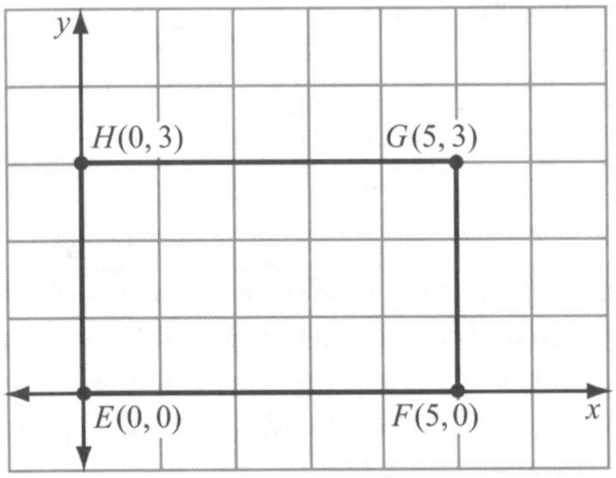

Step 1: Begin by plotting the vertices on a coordinate plane, then connect the vertices to draw the rectangle.

$A = \ell \times w$

Step 2: Recall that the area of a rectangle is $A = \ell \times w$.

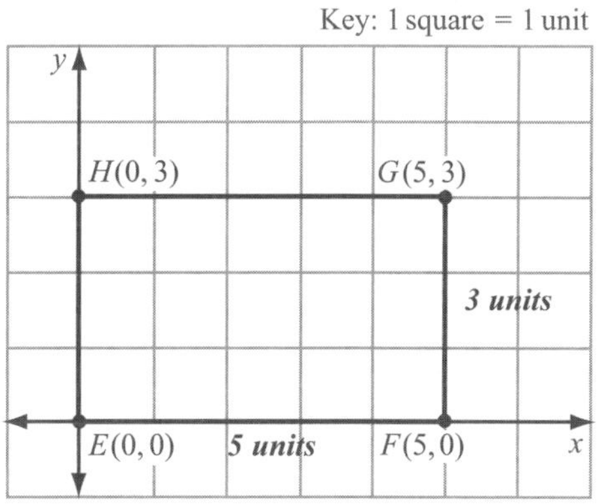

Step 3: The length of the rectangle is 5 units, and the width of the rectangle is 3 units.

$A = 5 \times 3$

$A = 15$

Step 4: Substitute these values into the formula and simplify.

Answer: The area of the rectangle is 15 square units.

Example 2 Find the area of triangle *ABC* if the vertices are *A* (2, 1), *B* (2, 4), and *C* (6, 1).

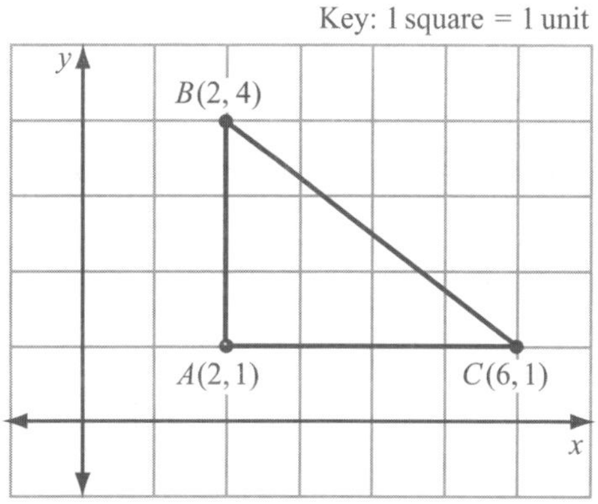

Step 1: Begin by plotting the vertices on a coordinate plane, then connect the vertices to draw the triangle.

$A = \frac{1}{2} \times b \times h$

Step 2: Recall that the area of a triangle is $A = \frac{1}{2} \times b \times h$.

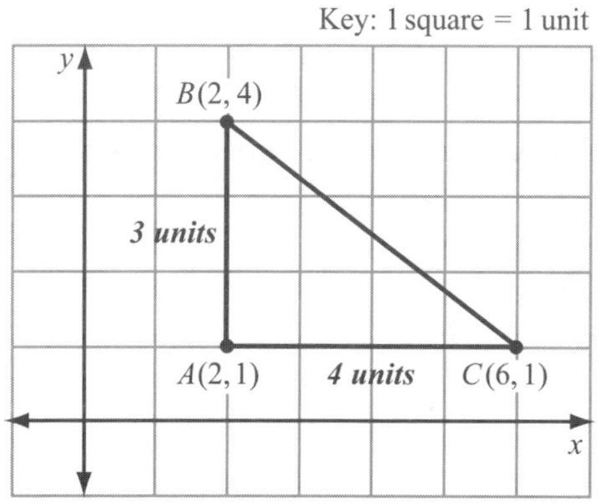

Step 3: The base of the triangle is 4 units, and the height of the triangle is 3 units.

$A = \frac{1}{2} \times 4 \times 3$

$A = \frac{1}{2} \times 12$

$A = 6$

Step 4: Substitute these values into the formula and simplify.

Answer: The area of the triangle is 6 square units.

Example 3 Find the area of parallelogram *JKLM* if the vertices are $J\,(-3,-4)$, $K\,(-2,-1)$, $L\,(2,-1)$, and $M\,(1,-4)$.

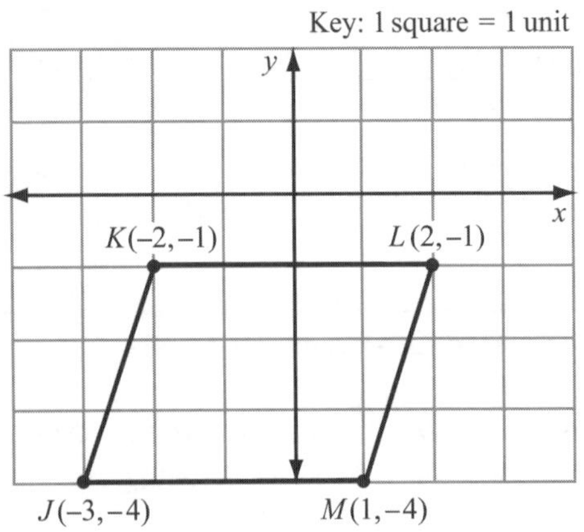

Step 1: Begin by plotting the vertices on a coordinate plane, then connect the vertices to draw the parallelogram.

$A = b \times h$

Step 2: Recall that the area of a parallelogram is $A = b \times h$.

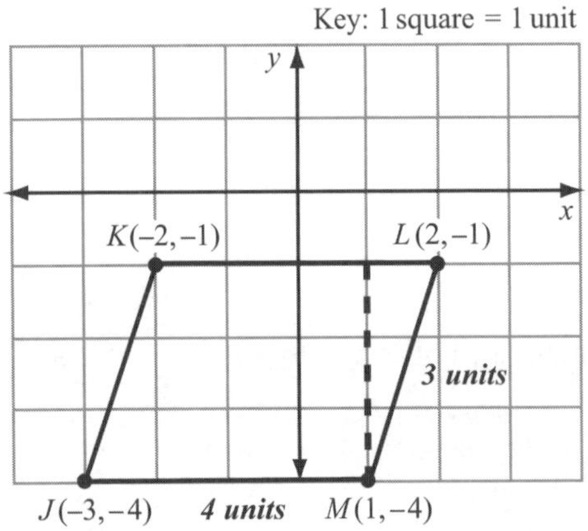

Step 3: The base of the parallelogram is 4 units, and the height of the parallelogram is 3 units.

$A = 4 \times 3$

$A = 12$

Step 4: Substitute these values into the formula and simplify.

Answer: The area of the parallelogram is 12 square units.

For irregular shapes, it may be necessary to subdivide the figure into two or more smaller sections. Find the area of each of these sections and then add them together.

Example 4 Estimate the area of the figure shown to the nearest square unit.

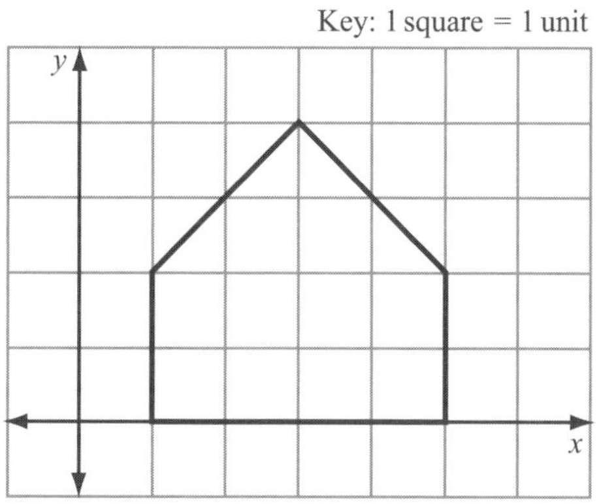

Step 1: Divide the figure into two sections – a triangle and a rectangle.

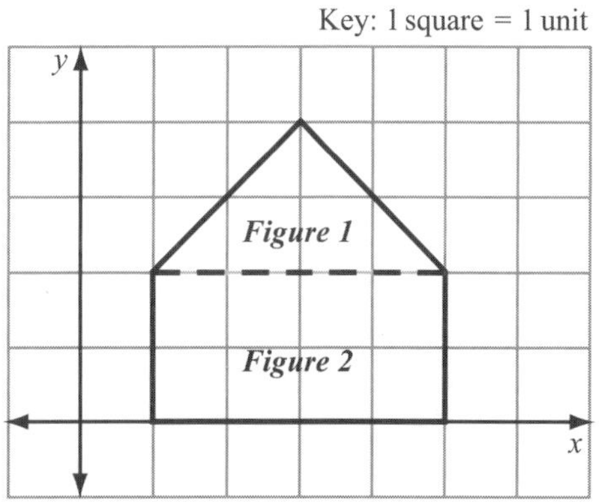

Step 2: Find the area of Figure 1, the triangle.

$A = \frac{1}{2} \times b \times h$

$A = \frac{1}{2} \times 4 \times 2$

$A = \frac{1}{2} \times 8$

$A = 4 \text{ units}^2$

Step 3: Find the area of Figure 2, the rectangle.

$A = \ell \times w$

$A = 4 \times 2$

$A = 8 \text{ units}^2$

Step 4: Add the areas of the two figures together to find the area of the irregular figure.

$4 \text{ units}^2 + 8 \text{ units}^2 = 12 \text{ units}^2$

Answer: The area of the figure is 12 units^2.

Example 5 Find the area of the irregular figure shown.

Key: 1 square = 1 unit

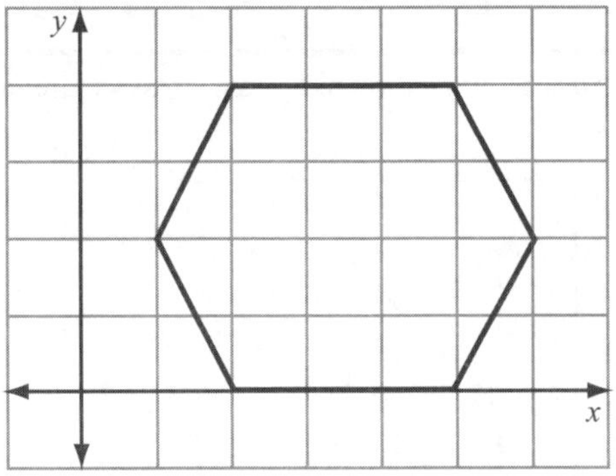

Key: 1 square = 1 unit

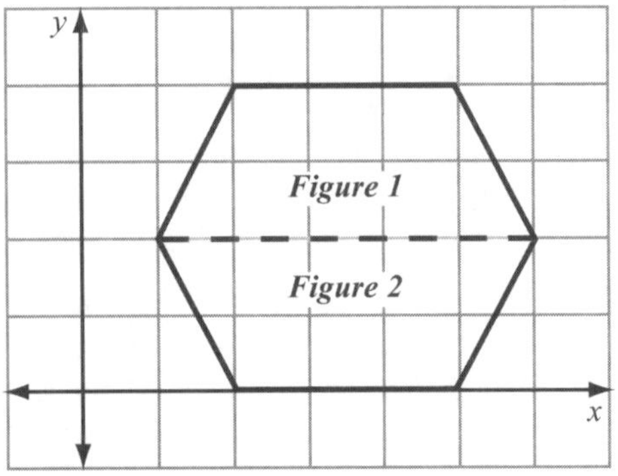

Step 1: Divide the figure into two sections – two trapezoids.

Step 2: Find the area of the trapezoid in Figure 1.

$$A = \frac{1}{2}(b_1 + b_2)h$$

$$A = \frac{1}{2}(5+3)2$$

$$A = \frac{1}{2}(8)2$$

$$A = \frac{1}{2}(16)$$

$$A = 8 \text{ units}^2$$

Step 3: Find the area of the trapezoid in Figure 2.

$$A = \frac{1}{2}(b_1 + b_2)h$$

$$A = \frac{1}{2}(5+3)2$$

$$A = \frac{1}{2}(8)2$$

$$A = \frac{1}{2}(16)$$

$$A = 8 \text{ units}^2$$

$8\text{units}^2 + 8\text{units}^2 = 16\text{ units}^2$	**Step 4:** Add the areas of the two figures together to find the area of the irregular figure.

Answer: The area of the figure is 16 units^2.

Problem Set

1. Find the area of rectangle *ABCD* if the vertices are *A*(0, 0), *B*(5, 0), *C*(5, 4), and *D*(0, 4).

2. Find the area of rectangle *ABCD* if the vertices are *A*(1, 0), *B*(3, 0), *C*(3, 6), and *D*(1, 6).

3. Find the area of rectangle *ABCD* if the vertices are *A*(0, 1), *B*(5, 1), *C*(5, 4), and *D*(0, 4).

4. Find the area of rectangle *ABCD* if the vertices are *A*(2, 0), *B*(7, 0), *C*(7, 5), and *D*(2, 5).

5. Find the area of triangle *PRQ* if the vertices are *P*(1, 0), *R*(5, 0), and *Q*(5, 5).

6. Find the area of triangle *XYZ* if the vertices are *X*(1, 1), *Y*(1, 4), and *Z*(8, 4).

7. Find the area of triangle *EFG* if the vertices are *E*(1, 6), *F*(7, 6), and *G*(7, 2).

8. Find the area of triangle *EFG* if the vertices are *E*(0, 6), *F*(0, 2), and *G*(0, 0).

9. Find the area of triangle *QRS* if the vertices are *Q*(–1, 2), *R*(4, 4), and *S*(4, –1).

10. Find the area of parallelogram *HIJK* if the vertices are *H*(–2, 0), *I*(–2, 3), *J*(3, 0), and *K*(3,–3).

11. Find the area of trapezoid *MNOP* if the vertices are *M*(–2,–6), *N*(–2,–1), *O*(3,–2), and *P*(3,–5).

12. Find the area of parallelogram *ABCD* if the vertices are *A*(–6, 1), *B*(–6, 3), *C*(1, 3), and *D*(1, 1).

13. Estimate the area of the figure shown to the nearest square unit.

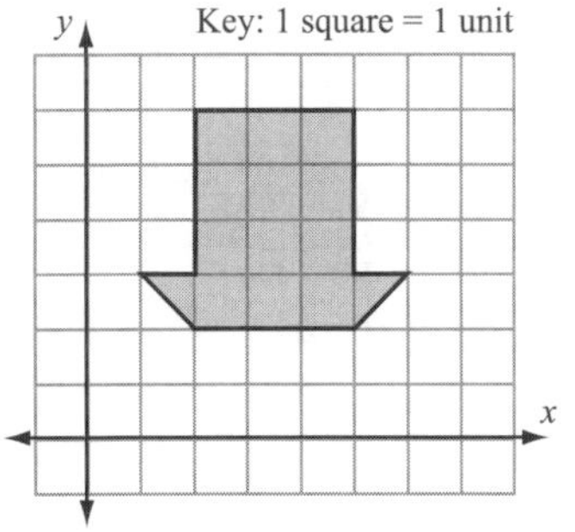

________units²

14. Estimate the area of the figure shown to the nearest square unit.

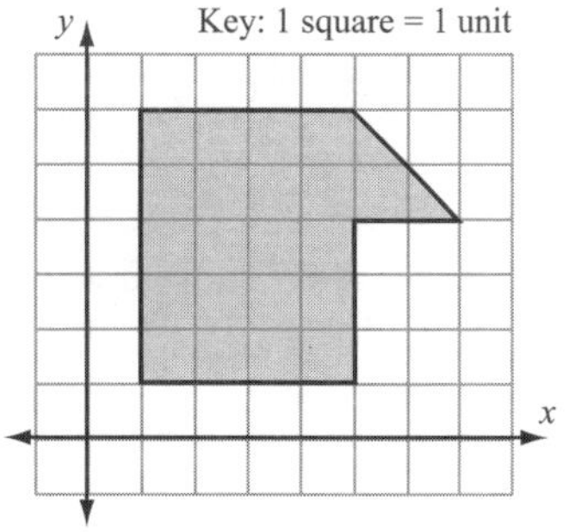

________units²

15. Estimate the area of the figure shown to the nearest square unit.

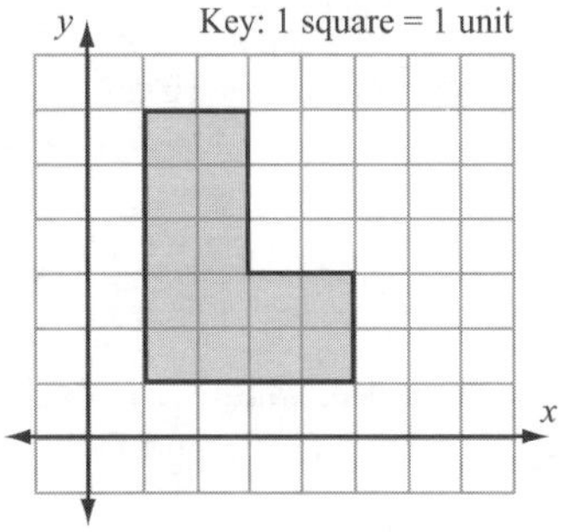

________units^2

16. Estimate the area of the figure shown to the nearest square unit.

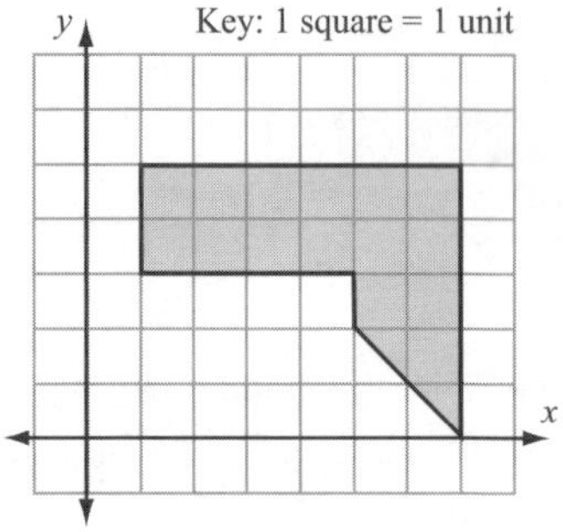

________units^2

17. Find the area of the irregular figure shown.

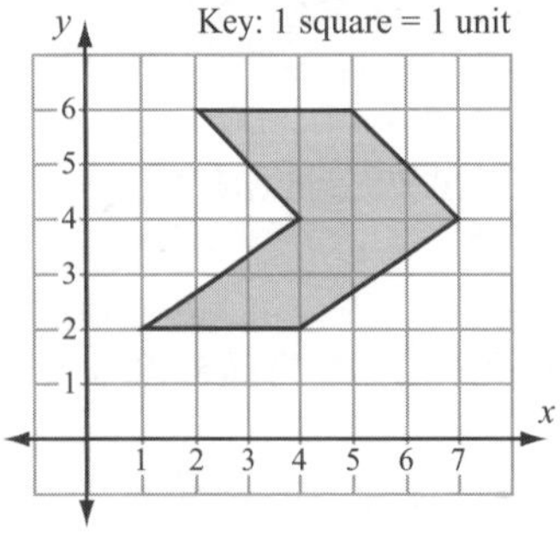

18. Find the area of the irregular figure shown.

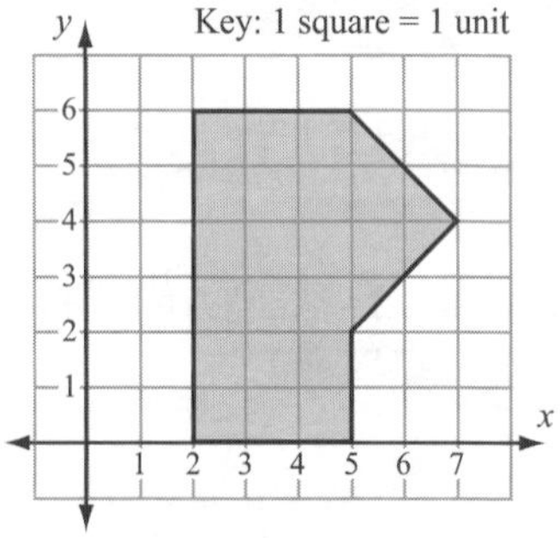

19. Find the area of the irregular figure shown.

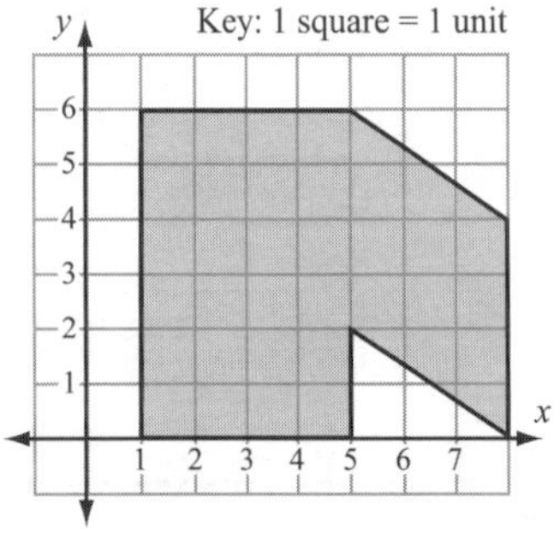

20. Find the area of the irregular figure shown.

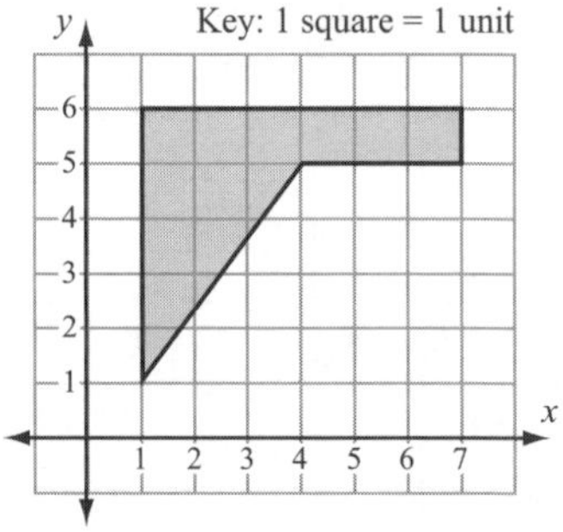

MPA-270: Generating Algebraic Expressions from Patterns of Models

Patterns of numbers, or sequences, are an important part of mathematics. Sequences can be written as a list of numbers. They can even be found in a variety of models and graphs. Since sequences follow patterns, algebraic expressions, or rules, can be written to represent the patterns.

Algebraic Expression	An expression that consists of at least one number and variable and contains one or more mathematical operations

If a rule, or algebraic expression, is given, it is quite simple to determine the sequence of numbers that follows that rule.

Example 1 Write a sequence of numbers that follows the rule $4n + 2$, where n represents the position of a term in the sequence.

n				
$4n + 2$				
Term				

Step 1: Create a table that will contain the step numbers, the expression, and the terms in the sequence.

n	1	2	3	4
$4n + 2$				
Term				

Step 2: Complete the first row of the table. The step numbers begin at 1 and increase by 1.

n	1	2	3	4
$4n + 2$	$4(1) + 2$	$4(2) + 2$	$4(3) + 2$	$4(4) + 2$
Term				

Step 3: Complete the second row of the table. Substitute the value of the step number into the rule.

n	1	2	3	4
$4n + 2$	$4(1) + 2$	$4(2) + 2$	$4(3) + 2$	$4(4) + 2$
Term	6	10	14	18

Step 4: Complete the last row of the table. Simplify each expression in the second row to obtain the terms in the sequence.

Answer: The sequence is 6, 10, 14, 18, …

Sequences are sometimes displayed in a table, such as in the following example.

Example 2 Using the data in the table, describe the relationship between a term and its position n in the sequence.

Position (n)	1	2	3	4	n
Term	5	10	15	20	

Position (n)	1	2	3	4	n
Term	5	10	15	20	

$$5 \cdot 1 = 5$$
$$5 \cdot 2 = 10$$
$$5 \cdot 3 = 15$$
$$5 \cdot 4 = 20$$

Step 1: Look for a relationship between the position of a term in the sequence and the value of the term. Each term in the sequence is obtained by multiplying its corresponding position value by 5.

Position (n)	1	2	3	4	n
Term	5	10	15	20	$5n$

Step 2: The nth term of the sequence is obtained by multiplying n by 5.

Answer: To obtain a term in the sequence, multiply n by 5.

Sequences can also be represented in models.

Example 3 Write a rule that describes the number of blocks. The rule should be in terms of n, where n represents the step number of the model.

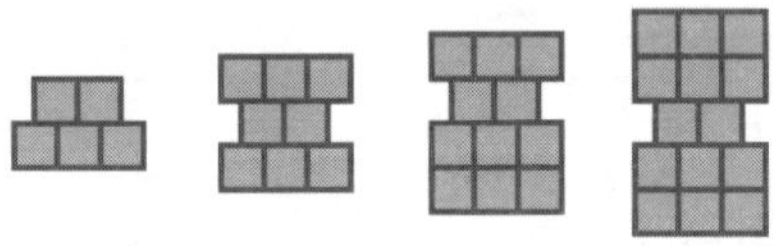

1 2 3 4

What is staying the same? One row that contains only two blocks.

Step 1: Identify what is staying the same between each step.

What is changing? The number of blocks above and below the row containing two blocks.

Step 2: Identify what is changing between each step.

Step Number	1	2	3	4
Number of blocks that are changing	3	6	9	12

Step 3: Create a table that shows the step number and the number of blocks that are changing.

The number of blocks that are changing is 3 times the step number ($3n$).

Step 4: Determine the relationship between the step number and the number of blocks that are changing.

Expression: $3n$

Number of blocks that are staying the same: 2

Rule: $3n + 2$

Step 5: Add the expression from Step 4 to the number of blocks that are staying the same to obtain the rule.

Answer: The rule that describes the number of blocks is $3n + 2$.

Example 4 Consider the sequence of numbers 2, 5, 8, 11, … Write a rule that describes this sequence. The rule should be in terms of n, where n represents the position of a term in the sequence.

Step 1: Create a table that contains the terms of the sequence and their corresponding positions in the sequence.

Position (n)	1	2	3	4
Term	2	5	8	11

Step 2: Observe the difference between each consecutive pair of position values and terms.

	+1	+1	+1

Position (n)	1	2	3	4
Term	2	5	8	11

	+3	+3	+3

Step 3: Divide the difference between the terms in the sequence by the difference between the values of the positions. This quotient is the coefficient of n in the rule.

Coefficient of $n = \frac{3}{1} = 3$

Step 4: Add a row to the table. This row should contain values for the expression obtained by multiplying n and the coefficient of n found in Step 3.

Position (n)	1	2	3	4
Term	2	5	8	11
$3n$	3	6	9	12

Step 5: Observe the difference between the terms in the sequence and the values in the third row. Each term is one less than the value of $3n$. Add a row to the table and write the values for the expression $3n - 1$.

Position (n)	1	2	3	4
Term	2	5	8	11
$3n$	3	6	9	12
$3n - 1$	2	5	8	11

Answer: The rule that describes the sequence is $3n - 1$.

Sequences can also be represented in graphs, such as the graph in the next example. Graphs often help determine patterns and relationships between two variables.

Example 5 The graph represents the number of phone calls an operator received during a 4-hour shift. In the graph, h represents the number of hours the operator worked. The variable n represents the number of calls received. Describe the relationship between the number of hours the operator worked and the number of calls she received.

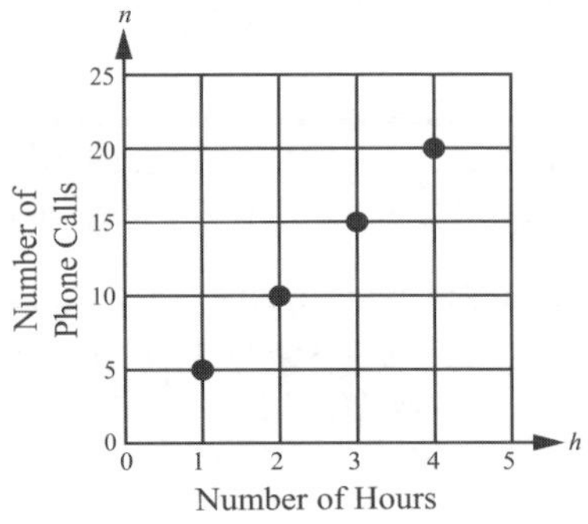

Number of Hours (h)	1	2	3	4
Number of Phone Calls (n)	5	10	15	20

Step 1: Create a table that contains values for the variables h and n.

The number of phone calls is obtained by multiplying the number of hours by 5.

Step 2: Determine the relationship between the two variables.

Answer: The relationship between the number of hours worked and the number of phone calls received is $n = 5h$.

Problem Set

1. Choose which sequence of numbers follows the rule $-10n$, where n represents the position of a term in the sequence:

10, 20, 30, 40, 50, ...

–9, –8, –7, –6, –5, ...

–10, –20, –30, –40, –50, ...

–11, –12, –13, –14, –15, ...

2. Choose which sequence of numbers follows the rule $12n$, where n represents the position of a term in the sequence:

0, 12, 24, 36, 48, ...

13, 14, 15, 16, 17, ...

12, 24, 36, 48, 60, ...

12, 13, 14, 15, 16, ...

3. Choose which sequence of numbers follows the rule $n - 8$, where n represents the position of a term in the sequence:

7, 6, 5, 4, 3, ...

–7, –6, –5, –4, –3, ...

–9, –10, –11, –12, –13, ...

9, 10, 11, 12, 13, ...

4. Choose which sequence of numbers follows the rule $5n - 8$, where n represents the position of a term in the sequence:

–2, –1, 0, 1, 2, ...

–35, –30, –25, –20, –15, ...

–3, 2, 7, 12, 17, ...

13, 18, 23, 28, 33, ...

5. Choose which sequence of numbers follows the rule $12 - n$, where n represents the position of a term in the sequence:

13, 14, 15, 16, 17, ...

11, 10, 9, 8, 7, ...

7, 8, 9, 10, 11, ...

−7, −8, −9, −10, −11, ...

6. Choose which sequence of numbers follows the rule $4 - 2n$, where n represents the position of a term in the sequence:

3, 2, 1, 0, −1, ...

2, 0, −2, −4, −6, ...

6, 8, 10, 12, 14, ...

−2, 0, 2, 4, 6, ...

7. Choose which sequence of numbers follows the rule $3n - 5$, where n represents the position of a term in the sequence:

8, 11, 14, 17, 20, ...

−12, −9, −6, −3, 0, ...

−2, 1, 4, 7, 10, ...

2, −1, −4, −7, −10, ...

8. Choose which sequence of numbers follows the rule $5n + 3$, where n represents the position of a term in the sequence:

20, 25, 30, 35, 40, ...

3, 8, 13, 18, 23, ...

15, 20, 25, 30, 35, ...

8, 13, 18, 23, 28, ...

9. Using the data in the following table, choose the answer that describes the relationship between a term and its position n in the sequence.

Position (n)	1	2	3	4	n
Term	15	16	17	18	

Multiply n by 15. Add 14 to n.

Subtract 14 from n. Divide n by 15.

10. Using the data in the following table, choose the answer that describes the relationship between a term and its position n in the sequence.

Position (n)	1	2	3	4	n
Term	21	22	23	24	

Add 20 to n. Multiply n by 11.

Divide n by 11. Subtract 20 from n.

11. Using the data in the following table, choose the answer that describes the relationship between a term and its position n in the sequence.

Position (n)	1	2	3	4	n
Term	−5	−4	−3	−2	

Multiply n by −5. Subtract 6 from n.

Add 6 to n. Divide n by −5.

12. Using the data in the following table, choose the answer that describes the relationship between a term and its position n in the sequence.

Position (n)	1	2	3	4	n
Term	−10	−9	−8	−7	

Multiply n by −10. Divide n by −10.

Subtract 11 from n. Add 11 to n.

13. Using the data in the following table, choose the answer that describes the relationship between a term and its position n in the sequence.

Position (n)	1	2	3	4	n
Term	2	4	6	8	

Add 1 to n. Divide n by 2.

Multiply n by 2. Subtract 1 from n.

14. Using the data in the following table, choose the answer that describes the relationship between a term and its position n in the sequence.

Position (n)	1	2	3	4	n
Term	−1	−2	−3	−4	

Subtract 2 from n. Add 2 to n.

Multiply n by −1. Divide n by 1.

15. Using the data in the following table, choose the answer that describes the relationship between a term and its position n in the sequence.

Position (n)	1	2	3	4	n
Term	$\frac{1}{8}$	$\frac{2}{8}$	$\frac{3}{8}$	$\frac{4}{8}$	

Subtract $\frac{7}{8}$ from n.　　Add $\frac{1}{8}$ to n.

Divide n by 8.　　Multiply n by 8.

16. Using the data in the following table, choose the answer that describes the relationship between a term and its position n in the sequence.

Position (n)	1	2	3	4	n
Term	$\frac{1}{2}$	1	$\frac{3}{2}$	2	

Subtract $\frac{1}{2}$ from n.　　Add $\frac{1}{2}$ to n.

Multiply n by 2.　　Divide n by 2.

17. The block structures follow a pattern. Write a rule that describes the number of blocks. The rule should be in terms of n, where n represents the step number of the model.

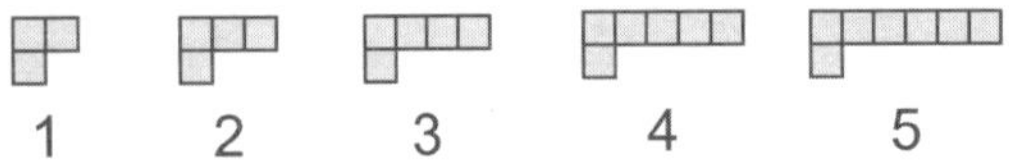

18. Lasaundra is creating several different block structures that follow a pattern. Write a rule that describes the number of blocks. The rule should be in terms of n, where n represents the step number of the model.

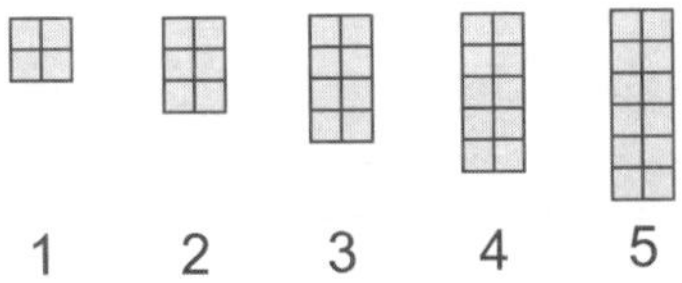

19. The blocks below follow a pattern. Write a rule that describes the number of blocks. The rule should be in terms of n, where n represents the step number of the model.

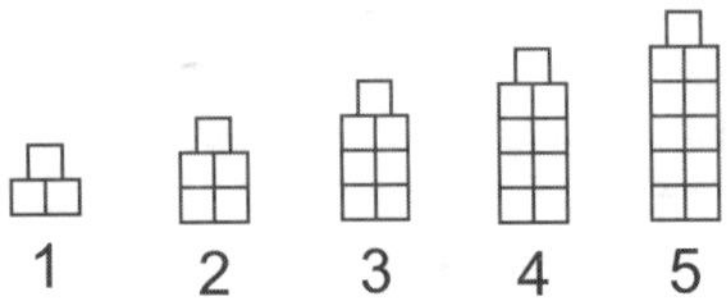

20. The yellow triangles below follow a pattern. Write a rule that describes the number of yellow triangles. The rule should be in terms of n, where n represents the step number of the model.

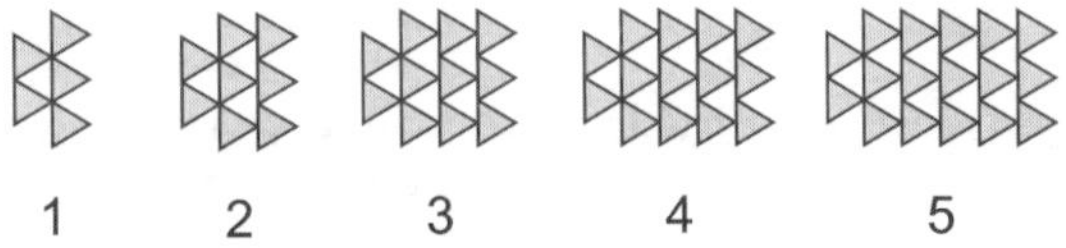

21. Jerene is creating several different block structures that follow a pattern. Write a rule that describes the number of blocks. The rule should be in terms of n, where n represents the step number of the model.

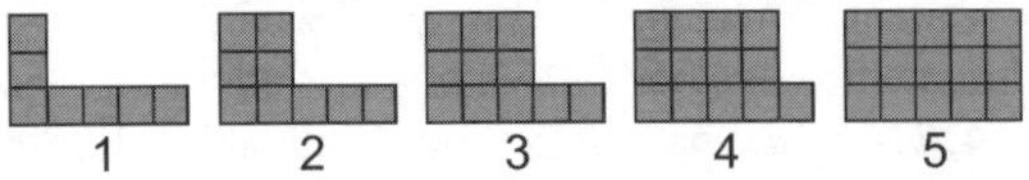

22. Nathan is creating several different block structures that follow a pattern. Write a rule that describes the number of blocks. The rule should be in terms of n, where n represents the step number of the model.

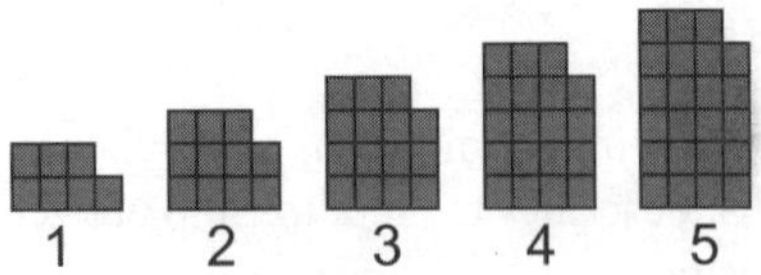

23. Tony is creating several different block structures that follow a pattern. Write a rule that describes the number of blocks. The rule should be in terms of n, where n represents the step number of the model.

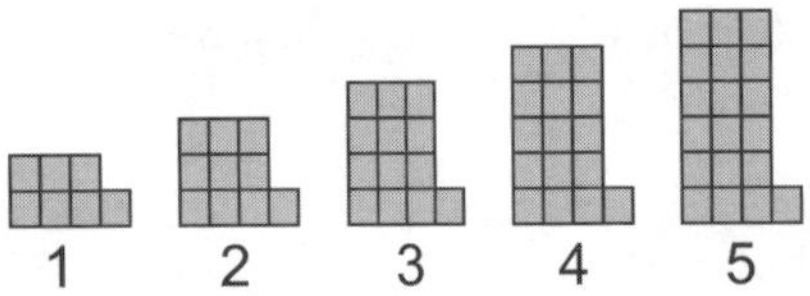

24. For the sequence of numbers 6, 8, 10, 12, ..., write a rule that describes this sequence. The rule should be written in terms of n, where n represents the position of a term in the sequence.

25. For the sequence of numbers 8, 11, 14, 17, ..., write a rule that describes this sequence. The rule should be written in terms of n, where n represents the position of a term in the sequence.

26. For the sequence of numbers 10, 17, 24, 31, ..., write a rule that describes this sequence. The rule should be written in terms of n, where n represents the position of a term in the sequence.

27. For the sequence of numbers –1, 3, 7, 11, ..., write a rule that describes this sequence. The rule should be written in terms of n, where n represents the position of a term in the sequence.

28. For the sequence of numbers 2, 5, 8, 11, ..., write a rule that describes this sequence. The rule should be written in terms of n, where n represents the position of a term in the sequence.

29. For the sequence of numbers 3, 8, 13, 18, ..., write a rule that describes this sequence. The rule should be written in terms of n, where n represents the position of a term in the sequence.

30. For the sequence of numbers 3, 9, 15, 21, ..., write a rule that describes this sequence. The rule should be written in terms of n, where n represents the position of a term in the sequence.

31. Which relationship is best represented by the data in the graph?

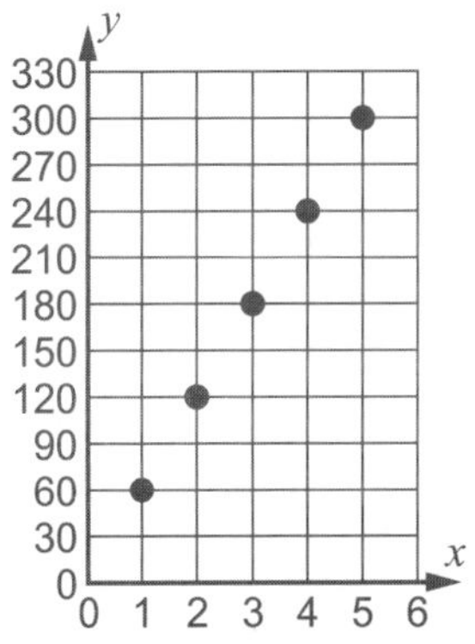

Conversion of days to months
Conversion of minutes to hours
Conversion of hours to days
Conversion of minutes to seconds

32. Which relationship is best represented by the data in the graph?

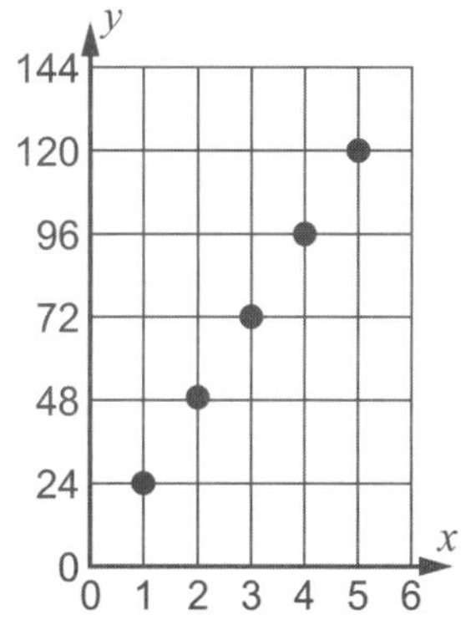

Conversion of days to hours
Conversion of days to months
Conversion of years to days
Conversion of minutes to hours

33. Which relationship is best represented by the data in the graph?

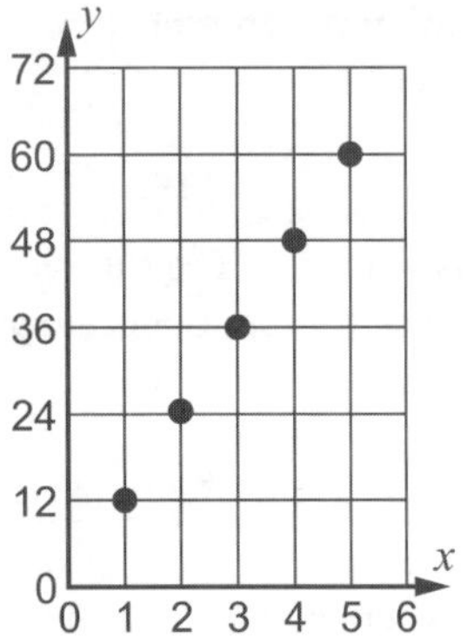

Conversion of feet to inches
Conversion of feet to centimeters
Conversion of feet to miles
Conversion of yards to feet

34. The graph represents the total number of customers that entered a store each hour during a 4-hour period. In the graph, h represents the number of hours the store was open. The variable n represents the number of customers that entered the store. Describe the relationship between the number of hours the store was open and the number of customers that entered the store.

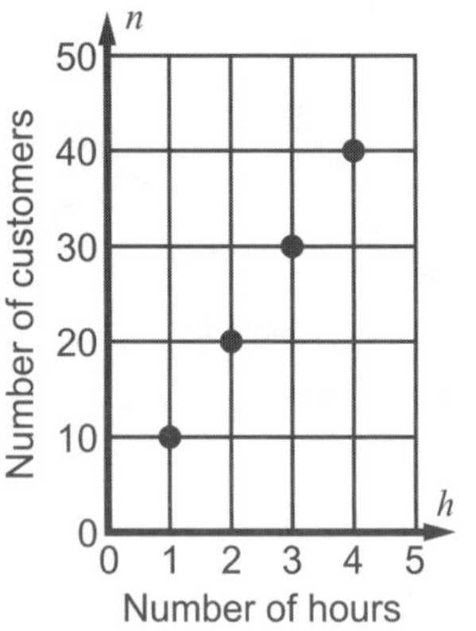

35. The graph represents the total number of customers that entered a store each hour during a 4-hour period. In the graph, h represents the number of hours the store was open. The variable n represents the number of customers that entered the store. Describe the relationship between the number of hours the store was open and the number of customers that entered the store.

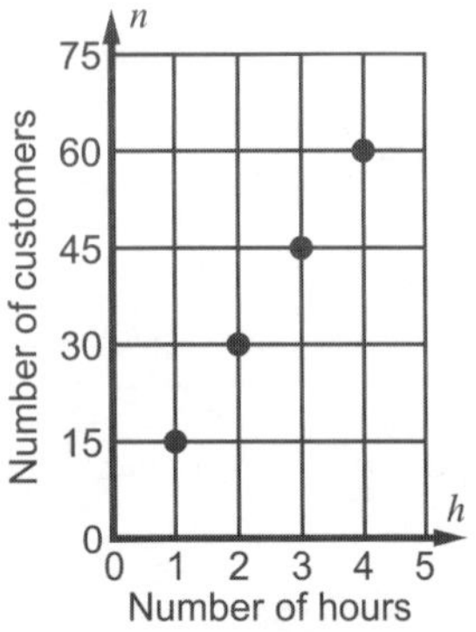

36. During a recent trip, the Matthews family recorded the number of miles they drove each hour. The graph represents the time t in hours and the distance d in miles they traveled. Describe the relationship between the time and the distance the family traveled.

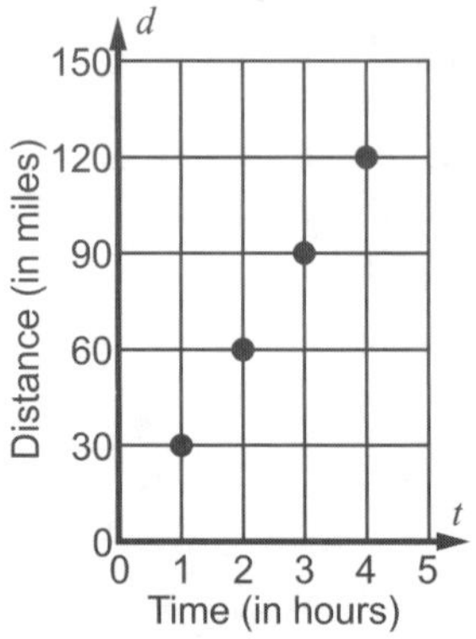

37. During a recent trip, the Turner family recorded the number of miles they drove each hour. The graph represents the time t in hours and the distance d in miles they traveled. Describe the relationship between the time and the distance the family traveled.

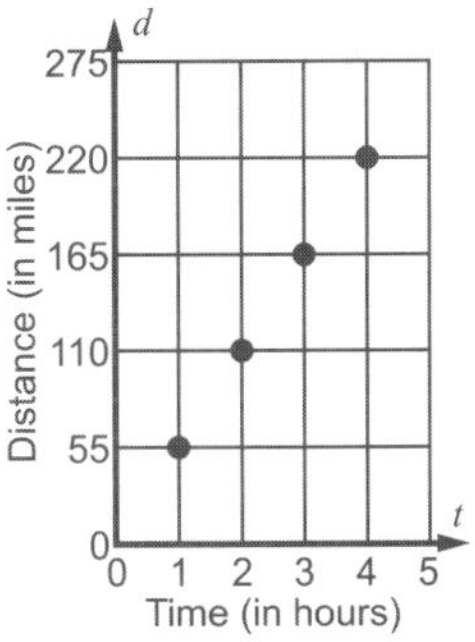

38. The graph shows the relationship between the cost to use a cellular phone and the amount of time the phone was used. In the graph, h is the number of hours the phone was used. The variable c is the cost to use the phone in dollars. Describe the relationship between the number of hours the phone was used and the cost to use the cellular phone.

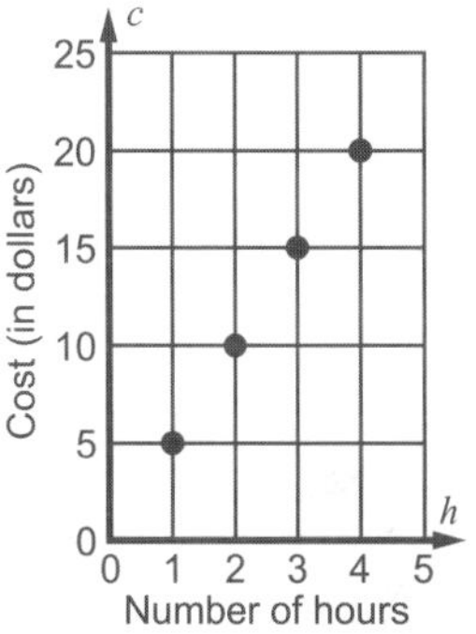

39. The graph shows the relationship between the cost to use a cellular phone and the amount of time the phone was used. In the graph, h is the number of hours the phone was used. The variable c is the cost to use the phone in dollars. Describe the relationship between the number of hours the phone was used and the cost to use the cellular phone.

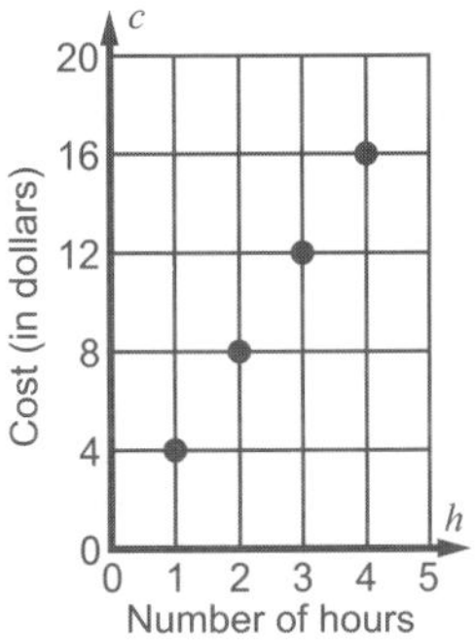

40. The graph shows the relationship between the cost to rent a car and the number of days the car is rented. In the graph, d is the number of days the car is rented. The variable c is the cost to rent the car in dollars. Describe the relationship between the number of days the car is rented and the cost to rent the car.

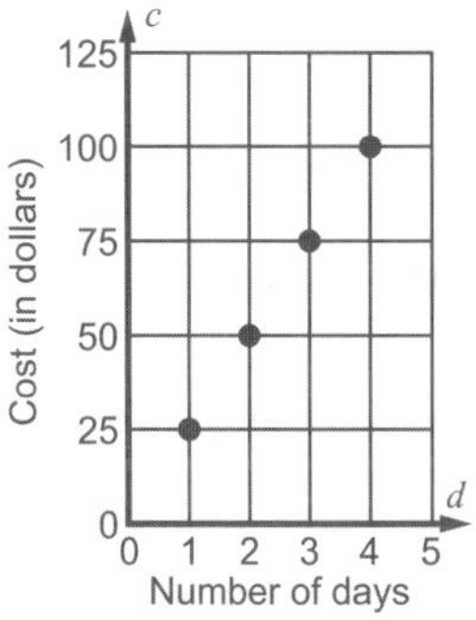

MPA-840: Interpreting Data

Data can be presented in many ways. Tables and graphs are useful when comparisons and analysis of collected data is necessary. The collecting, presenting, and analyzing of data is called **statistics**.

Surveys are conducted to study a population and samples are taken to represent the group when the population is very large. This **sample** represents the population.

When a comparison is needed between several data values, a bar graph is helpful. When a comparison of a data value to the sum of the data values is required, a pie chart works best.

The **average** or **mean** is often used to show the central tendency for a set of data. The **average** is a single value that summarizes a set of data.

Population	The collection of all objects that are of interest in a study.
Units or Subjects	The elements in a population.
Sample	A representation or model or example.
Data or Data Set	The collection of observations.
Bias	The tendency of a sample to misrepresent the population.
Average or Mean	The sum of the data values divided by the number of values.

The manner in which data values are presented can be misleading. Varying the range, chopping a chart, using objects instead of bars, and exploding a piece of a pie chart can skew data and affect the way the information is interpreted.

Example 1

The table to the right shows Kevin's four arithmetic test grades. Which statement can be inferred by the data presented in this table?

A. Kevin's test scores gradually improved.
B. Kevin's test scores gradually decreased.
C. Kevin's highest test score is the Chapter 2 test score.
D. Kevin's highest test score is 75.

Chapter	Test Score
1	87
2	86
3	75
4	73

Step 1: Examine each choice and determine from the table whether the statement is true or false.

A. False; the test scores are decreasing.
B. True.
C. False; the highest test score is the Chapter 1 test score.
D. False; the highest test score is 87.

Answer: Kevin's test scores gradually decreased.

Example 2 Ellen made crystal bracelets and sold them to her friends for $5.00 each. She sold 12 on Saturday and 10 on Sunday. What piece of information is needed to find the amount of profit she made from the sale of the bracelets on Saturday and Sunday?

Recall that profit is the difference between the amount earned and the cost of the materials to make the bracelets.

Step 1: Look up any definitions or formulas needed in solving the problem.

Answer: To find the profit, information on the cost of the materials to make the bracelets is needed.

Example 3 The church fair holds a charity raffle each year and gives away cash prizes. The table to the right shows the actual prizes awarded. They advertise that the average winner gets $1,025.00 in prize money. How is this misleading?

Prize Amount	Number Awarded
$10,000	1
$1,000	2
$500	17

Look at the table for prize awards.

Step 1: Examine the given information.

Only one person will win more than the average claimed in the advertisement.

Step 2: Nineteen prize winners will not earn the average advertised.

Answer: Nineteen people will not earn the average amount of prized money so the average is misleading. The one large prize skews the average.

Example 4 The bar graphs below show the number of games won by each of four soccer teams. Graph 1 shows the Jets as a superior team, since the height of the bar representing the Jets' wins is more than twice the height of the bar representing the Doves' wins. Did the Jets win twice as many games as the Doves?

Explain why these graphic representations of the same set of data give different impressions.

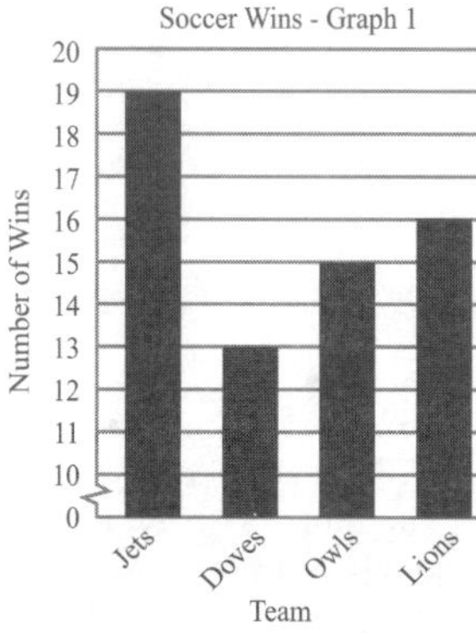

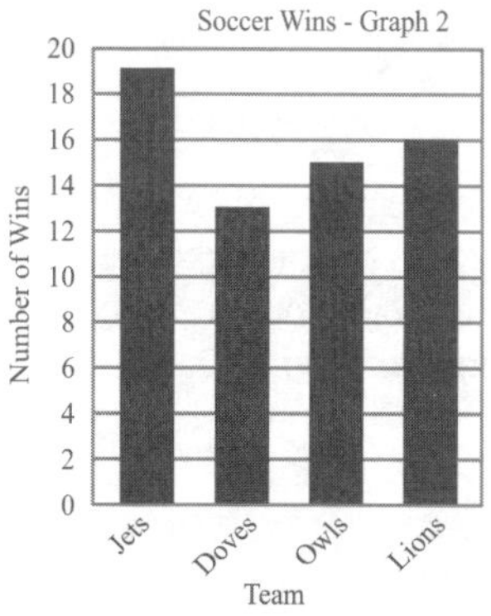

The range on the vertical axis of Graph 1 is $0 \rightarrow 20$ wins with a break in the vertical axis. The range for Graph 2 is $0 \rightarrow 20$. The scale is 1 in Graph 1 and the scale is 2 in Graph 2.

Step 1: Determine the range and scale on each graph by looking at the vertical axis. Varying the range or scale when representing data on a graph can cause misinterpretations.

The Jets won 19 games and the Doves won 13. The Jets did not win twice as many games as the Doves.

Step 2: Look at the actual data values.

Graph 1 makes it appear that the Jets won twice as many games as the Doves.

Graph 2 makes it appears that the six-game difference between the leading team and the trailing team is closer.

Step 3: Look at the vertical scale and the height of the bar for each team in Graph 1 and Graph 2.

Answer: The Jets did not win over two times as many games as the Doves. By altering the vertical scale (chopping off the bottom values) Graph 1 gives the impression that the Jets outplayed the other teams by a large margin.

Example 5 The table below shows the data for the number of soccer games won by four teams. From the table, Jim concluded that the Jets are the most skilled team. How is this table misleading?

Team	Jets	Doves	Owls	Lions
Soccer Games Won	19	13	15	16

Consider the conditions under which this data was collected. Are the Jets the most skilled team? Did the four teams play the same number of games? Is that information necessary?

Step 1: Analyze the data.

If the Jets played 38 games, they won 50% or one-half of their games.

$$\frac{19}{38} = \frac{1}{2} = 50\%$$

If the Doves played a total of 26 games, they won 50% or one-half of their games.

$$\frac{13}{21} = \frac{1}{2} = 50\%$$

Given this information, each team won one-half of the games played and appear to be equally skilled. If the teams did not play each other or played the same opponents, a comparison cannot be made among the four teams.

Step 2: Consider the different conditions under which competition occurs:
Did the teams play the same opponents?
Did they play the same number of games?
Did the teams play each other?
Was weather a factor in the outcome of some of the games?

Answer: The table is misleading because before the skill levels of the teams can be compared, the total number of games played is needed. If the four teams played the same opponents, the percentage of games won would be a better comparison of the skill level of each team.

Problem Set

Choose the statement that best answers the following questions from the given choices:

1. The data table shows the number of minutes Erin spent studying this week for each of her classes. Which statement can be inferred by the data presented in the table?

Class	Minutes
Math	60
English	75
History	45
Science	90
French	30

- Erin spent the most amount of time studying for English.
- Erin spent twice as much time studying for science than she did studying for history.
- Erin spent twice as much time studying for French than she did studying for math.
- Erin spent an average of 45 minutes studying for each class.

2. The data table shows the number of minutes Erin spent studying this week for each of her classes. Which statement *cannot* be inferred by the data presented in the table?

Class	Minutes
Math	60
English	75
History	45
Science	90
French	30

- Erin spent an average of 60 minutes studying for each class.
- Erin spent twice as much time studying for science than she did studying for history.
- Erin spent the most amount of time studying for English.
- Erin spent twice as much time studying for math than she did studying for French.

3. The data table shows Mark's test scores on each of his math tests. Which statement can be inferred by the data presented in the table?

Test	Score
Test 1	68
Test 2	74
Test 3	82
Test 4	89
Test 5	95

- Mark's test scores gradually decreased.
- Mark's average test score is 85.
- Mark's highest test score was on Test 3.
- Mark's test scores gradually improved.

4. The data table shows Mark's test scores on each of his math tests. Which statement can be inferred by the data presented in the table?

Test	Score
Test 1	68
Test 2	74
Test 3	82
Test 4	89
Test 5	95

- Mark scored 8 points higher on Test 3 than he did on Test 2.
- Mark scored 14 points higher on Test 3 than he did on Test 2.
- Mark scored 8 points lower on Test 3 than he did on Test 2.
- Mark scored 14 points lower on Test 3 than he did on Test 1.

5. The data table shows Mark's test scores on each of his math tests. Which statement can be inferred by the data presented in the table?

Test	Score
Test 1	68
Test 2	74
Test 3	82
Test 4	89
Test 5	95

- Mark scored 6 points lower on Test 2 than he did on Test 3.
- Mark scored 8 point lower on Test 1 than he did on Test 2.
- Mark scored 6 points lower on Test 4 than he did on Test 5.
- Mark scored 6 points lower on Test 3 than he did on Test 4.

6. A group of friends collected data on the number of siblings that each has. Their results are presented in the data table. Which statement can be inferred by the data presented in this table?

Friend	Number of Siblings
Ashley	5
Kristan	1
Will	3
Steven	5
Natasha	0

- Ashley and Steven have the same number of siblings.
- Kristan has the least number of siblings.
- Kristan and Natasha have the same number of siblings.
- Will has the most number of siblings.

7. A group of friends collected data on the number of siblings that each has. Their results are presented in the data table. Which statement can be inferred by the data presented in this table?

Friend	Number of Siblings
Ashley	5
Kristan	1
Will	3
Steven	5
Natasha	0

- Will has 2 fewer siblings than Steven.
- Will has the most number of siblings.
- Will has 2 more siblings than Ashley.
- Will has 2 more siblings than Natasha.

8. A group of friends collected data on the number of siblings that each has. Their results are presented in the data table. Which statement *cannot* be inferred by the data presented in this table?

Friend	Number of Siblings
Ashley	5
Kristan	1
Will	3
Steven	5
Natasha	0

- Kristan has the least number of siblings.
- Will has 2 fewer siblings than Ashley.
- Ashley has 5 more siblings than Natasha.
- Steven has 5 times as many siblings as Kristan.

9. The table shows the number of hours that five different people worked last week. What piece of information, if any, is needed to determine who earned the most money last week?

Person	Number of Hours Worked
Jamie	35
Alice	30
Olivia	15
Anna	5
Ronnie	40

- The number of years each person has worked at his or her job is needed.
- The number of days last week that each person worked is needed.
- No other information is needed.
- The hourly rate of each person is needed.

10. The table shows the amount of money that five different people earned last week. What piece of information, if any, is needed to determine who worked the most number of hours?

Person	Amount Earned
Pete	$300
Jackson	$500
Brad	$250
Thomas	$400
Pedro	$450

- The hourly rate of each person is needed.
- The number of days last week that each person worked is needed.
- No other information is needed.
- The amount of money that each person earned last month is needed.

11. The table shows the amount of money each of five different people earn per hour of work (hourly rate). What piece of information, if any, is needed to determine who earned the least amount of money last week?

Person	Hourly Rate
Michael	$8
Jacqui	$6.50
Darren	$12.50
Madison	$25
Clementine	$10

- No other information is needed.
- The number of days last week that each person worked is needed.
- The amount of money that each person earned last month is needed.
- The number of hours each person worked is needed.

12. The table shows the amount of money each of five different people earned last week as well as the number of hours each person worked. What piece of information, if any, is needed to determine who has the highest hourly rate?

Person	Amount Earned	Number of Hours Worked
Hannah	$300	30
Christi	$500	40
J.J.	$250	40
Carlos	$800	40
Byron	$650	25

- No other information is needed.
- The number of years that each person has worked at his or her job is needed.
- The number of days last week that each person worked is needed.
- The amount of money that each person earned last month is needed.

13. The table shows the number of points that each of five football teams scored last season. What piece of information, if any, is needed to determine which team scored the most points per game?

Team	Number of Points Scored
Bulldogs	210
Lions	172
Panthers	60
Patriots	135
Hawks	108

- The number of players on each team is needed.
- The number of hours each team practiced is needed.
- No other information is needed.
- The number of games each team played is needed.

14. The table shows the number of games that each of five football teams played last season. What piece of information, if any, is needed to determine which team scored the most points per game?

Team	Number of Games Played
Bulldogs	20
Lions	17
Panthers	18
Patriots	16
Hawks	16

- The number of hours each team practiced is needed.
- The number of players on each team is needed.
- No other information is needed.
- The number of points each team scored is needed.

15. Shameka painted pictures and sold them at an art fair. She sold 22 pictures for $25 each. What piece of information, if any, is needed to determine her profit?

- The number of days she sold paintings at the art fair is needed.
- The cost of supplies to paint the pictures is needed.
- No other information is needed.
- The size of the paintings is needed.

16. Each year, the county fair has a raffle to support local schools. The table shows the number of prizes that can be won and the value of each prize. It is advertised that the average price is approximately $1,000. Is this information misleading? If so, how is it misleading?

Prizes	Value
First prize (1 winner)	$25,000
Second prize (3 winners)	$10,000
Third prize (50 winners)	$50

- Yes, this information is misleading because 50 winners will win more than the average amount advertised.
- Yes, this information is misleading because 50 winners will win less than the average amount advertised.
- No, this information is not misleading.
- Yes, this information is misleading because the average prize is approximately $12,000.

17. Each year, the police department has a raffle to raise money for new cars. The table shows the number of prizes that can be won and the value of each prize. It is advertised that the average prize is approximately $500. Is this information misleading? If so, how is it misleading?

Prizes	Value
First prize (1 winner)	$15,000
Second prize (2 winners)	$300
Third prize (30 winners)	$50

- Yes, this information is misleading because only one winner will win at least the average amount advertised.
- Yes, this information is misleading because 32 winners will win more than the average amount advertised.
- No, this information is not misleading.
- Yes, this information is misleading because the average prize is approximately $5,000.

18. Each year, the police department has a raffle to raise money for new uniforms. The table shows the number of prizes that can be won and the value of each prize. It is advertised that the average prize is approximately $50. Is this information misleading? If so, how is it misleading?

Number of prizes	Value
10	$100
15	$50
25	$20

- Yes, this information is misleading because 40 winners will win less than the average amount advertised.
- No, this information is not misleading.
- Yes, this information is misleading because only 10 winners will win at least the average amount advertised.
- Yes, this information is misleading because the average prize is approximately $57.

19. A recent newspaper article listed the top 5 soccer players in the district. The article stated that Lance is first in the district because he scored the most goals. Using the data presented in the table, determine if this information is misleading. If so, how is it misleading?

Player	Goals	Number of Games Played
Lance	20	15
Danny	18	9
Bobby	16	10
Jermaine	16	16
Jesse	15	5

- Yes, this information is misleading because Danny scored the most goals per game.
- No, this information is not misleading.
- Yes, this information is misleading because Jesse scored the most goals per game.
- Yes, this information is misleading because Jermaine played the most number of games.

20. A recent newspaper article listed the top 5 soccer players in the district. The article stated that Bobbie and Jermaine have the same rank in the district because they each scored 16 goals. Using the data presented in the table, determine if this information is misleading. Is so, how is it misleading?

Player	Goals	Number of Games Played
Lance	20	15
Danny	18	9
Bobby	16	10
Jermaine	16	16
Jesse	15	5

- Yes, this information is misleading because Lance and Bobby have the same rank.
- No this information is not misleading.
- Yes, this information is misleading because Jermaine scored more goals per game than Bobby.
- Yes, this information is misleading because Bobby scored more goals per game than Jermaine.

21. A recent newspaper article listed the top 5 soccer players in the district. The article stated that Jesse is fifth in the district because he scored the fifth-most goals. Using the data presented in the table, determine if this information is misleading. If so, how is it misleading?

Player	Goals	Number of Games Played
Lance	20	15
Danny	18	9
Bobby	16	10
Jermaine	16	16
Jesse	15	5

- Yes, this information is misleading because Lance scored the fifth-most goals per game.
- Yes, this information is misleading because Jesse scored the second-most goals per game.
- No, this information is not misleading.
- Yes, this information is misleading because Jermaine scored the fifth-most goals per game.

22. A recent newspaper article listed the top 5 soccer players in the district. The article stated that Jesse is first in the district because he scored the most goals per game. Using the data presented in the table, determine if this information is misleading. If so, how is it misleading?

Player	Goals	Number of Games Played
Lance	20	15
Danny	18	9
Bobby	16	10
Jermaine	16	16
Jesse	15	5

- No, this information is not misleading.
- Yes, this information is misleading because Lance scored the most goals.
- Yes, this information is misleading because Danny scored the most goals.
- Yes, this information is misleading because Jermaine played the most number of games.

23. The bar graphs show the same data for the number of sales for each of 4 employees. Choose the statement that best compares the graphs and the data presented.

Graph 1:

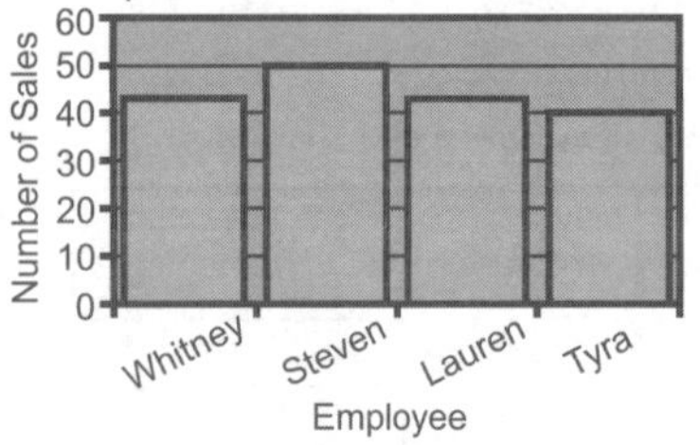

Graph 2:

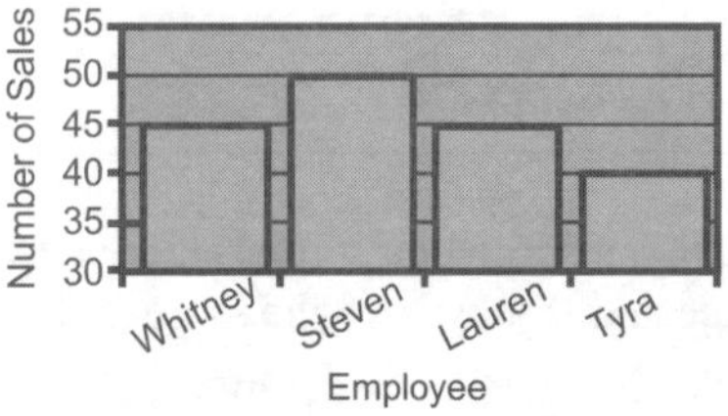

- The employee who made the least number of sales can be determined from Graph 2, but not Graph 1.
- Changing the vertical scale and range affects how the graph is interpreted.
- Changing the vertical scale and range does not affect how the graph is interpreted.
- The employee who made the most number of sales can be determined from Graph 2, but not Graph 1.

24. The graphs below show the number of students enrolled at a state university from the years 2000 to 2004. Choose the statement that best compares the graphs and the data presented.

Graph 1:

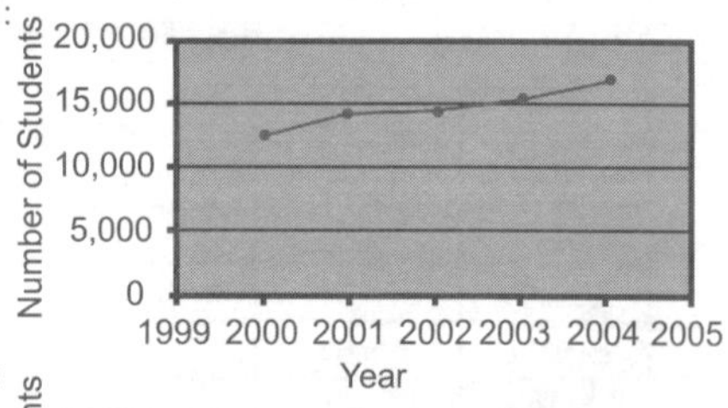

Graph 2:

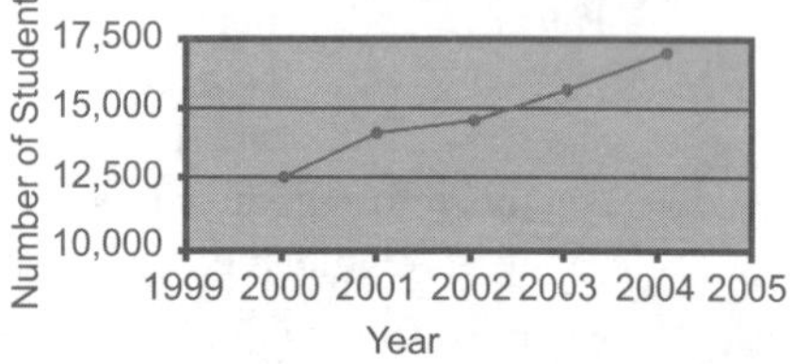

- The graphs represent the same data using different vertical scales and ranges.
- Graph 1 shows a more drastic increase in the number of students enrolled.
- Graph 2 shows a more gradual increase in the number of students enrolled.
- The graphs represent different data using the same vertical scale and range.

25. The graph below shows the number of houses in the same city in 1970 and 2000. Is the graph misleading? If so, how is the graph misleading?

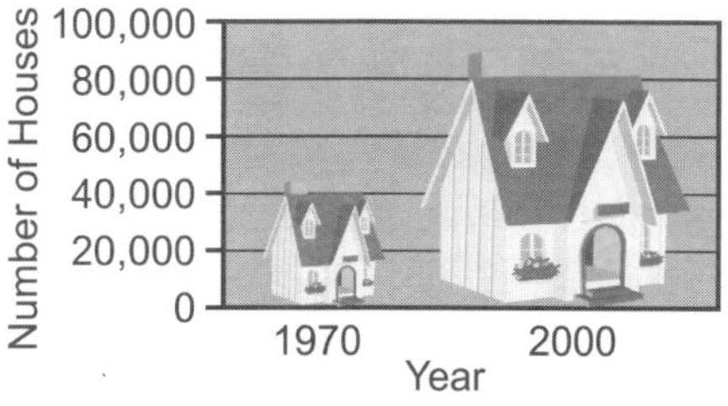

- No, the graph is not misleading.
- Yes, the graph is misleading because the size of the house representing the number of houses in 2000 is larger in height and width which makes the increase seem greater.
- Yes, the graph is misleading because it appears that there were half the number of houses in 1970 than there were in 2000.
- Yes, the graph is misleading because it appears that there were 2 times as many houses in 2000 than there were in 1970.

26. The graph below shows the number of houses in the same city in 1970 and 2000. Is the graph misleading? If so, how is the graph misleading?

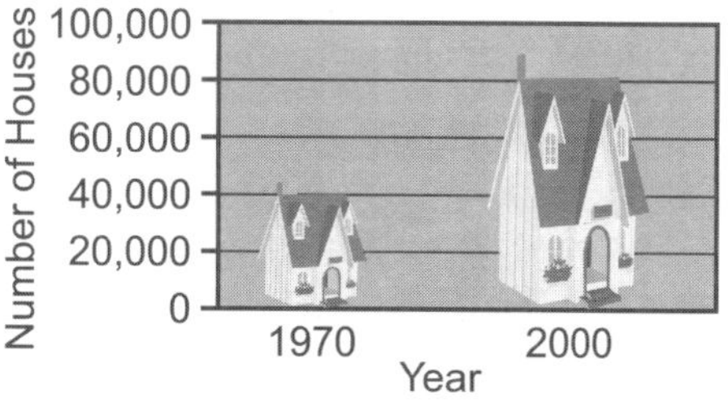

- Yes, the graph is misleading because it appears that there were 2 times as many houses in 2000 than there were in 1970.
- No, the graph is not misleading.
- Yes, the graph is misleading because it appears that there were 4 times as many houses in 2000 than there were in 1970.
- Yes, the graph is misleading because it appears that there were 2 times as many houses in 1970 than there were in 2000.

27. The school music club recently surveyed students about their music preferences. The choices were country, rock, R&B, or any other type of music. The results are shown in the pie chart below. Is the chart misleading? If so, how is it misleading?

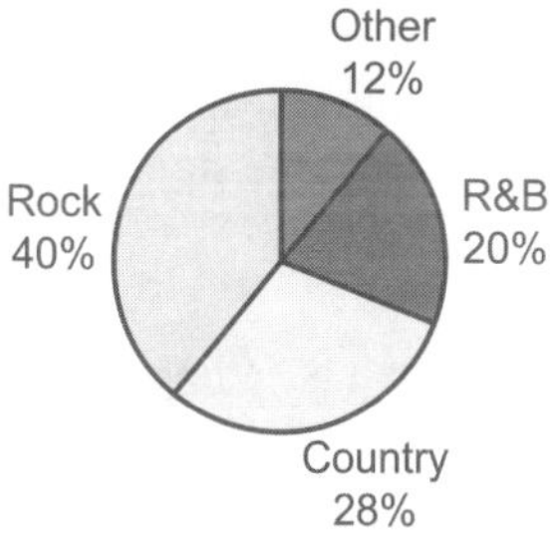

- Yes, the pie chart is misleading because Classical should also be a category on the chart.
- Yes, the pie chart is misleading because the section for rock music should be larger than half of the circle.
- No, the pie chart is not misleading.
- Yes, the pie chart is misleading because the section for R&B music should be larger than the section for country music.

28. The school music club recently surveyed students about their music preferences. The choices were country, rock, R&B, or any other type of music. The results are shown in the pie chart below. Is the chart misleading? If so, how is it misleading?

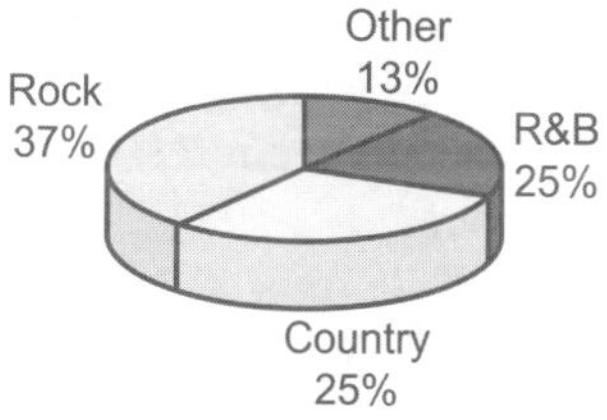

- No, the pie chart is not misleading.
- Yes, the pie chart is misleading because Classical should also be a category on the chart.
- Yes, the pie chart is misleading because the section for R&B music should be larger than the section for rock music.
- Yes, the pie chart is misleading because the section for Country appears to be larger than the section for R&B.

29. The school music club recently surveyed 275 students about their music preferences. The choices were country, rock, R&B, classical, and jazz. The results are shown in the bar chart below. Is the chart misleading? If so, how is it misleading?

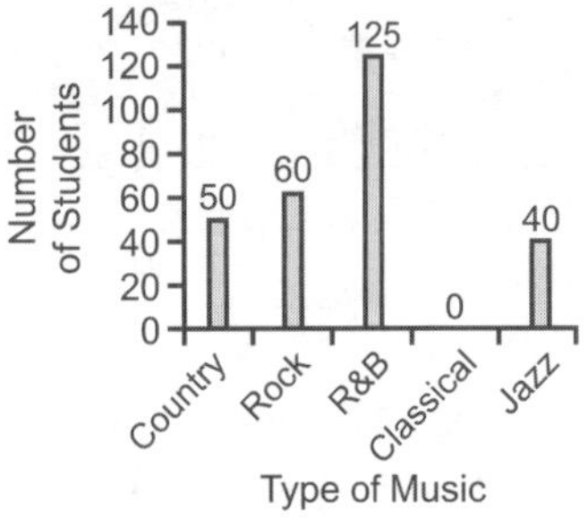

- Yes, the bar chart is misleading because there should be no category for classical music on the chart.
- No, the bar chart is not misleading.
- Yes, the bar chart is misleading because the bar for country music should be higher than the bar for rock music.
- Yes, the bar chart is misleading because the bar for R&B music should be twice as tall as the bar for jazz music.

30. The school music club recently surveyed 275 students about their music preferences. Their choices were country, rock, R&B, classical, and jazz. The results are shown in the bar chart below. Is the chart misleading? If so, how is it misleading?

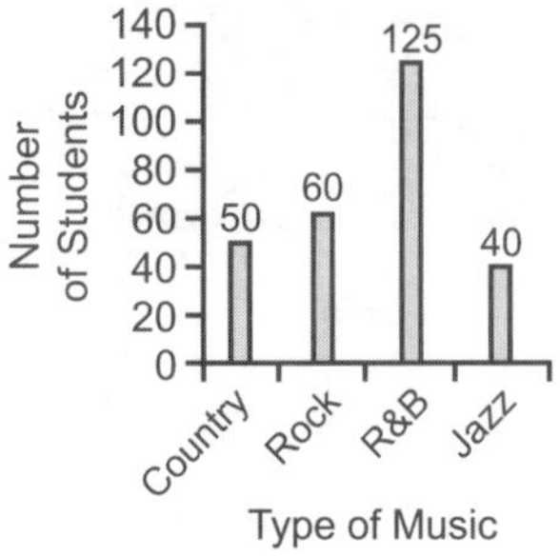

- Yes, the bar chart is misleading because there is no category for classical music.
- Yes, the bar chart is misleading because the bar for R&B music should be twice as tall as the bar for jazz music.
- No, the bar chart is not misleading.
- Yes, the bar chart is misleading because the bar for R&B music should be three times taller than the bar for country music.

31. Ten years ago, a survey was conducted and it was found that 120 people prefer regular cola to diet cola. A similar survey this year indicated that 400 people prefer regular cola to diet cola. The researchers concluded that more people now prefer regular cola to diet cola than they did ten years ago. Is the researchers' conclusion misleading? If so, how should the statistics be reported?

- The conclusion is misleading. The statistics should be reported as the increase in the number of people who prefer regular cola to diet cola.
- The conclusion is misleading. The statistics should be reported as the number of people who prefer diet cola to regular cola.
- The conclusion is not misleading.
- The conclusion is misleading. The statistics should be reported as the percent of people who prefer regular cola to diet cola.

32. Ten years ago, a survey was conducted and it was found that 80% of people prefer regular cola to diet cola. A similar study this year indicated that 60% of people prefer regular cola to diet cola. The researchers concluded that more people preferred regular cola to diet cola ten years ago. Is the researchers' conclusion misleading? If so, how should the statistics be reported?

- The conclusion is not misleading.
- The conclusion is misleading. The statistics should be reported as the percent of people who prefer diet cola to regular cola.
- The conclusion is misleading. The statistics should be reported as the decrease in the percent of people who prefer regular cola to diet cola.
- The conclusion is misleading. The statistics should be reported as the number of people who prefer regular cola to diet cola.

33. Ten years ago, a survey was conducted and it was found that 100 out of 125 people prefer regular cola to diet cola. A similar survey this year indicated that 600 people out of 750 people prefer regular cola to diet cola. The researchers concluded that more people now prefer regular cola to diet cola then they did ten years ago. Is the researchers' conclusion misleading? If so, how should the statistics be reported?

- The conclusion is misleading. The statistics should be reported as the percent of people who prefer regular cola to diet cola.
- The conclusion is not misleading.
- The conclusion is misleading. The statistics should be reported as the number of people who prefer diet cola to regular cola.
- The conclusion is misleading. The statistics should be reported as the increase in the number of people who prefer regular cola to diet cola.

34. Ten years ago, a survey was conducted and it was found that 100 out of 125 people prefer regular cola to diet cola. A similar survey this year indicated that 600 people out of 750 people prefer regular cola to diet cola. The researchers concluded that the same percentage of people prefer regular cola to diet cola now than they did ten years ago. Is the researchers' conclusion misleading? If so, how should the statistics be reported?

- The conclusion is misleading. The statistics should be reported as the increase in the number of people who prefer regular cola to diet cola.
- The conclusion is not misleading.
- The conclusion is misleading. The statistics should be reported as the number of people who prefer diet cola to regular cola.
- The conclusion is misleading. The statistics should be reported as the increase in the percentage of people who prefer regular cola to diet cola.

35. A city recently published a report about the number of deaths from car accidents that occurred over the past 50 years. Fifty years ago, there were only 15 deaths from car accidents. However, this year there were 235 deaths from car accidents. The report stated that car accidents result in more deaths this year than they did 50 years ago. Is the report's conclusion misleading? If so, how should the statistics be reported?

- The conclusion is not misleading.
- The conclusion is misleading. The statistics should be reported as the number of car accidents in which there was no death.
- The conclusion is misleading. The statistics should be reported as the increase in the number of deaths from car accidents.
- The conclusion is misleading. The statistics should be reported as the percent of car accidents that resulted in a death.

36. A city recently published a report about the number of deaths from car accidents that occurred over the past 50 years. Fifty years ago, 15 deaths resulted from 300 car accidents. However, this year there were 235 deaths resulting from 5,875 car accidents. The report stated that car accidents resulted in more deaths this year than they did 50 years ago. Is the report's conclusion misleading? If so, how should the statistics be reported?

- The conclusion is misleading. The statistics should be reported as the number of car accidents in which there was no death.
- The conclusion is not misleading.
- The conclusion is misleading. The statistics should be reported as the percent of car accidents that resulted in death.
- The conclusion is misleading. The statistics should be reported as the increase in the number of car accidents that resulted in a death.

37. A city recently published a report about the number of deaths from car accidents that occurred over the past 50 years. Fifty years ago, 3% of car accidents resulted in death. However, this year 5% of car accidents resulted in death. The report stated that car accidents resulted in more deaths this year than they did 50 years ago. Is the report's conclusion misleading? If so, how should the statistics be reported?

- The conclusion is misleading. The statistics should be reported as the increase in the number of car accidents that resulted in death.
- The conclusion is misleading. The statistics should be reported as the percent of deaths from car accidents.
- The conclusion is not misleading.
- The conclusion is misleading. The statistics should be reported as the decrease in the percent of car accidents that result in death.

38. A study was conducted twenty years ago and it was found that 65 out of 100 homes owned at least one television. A similar survey this year indicated that 40 out of 50 homes owned at least one television. The researchers concluded that fewer homes now own at least one television. Is the researchers' conclusion misleading? If so, how should the statistics be reported?

- The conclusion is misleading. The statistics should be reported as the decrease in the percent of homes that own at least one television.
- The conclusion is misleading. The statistics should be reported as the number of homes that do not own a television.
- The conclusion is misleading. The statistics should be reported as the percent of homes that own at least one television.
- The conclusion is not misleading.

39. A study was conducted twenty years ago and it was found that 80 out of 100 homes owned at least one television. A similar survey this year indicated that 40 out of 50 homes owned at least one television. The researchers concluded that fewer homes now own at least one television. Is the researchers' conclusion misleading? If so, how should the statistics be reported?

- The conclusion is misleading. The statistics should be reported as the percent of homes that own at least one television.
- The conclusion is misleading. The statistics should be reported as the number of homes that do not own a television.
- The conclusion is misleading. The statistics should be reported as the increase in the number of homes that own at least one television.
- The conclusion is not misleading.

40. A study was conducted twenty years ago and it was found that 75% of homes owned at least one television. A similar survey this year indicated that 85% of homes owned at least one television. The researchers concluded that more homes now own at least one television. Is the researchers' conclusion misleading? If so, how should the statistics be reported?

- The conclusion is misleading. The statistics should be reported as the decrease in the percent of homes that own at least one television.
- The conclusion is misleading. The statistics should be reported as the number of homes that own at least one television.
- The conclusion is not misleading.
- The conclusion is misleading. The statistics should be reported as the increase in the number of homes that own at least one television.

Answers to Selected Exercises

MPA-001: Identifying, Comparing, and Ordering Whole Numbers Through Billions

1. < **3.** < **5.** < **7.** 2,402; 2,489; 2,498; 2,561; 2,650 **9.** Ackerman **11.** 8 **13.** 3 **15.** 5 **17.** 0 **19.** Tens

MPA-002: Adding, Subtracting, Multiplying, and Dividing Whole Numbers

1. 133,017 **3.** 6,040 **5.** 173 r19 **7.** 39,660,802 **9.** 615,317,482 **11.** 82 r169 **13.** 6 **15.** 31,200

17. 3,993 **19.** 2,970

MPA-003: Using Four-Step Plan for Problem Solving

1. $52 **3.** $3,115 **5.** 63 pounds **7.** 42 chicken wings **9.** 23 hours and 46 minutes

11. 24 bananas **13.** $1.24 **15.** 16 gifts **17.** $9.00 **19.** $18.22

MPA-004: Using Rounding to Estimate

1. 5,000 **3.** 4,000 **5.** 18,000 **7.** 18,000 **9.** 10,000

11. 9,000 **13.** 29,000 **15.** 29,000 **17.** 16,000 **19.** 16,200

MPA-005: Estimating Products and Quotients Using Patterns

1. 37,800 **3.** 7,200 **5.** 73 **7.** $600,000 **9.** 1 book

11. 3,900,000 **13.** 80,000 **15.** 8 **17.** 27,500 **19.** 25

MPA-006: Determining Reasonableness of Answers and Appropriate Method of Computation

1. estimate **3.** exact **5.** exact **7.** no **9.** $50

11. estimate **13.** exact **15.** yes **17.** $470.00 **19.** $75,000.00

MPA-007: Solving Problems Using Logical Reasoning Skills

1. inductive reasoning **3.** inductive reasoning **5.** deductive reasoning **7.** younger **9.** smaller

11. less **13.** less **15.** less than **17.** Charles **19.** not enough info

21. John **23.** not enough info **25.** not enough info **27.** Paul **29.** James chose Group 1 and Blake chose Group 2

31. *Hate is Nothing* **33.** V. Flourney **35.** L. Erdrich **37.** S. Banks **39.** Rebecca Davis

MPA-008: Order of Operations

1. 15 **3.** 8 **5.** 2 **7.** 8 **9.** 21 **11.** 32 **13.** 8 **15.** 64 **17.** $23 **19.** $62

MPA-009: Solving One-Step Equations Using a Box

1. 5 **3.** 4 **5.** 20 **7.** $23.00 **9.** No **11.** No **13.** 39 **15.** No **17.** Yes **19.** 10

MPA-010: Solving One-Step Equations of Whole Numbers Using Addition and Subtraction

1. 54 **3.** $28.00 **5.** 836 **7.** 88 hair clips **9.** 5,104

11. subtracting 33 **13.** subtraction **15.** adding 544 **17.** adding 301 **19.** addition

MPA-011: Solving One-Step Equations of Whole Numbers Using Multiplication and Division

1. 13 weeks **3.** 7 **5.** 69 **7.** 154 **9.** 44 candy bars

11. dividing by 13 **13.** dividing by 917 **15.** divide by 102 **17.** multiply by 5 **19.** multiply by 3

MPA-012: Solving One-Step Equations of Whole Numbers Using All Operations

1. 1,630 **3.** 20,216 **5.** 603 **7.** 70 paintings **9.** 156

11. True **13.** False **15.** True **17.** False **19.** False

MPA-013: Using Powers and Exponents in Expressions

1. 125 **3.** 529 **5.** 196 **7.** 7 **9.** 13^5 **11.** 615 **13.** 876 **15.** 516 **17.** 1 **19.** 1,024

MPA-014: Evaluating Expressions for Given Variables

1. 11 **3.** 6 **5.** 35 **7.** 6 **9.** 15 **11.** 39 **13.** 16 **15.** 24 **17.** 8 **19.** 109

MPA-015: Identifying the Place Value of Decimals Through Thousandths

1. 0.28 **3.** 0.085 **5.** Thirty-eight hundredths **7.** 58.2 **9.** 0.892

11. Tenths **13.** Thousandths **15.** 0.11 **17.** 0.075 **19.** 12.085

MPA-016: Comparing and Ordering Decimals

1. > **3.** < **5.** (0.2; 0.26; 0.262; 0.263; 0.2632) **7.** (0.4; 0.47; 0.472; 0.473; 0.4732)

9. (6.26262; 62.6262; 626.262; 6,262.62; 62,626.2) **11.** Janet **13.** = **15.** (0; 0.21; 0.24; 0.241; 0.401)

17. Shaquita **19.** 6 hundredths

MPA-017: Rounding Decimals and Estimating Computations Using Decimals

1. 9.68 **3.** 6.292 **5.** 100 **7.** 26.2 **9.** 19.673 **11.** True **13.** False **15.** True **17.** 8 CDs **19.** 12 gifts

MPA-018: Adding and Subtracting Decimals

1. 110.914 **3.** 16.42 **5.** 10.79 **7.** 5.511 **9.** 20.0
11. 46.00 **13.** 1.923 **15.** $197.13 **17.** 40.93 meters **19.** 50.75 pages

MPA-019: Multiplying Decimals

1. 356.4 **3.** 78.3 **5.** 55.8 **7.** 332.621 **9.** $577.17
11. 1,231.2 **13.** 22.26 **15.** 73.12 **17.** 185.666 **19.** $51.17

MPA-020: Multiplying Decimals by Powers of Ten

1. 268.71 **3.** 43,865 **5.** 7,852.13 **7.** 54,674 **9.** 94.95
11. 979.8 **13.** 6.8543 **15.** 94.5065 **17.** 60 CDs **19.** $21,560

MPA-021: Converting Between Standard and Scientific Notation

1. 340 **3.** 1.2×10^8 **5.** 0.16×10^6 **7.** 215,000 **9.** 2.796×10^9 miles
11. 625 **13.** 78,660 **15.** 4.21×10^3 **17.** 8.99×10^8 **19.** 67,000,000 miles

MPA-022: Dividing Decimals

1. 60.2 **3.** divisor **5.** 5.9 **7.** 47.8 **9.** 4.1 oz
11. 45.78 **13.** 2.3 **15.** 104 **17.** $11 **19.** 16,135

MPA-023: Rounding Quotients Involving Decimals

1. 12.8 **3.** 1.7 **5.** 2.71 **7.** 9 **9.** $1.63 **11.** $0.22 **13.** $4.42 **15.** 1.68 **17.** 2.43 **19.** 1.83

MPA-024: Using Divisibility Rules

1. 5 **3.** 3 **5.** Yes **7.** Yes **9.** 6,642 **11.** No **13.** 5 **15.** Yes **17.** Yes **19.** 9 sectors

MPA-025: Identifying Prime and Composite Numbers

1. composite **3.** a number that has exactly two factors **5.** 21, 22, 24, 25, 26, 27, 28, 30 **7.** 39 **9.** False
11. 3, 5 **13.** False **15.** {19, 29} **17.** False **19.** composite

MPA-026: Using Prime Factorization

1. $2 \times 2 \times 3 \times 3$ **3.** $2 \times 2 \times 2 \times 2 \times 3$ **5.** $2 \times 3 \times 5$
7. 3×13 **9.** 83 **11.** $2^2 \times 3^2 \times 5$

13. 1, 2, 5, 10, 25, 50 **15.** $2 \times 2 \times 2 \times 5 \times 5 \times 5$ **17.** $3 \times 5 \times 7$

19. $2 \times 5 \times 17$

MPA-027: Finding the Greatest Common Factor

1. 12 **3.** 3 **5.** 3 **7.** 10 **9.** 2 **11.** 7 **13.** 9 pieces of bread **15.** 12 pads

MPA-028: Reducing Fractions to Lowest Terms / Simplest Form

1. $\frac{1}{2}$ **3.** $\frac{1}{4}$ **5.** $\frac{1}{7}$ **7.** $\frac{2}{7}$ **9.** $\frac{1}{3}$ **11.** $\frac{12}{16}$ **13.** $\frac{1}{2}$ **15.** $\frac{18}{19}$ **17.** $\frac{1}{5}$ **19.** $\frac{4}{12}$

MPA-029: Converting Fractions and Decimals

1. 0.17 **3.** 0.7 **5.** $\frac{12}{25}$ **7.** 0.875 **9.** 0.75

11. 0.25 **13.** $\frac{21}{100}$ **15.** $\frac{51}{200}$ **17.** $\frac{69}{100}$ **19.** $\frac{3}{15}$

MPA-030: Finding Least Common Multiple of Two or More Numbers

1. 35 **3.** 18 **5.** 36 **7.** 40 **9.** 60 seconds **11.** 75 seconds **13.** 60 **15.** 21

MPA-031: Comparing and Ordering Fractions and Decimals

1. $<$ **3.** $>$ **5.** $<$ **7.** $<$ **9.** $\frac{1}{8}, \frac{3}{8}, \frac{5}{8}, \frac{7}{8}$

11. $\frac{1}{4}, \frac{2}{6}, \frac{5}{12}, \frac{6}{8}$ **13.** Tabitha **15.** Plant A, B, D, C **17.** $\frac{1}{3}, \frac{4}{9}, \frac{3}{6}, \frac{13}{18}$ **19.** Stone's Discount Store

MPA-032: Converting Improper Fractions and Mixed Numbers

1. $\frac{7}{5}$ **3.** 6 **5.** $4\frac{3}{4}$ **7.** $\frac{12}{1}$ **9.** proper fraction **11.** $<$ **13.** $9\frac{2}{5}$ **15.** $\frac{31}{8}$ **17.** $5\frac{1}{4}$ **19.** $>$

MPA-033: Estimating Computations with Fractions and Mixed Numbers

1. $1\frac{1}{2}$ **3.** 18 **5.** 9 **7.** 9 **9.** 32 **11.** 7 **13.** 65 **15.** 39 **17.** No **19.** 33

MPA-034: Adding and Subtracting Fractions

1. $\frac{6}{7}$ **3.** $1\frac{1}{2}$ **5.** $1\frac{2}{25}$ **7.** $\frac{21}{22}$ **9.** $\frac{1}{11}$ **11.** $\frac{43}{60}$ **13.** $\frac{1}{2}$ cup **15.** $1\frac{1}{8}$ **17.** $\frac{7}{8}$ **19.** $\frac{31}{40}$

MPA-035: Adding and Subtracting Mixed Numbers with Unlike Denominators

1. $6\frac{5}{6}$ **3.** $8\frac{8}{21}$ **5.** $6\frac{16}{105}$ **7.** $11\frac{11}{18}$ **9.** $5\frac{1}{15}$

11. $4\frac{5}{42}$ **13.** $8\frac{7}{8}$ **15.** $47\frac{9}{10}$ miles **17.** $5\frac{1}{4}$ gallons **19.** $4\frac{80}{189}$

MPA-036: Multiplying Fractions and Mixed Numbers and Simplifying

1. 4 **3.** $\frac{5}{22}$ **5.** $\frac{1}{3}$ **7.** $\frac{1}{12}$ **9.** $\frac{63}{64}$ **11.** $\frac{1}{2}$ **13.** $1\frac{3}{5}$ **15.** $\frac{2}{11}$ **17.** 1 **19.** $\frac{5}{126}$

MPA-037: Dividing Fractions and Mixed Numbers and Simplifying

1. $\frac{1}{3}$ **3.** $1\frac{1}{9}$ **5.** $1\frac{5}{6}$ **7.** 26 **9.** 6 projects

11. $2\frac{1}{8}$ **13.** $\frac{2}{1}$ or 2 **15.** $\frac{1}{4}$ **17.** 32 packages **19.** $\frac{1}{12}$

MPA-038: Solving One-Step Equations with Fractions Using Addition and Subtraction

1. $z = 6\frac{1}{10}$ **3.** $x = 1\frac{1}{4}$ **5.** $m = 5\frac{2}{3}$ **7.** $x = 8\frac{5}{8}$ feet **9.** $x = 30\frac{2}{3}$

11. $r = \frac{1}{7}$ **13.** $f = 6\frac{4}{13}$ **15.** $x = \frac{1}{2}$ **17.** $b = \frac{3}{7}$ **19.** $d = \frac{4}{11}$

MPA-039: Solving One-Step Equations with Fractions Using Multiplication and Division

1. $\frac{1}{1}$ or 1 **3.** $\frac{32}{33}$ **5.** $\frac{4}{7}$ cup **7.** $\frac{14}{15}$ mile **9.** 13

11. $\frac{6}{13}$ **13.** 21 **15.** $\frac{3}{5}$ **17.** $\frac{6}{7}$ **19.** $\frac{8}{11}$

MPA-040: Solving One-Step Equations with Decimals Using All Four Operations

1. 0.6 **3.** 0.9 **5.** 7.2 **7.** 2.92 **9.** $4.02

11. $2.67 **13.** 8.5 minutes **15.** $0.26 **17.** 23 seconds **19.** 8.174

MPA-041: Writing Simple Algebraic Expressions from Phrases

1. $p + 9$ **3.** $\frac{b}{4}$ **5.** $17 - m$ **7.** $a + 14$ **9.** $52 - x$

11. $m - 16$ **13.** $\frac{b}{4}$ **15.** $b \times 49$ **17.** $20b - 8$ **19.** $17 + \frac{f}{3}$

MPA-042: Solving Problems Using an Equation

1. 34 **3.** 73 **5.** \$6.65 **7.** $f - 12 = 16$ **9.** $\frac{x}{5} = 11$

11. $3c = 24$ **13.** $\frac{n}{7} = 2$ **15.** $x - 34 = 7$ **17.** $d - 10 = 36$ **19.** $230 - k = 150$

MPA-043: Reading and Writing Integers

1. –977 **3.** –\$346 **5.** –85,452,000 **7.** Negative one hundred six **9.** –2 floors

11. –1,801 **13.** –5,526 **15.** –8,144 **17.** 1,760

19. Negative seven million, four hundred sixty-three thousand, two hundred eighty-five

MPA-044: Finding Opposite and Absolute Values of Integers

1. 150 **3.** 42 **5.** 15 **7.** –890 **9.** 7,578

11. 15 **13.** 89 **15.** False **17.** 6 feet **19.** \$800

MPA-045: Comparing and Ordering Integers

1. $<$ **3.** $>$ **5.** –5, 2, 12 **7.** today **9.** Ethel, Jamie, Danny

11. $>$ **13.** –13, –4, 0, 11, 14 **15.** –15, –14, –11, 5, 9 **17.** $>$ **19.** $=$

MPA-046: Graphing Points on a Coordinate Plane

1. (4, 6) **3.** (–2, –5) **5.** (7, 2) **7.** (–1, 8) **9.** Quadrant III **11.** Quadrant I

13. Quadrant III **15.** Quadrant II **17.** (7, 0) **19.** (0, 6) **21.** (0, 4)

23.

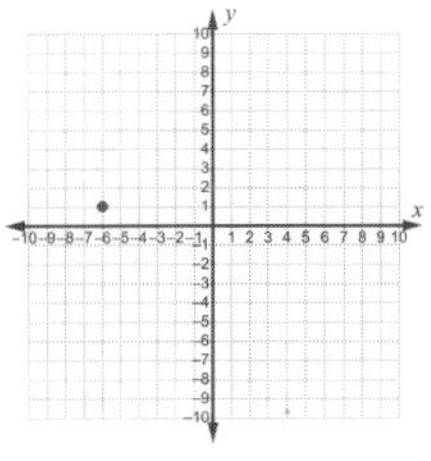

25.

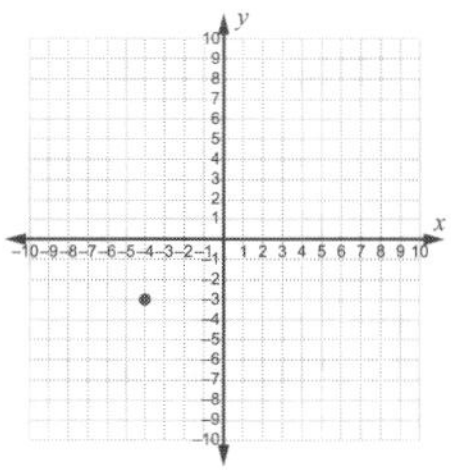

27.

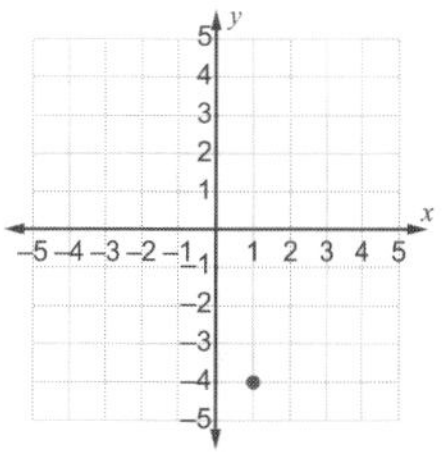

29.

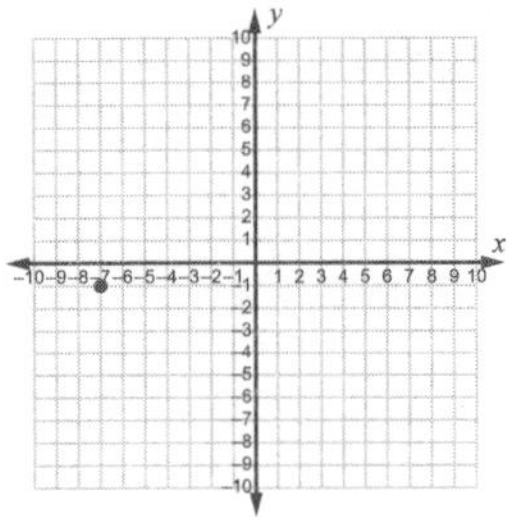

31.

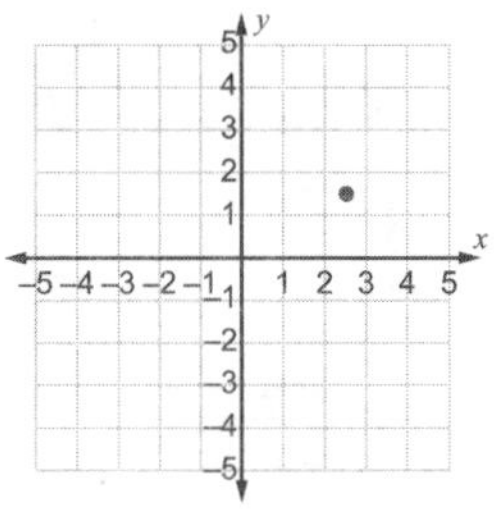

33.

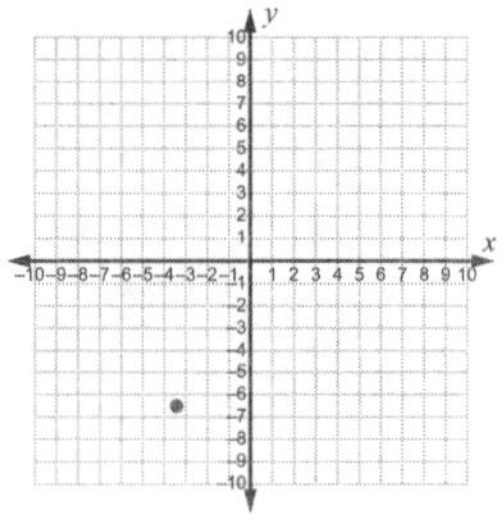

35.

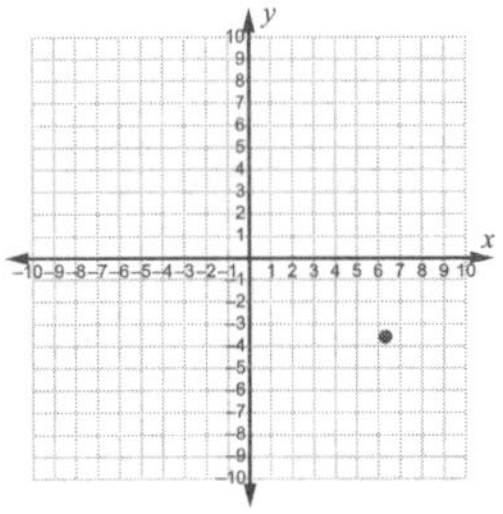

37.

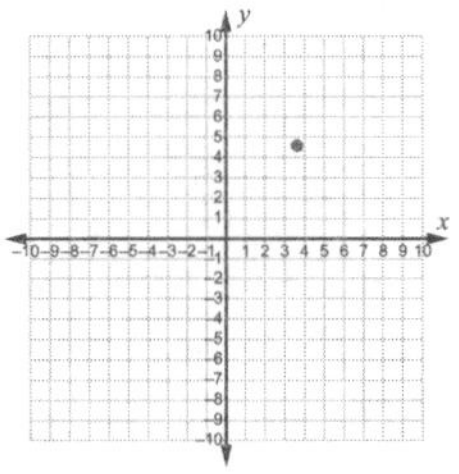

39.

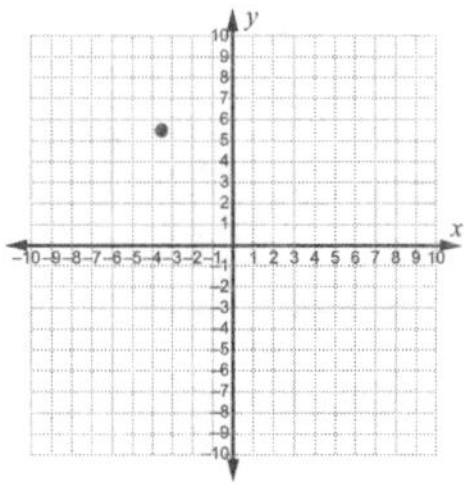

MPA-047: Adding Integers with Like Signs

1. 7 **3.** –10 yards **5.** 34 **7.** –200 **9.** 70

11. –7 **13.** 8 yellow counters **15.** 14 **17.** 107 **19.** greater than

MPA-048: Adding Integers with Unlike Signs

1. positive **3.** negative **5.** positive **7.** zero **9.** negative

11. –18 **13.** 31 **15.** 0 **17.** 8°C **19.** 3

MPA-050: Subtracting Integers with Unlike Signs

1. –16 **3.** –20 feet **5.** 320 **7.** –9 yards **9.** 16 **11.** –104 **13.** 28 **15.** 32 **17.** 6 **19.** –6

MPA-051: Multiplying Integers with Like and Unlike Signs

1. 42 **3.** –54 **5.** 54 **7.** 0 **9.** –868 **11.** 240 feet **13.** 630 **15.** –540 **17.** –8 **19.** –12

MPA-052: Dividing Integers with Like and Unlike Signs

1. 5 **3.** –3 **5.** 23 **7.** –7 **9.** 92 **11.** 2 **13.** 3 **15.** 10 **17.** 5 **19.** –8

MPA-053: Adding, Subtracting, Multiplying, and Dividing Integers

1. 36 **3.** –75 **5.** –1,638 **7.** 6 **9.** –4,248

11. –400 points **13.** 6 grandchildren **15.** positive **17.** $-2n+45$ **19.** $\frac{2n}{(-7)}$

MPA-054: Solving One-Step Equations with Integers Using all Four Operations

1. $n = -21$ **3.** $n = -12$ **5.** $x = 540$ **7.** $n = -12$ **9.** $n = 80$

11. $n = 14$ **13.** $n = 8$ **15.** $x = \$950$ **17.** $b = \$2$ **19.** $a = 11$ feet

MPA-055: Finding the Perimeter of a Figure

1. 208 **3.** 174 feet **5.** 6 in. **7.** 72 in. **9.** 25 ft

11. 20 yd **13.** 22 mi **15.** 39 **17.** 6 ft **19.** 30

MPA-056: Classifying Angles

1. B **3.** False **5.** True **7.** False **9.** Acute angle

11. $\angle K$ **13.** $\angle A$ **15.** $\angle U$ **17.** $\angle I$ **19.** Right

MPA-057: Identifying and Applying Supplementary and Complementary Angles

1. $\angle QRT$ or $\angle TRQ$ **3.** $\angle YXZ$ or $\angle ZXY$ **5.** $\angle NMO$ or $\angle OMN$ **7.** $\angle JIK$ or $\angle KIJ$ **9.** $\angle PQR$ or $\angle RQP$

11. $\angle JLK$ or $\angle KLJ$ **13.** $\angle MNO$ or $\angle ONM$ **15.** $\angle QRS$ or $\angle SRQ$ **17.** 135° **19.** 65°

21. 60° **23.** $m\angle A = 15°$ **25.** $m\angle E = 65°$ **27.** $m\angle S = 100°$ **29.** $m\angle X = 25°$

31. $m\angle CGD = 40°$ **33.** $m\angle 3 = 46°$ **35.** $m\angle 3 = 45°$ **37.** $m\angle 3 = 28°$ **39.** $m\angle 3 = 28°$

MPA-058: Identifying Polygons

1. not a polygon **3.** regular triangle **5.** True **7.** heptagon **9.** not a polygon

11. False **13.** Not a polygon **15.** Polygon **17.** Heptagon **19.** Hexagon

MPA-059: Classifying Triangles and Quadrilaterals

1. its angles and its sides **3.** True **5.** True **7.** trapezoids **9.** isosceles, right

11. obtuse scalene triangle **13.** rectangle **15.** quadrilateral **17.** obtuse **19.** scalene

MPA-060: Determining Which Figures Tessellate

1. No **3.** No **5.** Yes **7.** Yes **9.** Yes

11. Yes **13.** Yes **15.** No **17.** No **19.** regular hexagon

MPA-061: Converting Metric Units of Length, Capacity, and Mass

1. 834,000 centimeters **3.** 1,023 hectoliters **5.** meters **7.** Grams **9.** 100 milliliters; 99.9 centiliters; 99 deciliters; 1,000 liters

11. 23,000; 6,500 **13.** 930,000 milliliters **15.** $<$ **17.** 58, Yes **19.** 1.7 meters

MPA-062: Converting Units in Customary System

1. 9 pounds **3.** 5 gallons **5.** 72 quarts **7.** 108 inches **9.** 18 feet

11. 23.75 gallons **13.** 1.5 miles **15.** 20 gallons 1 quart **17.** 100 yards 2 feet **19.** 195 ounces

MPA-063: Converting Units Between Metric and Customary System

1. 118.4 milliliters **3.** 10.01 meters **5.** 59.2 milliliters **7.** No **9.** False

11. 3.071 ounces **13.** 8.036 ounces **15.** 2,940 grams **17.** 448 grams **19.** 38.46 yards

MPA-064: Finding Square Roots of Perfect Squares

1. 10 **3.** 12 **5.** 20 **7.** 48 in. **9.** 3

11. 6 **13.** 9 inches **15.** 3 cm **17.** 4 feet **19.** 64

MPA-065: Estimating Square Roots

1. 49, 64 **3.** 6, 7 **5.** 9.9 **7.** 5.7 **9.** 121, 144

11. 12 **13.** 15.17 **15.** 11 **17.** 4.6 **19.** 19.4 km

MPA-066: Solving Problems Using the Pythagorean Theorem

1. 13.6 **3.** 4.4 **5.** 8.5 **7.** 25.5 in. **9.** 45 **11.** 19.8 **13.** 93.6 **15.** 16.6 **17.** 11.3 in. **19.** No

MPA-067: Finding the Area of Rectangles and Parallelograms

1. 224 in.^{2} **3.** 12 in. **5.** 282.24 ft^2 **7.** $10\frac{11}{13}$ ft **9.** 49 ft^2

11. 6 ft **13.** $9\frac{16}{43}$ in. **15.** 12 **17.** 233.64 **19.** 93.5 in^2

MPA-068: Finding the Area of Irregular Figures

1. $57\frac{1}{2}$ square units **3.** 280 in.^{2} **5.** 72 in.^{2} **7.** 194 in.^{2} **9.** 56 cm^2

MPA-069: Finding the Area of Triangles and Trapezoids

1. 82.5 square units **3.** 300 square units **5.** 360 square units **7.** 10 square units **9.** 3 square units

11. 67.5 square units **13.** 46 square units **15.** 80 square units **17.** 12 in. **19.** 2 ft

MPA-070: Finding the Circumference of a Circle

1. 21.98 in. **3.** 98 in. **5.** 42 in. **7.** 288.88 in. **9.** 35 in.

11. False	**13.** 25.12 in.	**15.** 7.75 in.	**17.** 4.5 in.	**19.** 25.12 in.

MPA-071: Finding the Area of a Circle

1. 28.26 m^2	**3.** 7,850 $in.^2$	**5.** 1,962.5 yd^2	**7.** 3,629.84 cm^2	**9.** 4,775.94 cm^2
11. 31,400 mm^2	**13.** 19.63 $in.^2$	**15.** 45.34 cm^2	**17.** 14 ft	**19.** 40 mi

MPA-072: Identifying 3-D Figures

1. prism	**3.** sphere	**5.** 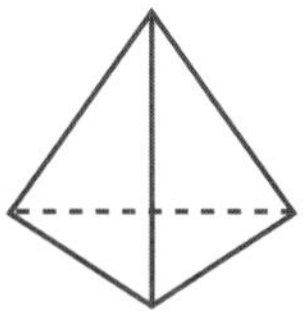	**7.** cylinder	**9.** 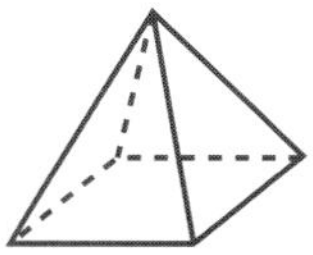
11. 12 edges	**13.** 5 faces	**15.** tetrahedron	**17.** None	**19.** 2 bases

MPA-073: Finding the Surface Area of Rectangular Prisms

1. 1,056 m^2	**3.** 112 $in.^2$	**5.** 112 ft^2	**7.** 92 cm^2	**9.** 158.52 cm^2
11. 88.8 $in.^2$	**13.** 0.54 cm^2	**15.** $\sqrt{6}$ in.	**17.** 22 in.	**19.** 96 $in.^2$

MPA-074: Finding the Surface Area of Cylinders

1. 489.84 ft^2	**3.** 282.6 $in.^2$	**5.** 37.68 $in.^2$	**7.** 282.6 ft^2	**9.** 397.9 cm^2
11. 163.22 $in.^2$	**13.** 1,444.4 $in.^2$	**15.** 678.24 cm^2	**17.** 37.7	**19.** 621.7

MPA-075: Finding the Volume of Rectangular Prisms

1. 27 dm^3	**3.** 5,832 cm^3	**5.** 43.7 $in.^3$	**7.** 2.744 km^3	**9.** $\frac{1}{22}$ $in.^3$
11. 42.875 $in.^3$	**13.** 1.331 cm^3	**15.** 64,000 ft^3	**17.** 756 cm^3	**19.** 468 cm^3

MPA-076: Finding the Volume of Cylinders

1. 175.84 ft^3	**3.** 2,826 $in.^3$	**5.** 50.24 ft^3	**7.** 2.355 $in.^3$	**9.** 14,243.04 cm^3
11. 254.34 m^3	**13.** 1,177.5 $in.^3$	**15.** 7,385.28 m^3	**17.** 3	**19.** 2.2

MPA-077: Solving Problems Using a Formula

1. 3 **3.** 6 **5.** 22 **7.** 176.4 **9.** $C = 3.14 \times D$

11. 110 miles **13.** 3 mph **15.** 5 hours **17.** 100.8 newtons **19.** $21\frac{m}{s^2}$

MPA-078: Expressing Ratios as Fractions and Determining Equivalency

1. $\frac{1}{2}$ **3.** $\frac{3}{1}$ **5.** No **7.** $\frac{3}{5}$ **9.** $\frac{5}{3}$ **11.** $\frac{7}{3}$ **13.** $\frac{2}{1}$ **15.** $\frac{20}{21}$ **17.** Yes **19.** $\frac{126}{21}$

MPA-079: Unit Rates

1. 4 miles per hour **3.** $0.12 per gram **5.** 143 miles per hour **7.** 76.5 miles per hour **9.** 0.11 cup

11. $1.64 per pound **13.** 9 miles per gallon **15.** True **17.** Job 2 **19.** Zesty

MPA-080: Solving Proportions

1. = **3.** = **5.** ≠ **7.** ≠ **9.** 11

11. Yes **13.** $d = 14$ **15.** 4,000 girls **17.** 45 hours **19.** 40 hours

MPA-081: Converting Fractions, Decimals, and Percents I

1. 0.86 **3.** 0.77 **5.** 29% **7.** 93% **9.** 60% **11.** 0.06 **13.** 80% **15.** $\frac{31}{100}$ **17.** 35% **19.** 36%

MPA-082: Converting Fractions, Decimals, and Percents II

1. 120% **3.** 580% **5.** 0.0045 **7.** 0.001 **9.** 267%

11. $1\frac{1}{4}$ **13.** 1.85 **15.** 3.1 **17.** 2.25 **19.** 275%

MPA-083: Finding Number Given Percent and Total

1. 9 homes **3.** 102 schools **5.** $1.88 **7.** 6,335.94 **9.** 5,600

11. 17,020 votes **13.** 676 immigrants **15.** 32 **17.** 119 **19.** 2.8

MPA-084: Finding Percent Given Number and Total

1. 12% **3.** 69% **5.** 31% **7.** 70% **9.** 50% **11.** 40% **13.** 90% **15.** 25% **17.** 25% **19.** 75%

MPA-085: Finding Total Given Number and Percent

1. 300 calories **3.** $36 **5.** $20 **7.** $250 **9.** 20

11. 40 **13.** 120 **15.** 140 **17.** 10 **19.** 120 pounds

MPA-086: Solving Problems Using Percent

1. 48 **3.** 18.8% **5.** 80 **7.** 140 **9.** 7 video games
11. $150 **13.** 20 miles **15.** 24% **17.** 38% **19.** 48 hours

MPA-087: Finding Percent Increase and Decrease

1. 21% **3.** 33% **5.** 11% **7.** 51% **9.** 16%
11. 21% **13.** 70% **15.** increase **17.** increase **19.** increase

MPA-088: Solving Real-World Problems Involving Percent

1. $39.79 **3.** $7.44 **5.** $64.80 **7.** $740.71 **9.** $69.07
11. $60.00 **13.** $15.48 **15.** $80.96 **17.** $436.30 **19.** $5.07
21. $4.72

MPA-089: Using Tree Diagrams

1. 6 **3.** 4 **5.** 9 **7.** 4 **9.** 18 **11.** 9 **13.** 12 **15.** 12 **17.** 8 **19.** 36

MPA-090: Finding the Probability of Simple Real-Life Events

1. $\frac{1}{5}$ **3.** $\frac{11}{25}$ **5.** $\frac{1}{3}$ **7.** $\frac{15}{46}$ **9.** $\frac{7}{29}$ **11.** $\frac{3}{14}$ **13.** $\frac{11}{38}$ **15.** $\frac{7}{18}$ **17.** $\frac{22}{29}$ **19.** $\frac{11}{12}$

21. $\frac{40}{63}$ **23.** $\frac{1}{3}$ **25.** $\frac{3}{7}$ **27.** $\frac{5}{9}$ **29.** $\frac{5}{12}$ **31.** $\frac{4}{9}$; not equally likely **33.** $\frac{1}{2}$; equally likely

35. $\frac{1}{2}$; equally likely **37.** $\frac{1}{2}$; equally likely **39.** $\frac{1}{3}$; equally likely

MPA-091: Finding the Number of Combinations of a Set of Objects

1. 3,024 **3.** 120 permutations **5.** 6 permutations **7.** 24 permutations **9.** 54,740
11. 6 **13.** 24 **15.** 3,628,800 **17.** 72 **19.** 18

MPA-092: Reading and Interpreting Bar, Line, and Circle Graphs

1. miscellaneous **3.** Group 1 **5.** August **7.** $2,496 **9.** $2,880
11. 70 degrees **13.** 50 students **15.** 72 girls **17.** 0 inches **19.** 48 students

MPA-093: Choosing Appropriate Scales and Intervals for Data

1. 67 **3.** 17 **5.** 69 **7.** 51 **9.** Yes
11. No **13.** Yes **15.** 9 tick marks **17.** Nothing **19.** The horizontal axis is inconsistent

MPA-094: Interpreting and Constructing Line Plots

1. 5 **3.** 66 **5.** 100 **7.** 12 **9.** yes

11. no **13.** no **15.** 32 and 35 **17.** 84 and 99 **19.** 83

MPA-095: Find the Mean, Median, and Mode

1. No **3.** 1 **5.** 29.1 **7.** 48 **9.** 56.5

11. 32 **13.** 22 **15.** 11 pounds **17.** 26, 77 **19.** $120

MPA-096: Constructing Stem-and-Leaf Plots

1. 2 **3.** 5 **5.** 9 **7.** No **9.** No **11.** 31 **13.** 73 **15.** 2.5 **17.** 6.25 **19.** 190

MPA-097: Constructing Box-and-Whisker Plots

1. 25 **3.** 32 **5.** 5 **7.** 1 **9.** 300

11. 12 **13.** 1,300 **15.** 99 **17.** 0% **19.** 500

MPA-098: Making Predictions from Graphs and Choosing the Correct Graph

1. pie graph **3.** pie graph **5.** blue **7.** 28 inches **9.** $1.00

11. 3 **13.** 29% **15.** pie graph **17.** bar graph **19.** Decreasing

MPA-099: Recognizing Misleading Statistics and Graphs

1. mean **3.** mode **5.** none of these **7.** No, because the president only called her friends, whose schedules are similar to hers. She should choose to call a variety of different people in the neighborhood

9. No **11.** Friday **13.** Not enough information **15.** False **17.** No **19.** Yes

MPA-100: Solving Two-Step Equations

1. 10 sheets **3.** 9 yards **5.** 14 **7.** $4 **9.** 6 hours

11. 5 **13.** 14 **15.** –11 **17.** –10 **19.** 4

MPA-101: Solving Equations by Combining Like Terms

1. 6 **3.** 8 **5.** 3.7 **7.** 9.6 **9.** 20 **11.** 5 **13.** –3 **15.** 4 **17.** 21 **19.** –26

MPA-102: Graphing Equations by Plotting Points

1. vertical **3.** horizontal **5.** 18 **7.** (3, 3) **9.** (11, –1)

11. 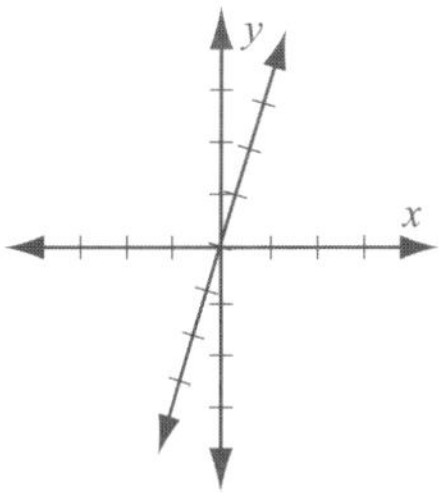

13. 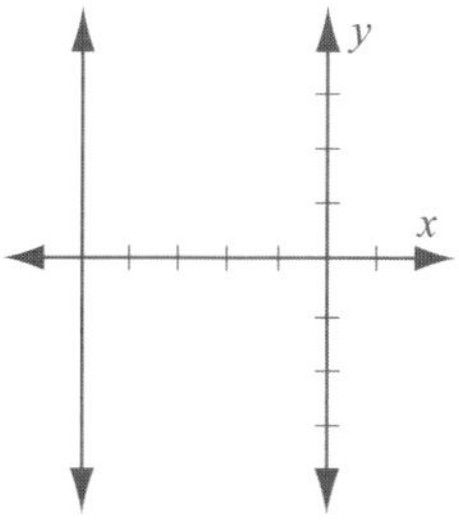

15. vertical **17.** horizontal **19.** neither

MPA-103: Distinguishing Between Relations and Functions

1. yes **3.** yes **5.** no **7.** no **9.** no **11.** no **13.** yes **15.** no **17.** no **19.** yes

MPA-104: Recognizing Patterns

1. addition **3.** addition **5.** subtraction **7.** 35.84 **9.** $6\frac{1}{2}$

11. 42 **13.** 32 **15.** –1 **17.** subtraction **19.** multiplication

MPA-105: Determining the Measure of Angles Made by Parallel Lines and a Transversal

1. $\angle 1$ and $\angle 5$ **3.** $\angle 3$ and $\angle 7$ **5.** $\angle 5$ and $\angle 1$ **7.** $\angle 1$ and $\angle 5$ **9.** $\angle 4$ and $\angle 8$

11. $\angle 1$ and $\angle 8$ **13.** $\angle 4$ and $\angle 7$ **15.** $\angle 1$ and $\angle 7$ **17.** 37° **19.** 130°

21. 130° **23.** 101° **25.** $m\angle 5 = 42°$ and $m\angle 6 = 77°$ **27.** $m\angle 7 = 83°$ and $m\angle 8 = 80°$ **29.** $m\angle 7 = 71°$ and $m\angle 8 = 69°$

31. 30° **33.** 45° **35.** 32° **37.** 44° **39.** 52°

MPA-106: Identifying a Solid Figure from a Net

1. 4, 4, 6 **3.** 8, 12, 18 **5.** 5, 5, 8 **7.** False

9. False **11.** False **13.** True **15.** triangular pyramid **17.** cone **19.** 54 square units

21. 20π units^2 **23.** 16 square units **25.** 33 units^2 **27.** 24 units^2 **29.** 38

31. 54 **33.** 16 **35.** 64 **37.** 304 **39.** 320

MPA-107: Constructing Three-Dimensional Figures and Examining Their Dimensions

1. cylinder **3.** sphere **5.** rectangular prism **7.** square pyramid **9.**

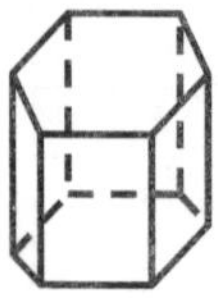

11.

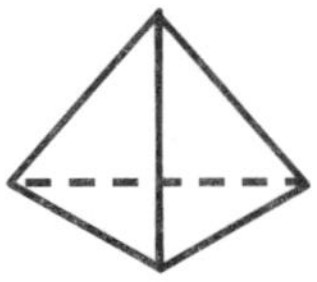

13.

15.

17. triangle **19.** square

21. octagon **23.** circle **25.** rhomboidal prism **27.** decagonal pyramid **29.** heptagonal prism

31. 81π square units **33.** 16π square units **35.** 121 square units **37.** 102 square units **39.** 48 square units

MPA-108: Graphing Translations and Reflections on the Coordinate Plane

1. No **3.** Yes **5.** Yes **7.** Yes **9.** True

11. False **13.** False **15.** False **17.** (1, –3) **19.** (4, 4)

21. (–5, 4)

23. (4, -3)
(7, -3)
(4, -5)
(7, -5)

25. (-4, 4)
(-8, 4)
(-9, 1)
(-5, 1)

27. (5, 3)
(4, 6)
(2, 4)

29. (-3, 0)
(-1, -1)
(1, 0)
(-1, 2)

31. (-2, 1)
(-6, 1)
(-2, 5)
(-6, 5)

33. (2, -1)
(4, -3)
(1, -4)

35. (2, 1)
(5, 1)
(1, 3)
(4, 3)

37. (-2, 4)
(-3, 5)
(-4, 4)
(-3, 2)

39. (-5, 0)
(-2, 0)
(-5, 3)
(-2, 3)

MPA-109: Solving and Graphing Linear Inequalities on a Number Line

1. $x > -3$ **3.** $x > 5$ **5.** $x < -2$ **7.**

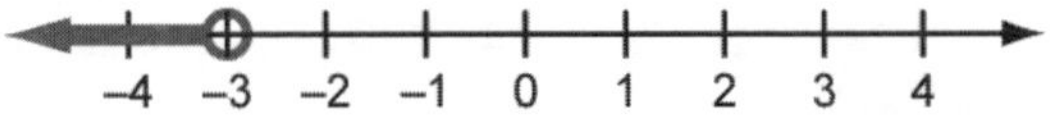

9. $x \geq 3$

11.

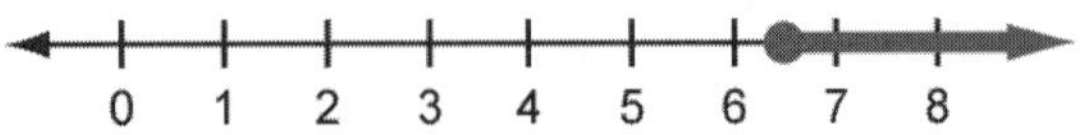

13.

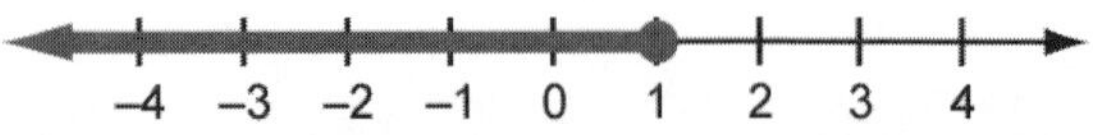

15. 

17. 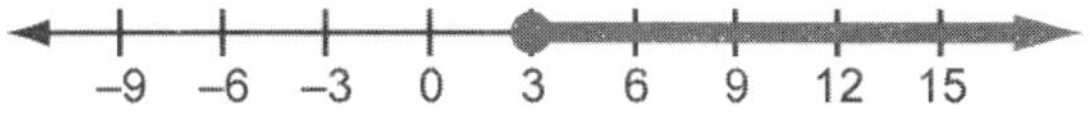

19. 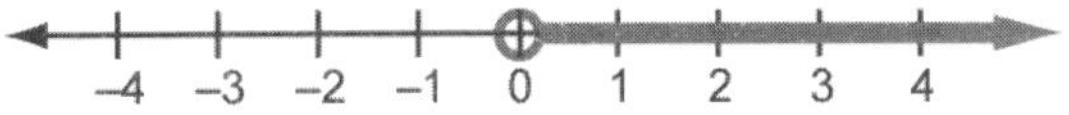

21. 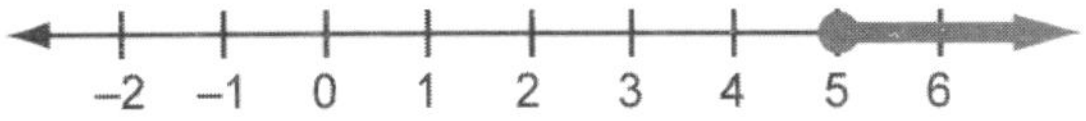

23. 

25. 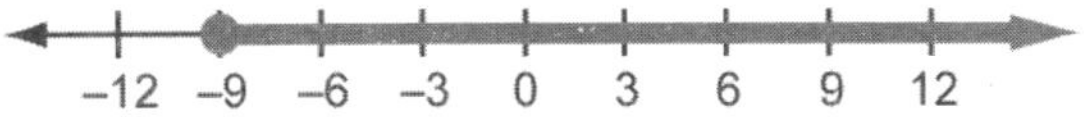

27. 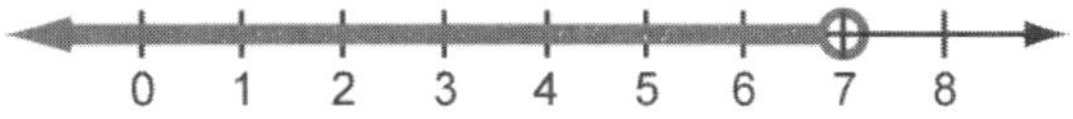

29. 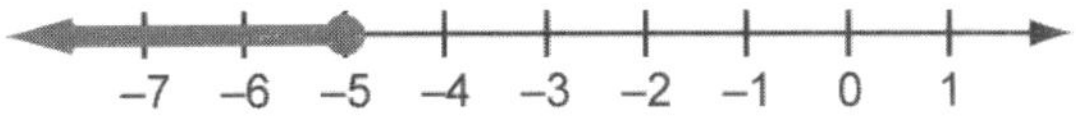

31. 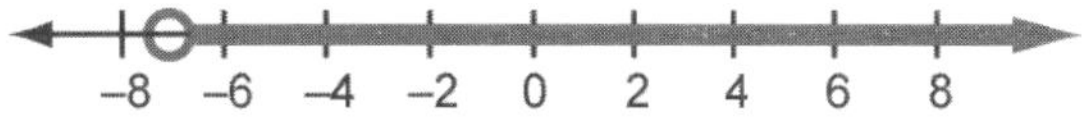

33. 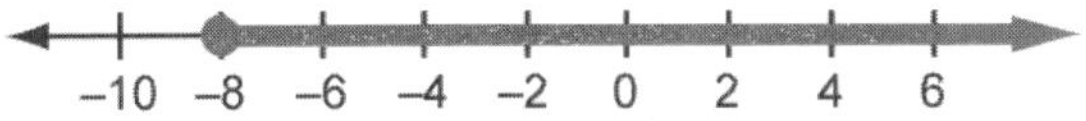

35. 

37. 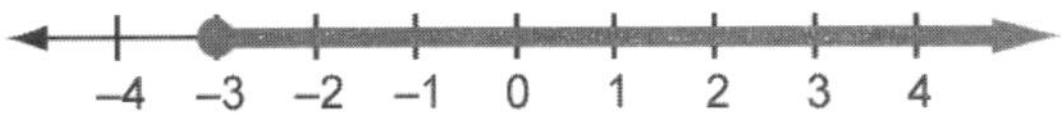

39. 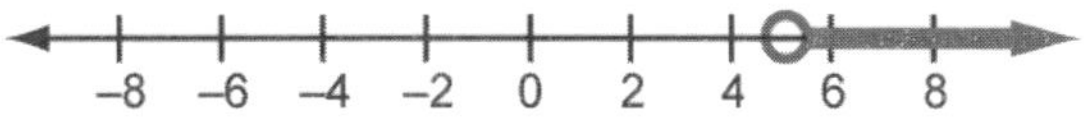

MPA-110: Solving Problems Using Proportions, Scale Drawings, Models and Maps

1. 1 inch : 40 miles **3.** 1 inch : 182 miles **5.** 1 inch : 352 kilometers **7.** 1 inch : 183 miles **9.** 1,640 meters

11. 2,250 meters **13.** 390 meters **15.** 550 meters **17.** 5 feet **19.** 40 feet

21. 71.4 feet **23.** 40 feet **25.** 60 feet **27.** 25 feet **29.** 28.6 feet

31. 1,000 miles **33.** 3,150 miles **35.** 900 miles **37.** 4 centimeters **39.** 11.09 centimeters.

MPA-111: Comparing Perimeters, Areas and Volumes of Similar Geometric Figures and Solids

1. 5 **3.** 8 **5.** 8 inches **7.** 12 centimeters **9.** 13 centimeters

11. 64 times **13.** 144 times **15.** 25 times **17.** 64 square centimeters **19.** 54 square inches

21. 32 square feet **23.** 4 square feet **25.** 4 times **27.** 25 times **29.** 4 times

31. 64 times **33.** 8 times **35.** 343 times **37.** 512 times **39.** 27 times

MPA-112: Constructing Sample Spaces for Compound Events (Dependent and Independent)

1. dependent **3.** dependent **5.** independent **7.** independent

9. {HHH, HHT, HTH, HTT, THH, THT, TTH, TTT} **11.** {CS, CL, CD, BS, BL, BD, PS, PL, PD}

13. {AX, AY, AZ, BX, BY, BZ, CX, CY, CZ} **15.** {RS, RP, GS, GP, BS, BP} **17.** $\frac{1}{10}$

19. $\frac{3}{16}$ **21.** $\frac{1}{5}$ **23.** $\frac{1}{7}$ **25.** $\frac{1}{4}$

27. $\frac{1}{5}$ **29.** $\frac{1}{4}$ **31.** $\frac{12}{23}$; dependent **33.** $\frac{11}{16}$; independent **35.** $\frac{12}{51}$; dependent

37. $\frac{15}{28}$; dependent **39.** $\frac{1}{52}$; independent

MPA-113: Finding the Probability of Compound Events Through Experimentation

1. $\frac{1}{8}$ **3.** $\frac{1}{8}$ **5.** 1 **7.** $\frac{3}{4}$ **9.** $\frac{19}{36}$ **11.** $\frac{7}{16}$ **13.** $\frac{1}{2}$ **15.** $\frac{7}{12}$ **17.** $\frac{1}{7}$ **19.** $\frac{1}{9}$

21. $\frac{1}{16}$ **23.** $\frac{1}{6}$ **25.** $\frac{1}{2}$ **27.** $\frac{5}{9}$ **29.** $\frac{8}{9}$ **31.** $\frac{1}{2}$ **33.** $\frac{1}{8}$ **35.** $\frac{1}{4}$ **37.** $\frac{1}{3}$ **39.** $\frac{1}{3}$

MPA-114: Finding the Odds of Events and Experimental Probability from a Math Model

1. 1:7 **3.** 1:3 **5.** 1:23 **7.** 1:18 **9.** 21:1 **11.** 7:1 **13.** 17:1 **15.** 9:1 **17.** 1:4 **19.** 3:8

21. 5:24 **23.** $\frac{3}{10}$ **25.** $\frac{3}{20}$ **27.** $\frac{7}{25}$ **29.** $\frac{1}{2}$ **31.** $\frac{13}{30}$ **33.** $\frac{17}{25}$ **35.** $\frac{19}{30}$ **37.** $\frac{5}{8}$ **39.** $\frac{33}{50}$

MPA-115: Finding the Volumes of Prisms, Cylinders, Pyramids, and Cones Using Models

1. 280 cm^3 **3.** 90 m^3 **5.** 336 ft^3 **7.** 24, 000 m^3 **9.** 320 yd^3

11. 20, 000 cm^3 **13.** 4, 500 ft^3 **15.** 780 in^3 **17.** 949.9 in^3 **19.** 2.4 ft^3

21. 137.4 in^3 **23.** 196 in^3 **25.** 56 in^3 **27.** 56.3 m^3 **29.** 36 in^3

31. 769.3 in^3 **33.** 602.9 ft^3 **37.** 56.5 m^3 **37.** 2, 198 ft^3 **39.** 5, 539 yd^3

MPA-116: Solving Real-Life Problems by Using Guess-and-Check and Working Backwards

1. 21 cups **3.** 12,334 miles **5.** $252 **7.** 18 people **9.** 36 bags

11. 57 pounds **13.** 35 yard line **15.** $25 **17.** 18 years old **19.** 18 points

21. three 10 dollar bills **23.** 9 of each coin **25.** 3 months **27.** 36 adults **29.** 36 people

31. 72 **33.** hot dog, salad, and water **35.** 16 questions **37.** 45 problems **39.** 36 cookies

MPA-117: Modeling Integer Arithmetic Using Cups and Counters

1. $2x - 3$

3. $(x + 1) + (2x + 4)$

5. 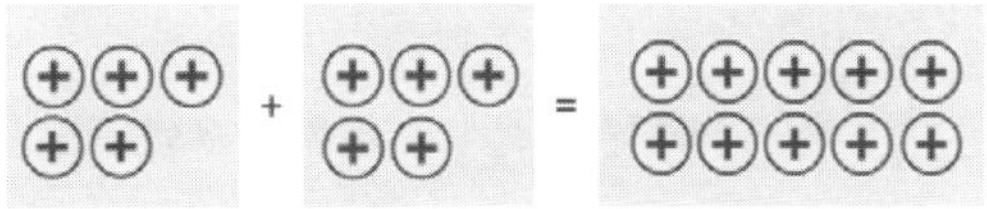

7. 

9. $2x - 3$

11. $-2 + 3 = 1$

13. 

15. 

17. $(2x - 2) - (2x + 2) = -4$

19. $x = 8$

21. $5x - 4$

23. $(2x + 1) + (x + 4)$

25. 

27. $-5 + 2 = -3$

29. $-3 + (-9) = -12$

31. 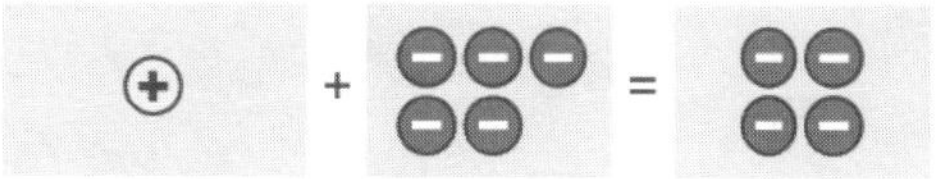

33. 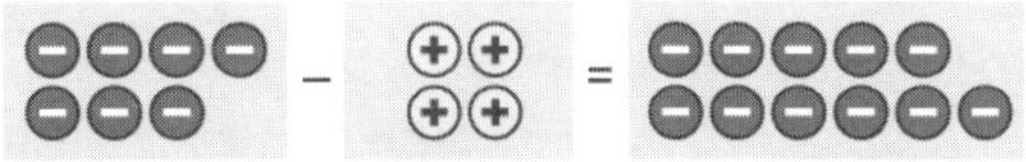

35. $-4 - 4 = -8$

37. $x = 6$

39. $x = -4$

MPA-118: Modeling Integer Arithmetic Using Algebra Tiles

1. $x^2 - 6x + 4$

3. $3x^2 - 4x + 2$

5. $5x^2 - 2x - 2$

7. $4x^2 + 4x - 3$

9. $(x + 5) + (4x - 1) = (5x + 4)$

11. $(5x + 2) + (x - 1) = (6x + 1)$

13. $(3x - 2) + (x - 2) = (4x - 4)$

15. $(2x + 4) + (x - 1) = (3x + 3)$

17. $x^2 - 4x + 3$

19. $6x^2 - x - 2$

21. $6x^2 + 5x + 1$

23. $4x^2 + 4x - 3$

25. 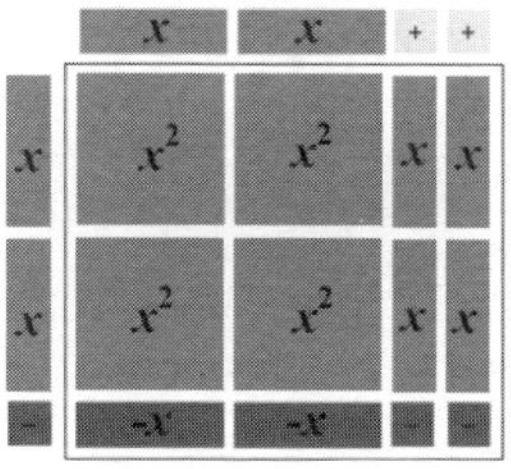

27. 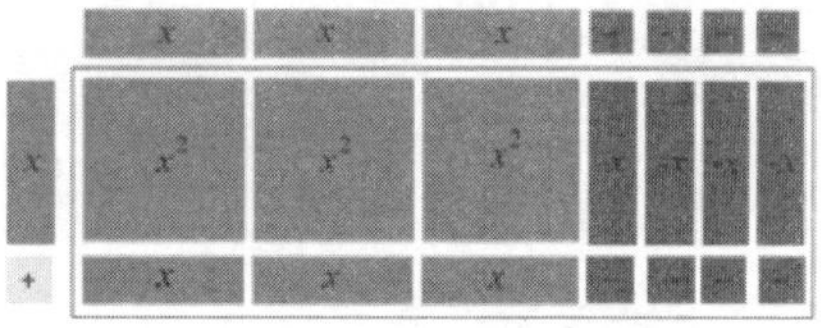

29. 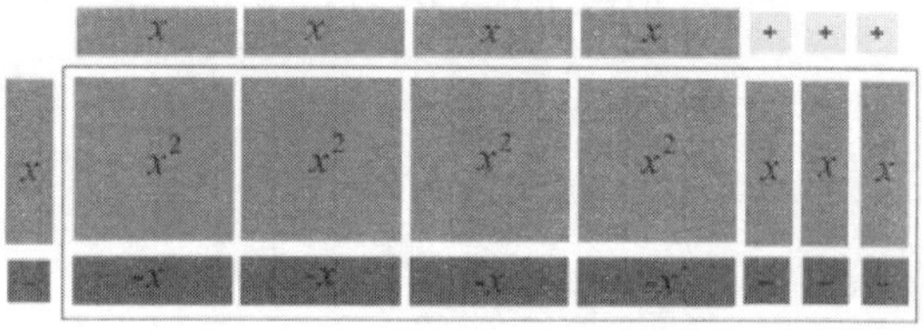

31. $x = 10$ **33.** $x = -8$ **35.** $x = 2$ **37.** $x = 5$ **39.** $x = 2$

MPA-119: Dividing Decimals (an Introduction)

1. 4.7 **3.** 9.7 **5.** 6.7 **7.** 9.7 **9.** 24 **11.** 33 **13.** 235 **15.** 65 **17.** 64.3 **19.** 56.36

21. 65.8 **23.** 15 **25.** 85 **27.** 75 **29.** 126

31. 8.4 milliliters **33.** 4.6 ounces **35.** 65.3 meters **37.** 1.6 pounds **39.** 2.2 miles

MPA-120: Applying Dilations in the Coordinate Plane

1. The scale factor is less than one.

3. The scale factor is greater than one.

5. The scale factor is less than one.

7.

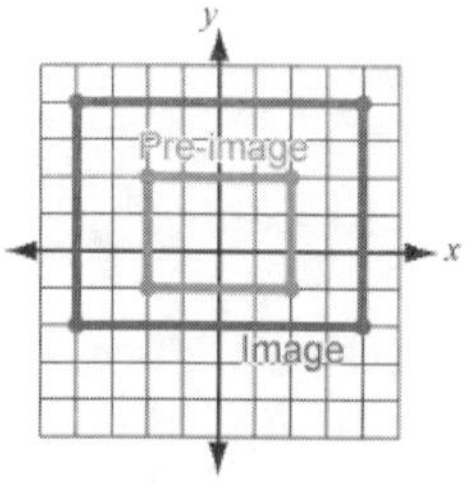

9.

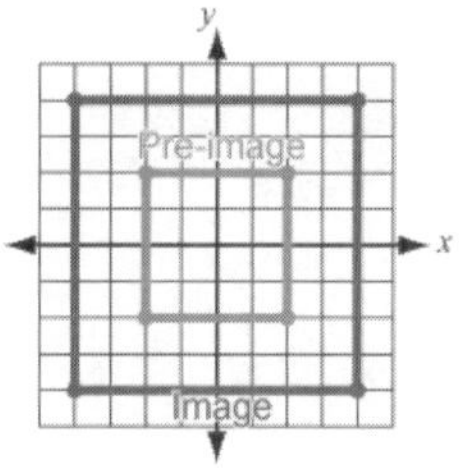

11.

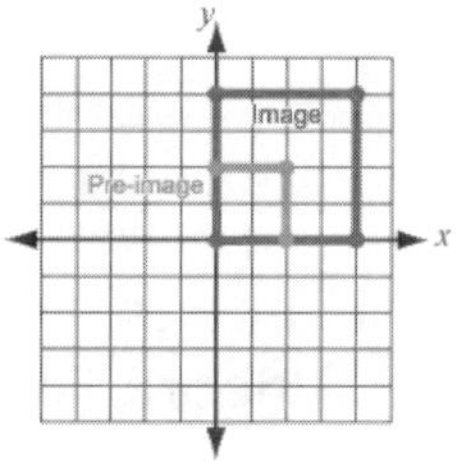

13.

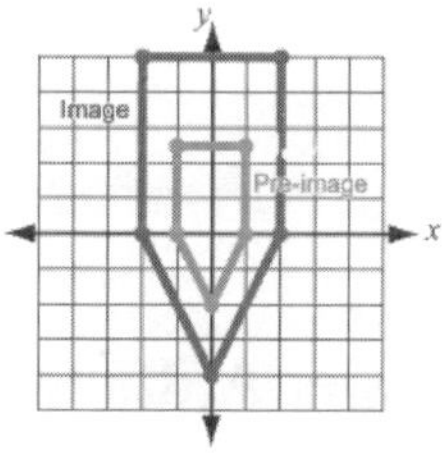

15.

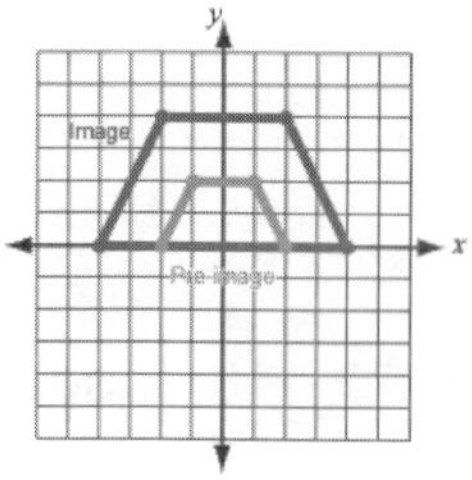

17.

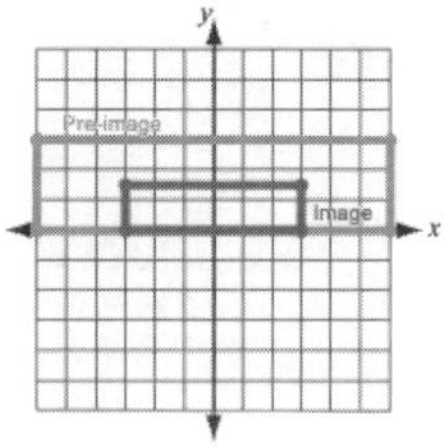

19.

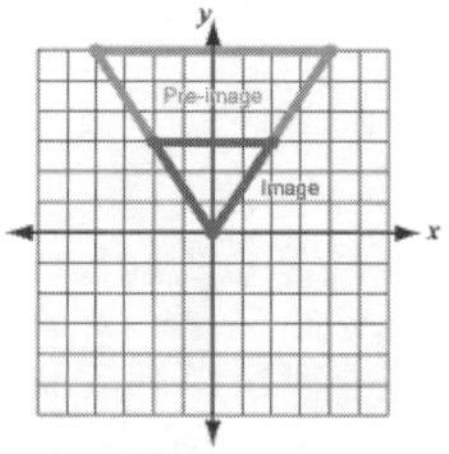

21.

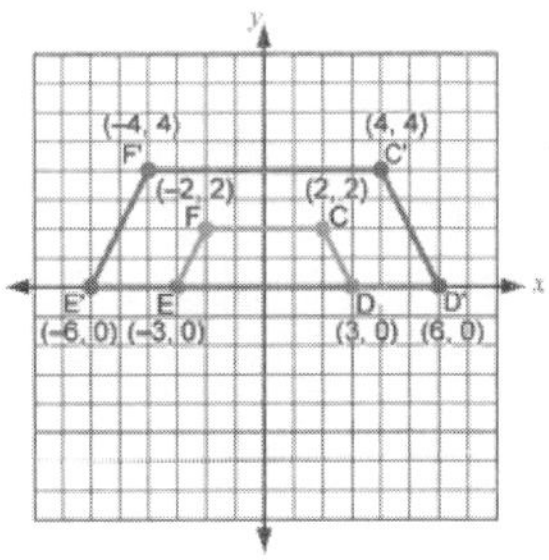

23.

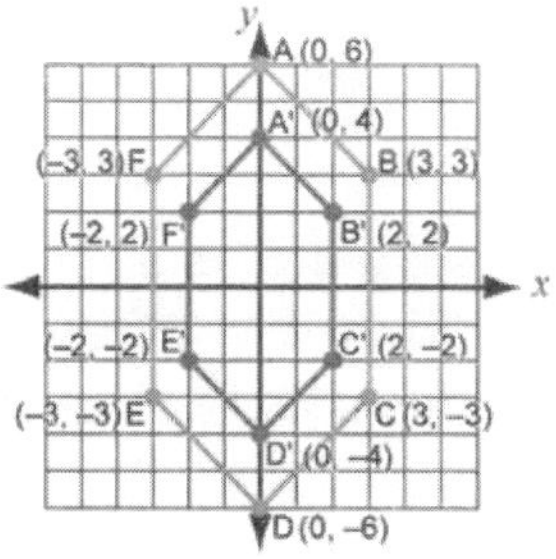

25.

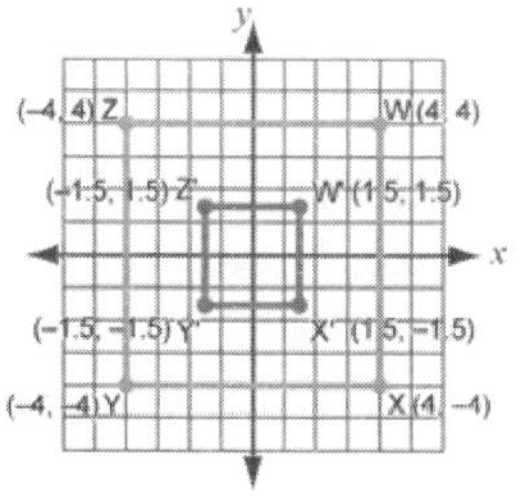

27. $\frac{3}{4}$

29. 4

MPA-121: Identifying Similar and Congruent Polygons Using Proportions

1.

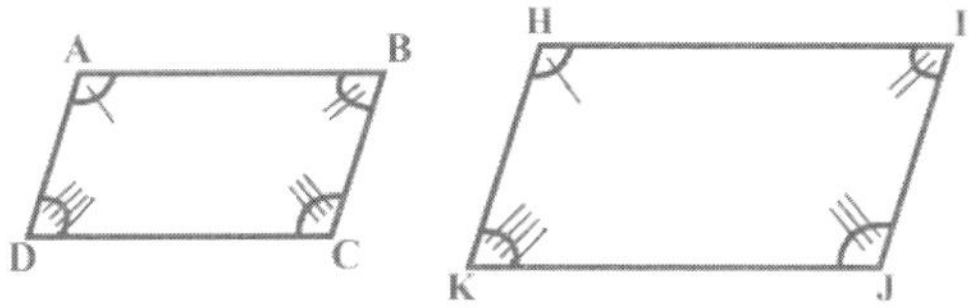

3.

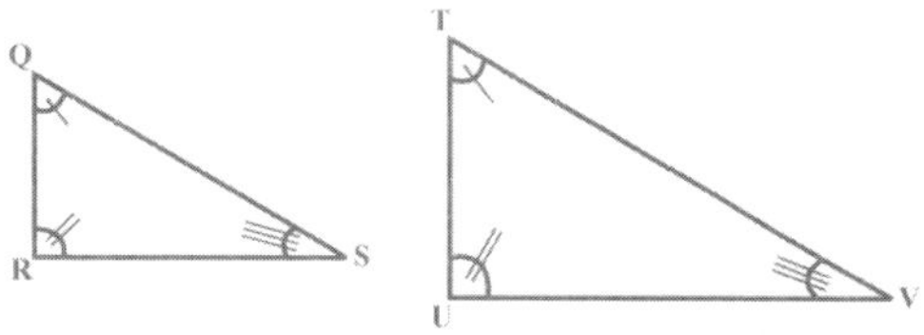

5.

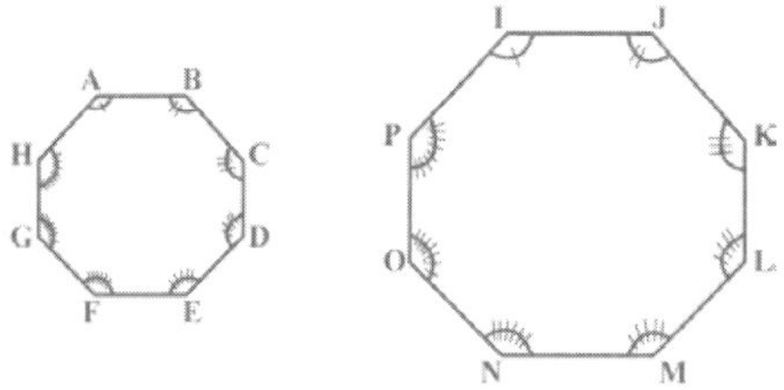

7. $\angle A \Leftrightarrow \angle X$ reads "Angle A corresponds with angle X."

9. $\overline{XY} \Leftrightarrow \overline{MN}$ reads "Segment XY corresponds to segment MN."

11. $\frac{AB}{HI} = \frac{BC}{IJ} = \frac{CD}{JK} = \frac{AD}{HK}$

13. $\frac{QR}{TU} = \frac{QS}{TV} = \frac{RS}{UV}$

15.

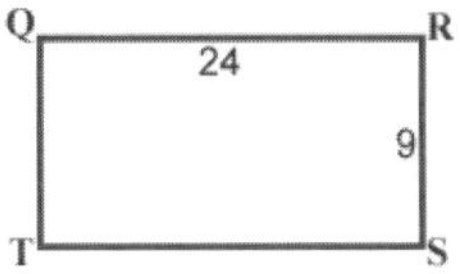

17. The unknown side has a length of 15.

19. The unknown side has a length of 20.

MPA-122: Modeling Multiplication and Division of Decimals

1.

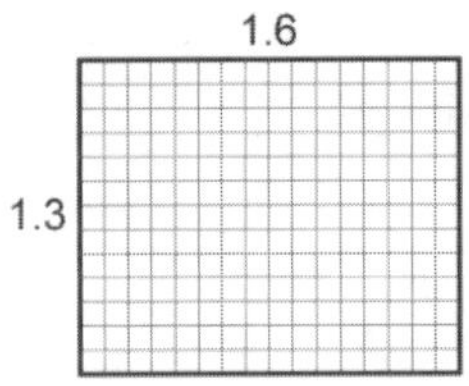

3.

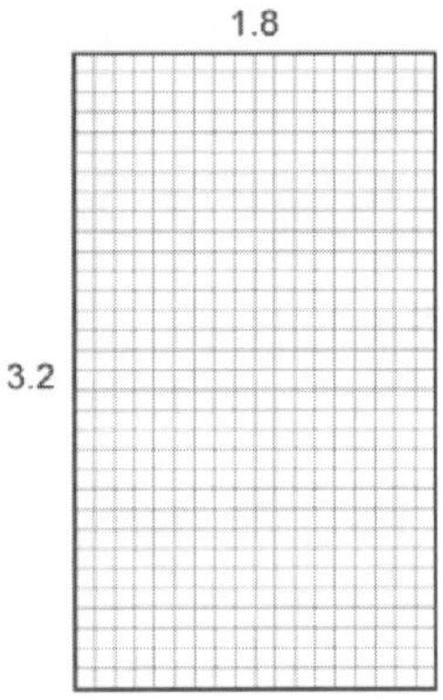

5.

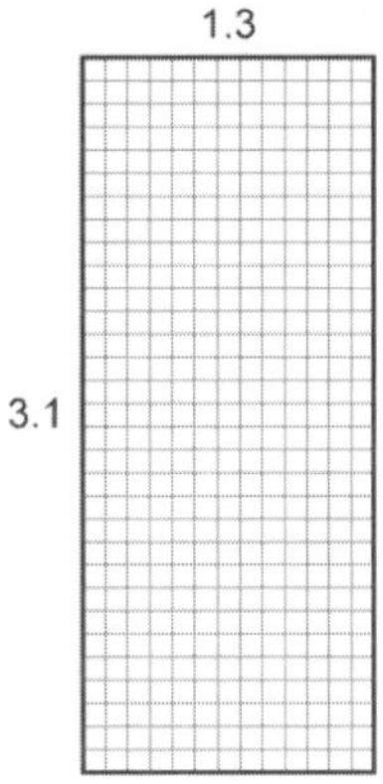

7.

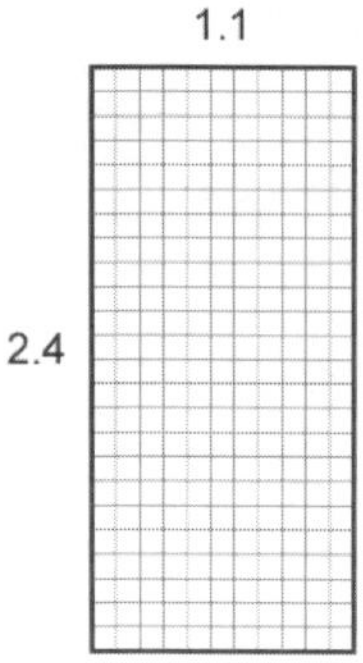

9. 32 squares long **11.** 29 squares long **13.** 22 squares wide **15.** 23 squares wide

17.

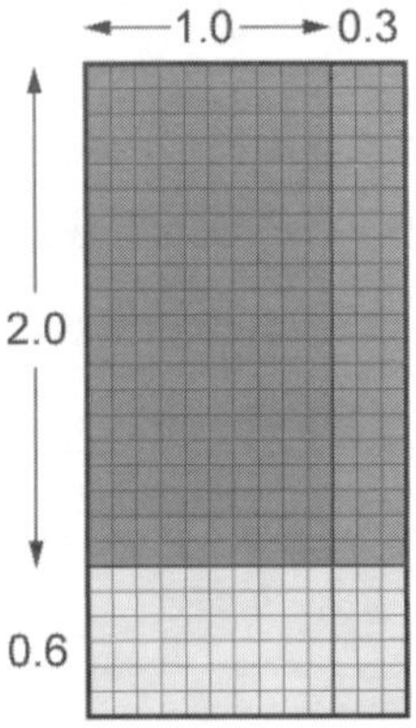

19.

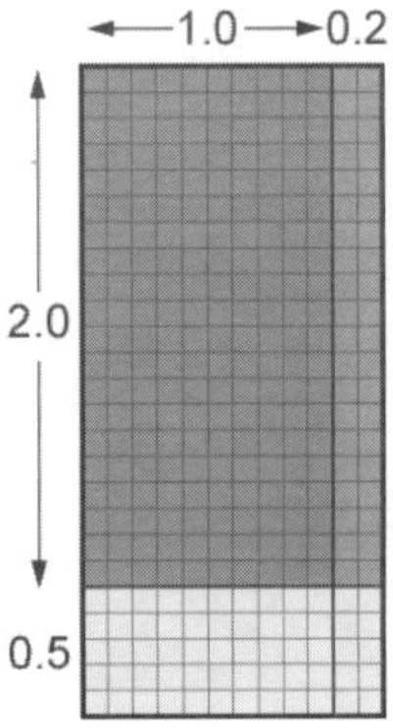

21.

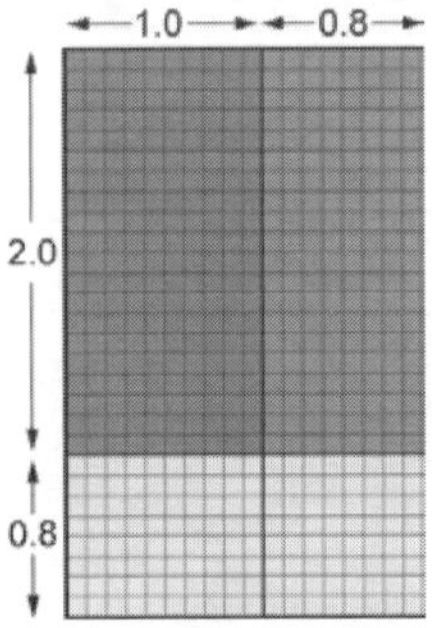

23.

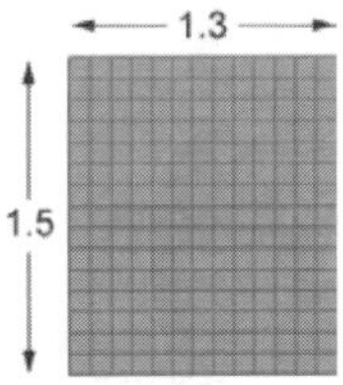

25.

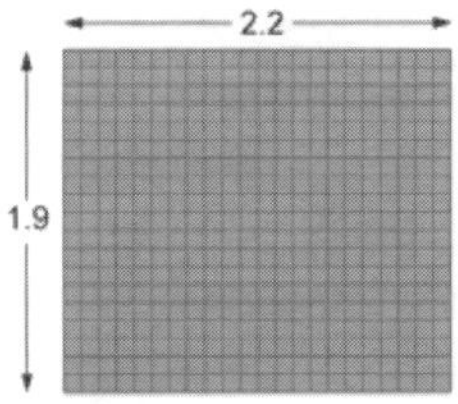

27.

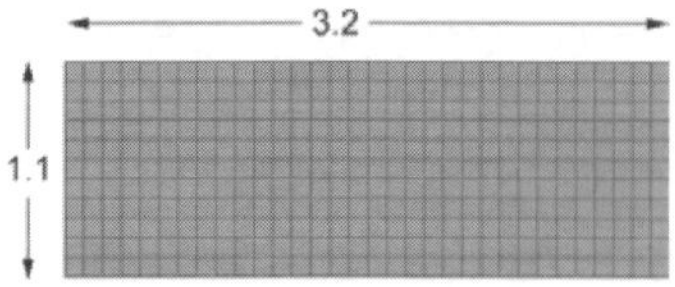

29.

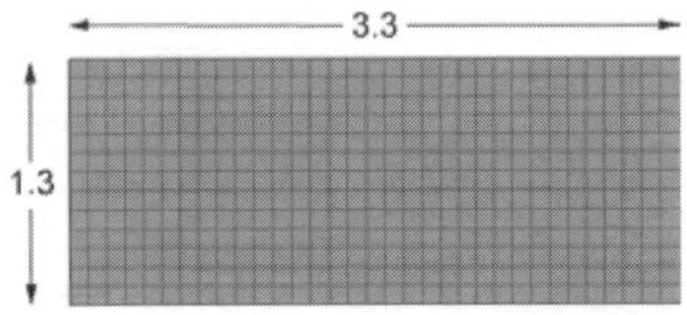

31. 1.6 × 1.7 **33.** 1.1 × 2.9 **35.** 1. 9 × 2.6 **37.** 1.2 × 2.7 **39.** 1.3 × 3.2

MPA-123: Modeling Multiplication and Division of Fractions

1. $\frac{7}{8}$ **3.** $\frac{15}{8}$ **5.** 5 **7.** 8

9.
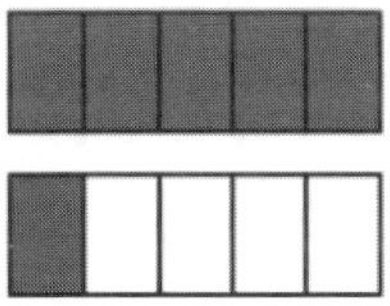

11.
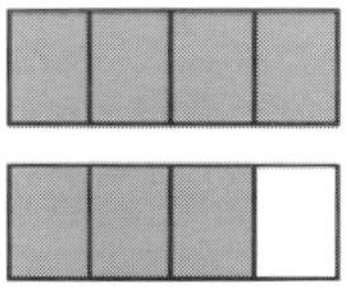

13.
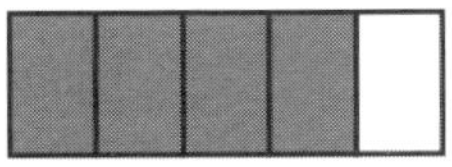

15.
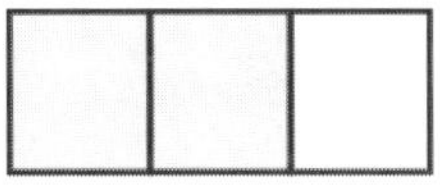

17.
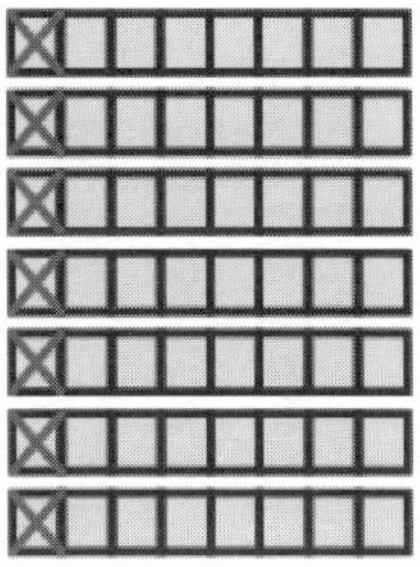

19.
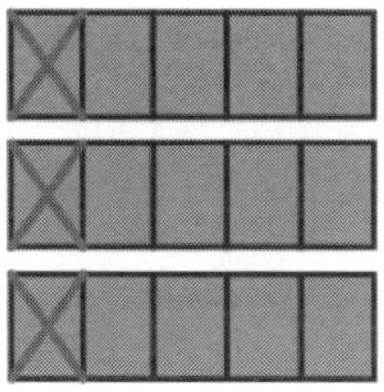

21.
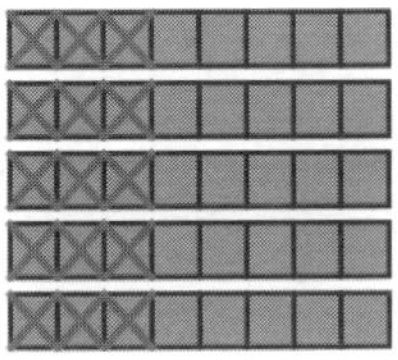

23.
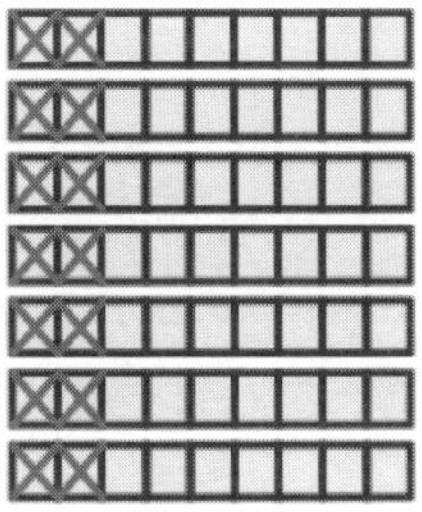

25.
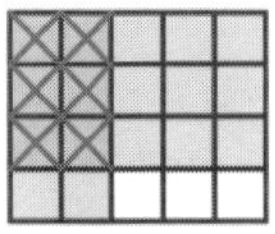

27.
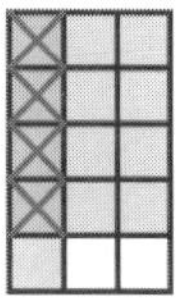

29.
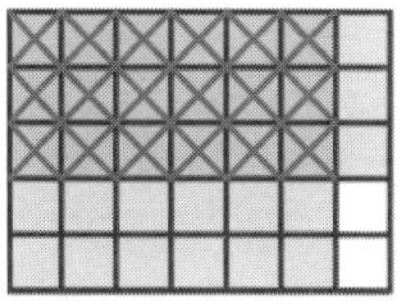

31.

33.
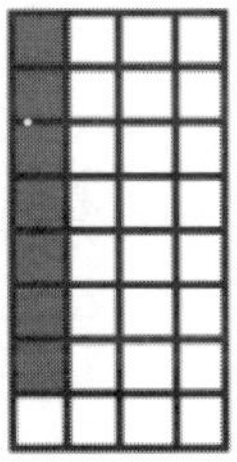

35.
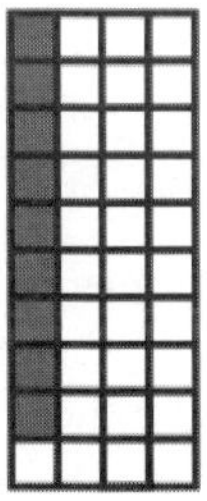

37.
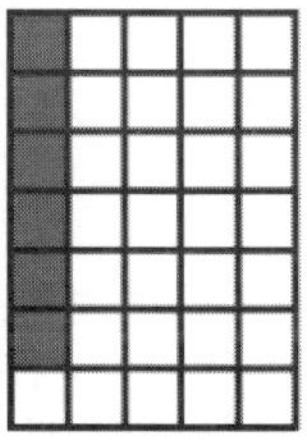

39.
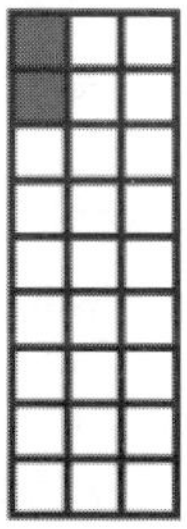

MPA-124: Classifying Numbers in the Real Number System

1. {1, 2, 3, 4, ...} **3.** irrational numbers **5.** {3.4545512..., 67.321234...} **7.** Real Numbers **9.** 0

11. 7 **13.** 6 **15.** 54 **17.** Natural Numbers **19.** Natural Numbers

21. Whole Numbers

23. $\sqrt{6}$

25. $\sqrt{56}$

27. 147

29. 0.35

31. It is an integer, a real number, a rational number, a whole number, and a natural number.

33. It is a real number, a rational number, an integer, and a whole number.

35. It is a real number and a rational number.

37. It is a real number and a rational number.

39. It is a real number and an irrational number.

MPA-125: Formulating a Possible Problem Situation Given an Equation

1. The cost of riding a taxi is $3.25 per mile.

3. The number of words typed is 60 times the number of minutes typed.

5. The time it takes to empty an auditorium is 0.1 minutes per person.

7. The total number of trees in a field is 40 times the number of rows of trees.

9. Franco has an aquarium that is 2 feet on each side. If he wants to fill it with water, how much water will it hold?

11. Hank has a tool box that measures 18 inches on each side. How many cubic inches of rock will it hold?

13. Nancy's room measures 14 feet on each side. How much tile does she need to cover her floor?

15. Pablo wants to paint a wall in his room that is 9 feet on each side. Find the area of the surface that he wants to paint.

17. An apartment complex charges a $400 deposit plus $800 per month.

19. A cable company charges a $125 installation fee plus $45 for each month of service.

21. At the beginning of the year gasoline was priced at $1.50 per gallon. Since then it has increased in price at about $0.05 per month.

23. Salene would like to cover a table with a piece of cloth. The table is 6 feet long and 3 feet wide. How much cloth does Salene need?

25. Uma is building a sidewalk that is 50 feet long and 3 feet wide. What is the surface area of the sidewalk?

27. Victor's room is 14 feet by 16 feet. He wants to put a new baseboard around the edges of his room. How much baseboard does he need?

29. Melissa's sunroom is 10 feet by 8 feet. She wants to place a string of lights along the edge of the room. How long will the string of lights be?

31. Problem: A kennel charges $15 per night for boarding a dog plus a $20 fee for bathing the dog. If the dog spends 7 nights at the kennel, what is the total coat of boarding?

    Answer: The cost of boarding the dog is $125.

33. Problem: The weekly income of a car salesman is $200 plus a $250 commission on each car sold. If he sells 5 cars in a week, what is his income for the week?

    Answer: The car salesman earned $1,450 for the week.

35. Problem: A workout facility charges a $75 membership fee plus $30 per month of membership. If Becky joined the facility 10 months ago, how much has she spent on her membership?

    Answer: Becky has spent $375 on membership.

37. Problem: An Internet CD club charges $50 to join plus $2.50 for each CD purchased. If Hamilton joined the club and then purchased 15 CD's, how much money did he spend?

    Answer: Hamilton spent $87.50 on the CD's.

39. Problem: A toll road offers a pass for $20 so that a driver pays a discount amount of $0.75 at each toll booth. If a driver purchased the pass and has paid tolls at 40 toll booths since he bought the pass, how much money has he spent on tolls?

    Answer: The driver has spent $50 on tolls.

MPA-126: Solving Real-World Problems Involving Sales Tax

1. \$1.98	**3.** \$3	**5.** \$0.39	**7.** $n = 0.13 \times 25$	**9.** $n \times 250 = 37.50$
11. \$24.53	**13.** \$4	**15.** \$15,000	**17.** 3%	**19.** 2%
21. \$0.91	**23.** $n \times 35 = 1.93$	**25.** $n = 0.0525 \times 52$	**27.** \$21.04	**29.** \$84.73
31. \$48	**33.** \$66.67	**35.** \$30	**37.** 9.25%	**39.** 8.5%

MPA-127: Solving Real-World Problems Involving Discounts

1. \$28.20 **3.** \$47.25 **5.** \$72.60 **7.** \$10.45 **9.** \$23 **11.** \$30 **13.** \$63 **15.** \$34 **17.** \$102.92 **19.** \$17.11

21. \$78.62 **23.** 15% **25.** 35% **27.** 18.5% **29.** 16%

31. \$22.54 **33.** \$49.98 **35.** \$42.08 **37.** \$282.19 **39.** \$65.92

MPA-128: Solving Real-World Problems Involving Simple Interest

1. \$40 **3.** \$210 **5.** \$150 **7.** \$189 **9.** 7% **11.** 1 year **13.** 5.5% **15.** \$6,250 **17.** \$630 **19.** \$4,812.50

21. \$3,750 **23.** \$2,145 **25.** \$5,250 **27.** \$125 **29.** \$240

31. 4 years **33.** 7 years **35.** 8% **37.** \$3,750 **39.** \$6,625

MPA-129: Choosing Appropriate Scales and Intervals for Data (an Introduction)

1. 41 **3.** 93 **5.** 188 **7.** 186 **9.** 10 **11.** 10 **13.** 50 **15.** 20

17.

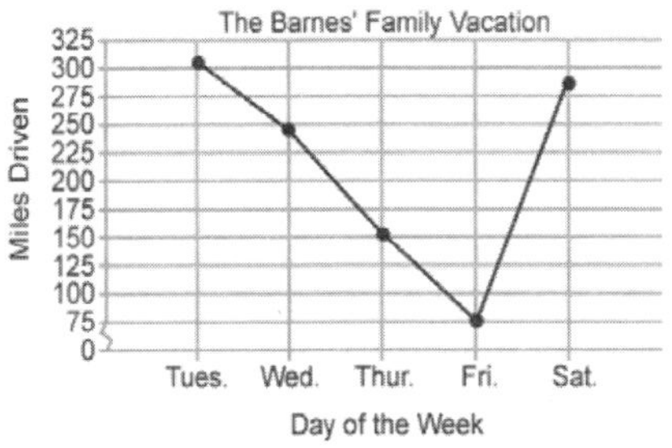

19.

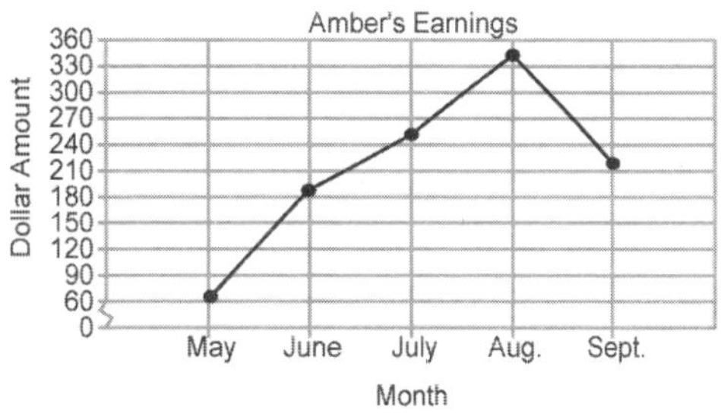

21.

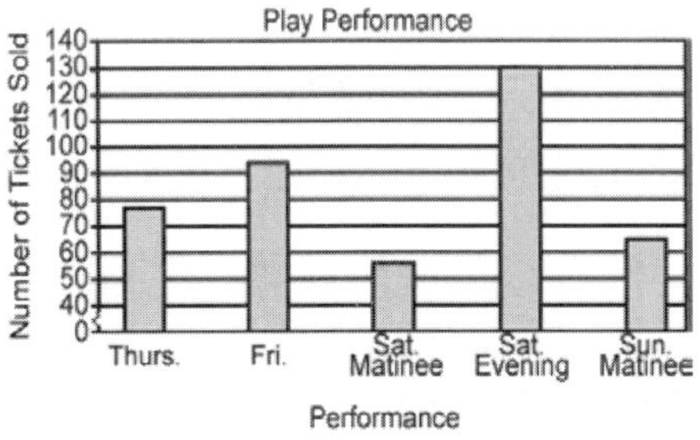

23.

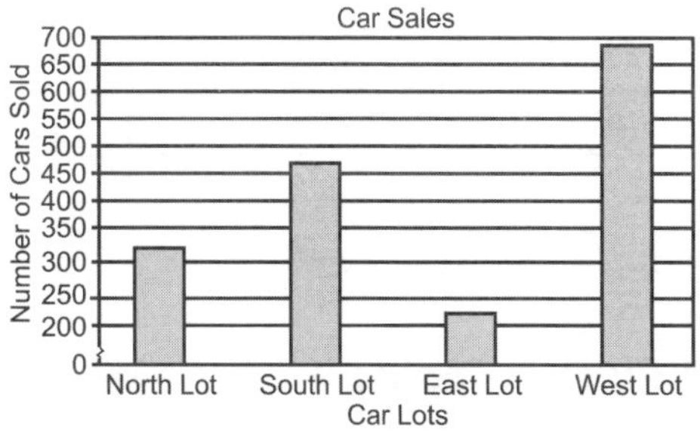

25.

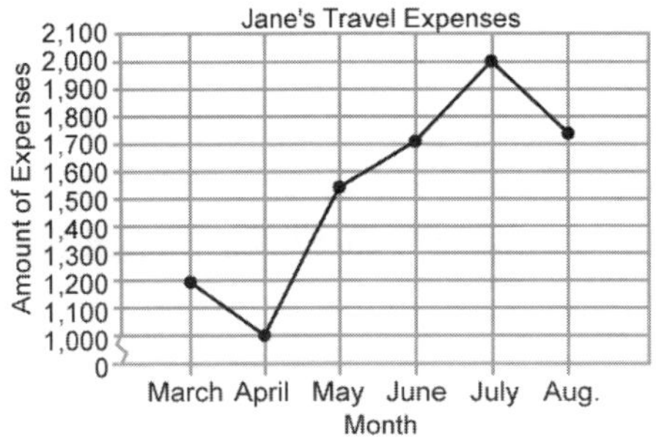

27.

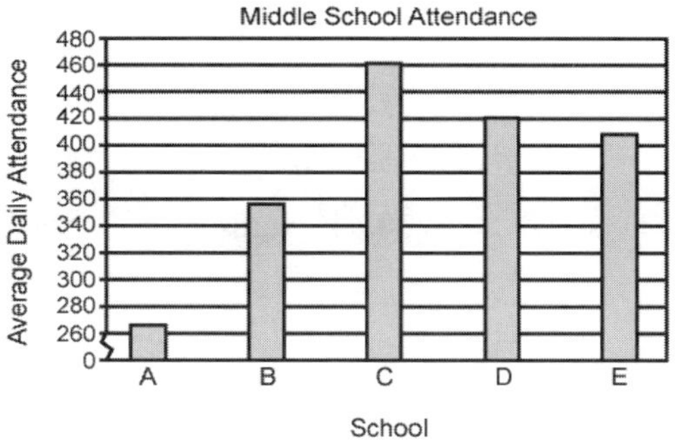

29.

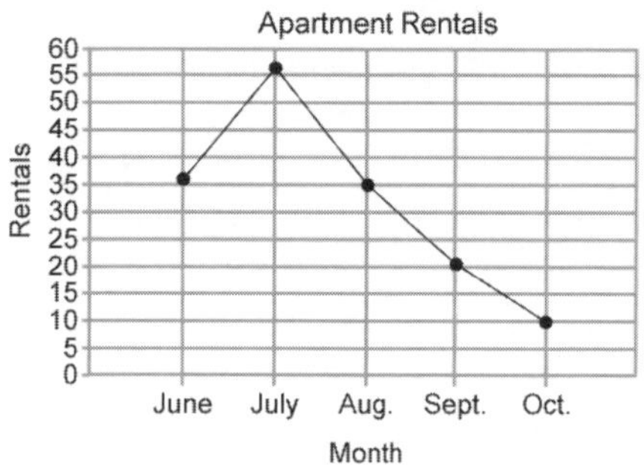

31.

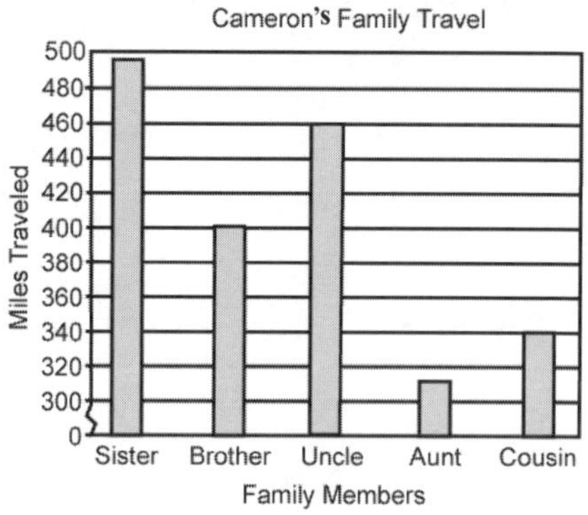

33.

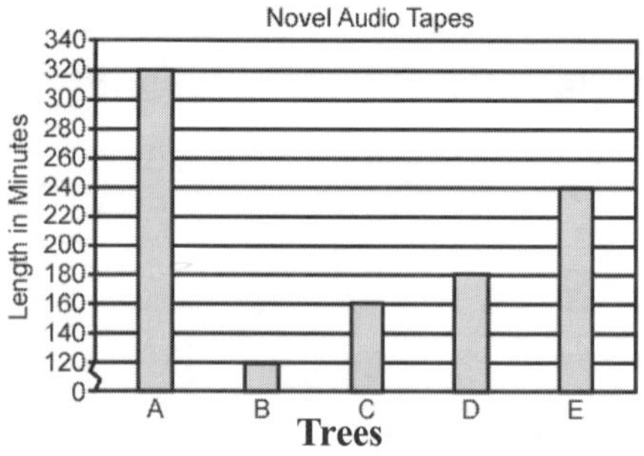

35.

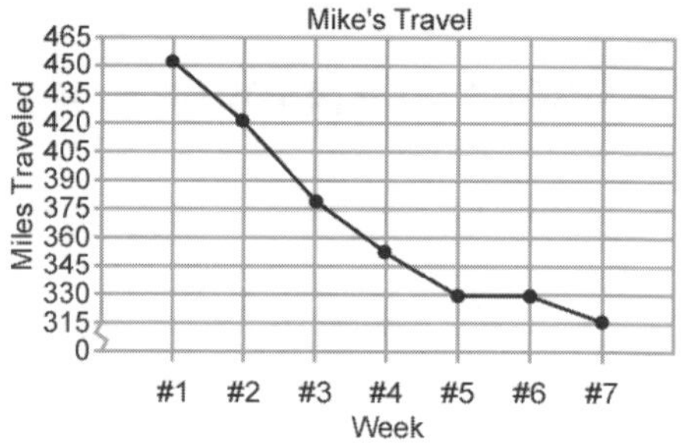

37.

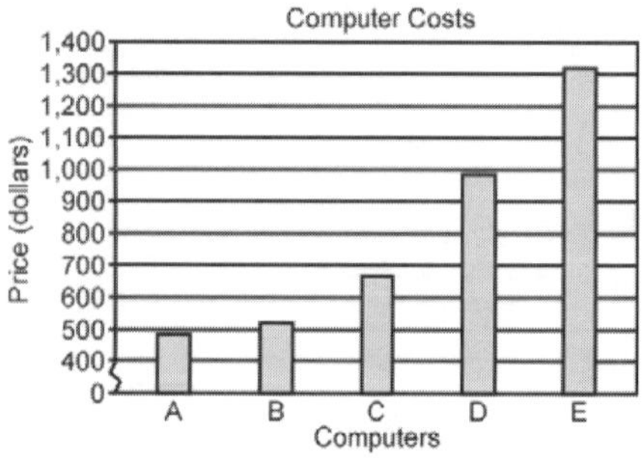

39.

MPA-130: Developing a Sense of Relative Sizes of Measures

1. meters **3.** ounces **5.** fluid ounces **7.** hectograms **9.** centimeters

11. centiliters **13.** grams **15.** meters **17.** mile **19.** hectogram

21. gallon **23.** $10\frac{1}{2}$ inches **25.** 1 kilogram **27.** 2 quarts **29.** 400 milliliters

31. 3 liters **33.** $4\frac{1}{2}$ yards **35.** 12 pounds **37.** 10 fluid ounces **39.** 6 meters

MPA-131: Interpreting and Creating Histograms

1. 19 **3.** 50 **5.** 48 **7.** 45

9.

Time Spent Watching Television Over the Weekend		
Number of Minutes	Tally	Frequency
30 – 59	III	3
60 – 89	卌	5
90 – 119	IIII	4
120 – 149	I	1
150 – 179	I	1
180 – 209	II	2
210 – 239	IIII	4

11.

New Magazine Subscriptions		
Number of Magazines	Tally	Frequency
190 – 239	卌	5
240 – 289	IIII	4
290 – 339	III	3
340 – 389	I	1
390 – 439	I	1
440 – 489	IIII	4
490 – 539	II	2

13.

Surface Area of Prisms		
Surface Area (ft^2)	Tally	Frequency
60 – 109	III	3
110 – 159	卌 I	6
160 – 209	IIII	4
210 – 259	II	2
260 – 309	III	3
310 – 359	II	2

15.

Weekly Pizza Deliveries		
Number of Pizzas	Tally	Frequency
75 – 94	IIII	4
95 – 114	II	2
115 – 134	IIII	4
135 – 154	I	1
155 – 174		0
175 – 194	III	3
195 – 214	I	1

17. 28 – 31 thousand; 45 employees **19.** 801 – 900 dollars; 40 apartments

21. 250 – 349 and 550 – 649 miles; 15 families **23.** 72 **25.** 41 **27.** 260 **29.** 32

31.

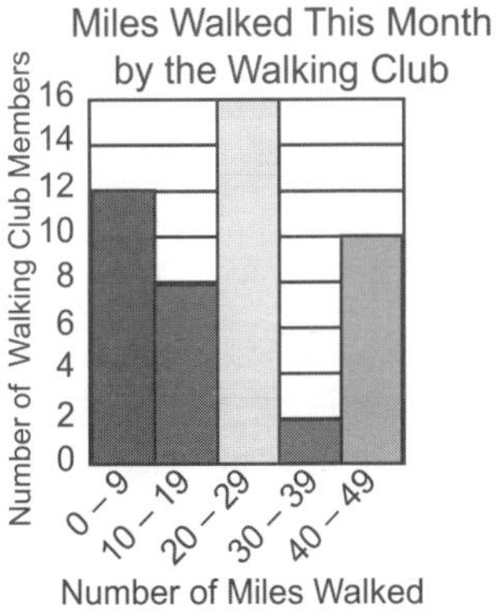

33.

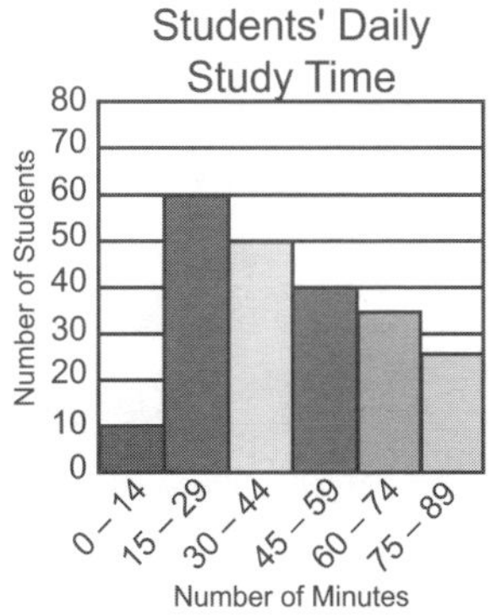

35.

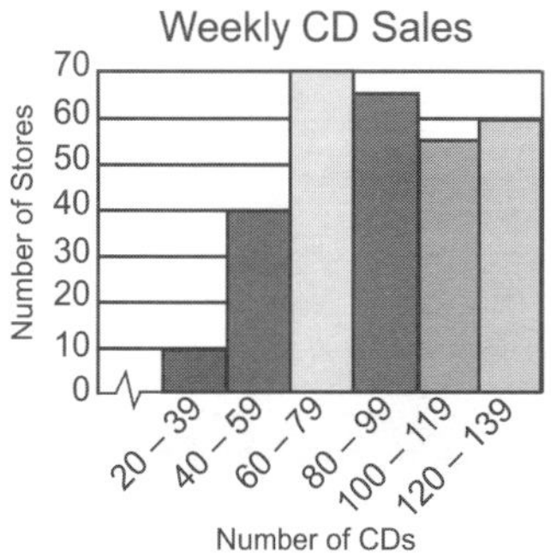

37.

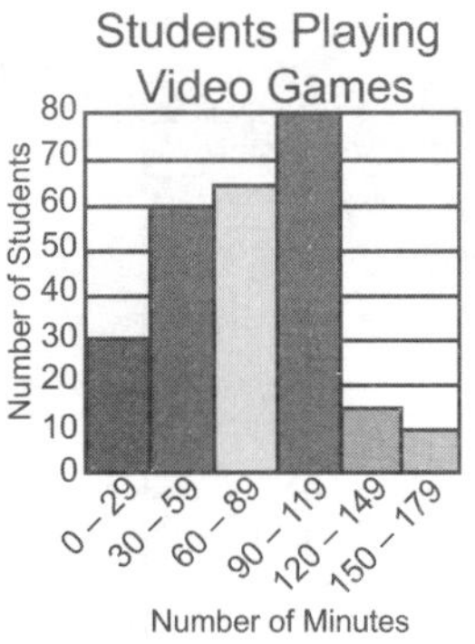

39.

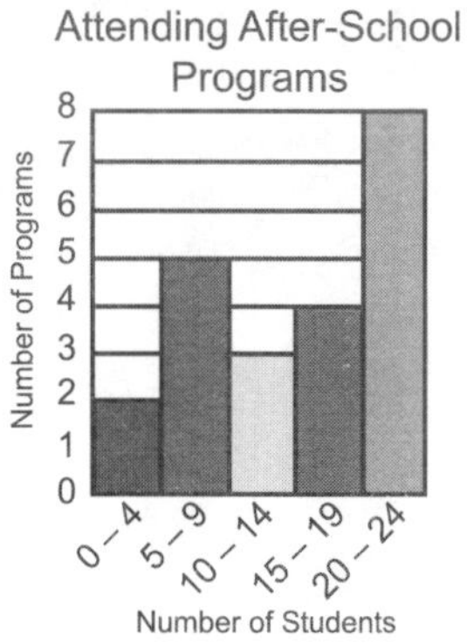

MPA-132: Interpreting and Creating Scatterplots

1. Positive correlation **3.** No correlation **5.** Negative correlation **7.** Negative correlation

9. As interest rates increase, the prices of homes decrease.

11. As the monthly rainfall increases, the number of mosquito bites increases.

13. As the amount of clouds in the sky increases, the number of cars in a lot neither increases nor decreases.

15. As the years of education increase, the salary increases.

17.

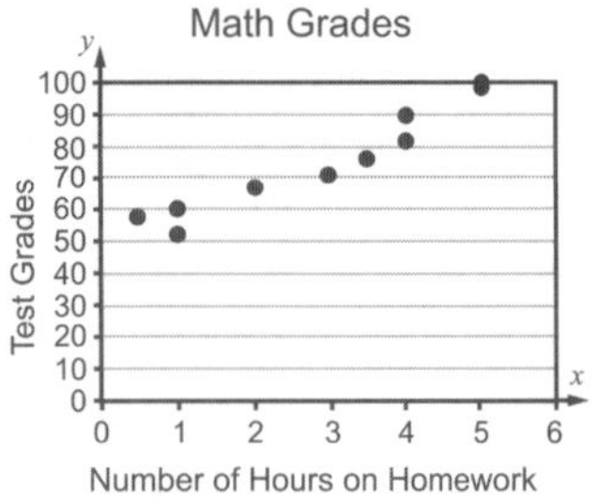

19.

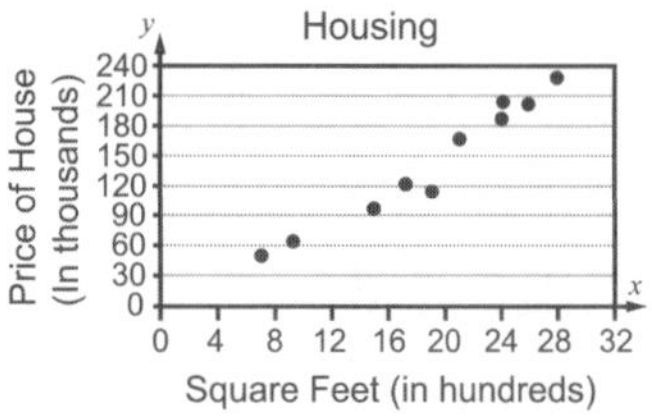

21.

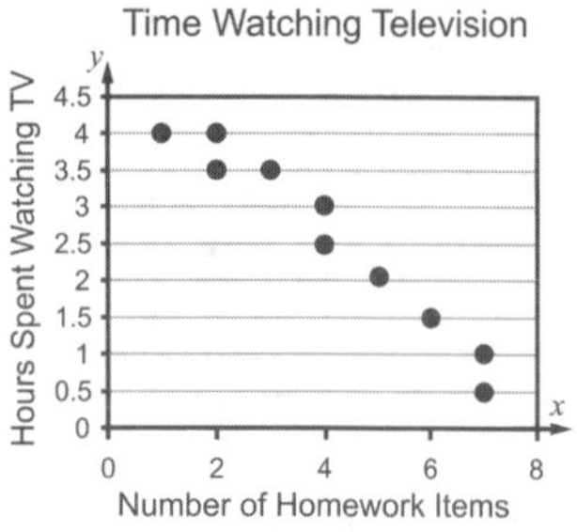

23. Positive correlation **25.** Positive correlation **27.** No correlation **29.** No correlation

31. No correlation **33.** $200 **35.** 85 **37.** $80 **39.** $7.00

MPA-133: Distinguishing Between Exact and Approximate Answers

1. Exact answer **3.** Exact answer **5.** Exact answer **7.** Approximate answer

9. George has to determine the amount of change that Mr. Martin should receive back from his purchases totaling $13.39 when Mr. Martin hands him a $20.00 bill.

11. Cameron must pay for her ticket to the basketball game.

13. A waitress must calculate the amount of a bill for dinner in a restaurant.

15. A bank teller must make change for a $20 bill using quarters.

17. Find the area of your hand using a tape measure.

19. Find the surface area of a large butterfly using a ruler.

21. Find the height of the triangle by comparing it to the base of the triangle.

23. ≈ 7.5 centimeters

25. $\approx 4\frac{1}{4}$ inches **27.** ≈ 18 centimeters **29.** ≈ 6.5 centimeters **31.** ≈ 64 square units

33. ≈ 75 square units **35.** ≈ 55 square units **37.** ≈ 135 square units **39.** ≈ 66 square units

MPA-134: Calculating with Precision, Accuracy, and Significant Digits

1. Accurate and precise **3.** Not accurate and not precise **5.** Not accurate but precise **7.** Accurate and precise

9. 4 significant digits **11.** 6 significant digits **13.** 5 significant digits **15.** 1 significant digit

17. 41.73 **19.** 370.4 **21.** 12.791 **23.** 75 **25.** 78.24 **27.** 2.5 **29.** 4.79

31. The students' measurements are accurate and precise. **33.** The students' measurements are precise.

35. The students' measurements are accurate and precise. **37.** Each measurement has 4 significant digits.

39. Each measurement has 4 significant digits.

MPA-140: Examining Linear Equations in Slope-Intercept Form

1. $y = 3x + 1$ **3.** $y = -\frac{1}{2}x + 3$ **5.** $m = 2$, $b = -5$ **7.** $m = \frac{3}{5}$, $b = 6$ **9.** $y = -3x - 7$

11. $y = \frac{5}{2}x + 2$ **13.** $m = 2$, $b = 3$ **15.** $m = -\frac{4}{3}$, $b = 4$ **17.** Company B **19.** Company C

MPA-142: Solving Problems with Linear Functions and Direct Variation

1. $y = 15x + 75$ **3.** $y = 1.5x + 36$ **5.** $y = 0.19x$ **7.** $y = 5x + 12$ **9.** –13

11. 5

13.

x	1	2	3	4
y	–8	–16	–24	–32

15.

x	–12	–9	–6	–3
y	–8	–6	–4	–2

17. 104 miles **19.** 6 hours

MPA-150: Identifying and Graphing Linear and Nonlinear Functions

1. $y = 7x^2$ **3.** $y = \frac{1}{8}x^2$

5.

x	y
1	3
2	12
3	27
4	48

7.

x	y
1	–2
2	–5
3	–8
4	–11

9. A parabola that opens downward and is narrower than $y = x^2$

11. An S-shaped curve that is narrower than $y = x^3$ with points in quadrants II and IV

13.

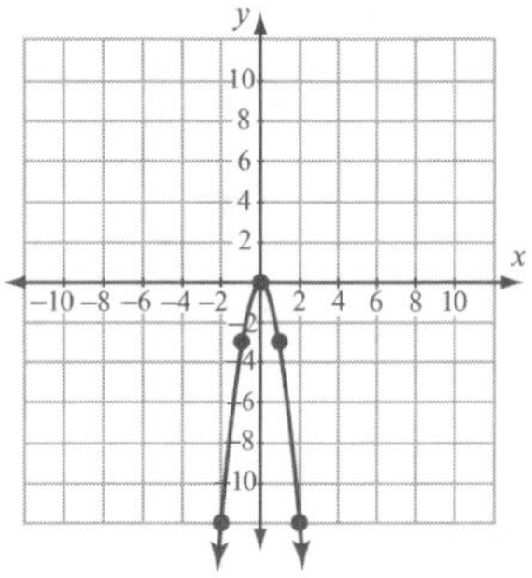

15.

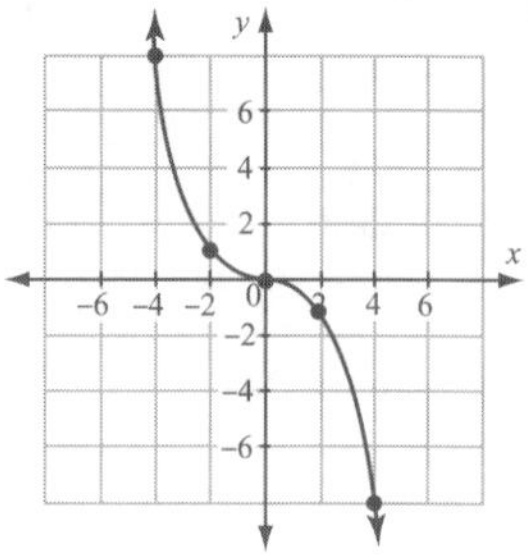

17.

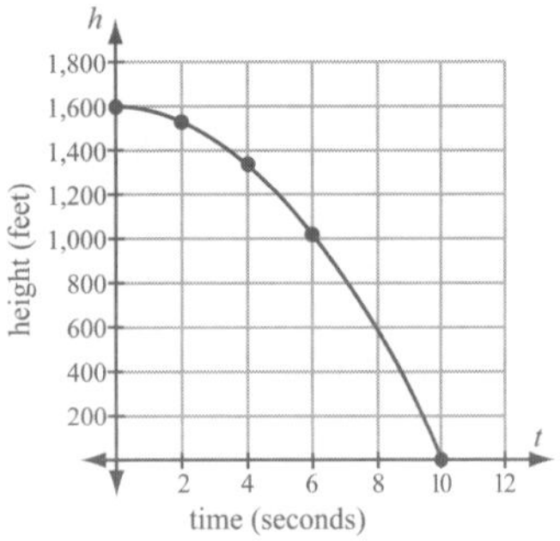

The ball strikes the ground in 10 seconds.

19.

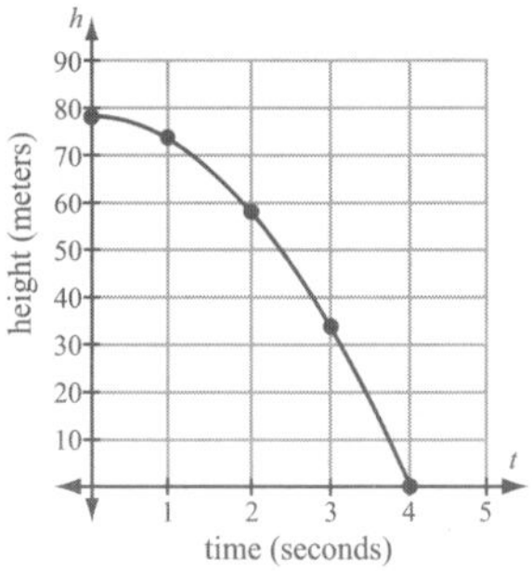

The ball strikes the ground in 4 seconds.

MPA-155: Comparing and Converting Rates

1. 65 miles per hour **3.** 22 miles per gallon **5.** Box D **7.** Package C **9.** 8,400 ft/min.

11. 3,600 gal/hr. **13.** \$15.75/sq yd **15.** \$2.75/sq ft **17.** 3.2 qt/min. **19.** 216,648 yen

MPA-160: Plotting Polygons and Finding the Area

1. 20 square units **3.** 15 square units **5.** 10 square units **7.** 12 square units **9.** 12.5 square units

11. 20 square units **13.** 13 units^2 **15.** 14 units^2 **17.** 12 units^2 **19.** 36 units^2

MPA-270: Generating Algebraic Expressions from Patterns of Models

Problem Set

1. −10, −20, −30, −40, −50, ... **3.** −7, −6, −5, −4, −3, ... **5.** 11, 10, 9, 8, 7, ...

7. −2, 1, 4, 7, 10, ... **9.** Add 14 to n. **11.** Subtract 6 from n.

13. Multiply n by 2. **15.** Divide n by 8. **17.** $n + 2$

19. $2n + 1$ **21.** $2n + 5$ **23.** $3n + 4$

25. $3n + 5$ **27.** $4n - 5$ **29.** $5n - 2$

31. Conversion of minutes to seconds **33.** Conversion of feet to inches **35.** $n = 15h$

37. $d = 55t$ **39.** $c = 4h$

MPA-840: Interpreting Data

Problem Set

1. Erin spent twice as much time studying for science than she did studying for history.
3. Mark's test scores gradually improved.
5. Mark scored 6 points lower on Test 4 than he did on Test 5.
7. Will has 2 fewer siblings than Steven.
9. The hourly rate of each person is needed.
11. The number of hours each person worked is needed.
13. The number of games each team played is needed.
15. The cost of supplies to paint the pictures is needed.
17. Yes, this information is misleading because only one winner will win at least the average amount advertised.
19. Yes, this information is misleading because Jesse scored the most goals per game.
21. Yes, this information is misleading because Jermaine scored the fifth-most goals per game.
23. Changing the vertical scale and range affects how the graph is interpreted.
25. Yes, the graph is misleading because the size of the house representing the number of houses in 2000 is larger in height and width which makes the increase seem greater.
27. No, the pie chart is not misleading.
29. No, the bar chart is not misleading.
31. The conclusion is misleading. The statistics should be reported as the percent of people who prefer regular cola to diet cola.
33. The conclusion is misleading. The statistics should be reported as the percent of people who prefer regular cola to diet cola.
35. The conclusion is misleading. The statistics should be reported as the percent of car accidents that resulted in a death.
37. The conclusion is not misleading.
39. The conclusion is misleading. The statistics should be reported as the percent of homes that own at least one television.

Index

E

Q

MARQUIS
Marquis imprimeur inc.
Québec, Canada
2007